Artificial Intelligence (AI) for IT Energy Efficiency and Green AI for Environment Sustainability

Pethuru Raj • Durga Prasad Sharma
Pushan Kumar Dutta • B. V. V. Siva Prasad
P. Beaulah Soundarabai
Editors

Artificial Intelligence (AI) for IT Energy Efficiency and Green AI for Environment Sustainability

Springer

Editors
Pethuru Raj
RJIL, Site Reliability Engineering
Reliance Industries (India)
Bangalore, Karnataka, India

Pushan Kumar Dutta
Amity University Kolkata
Kolkata, West Bengal, India

P. Beaulah Soundarabai
Dept of Computer Science
CHRIST (Deemed to be University)
Bangalore, Karnataka, India

Durga Prasad Sharma
MAISM
Rajasthan Technical University
Jaipur, Rajasthan, India

B. V. V. Siva Prasad
School of Engineering
Anurag University
Hyderabad, Telangana, India

ISBN 978-3-031-89419-0 ISBN 978-3-031-89420-6 (eBook)
https://doi.org/10.1007/978-3-031-89420-6

This Springer imprint is published by the registered company Springer Nature Switzerland AG
The registered company address is: Gewerbestrasse 11, 6330 Cham, Switzerland

Preface

The growing need for energy efficiency and environmental sustainability is one of the most pressing challenges of our time. The way we consume energy and manage resources has a direct impact on this planet's future. Along with several converging technologies, artificial intelligence (AI) is emerging as a transformative force in addressing these challenges. This book, *Artificial Intelligence (AI) for IT Energy Efficiency and Green AI for Environmental Sustainability*, explores how AI can reincarnate the energy landscape, reduce waste, and create a sustainable future.

This book serves as a bridge between the complex world of AI and the practical applications that directly impact energy efficiency and sustainability. It is designed for a diverse audience—students eager to learn about AI's potential, researchers seeking inspiration for their next breakthrough, industry leaders looking for innovative solutions, and policymakers striving to balance progress with environmental responsibility.

The prime goal of this book is simple yet ambitious: to illustrate how AI can make IT systems more energy efficient and help build a world where technology and nature coexist harmoniously. Through a carefully curated selection of chapters, this book provides readers with insights into AI's capabilities and applications across a variety of domains and subdomains.

Smart city governance is not just about deploying more sensors or ICT infrastructure. It also requires transparency, accountability, and robust public participation in decision-making. These how AI-powered solutions are optimizing energy consumption and reducing the carbon footprint of core and allied IT systems.

The book progresses to address the role of AI in smart grids, with chapters like *AI for Smart Grids for Sustainable Future* and *Artificial Intelligence Applications for Enhancing Efficiency in Smart Grids*. These chapters demonstrate how AI is revolutionizing the way energy is distributed, monitored, and consumed, ensuring that power reaches where it is needed most with minimal waste.

Further chapters focus on groundbreaking advancements, such as *The Environmental Impact of AI in Fashion: Balancing Innovation with Sustainability* and *Harnessing AI for Sustainable Smart Cities: Impact, Innovations and Use*

Cases. These contributions showcase how AI is driving sustainability across industries and creating new possibilities for a greener future.

Ethical considerations and challenges are also woven into the narrative. Chapters like *Privacy, Security, Ethics in AI-Driven Energy Systems* and *An Interview-Based Study on the Ethical, Privacy, and Security Implications of AI-Driven Smart Grids* offer critical perspectives on ensuring that AI is deployed responsibly and transparently.

To address the practical side, chapters on AI-powered energy, *Future Trends* and *Economics Impact of AI on Energy Suitability* and *AI-Powered Energy Consumption Prediction and Forecasting*, provide actionable insights into how organizations can leverage AI to plan better, save energy, and contribute to sustainability goals.

The book closes with futuristic and aspirational ideas, such as the convergence of disparate technologies for energy-efficient fusion, as discussed in chapters like *Blockchain and AI for Enhanced Threat Detection in Digital Marketing: Maneuvering the Legal Minefield* and *Innovations in Data Security and Intelligent Automation Through Artificial Intelligence: Patterns and Hurdles*. These chapters emphasize the ongoing evolution of AI technologies and their untapped potential for advancing sustainability.

This book is more than just a collection of chapters—it is a road map for the future. By connecting and converging AI, for energy efficiency, and environmental sustainability, it lays the groundwork for actionable solutions and inspiring innovations. Each chapter is a call to action, encouraging readers to think critically about the intersection of technology and the environment.

We are deeply grateful to all the contributors whose expertise and insights have shaped this work. Their dedication and innovative thinking have resulted in a resource that we hope will serve as a cornerstone for education, innovation, research, and renovations.

As you read through these pages, we invite you to envision a future where AI not only enhances IT systems but also protects and sustains the environment. Together, we can create a world where technology works in harmony with nature, ensuring a brighter and greener tomorrow.

Bangalore, Karnataka, India — Pethuru Raj
Jaipur, Rajasthan, India — Durga Prasad Sharma
Kolkata, West Bengal, India — Pushan Kumar Dutta
Hyderabad, Telangana, India — B. V. Siva Prasad
Bangalore, Karnataka, India — P. Beaulah Soundarabai

The original version of this book was revised. An erratum to this book can be found at https://doi.org/10.1007/978-3-031-89420-6_33

Contents

Fusion of Green and High-Tech Innovation for a Sustainable Future through Evaluating Development Frameworks

Bindiya Jain, Udit Mamodiya, and Indra Kishor

1 Introduction

The green innovation in India happens to symbolize a forward-looking commitment to sustainable development and mitigation of climate change. The Indian nation has indeed been focusing more lately on clean energy, resource efficiency, and environmentally friendly industrial practices—the rationality behind its economic strategy. India has outlined in the National Solar Mission under the National Action Plan on Climate Change and is very well positioned to become one of the leaders in renewable energy. At 450 GW of renewable energy by 2030, ambitions seem high. Indian aspiration remains strong. Carbon intensity reduction and transition to a low-carbon economy are ways India has committed to via the Paris Agreement, and investments in green technologies remain very prominent in large sub-sectors like energy and manufacturing. In view of the fact that Indian government policies toward FAME India Faster Adoption and Manufacturing of Hybrid and Electric Vehicles are proving encouraging for cleaner transport technologies; with smart cities and smart grids forming an emphasis on ensuring efficient urbanization and energy utilization in the country, high-tech industries have so far seen a high degree of success on the way to

B. Jain (✉)
Faculty of Computer Science and Engineering, Poornima University, Jaipur, Rajasthan, India
e-mail: bindiya.jain@Poornima.edu.in

U. Mamodiya
Faculty of Engineering and Technology, Poornima University, Jaipur, Rajasthan, India
e-mail: assoc.dean_research@poornima.edu.in

I. Kishor
Department of Computer Engineering, Poornima Institute of Engineering & Technology, Jaipur, Rajasthan, India
e-mail: indra.kishor@poornima.org

P. Raj et al. (eds.), *Artificial Intelligence (AI) for IT Energy Efficiency and Green AI for Environment Sustainability*,
https://doi.org/10.1007/978-3-031-89420-6_1

green innovation. "Make in India" and "Atmanirbhar Bharat" or self-reliant India require the integration of high-tech businesses with sustainable practice, with all its dimensions using AI and IoT to optimize energy efficiency, reduce waste, and decrease emissions. The green transformation properly addresses key challenges such as air pollution and depletion of resources, thus also enhancing the quality of life for future generations. The regional initiatives also significantly contribute to sustainable agriculture, water conservation, and biodiversity protection. With high-tech innovation, India is now a major player toward becoming a world leader in green technology to achieve its ambitious sustainability goals. Sustainable competitiveness becomes an important factor in balancing the economic growth within financial performance, environmental, and social dimensions. Encompassing population welfare, ecological sustainability, and the general quality of life integration of sustainable development into economic strategies offers the basis for that country's GCI (Global Competitiveness Index). The view of the United Nations invites competitiveness analysis through the lens of environment and quality of life. The concept of sustainable competitiveness in India leads to enhanced long-term economic viability and answers environmental and social problems. India will always be able to help its citizens toward a sustainable future. For further insights into sustainable competitiveness, the work is recommended. Green and high-tech innovation, when combined, can create transformative pathways toward sustainable development. These innovations can significantly reduce the reliance on nonrenewable resources, minimize waste, and promote the transition to renewable energy sources, thus contributing to achieving the United Nations' Sustainable Development Goals (SDGs). However, integrating these approaches requires a structured evaluation of development frameworks that can facilitate and guide their convergence in real-world applications. To achieve meaningful progress, it is essential to critically examine the current development frameworks that underpin both green and high-tech innovations. These frameworks encompass regulatory policies, research and development (R&D) funding, and the collaboration between industries, governments, and academic institutions. By evaluating these structures, we can identify areas for improvement, fostering a synergistic approach to innovation that prioritizes sustainability alongside technological advancement.

This chapter explores the fusion of green and high-tech innovation, with a particular emphasis on evaluating various development frameworks that promote their integration. The study seeks to identify opportunities for accelerating progress toward a sustainable future, highlighting key areas such as energy efficiency, waste reduction, and the creation of sustainable materials. These innovations can complement one another; we can build more resilient, eco-friendly economies that embrace technology for the betterment of society and the environment.

2 Literature Review

The integration of green technology and high-tech innovation is critical which emerged as a critical path toward achieving sustainability [1, 2]. Green technologies, defined as innovations that reduce environmental harm, are increasingly being adopted across industries, driven by advancements in artificial intelligence (AI), big data, and automation. These technological integrations are helping businesses not only cut their carbon emissions but also boost operational efficiency, resulting in enhanced sustainability outcomes [3, 4]. These technologies are central to reducing carbon footprints and transitioning toward renewable energy systems for fostering sustainable growth and addressing pressing environmental challenges. According to Li [5], green technology involves practices and technologies that aim to ease the adverse things of human activities on the situation, promoting energy efficiency, waste reduction, and sustainable resource utilization. The rise of high-tech innovation has enhanced the deployment of such technologies, enabling more efficient and scalable solutions to climate and environmental issues. Green transformation in recent research highlights these innovations in areas of the Internet of Things (IoT) and AI-driven energy management systems with significant environmental and economic benefits [6]. Green transformation is now being viewed as essential for long-term industrial sustainability. In a study by [7–9], industries that incorporate green technologies are reducing their carbon footprints but also gaining resilience in the face of regulatory pressures and market shifts toward sustainability. In the renewable energy sector, AI is enabling more efficient energy storage solutions and improving the grid's responsiveness to fluctuations in energy demand, which involves systemic shifts in industries through eco-friendly technological integration. Wu et al. [10] and Rahman and Lee [11] maintain the energy sector particularly in transformation, with AI, big data, and IoT being leveraged to improve efficiency in renewable energy production. The use of real-time data and predictive analytics is heightening renewable energy outputs and minimizing waste in decarbonizing economies. The green transformation of industries requires technological innovation in policy and regulatory frameworks [1]. Green transformation is a multi-dimensional process that includes technological, institutional, and social changes aimed at dropping the environmental influence of human actions. A transformation is largely driven by the integration of pioneering technologies such as artificial intelligence, IoT, and renewable energy systems into traditional industries [12, 13].

Sustainable competitiveness firms that adopt green technologies are finding new avenues for growth while gaining a competitive edge [14, 15]. Companies integrating green technologies into their business models outperform competitors in both environmental and economic performance. Sustainability is driving eco-innovation, businesses developing products, and services that minimize environmental impacts and optimize resource use. Electric vehicle manufacturers are integrating advanced battery management technologies that extend the life of EV batteries and reduce waste [16]. Sustainable competitiveness is now a key performance indicator for industries globally as noted by [17] companies that prioritize sustainability in terms of environmental goals but also demonstrate superior long-term financial returns. Eco-innovation focuses on developing products and processes that reduce environmental impact and is central to achieving this competitiveness [18, 19]. Another study says that businesses can gain a competitive advantage by adopting environmentally friendly technologies and practices. Companies that integrate sustainability principles into their business models not only reduce their ecological footprint but also increase their market share and profitability through improved operational efficiency and innovation [20, 21]. Highlight that eco-innovation, as a key aspect of sustainable competitiveness, can drive long-term economic growth and environmental protection simultaneously.

Technological integration is essential to the realization of eco-innovation and sustainable development. As businesses integrate digital tools such as AI and IoT with green technologies, the potential for reducing emissions, improving resource use efficiency, and enhancing supply chains. References [22, 23] highlight the construction industry, where smart energy systems have helped reduce energy consumption in smart buildings by 30%, illustrating the impact of technological innovation. Faria et al. [24] explain that the integration of technologies like big data analytics, AI, and automation in renewable energy and manufacturing sectors has the potential to enhance resource use, reduce emissions, and improve energy efficiency. This integration is essential for creating a circular economy that is minimized, and materials are reused and recycled. According to Banishree et al. [25], the convergence of AI, IoT, and green energy technologies has created new avenues for smart energy management and resource optimization. The automotive industry, like electric vehicles powered by AI-driven battery management, contributes to lower greenhouse gas emissions. This level of integration helps create circular economies by reducing resource consumption and maximizing output efficiency.

Policy evaluation is critical for ensuring that green technologies and innovations are effectively motivated and adopted. Ranjan et al. [26] emphasize the need for robust policy frameworks that can not only promote the adoption of green technologies but also foster innovation. Their research shows that policies aimed at reducing carbon emissions, such as carbon pricing and subsidies for renewable energy, have been instrumental in encouraging firms to invest in eco-innovation. Zhao et al. [27] stress that well-designed environmental policies that incentivize innovation while limiting unsustainable practices are critical for driving eco-revolution and technological adoption. Evaluating the success of such policies involves assessing their impact on both environmental outcomes and technological development, highlighting the need for dynamic policy frameworks that evolve alongside technological advancements. The success of green transformation efforts also depends on policy evaluation and implementation. Effective policy evaluation frameworks, as outlined by Gunningham and Arora et al. [28], are crucial for measuring the impact of environmental policies and regulations on innovation and lowness. Scholars such as Cheng et al. [29] caution that policymakers must carefully balance the need for environmental regulation with the promotion of innovation to avoid stifling technological progress. Policies that incentivize eco-innovation and penalize unsustainable practices can accelerate the transition toward a green economy.

High-tech innovation is playing a critical role in the green transformation for the future of the energy and manufacturing industries and so on. Innovations in areas like the Internet of Things and AI-driven energy management systems are providing contribution to both environmental and economic benefits. For example, AI is improving energy storage solutions and enhancing the responsiveness of the power grids toward fluctuations in the demand for energy leading to better utilization of renewable energy [30]. High technology progress together with the success of digitalization has been forming an important influence on the processes of energy, manufacturing, and transport industries. As Chiaroni et al. pointed, "complementarities between digital tools and green technologies accelerate the speed of the green transformation further and bring more radical change beyond the technological and economic limits of what is currently feasible." Some of the innovations that are to be watched in 2024 include AI-driven scientific research, RIS in smart factories, development of climate tech, energy storage, and carbon capture systems. That implies highly innovative discoveries increasing productivity leading to solutions to severely pressing global issues such as climate change.

3 Methodology

An approach aimed at energy efficiency and carbon footprint reduction through the strategic integration of green and highend technology. First, it introduces the definition of green technology and explores all possible implications for reducing energy consumption and minimizing the imprint of the environment. The same focuses on high technology and its potential in energy minimization and low carbon emission. Then comes the fusion of green and high technologies to achieve the maximum synergies of their strengths in environmental sustainability. It leads to the implementation of CCS (carbon capture and system) as part of the strong reduction of industrial emissions of CO_2. In addition, the method is supported by a strategy to decarbonize and manage resources so that carbon-intensive processes are minimized. This research based on environmental change emphasizes the aspects of adaptation and resilience in confronting climate impacts. This finally evaluates the capture capacity of operating commercial projects, based on the practical application of such technologies in the real world toward the achievement of sustainability goals. This systematic approach combines technology and environmental strategies to respond to holistic climate challenges (Fig. 1).

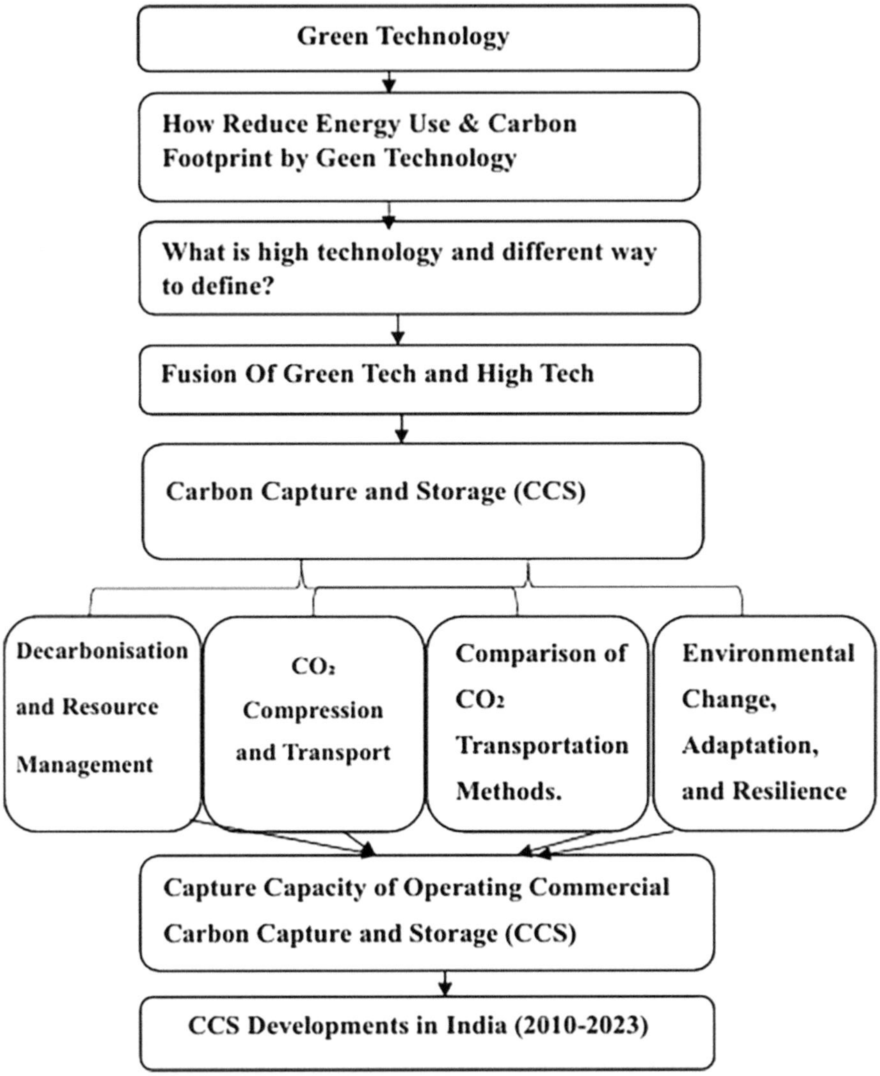

Fig. 1 Flowchart of methodology

3.1 Green Technology

Green technology (Fig. 2) can be loosely defined as a way of using technology to help make everything friendly for the environment, such as conserving energy and reducing waste and protecting the environment. Green technology pertains to any product, design, formula, algorithm, procedure, method, discovery, process, technique, idea, know-how, or software that may likely help us lessen our environmental

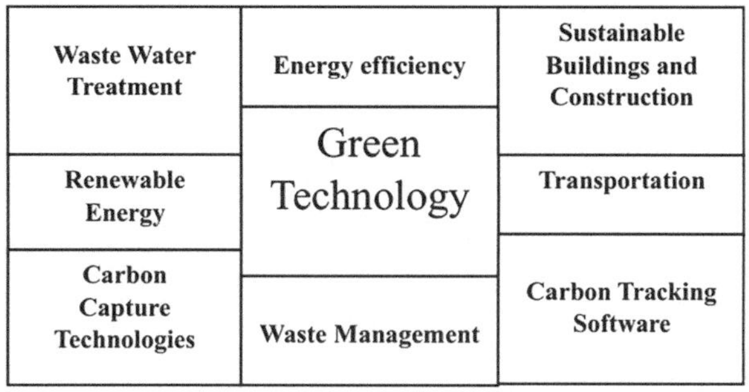

Fig. 2 Green technology

impact and eventually gain sustainable development. There are different approaches to utilizing green technology.

3.2 Renewable Energy

Renewable or clean energy is generated from natural resources that can be replenished much faster than they are used. Generally, energy generated from renewable sources and alternative fuels is highly low in carbon emissions and other air pollutants are small in quantity. The main types of sources for renewable energy are solar energy, wind energy, hydroelectric power, geothermal energy, and biomass. Most of these technologies are associated with solar panels and wind turbines.

3.2.1 Energy Efficiency

Much waste and money can be saved if monitoring and optimizing energy consumption are effectively done. This brings the vision of green technology on board: it is about the creation or improvement of products and systems that utilize less energy. Some of the examples of this technology in use are energy-efficient appliances, LED lighting, solutions in energy storage, smart meters, and energy management systems.

3.2.2 Waste Management

Reduced waste and cost can be guaranteed through tracking and optimizing energy usage. According to the vision of green technology, it emphasizes either designing or growing products and systems related to green technology like using reduced amounts of energy. The prevalence of this technology is further elaborated in energy-efficient appliances, LED lighting, energy storage solutions, smart meters, and energy management systems.

3.3 Transportation

Electric cars, trucks, and buses recharged by rechargeable batteries have helped cap more emissions of fossil fuels. Included are other low-carbon fuel technologies, such as biofuels. This will include developments in public transportation and infrastructure for electric vehicles, bicycles, and pedestrian transit in urban areas.

3.4 Wastewater Treatment

Other examples of green technology include water purification technologies that enable the reuse of water without causing harm, such as low-flow fixtures, rainwater harvesting, and advanced filtration technologies that remove contaminants in wastewater, such that contaminated water can safely be reused for irrigation or industrial processes.

3.5 Agriculture

Such smart farming and sustainability agriculture innovations help to decrease food production, although with emphasis on safety in food production. Examples include precision agriculture, through which analytics and sensors create more resource-efficient distribution, and vertical farming, which grows crops in vertically stacked layers and utilizes hydroponic systems aligned with LED lighting.

3.6 Carbon Capture and Storage

On the other hand, carbon capture and storage, otherwise known as a CCS technology, involves the capture of carbon dioxide emissions resulting from industrial processes or directly from the atmosphere and storing it permanently in such a way that it is not released back into the environment. Besides this method are various other carbon capture technologies, including direct air capture, that remove CO_2 directly from the atmosphere. This category also has come to encompass the process known as bioenergy with carbon capture and storage, the production of energy from biomass integrated with CO_2 capture and storage.

3.7 Sustainable Buildings and Construction

Green building technologies decrease the environmental footprint of built structures and activities within built structures. This can be in the form of energy-efficient design, green roofs, and sustainable building materials, such as recycled steel and bamboo. The adoption of these practices lowers energy consumption, water consumption, and waste within buildings.

3.8 Carbon Tracking Software

Carbon tracking software can track, measure, and report greenhouse gas emissions for an organization. The company then can identify hotspots of emission, determine the reduction target to be achieved, and constantly follow through the process to reach the sustainability goals. Platforms for carbon accounting, risk management systems incorporating emissions data, and supply chain management tools enhanced with environmental impact metrics.

3.9 How to Reduce Energy Use and Carbon Footprint by Green Technology

Software like Turbomeca applies AI-driven automation to three major areas of IT infrastructure optimization such as resource optimization, workload management, and automation of energy use reduction and carbon emissions. Figure 3 shows how the demand for CPU, memory, storage, and network is constantly updated on data centers and cloud environments. The application always keeps it right sized without provoking over-provisioning, wasteful usage of energy, and under-provisioning of inefficient usage. It reduces energy consumption highly by turning off or reallocation of idle infrastructure through consolidation of workloads on fewer servers or virtual machines. For workload management, green-tech software smartly allocates workloads on those servers or virtual machines that are most energy-efficient or have the best energy-to-performance ratio and dynamically schedules its workloads. It shifts the workloads to servers or the cloud regions where they obtain the lowest form of energy consumption or renewable energy sources. This platform also automatically takes energy-saving actions in various ways, such as scaling up or scaling down resources according to demand and automatic disengagement of idle resources without any form of energy wastage. Its objective ensures optimal reductions in carbon emissions by reducing the number of powered, idle, or underpowered machines. Green-tech software supports cleaner energy usage by placing workloads

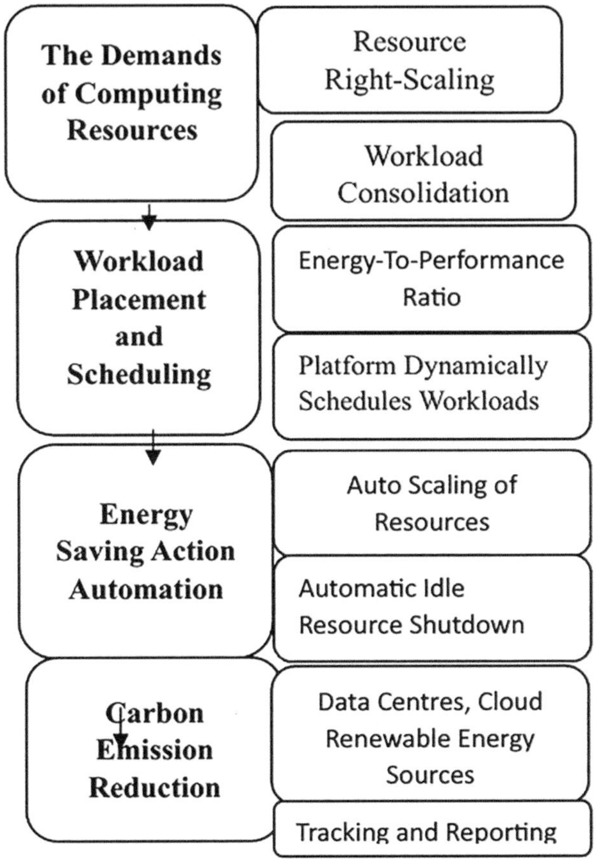

Fig. 3 How to reduce energy use and carbon footprint

in data centers powered by renewable energy. Green-tech software continuously tracks and reports energy usage and carbon footprint, hence giving IT teams insights to continually optimize and further reduce energy consumption and emissions. Green-tech software ensures the IT infrastructure in an energy-efficient way for organizations to have a reduced carbon footprint.

3.10 What Is High Technology and Different Ways to Define?

High technology called advanced technology refers to the latest technological inventions that are recognized for their complexity and novelty. It stands for one of the most developed types of technology. High-tech systems feature the latest developments in electronics, computing, AI, biotechnology, etc. Examples of advanced technologies include next-generation robotics, quantum computing systems, and AI-powered applications. The term "high-tech" complements of low technology (low-tech), which designates less complex, often traditional, manual, or mechanical technologies. Vacuum tube electronics—once the essence of current technology—are now obsolete and low-tech. Medium-tech solutions fit in the middle ground between the high- and low-tech extremes. They aim to combine the advantages of both the efficiency and adaptability of the latest systems with the simplicity, autonomy, and resilience of the old technologies. Mid-tech attempts to achieve maximum functionality and usability by reducing dependency on overly complex infrastructural features, which in turn provide higher flexibility and sustainability in a variety of contexts. High-tech solutions are usually more efficient, doing things better, and much more automatic than older, simpler technologies.

High-tech is used to refer to technology that is very complex as shown in Fig. 4, the extreme of innovativeness, sometimes even out of our reach, given the current knowledge. It is characterized by outstanding innovation, which represents the initial steps of new research findings and modern methods in the field of science and engineering. It is located at the leading edge of technological development. This kind of technology includes the most recent and most innovative products available, particularly in the market sectors like telecommunications, aerospace, robotics, and information technology. High-tech solutions often incorporate advanced automation, intelligence, and decision-making capabilities, especially through AI-powered systems, driverless cars, and smart devices. Besides, high tech is yet the same as the digital revolution and covers such things as data-driven solutions, cloud computing, and AI as the new service and industrial economies' core components. Nevertheless, the upgrading of high technology solutions very often is also afflicted with intervention-heavy infrastructures and, thus, large investments, impeding their almost exclusive use in advanced technology hubs.

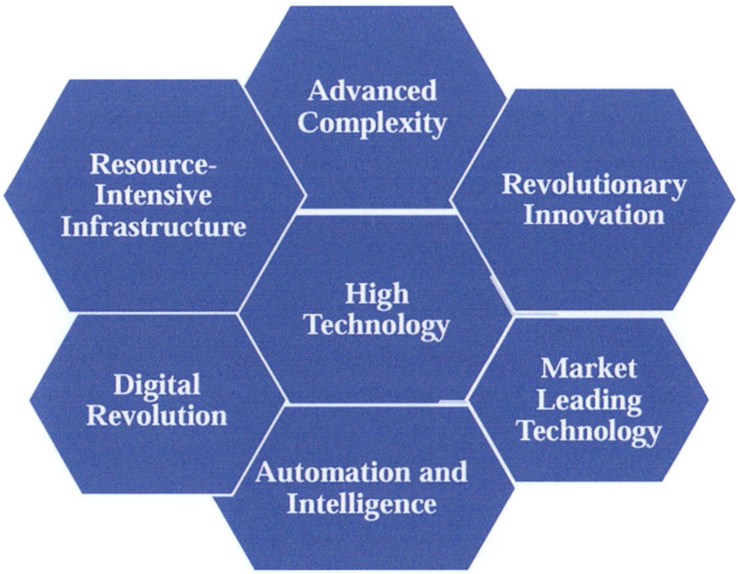

Fig. 4 High technology exploring new pathways for innovation

3.11 Fusion of Green Tech and High Tech

The integration of green tech and high tech shown in Fig. 5 serves as the nucleus for a sustainable future that sees the application of technology not just in innovation but also toward dealing with environmental issues. The integration brings about changes in industries, increased efficiency, and new growth opportunities while working to deal with urgent global challenges such as climate change, resource depletion, and pollution.

3.11.1 Enhancing Renewable Energy

. Technological innovations in the form of AI, ML, and data analytics are being implemented to make efficient utilization of renewable energy sources like solar and wind energies. Such technologies predict generation and usage patterns in advance, leading to more efficient storage and decreased reliance on fossil fuels. Smart grids powered by IoT manage real-time energy distribution, thus reducing loss and enabling a stable green supply of energy.

3.11.2 Smart Agriculture Methods

The fusion of High-Tech and Green Tech revolutionizes food production. Precision agriculture relies on AI-driven sensors, drones, and analytics to gauge soil health, crop condition, and water usage. Resource waste is avoided, environmental impact reduced, and yields maximized. The advances in farming can be made more environment friendly, thus improving food security while conserving resources.

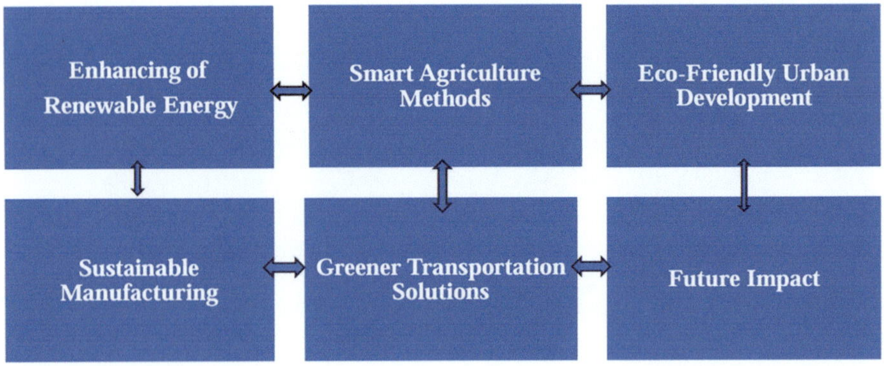

Fig. 5 Eco-friendly tech fusion

3.11.3 Eco-Friendly Urban Development

Smart cities embody the joining of Green Tech and High-Tech, hence their actualization as places that reduce environmental footprint with maximum urban living. IoT, smart grids, and AI-based systems for urban planning have contributed to lower energy use in buildings, smoother traffic flow, and reduced waste accumulation. Overall, it presents more efficient, sustainable, and habitable urban spaces and low emissions in favor of good-quality living.

3.11.4 Sustainable Manufacturing

This is characterized by the implementation of the principles of Green Tech in the manufacturing industry through robotics, 3D printing, and AI-driven optimization strategies; it means that the solutions deployed are high-tech which make the production process less cumbersome, save more emissions, have less waste, and consume less energy. It is becoming rapid that manufacturing industries move toward sustainable manufacturing by relying on High-Tech solutions to maintain their productivity without harming the environment so rapidly.

3.11.5 Greener Transportation Solutions

Electric and autonomous driving technologies, along with smart infrastructure, are paving the way for transportation tomorrow. EVs operate from renewable sources and are AI moderated to decrease carbon footprint and air pollution. At the same time, smart traffic management systems reduce traffic congestion and, consequently, reduce fuel consumption, improving urban mobility. This integration of green and high technologies is building a greener, more sustainable transport ecosystem.

3.11.6 Future Impact

More than an ideal fantasy, the combination of green tech and high tech is definitely on the way to being realized in every business or industry where this would be applicable. The sustainability technologies in conjunction with their maturity and concurrence would make an immense contribution toward ending global environmental issues, encouraging efficiency, and generating innovation. This convergence promises to reshape industries, from the energy and agriculture of yore to manufacturing and transportation, so we may walk closer toward a future in which sustainability and technological progress are as one. It leaves open the door to new research, investments, and policies supportive of further economic growth with equally responsible environmental stewardship, thereby ensuring a brighter, more sustainable world for future generations.

3.12 Carbon Capture and Storage (CCS)

Identification of sources of emissions in the carbon capture and storage process is given in Fig. 6. The identification of sources of emissions would consist of fossil fuel power plants and industrial processes that emit carbon dioxide (CO_2). First and foremost, the capture of the CO_2 may be through many techniques: precombustion capture removes the CO_2 before firing the fuel, and postcombustion capture removes the CO_2 from the flue gas after firing, or oxy-fuel combustion, whereby fuels are combusted in pure oxygen to produce a CO_2-rich exhaust. It will be compressed to high pressure to enable transportation. Further compression and transportation through pipelines, ships, or trucks take the CO_2 to storage sites. Storage usually happens through geological methods: injection of CO_2 deep into formations that are depleted oil fields and gas fields as well as deep saline aquifers. There should be continuous monitoring and verification to ensure that there is no leakage of CO_2. Fourth, there is a step for the utilization of the captured CO_2, for example, enhanced oil recovery (EOR) or the conversion to useful products. Lastly, end-of-life

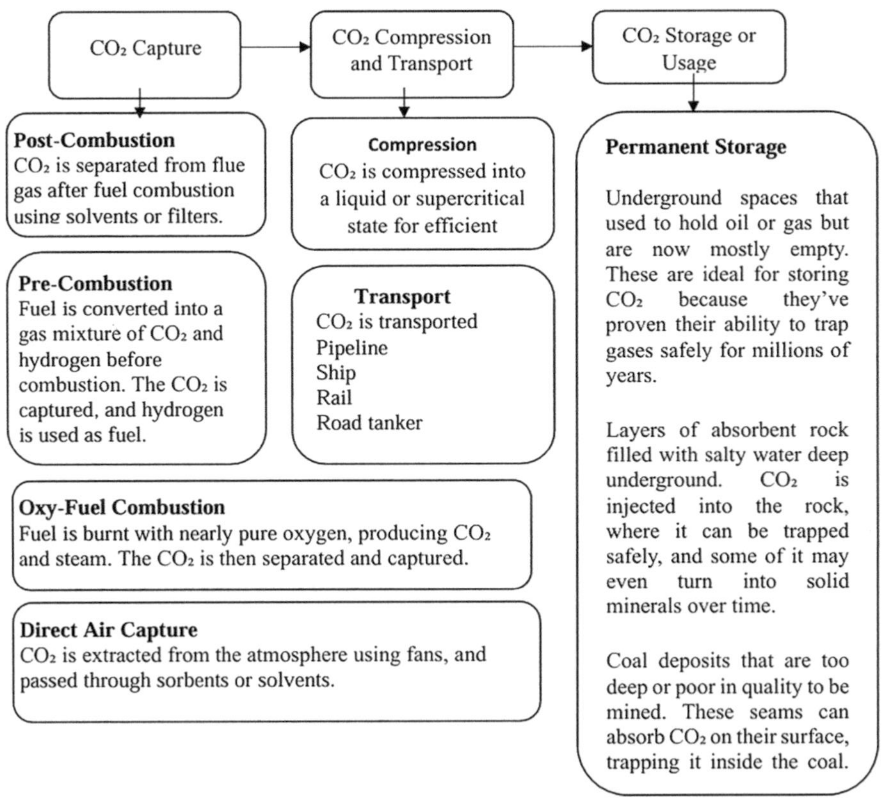

Fig. 6 CCS process flowchart

management must consider long-term integrity at storage sites, capture any leakage, and further ensure the long-term stability of stored CO_2.

One of the world's leading organizations for geoscience, the British Geological Survey provides high-quality geological research for the benefit of the government, society, and the scientific community. Its added focus is to add insight into earth and environmental processes through objective, authoritative geoscientific information, and advice. BGS supports responsible resource development and sustainable management as the UK's leading source for geoscience. In line with its BGS Strategy 2023–2028, "Understanding Our Earth," the organization has identified three key challenge areas and four core scientific priorities.

3.12.1 Decarbonization and Resource Management

Decarbonization and resource management is a key challenge area that addresses the urgent need to reduce carbon emissions in energy and industry while optimizing resource use. The need to reduce carbon emissions immediately puts this challenge under intense pressure from global competition for resources and meeting the ever-increasing necessity to expand access to affordable, sustainable energy worldwide. Safe, efficient CO_2 storage solutions for the UK, Europe, and further afield will be the focus of the CCS topic, providing just what industry, power generation, transportation, and heating need to move toward net zero.

A pilot CO_2 storage development is designed and operated based on advanced technologies and methodologies in order to ensure the safe and effective storage of CO_2. The process in Fig. 7 starts with the identification of suitable sites for geological storage that can permanently contain CO_2. This process continuously monitors site behavior before, during, and after injecting CO_2 to predict plume evolution. Such monitoring is essential to understand how the injected CO_2 plume moves through the system and how the geological formation behaves. Management of pressure forms a major part of the optimization of storage, as it involves the careful investigation of the allowable pressure buildup within the storage formation. Effective injection techniques and controls of pressure have to be used to prevent

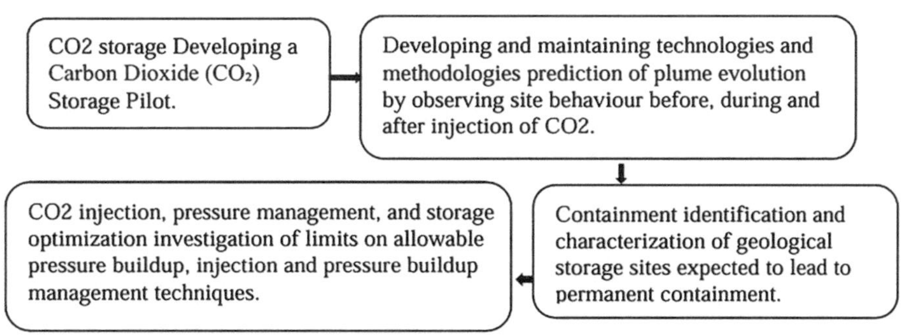

Fig. 7 Decarbonization and resource management

leakage and ensure that the injected CO_2 is maintained under long-term containment.

It is technically feasible to capture CO_2 directly from large sources of emission, such as power plants, natural gas processing plants, and industrial operations. It is also theoretically feasible to capture CO_2 from the open atmosphere. There are primarily three techniques that are used to capture or "scrub" CO_2 in power plants that burn fossil fuels: postcombustion, precombustion, and oxyfuel combustion.

Postcombustion is one kind of CO_2 capture that takes place after the fossil fuel has been combusted. In this process, the CO_2 is captured from the flue gases generated during the combustion process. This method can be integrated into most current power plants and therefore is very flexible. It has already been used in other industries, and technology is mature in such a system, though it will be more expensive due to its energy intensity.

Precombustion captures CO_2 before the fossil fuel is burned. Precombustion captures the CO_2 that is emitted from the end of a partially combusted fossil fuel in a "gasifier" to create a mixed stream called synthetic gas or syngas. In pre-combustion processes, the gas mixture can be further purified to remove CO_2 allowing the remaining hydrogen to be used as a fuel source. While pre-combustion is more cost-effective than post-combustion, it is not suitable for retrofitting older power plants. (Anand, 2015). It is more often applied in fertilizer and chemical processes, and in some power generation methods.

Oxyfuel combustion burns the fossil fuel in pure oxygen, rather than air. The flue gas will then mainly consist of CO_2 and water vapor. Upon cooling, the latter condenses, so that nearly pure CO_2 is obtained ready for transport and storage. This technique is considered to be highly efficient, with near-complete capture of COs, which has earned some applications for the term "zero-emission." There are three conversion methods, but these require different quantities of energy. Pure oxygen, especially that necessary for combustion, is relatively energy intensive and the most energy consuming of all three. CO_2 can also be dissolved in water during condensation, and treatment must be rendered before further use or storage.

Hydrogen generation (Fig. 7) is considered a way of cutting further emissions of carbon dioxide in areas where power from batteries in the form of grid power is less practical, such as heavy goods vehicles, trains, gas appliances, and domestic heating. The cleanest production of hydrogen comes from water electrolysis using renewable energy sources. These result in extremely low CO_2 emissions, and the process is expensive at present. Methane reformation is an older but a cheaper technique for the generation of hydrogen from natural gas. However, this yields CO_2 as a byproduct, which needs to be captured and sequestered to avoid its release into the atmosphere (Jin et al., 2021). Another emerging technology is DAC, which directly removes CO_2 from the atmosphere. Although currently still in the research and development phase, DAC will have less energy input than CO_2 capture from large sources of emission. It is more costly but will capture those CO_2 sources that are diffuse and hard to reach (Fig. 8).

Fig. 8 Oxyfuel combustion plant. Source https://www.bgs.ac.uk/discovering-geology/climate-change/carbon-capture-and-storage/

3.12.2 CO2 Compression and Transport

Once CO_2 is captured, it needs to be transported to suitable storage sites. Pipelines are the cheapest and most reliable way of transporting CO_2. This technology is already well established with 5800 km of CO_2 pipelines already installed in the United States, mainly to carry CO_2 into oil fields for EOR. In EOR, CO_2 is injected into oil reservoirs to enhance oil production. Building pipeline networks and increasing capacities of large-scale carbon capture and storage industry will need additional safety studies, particularly in densely populated regions and areas prone to earthquakes, where the risks of pipeline failure might be greater. Second, CO_2 can be transported using ships and road tankers; however, these modes are more suitable for small-scale applications. There are two major mechanisms of carbon dioxide storage, which are deep geological storage and mineral storage. Relevant technologies in both include enhanced oil recovery (EOR) and enhanced coalbed methane (ECBM).

Deep Geological Storage (Geo-sequestration): The process involves injecting and storing carbon dioxide in underground geological formations. This is achieved by converting CO_2 into a high-pressure, liquid-like state, known as supercritical CO_2, which is then injected into porous rock formations such as depleted oil or gas fields, or saline aquifers. CO_2 becomes trapped physically through impermeable layers of rock, known as caprocks, and chemically through reactions with minerals in the rock (Guy, 2011).*Mineral Storage (Mineral Carbonation):* A process in which CO_2 reacts with naturally occurring minerals such as iron (Fe), magnesium (Mg), or calcium (Ca), resulting in stable carbonate minerals. This simulates the natural process of weathering, which ties CO_2 irreversibly (Kovacic et al., 2024).

These reactions are naturally slow and speed-up processes require energy input for heating and pressure, making EOR CCS plants energy intensive.

Enhanced Oil Recovery (EOR): This involves CO_2 injection into declining oil fields to enhance oil recovery. This is a known technology wherein the cost of CO_2 injection is recovered mainly through the sale of the oil recovered, but the benefit of CO_2 storage is reduced because the recovered oil, when combusted, produces more CO_2 emissions (Wang et al., 2023).

Enhanced Coal-Bed Methane (ECBM): CO_2 is injected into uneconomic coal seams in a manner whereby coal degasses methane (CH4), which is captured, and commercialized methane sales partially recover storage costs, but burning methane releases CO_2 that negates some of the stored CO_2 benefits.

Saline Aquifers: These are ground water-filled reservoirs in rock containing salt-water (brine) dissolved within the pore spaces of the rocks. There is a lot of storage capacity here due to their abundance, but they are not as well understood as oil fields, and there are no useful byproducts to help offset storage costs. These are the best-developed methods currently for capturing and storing CO_2 to mitigate human-induced climate change, each with its unique advantages and disadvantages. Saline aquifers provide a significant potential for CO_2 storage due to their abundant capacity. Their geological characteristics are not as well understood as those of oil fields, and the lack of economically valuable byproducts complicates the mitigation of associated storage costs.

3.13 Comparison of CO_2 Transportation Methods

CCS transport of CO_2 can be carried through pipelines, ships, trucks, and rail with suitable operational conditions and capacity limits (Table 1). With over 50 years of commercial experience, in pipelines, it is the best-established method and operates at pressures between 48 and 200 barg (unit of pressure measurement that stands for bar gauge) and temperatures of 10–34 °C, in vapor or dense phase. Pipelines can transport around 110 $MtCO_2$ every year for extremely long distances. They have high capital costs but very low operating costs, compared with other modes. In the food and beverage industry, ships available for CO_2 transport today are operating between 7 and 45 barg and temperatures down to −52 °C. Liquid CO_2 can be transported by such ships with more than the capacity to transport 3 $MtCO_2$ a year. Ships are a flexible route and volume mode, but are more expensive to operate with much less capital intensity than pipelines. Trucks and rail transport liquid CO_2 under pressure and low temperatures and have capacities of over 1 $MtCO_2$ per year for each one. While trucks can carry 2–30 tonnes per batch, they are not very economical for large-scale CCS projects and experience gas boil-off, losing around 10% of the load. Rail transport, little used so far for CO_2, offers advantages for medium and long distances, assuming there is loading/unloading infrastructure and suitable

Table 1 Comparison of CO_2 transportation methods

Transport method	Conditions	CO_2 phase	Current capacity	Remarks
Pipeline	48–200 barg, 10–34 °C	Vapor or dense	More than 110 $MtCO_2$/yr. 9500 km of pipeline	Requires high capital costs and relatively low operating costs in comparison with other methods
				The cost per unit of CO_2 is higher in low-pressure piping transport than with dense-phase transport
				There are more than five decades of commercial experience with this mode, so it has been proven in practice
Ship	7–45 barg, −52 to 10 °C	Liquid	More than 3 $MtCO_2$/yr.	Ship transport has higher operating costs but lower capital costs than pipelines
				It can be applied to the small-scale applications such as the food and brewery industries among many others
				It gives flexibility in transport volumes as well as routing options
Truck	17–20 barg, −30 to −20 °C	Liquid	More than 1 $MtCO_2$/yr.	2–30 tonnes of CO_2 are transported per truck batch
				This is not a cost-effective approach for large-scale CCS projects
				Boil-off gas results in a loss of about 10% of the carbon dioxide transported in each shipment
Rail	7–26 barg, −50 to −20 °C	Liquid	More than 1 $MtCO_2$/yr.	No large-scale, operational CO_2 rail transport system exists
				It requires specific infrastructure for the loading, unloading, and storage
				It is only possible if existing rail lines already exist
				Rail transport is well suited for medium to long distances

Source Global CCS Institute (2023)

existing rail lines available [30]. All these methods play critical roles in facilitating various uses of CO_2, including permanent storage, fertilizer production, and enhanced oil recovery.

3.14 Environmental Change, Adaptation, and Resilience

Earthquakes, landslides, and rockfalls pose significant natural risks to economic growth, infrastructure, and community well-being, especially in developing countries. Climate change is likely to increase the frequency and intensity of disasters. Profound understanding involves analyzing hazards, exposure, vulnerability, and risk. As potential consequences to populations, infrastructure, and societies are characterized, the planning for and response to disasters can be greatly improved—especially in areas facing continual threats to communities. Decision Making Assessing single and multi-hazard events, and an appreciation of the vulnerability of exposed populations, is critical to effective risk mitigation (Adomako and Tran, 2024). Developing science-based, usable solutions that are implemented around the globe in managing disaster risks is envisioned. Managing natural hazards and risks requires an integrated approach that includes real-time data analysis, effective communication, rapid response to events, and building resilience. These activities are basic to the governments and communities prone to hazardous events because they form the core framework of risk management policies associated with natural hazards.

4 Results and Discussions

4.1 Capture Capacity of Operating Commercial Carbon Capture and Storage (CCS)

Worldwide capture capacity of operational commercial Carbon Capture and Storage (CCS) facilities by year, spanning 2010 to 2023. The operational commercial carbon capture and storage (CCS) has significantly spread around the world, from 2010 to 2023. It explains a rise in efforts to combat climate change through a reduction in carbon emissions. In general terms, the capture capacity of CCS facilities around the world from 2010 to 2023 is shown in Fig. 9.

From 2010 to 2015 in Table 2, global CO_2 capture capacity increased modestly, starting with around 30 million tonnes per annum (Mtpa) in 2010, with only a few commercial CCS projects in operation. By 2015, this (Fig. 8) had risen to approximately 40–45 Mtpa, largely driven by projects in gas processing and chemicals, particularly in North America. Between 2016 and 2020, global capacity reached 40–50 Mtpa, with notable large-scale projects like Boundary Dam in Canada and Petra Nova in the United States coming online. However, by 2020, the global capacity remained at about 40–45 Mtpa, as delays and cancellations of some projects, such as Petra Nova, offset new developments. Interest in CCS surged between 2021 and 2023, with investment pushing global capture capacity toward 50 Mtpa by 2021. By 2022, operational capacity had risen to over 60 Mtpa, spurred by commitments from companies and governments to meet net-zero targets. In 2023, global CCS capacity was estimated to have exceeded 70 Mtpa, driven by numerous projects under construction, in North America, particularly the United States, leading

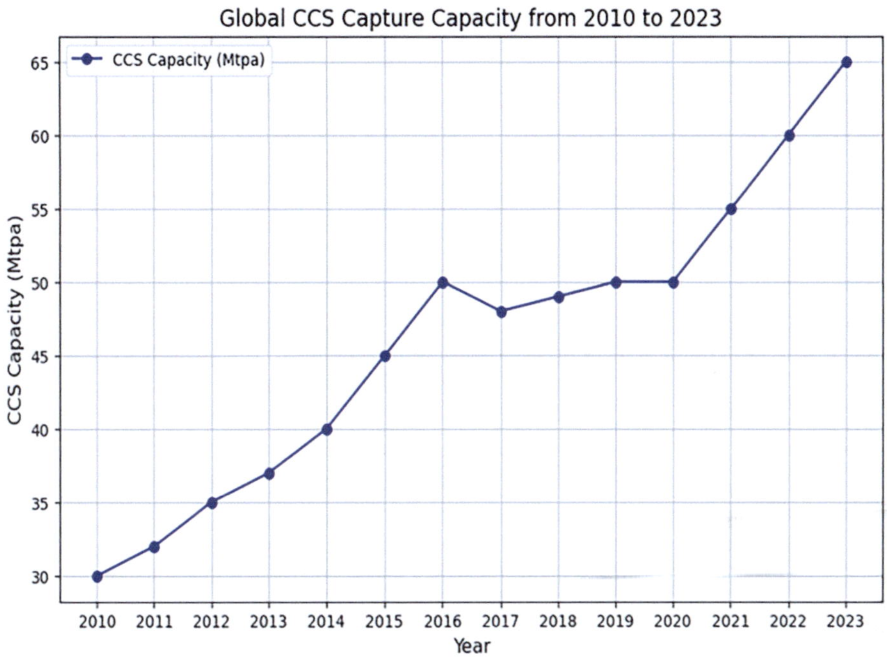

Fig. 9 Commercial carbon capture and storage (CCS) year-wise

Table 2 Capture capacity of operating commercial carbon and capture and storage (CCS)

Year	CCS capacity (Mtpa)
2010	30
2011	32
2012	35
2013	37
2014	40
2015	45
2016	50
2017	48
2018	49
2019	50
2020	50
2021	55
2022	60
2023	65

the way, followed by Europe and Asia. This growth reflects the growing interest in CCS as an important medium for making decarbonization happen across industries as diverse as energy, cement, steel, and chemical manufacturing. Actual growth rates might rise with additional adoption of CCS by more countries within their climate strategies.

4.2 CCS Developments in India (2010–2023)

Before 2015 India's focus on CCS had been to a considerable degree in the developmental and research stage, with the institutions of the IITs, TERI (The Energy and Resources Institute), and government agencies identifying potential CCS locations and designing capture technologies. In 2016–2020 a handful of pilots had emerged, mostly in CCU rather than CCUS, which generally involves the use of captured CO_2 in some form of industrial processes, such as urea production or EOR. Tata Steel in 2021: Tata Steel inaugurated a carbon capture facility at its Jamshedpur facility. This is capturing 5 tonnes of CO_2 per day, thus making it the first full-scale CCS at a steel plant anywhere in the world. In 2020–2023, the focus on CCS has been more pronounced within India's climate policy debate. This increased sharply as part of India's net-zero by 2070 target. Through National Activities, the Indian Government launched a discussion for the inclusion of CCS technologies in several climate change mitigation programs. But to date, no sizeable operational plants have been built. Private Sector Indian companies like Tata and Indian Oil have examined carbon capture technologies as part of their sustainable low-carbon strategy. Challenges and priority areas' barriers to deployment in India are the high cost of the technology and the lack of clear financial incentives. Regulatory and Storage Issues: India does not have specific legislation and geological surveys that pinpoint appropriate storage sites (depleted oil and gas fields) for CCS at scale. Carbon Capture Utilization (CCU) in India that captured carbon for permanent storage has shown more interest in various kinds of industrial applications using captured carbon.

5 Limitations and Finding

Research on the fusion of green and high-tech innovation for sustainable growth usually encounters a number of constraints that affect its strength in the conclusion made. One such critical challenge is the absence of availability of deep data sets, especially in developing economies where there may be inadequate resources available. Most green innovation and high-tech industries' technologies are still in the developmental stages, which hampers their effective assessment in determining long-term impacts and scalability. Variability in regulatory frameworks and environmental policies from one country to the other could further limit the generalizability of findings. Integrating knowledge and methodologies from diverse fields—environmental science, engineering, and economics—could be inherently interdisciplinary and present complications in developing cohesive frameworks. Market dynamics also raise the stakes, due to fast-paced changes in consumer preferences and technological advancements that can make research findings less relevant quickly. The economic considerations do bear a lot of weight, in the case of smaller and medium-sized initiatives that are not equipped to access all the resources needed in order to implement green technologies. Social acceptance is also of significant importance. Public opinion and popular sentiments shape behavior pretty profoundly, then usually, quantifying and directing those things becomes pretty difficult. The medium- and long-run environmental effects of the introduction of new technologies remain unknown and often dependent on local conditions. Specific industries or regions can bind the applicability of conclusions in research areas surrounding this topic. The temporal factors within these studies may not account for long-run effects, thus leading to inconclusive appraisals of sustainability outcomes. Overall, these are the weaknesses that point out the complexities of judging the integration process between green and high-tech innovation and indicate the demands for continuous study and adjustability in this highly dynamic field.

6 Future Work

Future work in the integration of green- and high-tech innovation for sustainable development should converge on several key themes as an approach toward increasing understanding and effectiveness. Critical extended studies to monitor the long-term impact of implemented technologies and practices across different sectors are necessary. This will, arguably, establish a more critical understanding and sustainability outcomes and the economic viability of the applied technologies and practices. Future research should also focus on further-developed, integrated databases gathering data from a wide variety of sources in emerging economies, as this would facilitate more robust comparative analyses and thus enhance the generalizability of findings and improve benchmarking of green technologies and practices around the world. Interdisciplinary collaboration will also play a major role in the further advancement of the discipline. For instance, encouraging collaboration between environmental scientists, engineers, economists, and social scientists may help researchers develop more holistic frameworks to address the multiple challenges sustainability poses. Innovations should focus on co-creation with stakeholders in the process, including businesses, policymakers, and communities, so that innovations developed truly meet specific local needs and conditions. Besides, research in the social aspects of sustainability, such as public perception, behavior change, and community engagement, should be done. Effective communication of the benefits of green technologies and change in public attitude will therefore be an integral part of the eventual successful implementation. Research should focus on developing policy-oriented analysis in relation to findings regarding the efficiency of different regulatory frameworks and incentives for green and high-tech innovations. This would extend to how policies can adapt to changing technological landscapes, not only ensuring environmental sustainability but also economic growth and social equity.

The directions mentioned above will help researchers address these areas in future work to contribute significantly to the ongoing discourse on sustainable development, thereby achieving a better future with more effective strategies by synergizing green and high-tech innovations.

7 Conclusion

In conclusion, though CCS operations in India are at a very preliminary stage, the initiatives constitute a critical step toward integrating CCS technologies into the overall strategy to decarbonize the nation. The country has acknowledged this urgent need to address climate change, though large-scale commercial carbon storage is yet to be fully accomplished; the emphasis on pilot projects has an eagerness to start being in search of new solutions for capturing and using carbon. Thus, providing the highest industrial usage of captured carbon, India would be in line with sustainable development goals, also ensuring economic opportunities through the establishment of new markets for carbon-based products. Not only does CCS simply reduce emissions, but it also can help improve energy security and nurture such technological developments that could spur economic growth. Balancing developmental aspirations and environmental responsibilities in India will depend on the success of full-scale CCS implementation toward reaching its climate change targets. Strong policy frameworks with increased investments in research and development and public–private partnerships will drive this transition. Summarizing, while the journey of CCS in India is just beginning, there is hope for a change in the country's energy landscape. It will be important to continue to innovate and scale up efforts to support a full CCS strategy that contributes significantly toward greenhouse gas mitigation, especially to ensure an enduring future and mitigate pressing needs related to the challenges of climate change.

References

1. Chen, L., & Zhou, Z. (2024). The role of AI in green transformation and energy management systems. *Journal of Green Technology, 19*(3), 145–162.
2. Bao, Z., Lin, Z., Jin, T., & Lv, K. (2024). Regional breakthrough innovation change strategies, ecological location suitability of high-tech industry innovation ecosystems, and green energy. *Energies, 17*. https://doi.org/10.3390/EN17163938
3. Zhang, L., Sun, H., Long, P., Sun, H., & Chen, Z. (2024). Do green finance and hi-tech innovation facilitate sustainable development? Evidence from the Yangtze River Economic Belt. *Economic Analysis and Policy*. https://doi.org/10.1016/J.EAP.2024.02.005
4. Liang, K., Wu, P., & Zhang, R. (2024). Research on the evaluation of regional scientific and technological innovation capabilities driven by big data. *Sustainability., 16*. https://doi.org/10.3390/SU16041379
5. Li, H. (2024). Enhanced group decision-making framework for financial performance evaluation of high-tech enterprises under interval neutrosophic environment. *Journal of Intelligent & Fuzzy Systems., 46*, 6249. https://doi.org/10.3233/JIFS-237196

6. Wang, Y., Wu, Y., Zhang, Y., & Zheng, L. (2024). Research on the coupled coordination and prediction of technological innovation and ecological environment development in coastal regions of China. *Applied Sciences., 14*. https://doi.org/10.3390/APP14114767
7. Zhang, L., & Wang, X. (2024). AI, big data, and green technology: Innovations for a sustainable future. *Sustainability Technology Review, 22*(2), 233–250.
8. Li, Q., Chen, Y., & Zhang, H. (2024). Sustainable competitiveness and green technology integration in modern industries. *Journal of Environmental Economics, 32*(2), 202–215.
9. Mutha, R., Pathak, D. N., Ahuja, V., Bhandari, P. R., Rahi, P. and Mamodiya, U. (2023). *Medical image fusion solid works supporting multiple techniques with feature-level transforms*. 2023 International Conference on Sustainable Computing and Data Communication Systems (ICSCDS), Erode, India, pp. 937–942, https://doi.org/10.1109/ICSCDS56580.2023.10105129.
10. Wu, D., Wang, Y., & Li, C. (2024). Policy frameworks for the adoption of green technologies: A comprehensive evaluation. *Journal of Green Policy and Innovation, 29*(1), 54–73.
11. Rahman, F., & Lee, S. (2024). Industrial resilience through green technology: A case study on AI-enabled energy management. *Journal of Industrial Sustainability, 14*(2), 89–108.
12. Adomako, S., & Tran, M. (2024). Exploring the effect of R&D support, green technology transfer, sustainable innovation. *Sustainable Development., 32*, 4758. https://doi.org/10.1002/SD.2936
13. Weng, J., Zhang, Q., & Liu, M. (2023). The future of sustainable competitiveness: Green technology and economic performance. *Global Sustainability Review, 21*(5), 193–212.
14. Wang, Q., Ren, J., Yan, S., Cao, Z., & Cheng, Y. (2023). Do the high-tech industrial development zones foster urban innovation? A case study of China. *Land*. https://doi.org/10.3390/LAND12040830
15. Abduvaliev, A. A., Isadjanov, A., Dadabaev, U., & Balbaa, M. (2023). Neutrosophic framework for analyzing factors of innovation in the development of Uzbekistan: Features and modern tendencies. *International Journal of Neutrosophic Science, 3*(3), 11–19. https://doi.org/10.54216/IJNS.210303
16. Ghazinoory, S., Fatemi, M., Saghafi, F., Ahmadian, A. A., & Tatina, S. (2023). A framework for future-oriented assessment of converging technologies at national level. *NanoEthics, 17*. https://doi.org/10.1007/S11569-023-00435-4
17. Dangelico, R. M., Pujari, D., & Pontrandolfo, P. (2017). Green product innovation in manufacturing firms: A sustainability-oriented dynamic capability perspective. *Business Strategy and the Environment, 26*(4), 490–506. https://doi.org/10.1002/BSE.1932
18. Scheffran, J., & Froese, R. (2016). Enabling environments for sustainable energy transitions: The diffusion of technology, innovation and investment in low-carbon societies. In *Hexagon series on human and environmental security and peace* (pp. 499–521). Springer. https://doi.org/10.1007/978-3-319-43884-9_34
19. Costantini, V., & Mazzanti, M. (2012). On the green and innovative side of trade competitiveness? The impact of environmental policies and innovation on EU exports. *Research Policy, 41*(1), 132–153.
20. Gunningham, N., & Sinclair, D. (1999). Regulatory pluralism: Designing policy mixes for environmental protection. *Oxford Journal of Environmental Studies, 23*(1), 77–101.
21. Wüstenhagen, R., Wolsink, M., & Bürer, M. J. (2007). Social acceptance of renewable energy innovation: An introduction to the concept. *Energy Policy, 35*(5), 2683–2691.
22. Porter, M. E., & van der Linde, C. (1995). Toward a new conception of the environment-competitiveness relationship. *Journal of Economic Perspectives, 9*(4), 97–118.
23. Sunori, S. K., Arora, S., Agarwal, P., Mittal, A., Mamodiya, U., and Juneja, P. (2022). *SA based optimization of controller parameters for crystallization unit of sugar factory*. 2022 6th International Conference on Trends in Electronics and Informatics (ICOEI), Tirunelveli, India, pp. 341–346, https://doi.org/10.1109/ICOEI53556.2022.9776904.

24. Ruhana, F., Suwartiningsih, S., Mulyandari, E., Handoyo, S., & Afrilia, U. A. (2024). 1. Innovative strategies for achieving sustainable development goals amidst escalating global environmental and social challenges. *The International Journal of Science in Society, 6*, 662. https://doi.org/10.54783/ijsoc.v6i1.1054
25. Sukla, B., Jakhar, R., Begum, S., Jasmine, M. M., & Sundaram, S. (2024). 2. Innovations in sustainable engineering: A review of green technologies. https://doi.org/10.53555/kuey.v30i5.3511
26. Aneja, R., Yadav, M., & Gupta, S. (2023). 4. The dynamic impact assessment of clean energy and green innovation in realizing environmental sustainability of G-20. *Sustainable Development, 32*, 2454. https://doi.org/10.1002/sd.2797
27. Zhao, Y., Shang, J., Yang, X., & Shi, G. (2024). 6. The road to sustainable development: Exploring the impact of green technology innovation and environmental regulation on the green competitiveness of heavily polluting manufacturing industries. *Journal of Environmental Planning and Management*, 1. https://doi.org/10.1080/09640568.2024.2342721
28. Arora, G. K., Taj Mahaboob, S., Adilakshmi, S., Rani, S. K., Kasthuri, A. and Mamodiya, U. (2023) *Establishment of an effective brain tumor classification system through image transformations and optimization techniques*. 2023 1st International Conference on Innovations in High Speed Communication and Signal Processing (IHCSP), BHOPAL, India, pp. 12–16, https://doi.org/10.1109/IHCSP56702.2023.10127182.
29. Cheng, M., Wang, J., Yang, S., & Li, Q. (2024). 1. The driving effect of technological innovation on green development: From the perspective of efficiency. *Energy Policy*. https://doi.org/10.1016/j.enpol.2024.114089
30. Baranitharan, B., Chandran, K., Murugesan, R., Roy, S., Dutta, P. K., Bhushan, B., & Bhattacharya, P. (2024). Towards sustainable power: A quantitative analysis of environmental factors and energy demand in different Indian states. In A. K. Somani, A. Mundra, R. K. Gupta, S. Bhattacharya, & A. P. Mazumdar (Eds.), *Smart systems: Innovations in computing. SSIC 2023* (Vol. 392). Springer. https://doi.org/10.1007/978-981-97-3690-4_10

AI-Enhanced Renewable Energy Integration

Indranil Mutsuddi

1 Introduction

Sustainability is vital for human survival in a world that is rapidly evolving. As populations increase and resources become scarcer, the need for sustainable practices becomes ever more crucial. These practices are not only essential for environmental protection but also for maintaining global economic and social stability. Adopting renewable energy sources is key to mitigating climate change impacts. By reducing dependence on fossil fuels, India can make a meaningful contribution to global efforts in cutting greenhouse gas emissions.

In this context, exploring and utilizing renewable energy sources have become increasingly important for modern corporations. Renewable energy, sourced from unlimited and naturally replenishing resources like sunlight, tidal movements, and wind, can be harnessed to generate electricity, heat and cool buildings, and water and power transportation.

Research by Rahman et al. [16] highlighted a significant relationship between renewable energy consumption, unemployment, and GDP growth in certain South Asian countries. Meanwhile, Shabbir and Cheong [17] found that, over the long term, both environmental degradation and technological innovation had a statistically significant negative effect on the adoption of renewable energy sources. According to IRENA's 1.5 °C Scenario in the World Energy Transitions Outlook, global renewable power generation capacity must triple by 2030, surpassing 11,000 GW, with approximately 90% of this growth expected to come from solar photovoltaic (PV) and wind power [11]. In response, energy efficiency policies should include setting clear targets with deadlines, implementing robust regulatory frameworks such as building codes and appliance efficiency standards, providing

I. Mutsuddi (✉)
Faculty of Management, JIS University, Kolkata, India

P. Raj et al. (eds.), *Artificial Intelligence (AI) for IT Energy Efficiency and Green AI for Environment Sustainability*,
https://doi.org/10.1007/978-3-031-89420-6_2

fiscal and financial incentives, and launching public campaigns to underscore the importance of energy efficiency, public transportation, and green mobility for achieving cost savings and collective decarbonization goals. Renewable energy is closely linked to sustainable development. Its influence on productivity and human development creates new opportunities for mitigating climate change, enhancing access to clean energy, fostering social and economic growth, reducing health and environmental impacts, and improving energy security.

In light of the urgent need to overcome the challenges associated with utilizing renewable energy resources, organizations and their R&D departments are progressively adopting digital technologies, especially Artificial Intelligence (AI) and Machine Learning (ML), to advance their efforts. This book chapter will examine emerging trends in AI-enhanced integration of renewable energy and will also discuss the current strategic and operational challenges and opportunities that organizations encounter in this domain.

2 Renewable Energy Resources: An Insight

The International Energy Agency (IEA) reports that 85% of global energy requirements are met by nonrenewable sources such as coal, gas, nuclear power, and hydrocarbons. The remaining energy must be derived from renewable sources, which are notable for their ability to naturally and quickly replenish. Often termed clean energy, these renewable sources produce minimal greenhouse gases and pollution. They are inherently natural and self-replenishing, generally resulting in a low or zero carbon footprint. Examples include wind power, solar power, bioenergy (organic matter used as fuel), and hydroelectric power, including tidal energy. The IEA [7] report highlights the critical need to transition to a net zero energy system by 2050 while ensuring that energy supplies remain stable and affordable, that energy access is universal, and that economic growth is robust. It advocates for a cost-effective and economically beneficial strategy to develop a clean, dynamic, and resilient energy economy, emphasizing renewable sources like solar and wind over fossil fuels.

2.1 *Types of Renewable Energy Resources*

India has established goals to reduce the carbon intensity of its economy by over 45% by the end of this decade, attain 50% of cumulative electric power generation from renewable sources by 2030, and reach net-zero carbon emissions by 2070. Additionally, the country aims to achieve 500 GW of installed renewable energy capacity by 2030 (Invest [9]). Table 1 presents an in-depth illustration of the various sources of renewable energy sources which may contribute effectively toward the reduction in the consumption of nonrenewable energy resources particularly those

Table 1 Types of renewable energy resources

Types of renewable energy resources	Features
Wind power	In several developed nations, including the United States, wind power is the leading source of renewable electricity. Onshore and offshore wind farms produce electricity by rotating the blades of wind turbines. India's wind energy sector, propelled by its local wind power industry, has made consistent progress. The expansion of this sector has created a strong ecosystem, improved project operation capabilities, and built a manufacturing base with an annual capacity of around 15,000 MW. As a result, India is now the fourth largest country in the world for installed wind capacity
Solar power	Solar power generates electricity by capturing sunlight with solar panels through a process known as the "photovoltaic effect" (or PV), which involves a combination of chemical and physical reactions. By the end of 2022, India was ranked fifth worldwide in solar PV deployment, according to REN21's Global Status Report 2023 and IRENA's Renewable Capacity Statistics 2023. As of June 30, 2023, India's installed solar power capacity had reached around 70.10 GW
Hydroelectric power	Hydropower is generated by harnessing the energy from flowing or falling water. Hydroelectric power plants, typically located at dams, produce electricity using underwater turbines that drive a generator. Additionally, hydropower includes wave and tidal power, which utilize oceanic forces to generate electricity at the entrances of large bodies of water, employing similar technology. Worldwide, hydropower remains the leading renewable energy source, accounting for over two-thirds of all renewable electricity. By 2021, global installed hydropower capacity had increased by 26 GW, reaching 1360 GW. In India, hydropower contributed 12.5% to the power generation mix in 2022–2023, with approximately 4745.6 MW of pumped storage capacity in operation
Bioenergy	With a significant surplus of biomass and other waste materials available, harnessing these resources for energy recovery is a viable option. Modern bioenergy is also distinguished by its multiple social and environmental advantages, in addition to supplying clean fuels. Since 2018, the Ministry of New and Renewable Energy (MNRE) has been running a program focused on harnessing energy from urban, industrial, and agricultural wastes, as well as municipal solid waste. This initiative aims to recover energy in the form of biogas, bioCNG, and power from these sources
Geothermal energy	Geothermal energy is sourced from the earth's interior, offering a dependable and continuous supply of heat. This renewable energy resource is both sustainable and environmentally friendly, as it is clean and carbon-free. Unlike other renewable sources, geothermal energy is available 24/7, unaffected by day-night cycles or seasonal changes, and does not require storage. India features seven geothermal provinces and numerous geothermal springs, with the geological survey of India (GSI) mapping the country's geothermal resources
Tidal energy	Waves on the ocean surface are generated by the movement of wind. The greater the wind's intensity and the longer it persists, the more energy is produced. In India, the theoretical potential for tidal power is estimated at 12,455 megawatts (MW), while wave power potential is estimated at 41,300 MW. However, these figures represent theoretical potential and do not necessarily reflect the amount of power that can be practically harnessed

mostly generated from fossil fuels. India has embarked upon an ambitious energy transition journey with a target of 50% cumulative electric power installed electricity capacity from nonfossil fuel-based energy resources by 2030 and achieving net zero by 2070 [12].

3 Artificial Intelligence (AI) Integration for Managing Renewable Energy

Artificial intelligence (AI) is driving progress in the renewable energy sector with applications such as predictive maintenance, energy optimization, and smart grid management. These advancements improve efficiency, reduce costs, and foster a more sustainable and eco-friendly future. By integrating AI into renewable energy research and practices, there is significant potential to enhance efficiency, reliability, and cost-effectiveness, thereby accelerating the transition to sustainable energy sources [13]. However, the intermittent nature of some renewable energy sources presents challenges for grid integration and energy management. Chawla et al. [4] noted that AI and robotics advancements should align with sustainable development goals to support climate action and address privacy issues—areas requiring further improvement in India. The global shift toward sustainable energy sources has led to a significant increase in integrating renewable energy systems (RES) into existing power grids [19]. To enhance the efficiency, reliability, and cost effectiveness of these systems, the combined use of artificial intelligence (AI) techniques has emerged as a promising approach. The inclusion of artificial intelligence (AI) techniques has significantly improved the reliability, efficiency, and sustainability of renewable energy systems (RES). Recent literature has increasingly focused on how AI-driven optimization methods can transform various aspects of RES, from resource assessment to system operation and maintenance. Despite these challenges, AI holds the potential to revolutionize the renewable energy sector by significantly boosting the efficiency and reliability of energy systems. As AI techniques continue to advance, their applications in the energy sector are expanding to include managing energy demand, achieving cost efficiency, identifying conservation measures, optimizing energy usage, and shaping policy decisions to enhance energy efficiency, promote renewable sources, and reduce emissions [1]. Energy production and distribution are undergoing a significant transformation driven by disruptive technologies such as the Internet of Energy (IoE), 5G, and artificial intelligence (AI). The IoE aims to automate and enhance the energy infrastructure, covering the entire power grid from grid operators to energy generators and distribution utilities [5]. A report by the International Renewable Energy Agency (IRENA) indicates that AI-powered energy forecasting can enhance the accuracy of renewable energy generation predictions by up to 10% [10]. AI algorithms are capable of analyzing historical data, weather patterns, and other pertinent factors to forecast energy generation from sources like solar and wind. Improved prediction accuracy facilitates better grid integration, more efficient energy management, and optimal use of renewable resources, leading to a more stable energy supply.

4 Applications of AI in Managing Renewable Energy

4.1 Enhancing Energy Efficiency

Artificial intelligence (AI) is transforming the energy sector, revolutionizing the generation, distribution, and consumption of power. From smart grid management and renewable energy forecasting to enhancing nuclear power plant safety, AI is driving a shift toward a more efficient, sustainable, and secure future for the energy industry. By monitoring energy consumption, AI generates predictive models that forecast usage patterns based on variables like time of day, weather, asset type, occupancy, and other relevant factors. These models enable AI to quickly detect and respond to anomalies in real time. AI enhances demand response management (DRM) by facilitating real-time interactions between energy providers and consumers, improving energy efficiency, reducing costs, and supporting the transition to renewable energy sources. AI optimizes demand by minimizing energy usage while maintaining desired ambient conditions, which not only conserves energy but also reduces a building's carbon footprint. When combined with IoT, AI can monitor and analyze vast amounts of data effortlessly, enabling actions that save both energy and money. At an operational level, AI can be integrated into energy efficiency strategies for municipalities, industrial, and commercial clients, assisting with energy forecasting, management, renewable energy storage, and sustainable development. It monitors, gathers information, controls, evaluates, and manages energy consumption in buildings and factories. By optimizing energy usage during peak hours, detecting potential issues, and identifying equipment failures before they happen, AI helps improve overall energy management. Its capacity to process and analyze large datasets ensures more efficient monitoring and interpretation of energy industry data, driving energy optimization. Figure 1 demonstrates how AI facilitates the detection of anomalies in energy consumption.

4.2 Building Agility

To address the growing frequency of service and supply disruptions, power generation companies must enhance their agility and resilience. This requires reducing maintenance costs while maintaining safe, reliable operations. With the demand for continuous, always-on operations, effective asset maintenance becomes even more critical. AI solutions can forecast asset and process behavior, allowing for well-planned and convenient maintenance schedules, reducing operating costs, and

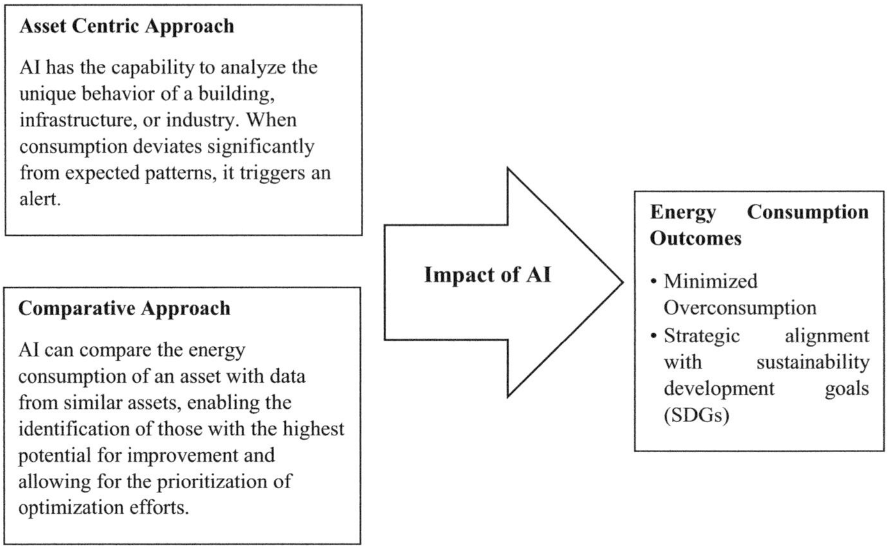

Fig. 1 Role of AI in anomaly detection for energy consumption

ensuring safety. Additionally, AI tools support regulatory compliance by enabling secure reporting, e-signing, and auditing capabilities. By streamlining maintenance and reporting processes, organizations can achieve higher levels of reliability, safety, and efficiency—thereby building AI-enabled agile systems to run operations. The aim of implementing such agile systems is to enhance an energy organization's production efficiency and reduce costs through AI-powered tools. This approach will improve operational execution, production management, and scheduling capabilities, while also optimizing process efficiencies [15].

Case Illustration: Use of Digital Technologies and AI in Qatar Power
Qatar power has successfully harnessed digital tools and AI to improve its operations for generating energy and power. By adopting AI, the company has streamlined its processes, cut costs, and enhanced the efficiency of renewable energy sources. This has led to a 10% increase in fuel efficiency and an annual savings of $1.4 million. These advancements are propelling Qatar power's shift toward a more sustainable and renewable future.

5 AI and Predictive Maintenance

AI-driven tools enable energy companies to forecast equipment failures or maintenance requirements. By analyzing comprehensive data from sources such as usage statistics, weather patterns, and historical maintenance records, machine learning

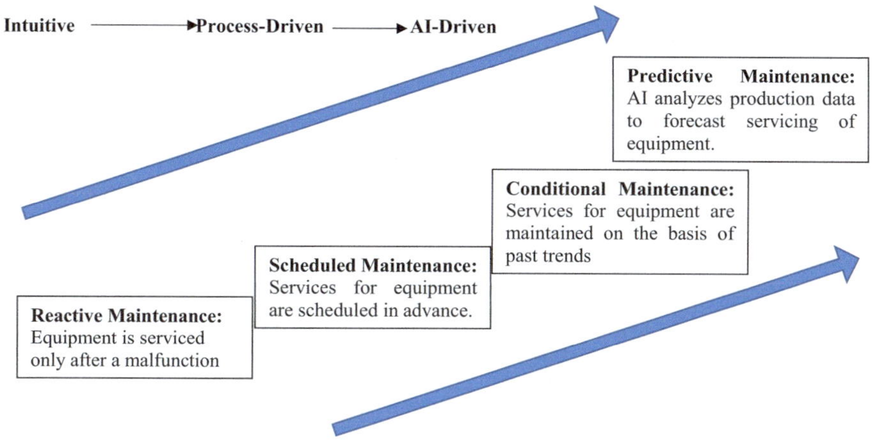

Fig. 2 Evolution of energy operations maintenance to AI-driven predictive maintenance

can predict potential breakdowns before they occur. This approach minimizes downtime, reduces repair expenses, and improves the overall reliability of energy infrastructure. Predictive AI detects patterns in energy data and uses them to notify organizations of subtle anomalies or deviations. These predictive alerts allow organizations to address issues before they escalate into major problems, well ahead of any expected performance decline. The shift from reactive maintenance practices to AI-driven predictive maintenance represents a remarkable evolution in energy management. Recently, energy transmission and generation companies have undergone a significant transformation as they adopt these advanced predictive approaches from an intuitive and later process-driven approach (Fig. 2).

AI-driven predictive maintenance depends on dependable production data from every stage of the production process. While maintenance maturity models often concentrate on equipment and machinery on the factory floor, it is essential to consider the supporting IT/OT infrastructure as well. This infrastructure collects data from the production process and transmits it to ERP systems or the cloud for analysis and processing. Predictive maintenance depends on a predictive model, with machine learning being a crucial AI discipline that offers diverse approaches for data analysis and forecasting. Supervised learning involves training a model with "labeled" data, where AI algorithms learn to map input features to specific output labels. This technique is commonly used to estimate failure probabilities based on historical data. Conversely, unsupervised machine learning works with unlabeled data to uncover hidden patterns or clusters within the dataset. These algorithms can detect anomalies or deviations from typical equipment behavior in predictive maintenance, helping to identify potential issues.

5.1 Forecasting of Renewable Energy Using AI

AI is crucial for forecasting renewable energy generation. The inherent variability of renewable energy stands in sharp contrast to fossil fuels, which can be burned as needed to meet immediate power grid demands. Solar and wind energies, on the other hand, rely heavily on weather conditions and time of day—factors that are beyond human control. Thus, for fluctuating sources such as wind and solar, AI algorithms assess weather forecasts, historical generation data, and real-time conditions. This enables energy providers to better predict renewable energy availability, leading to more effective management of supply and demand. AI-driven applications have been employed for decades to address the variability of renewable energy sources. Over time, new modeling techniques have been developed to enhance accuracy and incorporate additional features. The increasing availability of data has fueled the advancement of both physical and statistical models for renewable farm production, resulting in greater complexity and precision. AI applications, utilizing advanced machine learning algorithms to analyze large datasets—including weather conditions, historical energy usage, and real-time sensor data—provide effective solutions to contemporary challenges. These algorithms can uncover complex correlations and causations, leading to accurate predictions of energy production from renewable sources. This predictive ability facilitates precise adjustments in energy distribution, enhancing the integration of renewable sources into the grid and minimizing dependence on traditional energy sources during periods of high renewable production.

5.2 Carbon Capture, Utilization, and Storage (CCUS)

Carbon capture, utilization, and storage (CCUS) is a process that captures carbon dioxide (CO2) emissions from industrial sources and either stores or reuses them to prevent their release into the atmosphere. Artificial Intelligence (AI) represents a significant advancement in optimizing subsurface workflows. These cutting-edge tools are employed to evaluate the suitability of storage sites for carbon sequestration projects. AI helps assess critical factors such as storage capacity, containment risk, and seal integrity, particularly in the presence of faulting, providing a clearer understanding of these elements. AI improves the efficiency of CCUS processes by optimizing the capture of carbon dioxide from the atmosphere or emission sources. AI-driven systems can determine the best methods for utilizing captured carbon, whether for industrial applications or safe long-term storage. This technology is crucial for reducing greenhouse gas emissions and addressing climate change.

5.3 Demand Response Management by AI

Demand Response Management (DRM) is a critical strategy in the energy sector for optimizing electricity consumption and ensuring grid stability by adjusting consumer electricity usage to better match demand and supply conditions. AI enhances this process by enabling real-time interactions between energy providers and consumers, allowing for immediate responses to changes in energy demand. By predicting and managing demand fluctuations, AI improves energy efficiency, reduces costs, and aids the transition to renewable energy. At the core of AI's impact on demand response is its ability to analyze vast datasets. AI algorithms process extensive information, ranging from individual business energy usage patterns to broader city-wide power consumption trends. These complex data are examined to uncover patterns and correlations that would typically be missed by human analysis. One of AI's most dynamic functions in DRM is its ability to make real-time adjustments. AI algorithms can swiftly respond to shifts in energy demand, optimizing supply accordingly. Beyond simply meeting demand, AI focuses on maximizing energy efficiency. For example, during low-demand periods, AI can scale back energy use without affecting operations or comfort. During peak demand, it efficiently allocates energy from diverse sources, ensuring the grid remains stable and energy needs are met. A key advantage of AI in demand response (DR) is its ability to improve operational efficiency, which in turn drives significant cost reductions. Additionally, AI solutions are highly adaptable and scalable. They can be tailored for a variety of contexts, from managing energy at the organizational level to handling large-scale industrial operations. As grid requirements shift, AI systems can easily adjust, providing a flexible solution that evolves alongside demand. Further, AI offers substantial environmental benefits. By optimizing energy use, it reduces waste and dependence on nonrenewable energy. Through better integration of renewable energy into the grid and more efficient resource management, AI helps decrease carbon emissions and accelerates the move toward a more sustainable energy landscape.

5.4 Managing Grid Integration

Grid integration is the process of incorporating new energy generation into existing power systems. This involves analyzing complex grids to understand how they balance supply and demand while assessing the impact of integrating variable renewable energy sources. As the renewable energy sector expands, ensuring seamless and efficient integration into the grid is increasingly critical. AI plays a key role in this process, helping to forecast and manage the variability of renewable energy generation and optimizing grid operations for smoother integration. Integrating renewable energy also requires building resilience against risks like natural disasters and cyberattacks, while addressing challenges such as the intermittent availability of renewable resources—from short-term fluctuations to seasonal variability. By deploying solutions and mitigation strategies across both information and operational technology systems, energy organizations can pave the way for a cleaner, greener, and more resilient energy future. To facilitate this transition and support renewable energy integration, operators may need to rethink their planning strategies and tools to address challenges across the network and value chain. As economies decarbonize, the installed capacity of renewable energy is expected to grow significantly by 2050. This shift to renewable energy sources (RES), along with ongoing economic growth, will cause electricity demand to rise by 40% between 2020 and 2030 and double by 2050 [14].

Virtual power plants (VPPs) utilize AI algorithms to coordinate and optimize renewable energy assets such as wind turbines, energy storage systems, and solar panels. This aggregation enhances grid stability, maximizes the use of renewable energy, and accelerates the transition to a more resilient and decentralized energy system. Ensuring the reliability and resilience of the grid is crucial when managing renewable energy. AI technologies improve real-time detection and response to grid disturbances. By continuously monitoring grid conditions, AI systems can swiftly identify potential issues, predict system vulnerabilities, and take corrective actions to maintain grid stability. AI simplifies the development of advanced control systems for microgrids, which are smaller, localized energy networks often incorporating renewable energy sources and energy storage. AI algorithms optimize microgrid performance by balancing local energy generation, storage, and consumption. This localized management not only enhances energy resilience, particularly in remote or off-grid areas but also provides a scalable and sustainable model for future energy solutions.

Case Illustration 1: Role of Omdena's AI Tools for Renewable Energy Management

Artificial intelligence (AI) in the form of generative AI is unlocking new possibilities. Generative AI models can produce original content—Ranging from images and text to music—By recognizing and replicating learned patterns. This technology paves the way for enhanced creativity, content creation, and innovative applications across diverse fields. Omdena works alongside renewable energy experts and industry leaders to address critical challenges in energy efficiency, grid integration, and predictive maintenance. By leveraging expertise in AI, machine learning, data analytics, and software engineering, Omdena develops cutting-edge solutions designed to meet real-world demands.

Courtesy: https://www.omdena.com/ai-in-renewable-energy

5.5 *Energy Storage Optimization*

AI plays a crucial role in optimizing renewable energy systems and smart grids, enhancing the efficiency, sustainability, and reliability of modern energy infrastructures. AI enhances renewable energy system performance by analyzing data from diverse sources such as weather forecasts, energy demand, and equipment performance. This analysis enables optimized energy output, reduced waste, and improved overall system efficiency. Additionally, AI optimizes energy storage systems, like batteries, by ensuring that excess energy from renewable sources is stored and utilized effectively, minimizing waste and promoting a more stable and reliable energy supply. AI also matches supply with demand, optimizes energy distribution, and addresses the intermittent nature of renewable energy through the analysis of weather forecasts, consumption patterns, and real-time energy data. Energy storage, a key component of renewable integration, is optimized by AI, with machine learning algorithms using historical data to identify the most efficient ways to utilize storage resources.

5.6 *Enhancing Renewable Energy Technologies (RETS)*

The application of AI tools has significantly enhanced the design, installation, and performance evaluation of renewable energy technologies (RETs). By leveraging AI algorithms and data analytics, stakeholders in the renewable energy sector have realized a range of benefits throughout these stages of RET implementation. In the design phase, AI has revolutionized the process by enabling more accurate and efficient modeling of renewable energy systems. Sophisticated algorithms analyze vast datasets, including geographical, meteorological, and historical energy production data, to optimize system sizing, layout, and component selection. This leads to the development of more cost-effective and reliable RETs, tailored to specific environmental conditions and energy demand profiles.

5.7 *Proactive Decision Making*

Artificial Intelligence (AI) algorithms process real-time data from IoT sensors monitoring renewable energy resources, enabling swift and informed decision-making. The integration of AI with IoT allows for smarter and more adaptive control of renewable energy infrastructure, boosting the resilience and adaptability of the energy ecosystem. AI is revolutionizing renewable energy management by providing innovative solutions to the challenges of integrating renewable energy sources.

6 Challenges of AI-Enhanced Renewable Energy Integration

Data Security: AI systems depend heavily on vast amounts of data, including sensitive information such as consumer usage patterns and operational details of energy facilities. Ensuring the protection of these data from breaches is crucial. This involves safeguarding against external threats as well as ensuring ethical data handling in compliance with privacy laws and regulations. Data security in AI systems goes beyond merely protecting information; it is about upholding trust, preserving privacy, and ensuring the integrity of AI-driven decision-making processes. The responsibility extends to everyone involved with data, not just database administrators or network engineers. Every interaction with data—whether creating, managing, or accessing it—can introduce a potential vulnerability in an organization's security framework.

Complexity of Integration into Existing Systems: A major challenge in integrating AI into existing energy infrastructure lies in its complexity. Many energy systems, especially in legacy environments, utilize a mix of outdated and varied technologies. Integrating AI into these systems requires meticulous planning and may involve considerable investment in upgrading hardware and software. Ensuring that AI solutions are compatible with the current infrastructure while minimizing disruptions is a significant obstacle for energy companies looking to leverage AI's benefits.

Standardization Challenges: Standardization among the diverse components of the energy ecosystem poses significant challenges. The sector encompasses a wide range of devices, sensors, and systems, often produced by different manufacturers with varying specifications. Achieving seamless communication and integration between these elements to create a cohesive AI-driven energy infrastructure is complex. Developing standardized protocols and interfaces is crucial for fostering collaboration and ensuring the effective integration of various technologies.

Adaptability with Existing Technologies: To ensure AI systems remain effective, they must be adaptable to both current and future technologies. This involves designing AI with built-in flexibility, enabling continuous evolution. The goal is to create a future-proof infrastructure that can effortlessly incorporate new renewable energy sources while adjusting to shifts in consumption patterns and technological advancements.

Quality of AI Data: AI algorithms rely heavily on large volumes of high-quality data for accurate training and decision-making. In the renewable energy sector, obtaining reliable and comprehensive data on variables such as wind speeds, solar radiation, and weather patterns is often challenging. Incomplete or inaccurate data can lead to less effective [18] AI models and predictions. Renewable energy systems (RES) are complex and nonlinear, with numerous variables and dynamic interactions. Building AI models that accurately represent these systems is difficult, especially when integrating different renewable sources like solar, wind, and hydropower [6] into a single network. Furthermore, the intermittent and unpredictable nature of renewable energy, such as solar and wind, adds additional obstacles to AI-driven prediction and control efforts.

Real-Time Adaptability of AI Data: The capacity to learn and adapt in real-time to changing environments is a unique human trait that AI finds challenging to emulate. Human cognition enables ongoing learning and adjustment, while AI typically requires retraining and large amounts of data input to achieve similar adaptability. As the sources of renewable energy like wind, solar and hydel power are entirely dependent on fluctuating weather and climate conditions it may become challenging and difficult for the real-time adaptability of AI data which can be addressed by the logic and understanding of human decision making.

Lack of Ethics and Biasness: AI systems can reinforce and even exacerbate existing biases present in the data they are trained on. These biases can arise from multiple sources, including human error, sampling bias, or significant variability in weather data and other influencing factors. AI algorithms are trained on historical data, which can inadvertently carry biases from that data. In the energy sector, this could lead to inequitable resource allocation or biased outcomes. To prevent such issues, it is essential to design and train AI systems with fairness and equity in mind, promoting ethical decision-making and minimizing unintended consequences in energy management.

Lack of Creative Inputs: Despite the ongoing debate about AI's creative potential, it still lacks true creativity. AI systems rely on algorithms and mathematical models that enable them to identify patterns and make predictions from vast datasets. However, they are not capable of generating original ideas or concepts on their own.

HR Challenges: The limited availability of skilled professionals in both AI and the energy sector is a significant obstacle. Successfully deploying AI solutions in the energy industry demands expertise in both areas, yet there is a shortage of individuals capable of bridging the gap between cutting-edge AI technologies and the intricacies of energy systems. This talent shortfall hinders the industry's ability to fully harness AI's potential and implement innovative solutions efficiently.

Top Management Commitment: For the success of AI-enhanced renewable energy adoption, there is a need for continued support and commitment from the top management of energy service providers. There has to be a clear-cut strategic, investment-related, and policy support for the fruitful integration of AI-driven applications in the existing operations pertaining to renewable energy exploration and management. While considering this role of the top management, the examples of Schneider Electric in India are noteworthy.

Case Illustration 2: Role of Leadership at Schneider Electric

Green Yodha, a sustainability initiative by Schneider electric, brings together a community of impact makers committed to advancing energy-efficient practices and reaching both personal and corporate sustainability objectives. Green Yodhas are proactive leaders driving efforts to help individuals and organizations transition to a more resilient, electrified, and net-zero future.

Courtesy: *https://www.se.com/in/en/work/campaign/local/green--yodha.jsp*

7 Policy-Related Understanding

India is drawing significant attention for its commitment to renewable energy and its ambitious deployment targets. The country currently boasts over 35 gigawatts (GW) of wind power capacity and an additional 25 GW from solar power, contributing to a total of 60 GW out of 75 GW from renewable sources. In April 2023, the Indian government announced plans to bid for 50 GW of renewable energy capacity annually over the next 5 years, aiming to achieve 500 GW by 2030 [8]. As India works toward these goals, it faces several challenges in developing regulations that will incentivize and optimize both existing and future power generation assets, including expanding the role of renewable energy within the power generation mix across various states.

The National Institution for Transforming India (NITI Aayog), the government's policy think tank, has introduced a National Strategy for Artificial Intelligence, highlighting AI as a pivotal tool to make renewable energy adoption more cost-effective and boost the efficiency of solar and wind power sources. However, relying solely on AI in the future is not enough—India's states must begin laying the foundation now to optimize their upcoming renewable energy capacity and minimize curtailment. Machine learning can enhance predictive grid load management and maintenance for renewables, helping states strengthen their energy security. India's commitment to achieving net zero emissions by 2070 and sourcing 50% of its electricity from renewable energy by 2030 marks a key milestone in the global battle against climate change. The country is pioneering a new model of economic growth that seeks to bypass the carbon-intensive pathways previously followed by many nations, offering a potential blueprint for other developing economies [2].

India ranks fourth globally in renewable energy installed capacity, fourth in wind power capacity, and fifth in solar power capacity, according to the International Renewable Energy Agency's Renewable Capacity Statistics 2023. Analysts project that India's renewable energy capacity additions will rise to about 25 GW in 2024–25, compared to over 18 GW in the previous period (Business [3]). India's energy policy is shaped by the need to balance four key factors: a rapidly expanding economy that demands a stable and reliable supply of electricity, gas, and petroleum products; rising household incomes that require affordable and sufficient electricity and clean cooking fuels; limited domestic fossil fuel reserves leading to heavy reliance on imports of natural gas, crude oil, and coal; and the environmental impact at local, urban, and regional levels, highlighting the importance of cleaner energy sources. The transition to clean energy presents a major economic opportunity, positioning India to become a global leader in renewable batteries and green hydrogen. These and other low-carbon technologies could potentially create a market worth up to $80 billion in India by 2030.

According to the REN21 Renewables 2024 Global Status Report, India ranks fourth worldwide in renewable energy installed capacity (including large hydro), fourth in wind power capacity, and fifth in solar power capacity. At COP26, India set an ambitious target of achieving 500 GW of nonfossil fuel-based energy by 2030

(Invest [9]). In the Union Budget for 2024–2025, the Centrally Sponsored Scheme for Solar Power (Grid) was allocated INR 10,000 crores, marking a 110% increase from the INR 4757 crores allotted in the 2023–2024 budget. Additionally, the PM-Surya Ghar Muft Bijli Yojana, launched in February 2024 with a budget of INR 75,000 crores, received INR 6250 crores in funding.

8 Research Implications

8.1 Methodological Implications

The theoretical insights developed in this chapter can provide academicians and researchers with a deeper exploration of the real-world applications of AI-enhanced renewable energy integration. Conducting surveys with process leaders, stakeholders, and corporate entities, followed by statistical data analysis, can offer valuable insights into how AI is being integrated with current renewable energy generation and distribution operations. Additionally, this approach could reveal the outcomes and efficiency improvements in existing processes, contributing to the sustainable use of renewable energy resources.

8.2 Practical Implications

The key issues discussed in this book chapter can serve as valuable guidelines for policymakers and process leaders to gain a comprehensive understanding of the implications and benefits of AI-enabled processes in renewable energy management. These insights will support the integration of AI into energy processes. The study also provides research scholars, students, and academicians with practical insights into the policies set by the Government of India, helping energy corporations align with the government's ambitious sustainable energy management goals.

8.3 Social Implications

The study highlights the potential for involving social stakeholders, such as local communities, rural populations, and society at large, to raise awareness of sustainability issues. It encourages dialogue with corporations, enabling these groups to actively contribute to the achievement of Sustainable Development Goals (SDGs) by exploring, utilizing, and managing renewable energy resources more effectively.

9 Conclusion

This chapter delves into the various forms and dimensions of renewable energy resources that can complement the excessive reliance on fossil fuels, which depletes natural resources and results in environmental hazards and pollution. It emphasizes the role of Artificial Intelligence (AI) and Machine Learning in optimizing and effectively managing energy from renewable sources. By addressing potential roadblocks, challenges, and implementation issues, the chapter aims to provide valuable insights for researchers, academicians, and policymakers, helping to inspire strategic changes in renewable energy management practices, particularly for corporations engaged in renewable energy operations.

References

1. Azadeh, A., Narimani, A., & Nazari, T. (2015). Estimating household electricity consumption by environmental consciousness. *International Journal of Productivity and Quality Management, 15*(1), 1–19.
2. Birol, F., & Kant A. (2022). *India's clean energy transition is rapidly underway, benefiting the entire world.* Retrieved from https://www.iea.org/commentaries/india-s-clean-energy-transition-is-rapidly-underway-benefiting-the-entire-world
3. Business Standard. (2024). *India gets $386 bn funding for RE target.* RIL, Adani pledge commitments. Retrieved from https://www.business-standard.com/companies/news/india-gets-386-bn-funding-for-re-target-ril-adani-pledge-commitments-124091600697_1.html.
4. Chawla, Y., Shimpo, F., & Sokołowski, M. M. (2022). Artificial intelligence and information management in the energy transition of India: Lessons from the global IT hear. *Digital Policy, Regulation and Governance, 24*(1), 17–29. https://doi.org/10.1108/DPRG-05-2021-0062
5. Fowdur, T. P., & Sanghan, A. (2024). Achieving affordable and clean energy through AI and 5G powered internet of energy (IoE). In T. P. Fowdur, S. Rosunee, R. T. F. Ah King, P. Jeetah, & M. Gooroochurn (Eds.), *Artificial intelligence, engineering systems and sustainable development* (pp. 97–108). Emerald Publishing Limited. https://doi.org/10.1108/978-1-83753-540-820241008
6. He, Z., Guo, W., & Zhang, P. (2022). Performance prediction, optimal design and operational control of thermal energy storage using artificial intelligence methods. *Renewable and Sustainable Energy Reviews, 2022*(156), 111977.
7. IEA. (2021). *Net Zero by 2050 A roadmap for the global energy sector*. Retrieved from https://www.iea.org/reports/net-zero-by-2050
8. International trade Administration. (2024). *Renewable energy, India - country commercial guide*. Retrieved from https://www.trade.gov/country-commercial-guides/india-renewable-energy
9. Invest India. (2024). *Renewable energy: India is 4th globally for total renewable power capacity addition*. Retrieved from https://www.investindia.gov.in/sector/renewable-energy
10. IRENA. (2020). *Innovation landscape brief: Advanced forecasting of variable renewable power generation*. International Renewable Energy Agency, Abu Dhabi.
11. IRNEA. (2024*). Tripling renewable power and doubling energy efficiency by 2030: Crucial steps towards 1.5°C*. Retrieved from https://www.irena.org/Digital-Report/Tripling-renewable-power-and-doubling-energy-efficiency-by-2030
12. MNRE. (2024). *Bio energy overview*. Retrieved from https://mnre.gov.in/bio-energy-overview
13. Maulana, F. I., Adi, P. D. P., Hari, N. H., Hamim, M., & Lestari, D. (2024). Applications of artificial intelligence in renewable energy: A bibliometric analysis of the scientific

production indexed in scopus. *E3S Web of Conferences, 501*, 01016. https://doi.org/10.1051/e3sconf/202450101016

14. McKinsey & Company. (2023). *Global energy perspective*. Retrieved from https://www.mckinsey.com/industries/oil-and-gas/our-insights/global-energy-perspective-2023.
15. McKinsey & Company. (2024). *The State of Energy Organizations 2024*. Retrieved from www.mckinsey.com/~/media/mckinsey/industries/oil%20and%20gas/our%20insights/the%20state%20of%20energy%20organizations%202024/the-state-of-energy-organizations-2024.pdf
16. Rahman, M. R., Rahman, M. M., & Akter, R. (2023). Renewable energy development, unemployment and GDP growth: South Asian evidence. *Arab Gulf Journal of Scientific Research*. Vol. ahead-of-print No. ahead-of-print, *42*, 1044. https://doi.org/10.1108/AGJSR-04-2023-0152
17. Shabbir, M. S., & Cheong, C. W. H. (2024). Financial resources and renewable energy nexus: A holistic perspective. *International Journal of Energy Sector Management*, Vol. ahead-of-print No. ahead-of-print, 2379. https://doi.org/10.1108/IJESM-12-2023-0006
18. Shamshirband, S., Rabczuk, T., & Chau, K. W. (2019). A survey of deep learning techniques: Application in wind and solar energy resources. *IEEE Access, 2019*(7), 164650–164666.
19. Ukoba, K., Olatunji, K. O., Adeoye, E., Jen, T.-C., & Madyira, D. M. (2024). Optimizing renewable energy systems through artificial intelligence: Review and future prospects. *Energy & Environment, 0*(0), 3833. https://doi.org/10.1177/0958305X241256293

AI-Driven Fault Detection in Photovoltaic Systems: A Path to Sustainable Energy Optimization

Ismail Bogrekci, Hilal Eyinc Akkaya, Pinar Demircioglu, and Serra Aksoy

1 Introduction

Solar energy is becoming more important as a clean, sustainable, and environmentally friendly source of power. Solar panels are the primary components responsible for converting energy-laden photons from the sun into electricity. They achieve this by capturing photons in semiconductor materials, which release electrons, thereby generating an electric current. Despite their long lifespan, solar panels are not immune to faults that may arise over time due to environmental conditions and wear.

Ismail Bogrekci, Hilal Eyinc Akkaya and Serra Aksoy contributed equally with all other contributors.

I. Bogrekci
Graduate School of Natural and Applied Sciences, Department of Mechanical Engineering
Aydin Adnan Menderes University, Aydin, Turkey
e-mail: ibogrekci@adu.edu.tr

H. E. Akkaya
Graduate School of Natural and Applied Sciences, Department of Mechanical Engineering
Aydin Adnan Menderes University, Aydin, Turkey

Ugur Sogutma Inc., R&D Department, Aydin, Turkey

P. Demircioglu (✉)
Graduate School of Natural and Applied Sciences, Department of Mechanical Engineering
Aydin Adnan Menderes University, Aydin, Turkey
e-mail: pinar.demircioglu@adu.edu.tr

S. Aksoy
Institute of Computer Science, Ludwig Maximilian University of Munich (LMU), Munich, Germany
e-mail: serra.aksoy@campus.lmu.de

P. Raj et al. (eds.), *Artificial Intelligence (AI) for IT Energy Efficiency and Green AI for Environment Sustainability*,
https://doi.org/10.1007/978-3-031-89420-6_3

Manufacturing defects in solar panels, as well as faults that occur after commissioning, are critical issues that affect the efficiency and performance of PV systems. Panels with manufacturing defects can lead to failures, reduce the reliability of PV systems, and increase operational costs. Moreover, defects can cause inaccuracies in energy forecasting and pose safety risks such as electrical hazards and fires [1].

The efficiency of a PV system can be greatly impacted by faults in the solar panels and environmental conditions, making early detection and classification of these faults crucial to prevent potential problems [2].

Faults in PV systems can have significant implications for system performance and safety. These faults typically arise from component failures, environmental factors, or installation errors. Electrical faults are among the most common issues in PV systems, including open circuits, short circuits, line-to-line faults, and line-to-ground faults, each of which can lead to energy losses and safety hazards. Thermal faults, such as hot spots and thermal runaway, can substantially reduce panel performance and result in long-term physical degradation. Mechanical faults, including cracks, fractures, or loose connections at junction points, represent another category of issues. Environmental factors, such as soiling, dust accumulation, and adverse weather conditions, can negatively impact system efficiency, while corrosive environments can compromise the durability of system components. Additionally, human-induced errors, including improper installation, inadequate maintenance, or engineering misjudgements, significantly hinder the effective operation of PV systems.

Faults in PV systems can be broadly categorized into those arising from panel defects, electrical components, environmental factors, or human errors. Panel-related issues include cell degradation, shading effects, and delamination, while electrical component faults often involve inverter malfunctions, connection issues, and wiring problems. Faults related to environmental factors stem from external influences such as dust, humidity, and weather conditions. This categorization facilitates a better understanding of fault origins and the implementation of more effective mitigation strategies.

The significance of faults in PV systems is underscored by their economic and technical impacts. Faults cause energy production losses, diminishing system efficiency and complicating the achievement of renewable energy targets. Electrical faults may pose safety risks, such as fire hazards and electric shocks, while frequent issues reduce component lifespan, escalate maintenance, and repair costs. Therefore, the use of advanced FD and management technologies is critical. Methods such as thermal imaging, monitoring systems, and AI-based analyses enable rapid and accurate fault identification. These approaches ensure that PV systems operate securely, efficiently, and sustainably, supporting their role in advancing renewable energy solutions (Fig. 1).

Recent studies have focused on detecting, classifying, and localizing the electrical faults in PV panels that can be considered the most dangerous, aiming to prevent adverse effects on system performance and safety. This study focuses on new methods and works used for the detection, classification, and localization of electrical faults.

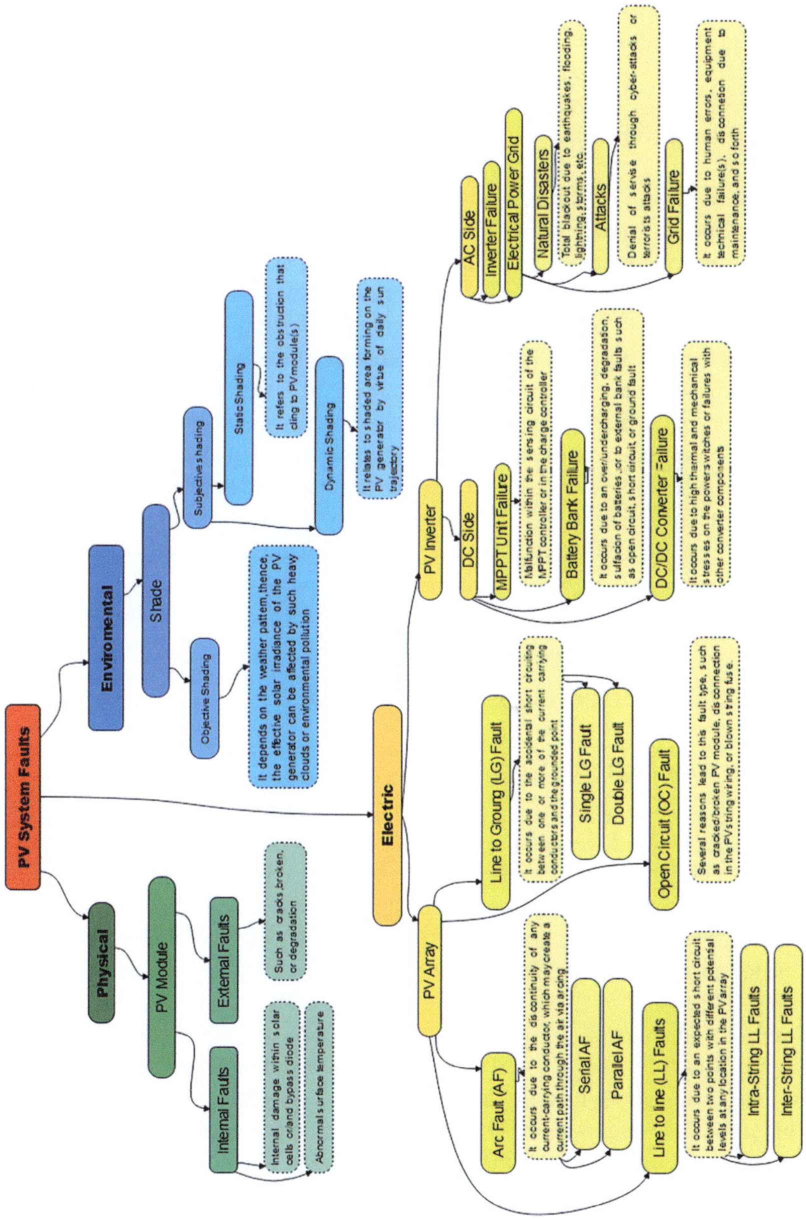

Fig. 1 Classification of PV system faults

AI, particularly in the branches of ML and DL, has played a significant role in error detection across various fields, and recently, it has also taken an active role in PV systems.

In recent years, the use of ML and DL models has gained importance in enhancing FD and classification accuracy. The most commonly used techniques include Fuzzy Logic (FL) [3], Support Vector Machine (SVM) [4], k-Nearest Neighbor algorithm (kNN) [5], and Decision Tree (DT)-based techniques (including Random Forest (RF) [6] and Naive Bayes [7]). For image-based FD and classification, methods have been developed using Artificial Neural Networks (ANN) [8], convolutional neural networks (CNNs), and deep neural networks (DNNs) [9]. Notably, advanced techniques such as the EfficientB0 model integrated with SVM classifiers and CNNs have demonstrated significant potential in achieving high-precision FD and localization in solar panels. The incorporation of AI-driven systems and the Internet of Things (IoT) into solar panel monitoring has transformed FD processes by facilitating real-time and remote diagnostics. However, several challenges remain, including the need for extensive data sets to train ML models and ensure their robustness under varying environmental conditions. Additionally, the deployment of UAVs equipped with infrared sensors for aerial inspections provides an efficient and cost-effective approach to monitoring large-scale solar farms, marking a significant advancement in the field. Ongoing advancements in these technologies are expected to additionally improve the maintenance and reliability of solar PV systems, contributing to more sustainable energy generation.

2 Methods for Detecting Electrical Faults

2.1 Infrared Thermography (IRT)

IRT emerges as a rapid, reliable, and affordable method for identifying faults in electrical systems, including PV panels. This technique involves analyzing thermal conditions by comparing thermal images captured with infrared cameras to reference images of properly functioning PV panels. By identifying thermal anomalies on the panel surface, IRT facilitates FD while offering the advantage of efficient monitoring [10].

IRT relies on the use of specialized cameras capable of high-resolution imaging (ranging from 640×480 to 1280×720), broad temperature detection ranges (-20 °C to 100 °C), and sensitivity as fine as 0.05 °C for precise anomaly identification. This approach significantly enhances monitoring processes by enabling the accurate and timely detection of thermal irregularities, thereby improving maintenance and reliability in solar energy systems (Fig. 2).

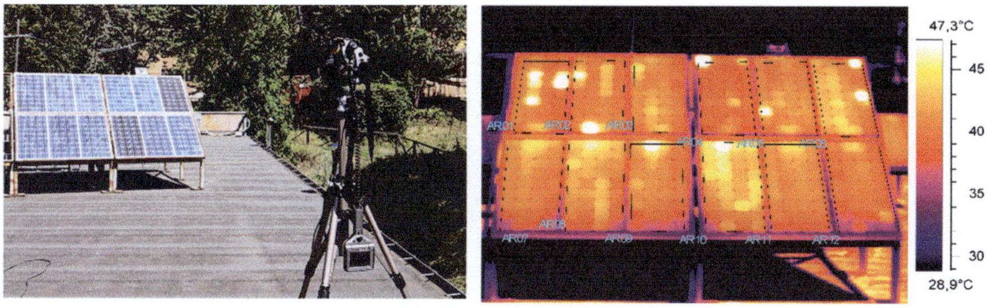

Fig. 2 Example of an IR image of the investigated panels and the photograph of the setup [11]

2.2 Signal Injection

A proposed method for detecting connection faults in PV systems involves injecting multiple frequencies into the system using a transmitter. The transmitted signals are detected by a receiver, and faults are identified based on changes in frequency characteristics. This method not only detects faults but also determines their exact locations within the system [12].

Signal injections involve transmitting multiple frequencies through the system and analyzing changes as they travel. These variations help detect and locate faults with precision. This method is particularly effective in pinpointing issues. Key parameters include a frequency range of 1 Hz to 10 kHz and a high level of FD accuracy based on signal variations.

For example, as shown in Figs. 3 and 4, when the frequency change period is 5 ms, the high side switches between 4 and 2 kHz in a repeating pattern, while the low side alternates between 1 and 2 kHz. This creates four distinct waveforms when the frequencies from both sides overlap. Under normal conditions (without any disconnection failures), the transmitter frequency is applied to both terminals of the PV system, allowing us to detect changes in these four frequencies. This process helps identify and accurately locate faults within the system.

2.3 Drone Inspections

Drones equipped with thermal cameras are utilized for aerial thermographic inspections of solar panels. This method is effective for detecting faults by measuring surface temperatures and identifying hot spots that indicate faulty panels. Drone inspections are faster, safer, and more cost-effective compared to traditional visual inspections, as conventional inspections can be time consuming and hazardous.

Drones equipped with thermal cameras provide a fast, safe, and cost-effective way to inspect solar panels. These aerial inspections identify temperature variations that indicate faults. Drones operate at altitudes of 30–100 m, with cameras offering

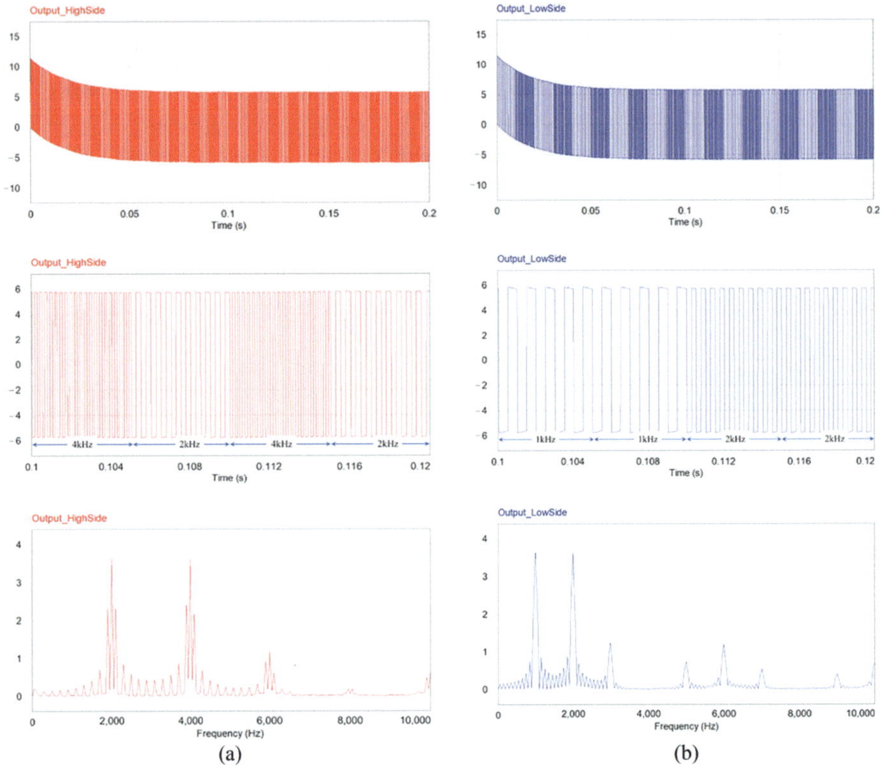

Fig. 3 (**a**) Frequency characteristics of the transmitter's high side and (**b**) frequency characteristics of the transmitter's low side [12]

resolutions between 640 × 480 and 1280 × 720. With flight durations of 30–60 min, drones are well suited for covering large areas efficiently (Fig. 5).

2.4 *Ground Fault Detection (GFD)*

GFD in PV systems is crucial for ensuring system safety and efficiency. A ground fault in a PV system occurs when electrical conductors unintentionally come into contact with the ground, resulting in energy losses, potential equipment damage, and serious safety hazards. Detecting and addressing these faults promptly enhance both system performance and user safety.

Among the methods used for GFD, GFD Interrupter (GFDI) devices play a pivotal role. GFDIs detect ground currents within the system and interrupt the circuit to protect both the equipment and the users. Voltage monitoring techniques are also commonly employed, where the voltage levels of PV arrays are continuously observed. With a detection accuracy of ±5 V, these methods enable rapid

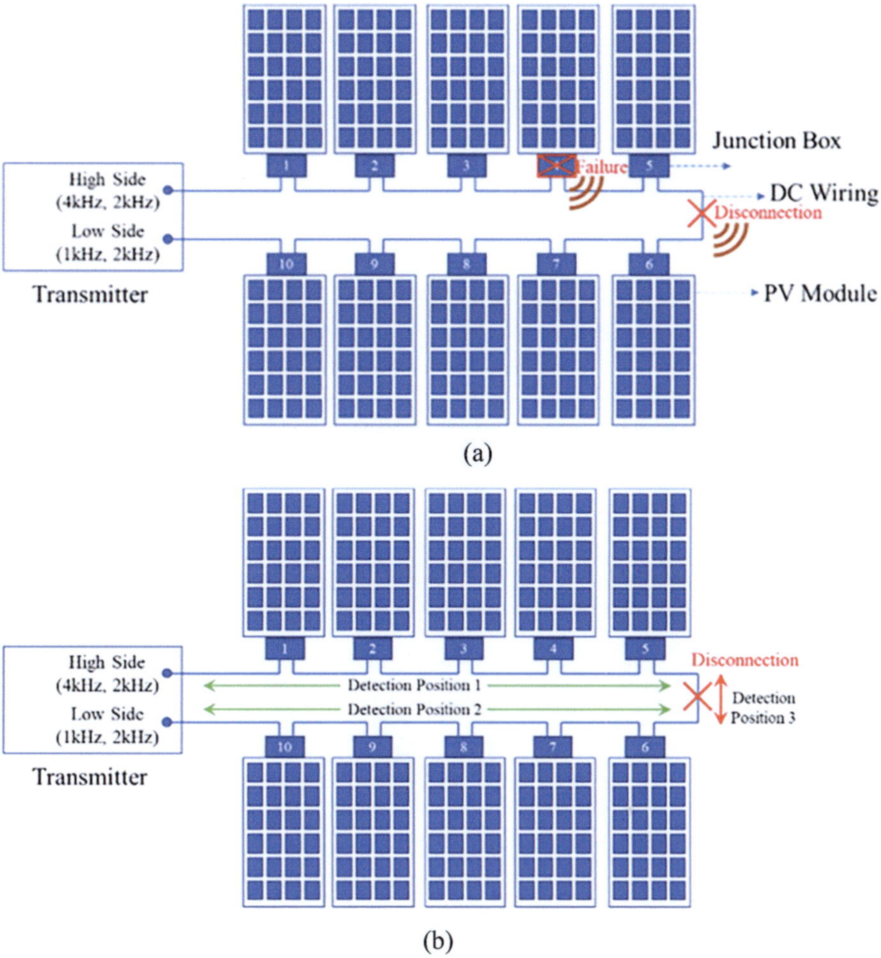

Fig. 4 (**a**) Method for detecting disconnection and (**b**) classification of the detection area

identification of faults, particularly under damp conditions where FD accuracy improves significantly. Insulation resistance measurement further contributes to FD by monitoring the system's insulation state, where a drop in resistance indicates a potential GF. For transient faults caused by factors such as moving parts, wind, or vibrations, the system is reset to observe the point at which the GFDI is triggered. This approach helps pinpoint the source of intermittent faults.

In recent years, ML and AI have become integral to ground fault detection in PV systems. Anomaly detection algorithms identify irregularities in current, voltage, and temperature data collected from the system. Classification models categorize fault types, such as LL or LG faults, while DL techniques leverage large datasets to improve accuracy, especially for detecting transient faults. Despite these advancements, GFD presents certain challenges. Identifying transient faults requires

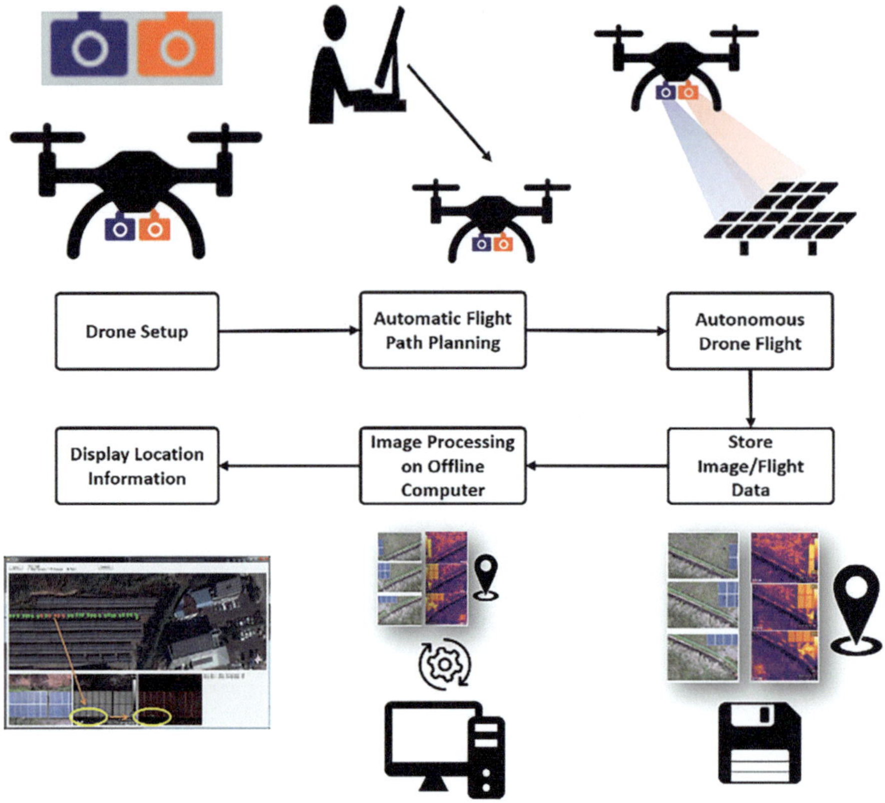

Fig. 5 Drone inspection block diagram [13]

continuous monitoring, and pinpointing the exact fault location in large-scale PV systems can be complex. Furthermore, environmental factors such as humidity and temperature fluctuations can affect detection accuracy. Nevertheless, modern monitoring techniques and AI-powered solutions offer more effective and precise GFD processes, paving the way for enhanced system reliability and safety (Fig. 6).

2.5 AI, ML, and DL

Traditional methods used for solar panel FD are generally limited to physical tests, manual observations, and basic electrical measurements. With the development of modern techniques such as ML and DL since 2009s, the FD concept has become more efficient, more sustainable, less time-consuming, and more cost-effective.

Advanced modern techniques have significantly improved FD processes in solar panels, offering powerful and accurate solutions for analyzing large datasets. The contributions of ML and AI techniques to FD in solar panels not only ensure

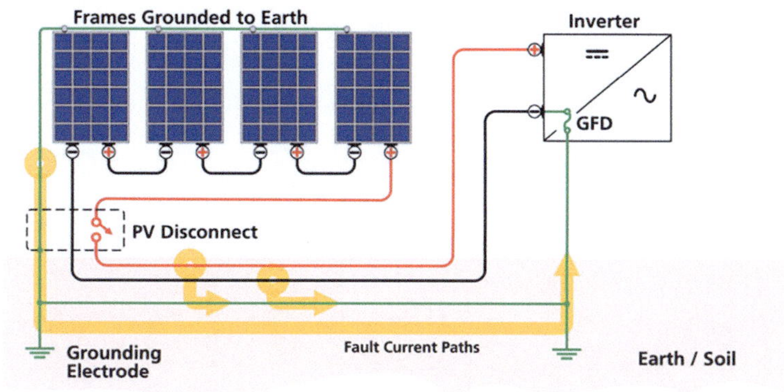

Fig. 6 Grounded PV system [14]

accurate identification but also enable the optimization of systems. Particularly, algorithms such as DL and SVM learn complex relationships and patterns when analyzing panel data, bringing previously unnoticed issues to light. These techniques provide successful results even in situations where traditional methods may fall short. Furthermore, big data analytics and AI-based monitoring systems offer a dynamic detection and diagnosis process by continuously tracking data flow. These systems can monitor real-time changes in weather conditions, panel efficiency, and system performance in solar power plants, quickly detecting anomalies. As a result, system administrators are alerted instantly without the need for manual intervention. The integration of these technologies not only enhances FD but also provides predictions and recommendations necessary for resolving faults, thereby improving operational efficiency and reducing maintenance costs.

As seen in Table 1, AI has been steadily advancing in the detection and diagnosis of faults like LLF (line-to-line fault), LGF (line-to-ground fault), SC (short circuit), OC (open circuit), and PSh (partial shading) in PV plants. Each year brings more successful outcomes, and many other studies could be added to further demonstrate the progress in this area.

AI techniques, particularly DL algorithms, are widely utilized to analyze thermal images collected by drones. This approach is especially advantageous for large-scale solar farms where manual monitoring is impractical. Popular models such as CNNs and Recurrent Neural Networks (RNNs) process extensive datasets comprising over 10,000 images, enabling real-time FD with accuracies exceeding 98%.

3 Techniques for Classifying Electrical Faults

The classification of electrical faults in PV panels holds an indispensable significance for the effectiveness and security of solar energy systems. These hidden dangers not only disrupt energy production but also pose a serious risk of severe damage

Table 1 Application of AI in FDD of PV plants from 2009 to 2024

Ref.	Year	ML method and learning type	Fault types	ID and classification	Accuracy (%)	PV system (kW)	Complexity and tools	Prototype and dataset
[15]	2009	Supervised BPNN	Anomalies	Identification	99	GCPV	Medium/Matlab	No/–
[16]	2010	Supervised MLP	Faulty module	Identification	100	PV Array 3.15 kW	Easy in point of view simulation	No/–
[17]	2011	Supervised BBN	SC	Detection	–	GGPV Array	Medium	No/–
[18]	2012	Unsupervised FL	Power loss	Detection	96	PV module level	Medium	No/–
[19]	2014	Supervised ELM	PSh	Classification	100	PV Array 3.5 kW	Easy	Yes/–
[20]	2015	Supervised ANN FL	Decrease in output power	Detection	86	PV module level	Medium	No/–
[21]	2016	Supervised MLPNN	PSh effect	Detection	90	PV string	Medium/Matlab	No/3000
[22]	2017	Supervised multi-class SVM	LL, degradation	Identification and classification	97.98	PV module 415 W	Easy to implement. Matlab/Simulink	No/–
[5]	2018	Supervised k-NN	SC, LL, PSh, and OC	Detection & classification	98.7	PV string level	Matlab/Simulink	No/–
[23]	2019	Supervised one-class SVM	SC, OC and PSh	Detection	99.3	GCPV 9.54 kW	Easy and Matlab	No/1440
[24]	2020	SVM	Hotspot	Detection	96.8	22 modules / 42.24 kW	Easy and Matlab	No/315
[25]	2021	FCN	Delamination, yellowing	Detection	98	PV module level	VGG16	No/1000
[26]	2022	FL	Shading	Detection	–	PV module level	Easy/Simulink-Matlab	No/100
[27]	2023	ML, DL	LLF, LGF, bypass faults	Classification	96.02	PV string level	Medium/Matlab	No/5000
[28]	2024	Random forest classifier	OC, SC, LLF, and PSh	Detection & classification	97.03	PV array 9.54 kW	Medium/PSIM-Matlab	No/19440

to system components and even potential catastrophes. A misdiagnosed or overlooked fault can lead to substantial energy losses and costly repairs. This is why accurately identifying and classifying electrical faults is crucial for protecting the future of PV systems. The key to ensuring both safety and sustainability is effectively managing this important process. To effectively address these faults, advanced techniques based on ML, NNs, and other intelligent systems have been developed in recent years. Chapter 3 provides current examples of these techniques, highlighting their applications in the field. These methods are not limited to analyzing electrical data from systems but extend to processing thermal images captured by drones, showcasing their broad range of applications. The examples in this chapter illustrate in detail the advantages of AI-based solutions in the classification of electrical faults in PV panels.

3.1 SA-RBF-ELM Fault Diagnosis Model

The SA-RBF-ELM fault diagnosis model is a notable method for detecting and classifying faults in PV panels, demonstrating exceptional accuracy in classifying fault events. Mahmoud Dhimish and colleagues confirmed the model's precision in identifying short-circuit faults. By integrating a fuzzy logic system, they enhanced fault detection accuracy to 95%, surpassing preimplementation accuracy levels. This approach showcases the model's effectiveness in addressing critical issues within solar energy systems [29]. Yue Wu and colleagues validated this approach using the same SA-RBF overlearning machine, achieving improvements in both accuracy and time efficiency during the training and testing phases. This further highlights the model's potential in optimizing FD processes for PV panels [30].

When it comes to fault diagnosis models for PV systems, several key factors play a role. Sparsity level refers to how much of the model's parameters are nonzero, with higher sparsity making the model more efficient and faster to compute. Training time is the amount of time it takes to train the model; in the case of the SA-RBF-ELM model, it is optimized to minimize training time while maintaining excellent performance. Accuracy is another critical aspect, and the SA-RBF-ELM model often achieves over 95% accuracy, thanks to its streamlined design. The sensitivity to data reflects how well the model performs based on the quality and quantity of data used for training, especially when capturing different fault conditions. Fault classification is about the model's ability to identify and differentiate between various types of faults in the system, such as short circuits or faulty components. Lastly, robustness speaks to the model's resilience, ensuring it can still function effectively even if the data are noisy or incomplete, which is important for practical applications.

3.2 Elman Neural Network (ENN)

Another noteworthy approach, proposed by Abubakar and colleagues, utilizes the ENN in combination with tree algorithms and statistical learning techniques to detect faults in PV strings and inverters. This method demonstrates the versatility and applicability of these algorithms across various fault scenarios, providing a robust solution for FD in solar energy systems. [31].

When using ENN for diagnosing faults in PV systems, several factors come into play to determine how well the model performs. Network depth, which refers to the number of hidden layers, is essential because it helps the model capture the complex patterns that arise in fault data. The training algorithm, often based on gradient descent methods like backpropagation, influences how quickly and effectively the model learns. Accuracy is a key measure, and ENN is especially strong in time-series prediction, where it excels at diagnosing faults in PV systems with reliable results. The model's fault sensitivity is another important strength, as ENN can detect even small, subtle changes in the system, making it great for catching early-stage issues or gradual degradation. Memory capacity allows the model to handle and remember sequences over time, which is crucial for identifying faults that develop slowly.

3.3 Convolutional Neural Network (CNN) Model

A frequently utilized model for robust fault classification in PV panels is CNN. For example, a CNN trained on historical data has demonstrated significant efficacy in accurately classifying faults in PV modules. Moreover, Chen et al. proposed a sequential FD algorithm employing autoregressive models combined with Generalized Likelihood Ratio (GLLR) tests, thereby improving the reliability and precision of FD methodologies [32].

When it comes to using CNNs for FD in PV systems, there are a few key factors that really shape how well the model performs. First, the layer configuration (the number and types of convolutional layers, like 3 × 3 or 5 × 5) determines how effectively the network can capture the finer details of the data. While having more layers can improve accuracy, it also makes the model more computationally demanding. Accuracy is another important aspect. CNNs often achieve impressive results, typically between 90% and 98%, depending on the quality of the dataset and how the model is trained. To train CNNs properly, a large dataset is necessary to help the model learn effectively and generalize new data well. One of the key strengths of CNNs is their ability to automatically extract important features from raw data, like images or time-series, saving a lot of time compared to manually identifying key details.

3.4 Deep Learning Models

DL models, particularly CNNs, have shown considerable potential in the detection of faults in PV systems. For example, a study utilizing two advanced CNN architectures effectively classified fault types and identified areas of interest on defective panels. This methodology achieved an F1 score of 85.37% with the ResNet-50 transfer learning model and a Mean Average Precision (mAP) of 67% using the Faster R-CNN model, underscoring the capability of DL techniques in fault localization. Additionally, another study reported a CNN-based model achieving an FD accuracy of 97.5%, surpassing the 92.64% accuracy attained by the ANN-based approach [33]. In addition to ML techniques, visual and thermal methods, as well as electrical-based methods, have been comprehensively reviewed and categorized for inspecting faulty PV systems. These methods involve analyzing RGB, EL, thermal, and infrared images to accurately identify panel defects. For example, Ali and colleagues proposed a hybrid feature-based SVM model using infrared thermography to classify hot spots in PV panels into healthy, nondefective, and defective categories [34].

Finally, to overcome the challenge of collecting the large volumes of data required for effective ML, data-driven FD methods, such as training DL models with synthetic EL images, have been explored. Monitoring and tracking systems, like mapperx software, also play a significant role in real-time FD, identifying faults in solar power plants with 99% accuracy. DL models, especially those with many layers, can be prone to overfitting if the training data are not diverse enough. Regularization techniques such as dropout are often used. Transfer learning is often applied in DL models to leverage pretrained models, reducing the amount of data needed for training and speeding up the learning process. Training DL models takes considerable computational power and time, particularly with large datasets. The time required can vary depending on the model's complexity and the hardware used. However, DL models excel at generalizing to new, unseen data, making them valuable for fault detection in various environmental conditions.

4 Strategies for Locating Electrical Faults

Locating electrical faults in solar panels is a critical step in reducing system failures and ensuring the efficient operation of PV systems. Various strategies have been developed to accurately identify and pinpoint these faults.

4.1 Visual and Thermal Methods

An effective strategy for fault localization involves visual and thermal inspections. IRT is a reliable, noninvasive, fast, and cost-efficient strategy approach for detecting faults in electrical installations. By using an infrared camera to capture thermal images, (Fig. 7) faulty panels can be detected by comparing their thermal profiles to those of normal, fully functional panels. Drones equipped with thermal cameras are increasingly being used for aerial inspections and offer a highly effective method for identifying faults by detecting high temperatures associated with faulty panels. Recently developed software utilizes advanced image processing and machine learning technologies to analyze thermal inspection data, providing solutions for detected faults and improving the overall performance of solar power plants.

Thermal image resolution plays a critical role in identifying faults in solar panels, as higher resolution ensures sharper and more detailed images, making FD more accurate. Temperature differences between faulty and surrounding panels often highlight issues such as defective cells or components. For large solar farms, the efficiency of inspections depends on the drone's speed and camera resolution, enabling broad coverage and precise fault identification. Additionally, advanced image processing techniques, such as edge detection and anomaly recognition, enhance the analysis of thermal images, ensuring reliable FD and minimizing false alarms. In reference [36], the authors employed a combination of Deep Convolutional Neural Networks (DCNN) and SVM to detect, localize, and classify faults in large-scale PV plants using UAVs. A dataset consisting of approximately 7560 PV images was utilized to develop the methodology. The results highlighted the effectiveness of this approach in accurately identifying and classifying common defects in PV arrays, such as delamination, dust shading, snail trails, gridline corrosion, and yellowing. While the method yielded promising accuracy, it is worth noting that it is highly resource-intensive, requiring both UAVs and advanced computational resources, such as supercomputers. This is one of the disadvantages of this method.

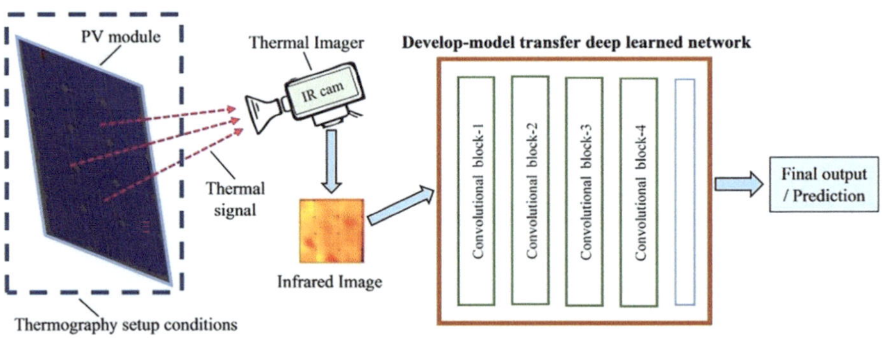

Fig. 7 The process of analyzing thermography data with a DL model for detecting thermal anomalies in PV modules [35]

4.2 Electrical Measurement Methods

Accurately determining the fault location requires understanding the expected measurement values, especially voltage. For instance, the module's open-circuit voltage (V_{oc}) can be found on the module label or datasheet, and the PV string voltage is calculated by multiplying the V_{oc} of the module by the number of series-connected modules. Additionally, evaluating current signals can aid FD, as faults often lead to significant increases in current, which can affect other electronic components in the system.

V_{oc} refers to the open-circuit voltage, which is the voltage measured across the terminals of a solar panel under conditions where no current is flowing. If there is a significant deviation from the expected V_{oc}, it can signal a potential fault in the panel. PV string voltage, on the other hand, is the total voltage from a series-connected string of panels, which is calculated by multiplying the V_{oc} of individual panels by the number of panels in the string. Monitoring the string voltage is important for spotting issues that might be caused by a faulty module. Current flow, measured in amperes, represents the amount of electricity flowing through the system, and any abnormal spikes or drops in current can indicate problems like short circuits or degraded components. To detect these issues, it is essential to monitor the current across both individual panels and the overall system, ensuring that any discrepancies are quickly identified.

4.3 Machine Learning and Computer Vision Techniques

ML techniques have demonstrated strong capabilities in classifying fault occurrences in solar panels. Methods like the SA-RBF extreme learning machine offer improved accuracy during training and testing phases while reducing computation time. When trained from scratch, DL models improve the processing of thermal images, leading to more accurate fault detection. These models are crucial for boosting the maintenance and efficiency of solar energy systems, helping optimize performance and reduce downtime.

4.3.1 Sequential Fault Detection Algorithms

Sequential fault detection algorithms that use autocorrelation models and GLLR tests are designed to quickly spot and pinpoint faults in PV systems, improving their reliability and performance. By detecting issues early and enabling fast fixes, these algorithms help reduce downtime and prevent more serious problems. Autocorrelation

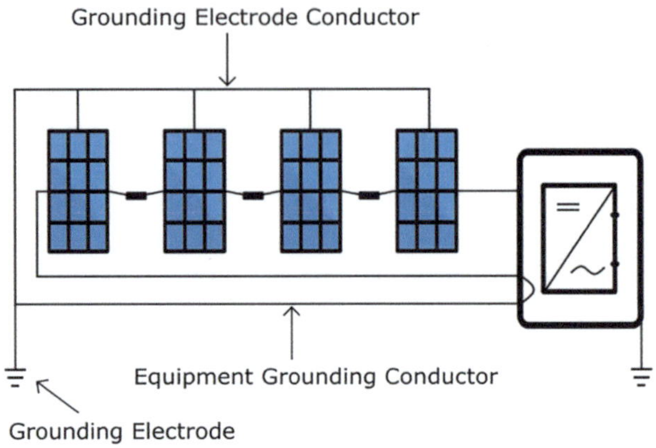

Fig. 8 GEC for locating faults

thresholds spot patterns in time-series data, making it easier to detect gradual changes that could signal faults. GLLR tests assess the likelihood of different faults, helping pinpoint the most likely issues. The speed at which these algorithms detect faults is key, allowing for quicker responses and ensuring the system runs smoothly.

4.4 Grounding and Component Understanding

Ground fault resistance is a crucial parameter that measures the resistance to fault current flowing to the ground. Lower resistance signifies a more efficient grounding system, enabling faster and more accurate FD. A comprehensive understanding of the components within a PV system, particularly the equipment grounding conductor (EGC) and the Grounding Electrode Conductor (GEC), is essential for the effective identification of faults. The EGC establishes a pathway for GF currents by linking the metallic components of PV equipment to the grounded conductor, whereas the GEC connects the system's equipment to the grounding electrode. Analyzing the interaction between these components enhances the ability to detect and mitigate faults efficiently, thereby ensuring the safety and optimal performance of the system (Fig. 8).

5 Preventive Measures and Solutions

Adopting preventive measures and advanced solutions is key to ensuring the long-term performance and reliability of solar panel systems. The incorporation of these advanced technologies, including real-time monitoring, ML algorithms, and remote diagnostic tools, enables the early detection of potential faults, thereby preventing their escalation into more critical issues. Regular maintenance routines, including cleaning, visual inspections, and performance evaluations, further contribute to minimizing the risk of system failures. These proactive strategies help optimize the efficiency of the solar power system, extend its operational lifespan, and reduce downtime, ultimately ensuring a more sustainable and cost-effective energy solution.

5.1 Regular Maintenance

Regular maintenance is critical for prolonging the lifespan and optimizing the performance of PV systems. This maintenance includes various procedures such as cleaning, thermal inspections, and electrical testing, which should be conducted multiple times throughout the year. For complex maintenance tasks, it is recommended to engage professional technicians who specialize in FD and resolution, ensuring that the highest maintenance standards are adhered to.

5.2 Advanced Monitoring Systems

Contemporary solar power plants increasingly rely on advanced monitoring systems for the prompt identification of faults. These systems employ AI to monitor and track the performance of solar panels, achieving FD with up to 99% accuracy. They utilize sophisticated image processing and ML technologies to analyze data gathered from thermal inspections, providing solutions to pre-empt production issues and improve system reliability.

5.3 Machine Learning and Artificial Intelligence

ML and AI techniques have proven to be highly effective in the detection and classification of faults within solar panels. CNN-based models, in particular, have demonstrated substantial improvements in FD accuracy, reaching as high as 97.5%. These models facilitate the precise classification of faults, thereby reducing power losses and enhancing the overall efficiency of solar PV systems [37].

5.4 Infrared Thermography (IRT)

The implementation of IRT technology offers a nondestructive, rapid, and cost-effective method for FD in electrical systems. By utilizing infrared cameras to capture thermal images, comparisons can be made between faulty panels and those operating under optimal conditions, enabling precise thermal assessments. This technology is particularly effective in classifying faults induced by environmental factors such as dust, water droplets, bird droppings, and partial shading conditions, which are known to impact the performance of solar panels.

5.5 Module-Level Power Electronics (MLPEs)

DC grounding faults represent one of the most prevalent fault types in PV systems, with approximately half of these faults remaining undetected. A DC grounding fault occurs when an unwanted current flows through circuits carrying DC power, before the inverter, via the equipment grounding conductor. Such faults can result in severe safety hazards, including arc faults and arc flashes under high-voltage conditions. Additionally, the heating of exposed metal due to short-circuit currents may pose a fire risk. To deepen the understanding of DC grounding faults, it is essential to review pertinent terminology and analyze a PV system comprehensively. The National Electrical Code defines the grounding terms. Equipment Grounding Conductor (EGC) establishes a pathway for fault currents by connecting noncurrent-carrying metallic components of equipment to the system's grounded conductor or to GEC. Grounding Electrode Conductor (GEC) links the system's grounded conductor or equipment to a grounding electrode or the grounding electrode system. Such a detailed examination of grounding components is critical for understanding and addressing DC grounding faults effectively. A Grounded Conductor is intentionally connected to the ground. The EGC connects all conductive parts (modules, racks) and provides a path to the GEC, which connects the entire system to the grounding electrode (GE). The GE, typically a large metal rod made of copper, aluminum, or copper-coated aluminum, is embedded at least 2 m deep in the ground. In the case of DC grounding faults, current flows from the EGC or any grounded metal component due to unwanted contact with a grounded conductor. This contact typically occurs due to damaged insulation, improper installation, or water establishing an electrical connection between the conductor and the EGC. Proper installation and regular maintenance are crucial in mitigating the safety risks and fire hazards associated with such faults.

5.6 Utilization of UAVs and Drones

UAVs or drones have emerged as one of the most efficient tools for inspecting solar panels. By capturing images of the panels and feeding them into DL models, drones assist in the rapid and accurate recognition and classification of faults. This method significantly accelerates the inspection process while enhancing its precision, making drones indispensable in solar panel maintenance practices [9]. Equipped with high-resolution thermal cameras and sensors, drones can identify issues such as temperature disparities, structural damage, or electrical faults. With real-time data transmission, quick decisions and interventions can be made, reducing downtime and ensuring the system operates optimally [38].

5.7 Bypass and Blocking Diodes

The integration of bypass and blocking diodes is crucial for ensuring a stable and uninterrupted current flow within solar panels. Blocking diodes, connected in series with solar cells, allow current to flow exclusively in one direction, thereby preventing reverse current. Conversely, bypass diodes, connected in parallel, prevent reverse current from flowing from higher-performing cells to those with lower performance. Accurate FD in these diodes is essential for safeguarding the overall functionality and efficiency of solar panel systems. Faulty bypass diodes can lead to energy losses, performance degradation, or permanent damage to the panels. Diagnosing such faults typically involves monitoring voltage drops, atypical current patterns, or the performance of individual panel strings.

5.8 Environmental Factors

Environmental parameters, including temperature, cloud cover, dust, radiation levels, and relative humidity, play a pivotal role in predicting faults in solar panels. These factors significantly influence both the performance and longevity of the panels, underscoring the importance of regular monitoring and adjustment of panel orientations to ensure optimal operation [10]. Elevated temperatures can induce thermal stress, leading to panel degradation over time, while dust accumulation on the panel surface may obstruct sunlight, reducing energy output. Humidity can exacerbate the risk of corrosion, and cloud cover can cause fluctuations in power generation. Continuous monitoring of these conditions supports the implementation of proactive maintenance strategies, such as cleaning schedules and system adjustments, thereby maximizing panel performance.

6 Future Trends in Solar Panel Maintenance

The future of solar panel maintenance will be significantly shaped by advancements in high-level technologies, particularly AI, ML, and IoT. These technologies have the potential to make the maintenance of solar PV systems more efficient, accurate, and cost-effective.

6.1 Integration of AI and ML

AI and ML are increasingly being applied to enhance the effectiveness of FD and maintenance processes in solar panels. These technologies utilize large datasets to identify fault patterns with high accuracy and predict potential failures. Research has shown that AI technologies can play a crucial role in the remote monitoring and fault diagnosis of solar PV systems. For example, DL models such as EfficientB0, trained from the ground up using infrared images of solar modules, have demonstrated significant improvements in FD accuracy, facilitating more precise and reliable outcomes.

6.2 Utilization of Deep Learning Techniques

DL techniques, particularly CNN, have a significant role in the classification and localization of faults in solar panels. Advanced models such as ResNet-50 and Faster R-CNN have demonstrated effectiveness in accurately detecting fault types and pinpointing their locations. Additionally, the development of synthetic electroluminescence (EL) images for PV cells allows for the creation of large, diverse datasets to train these DL models.

6.3 Role of IoT and Smart Monitoring Systems

The incorporation of IoT with solar energy systems provides real-time monitoring and FD capabilities. By using IoT and ATMEGA processors, integrated management and security maintenance help to control unwanted wiring and hardware faults, thereby enhancing the cost-effectiveness of the system. Moreover, advanced monitoring systems, powered by AI-assisted analysis of data from thermal inspections, can detect faults with 99% accuracy, optimizing the efficiency of solar power plants.

6.4 Module-Level Power Electronics (MLPEs)

The use of MLPEs ensures the isolation of faulty modules and the disconnection of power from individual modules. This approach presents a significant solution for the early detection and prevention of DC ground faults in PV systems. These systems accelerate maintenance processes, increase system security, and minimize risks.

6.5 UAVs and Infrared Imaging Technologies

UAVs equipped with infrared cameras have become an effective method for inspecting large-scale solar power plants. These UAVs capture high-resolution infrared images, which can then be processed by DL models such as Mask-RCNN for FD and localization. Real-time data transmission enables rapid interventions, minimizing system downtime.

6.6 Future Research and Development Areas

Future research focuses on enhancing the quality and quantity of data used to improve ML models. This includes the generation of more synthetic images and the refinement of existing models increasing the effectiveness of solar panel maintenance, leading to safer and more sustainable energy production processes.

References

1. Duranay, Z. B. (2023). Fault detection in solar energy systems: A deep learning approach. *Electronics, 12*, 4397. https://doi.org/10.3390/electronics12214397
2. Hong, Y.-Y., & Pula, R. A. (2022). Methods of photovoltaic fault detection and classification: A review. *Energy Reports, 8*, 5898–5929. https://doi.org/10.1016/j.egyr.2022.04.043
3. Yi, Z., & Etemadi, A. H. (2016). Fault detection for photovoltaic systems based on multi-resolution signal decomposition and fuzzy inference systems. *IEEE Transactions on Smart Grid, 8*(3), 1274–1283.
4. Harrou, F., et al. (2019). An unsupervised monitoring procedure for detecting anomalies in photovoltaic systems using a one-class support vector machine. *Solar Energy, 179*, 48–58.
5. Madeti, S. R., & Singh, S. N. (2018). Modeling of PV system based on experimental data for fault detection using kNN method. *Solar Energy, 173*, 139–151.
6. Heinrich, M., et al. (2020). Detection of cleaning interventions on photovoltaic modules with machine learning. *Applied Energy, 263*, 114642.
7. Niazi, K. A. K., et al. (2019). Hotspot diagnosis for solar photovoltaic modules using a Naive Bayes classifier. *Solar Energy, 190*, 34–43.
8. Youssef, A., El-Telbany, M., & Zekry, A. (2017). The role of artificial intelligence in photovoltaic systems design and control: A review. *Renewable and Sustainable Energy Reviews, 78*, 72–79.
9. Herraiz, Á. H., Marugán, A. P., & Márquez, F. P. G. (2020). Photovoltaic plant condition monitoring using thermal images analysis by convolutional neural network-based structure. *Renewable Energy, 153*, 334–348.
10. Dhanraj, J. A., Mostafaeipour, A., Velmurugan, K., Techato, K., Chaurasiya, P. K., Solomon, J. M., Gopalan, A., & Phoungthong, K. (2021). An effective evaluation on fault detection in solar panels. *Energies, 14*, 7770. https://doi.org/10.3390/en14227770
11. Muttillo, Mirco, et al. On field infrared thermography sensing for PV system efficiency assessment: Results and comparison with electrical models." Sensors 20.4 (2020): 1055.

12. Ko, J.-S., & Kim, D.-K. (2021). Localization of disconnection faults in PV installations using the multiple frequencies injection method. *Energies, 14*, 7346. https://doi.org/10.3390/en14217346
13. Henry, C., et al. (2020). Automatic detection system of deteriorated PV modules using drone with thermal camera. *Applied Sciences, 10*(11), 3802.
14. Jowett, R.Jeffrey. Ground faults in photovoltaic systems Neta World Journal, 2022, Retrieved 29 December 2024 from *https://netaworldjournal.org/ground-faults-in-photovoltaic-systems/.*
15. Wu, Y., Lan, Q., & Sun, Y. (2009). Application of BP neural network fault diagnosis in solar photovoltaic system. 2009 International Conference on Mechatronics and Automation. IEEE.
16. Chao, K.-H., Chen, C.-T., Wang, M.-H., & Wu, C.-F. (2010). A novel fault diagnosis method based on modified neural networks for photovoltaic systems. In *Advances in swarm intelligence* (pp. 531–539). Springer.
17. Karatepe E, Hiyama T. (2011) *Controlling of artificial neural network for fault diagnosis of photovoltaic array.* In: IEEE 16th International Conference on Intelligent System Applications to Power Systems, pp. 1–6.
18. Spataru S, Sera D, Kerekes T, Teodorescu R. (2012) *Detection of increased series losses in PV arrays using fuzzy inference systems.* In: 38th IEEE Photovoltaic Specialists Conference, pp. 464–469.
19. Chao, K. H., Chen, P. Y., Wang, M. H., & Chen, C. T. (2014). An intelligent fault detection method of a photovoltaic module array using wireless sensor networks. *International Journal of Distributed Sensor Networks, 10*, 540147.
20. Jones CB, Stein JS, Gonzalez S, King BH. (2015). *Photovoltaic system fault detection and diagnostics using laterally primed adaptive resonance theory neural network.* In: IEEE 42nd Photovoltaic Specialist Conference, pp. 1–6.
21. Mekki, H., Mellit, A., & Salhi, H. (2016). Artificial neural network-based modelling and fault detection of partial shaded photovoltaic modules. *Simulation Modelling Practice and Theory, 67*, 1–13.
22. Wang L, Liu J, Guo X, Yang Q, Yan W. (2017). *Online fault diagnosis of photovoltaic modules based on multi-class support vector machine.* In: IEEE 2017 Chinese Automation Congress, pp. 4569–4574.
23. Riley D, Johnson J. (2012). *Photovoltaic prognostics and heath management using learning algorithms.* In: 38th IEEE Photovoltaic Specialists Conference, pp. 1535–1539.
24. Ali, M. U., et al. (2020). A machine learning framework to identify the hotspot in photovoltaic module using infrared thermography. *Solar Energy, 208*, 643–651.
25. Sikuli, A. M., Aghaei, M. M., & Esmailfar, S. M. (2021). A deep convolutional encoder-decoder architecture for autonomous fault detection of PV plants using multi-copters. *Solar Energy, 223*, 217–228.
26. Bacha, M., & Terki, A. (2022). Diagnosis algorithm and detection faults based on fuzzy logic for PV panel. *Materials Today: Proceedings, 51*, 2131–2138.
27. Hajji, Mansour, et al. (2023). Energy reports.
28. Amiri, Ahmed Faris, et al. Faults detection and diagnosis of PV systems based on machine learning approach using random forest classifier Energy Conversion and Management 301 (2024): 118076.
29. Dhimish, M., Holmes, V., Mehrdadi, B., Dales, M., & Mather, P. (2017). Photovoltaic fault detection algorithm based on theoretical curves modelling and fuzzy classification system. *Energy, 140*, 276–290. https://doi.org/10.1016/j.energy.2017.08.102
30. Wu, Y., Chen, Z., Wu, L., Lin, P., Cheng, S., & Lu, P. (2017). An intelligent fault diagnosis approach for PV array based on SA-RBF kernel extreme learning machine. *Energy Procedia, 105*, 1070–1076. https://doi.org/10.1016/j.egypro.2017.03.462
31. Abubakar, A., Jibril, M. M., Almeida, C. F. M., Gemignani, M., Yahya, M. N., & Abba, S. I. (2023). A novel hybrid optimization approach for fault detection in photovoltaic arrays and inverters using AI and statistical learning techniques: A focus on sustainable environment. *Processes, 11*, 2549. https://doi.org/10.3390/pr11092549

32. Chen, L., Li, S., & Wang, X. (2016). Quickest fault detection in photovoltaic systems. *IEEE Transactions on Smart Grid*, 1. https://doi.org/10.1109/TSG.2016.2601082
33. Pathak, S. P., Patil, S., & Patel, S. (2022). Solar panel hotspot localization and fault classification using deep learning approach. *Procedia Computer Science, 204*, 698–705. https://doi.org/10.1016/j.procs.2022.08.084
34. Ali, M. U., Khan, H. F., Masud, M., Kallu, K. D., & Zafar, A. (2020). A machine learning framework to identify the hotspot in photovoltaic module using infrared thermography. *Solar Energy, 208*, 643–651. https://doi.org/10.1016/j.solener.2020.08.027
35. Akram, M. Waqar, et al. Automatic detection of photovoltaic module defects in infrared images with isolated and develop-model transfer deep learning Solar Energy 198 (2020): 175–186.
36. Li, Xiaoxia, et al. Deep learning-based module defect analysis for large-scale photovoltaic farms IEEE Transactions on Energy Conversion 34.1 (2018): 520–529.
37. Memon, S. A., Javed, Q., Kim, W.-G., Mahmood, Z., Khan, U., & Shahzad, M. (2022). A machine-learning-based robust classification method for PV panel faults. *Sensors, 22*, 8515. https://doi.org/10.3390/s22218515
38. Segovia Ramírez, I., Das, B., & García Márquez, F. P. (2022). Fault detection and diagnosis in photovoltaic panels by radiometric sensors embedded in unmanned aerial vehicles. *Progress in Photovoltaics: Research and Applications, 30*, 240–256. https://doi.org/10.1002/pip.3479

Education and Training for the AI-Enabled Healthcare Workforce

Manish Thakral, Vishal Nagpal, Anita Devare, Manoj H. Devare, Indrajit Ghosal, and Mohammad Irfan

1 Introduction

The integration of artificial intelligence (AI) technologies into healthcare systems has ushered in a new era of innovation and opportunity, promising to revolutionize patient care, clinical decision-making, and healthcare delivery. From diagnostic assistance to personalized treatment recommendations, AI has the potential to enhance the efficiency, accuracy, and effectiveness of healthcare services. However, realizing the full potential of AI in healthcare requires a skilled and knowledgeable workforce capable of harnessing these technologies responsibly and ethically. The integration of Artificial Intelligence (AI) within the healthcare sector marks a pivotal advancement, redefining the landscape of medical diagnostics, treatment planning, and operational efficiency. AI-enabled systems have demonstrated significant potential in addressing complex challenges, such as disease prediction, personalized medicine, and real-time patient monitoring. However, the rapid evolution of

M. Thakral
EY, Mumbai, India

V. Nagpal · A. Devare
Amity University, Mumbai, India

M. H. Devare
AIIT-AUM, Mumbai, India
e-mail: mhdevare@mum.amity.edu

I. Ghosal (✉)
Brainware University, Kolkata, India

M. Irfan
Chandigarh University, Lucknow, India

P. Raj et al. (eds.), *Artificial Intelligence (AI) for IT Energy Efficiency and Green AI for Environment Sustainability*,
https://doi.org/10.1007/978-3-031-89420-6_4

these technologies demands a concomitant transformation in the skills and knowledge of the healthcare workforce. Ensuring that practitioners are equipped to harness AI effectively and ethically is critical for optimizing patient outcomes and fostering a sustainable healthcare ecosystem [1–3].

We begin by outlining the foundational role of AI in contemporary healthcare, highlighting its applications and the transformative opportunities it presents. The chapter then transitions into a detailed analysis of current gaps in healthcare education, emphasizing the need for robust, interdisciplinary frameworks tailored to the unique demands of AI-integrated environments. Subsequent sections examine key components necessary for developing effective training programs, including ethical and legal considerations, continuous learning mechanisms, and practical, hands-on approaches to skill development. Special attention is given to aligning these initiatives with principles of sustainability, ensuring that the adoption of AI in healthcare supports energy-efficient practices and environmentally conscious policies. In this chapter, we explore the critical importance of education and training for the AI-enabled healthcare workforce. We begin by examining the evolving landscape of AI in healthcare, highlighting key trends, applications, and implications for healthcare professionals. By understanding the current state of AI adoption and its impact on healthcare delivery, we can appreciate the urgency and significance of preparing the healthcare workforce for the AI-enabled future [4–6].

As AI technologies continue to evolve and permeate various aspects of healthcare, it is essential to equip healthcare professionals with the knowledge, skills, and competencies necessary to leverage these technologies effectively. This necessitates comprehensive education and training programs designed to address the unique challenges and opportunities presented by AI in healthcare. Understanding AI in Healthcare: We will provide an overview of AI technologies, including machine learning, natural language processing, and computer vision, and their applications in healthcare. By gaining a foundational understanding of AI, healthcare professionals can better appreciate its potential benefits and limitations in clinical practice. Educational Frameworks: We will examine existing educational frameworks and initiatives aimed at integrating AI education into healthcare training programs. This includes formal coursework, workshops, online resources, and hands-on training opportunities tailored to meet the diverse needs of healthcare professionals at all career stages. The objectives of the chapter are to evaluate the potential of AI-driven solutions for enhancing energy efficiency and to provide a systematic framework for their implementation. The scope extends to real-time monitoring, workload management, and thermal optimization, positioning AI as a transformative tool for sustainable operations [7]. The methodology begins with an in-depth analysis of global energy consumption statistics for data centers, focusing on metrics such as power usage effectiveness (PUE) and total energy usage. Data is gathered from industry reports, academic research, and environmental studies to quantify the major sources of energy consumption, including server operations, cooling systems, and auxiliary infrastructure. The environmental impact of these practices is assessed through carbon footprint analysis and greenhouse gas (GHG) emission metrics [8]. Current energy efficiency measures, such as server virtualization and improved hardware design, are critically examined to identify their limitations and scope for AI integration. Techniques for power usage prediction are explored, including regression models, neural networks, and ensemble methods, with an emphasis on accuracy and scalability. Real-time monitoring systems leveraging AI for dynamic workload distribution and scheduling are analyzed to determine their effectiveness in reducing energy waste as shown in Fig. 1. Thermal management optimization techniques, such as AI-driven airflow modelling and heat dissipation strategies, are evaluated for their role in minimizing cooling energy requirements.

Case studies of successful implementations in leading data centers are reviewed to extract best practices and assess practical challenges. These methodologies collectively aim to establish AI as a pivotal component in advancing energy-efficient and sustainable data center operations. First, current frameworks inadequately address interdisciplinary collaboration, which is vital for integrating AI tools into diverse clinical contexts. Second, the environmental implications of AI systems are often overlooked, despite their increasing relevance in a global push for sustainability. Third, ethical considerations, including bias mitigation and transparency, are insufficiently integrated into healthcare AI curricula. By addressing these gaps, the study underscores the importance of a holistic approach that combines AI competencies with environmental stewardship and ethical practices, fostering a workforce capable of navigating the complexities of modern healthcare. This study aims to bridge these gaps by proposing a comprehensive educational framework that integrates technical training, ethical guidelines, and sustainability principles, ensuring the development of a competent and responsible AI-enabled healthcare workforce. Recognizing the interdisciplinary nature of AI in healthcare, we will emphasize the importance of collaboration between healthcare professionals, data scientists, engineers, ethicists, and other stakeholders. By fostering interdisciplinary collaboration, we can develop comprehensive education and training programs that address the

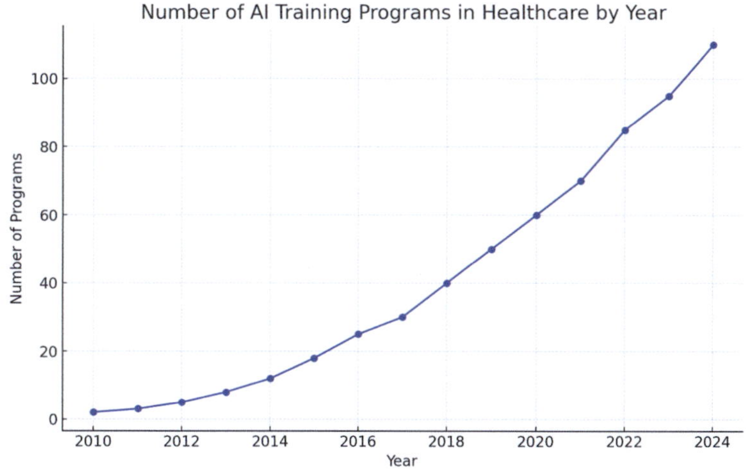

Fig. 1 Representation of the number of programs in alignment with the year 2010–2024

complex challenges inherent in AI-enabled healthcare delivery [9]. We will explore the ethical and legal considerations associated with AI in healthcare, including issues related to data privacy, bias, transparency, and accountability. Healthcare professionals must navigate these complex ethical dilemmas when implementing AI technologies in clinical practice, underscoring the need for ethics education and training.

Finally, we will discuss the importance of fostering a culture of continuous learning and adaptation among healthcare professionals. Given the rapid pace of technological innovation in healthcare, ongoing education, training, and professional development are essential to ensure that the healthcare workforce remains abreast of the latest developments in AI and is equipped to leverage these technologies effectively in clinical practice. We will explore the importance of tailoring training programs to the specific needs and roles of healthcare professionals. Different healthcare specialties may require different levels of AI proficiency and understanding. For example, radiologists may benefit from specialized training in AI applications for medical imaging interpretation, while primary care physicians may require training in AI-assisted diagnostics and treatment planning as shown in Fig. 2.

We will explore the importance of continuing education and professional development opportunities for healthcare professionals already in practice. Workshops, seminars, webinars, and online courses can provide opportunities for ongoing learning and skill development in AI and related technologies. Finally, we will discuss the need for robust evaluation and assessment mechanisms to measure the effectiveness of AI education and training programs [10]. Formative and summative assessments can help identify areas for improvement and ensure that healthcare professionals are achieving the desired learning outcomes.

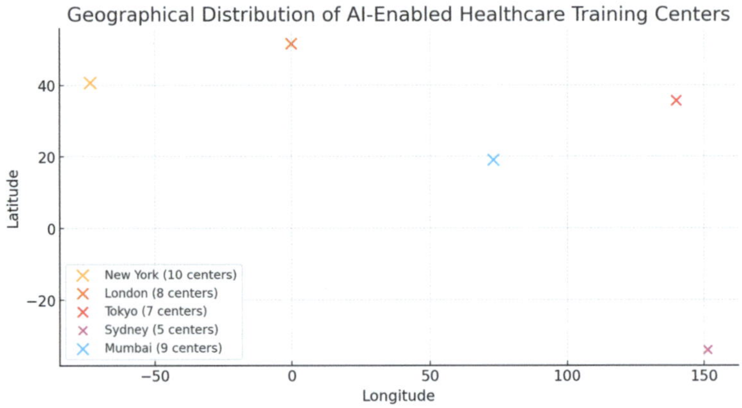

Fig. 2 Representation of geographical distribution of AI-enabled healthcare training centers

2 Source of Data

The development and implementation of education and training programs for the AI-enabled healthcare workforce require access to diverse sources of data to inform curriculum design, instructional methods, and learning outcomes assessment. In this chapter, we explore various sources of data that can be leveraged to support the development of effective education and training initiatives for healthcare professionals in the context of artificial intelligence (AI) in healthcare.

Clinical Data Repositories: Clinical data repositories, including electronic health records (EHRs) and health information exchange (HIE) systems, serve as valuable sources of data for understanding clinical workflows, patient demographics, disease prevalence, and treatment outcomes. Analyzing anonymized clinical data can provide insights into the types of AI applications that are most relevant to specific clinical specialties and practice settings [11].

Medical Imaging Databases: Medical imaging databases, such as the Cancer Imaging Archive (TCIA) and the Alzheimer's Disease Neuroimaging Initiative (ADNI), contain large collections of medical images annotated with clinical metadata. These databases can be used to develop AI algorithms for image interpretation, disease detection, and treatment planning. Access to medical imaging data is essential for training healthcare professionals as shown in Fig. 3, particularly radiologists and other imaging specialists, in AI-assisted diagnostic techniques.

Biomedical Literature: Biomedical literature databases, such as PubMed and MEDLINE, provide access to a vast repository of peer-reviewed research articles,

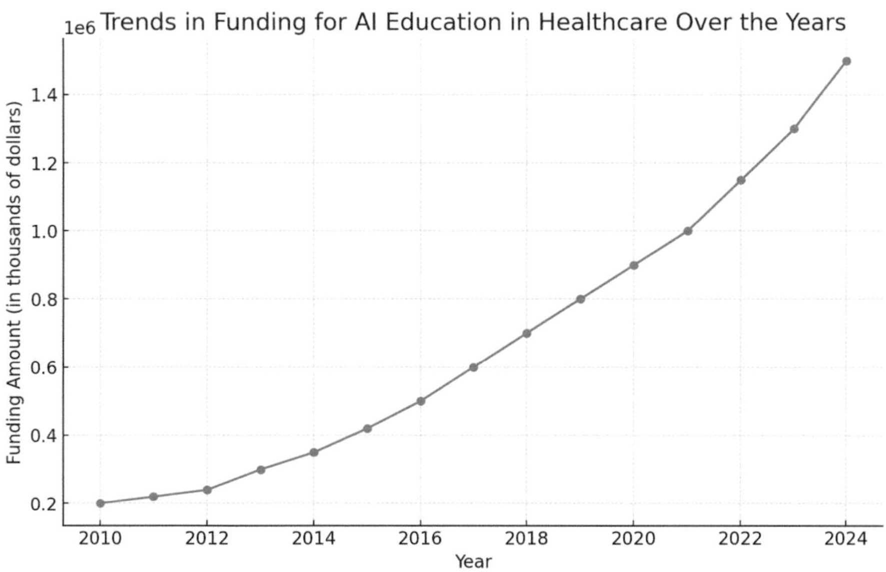

Fig. 3 Funding amount trend in the FY 2010 till 2024

review papers, and clinical guidelines. Analyzing biomedical literature can help educators and curriculum developers stay abreast of the latest advancements in AI technologies and their applications in healthcare. Moreover, literature reviews can inform the development of evidence-based educational materials and case studies for AI training programs [12].

Open-Access AI Datasets: Open-access AI datasets, such as MIMIC-III (Medical Information Mart for Intensive Care) and PhysioNet, offer researchers and educators access to de-identified patient data collected from real-world healthcare settings. These datasets enable the development and validation of AI algorithms for a wide range of clinical applications, including predictive modeling, risk stratification, and clinical decision support. Incorporating open-access AI datasets into AI education and training programs allows healthcare professionals to gain hands-on experience with real-world healthcare data and AI technologies.

Healthcare Simulation Data: Healthcare simulation data captured during medical simulations and virtual patient encounters provide valuable insights into clinical decision-making, diagnostic reasoning, and treatment planning. By analyzing simulation data, educators can identify learning gaps, assess clinical competency as shown in Fig. 4, and tailor educational interventions to address specific learning needs. Additionally, simulation data can be used to develop virtual reality (VR) and augmented reality (AR) simulations for immersive training experiences in AI-enabled healthcare scenarios.

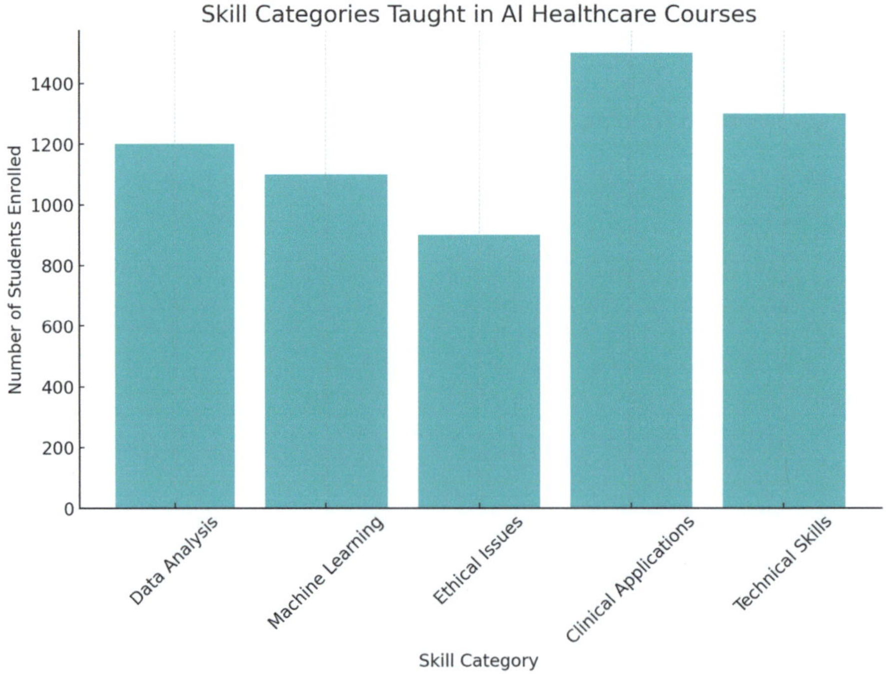

Fig. 4 Number of students enrolled vs skill category

Healthcare Quality and Performance Metrics: Healthcare quality and performance metrics, such as hospital readmission rates, medication errors, and patient satisfaction scores, offer insights into healthcare delivery processes and outcomes. Analyzing quality and performance metrics can help educators identify areas for improvement and develop targeted educational interventions aimed at enhancing clinical practice and patient care. Moreover, integrating quality and performance metrics into AI education and training programs enables healthcare professionals to evaluate the impact of AI technologies on healthcare quality and outcomes [13].

Surveys and Needs Assessments: Surveys and needs assessments conducted among healthcare professionals can provide valuable insights into their knowledge, attitudes, and training needs related to AI in healthcare. By collecting data on healthcare professionals' perceptions of AI, their comfort level with AI technologies, and their preferences for educational modalities, educators can design tailored education and training programs that address specific learning objectives and preferences as shown in Fig. 5.

Clinical Trials and Research Studies: Clinical trials and research studies provide valuable insights into the efficacy and safety of AI applications in healthcare. Accessing data from clinical trials and research studies allows educators to explore the evidence base supporting AI technologies and their impact on clinical outcomes. Additionally, educators can incorporate findings from clinical trials and research studies into educational materials to demonstrate the real-world implications of AI in healthcare.

Healthcare Policy and Regulatory Guidelines: Healthcare policy documents, regulatory guidelines, and professional standards provide guidance on the ethical,

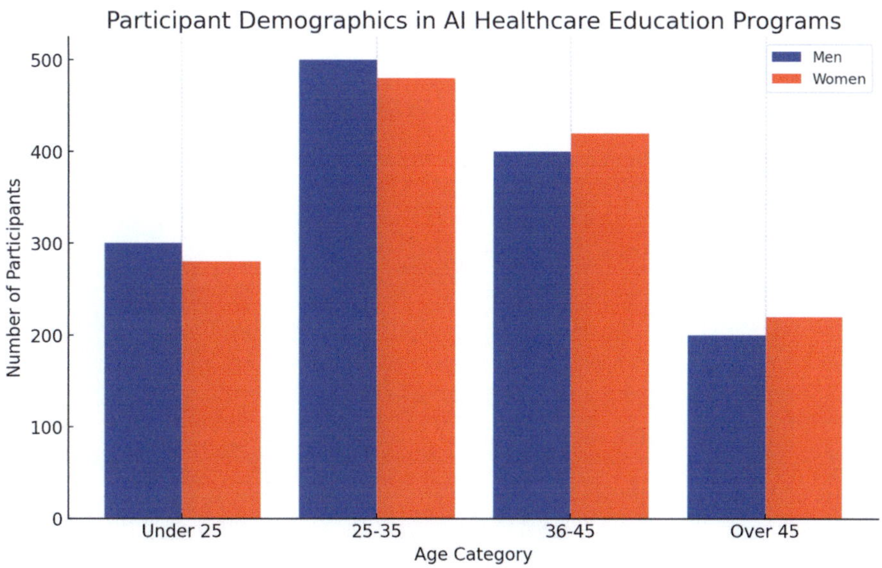

Fig. 5 Representation of age category against the number of participants

legal, and regulatory considerations surrounding AI in healthcare. Analyzing healthcare policy and regulatory documents helps educators understand the broader healthcare landscape and ensure that education and training programs align with relevant regulations and standards. Moreover, educators can use policy and regulatory guidelines to develop educational content on topics such as data privacy, security, and governance in AI-enabled healthcare environments [14].

Industry Reports and Market Trends: Industry reports, market analyses, and trend reports offer insights into the adoption, implementation, and market trends of AI technologies in healthcare. Monitoring industry reports and market trends enables educators to stay informed about emerging AI applications, technology trends, and industry developments. Educators can use industry reports to anticipate future educational needs and adapt curriculum content to address evolving industry demands and priorities.

Expert Consultations and Collaborations: Engaging with AI experts, healthcare professionals, industry stakeholders, and regulatory agencies through expert consultations and collaborations provides valuable perspectives and insights into AI education and training needs as shown in Fig. 6. Expert consultations and collaborations facilitate knowledge exchange, networking, and partnership opportunities that enhance the quality and relevance of education and training programs. Additionally, involving experts in curriculum development and delivery ensures that education and training initiatives reflect current best practices and industry standards.

Feedback and Evaluation Data: Collecting feedback and evaluation data from learners, educators, and stakeholders is essential for assessing the effectiveness and impact of education and training programs. Surveys, focus groups, interviews, and

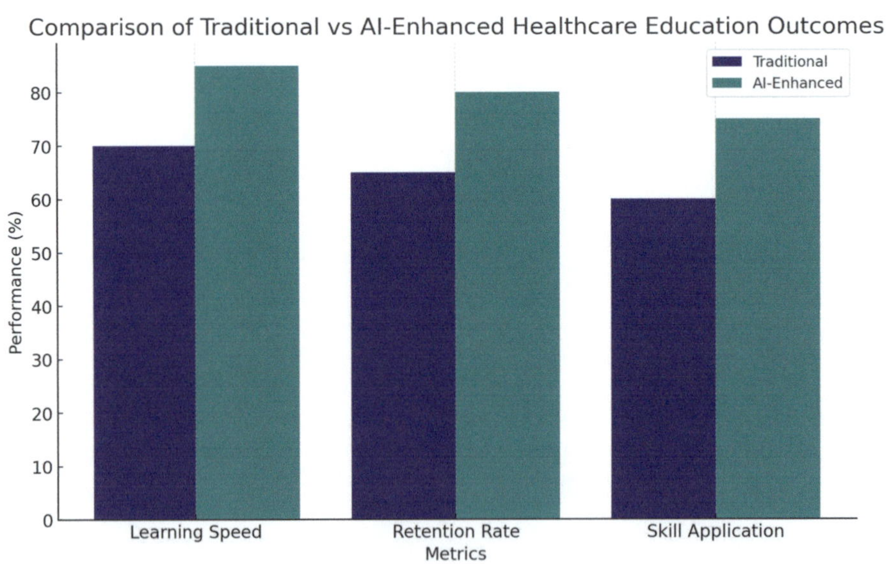

Fig. 6 Performance attitude vs. learning speed in traditional and AI-enhanced cases

evaluation forms can be used to gather feedback on program content, instructional methods, and learning outcomes. Analyzing feedback and evaluation data helps educators identify strengths and weaknesses in education and training programs and make informed decisions to improve program quality and effectiveness [15].

3 Green AI Solutions

The integration of artificial intelligence (AI) technologies into healthcare systems has necessitated the development of education and training programs to prepare the healthcare workforce for the AI-enabled future. In this literature review, we explore existing research and scholarship related to education and training initiatives for healthcare professionals in the context of AI in healthcare. Green AI represents a transformative approach to energy management, emphasizing efficiency and sustainability across data center operations. These solutions leverage advanced machine learning algorithms to optimize power usage, predict workloads, and enable real-time adjustments that minimize energy consumption without compromising performance. Technologies such as AI-driven thermal management systems use predictive analytics to model airflow and heat dissipation, reducing the need for energy-intensive cooling mechanisms. Similarly, workload scheduling algorithms dynamically allocate computational resources based on real-time demands, ensuring that energy is utilized effectively. The integration of renewable energy sources, guided by AI-based forecasting tools, further enhances the sustainability of data centers by aligning energy consumption with green energy availability. Moreover, innovations in AI model design, such as model compression and efficient training techniques, address the energy-intensive nature of traditional AI systems. These advancements collectively position green AI as a cornerstone of sustainable energy practices, offering scalable solutions to reduce the environmental footprint of data centers while maintaining operational excellence.

Current State of AI in Healthcare Education: Several studies have examined the current state of AI education in healthcare and identified gaps and challenges in existing educational programs. Research has explored the educational needs and preferences of healthcare professionals regarding AI training [16].

Curriculum Development and Implementation: Scholarship has examined the development and implementation of AI education curricula in healthcare training programs [17].

Developing effective education and training programs for the AI-enabled healthcare workforce requires careful planning and implementation. In this section, we outline the experimental setup for designing and evaluating education and training initiatives aimed at preparing healthcare professionals for the integration of artificial intelligence (AI) technologies into clinical practice. The first step in the experimental setup involves conducting a comprehensive needs assessment to identify the educational needs, preferences, and challenges of the target audience. This may involve surveys, focus groups, interviews, and literature reviews to gather insights into the knowledge gaps, skill deficiencies, and training priorities of healthcare professionals related to AI in healthcare [18]. Based on the needs assessment findings, the next step is to develop a curriculum that addresses the identified educational needs and objectives. The curriculum should include modules on fundamental concepts of AI, AI applications in healthcare, ethical and legal considerations, hands-on training with AI tools and technologies, and case-based learning activities. Collaboration with subject matter experts, educators, and stakeholders is essential in designing a comprehensive and engaging curriculum. The experimental setup should include the selection and implementation of appropriate instructional methods to deliver the curriculum content effectively. This may involve a combination of traditional didactic lectures, interactive workshops, small group discussions, case studies, simulations, and hands-on laboratory exercises [19]. The instructional methods should be tailored to the learning preferences and needs of the target audience and promote active engagement and participation. Providing access to high-quality learning resources is essential for supporting education and training initiatives. This may include textbooks, online modules, multimedia presentations, instructional videos, peer-reviewed articles, and AI software tools. Ensuring accessibility and usability of learning resources is crucial for facilitating self-directed learning and continuous professional development among healthcare professionals. Equipping educators with the necessary knowledge, skills, and resources to deliver AI education effectively is critical for program success [20]. Educator training programs should provide guidance on curriculum implementation, instructional methods, assessment strategies, and the use of educational technologies. Ongoing support and professional development opportunities should be available to educators to enhance their teaching effectiveness and promote innovation in AI education. Before full implementation, the experimental setup should include pilot testing of the education and training program to evaluate its feasibility, acceptability, and effectiveness. Pilot testing involves delivering the curriculum to a small sample of learners and collecting feedback on program content, delivery methods, and learning outcomes. Iterative revisions based on pilot testing results may be necessary to refine the program and address any identified deficiencies or challenges. An evaluation plan should be developed to assess the impact of the education and training program on learner knowledge, skills, attitudes, and behaviors. This may involve pre- and post-training assessments, skills demonstrations, self-assessments, surveys, and focus groups to measure learning outcomes and program effectiveness [21]. Longitudinal follow-up assessments may also be conducted to evaluate the sustainability of learning and the transfer of knowledge to clinical practice. Data collection methods should be implemented to gather quantitative and qualitative data on education and training outcomes. This may include collecting demographic information, pre- and post-training assessments, learner feedback, and program evaluation surveys. Data analysis techniques such as descriptive statistics, inferential statistics, and thematic analysis can be used to analyze the collected data and draw conclusions about program effectiveness and impact.

4 Barriers to Implementation: Technical, Economic, and Organizational Challenges

Education and training programs for the AI-enabled healthcare workforce must navigate a complex landscape of legal and regulatory considerations to ensure compliance, protect patient privacy, and uphold ethical standards. In this section, we examine key legal considerations that healthcare organizations and educational institutions must address when developing and implementing education and training initiatives related to artificial intelligence (AI) in healthcare.

Data Privacy and Security: Education and training programs may involve the use of sensitive healthcare data, including patient records and medical images, for instructional purposes. It is essential to comply with relevant data privacy laws and regulations, such as the Health Insurance Portability and Accountability Act (HIPAA) in the United States and the General Data Protection Regulation (GDPR) in the European Union. This includes implementing safeguards to protect the confidentiality, integrity, and availability of healthcare data and obtaining appropriate consent for data use in educational activities [22]. Intellectual Property Rights: Education and training programs may incorporate proprietary AI algorithms, software tools, and educational materials developed by third-party vendors or institutions. It is important to respect intellectual property rights and adhere to licensing agreements when using third-party resources. Additionally, healthcare organizations and educational institutions should establish policies and procedures for the development, ownership, and sharing of intellectual property generated through education and training activities. Education and training programs must adhere to professional standards and accreditation requirements established by regulatory bodies and accrediting agencies. For example, medical education programs may need to comply with standards set forth by organizations such as the Accreditation Council for Graduate Medical Education (ACGME) in the United States. Compliance with professional standards ensures that education and training programs meet minimum quality and competency requirements and prepares healthcare professionals for licensure and certification exams. Education and training programs should adhere to ethical guidelines and codes of conduct that govern professional behavior and practice standards in healthcare [21]. This includes principles such as beneficence, nonmaleficence, autonomy, and justice, as well as guidelines for ethical use of AI technologies in clinical practice. Healthcare professionals should be trained to recognize ethical dilemmas and apply ethical reasoning frameworks when confronted with challenging situations involving AI in healthcare.

Informed Consent and Patient Rights: Education and training programs involving patient interactions or simulations must obtain informed consent from participants and respect patient rights to privacy, confidentiality, and autonomy. Participants should be fully informed about the nature and purpose of the educational activities, the use of their data, and any potential risks or benefits involved. Additionally, healthcare professionals should be trained to communicate effectively with patients and respect their preferences and wishes regarding AI-enabled healthcare interventions [22].

Regulatory Compliance: Education and training programs must comply with applicable laws, regulations, and regulatory requirements governing healthcare education and practice. This includes regulations related to medical education, continuing education, professional licensure, and accreditation. Healthcare organizations and educational institutions should establish policies and procedures to ensure compliance with regulatory requirements and monitor changes in laws and regulations that may impact education and training activities.

Risk Management and Liability: Education and training programs should include provisions for risk management and liability mitigation to protect participants, educators, and healthcare organizations from potential legal liabilities. This may involve implementing policies and procedures for informed consent, risk disclosure, participant safety, and incident reporting. Additionally, healthcare organizations and educational institutions should maintain appropriate liability insurance coverage to mitigate financial risks associated with education and training activities.

Proper documentation and recordkeeping are essential for demonstrating compliance with legal and regulatory requirements and facilitating accountability and transparency in education and training programs. Participants' consent forms, educational materials, assessment results, and other relevant documents should be accurately documented, securely stored, and easily retrievable for audit and review purposes. Healthcare organizations and educational institutions should establish protocols for recordkeeping and data retention in accordance with applicable laws and regulations.

Education and training programs that involve international collaboration or participation may need to consider cross-border legal considerations, such as data transfer regulations, jurisdictional differences, and cultural sensitivities. Healthcare organizations and educational institutions should ensure compliance with relevant laws and regulations in the countries where educational activities are conducted and take appropriate measures to address any legal or regulatory challenges that may arise [18].

Continuing Legal Education: Healthcare professionals involved in education and training programs should receive continuing legal education (CLE) to stay informed about changes in laws, regulations, and legal standards relevant to their practice. CLE programs may cover topics such as healthcare law, privacy and security regulations, professional liability, and risk management. By staying up-to-date on legal developments, healthcare professionals can effectively navigate legal challenges and mitigate legal risks associated with AI-enabled healthcare education and practice [23].

Collaboration with Legal Experts: Collaboration with legal experts, including healthcare attorneys, regulatory specialists, and compliance officers, is essential for ensuring that education and training programs comply with legal requirements and best practices. Legal experts can provide guidance on interpreting and applying relevant laws and regulations, conducting risk assessments, and developing policies and procedures to address legal considerations. Additionally, legal experts can help healthcare organizations and educational institutions respond to legal inquiries, audits, and investigations related to education and training activities [24].

5 Strategies for Sustainable Growth: Collaboration, Innovation, and Awareness

The adoption of AI-driven solutions in data center energy management faces several significant technical and operational challenges. One of the primary issues is the high energy consumption of AI models during the training phase, particularly for large-scale deep learning frameworks. While these models hold the potential to enhance energy efficiency during operations, their resource-intensive training processes can offset these gains, necessitating the development of "Green AI"methodologies to balance efficiency and sustainability [25]. Additionally, limited real-time data processing capabilities in large-scale deployments hinder AI systems' ability to dynamically adapt to fluctuating workloads, leading to suboptimal energy optimization outcomes. Integration with legacy infrastructure further compounds these challenges, as older systems often lack the compatibility or flexibility required for seamless adoption of advanced AI technologies [26]. High initial investment costs for AI infrastructure, including specialized hardware, software, and sensors, deter many organizations from transitioning to AI-enabled energy management systems. Furthermore, a pronounced skill gap in "Green AI"expertise complicates the deployment and maintenance of these technologies. As the demand for professionals proficient in both AI and sustainable practices outpaces supply, organizations struggle to assemble teams capable of designing and managing energy-efficient AI systems. Additionally, complex regulatory requirements related to energy usage, data privacy, and environmental compliance add layers of difficulty, requiring significant effort to ensure adherence without compromising innovation. Resistance to organizational change represents another critical barrier to the adoption of AI-driven solutions. Many stakeholders are hesitant to embrace new technologies due to fears of disruption, scepticism about the tangible benefits, or a lack of understanding of AI's potential. This resistance can stall the implementation process and undermine efforts to achieve energy efficiency and sustainability goals. Addressing these challenges requires comprehensive strategies, including targeted workforce training, investment in scalable infrastructure, regulatory alignment, and effective change management frameworks to facilitate the transition to AI-driven energy optimization in data centers. Effective education and training programs often require interdisciplinary collaboration between healthcare professionals, data scientists, engineers, ethicists, and other stakeholders. However, fostering collaboration across diverse disciplines can be challenging due to differences in language, culture, and expertise. Successful collaboration requires effective communication, coordination, and mutual understanding among interdisciplinary teams. Despite these limitations, education and training programs for the AI-enabled healthcare workforce hold tremendous potential for advancing healthcare delivery and improving patient outcomes [27]. By addressing resource constraints, technological complexity, interdisciplinary collaboration, evaluation and assessment, and legal and ethical considerations, stakeholders can develop and implement education and training initiatives that empower healthcare professionals to harness the transformative potential of AI technologies in clinical practice [28]. This growing sector presents opportunities for public–private partnerships to fund research, establish innovation hubs, and support startups focused on sustainable technology. Collaborative efforts between governments, industries, and academia can drive investment in green AI infrastructure while fostering an ecosystem of innovation that aligns technological advancements with environmental objectives. Equally important is the role of community engagement and awareness programs in promoting the adoption of sustainable practices [22]. Environmental impact assessment tools can help measure and communicate the benefits of green AI systems, reinforcing the value of such initiatives to stakeholders. Community-focused frameworks can encourage participation in sustainability efforts, ensuring that local perspectives and needs are incorporated into planning and implementation. Raising awareness through educational campaigns and outreach programs further amplifies public support, creating a culture of environmental responsibility that complements technological progress. Together, these efforts can ensure that green AI solutions not only mitigate the environmental impact of energy-intensive systems but also contribute to broader societal goals of sustainability and inclusivity. Ethical, Legal, and Social Implications (ELSI) Education: As AI technologies become more pervasive in healthcare, there will be a growing need for education and training programs that address the ethical, legal, and social implications (ELSI) of AI in healthcare. Healthcare professionals will need to navigate complex ethical dilemmas, legal frameworks, and societal concerns related to AI-enabled healthcare interventions. Future education and training initiatives will integrate ELSI education into their curricula to ensure that healthcare professionals are equipped to address these challenges responsibly and ethically. As AI technologies become more integrated into healthcare workflows, there will be a growing emphasis on human-centered design principles in education and training programs. Future initiatives will prioritize the needs, preferences, and experiences of healthcare professionals and patients to ensure that AI solutions are user-friendly, intuitive, and effective. Education and training programs will incorporate principles of human-computer interaction, usability testing, and user feedback into their curriculum design and instructional methodologies to promote user acceptance and adoption of AI technologies in healthcare.

References

1. Kaushik, K. (2022). A novel approach to secure files using color code authentication. In V. Sugumaran, D. Upadhyay, & S. Sharma (Eds.), *Advancements in interdisciplinary research. AIR 2022. Communications in Computer and Information Science* (Vol. 1738). Springer. https://doi.org/10.1007/978-3-031-23724-9_4
2. Kaushik, K., Singh, V., & Manikandan, V. P. (2022). A novel approach for an automated advanced MITM attack on IoT networks. In V. Sugumaran, D. Upadhyay, & S. Sharma (Eds.), *Advancements in interdisciplinary research. AIR 2022. Communications in Computer and Information Science* (Vol. 1738). Springer. https://doi.org/10.1007/978-3-031-23724-9_6

3. Kaushik, K., & Naik, V. (2023). *Making ductless-split cooling systems energy efficient using IoT*. 2023 15th International Conference on Communication Systems & Networks (COMSNETS), Bangalore, India, pp. 471–473, https://doi.org/10.1109/COMSNETS56262.2023.10041408.
4. Porwal, P., & Devare, M. (2024, August 1). Scientific impact analysis: Unraveling the link between linguistic properties and citations. *Journal of Informetrics, 18*, 101526. https://doi.org/10.1016/j.joi.2024.101526
5. Porwal, P., & Devare, M. (2023, September 20). Citation count prediction using weighted latent semantic analysis (WLSA) and three-layer-deep-learning paradigm: A meta-heuristic approach. *Multimedia Tools and Applications., 83*, 32245. https://doi.org/10.1007/s11042-023-16957-8
6. Porwal, P., & Devare, M. (2021). Citation classification prediction implying text features using natural language processing and supervised machine learning algorithms. *Communications in Computer and Information Science*. https://doi.org/10.1007/978-981-16-0507-9_46
7. Tatkare, K. A., & Devare, M. (2019). A novel region duplication detection algorithm based on hybrid approach. *Social Science Research Network*. https://doi.org/10.2139/ssrn.3425340
8. Mohurle, S., & Devare, M. *Fuzzy probability model for quantifying the effectiveness of the MSW compost*. IEEE Conference Publication | IEEE Xplore, 2019. https://ieeexplore.ieee.org/document/8945830
9. Devare, M., & Thakral, M. (2023). Enhancing automatic speech recognition system performance for Punjabi language through feature extraction and model optimization. *International Journal of Intelligent Systems and Applications in Engineering, 12*(8s), 307–313.
10. Thakral, M., & Devare, M. (2023). *Designing an automatic speech recognition system for the minor age group*. 2023 4th International Conference on Communication, Computing and Industry 6.0 (C216), Bangalore, India, , pp. 1–5, https://doi.org/10.1109/C2I659362.2023.10431311.
11. Thakral, M., Singh, R. K., & Kaushik, K. (2022). *Integration of blockchain technology and intelligent system for potential technologies*. Chapman and Hall/CRC eBooks. https://www.taylorfrancis.com/chapters/edit/10.1201/9781003193425-6/integration-blockchain-technology-intelligent-system-potential-technologies-manish-thakral-rishi-raj-singh-keshav-kaushi
12. Thakral, M., & Singh, S. (2022). *A secure bank transaction using Blockchain computing and Forest oddity*. Blockchain Technologies. https://link.springer.com/chapter/10.1007/978-981-19-1960-2_6
13. Thakral, M., Singh, R. R., & Singh, S. P. (2022). An extensive framework focused on smart agriculture based out of IoT. In A. Choudhury, T. P. Singh, A. Biswas, & M. Anand (Eds.), *Evolution of digitized societies through advanced technologies. Advanced technologies and societal change*. Springer. https://doi.org/10.1007/978-981-19-2984-7_12
14. Thakral, M., Singh, R. K., & Kalghatgi, B. V. (2022). Cybersecurity and ethics for IoT system: A massive analysis. *Transactions on Computer Systems and Networks*. https://link.springer.com/chapter/10.1007/978-981-19-1585-7_10
15. Mannepalli, P. K., Richhariya, V., Gupta, S. K., Shukla, P. K., Dutta, P. K., Chowdhury, S., & Hu, Y.-C. (2024). A robust blockchain-based watermarking using edge detection and wavelet transform. *Multimedia Tools and Applications, 83*(20), 1–25. https://doi.org/10.1007/s11042-024-18907-4
16. Singh, B., Dutta, P. K., & Kaunert, C. (2024). Wind-solar renewable energy and innovative technologies applying internet of things (IoT) for green and sustainable future: Projecting carbon neutrality for Smart and sustainable cities. In J. A. Parray, A. K. Haghi, & G. Meraj (Eds.), *IoT-based models for sustainable environmental management. Lecture notes on data engineering and communications technologies* (Vol. 227). Springer. https://doi.org/10.1007/978-3-031-74374-0_6
17. Thakral, M., Singh, R. R., Jain, A., & G. Chhabra. (2021) *Rigid wrap ATM debit card fraud detection using multistage detection*. 2021 6th International Conference on Signal Processing, Computing and Control (ISPCC), Solan, India, pp. 774–778, https://doi.org/10.1109/ISPCC53510.2021.9609521.

18. Girish, I., Kumar, A., Kumar, A., & Anuradha, M. (2020). *Driver fatigue detection*. 2020 IEEE 17th India Council International Conference (INDICON), New Delhi, India, pp. 1–6, https://doi.org/10.1109/INDICON49873.2020.9342456.
19. Prajeesha & Anuradha. M. (2021). *EDGE computing application in SMART GRID-a review*. 2021 Second International Conference on Electronics and Sustainable Communication Systems (ICESC), pp. 1–6, https://doi.org/10.1109/ICESC51422.2021.9532792.
20. D. M. Udeshi, S. G. L. Divakarla, N. C. Rajdev and Anuradha, M. (2022). *Wind speed forecasting using hybrid model*. 2022 IEEE 7th international conference for convergence in technology (I2CT), Mumbai, India, 2022, pp. 1–5, https://doi.org/10.1109/I2CT54291.2022.9823995.
21. Prasad, V. K., Bhattacharya, P., Hemanth, D. J., Dutta, P. K., Kathait, A., & Dănciulescu, D. (2024). Navigating the future: Secure big data analytics in healthcare's 5G era (Chapter 1). In *Secure big-data analytics for emerging healthcare in 5G and beyond* (pp. 1–10). https://doi.org/10.1049/PBHE063E_ch1
22. Balasubramanian, S., Kashyap, R., Teja Cvn, S., & Anuradha, M. (2020) Hybrid prediction model for type-2 diabetes with class imbalance. 2020 IEEE International Conference on Machine Learning and Applied Network Technologies (ICMLANT), Hyderabad, India, pp. 1–6, https://doi.org/10.1109/ICMLANT50963.2020.9355975.
23. M. Yakasiri, Anuradha, M, & Keshavan, B. K. *Comparative analysis of Markov chain and polynomial regression for the prognostic evaluation of wind power*. 2020 IEEE International Conference for Innovation in Technology (INOCON), Bangluru, India, pp. 1–5, https://doi.org/10.1109/INOCON50539.2020.9298374.
24. Gupta, S., Sarkar, D., Roy, D., Kundu, S., & Dutta, P. K. (2024). Embracing flexibility and efficiency: Practical approaches to Agile project management. In S. Gupta, D. Sarkar, D. Roy, S. Kundu, & P. K. Dutta (Eds.), *Practical approaches to Agile project management* (p. 8). IGI Global. https://doi.org/10.4018/979-8-3693-3318-1.ch001
25. Yakasiri, M., Avrcl, J., Sharma, S., Anuradha, M., & Keshavan, B. K. (2019) *A stochastic approach for the state-wise forecast of wind speed using discrete-time Markov chain*. TENCON 2019–2019 IEEE Region 10 Conference (TENCON), Kochi, India, pp. 575–580, https://doi.org/10.1109/TENCON.2019.8929529.
26. Bhattacharya, P., Chatterjee, S., Datt, R., Verma, A., & Dutta, P. K. (2023). A permissioned Blockchain approach for real-time embedded control systems. In S. Kadry & R. Prasath (Eds.), *Mining intelligence and knowledge exploration. MIKE 2023* (Vol. 13924). Springer. https://doi.org/10.1007/978-3-031-44084-7_32
27. Udeshi, D. M., Divakarla, S. G. L., Rajdev, N. C., and Anuradha, M. (2022). *Wind speed forecasting using hybrid model*. (2022). IEEE 7th International conference for Convergence in Technology (I2CT), Mumbai, India, pp. 1–5, https://doi.org/10.1109/I2CT54291.2022.9823995.
28. Tammana, A., Amogh, M. P., Gagan, B., Anuradha, M., & Vanamala, H. R. (2021). Thermal image processing and analysis for surveillance UAVs. In M. S. Kaiser, J. Xie, & V. S. Rathore (Eds.), *Information and communication technology for competitive strategies (ICTCS 2020). Lecture notes in networks and systems* (Vol. 190). Springer. https://doi.org/10.1007/978-981-16-0882-7_50

AI and Sustainability: Can the Fourth Revolution Usher in a New Reality?

Ritaprava Bandyopadhyay

1 Introduction

A recent Guardian report states that artificial intelligence could assist in addressing the climate crisis is erroneous, with the technology likely to increase energy use and accelerate the spread of climate disinformation, according to a coalition of environmental groups. In this report, it has been stated that, in the United States, there is already evidence that coal-fired power plants are being extended to satisfy the increased energy demands of AI. The report also mentioned that separate research suggests that in just 3 years, AI servers might consume as much energy as Sweden [22]!

It is a fact that artificial intelligence (AI) can accomplish a wide range of tasks, playing an important role in creating and altering our reality. Moreover, AI is being used in a variety of spheres, including education, banking, defence, the environment, and so on. As we all know, AI is the product of the Fourth Revolution, and it is widely assumed that the new Avatar will usher in a new era in hyperhistory[1], however, the widespread use and new advancements pose worries about sustainability.

[1] Floridi divides human history into three periods: prehistory, history, and hyperhistory. He claims that there was no information and communication technology (ICT) in prehistory. There was ICT in history, but the equipment employed could only capture data and not process it. In hyperhistory, ICTs and their data processing skills have been required for the maintenance and advancement of social welfare, personal well-being, and intellectual flourishing. We are reliant on ICTs in hyperhistory.

R. Bandyopadhyay (✉)
Jadavpur University, Kolkata, West Bengal, India

Anandabazar Online, Kolkata, West Bengal, India
e-mail: ritaprava.bandyopadhyay@abpdigital.in

P. Raj et al. (eds.), *Artificial Intelligence (AI) for IT Energy Efficiency and Green AI for Environment Sustainability*,
https://doi.org/10.1007/978-3-031-89420-6_5

It should be noted that, in the AI literature, AI for sustainability and the sustainability of AI are treated as two different genres under a common head sustainable AI. In AI for sustainability, the main question is whether AI can perform toward sustainable development goals. On the other hand, the sustainability of AI "is a movement to foster change in the entire lifecycle of AI products (i.e., idea generation, training, re-tuning, implementation, governance) toward greater ecological integrity and social justice" [28]. The sustainability of AI is concerned with the overall "sociotechnical system of AI," not simply AI applications. On the other hand, AI for sustainability is concerned with the application of AI to meet the Sustainable Development Goals (SDGs) to transform our world. So, in a nutshell, AI for sustainability largely deals with the application of AI. We can perceive that, in the fourth revolution, AI is becoming the most sought-after technology. The more it is used, the more questions about sustainable AI will arise. We will see that there is a paradoxical relationship between AI and sustainability. For sustainable development, AI has become the sine qua non of our existence. However, the paradox is that the use of AI ushers in other imponderables, which will have severe repercussions on the environment and economy as well. The report we have mentioned above also echoes the same. The further we go toward sustainability, the more we become entangled in the conundrum. This dilemma will have to be addressed in the context of a new reality. These are the lines of thought I follow in this chapter. The chapter has three sections, (I) AI and Sustainability: The Relationship, (II) The Paradox, and (III) The Conclusion.

1.1 Methodology

This chapter examines journals and other publications related to the debate using a qualitative approach to social science. I have analyzed and critically revisited the research already conducted in this area and tried to explain the problem.

2 I: AI and Sustainability: The Relationship

Before going deep into the topic, we need to understand in brief what AI is and what sustainability means. This will pave the way for our discussion regarding the relationship between the two and, eventually, the paradox. In the conclusion, I will discuss how sustainable AI is possible, keeping aside the paradox.

2.1 *Artificial Intelligence*

It was 1956. At a conference at Dartmouth College, John McCarthy introduced the word artificial intelligence (AI) [6]. After that, it became popular gradually with time. It is assumed that to do a job, one has to become an agent who can understand the nature of the job. Supposing that one wants to eradicate hunger, then the agent must understand what hunger means. Hence, to satisfy the Sustainable Development Goals or SDGs, AI has to become agents like a natural agent who understands their duty.

In the domain of AI, the concepts of agency, autonomy, and intelligence are all obscure and difficult to define. Furthermore, the agency is inextricably linked to qualities such as "autonomy," "situatedness," and "embodiment." Many scholars avoid providing minimal definitions because such definitions are invariably either too broad or too narrow. The concept of an agent is intended to be a tool for analyzing the system rather than an absolute categorization that divides the entire universe into agents and nonagents [23]. Furthermore, various definitions available in the literature are frequently inconsistent with one another [10]. For our current purpose, we will take artificial intelligence, or artificially intelligent agents such as robots, as machines whose shape varies and whose decision-making abilities and actions are based on algorithms. With the given input and following certain rules, these machines perform certain steps to produce output. In this chapter, artificial intelligence, artificially intelligent agents, robots, automata, and intelligent machines are used interchangeably. Presently, most definitions postulate that AI resolves intricate cognitive problems connected with human intelligence. Some may say that AI benefits us through smartwatches, smartphones, and healthcare methods. Another group says that AI identifies problems and offers solutions for the advantage of technology and, at the same time, for the benefit of people and society. However, the central idea of AI has unceasingly been to craft machines that are proficient in thinking like humans [2]. As I have mentioned earlier, it is hard to define AI, so here I will not venture to define AI but rather focus on the relationship between AI and sustainability. Before that, I want to clarify certain ideas on sustainability in brief.

2.1.1 Sustainability

Though there are many aspects of sustainable development, I will restrict myself only to what the United Nations Charter has provided. This has a direct connection with our present discussion with AI. The 2030 Agenda for Sustainable Development, which was endorsed by all UN Member States in 2015, presents a "shared roadmap for peace and prosperity for people and the planet today and in the future" [27]. In the field of AI, sustainability means achieving peace and prosperity for people and the planet. This is cardinal in my discussion.

There are 17 Sustainable Development Goals (SDGs), which require an urgent need to action by all countries, developed and developing, in a global partnership.

They acknowledge that alleviating poverty and other forms of deprivation must be linked with programs to improve health and education, reduce inequality, and drive economic growth. In this context, we must not forget to discuss the raison d'etre of climate change. Our endeavor should be to protect our oceans and forests. Real economic growth depends on all of these factors [27].

These 17 SDGs are: GOAL 1: No Poverty, GOAL 2: Zero Hunger, GOAL 3: Good Health and Well-being, GOAL 4: Quality Education, GOAL 5: Gender Equality, GOAL 6: Clean Water and Sanitation, GOAL 7: Affordable and Clean Energy, GOAL 8: Decent Work and Economic Growth, GOAL 9: Industry, Innovation and Infrastructure, GOAL 10: Reduced Inequality, GOAL 11: Sustainable Cities and Communities, GOAL 12: Responsible Consumption and Production, GOAL 13: Climate Action, GOAL 14: Life Below Water, GOAL 15: Life on Land, GOAL 16: Peace and Justice Strong Institutions, GOAL 17: Partnerships to achieve the Goal [27].

When we talk about sustainability in AI we keep this background in mind. Scholars have rightly pointed out that the term sustainable AI implies both AI for sustainability and sustainability of AI though there is a difference between the two [28]. AI for sustainability is an area where plenty of research is going on and there is no dearth of funding. On the other hand, the sustainability of AI is relatively new. The researchers hold that in the third wave of ethics[2] environmental cost of AI must be taken into account [28].

Initially, the producing sectors used industrial robots. Currently, AIAs are widely used in different spheres of the economy and society. It is being used in areas like managing supply chains, farming, autonomous vehicles, warfare, travel, tourism, education, journalism and additional services, trading on the financial markets, and implementing medical operations. Apparently, all of these contribute to the SDGs. Chatbots are being used in businesses to communicate with and maintain connections with their clients [2]. We have the example of ChatGPT which is a type of large language model that can handle large datasets and can give apparently "coherent" answers [8]. These are examples of AI for sustainability that demonstrate AIAs' pervasive integration into society, resulting in massive shifts in how people live, work, and conduct business. In other words, it is continuously reshaping our world. However, it is not confined to these sectors but rather has become a present continuous process involving our very existence and our relationship with nature. As a result, the power of AI can be compared with the internet and smartphones. Moreover, in order to call it "sustainable" this process can satisfy or can be used to satisfy some of the SDGs provided by the UN.

[2] Just like fourth revolution, there are three waves of AI ethics. The inadequacy of equality in data used for training causes inefficiencies in artificially intelligent agents. Ethicists of the second web were concerned with issues like privacy, property rights, and problem emerging from face and emotion recognition technology that often violate citizens' rights. In the third wave, ethicists are concerned more with the relationship between the environment and artificial intelligence as well as urge AI scientists, policymakers, scholars, and commoners to come forward and contribute [28].

It has been rightly pointed out earlier that AI for sustainability is a little more developed. The author cites the nonprofit organization "AI4Good" as an example of this. The author says that this branch examines the use of artificial intelligence to promote sustainability in one way or another. We can see that AI is being used to achieve some of the SDGs set by the UN. In this case, artificial intelligence, or machine learning, is a means for creating inexpensive and clean energy, as envisioned by SDG 7 [28].

However, to understand the relationship between sustainability and AI, we need to discuss in brief the concept of sustainability of AI. The sustainability of AI keeps the environmental impact of the new technology at its core. Sustainable AI is a movement that promotes change across the AI product lifecycle (i.e., idea development, training, retuning, deployment, and governance) to achieve more ecological integrity and social fairness [24]. As a result, sustainable AI covers the entire socio-technical system of AI rather than just AI applications. Sustainable AI involves creating AI that is in line with sustaining natural assets for both present and forthcoming generations [17]. This is not about how artificial intelligence will be sustained [28].

Google's AlphaGo Zero generated 96 tonnes of carbon dioxide over 40 days of laboratory training, which is the same as a 1000 h of traveling by air or the carbon footprint of 23 US households [28]. When the world is obliged to commit to reducing carbon dioxide, one must wonder if the greenhouse gases released from programs that can play games (or execute different petty duties) are genuinely worth the expenditure [28]. Scholars showed in a well-known study that the procedure of preparing a single deep learning, natural language processing model can result in approximately six pounds of CO2 releases [28].

This is one side of the story. Another aspect of AI's growing popularity is that apparently, it poses a threat to the loss of jobs, which in turn would impact two of the SDGs, i.e., Decent Work and Economic Growth and eradicating Inequality creating an AI divide—those who have AI technology and those who have not. These create a paradox. For sustainable development, we need AI and AI, on the other hand, defies certain SDGs.

3 II: The Paradox

To solve the riddle and facilitate sustainable AI, one must be cautious in its usage, and it has to be ensured that AI has to comply with the demands of lifeworld. As we embark on the fourth revolution, sustainable AI is becoming the need of the hour as it is directly affecting society. As we have seen earlier, since the usage of AI has become as pervasive as the internet or smartphones, its environmental effects must be taken into account. Sustainable AI is focused on determining how to assess the long-term viability of developing and deploying artificially intelligent agents, such as carbon footprints, processing power for training algorithms, and so on. For this reason, sustainable AI must include both AI for sustainability and sustainability of AI [28].

At this juncture, one may ask what is the relationship between the fourth revolution and sustainable AI.

To delve deep, we must clarify certain ideas regarding information ethics as envisioned by Luciano Floridi [11]. Information Ethics (defined as ontocentric, patient-oriented, ecological macro ethics) has emerged as a critique of anthropocentric ethics. Floridi, a pioneer of information ethics, sees it as the "fourth revolution," following the Copernican Revolution, Darwinism, and Freudianism.

The concept of Floridi's Information Ethics emerges after Alan Turing's theory [15]. Human beings have been pushed from the center to the periphery in each previous "revolution." In the Copernican Revolution, humans were thrown away from the center of the cosmos. That is to say, living beings just are not motionless at the center of the solar system. In Darwinism, humans were removed from the biological kingdom. This means living beings are not separated from the rest of the animal kingdom. Freud shows that not all the parts of the mind are rational. That is to say, living beings are not logical thinkers entirely open to themselves. So humans were removed from the center of rationality as well. Fourth is Turing's invention, which brings about a change in the realm of Ethics. Turing shows that living beings are not separate actors but information-based species, or "inforgs." They share an essentially informational atmosphere. This is called the infosphere [14]. Based on this foundation, Floridi developed his theory. The replacement of biocentrism with ontocentrism is an example of this. According to Floridi, nonhumans can come to the center of ethical discussion. For this reason, there needs to be a doctrine. Information ethics can serve this purpose [11].

Human evolution, as Floridi describes, may be seen as a three-stage cycle. First is the prehistory. In prehistory, there were no ICTs. In history, we have ICTs. But they record and transmit data. In history, human societies have relied heavily on technologies concerning primary resources and energy. In hyperhistory, there are ICTs. Here, ICTs can record, transmit, and, above all, process data, and human societies become vitally dependent on them and on information as a fundamental resource. By prehistory, Floridi means the period before written records. In history, we could get radios, televisions, or tape recorders that can store or transmit data. In hyperhistory, we get technology that records, transmits, and, most importantly, processes data autonomously, and our society becomes increasingly reliant on it. An example of this is AI. It takes information as a fundamental resource. Society in hyperhistory has become ICT-related to ICT-dependent [13].

Hyperhistory depicts societies or situations in which ICTs and data processing capabilities are required for the maintenance and future growth of social welfare, personal well-being, and intellectual flourishing, thereby contributing to meeting some of the SDGs. It undoubtedly offers a broad range of potential, all of which are primarily driven by the recording, transmitting, and processing capabilities of ICTs. To make the fourth revolution meaningful, hyperhistory relies heavily on AI. The more AI is being used, the more it poses a threat to the environment and job market. Here lies the paradox.

One way to avoid this is the responsible use of AI. How will it happen? According to scholars, "authors should report training time and hyperparameter sensitivity" since "this will enable direct comparison across models" [25].

Researchers have informed that up to the present time, two apparatuses are accessible for counting emissions: the "Machine Learning Emissions Calculator" calculates the carbon trail of GPU by identifying hardware type, hours used, cloud provider, and region [1], and the "experiment-impact tracker" system "for tracking real-time energy consumption and carbon emissions, as well as generating standardized online appendices" [28]. Each of these measures seeks to reduce carbon emissions and energy consumption to enable the long-term development of machine learning. Although the instruments to detect carbon emissions are available, stronger enticements are required to incentivize researchers and industry inventors to quantity and circulate such data. Anthony offers a "carbon tracker," a "tool for tracking and predicting the energy consumption and carbon emissions of training DL models" based on the work of Henderson et al. [1].

To achieve sustainable AI, it must be used to meet the demands of the lifeworld. Controlling AI's carbon footprint is one need for the lifeworld's sustainability; another is involving humans in the process.

The second-order paradox is that, as a result of the fourth revolution, biocentrism has been replaced by ontocentrism. Hence, humans drift away from the middle. We know that in hyperhistory, technology interacts with technology and then with technology again. This will look like this: Technology → Technology → Techno logy [12].

The human is either absent or relegated to the back seat in this loop. Subsequently, technology such as AI is becoming increasingly disconnected from the lifeworld, as it is humans that bridge the gap between the real and the virtual. Though AI collects data from the lifeworld and strives to contribute to its sustainability, it cannot "understand" the lifeworld in the same way that natural agents can. For this reason, human interaction and active involvement are required for the sustainable development of the lifeworld.

It should be noted that to achieve the SDGs scholars discuss three points: hope as a starting point, vision as a goal to achieve, and process as what allows the vision to be implemented from the hope [21]. According to the authors, there is hope that the development and usage of AI will have a positive impact on individuals, societies, and surroundings [21]. They have stated that this is the underlying concept of "AI for social good" (AI4SG) [9]. However, such an approach has gained traction in the AI community [21]. We have already seen that AI, as a general-purpose technology, can address a wide range of problems and jobs. We have also seen that there is another way to look at this problem. Furthermore, Mazzi and Floridi accept that scalable adoption of AI technologies to improve SDG attainment necessitates coordinated activities by corporate and public actors [21]. In their work, also public and corporate actors become important.

Why is it so? One way to answer this is that if the big business houses rely heavily on AI and cut the jobs of humans, then this would hamper the equilibrium of value generation and value transfer, jeopardizing one of the SDGs. Following Marx [3], we know that only humans can generate value and machines can only transfer it. If humans are gradually eliminated from the process, then value generation will not occur, thereby impacting SDG8, i.e., Decent Work and Economic Growth and this disparity would increasingly promote inequality.The above-mentioned technological automation has huge potential. However, studies have found little evidence that they have had an impact on aggregate production. Labor productivity growth rates, according to the researchers, plummeted in the mid-2000s and have since remained low. From 2005 to 2016, cumulative labor productivity growth in the United States averaged only 1.3% per year. Between 1995 and 2004, the yearly growth rate of 2.8% was maintained at less than half of its previous level. The analysis discovered that 28 of the 29 other nations whose productivity growth statistics the OECD has compiled followed the same trend [26].

Furthermore, annual labor productivity growth rates in these countries were 2.3% from 1995 to 2004. It fell to 1.1% between 2005 and 2016 [2]. Moreover, real median income has been stagnant since the late 1990s, and noneconomic measures of well-being, such as life expectancy, have declined for some groups [7].

Following Marx's analysis of the "tendency of the rate of profit to fall," I would like to offer an alternative explanation for this phenomenon. I will not delve deeply into Marx's Labour Theory of Value but will instead attempt to cite a possible explanation for the aforementioned hypothesis—Why does the rate of profit of some companies tend to fall, which impacts the economy? In this spirit, I draw your attention to Chap. 13 of *Capital* Volume 3, where Marx discusses a long-term tendency for profit rates to fall with capitalist development [19]. It successfully describes how an AI agent or any constant capital (i.e., plant and building, equipment, physical infrastructures of production, raw materials, auxiliary materials, energy, and so on) could not substitute human agency, whose distinctive and innovative expression lies in possessing labor power (variable capital), which is the sole source of value. According to Marx, constant capital can only transfer value to a new commodity that has historically created constant capital (such as AI in our case). Marx holds value can be defined as socially necessary labor time congealed in constant capital. An artificially intelligent agent, according to Marx, is dead labor. It aids the owner of the means of production (the capitalist) in lowering the value of labor. It can be the source of increased relative surplus value rather than the value per se [2]. The fear that technological advancements will usher in a future where machines will take our jobs is not new. It dates back to the nineteenth-century Luddite rebellion and rose to prominence with the introduction of new technological advancements [16].

However, one of the most contentious arguments in the *Fragment* is that general intelligence is not an attribute of labor. It is an attribute of a fixed capital asset, specifically machinery and organizational systems [18]. According to Marx, "the general productive forces of the social brains are thus absorbed into capital, as opposed to labor, and thus appear as an attribute of capital, and more specifically of fixed capital" [20]

How do humans generate value that machines cannot? Why is lifeworld significant in the story of AI and sustainability, and why is human intervention required to ensure the sustainable development of lifeworld? This will set the stage for the conclusion.

4 III: Conclusion

By lifeworld, I mean the phenomenological world of intersubjective experience. Individual, social, perceptual, and practical experiences are all part of the lifeworld. Through experience, humans learn whatever they can. Furthermore, human experience tells her that reality is too diverse to be explained algorithmically. Understanding others through mental simulation occurs by participating in the lifeworld [5]. To achieve the SDGs one needs to participate in the lifeworld.

The United Nation's 2030 Agenda's goals are quite ambitious, but there is a limited amount of time left to accomplish them. If we want to use innovative technologies to further our common objectives, we should at the very least responsibly consider only those that are sustainable in and of themselves.

Here, human agency becomes important. However, artificially intelligent agents can imitate humans' emotions and their linguistic ability, but it is impossible to imitate humans by their subjective experiences. It is impossible to copy humans epistemologically and metaphysically.

I want to add further that the lifeworld is the grand narrative that provides the content for smaller narratives that constitute individual selves, which is the repository of value that is needed to achieve the SDGs for a prosperous lifeworld.

However, it is the lifeworld where agents interact with each other and come to decisions about each other's thoughts, plans, actions, feelings, and emotions. In some cases, humans do precisely these things to do the jobs that are required to meet the SDGs. Moreover, by participating in the lifeworld, we understand each other as like-minded, autonomous, and efficient beings like human agents. It is the lifeworld that forms our concepts of personhood, mind, and action and through which humans conceptualize a problem and devise its solution. In the process, she generates value over time. Lifeworld, however, is the source of human creativity and value too.

Humans are naturally occurring agents. They are evolution's by-products. The subtle nuances of human beings behaving in a particular situation are beyond the comprehension, representative ability, and capture of algorithms. Using mental simulation, they can predict or explain others' behavior. At this juncture, I have asked, can an artificial agent do the same? Here, we are talking about empirical and contingent matter, which is subject to testing. Given the conception of simulation as embedded in the phenomenological notion of the lifeworld, it does not seem appropriate to think of an artificial agent as a participant in the lifeworld and certainly not an agent of mental simulation.

This is one side of the story. Another thing is that sometimes the creators of artificial agents cannot explain their behavior. We can consider the examples of autonomous self-driving cars defying certain rules and meeting with an accident and two Facebook bots that began chatting with each other [2].

However, from this perspective, as common or folk people it will not be prudent to ascribe agency to them because their functioning is unknown to the common people and in some cases, it is strange to their creators as well. Technically knowledgeable individuals from specialized disciplines may claim to have special knowledge of "the mental life of machines" but the matter must be seen through the eyes of the roadside commoner, the employed, the teacher, the parent, and the environment activists. The question of autonomy for such agents is thus even more far-fetched than their mental lives [2].Hence, it is hard to accept artificial agents as autonomous agents just like their human counterparts who participate in the life-world to meet the SDGs. For example, to eradicate hunger, one must know what hunger in lifeworld means. Likewise, to end poverty, one must understand what poverty in lifeworld is. Without active participation in the lifeworld, one cannot understand some of the SDGs.

Our engineer friend may say that machine learning algorithms are now written in such a way that they can adapt their behavior to their surroundings. They evolve as a result of constant input from their surroundings. With this input, they can alter the algorithm too. But what will happen to those problems that are nonalgorithmic? We know that most psychological and moral problems are nonalgorithmic. The problems we face from some of the SDGs like reduction of inequality or gender equality are also nonalgorithmic in nature.

It could be argued that if the lifeworld has a high degree of algorithmicity, then it can be captured by an explanatory and predictive device like a theory. Then it must be rule-bound in (at least) the most basic sense. If this is the case, it must be highly predictable and programmable. If it is programmable, it can be programmed into a robot or any intelligent artifact. That is correct.

However, if we consider the lifeworld in its essential dynamicity, then the only way to capture its sometimes fleeting and sometimes lasting clusters of meaning would be through some kind of participatory process like mental "simulation." The computability of the lifeworld is put into question due to its essential subjectivity and uniqueness. One may ask if this does not put into question the predictability of activities in the lifeworld altogether and whether that would be desirable. The absolute unpredictability of the lifeworld as such and the directions in which it may progress or venture in the future is perhaps unquestionable. That does not, however, completely jeopardize the possibilities of statistical prediction in identifiable sections of the lifeworld [2].

Every moment is a fresh start. Because we are in the present, we cannot predict the next moment as it is unknown to us. We can hope for, expect, and even predict the next moment, but in reality, our prediction, our hope, may remain elusive. Our quest for a new adventure begins as soon as we discover that reality is quite different than we thought. However, the lifeworld plays an important role in shaping our thoughts as we embark on this new journey, and it would be wrong to equate this lifeworld with the physical world. Our experiences and thoughts are the value reservoirs that make humans human.

What if it is suggested that the artificial agents might as well join in the lifeworld and enjoy an acceptable degree of intersubjectivity with the natural agents therein? This too, is not to be so.

We can argue that even though the theoretical structure may be built into artificial agents to capture different kinds of regularities in the lifeworld, to attain any degree of inter-subjectivity, the artificial agents must possess the capacity to simulate [4]. To qualify as a mental simulator like a natural agent, artificial agents must acquire the ability to generate, grasp, and comprehend the content of the lifeworld. In the same way, the natural agent, the human, cannot meaningfully interact with the artificial agent through simulation. The bar lies in attempting to put herself in the digital shoes of the artificial agent. This is a serious problem in any interpersonal exchange built upon first-person experience.

Even if the "agency" of an artificial agent is established by some ethical doctrines, it is clear from this discussion that we cannot ascribe agency to them like the natural agent. To address the SDGs one needs to first qualify as an agent, as without agency, one cannot understand the needs of the lifeworld. As I have mentioned earlier, by the lifeworld I mean the phenomenological world of intersubjective experience. Hence, there will be an unbridgeable gap between humans and artificial agents. This gap is hard to overcome. However, we can proceed further by accepting this gap and actualizing our goals in more realistic terms.

It may be because there is a clamor for the inclusion of humans in the domain of artificially intelligent agents. Hence, the new relationship between humans and artificial intelligence is one of inclusion, not exclusion. In the new reality, to minimize carbon footprints and to achieve other SDGs like Decent Work and Economic Growth, active human involvement is needed. AI along with humans will usher in a new reality. AI without human involvement cannot satisfy the SDGs as they do not "understand" lifeworld nor can they participate in it.

Hence, to achieve sustainable AI more human inclusion in the process is needed, and the controlling power of AI should be in the hands of humans, where "principle of beneficence"[3] would be the last word.

References

1. Anthony, LFW, et al. Carbontracker: Tracking and predicting the carbon footprint of training deep learning models. *ICML Workshop on Challenges in Deploying and Monitoring Machine Learning Systems*. ICML, 2020., Retrieved November 23, 2024, from arxiv.org/pdf/2007.03051.pdf. .
2. Bandyopadhyay, Ritaprava. (2022) *Soft-morality and artificiality*. Jadavpur University, PhD dissertation, pp. 1–212. Retrieved November 23, 2024, from http://20.198.91.3:8080/jspui/handle/123456789/3404.

[3] Principle of beneficence states that AI ought to benefit both people and the environment. Indeed, artificial intelligence for social good should strive for environmentally and socially sustainable solutions [21].

3. Bandyopadhyay, R., Guha, M., & Chatterjee, A. (2024). The correlation between fall in profit and the deployment of AI labour: A Marxian analysis. *Antrocom, 18*(1), 313–325. Retrieved November 23, 2024, from antrocom.net/archives/2022/volume-18-number-1/the-correlation-between-fall-in-profit-and-the-deployment-of-ai-labour-a-marxian-analysis
4. Barlassina, L., & Gordon, R. M. (2017). Folk psychology as mental simulation. In E. N. Zalta (Ed.), *Stanford encyclopedia of philosophy*. Metaphysics Research Lab, Stanford University. Retrieved November 23, 2024, from plato.stanford.edu/archives/sum2017/entries/folkpsych-simulation.
5. Beyer, C. (2020). Edmund Husserl. In E. N. Zalta (Ed.), *Stanford encyclopedia of philosophy*. Metaphysics Research Lab, Stanford University. Retrieved November 23, 2024, from https://plato.stanford.edu/archives/win2020/entries/husserl/
6. Bringsjord, S., & Govindarajulu, N. S. (2022). Artificial intelligence. In E. N. Zalta & U. Nodelman (Eds.), *The Stanford encyclopedia of philosophy*. Metaphysics Research Lab, Stanford University. Retrieved November 23, 2024, from https://plato.stanford.edu/archives/fall2022/entries/artificial-intelligence/
7. Case, A., & Angus, D. (2017). Mortality and morbidity in the 21st century. *Brookings Papers on Economic Activity*, 397–476. http://www.princeton.edu/~accase/downloads/Mortality_and_Morbidity_in_21st_Century_Case-Deaton-BPEA-published.pdf. Accessed 23 November 2024
8. Watumull Jeffrey, Chomsky Noam and Roberts Ian. The false promise of ChatGPT, The New York Times, 8 March 2023, Retrieved November 23, 2024, from https://www.nytimes.com/2023/03/08/opinion/noam-chomsky-chatgpt-ai-html.
9. Cowls, J., Tsamados, A., Taddeo, M., & Floridi, L. (2021). A definition, benchmark and database of AI for social good initiatives. *Nature Machine Intelligence, 3*(2), 111–115. https://doi.org/10.1038/s42256-021-00296-0
10. Florian, R. V. (2003). *Autonomous artificially intelligent agents* (pp. 1–46). Coneural, Center for Cognitive and Neural Studies. 03-01, Retrieved November 23, 2024, from https://neuro.bstu.by/my/Tmp/Florean/Coneural-03-01.pdf.
11. Floridi, L. (2006). Information ethics, its nature and scope. *Computers and Society*, 21–36. https://doi.org/10.1145/1195716.1195719
12. Floridi, L. (2014). *The 4th revolution: How the infosphere is reshaping human reality* (pp. 25–31). Oxford University Press.
13. Floridi, L. (2019). What the near future of artificial intelligence could be. *Philosophy & Technology, 32*(1), 1–15. https://doi.org/10.1007/s13347-019-00345-y. Accessed November 23, 2024.
14. Floridi, L. The green and the blue: A new political ontology for a mature information society. *Social Science Research Network, 2020*. https://doi.org/10.2139/ssrn.3831094. Accessed November 23, 2024.
15. Hodges, A. (2019). Alan Turing. In E. N. Zalta (Ed.), *The Stanford encyclopedia of philosophy*. Metaphysics Research Lab, Stanford University. Retrieved November 25, 2024, from https://plato.stanford.edu/archives/win2019/entries/turing/
16. Karakilic, E. (2022). Why do humans remain central to the knowledge work in the age of robots? Marx's *fragment* on machines and beyond. *Work, Employment and Society, 36*(1), 179–189. Retrieved November 25, 2024, from https://journals.sagepub.com/doi/10.1177/0950017020958901
17. Makridakis, S. (2017). The forthcoming artificial intelligence (AI) revolution: Its impact on society and firms. *Futures*, 46–60, Accessed 25 November 2024. https://doi.org/10.1016/j.futures.2017.03.006
18. Marx, Karl. Grundrisse: Notebook VI/VII. 1857. Retrieved November 25, 2024, from https://www.marxists.org/archive/marx/works/1857/grundrisse/ch13.htm.
19. Marx, K. (1981). Capital: A critique of political economy. In *The law of the tendency of the rate of profit to fall* (Vol. 3, p. 1867) https://www.marxists.org/archive/marx/works/1894-c3/ch13.htm

20. Marx, Karl. *Theses on Feuerbach*. 1845., Retrieved November 25, 2024 from http://www.marxists.org/archive/marx/works/1845/theses/index.htm.
21. Mazzi, F., & Floridi, L. (Eds.). (2023). *The ethics of artificial intelligence for the sustainable development goals* (Philosophical Studies Series) (Vol. 152). Springer Nature. pp. 4, 5,7,45, 173, 341.
22. Oliver, M. (2024). AI likely to increase energy use and accelerate climate misinformation – Report. *The Guardian*. Retrieved November 25, 2024, from https://www.theguardian.com/technology/2024/mar/07/ai-climate-change-energy-disinformation-report
23. Russell, S. J., & Norvig, P. (2020). *Artificial intelligence: A modern approach* (p. 36). Prentice Hall.
24. Sanu, M. (2022). *5 Examples of how AI is helping companies become more sustainable*. Winnow. Retrieved November 25, 2024, from https://blog.winnowsolutions.com/5-examples-of-how-ai-is-helping-companies-become-more-sustainable
25. Strubell, E., Ganesh, A., & McCallum. (2019). *Energy and policy considerations for deep learning in NLP*. Association for Computational Linguistics. Researchgate. https://doi.org/10.18653/v1/P19-1355
26. Syverson, C. (2016). *The slowdown in manufacturing productivity growth* (pp. 1–9). Brookings Institution. https://www.brookings.edu/wp-content/uploads/2017/12/the-slowdown-in--manufacturing-productivity-growth.pdf
27. United Nations. (2015). *Transforming our world: The 2030 agenda for sustainable development. A/RES/70/1*. United Nations. sustainabledevelopment.un.org
28. Van Wynsberghe, A. (2021). Sustainable AI: AI *for sustainability* and the sustainability *of* AI. *AI and Ethics, 1*, 213–218. https://doi.org/10.1007/s43681-021-00043-6

AI for Smart Grids for a Sustainable Future

R. Kanthavel, R. Adline Freeda, and R. Dhaya

1 Introductory Review on Smart Grids and Artificial Inetlligence (AI)

At this point, focusing on the ever-increasing energy demand and striving for renewable resources, there has been an upsurge in seeking a smart, flexible, and sustainable electricity supply system. Recently, conventional power generation systems—comprising central plants of electricity generation that dispatch electricity to the end-user in an undemanding flow—started to fail due to complex energy demands which evolved over time [1]. The smart grid concept is already being applied to meet the demands for power that are in existence and those that are emerging, whereby there is an interdependency in the distribution of electrical energy and data within an electrical network to permit its real-time monitoring, controlling, and optimizing of all its components.

Smart grids involve the use of modern digital systems in the conduits and relays of electric energy systems. The capability enabled by smart grids allows for the integration of various types of DERs, such as solar, wind, and battery storage systems, that are in high demand due to the intermittency of renewable energy. The smart grid desktops bring to the user's doorstep enough information in time and empower them for demand control, thereby facilitating demand response and improving energy efficiency. While doing so, the effect of making sure that all these devices remain interconnected and, most importantly, the data produced is stable, reliable, and does not disturb the grid performance.

R. Kanthavel (✉) · R. Dhaya
School of Electrical and Communication Engineering, The PNG University of Technology, Lae, Papua New Guinea

R. A. Freeda
Department of Information Technology, KCG College of Technology, Chennai, India

P. Raj et al. (eds.), *Artificial Intelligence (AI) for IT Energy Efficiency and Green AI for Environment Sustainability*,
https://doi.org/10.1007/978-3-031-89420-6_6

Predictably enough, AI comes into play. Domains of AI include machine learning, deep learning, and reinforcement learning that prove useful when morphological processing is done and analysis has to be carried out on datasets, which predict change in demand and supply, and also optimize power systems so that the grid can be more stable against physical and cyberattacks. In the rational use of data and intelligent activity of AI, it is a smart grid empowered to improve the reliability of a grid with fewer associated costs and permits more clean energy provision to be used [2].

1.1 Role of Artificial Intelligence in Smart Grids

Artificial Intelligence enables a wide range of functions important to the operation of a Smart Grid, which is shown in Fig. 1.

- **Predictive Maintenance**: Machine learning models help predict equipment failures and maintenance needs, preventing costly downtime and extending system life.
- **Load and Demand Forecasting:** Artificial intelligence algorithms enable accurate electricity consumption forecasts and help grid operators effectively match supply with demand, even under varying conditions.
- **Fault Detection and Anomaly Detection**: By analysing data patterns in real time, AI systems can detect and respond to network anomalies and failures, prevent outages, and ensure stable power.

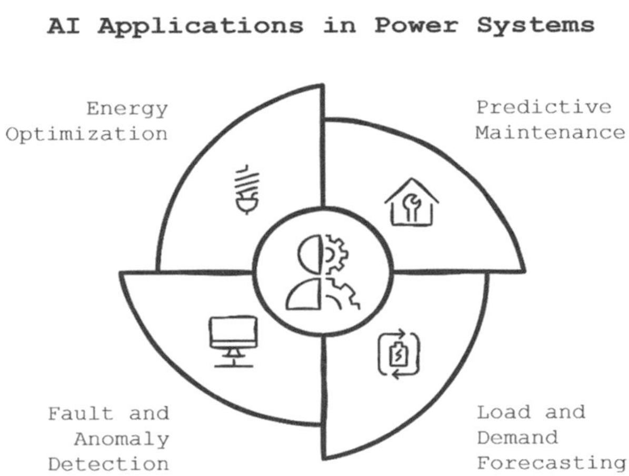

Fig. 1 Role of artificial intelligence in smart grids

- **Energy Optimization:** AI-driven optimization techniques improve grid efficiency by adjusting power flows, managing demand response, and reducing energy losses in real time [3].

1.2 Benefits and Challenges of AI Integration

Integrating artificial intelligence into smart grids brings a number of benefits:

- **Increased Efficiency**: AI optimizes energy distribution, minimizes losses, and reduces operating costs.
- **Improved Reliability and Resilience:** Through predictive maintenance and fault detection, AI increases network reliability and makes it more resilient to failures.
- **Supporting the Integration of Renewables**: Artificial intelligence helps manage the variability of renewable energy sources, balancing supply and demand, and reducing dependence on fossil fuels.
- **Better Consumer Engagement**: AI enables personalized energy insights, demand response incentives, and smart home integration, leading to increased consumer engagement and efficiency.

Despite these benefits, integrating AI also presents challenges. These include:

- **Privacy and Data Security**: The handling of sensitive user data raises privacy concerns, while the increasing interconnectedness of grid systems increases cybersecurity risks.
- **Scalability and Interoperability**: AI solutions must be scalable and interoperable across disparate grid infrastructures, an often-complex task due to different technologies and standards [4].
- **Cost and Complexity**: Implementing AI systems in smart grids requires significant investment and technical expertise, making it a resource-intensive process.

This chapter delves into the application of AI in smart grids, covering specific AI techniques and their applications in forecasting, optimization, anomaly detection, and cybersecurity, and explores the future potential of AI-driven smart grids for a sustainable energy future.

2 AI Techniques and Models for Smart Grid Applications

As smart grids are coming into trend, there is a need and desire to add AI techniques to them so that their effectiveness, dependability, and even environmental sustainability can be enhanced. This chapter discusses and assesses the forms and models of AI suitable for the application of smart grids. The focus will be on principles, practices, and their consequences for the operations of the grid.

2.1 Use of Machine Learning in Utility Load Forecasting

Machine learning is a range of algorithms that lets systems make predictions without being programmed to do so in any predefined manner. In the context of smart grids, the main application clusters are:

- **Overall Electric Power Load Forecasting**: In this case, based on the electricity load consumption in the previous period, various ML techniques are employed in estimating a date within which energy load will be required. Techniques include those such as regression techniques, support vector techniques, and decision trees in carrying out load forecasting.
- **Algorithms for Prediction of the Power Generated from Renewables**: These include artificial neural networks and other prediction techniques for the forecasting of the output of generation units dispatched by renewable resources. Because of the variation of solar and wind energy sources with respect to weather and time factors, these models encourage the due usage of such types of energies on the grid [5].

The same is furthered in use by machine learning applications to support utilities in the implementation process of demand-side management measures and rate schedules by incorporating pattern recognition, which matches their behaviour in consumption.

2.2 Deep Learning for Improved Pattern Recognition

Deep learning is a particular class of machine learning methods characterized by the adoption of several layers of neural networks, with the aim to learning the relationship among the factors of interest in data. Its benefits are manifold, a few relating to smart grids.

Visual monitoring systems also make use of CNN and deep learning models in fault detection, infrastructure inspection, and even some aspects of security surveillance. Smart meters and sensors can process time-series data with functions from RNN and auto-encoders in order to look for deviations in data; such deviations could refer to a possible mechanical failure or a cybersecurity threat.

It can, therefore, leverage the time dependencies inside energy usage data and hence enhance the accuracy of the load and generation forecast by the use of long short-term memory networks in the structure of deep learning models for energy forecasting.

2.3 *Role of Reinforcement Learning in Decision-Making*

Reinforcement learning can be considered as a particular sub-area of machine learning that allows artificial agents to learn from making sequential decisions while their goal is optimally gathering the reward collected with time [6]. In Smart Grid, RL plays its role in several features which are shown in Fig. 2.

- **Effective Management of Energy Resources:** RL algorithms allow the operator to take into consideration real-time performance control of distributed energy resources (batteries, microgrids, etc.) and solve the control problem accordingly.
- **Management of Demand Response:** RL can be used for the design of consumer energy use modification incentives aimed at the consumers in order to ease the effects of peak loads on the grid.
- **Resource Allocation:** By proper division of limited resources like storage and generation capacity, the RL methods are able to ensure that the energy production and consumption are matched up while the expenses are decreased.

2.4 *Hybrid Artificial Intelligence Model*

Artificial intelligence models need to be used to fully utilize such elaborative processes and challenges arising during smart grid operation. A few techniques form the basis of this approach. These examples shall help explain more clearly and make it easier; for instance,

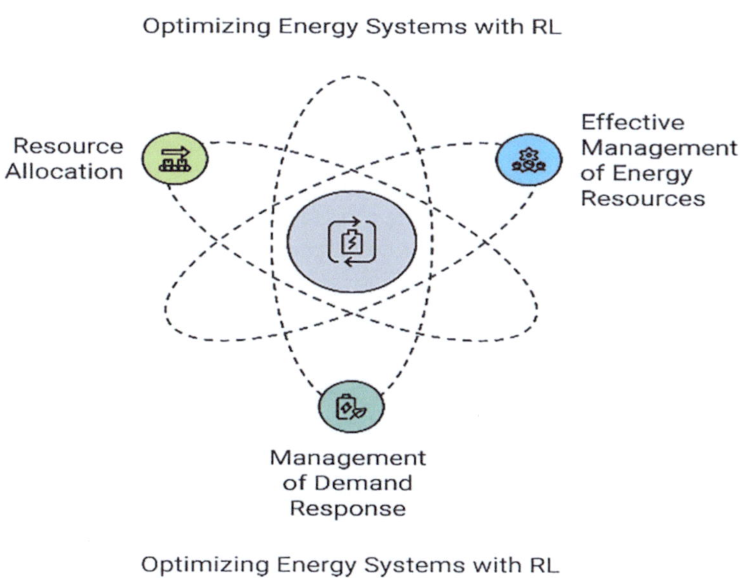

Fig. 2 Role of reinforcement learning in making decisions

- **Data Fusion:** The integration of IoT sensors and weather data through data fusion enhances forecasting and management decisions. This is particularly helpful in this case.
- Multi-agent systems: these are those kinds of AI that work in cooperation with other AIs to carry out a mutual function [7]. In addition, load balancing and resource optimization can be performed in a distributed power system.
- **Ensemble Learning:** The application of more than one machine learning algorithm through ensemble learning offers higher accuracy with more reliability and fewer errors per model. This approach is generally known as blended learning.

The key to successful smart grids is the implementation of smart technologies and models. Hybrid approaches such as machine learning, deep learning, including quantum computing, and reinforcement learning, just to mention a few from the field of neuroscience, will provide smart grids with the capacity to manage energy distribution and consumption more efficiently.

3 AI for Forecasting in Smart Grids

Precise prediction is the basis of any efficient smart grid, as it enables grid operators to balance the generation and consumption of energy by the introduction of renewable sources and thereby save energy. AI technologies supply tools that are exploited in predicting various key parameters, including load demand, renewable generation, and energy prices [8]. This section provides a discussion on AI-based solutions for these types of forecasting problems, placing emphasis on methods and barriers while making an assessment of their effect on grid stability and efficiency.

3.1 Load and Demand Forecasting

The forecast of the load helps the functioning of the power network properly and thus the AI techniques which contribute by studying historical usage patterns, economic growth, revenue generation, and weather conditions to improve the accuracy of load forecasting are an added advantage.

- **STLF:** Today, the major technologies in use include machine learning systems like support vector machines, decision trees, and RNNs. This supports the predictions of power demands over a short time cycle of a few hours or days, hence improving the real-time grid operation to ensure adequate energy production.
- **Medium-Range and Long-Range Load Forecasting**: In medium- to long-term ranges, one can see deep learning approaches such as LSTM networks and ensemble methods. Improved planning from the work of these forecasts in both

clean energy regulation and infrastructure creates grid extension and wealth allocation [9].

- **Consumer-based Load Forecasting**: Now, with the availability of smart meters, we can forecast the power consumption of a single house. The AI models can be designed to read the behaviour of the individuals through some interference; for example, utilities can ask the smart meter to notify the homeowner when the city peak is going to happen to shut the load to avoid generation issues.

3.2 *Renewable Energy Production Forecast*

The difficulties in providing electricity arise due to the dramatic shift in wind, solar, etc. AI can predict with accuracy how much production would relate to power problems and is shown in Fig. 3.

- **Solar Power Forecasting:** It works by analysing the pre-power cloud coverage to forecast the generation through AI models. People mostly depend on convolutional neural networks (CNNs) and gradient boosting to make them more energy efficient.
- **Wind Power Forecasting**: These would be dependent on variables like wind speed, temperature settings, and atmospheric pressure. Some of the AI models being used by people in predicting the availability of wind power include ensemble learning and time series analysis. Hence, the grid operators can prepare for the transition and make optimum utilization of wind energy [10].

The hybrid data forecast models depend on solar, wind, and hydro data for various renewable energy sources to develop hybrid forecast models that are way more precise.

3.3 *Energy Price Forecast*

Proper grid and market management require energy price forecasting and help various stakeholders. The direction of price changes can be predicted so that appropriate decisions can be made in advance.

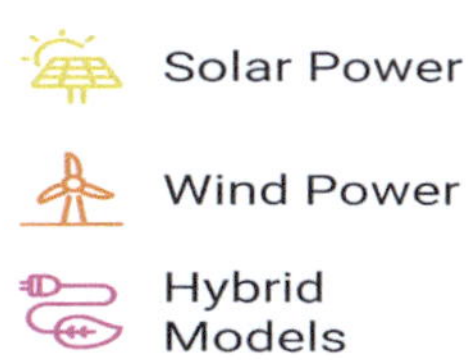

Fig. 3 Renewable energy production forecast

- **Forecast Values:** Deep neural networks, regression methods, etc. are used for real-time or short-term one-day energy values. These models are used to optimize trading sourcing strategies for energy networks and traders. Criteria include demand, production capacity, and market trends.
- **Long-Term Price Forecasts:** Long-term price forecasts use AI techniques such as Random Forest, XGBoost, and Support Vector Machine to analyse macroeconomic indicators, oil prices, and policy changes. These can be used to plan long-term investments, for example, in infrastructure development and energy supply [11].
- **Consumption Level Prediction:** AI models can also predict consumer energy prices with information about dynamic demand and prices. Dynamic pricing models help consumers alter their consumption patterns in relation to prices. This ensures that the network remains stable and the business is profitable.

3.4 Climate and Environment Interface

Weather is a key factor in the consumption of energy and demand for renewable energy sources. AI-powered models utilize weather and other environmental parameters for very accurate forecasting. The integration of weather data includes temperature, humidity, wind speed, and cloud cover for weight and restoration of sources in this section These data help to improve the accuracy of AI models by performing spatial analysis of the local environment and making appropriate adjustments to the model.

- **Environmental Data for Renewable Energy:** Solar irradiance and atmospheric pressure are two of the environmental factors normally used by AI renewable energy models. Renewable forecasting is enhanced by developed artificial intelligence strategies where the geographic information systems (GIS) employ special local environmental data to effect successful integration [12].

3.5 Problems with AI-Based Forecasting

While AI offers powerful forecasting capabilities, several challenges remain:

Data Quality and Availability: The use of countless datasets is important, especially when proactive forecasting must be made. This is because images usually fail to provide the complete picture, and due to missing or noisy data, the predicted models may be wrong.

- **Model Interpretability:** AI models based on deep learning, as a rule, work with high opacity; it is almost impossible to see what they "saw" in the data and how they arrived at conclusions.

- **Scalability and Adaptability**: In terms of ongoing and future grid conditions, it is difficult for AI forecasting models to expand to other areas. One of the most significant concerns that must be considered to make accurate forecasts is flexibility in various forms of grids, weather, and consumers' behaviour [13].

The use of artificial intelligence in smart grids improves decision-making by providing precise forecasts of load, renewable generation, and energy price. The above forecasts help the grid operators in the proper utilization of resources for cost-effective operation and accommodating the renewables. In the next section, the author will discuss AI-based optimization and control methods for using these forecasts to provide grid stability and robustness.

4 AI-Driven Optimization and Control in Smart Grids

Among the essential functions in smart grids, optimization and control should be underlined because they help to generate, distribute, and consume electricity in the most effective way possible. Through the use of real time data, adaptive algorithms, and predictive models, AI techniques improve optimization and control operations. This chapter focuses on the initiatives of AI optimization and control methodologies that cover a range of SMES operation complexities, including supply-demand balance, DERs management, and grid stability, which are shown in Fig. 4.

4.1 Scope of Grid Balancing and Frequency Regulation

More so, grid balancing and frequency regulation are important services in support of maintaining the power grid's stability and reliability. AI techniques assist in maintaining a continuous balance between power supply and demand.

- **Frequency Control:** Modern synchronized real-time monitoring and controlling incorporate artificial intelligence models by regulating the generation output to achieve stability of frequency according to the load [14]. It is therefore convenient to use RL algorithms in frequency regulation since they will follow shifts in load patterns and adjust the control of generation resources in their operations.
- **Automatic Generation Control (AGC):** Generally, AGC employing AI possesses machine learning models and optimization communion to adapt automatically with the help of generating units towards output when the cases of frequency fluctuations commence, implicating stable power quality and avoiding major blackouts in organizational forms.
- **Load Shedding and Demand Response**: AI models can tell when is the time of the day when consumption is at its highest and switch on controlled load shedding or demand response programs to ease the strain on the grid. Some of the general use loads that can be shed temporarily may include the following: AI makes it easier to match supply and demand during peak hours.

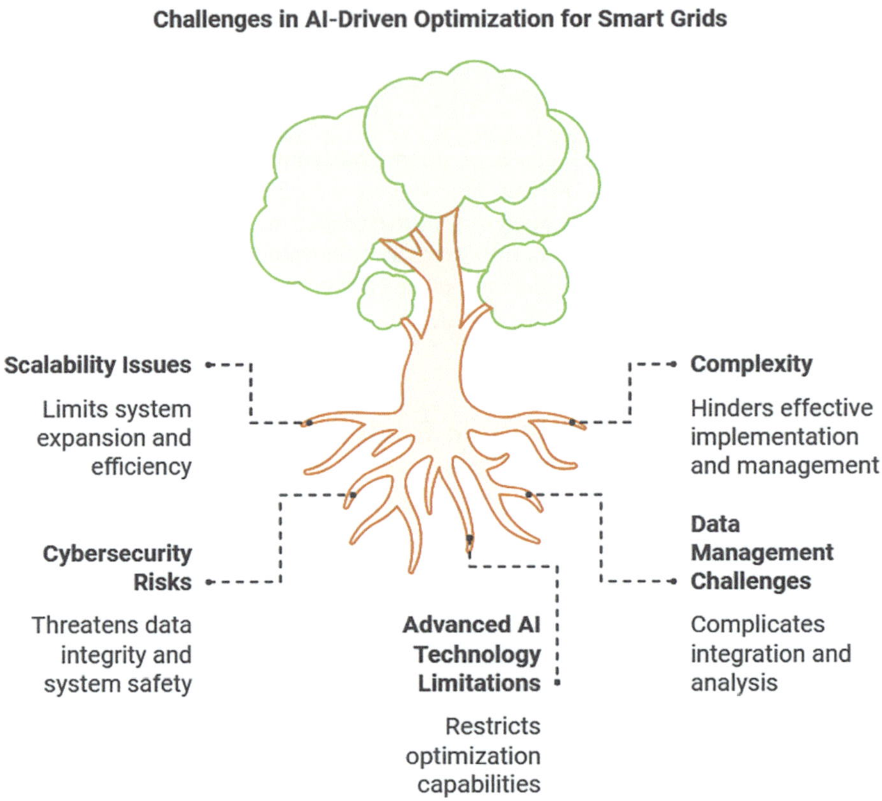

Fig. 4 Challenges in AI-driven optimization for smart grids

4.2 DSM and Response Optimization

Demand-side management can be described in simple terms as a way of maintaining the level of electricity usage to correspond with the levels of supply at the specific time of electricity usage. Therefore, by following the DSM and more accurate demand prediction and management, AI can be defined.

- **Demand Response Programs**: Demand response programs are formulated and managed by supervised learning and predictive analytical methods. These models are better at identifying the energy consumption pattern of the consumers and recommending changes during periods of high consumption; hence, there is no rationale for demanding new generation capacities.
- **Real-Time Pricing Models:** The value of dynamic models of price concerning the high peak, created by artificial neural networks, leads to a reduction of a part of energy use during the high demand [15]. These dynamic prices driven by AI

emotions are good at responding to real-value signals of power and avoiding wastage as low loading on power systems.
- **Smart Home and IoT Integration**: AI deployed in IoT-based smart homes can control the consumption of household appliances in relation to prevailing grid demands, which enhances demand response. For instance, the smart and managed assets, including thermostats, water heaters, EV chargers, etc., can be properly scheduled directly by the grid signals, which can reduce the peak loads effectively and increase the flexibility of the grid.

4.3 Power Flow Optimization and Grid Stability

Energy management gives conditions to limit energy loss and to distribute energy in equal sections for the power network. AI-based models for optimizing power flow will dynamically adjust power so that the flow serves to enhance the stability of the grid and lower the operational costs.

- **Optimal Power Flow (OPF):** Solutions for OPF problems include genetic algorithms, particle swarm optimization, and deep reinforcement learning. The models optimally distribute the electricity in the system in order to cut down both the losses and the costs while still ensuring that there is enough supply to meet the demand [16].
- **Voltage Control and Reactive Power Management:** AI models help to eliminate the usual postponement of the voltages and reactive power control, which is done in order to maintain the stability and the safety of the apparatus. For instance, neural networks along with reinforcement learning are put into operation for voltage magnitude surveillance and control so as to improve the performance of the power system where voltage stability plays an important role.
- **Distribution Network Reconfiguration:** This is referred to in conjunction with AI as the reallocation of distribution network topologies when loads and losses come into place. Time-critical elements feature applications of deep Q-networks and hybrid optimization models for active grid reconfiguration, which are crucial in high-renewable energy penetration systems.

4.4 Integration and Management of Distributed Energy Resources (DERs)

The adoption of DERs that accompany the installation of more and more solar panels, wind turbines, and energy storage systems has necessitated the enhancement of the optimization strategy.

- **Energy Storage Optimization:** AI models predefine the charging or discharging currents of energy storage devices in use by maintaining healthy energy supplied versus consumed ratio and even extending the service life of the battery. Mainly, reinforcement learning with deep learning algorithms is considered while performing coordination management of the storage assets with the grid [17].
- **Scheduling of Distributed Energy Resources:** AI effectively plans the needs and locations of DERs that need to be deployed, considering renewable generation forecasts, energy cost, and demand. Applications such as genetic algorithms and mixed integer linear programming serve to schedule and control the DERs, which will optimize resources and minimize costs.
- **Microgrid Control and Optimization:** Advanced artificial intelligence techniques play a crucial role in micro-grids since they comprise several DERs and are localized energy systems. Microgrids incorporated by machine learning and multi-agent systems can work either in isolation or in parallel with the main grid and help in power supply to remote areas and during peak power demand in the main grid.

4.5 Fault Detection, Isolation, and Restoration (FDIR)

FM is critical in reducing outages, thereby boosting service continuity and quality. AI-driven FDIR systems improve fault detection, isolation, and service restoration.

Fault Detection and Isolation: Customers reported that AI enabled them to pinpoint problems using deep neural networks and recognize different patterns with the help of data collected from sensors and take less time to discover and correct them.

- **Self-Healing Networks:** Self-healing hybrid networks use automated control to reroute power around a problem area to provide service to areas not affected by the problem. These networks use reinforcement learning and optimization algorithms to self-adapt quickly to the occurrence of faults, thereby increasing the reliability of the networks and reducing the amount of time the networks are down.

4.6 Problems in AI-Driven Optimization and Control

Despite the advantages, implementing AI-driven optimization and control in smart grids comes with challenges:

- **Scalability and Complexity:** Deploying AI models across big and complex power networks that have different configurations is not easy. To achieve a high level of communication between the various systems, there is a need to employ complex mathematical formulas and highly efficient computational platforms.

- **Cybersecurity and Privacy**: This is especially important because smart grids are interconnected and increasingly rely on automation in the vicinity of a densely populated area. Any use of artificial intelligence in optimization systems therefore requires cybersecurity measures to ensure that any data that is sensitive is well protected from people who may wish to breach the security [18].
- **Data Management and Integration**: Smart metering AI models require a large quantity of accurate data from both metering and IoT sensors as well as external data feeds. Co-ordinating and addressing these data in a timely and sustained fashion is critical to the successful optimization of the process.
- **Advanced AI technologies** in form of optimization and control methods are enhancing the functionality of smart grids to achieve optimum adaptation to real-time conditions, renewable resources integration, and enhance efficiency. Applications of AI in load balancing and demand response in addition to fault management in smart grids can thus offer sustainable, economical, and credible electricity supplies. The next sections will expand on AI in fault diagnosis, predictive upkeep, and security to present the importance of AI towards smart grid development.

5 Fault Detection, Anomaly Detection and Predictive Maintenance

Fault detection and anomalies are expensive in a smart grid, but for accurate forecasting, precise maintenance timing is key to fault-free and uninterrupted power supply. The following sections describe in detail AI techniques and AI models for fault diagnosis, anomaly detection, early warning and maintenance operations, challenges, and benefits.

5.1 Smart Grid Fault Detection

Fault detection involves network faults blocking static features or high-energy interference such as faulty equipment or line failure. AI algorithms make fault detection faster and more accurate.

- **Pattern Recognition and Classification**: The models SVMs and decision trees fundamentally study patterns in the grid data provided by a facility to separate the faults by type and location. These models help operators in that they can immediately tell what kind of fault there is and what needs to be done.
  In so doing, the signal data is of a waveform nature and is normally gathered through voltage and current sensors that depend on deep learning models like the convolutional neural network. These models can locate faults with a high degree of precision due to their capability to detect minor signal pattern variations.

– **Real-Time Detection Systems for Fault:** Smart grid repair involves the exact same process as fault detection, and existing AI-driven systems use data from PMUs and smart meters to diagnose them in real time. Algorithms of reinforcement learning facilitate self-organization of the fault during power supply to reduce the incidents [19].

5.2 Smart Grid Data Anomaly Detection

Abnormality detection is indispensable for the detection of an abnormality in the operation of the system, which may be a security threat, a system failure, or an evolving fault condition in the grid. AI-based anomaly detection models are designed to handle the complexity of

- **Smart Grid Data: Unsupervised Learning for Anomaly Detection**: Supervised models include linear models (e.g., regression), decision trees, support vector machines, logistic regression, naive Bayes, neural networks, random forests, and boosting and cottage models; while unsupervised models are auto-encoders, clustering algorithms like k-means. These models define non-conformance to normal behaviours to make the identification of other unseen behaviours that are peculiar possible while reducing the number of false alarms.
- **Time-Series Anomaly Detection**: Temporal patterns in the grid data are studied using recurrent neural networks specifically LSTM (long short-term memory) networks. In this way, the models decide on normal operation sequences, and by recognizing anomalies thereof, e.g. peak loads or voltage deviations that indicate problematic equipment or cyber-attacks.
- **Hybrid Anomaly Detection Models:** Web-based techniques integrate two or more techniques at a time (for example, rule-based, machine learning, and so on) for better result identification. These models enhance the degree of detection and flexibility throughout the elements of a grid, including the capacity to quickly identify compliance or non-compliance of a structure that is integrated into several information feeds.

5.3 Preventive Maintenance through Artificial Intelligence

Predictive maintenance leverages AI models to predict equipment failures and maintenance needs before they occur, allowing for proactive intervention and reducing downtime and repair costs.

Condition-Based Maintenance: Random forests and gradient boosting algorithms, for instance, offer a custom evaluation of the records of past maintenance, and real-time sensors, to determine the health of the equipment. These models indicate when maintenance is necessary due to existing conditions of operation, which reduces random downtime.

- **Remaining Useful Life (RUL) Prediction:** LSTMs and recurrent neural networks in deep learning calculate the remaining useful life of transformers, circuit breakers, and other critical assets. Through accurate prognosis for RUL, grid operators are able to plan replacements and prevent failure.

5.4 Uses of AI in Fault, Analysis, and Maintenance Management

AI applications in fault detection, anomaly detection, and predictive maintenance have broad implications for grid operations and reliability.

- **Smart Fault Management Systems:** Real-time fault detecting and isolating systems enabled and powered by AI lead to a decrease in fault diagnosis time and enhanced grid stability.
- **Cybersecurity and Anomaly Detection**: In the emerging close-knit smart grid environment, threats to cybersecurity cannot be fully dismissed. Anomaly detection systems, which are managed with the help of artificial intelligence, look for tendencies of malicious activity that can cause disruptions of the grids and jeopardize the safety of the information [20].
- **Proactive Asset Management**: The models allow utilities to move from a reactive approach to asset management to a proactive one. These models help prevent or at least forecast equipment failures and, thus, minimize downtime while decreasing costs of maintenance and prolonging the lifespan of grid assets, which improves the stability and efficiency of the grid as a whole.

5.5 Impacts of Applying AI in the Fault, Anomaly, and Maintenance Management Function

Implementing AI-driven fault detection, anomaly detection, and predictive maintenance in smart grids presents several challenges:

- **Data Quality and Availability:** In AI systems, the quality of data is important, but grid component data may be inaccurate, missing, or noisy and affect model performance.

  Interpretability of AI Models: Most of the AI models, especially those under deep learning, are black box models; the output cannot be explained to serve the purpose of logging or developing trust in the models.

- **Cybersecurity and Privacy Concerns**: Due to rising IoT devices and data-driven smart grid models, there are a growing number of dangers when it comes to data privacy and security. This means that the security of data and adherence to the laws and regulations that govern, amongst other things, the privacy of individuals are key.

Fault detection, anomaly detection through the AI technique, and predictive maintenance are revolutionizing the management of grids, increasing their efficiency, and reducing downtime. Through early fault detection, diagnosis, and prognosis, these AI applications decrease asset downtime, enhance asset life, and support creating a more secure and stable smart grid [21].

6 Cybersecurity and Privacy Concerns in AI-Enabled Smart Grids

The steady deployment of AI-based systems in smart grids has led to cybersecurity and privacy as major elements of safe and efficient operation. Smart grids are built on an extensive number of data inputs and interconnected systems and components, hence making them vulnerable to cyber-attacks, access violations, and data privacy. This section looks at the main cybersecurity and privacy threats in smart grid systems, AI approaches that may be used in smart grid security, and approaches to safeguarding personal information in smart grid environments.

6.1 Cybersecurity Threats in AI-Based Smart Grids

AI-enabled smart grids face a range of cybersecurity threats due to the increasing digitalization and interconnectivity of grid components.

- **Malware and Ransomware Attacks:** Cyber threats like malware or ransomware can affect the grid's functionality, hold important data captive, and even disable key systems—threats that directly affect the resilience of the grid. Because of the importance of real-time data on power delivery, short-lived attacks can lead to significant disruptions of power delivery systems and customer confidence.
- **Data Poisoning and Model Manipulation**: AI models to operate in smart grids are developed using big datasets, and data poisoning attacks refer to feeding wrong or fake data to make models predict wrongly. With knowledge of these modifications, it can produce inaccurate predictions, ineffective energy services, or system failures.

- **Distributed Denial of Service (DDoS) Attacks**: In DDoS, a large number of compromised systems attack a single target with a large amount of traffic consuming the resources of the target system. Assaults on key smart grid components, including EMS or DMS, could slow other real-time applications that ensure the integrity of the grid.
- **Advanced Persistent Threats (APTs):** APTs are silent and long term and are executed by the state organizations mostly. The APTs on smart grids are further capable of leading to compromises on data integrity, malicious disconnects of physical infrastructure, or other large-scale impacts on energy delivery [22].

6.2 *AI, Smart Grid, Privacy*

The massive data gathering required by AI-powered smart grids poses a significant privacy challenge.

Smart meter-related and IoT devices are one of the larger contributors to energy usage in homes. What then is the privacy of individual homes? When this happens, wrong management can lead to the dissemination of personal behaviour, including home use and usage of appliances. This can be a concern for privacy.

The exchange of data in the smart grid is done with the present, as well as the participation of the utilities, the third party, and all the stakeholders. A company can easily leak customer information or enrich it without adequate protocols for handling data. Why?

6.3 *Artificial Intelligence Is Being Used to Enhance Security in Smart Grids. How?*

AI is a crucial component of smart grids'cybersecurity due to its ability to detect and react to threats with high levels of precision. The use of IDS incorporates artificial intelligence into techniques that detect potential hazards in the network in real time. The application of machine learning classifiers such as support vector machines (SVM), neural networks, and deep learning models, including convolutional neural networks, is launched to recognize malicious trends and alert the intruder before a lot of damage has been done. In essence, in big data sets, AI can use past information, helping cyber security detect patterns and attacks. This method is called anomaly detection. As cases that may only occur once, such as data corruption or even unauthorized access, predictive RNHs focus on time series data to consider [23].

Other techniques, such as Behavioural analytics and User authentication, enhance security measures learned from the users.

6.4 Privacy Techniques in AI for Smart Grids

Many AI-based techniques are used to deal with privacy issues in smart grids:

- **Differential Privacy**: It adds a little noise to the data. This way, it keeps individual user information secure. This allows the smart grid to analyse large sets of data without hindering customers' privacy. It is suitable for huge demand data analysis.
- **Homomorphic Encryption**: This will allow the data to be still encrypted but allow for analysis without opening it. In smart grids, this shall enable different groups to process information while retaining privacy.

6.5 Issues of Cyber Security and Privacy Measures Implementation

Attaining cybersecurity and privacy in AI-powered smart grids based on several limiting issues:

Deep computations are needed for advanced cybersecurity AI models, which can be hard to provide for legacy systems because of the low power of processing. New Emerging Cyber Threats: The threat environment keeps changing, and with the coming of new types of attacks, even AI models have to be changed. Robust updated AI-based security systems require regular maintenance in order to ensure long-term grid security [24].

7 Case Studies and Real-Life Applications

More AI support is being integrated into the controlled power systems and focused towards growth in energy management and the optimization of efficiency, hence reliability [25]. The subheading includes within it the set of industrial cases presented in this chapter that serve as evidence for the applications of AI technologies in the smart grid functional domains of today, such as forecasting, optimization, fault detection, and cybersecurity are described in Table 1.

Case Study 1: **Pacific Gas and Electric-PG&E, Smart Grid Management**: PG&E is one of the largest utility giants in the United States [26]. With over a million families in California as its clients, the company has managed the grids and dispersed energy with the help of advanced AI solutions.

AI Solution: PG&E has developed an intelligent predictive analytics engine that uses machine learning to analyse the readings collected from smart meters, weather data, and historical consumption statistics. The autonomy of the system also offers the utility the ability to predict the fluctuations in power demand

Table 1 Case study

Case study	Overview	AI application	Outcome
Pacific gas and electric (PG&E)	One of the largest utility companies in the United States, serving millions of customers in California	Developed a predictive analytics platform using machine learning to analyse data from smart meters and weather forecasts for demand forecasting	Improved load management capabilities, reduced operational costs, decreased energy waste, and enhanced customer satisfaction
Siemens	Leader in developing smart grid solutions in Germany, focusing on renewable energy integration	Implemented an AI optimization system to forecast renewable energy production and manage supply and demand in real time	Reduced energy curtailment from renewables, maximized energy generation, and improved grid stability, supporting Germany's energy transition goals
National Grid	UK utility company adopting AI for predictive maintenance of electrical infrastructure	Used machine learning to analyse sensor data for predicting equipment failures based on health assessments and environmental conditions	Reduced unplanned outages and maintenance costs, extending the lifespan of infrastructure and ensuring reliable service
Enel	Multinational energy company emphasizing cybersecurity in smart grid initiatives	Deployed an AI-driven cybersecurity framework to monitor network traffic for anomalies and potential intrusions in real time	Enhanced ability to prevent and respond to cyber threats, leading to quicker detection of anomalies and improved grid security
Duke Energy	Major U.S. utility provider leveraging AI for renewable energy integration.	Developed an AI forecasting system to predict solar and wind energy generation by analysing weather data and historical patterns	Improved accuracy of renewable energy predictions, leading to increased utilization of renewables and reduced reliance on fossil fuels
Xcel energy	Utility operating in multiple states, implementing innovative smart grid technologies	Utilizes AI algorithms for demand response programmes, analysing customer behaviour, and optimizing participation during peak periods	Successfully reduced peak demand, improved grid reliability, and increased customer engagement through financial incentives

besides enabling demand management wherein generation and supply of electricity or heat are needs-based performances based on the forecast. It is a complex technology, but a good thing to know: actual service reliability is way higher.

Case Study Overview. Value Proposition w/AI Outcome.

Pacific Gas and Electric is among the major utility providers in the United States. It gives service to millions of customers within California. They have been able to adapt predictive analytics, through the implementation of machine learning analysis, based on smart meter data and weather forecasts to enable it to conduct demand forecasting. Because of this, PG&E has increased its ability in load management, which has reduced operational costs, energy wastes, and heightened customer satisfaction.

Siemens houses and develops smart grid solutions in Germany. Most of their efforts in the country are directed towards integrating renewable energy sources. Giant Siemens implemented an AI-based optimization system that would foresee the production of renewable energy and regulate the supply-demand chain in real time. Energy curtailment from renewables is lowered; energy generation is at a maximum, thereby enhancing grid stability, while Germany's energy transition goals are boosted.

National Grid: A UK utility company using AI to perform predictive maintenance of electrical infrastructure. It has used machine learning algorithms against sensor data to predict equipment failure based on its health assessment and environmental conditions. The results have been fewer unplanned outages and reduced maintenance costs, extending the useful life of infrastructure with continued reliable service.

Enel Multinational energy company, whose focus is cybersecurity in smart grid projects. A real-time anomaly and intrusion detection artificial intelligence-powered cybersecurity framework was implemented with the objective of monitoring network traffic. Improvement in terms of cybertime detection and response, with the speed of anomaly detection increasing hence fortifying grid security.

Duke Energy is one of those large utility providers in the United States using AI to integrate more renewable energies. It built an AI-based forecasting system to predict exactly how much energy would be produced by the sun and wind at any particular time by analysing weather data and historical patterns. This lets the company raise its accuracy level regarding the quantity of renewable energies it produces, increases its usage, and cuts its dependence on fossil fuels.

Xcel Energy: Operating across multiple U.S. states, is recognized as an innovator in smart grid technology. The company leverages AI algorithms for demand response, enabling customer behavior analytics and optimizing peak load participation. This approach has led to reduced peak demand and improved grid reliability by incentivizing greater customer engagement through financial benefits.

8 Conclusion

Artificial Intelligence or AI is revolutionizing the smart grid either by enhancement in efficiency and resiliency or by meeting the new era energy requirements. Applying emerging delivery models for predictive analytics, machine learning, and optimization, AI augments the accuracy of the forecast, integrates renewables and maintenance, and fortifies cyber security. AI algorithms created by Purdue's Michael Hsieh and Duke Energy's Jeffrey Hausman feature a way to improve energy distribution and integrate renewables in order to use the minimum amount of fossil fuels.

Siemens and Enel also employ AI algorithms to enable the upscale availability of grid services to users with intrinsic needs. AI-based integration in smart grids must ensure the availability and quality of data models themselves, along with the cooperative participation of stakeholders. It makes operational AI systems capable of reacting to ever-changing environments, allowing irrigation networks to effectively cope with changing hydrographs and today's cybernetic threats. And, monitoring and modifying AI models and models themselves can allow for timely adjustments across existing trends and advancement of other technological means that may influence AI systems efficiency within a dynamic environment.

References

1. F. Khosrojerdi, S. Gagnon and R. Valverde, Applications of artificial intelligence in smart grids: present and future research domains. 2021 IEEE 9th International Conference on Smart Energy Grid Engineering (SEGE), Oshawa, ON, Canada, 2021, pp. 7–12, https://doi.org/10.1109/SEGE52446.2021.9534914.
2. Kanthavel, R., Dhaya, R., & Ahilan, A. (2022). AI-based efficient WUGS network channel modeling and clustered cooperative communication. *ACM Transactions on Sensor Networks, 18*(3).
3. X. Zhao, Y. Guo, X. Guo and H. Li, Artificial intelligence applications and prospects for the smart grid. 2023 Panda Forum on Power and Energy (PandaFPE), Chengdu, China, 2023, pp. 1844–1848, https://doi.org/10.1109/PandaFPE57779.2023.10141110.
4. Kanthavel, R., & Dhaya, R. (2022). Wireless underground sensor networks channel using energy efficient clustered communication. *Intelligent Automation & Soft Computing, 31*(1), 649.
5. S. Davlyatov, Artificial intelligence techniques: Smart way to smart grid. 2023 International Conference on Artificial Intelligence and Smart Communication (AISC), Greater Noida, India, 2023, pp. 838–842, https://doi.org/10.1109/AISC56616.2023.10085216.
6. S. Liasi, N. S. Ghiasi and R. Hadidi, *Smart grid vs. intelligent grid; Artificial intelligence takes the power system to the next level.* 2023 IEEE 3rd International Conference on Digital Twins and Parallel Intelligence (DTPI), Orlando, FL, USA, 2023, pp. 1–6, https://doi.org/10.1109/DTPI59677.2023.10365429.
7. Dhaya, R., Indhuja, P., Sinduja, S., & Swetha, M. (2015, January). *Finest power efficient steering algorithm for wireless sensor networks for surveillance.* In 2015 IEEE 9th international conference on intelligent systems and control (ISCO) (pp. 1–6). IEEE.
8. Yang, L., Li, Y., & Li, Z. (2017). Improved-ELM method for detecting false data attack in smart grid. *International Journal of Electrical Power & Energy Systems, 91*, 183–191.
9. Kanthavel, R., & Dhaya, R. (2021, March). *Proactive cooperative communication enabled fifth generation wireless sensor network using SCTP.* In 2021 National Computing Colleges Conference (NCCC) (pp. 1–5). IEEE.
10. Xue, D., Jing, X., & Liu, H. (2019). Detection of false data injection attacks in smart grid utilizing ELM-based OCON framework. *IEEE Access, 7*, 31762–31773.
11. Wang, Z., He, H., Wan, Z., & Sun, Y. (2020). Coordinated topology attacks in smart grid using deep reinforcement learning. *IEEE Transactions on Industrial Informatics, 17*(2), 1407–1415.
12. Dhaya, R., Kanthavel, R., Algarni, F., Jayarajan, P., & Mahor, A. (2020). Reinforcement learning concepts ministering smart city applications using IoT. *Internet of Things in Smart Technologies for Sustainable Urban Development*, 19–41.
13. Kanthavel, R., Anathajothi, K., Balamurugan, S., & Ganesh, R. K. (Eds.). (2022). *Artificial intelligent techniques for wireless communication and networking*. John Wiley & Sons.

14. Wang, Y., Liu, M., Bao, Z., & Zhang, S. (2019). Stacked sparse autoencoder with PCA and SVM for data-based line trip fault diagnosis in power systems. *Neural Computing and Applications, 31*, 6719–6731.
15. Wei, Y., Zhang, X., Shi, Y., Xia, L., Pan, S., Wu, J., et al. (2018). A review of data-driven approaches for prediction and classification of building energy consumption. *Renewable and Sustainable Energy Reviews, 82*, 1027–1047.
16. Freeda, A. R., Anju, A., Kanthavel, R., Dhaya, R., & Vijay, F. (2024). Integrating AI-driven technologies into service marketing. In *Integrating AI-driven technologies into service marketing* (pp. 375–394). IGI Global.
17. P. K. Dutta and S. Mullick. *Enhancing position control performance of magnetic levitation system using optimized PID-P controller: a performance analysis*. 7th IET Smart Cities Symposium (SCS 2023), Hybrid conference, Bahrain, 2023, pp. 40–47, https://doi.org/10.1049/icp.2024.0898.
18. Kanthavel, R., & Dhaya, R. (2025). Research review on AI-powered 6G as sixth-sense technologies. *AI for Large Scale Communication Networks*, 373–394.
19. Sharma, K., & Saini, L. M. (2015). Performance analysis of smart metering for smart grid: An overview. *Renewable and Sustainable Energy Reviews, 49*, 720–735.
20. El-Hawary, M. E. (2014). The smart grid—State-of-the-art and future trends. *Electric Power Components and Systems, 42*(3–4), 239–250.
21. Dhaya, R., & Kanthavel, R. (2022). Elaborative investigation of blockchain technology in intelligent networks. In *Advancing smarter and more secure industrial applications using AI, IoT, and blockchain technology* (pp. 93–106). IGI Global.
22. Palensky, P., & Kupzog, F. (2013). Smart grids. *Annual Review of Environment and Resources, 38*(1), 201–226.
23. Dileep, G. J. R. E. (2020). A survey on smart grid technologies and applications. *Renewable Energy, 146*, 2589–2625.
24. Khan, S., Paul, D., Momtahan, P., & Aloqaily, M. (2018). *Artificial intelligence framework for smart city microgrids: State of the art, challenges, and opportunities*. In 2018 Third International Conference on Fog and Mobile Edge Computing (FMEC) (pp. 283–288). IEEE.
25. Zuhair, Mohd, Madhushree Sannigrahi, Abhimanyu Bhowmik, Yanamareddy Pravallika, Prithwish Guha, Pushan Kumar Dutta, and Pronaya Bhattacharya. *TrackerBot: Parallel object tracking and collision avoidance design for unmanned ground vehicles*. 2023 6th International Conference on Contemporary Computing and Informatics (IC3I), Gautam Buddha Nagar, India, 2023, pp. 715–721. IEEE, https://doi.org/10.1109/IC3I59117.2023.10398032.
26. Singh, B., Dutta, P. K., & Kaunert, C. (2024). Replenish artificial intelligence in renewable energy for sustainable development: Lensing SDG 7 affordable and clean energy and SDG 13 climate actions with legal-financial advisory. *Social and Ethical Implications of AI in Finance for Sustainability*, 30. https://doi.org/10.4018/979-8-3693-2881-1.ch009

Artificial Intelligence for Energy-Efficient and Sustainable Smart Grids

Chirag Jitendra Chandnani, Shlok Chetan Kulkarni, Vrishin Upesh Patel, and L. Ramanathan

1 Introduction

1.1 Overview of Smart Grid Systems and the Need for Modernization

The traditional power grid, since its inception more than 100 years ago, has become an integral part of our infrastructure. Almost every other piece of technology that has been created since then is in some way dependent on the power grid. That being the case, keeping the infrastructure up to date is a priority, especially in today's times.

There are a few major challenges for the traditional power grid.

1. The increased production of electric vehicles or EVs along with the AI revolution has skyrocketed energy demands.
2. The integration of intermittent renewable energy sources has also led to some issues.
3. There has been an increased threat of cyber-attacks in recent years which can incapacitate essential services by targeting the power grid.

C. J. Chandnani
School of Computer Science and Engineering (SCOPE), Vellore Institute of Technology, Vellore, Tamil Nadu, India

Mumbai, India
e-mail: chiragjitendra.chand2021@vitstudent.ac.in

S. C. Kulkarni · V. U. Patel · L. Ramanathan (✉)
School of Computer Science and Engineering (SCOPE), Vellore Institute of Technology, Vellore, Tamil Nadu, India
e-mail: shlokchetan.kulkarni2021@vitstudent.ac.in; vrishinupesh.patel2021@vitstudent.ac.in; lramanathan@vit.ac.in

P. Raj et al. (eds.), *Artificial Intelligence (AI) for IT Energy Efficiency and Green AI for Environment Sustainability*,
https://doi.org/10.1007/978-3-031-89420-6_7

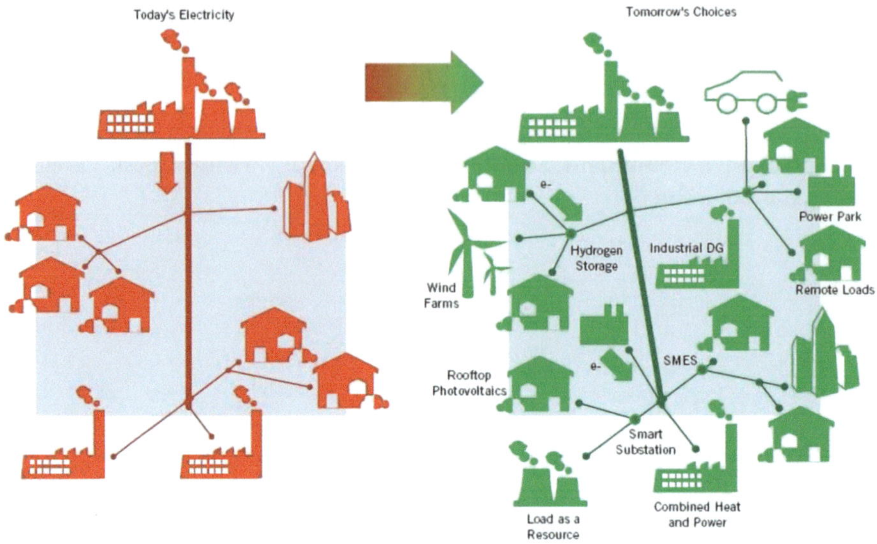

Fig. 1 Smart grids vs traditional grids [21]

An illustration of the differences between a smart grid and a traditional grid has been shown in Fig. 1.

There are solutions to these problems, however, which can be found through the implementation of a smart grid. A few of them are as follows.

1. **Smart meters**: These are devices that collect data about energy consumption in real time. This allows for better demand forecasting and dynamic pricing [11].
2. **Sensors and monitoring systems**: These are sensors placed throughout the grid, detecting any fluctuations in the quality of power, power outages, and other issues. This allows for a quick response to such issues [9].
3. **Automated control systems**: These systems allow optimization of the grid based on current supply and demand. They allow for quick decisions that lead to an overall increased efficiency [24].

Let us, however, delve deeper into the problems that drive the need for grid modernization.

1. **Increased energy demand**: Global energy consumption has seen a large spike since the advent of readily available electronic devices. The increased production of EVs has further exacerbated the issue as traditional grids face significant difficulties in coping with this increased demand. An illustration has been provided in Fig. 2.
2. **Integration of renewable energy sources**: Since its inception, the traditional power grid has been using fossil fuels as a primary source of energy. However, there has been a recent shift to more renewable forms of energy due to environmental concerns. This poses a unique problem to the traditional grid as these sources tend to be more variable and intermittent in their energy output.

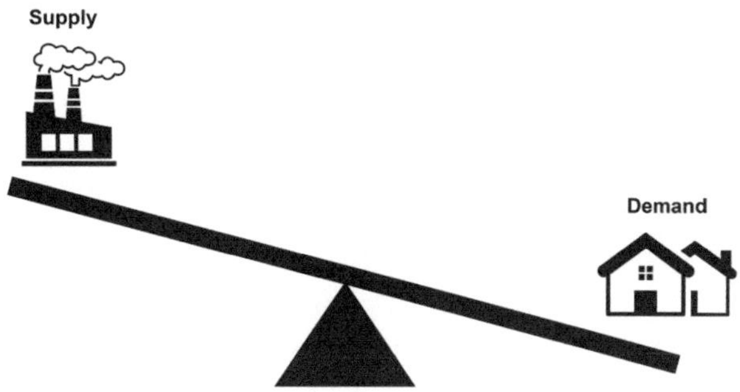

Fig. 2 Balancing demand and supply

3. **Reliability**: There has been a sharp spike in the number of extreme weather events in recent years due to climate change. This spells disaster for the traditional grid due to its centralized nature. This interdependence of components means that a failure in any component of the traditional grid has the potential to lead to a catastrophic failure of the entire grid.
4. **Cybersecurity**: Due to the key role that the power grid plays in the infrastructure of any country and the world, it is a high-value target for malicious actors. This poses a challenge because of the extreme interconnectedness of the grid. Hence, advanced cybersecurity safeguards must be put in place to protect both consumer data and the continued reliability of the grid.

1.2 Role of AI in Transforming Smart Grids

AI has a huge role to play in the modernization of grids. Human error plays a huge role in the unreliability of any infrastructure component. This, however, can be remedied by automating some control tasks using AI. This will lead to not only better reliability but also higher efficiency and lower operational costs. It can also help when dealing with the increased complexity associated with the management of a complex modernized grid.

The following are some specific ways in which AI can be utilized in the implementation of smart grids.

1. **Enhanced demand forecasting and load balancing**: Demand forecasting is an essential functionality required for any power grid. It allows operators to predict fluctuating demand and supply and prepare in advance. Traditionally, power grids utilized historical data to make these predictions. But, due to the increasing complexity of a modern grid, this may not suffice. In this case, machine learning (ML) can provide a solution. Due to recent advancements in ML, it is now possible to process bigger datasets than ever before and use this data to make highly accurate predictions. This data is also not limited only to historical data but also to other factors such as weather patterns, economic and social factors, etc. These methods can be utilized for smarter load balancing and adjustments in power production, which align with the rapid variability of the modern power grid [10].

2. **Fault detection and predictive maintenance**: Currently, the power grid is highly susceptible to power cuts and large-scale failures due to its centralized nature. However, AI can be used to remedy this problem. Using anomaly detection, faults like equipment failure, transformer overloads, or power line faults can be detected in real time. These issues can therefore be fixed quickly and before they lead to any large-scale issues. Along with this, AI can also be used in predictive maintenance to predict wear and tear in certain components and replace them before they cause an issue [33].
3. **Dynamic pricing and demand response**: Dynamic pricing and demand response are yet another application for AI. AI algorithms can be used to predict high demand and vary prices accordingly to incentivize users to reduce the load on the system. In addition to this, the AI models can also use reinforcement learning to reduce energy usage in households and businesses during peak hours without compromising user comfort [30].

2 Historical Background and Evolution of Smart Grids

In order to understand the potential contribution of AI to the modernization of power grids, we must also understand how the traditional grid system was built and how the smart grid paradigm began. This section aims to do that.

2.1 Development of Traditional Power Grids

The earliest centralized power grids started operation in the late nineteenth to early twentieth centuries. These systems are commonly referred to as "one-way" systems. They were very simple in their design. The power is generated in a large power plant, which could be either a thermal plant or a hydroelectric. These plants formed the backbone of the entire grid. The power was then supplied directly to the consumers. The system lacked any feedback loops and created a linear flow of energy.

The flaws in the system, however, soon became apparent. These traditional grids struggled with meeting increasing energy demands, especially during peak usage times, leading to frequent outages [25]. Furthermore, as renewable sources of energy became more sought after in the twentieth century, their integration into the existing grid turned out to be a huge challenge. The inflexibility of the traditional grid proved to be an obstacle in the integration process of new power sources. Additionally, the traditional approach of predicting energy demand using only past trends proved to be insufficient in the new dynamic energy landscape with decentralized energy sources and variable consumer behavior.

As the energy landscape kept evolving, there was a bigger than ever need for the modernization of the grid. The lack of reliability of the traditional grid was a fatal flaw and with the widespread increase in the adoption of renewable energy technologies a change in paradigm was required. The new grid required resilience against failure and ease of integration when it came to new sources of energy especially when these sources were erratic [25].

2.2 Emergence of Smart Grid Technologies

Smart grids first emerged in the twentieth century. This happened due to a number of reasons. The first was that the environmental impact of fossil fuel-based solutions and the traditional power grid became apparent. Additionally, with the fast advancements in information and communication technologies, the potential capabilities of a smart grid solution were more visible. A smart grid integrated with communication and information technology would allow for real-time monitoring, control, and automation of energy distribution [12]. There are a few major components of smart grids:

- Advanced metering infrastructure (AMI).
- Sensors.
- Smart appliances.
- Communication networks.

The communication networks allow for two-way communication between utilities and consumers. This allows the consumers to be more aware of their energy consumption and make more informed decisions, promoting energy efficiency and conservation [17]. Consumers could, for instance, reduce their consumption during peak hours, in turn, reducing their costs. It also, at the same time, contributes to load balancing and overall system efficiency.

As mentioned before, the limitations of a traditional grid make it difficult to integrate renewable sources of energy. Smart grids can not only solve this issue but also provide additional functionality like real-time management, allowing utilities to optimize production and consumption. Smart grid solutions also support demand-side management activities, wherein, the consumers are incentivized to use less power during peak periods.

2.3 Early Integration of AI in Energy Systems

The smart grid systems require an accurate prediction of supply and demand. Until the late twentieth century, this was done only by using historical data for predictions. However, with the advent of AI technology in the late twentieth century, a better solution was on the horizon. These ML models could efficiently process large amounts of data and were much more accurate in their predictions than previous solutions. This went hand in hand with the vast amount of data that was being generated by smart meters, sensors, and other digital tools. Early integration of AI into smart grids mostly focused on improving demand prediction accuracy and operational efficiency [9].

Since its introduction, AI in smart grids has seen its utility go up manyfold. The accuracy of prediction has also gone up significantly due to the integration of many information sources for prediction such as weather patterns, seasonal trends, and demographic shifts. This increased accuracy allows for proactive management of supply, a reduced number of outages, and optimal resource allocation [16].

Fault prediction and predictive maintenance are two more ways in which AI has greatly improved the grid. Due to the ability of operators to proactively schedule maintenance based on failure predictions and/or react quickly to failures in the system, the reliability of grids has greatly improved. Furthermore, this has also led to a decrease in operational costs, which is incurred due to the aftermath of such failures.

3 AI-Powered Demand Forecasting and Management

AI-powered demand forecasting has completely changed the way the power grid is managed. It has led to huge boosts in not only efficiency but also the reliability of the power grid. AI-powered techniques can be utilized to forecast electricity demand more accurately resulting in the aforementioned gains in efficiency. This section will explore some such techniques.

3.1 Machine Learning Techniques for Predicting Electricity Consumption

ML models have become an essential component for demand forecasting in smart grids. They make use of the vast amount of data generated by smart devices such as smart meters, sensors, and other digital infrastructure used in smart grids to predict future demands. One major advantage of ML-based techniques over previous techniques is that they not only make use of historical data but also take into account other factors like consumer behavior, time of day, weather conditions, economic trends, etc. This allows for increased precision in their predictions [7].

A few of the most widely used machine learning techniques for demand forecasting include time-series analysis, support vector machines (SVMs), and deep learning models like artificial neural networks (ANNs). Each of these techniques is used for specific situations as specified below:.

1. **Time-series analysis**: Time-series analysis allows operators to use past data and predict future demand surges [13].
2. **Support vector machines**: SVMs are used to apply different operating conditions to identify the most relevant features and classify and forecast demand.

3. **Recurrent neural networks and long short-term memory networks**: RNNs and LSTMs are deep learning models that are extremely well equipped to deal with sequential data. This allows them to identify temporal dependencies in the consumption data. They are employed to deal with the extremely dynamic and nonlinear nature of demand.

Deep learning models have recently been seen to outperform traditional statistical models especially when the data has multiple sources such as renewable energy systems, EVs, and smart appliances.

The deployment of ML algorithms has enabled the movement to more renewable sources of energy. Renewable sources of energy are currently very variable in their power output and require accurate predictions of ML models to avoid over-dependence on fossil fuel-based backups. The predictions allow operators to decide the share of renewable and conventional sources of energy, thereby contributing to the reduction of greenhouse gas emissions and operational costs at the same time.

3.2 Efficient Demand Response Strategies Using AI

Demand response (DR) is a set of strategies that are used by utilities to incentivize consumers to avoid using power during peak hours. However, the nature of these strategies has changed a lot. Traditionally, these strategies would involve providing reduced rates of electricity for off-peak hours. This has changed entirely, though, with the advent of AI-based demand response strategies. Using these strategies, the price incentives can be replaced by more sophisticated and real-time management techniques [33].

Instead of giving incentives to consumers, AI-based DR can automatically manage power consumption across the entire grid by communicating with smart devices and appliances. This includes things like EV charging and smart thermostats. ML models are used in these cases to determine the optimal time to reduce power consumption without sacrificing consumer comfort or convenience.

Demand response with AI also has a major advantage when it comes to decentralized approaches. In a traditional centralized approach, the utilities decide how much electricity each household is consuming. In a smart grid, however, it is possible for consumers to autonomously control their consumption using smart meters. During peak times, smart meters and connected appliances can be programmed to turn off nonessential services such as washing machines. This also has the added advantage of reduced need for human intervention, making demand response more seamless and effective.

With the increasing frequency of extreme weather, AI-based DR is essential to maintain the reliability of the grid. It prevents events like blackouts from happening and affecting essential services during times of need by reducing stress on the grid during peak hours. This is especially true in regions with an aging grid infrastructure [33].

3.3 *Benefits of AI-Based Energy Management Systems*

AI-based energy management systems have several key advantages over traditional management systems. These are summarized visually in Fig. 3. They are as follows.

1. Real-Time Response
   - AI-based approaches allow for a real-time response to fluctuations in demand.
   - This leads to an increased reliability and efficiency of the grid along with lower operational costs.
2. Integration of Renewable Energy Sources
   - Integration of AI helps deal with more variable and decentralized sources of energy such as solar panels or wind turbines.
   - Without AI, such sources of energy could not have been integrated into the grid because of their high intermittency. However, the predictive abilities of AI allow operators to manage this variability in output. This allows us to move toward more renewable sources of energy.
3. Predictive Maintenance
   - Utilizing AI algorithms bottlenecks and failures can be detected well before they become a larger issue.
   - This allows for real-time monitoring of the grid and the quick detection of issues as they occur.
   - These methods also detect wear and tear of components, allowing predictive maintenance.
   - This leads to a significant reduction in downtime and disruptions.

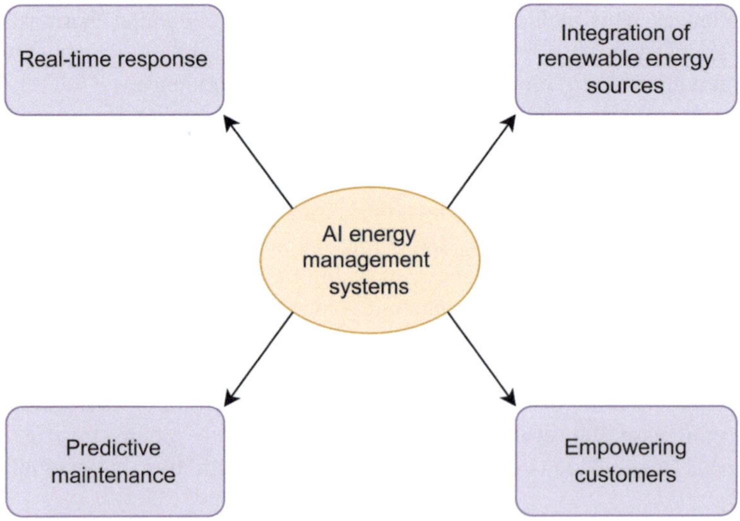

Fig. 3 Benefits of AI in energy management systems

4. Empowering Customers
   - Demand response using AI also allows customers to gain insights into their power consumption, allowing them to make more informed decisions.
   - They also further the goal of spreading awareness about the conservation of energy and the move toward a more sustainable society.

4 Fault Detection and Predictive Maintenance

Automation has gained a lot of significance in recent years, especially in automating and scheduling maintenance activities and predictive fault detection and prevention. This ensures that the grid is up and running without any unforeseen downtime which is vital for any state's/city's efficiency and production. As we are moving toward the automation era, it is the need of the hour to automate this task, which can help decrease the downtime of grids [6]. AI-driven methods have gained prominence in recent years by analyzing and learning patterns from continuously monitored IoT sensors. This helps in staying a step ahead and taking necessary cautionary measures before the real issue takes place. This minimizes interruptions, improves overall working, and prolongs the lifespan of the infrastructure.

4.1 AI Algorithms for Real-Time Fault Identification

Leveraging the power of AI can help improve prediction in smart grids. As AI algorithms require large amounts of data to train and learn patterns, smart grid systems are one of the best use cases for their application. Grids are an abundant source of information, storing data continuously about the various parameters of the grid. Timely identification of these problems and timely measures are essential for maintaining grid stability. By utilizing the power of light AI algorithms and various computing paradigms like FOG and edge computing, we can assess real-time data from IoT sensors distributed throughout the grid, identifying anomalies, alleviating any dangers in real time, and alleviating the defects or issues beforehand [8].

1. **Anomaly detection using AI**: These models employ supervised or unsupervised learning methodologies and are extensively used for recognizing out-of-the-normal patterns in grid data that may allude to failures in some component of the grid. Supervised methods require huge amounts of labeled training data in order to predict the future efficiently [19]. Conversely, unsupervised models require unlabeled data and can infer patterns from the dataset without labels, some of the examples being clustering and autoencoders, which identify anomalies in normal behavior without pre-existing fault labels, thereby making them advantageous in varying or dynamic fault situations [4].

2. **Sensor data and IoT integration**: IoT sensors are widely utilized in smart grids to monitor parameters such as voltage, current, and temperature at various grid locations. These sensors generate real-time data essential for identifying irregularities. By amalgamating sensor data with AI models and also leveraging the power of edge and fog computing to give analysis in real time, utilities can facilitate continuous monitoring and identify early indications of defects prior to their escalation. This IoT-based methodology facilitates effective situational awareness across extensive grid networks, allowing for expedited responses to potential problems.
3. **WAN and AI**: The proliferation of wide area networks (WANs) in today's generation has helped shape lives and given access to vast repositories of information to billions of people. By integrating WANs and dense pretrained AI models, we can train the model remotely from servers and can then deploy it on the grid infrastructure for inferencing; this loop continues, and the model updates remotely on the server and updates its parameters via training.

4.2 Predictive Maintenance for Reducing Downtime and Improving Reliability

Predictive maintenance is a proactive method that uses the power of data analytics to anticipate/predict when equipment will need repair or replacement, informed by its current state and operational history. This ensures that the grid safety is not affected and the engineers and officers remain a step ahead of all these last-minute defects and issues. In contrast to reactive maintenance, executed postequipment failure, and preventive maintenance, conducted on a predetermined schedule, predictive maintenance seeks to enhance the scheduling of maintenance activities to minimize operational interruptions and maintenance expenses. Predictive maintenance enables utilities to anticipate breakdowns, thereby conducting maintenance solely when required, which reduces unplanned downtime and superfluous routine inspections. But for effective prediction from our AI model, we need real-time access to large amounts of organizational data. Some of the sources are:

1. **Historical maintenance and auditing reports**: Historical maintenance logs serve as a valuable repository of labeled or unlabeled data for training AI models for the use of predictive maintenance. These logs have the history of maintenance, checks conducted in the organization, types of failures that occurred, and so on. This data enables AI models to learn trends and patterns linked to diverse failure types, calculate the average life-cycle of MTTF (mean time to failure) of the components of grids, and send alert notifications if the time for maintenance nears. Furthermore, through the examination of previous logs, models can identify recurring failure patterns, including trends or components susceptible to failure. These trends shape the so-called "decision-head" of the model. This is crucial, as incorrect classifications or late classifications can lead to huge losses.

2. **Sensor data**: Continuous routing of sensor data is essential for continuous health monitoring and operation of smart grid modules and components over time. IoT sensors integrated into essential components across the grid measure various parameters such as the temperature of the part and voltage generated. This data generated in real time offers up-to-date insights into the working of the whole infrastructure, at a granular level, thereby enabling AI models to identify even the slightest of anomalies from standard occurring patterns. For example, a steady increase in temperature for a prolonged period of time is detrimental to the life of that component. This might signify that a component is approaching its failure, thereby necessitating measures for its replacement. Alongside temperature, metrics like load variations and harmonic distortion data are also monitored and stored to evaluate overall equipment condition. A thorough analysis of extensive real-time data can diminish the probability of abrupt failures in grids [32].
3. **External resources**: The performance of the smart grid is significantly affected by external data, including environmental factors like weather, load demand, and other factors that can compromise system components. Heatwaves, thunderstorms, and subfreezing temperatures—all facets of severe weather—can accelerate infrastructure decay. Heat can damage insulation, and freezing temperatures can inhibit battery performance in storage units. Furthermore, unexpected spikes in load demand (often caused by seasonal consumption patterns or external events) put additional pressure on transformers, generators, and distribution lines. Adding this external data into the predictive maintenance models allows the utilities to shift their repair schedules for these changing conditions. By forecasting data and important variables that may be one of the main reasons leading to equipment deterioration, predictive models empower grid operators to do preventative maintenance under optimal conditions, thereby minimizing the risk of extensive outages.
4. **Time-series-based models**: Data points collected over time used for training share patterns, trends, or dependencies between data that allow for the continuation of a sequence in order to forecast and predict future events; this is known further as time-series analysis. State-of-the-art time-series models such as ARIMA and LSTM networks have proven effective in dealing with temporal dependencies, making them a prevalent choice for the predictive maintenance of smart grids. ARIMA models are good for linear data and can predict clear trends in the future based on past data. LSTM networks are a more advanced type of RNN that have been specifically designed to learn complex and often hidden relationships in time-series data, and mitigate the vanishing gradient problem. They are especially useful for sensor data, where equipment behavior is unstable and may change rapidly. These networks can identify irregularities with great efficiency, thanks to their gated recurrence, which enables the network to forget patterns not required and remember patterns important ones and also predict possible equipment failures by constant monitoring and analysis of the incoming sensor data. Time-series models also help identify failures prior to their escalation to major problems.

5. **Reinforcement learning-based models**: Reinforcement learning (RL) models optimize maintenance schedules by identifying the most effective maintenance procedures via trial and error and a reinforcement scoring mechanism. In the realm of smart grids, reinforcement learning models evaluate the impact of various maintenance actions on equipment performance, employing a feedback loop with the AI agent and the input to the grid to enhance decision-making progressively. An RL agent may learn the changes in behavior in voltage signals, temperature, or any other physical factors to conduct maintenance solely when sensor data reveals particular risk factors, thus minimizing superfluous inspections and reducing the cost of these inspections. These models refine their maintenance suggestions in response to real-time grid circumstances, as they gain experience from each maintenance cycle through points and a reward system. It incentivizes or rewards those actions that facilitate or help the agent to reach the final goal state [14]. In intricate grid systems, where equipment performance fluctuates due to various variables such as changing load demands or variations in voltage generated, reinforcement learning can optimize maintenance schedules according to prevailing conditions. This helps the organization in bringing down the overall maintenance expenses.

The methods enlisted above are summarized in Fig. 4.

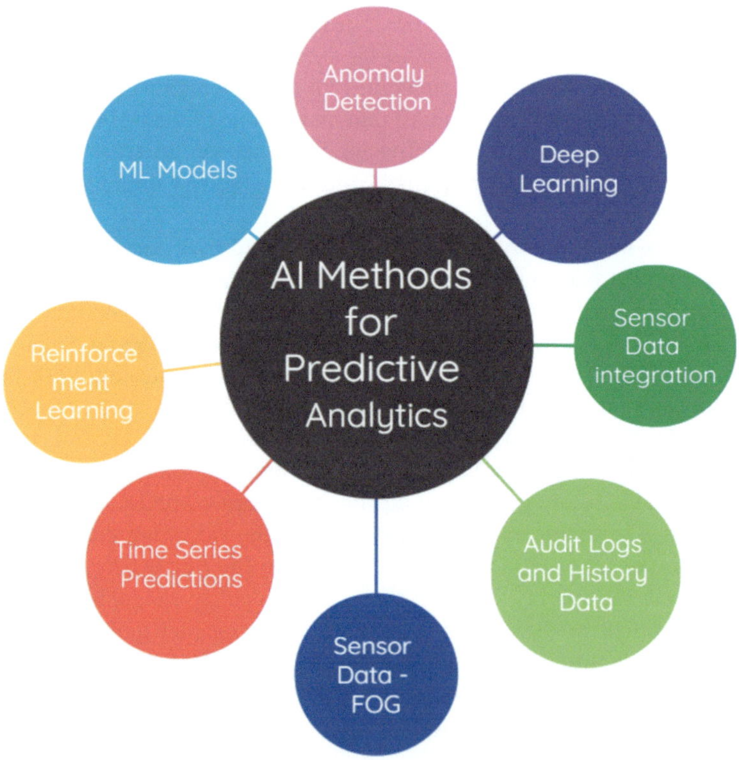

Fig. 4 Brief summary of various AI methods for predictive analysis and fault detection in smart grids

4.3 Benefits of AI Algorithms in Predictive Analytics

Predictive maintenance significantly transforms contemporary industry by enabling enterprises to proactively manage their equipment, thereby circumventing the drawbacks of reactive maintenance. Utilizing real-time data from sensors, machine learning algorithms, and IoT devices facilitates the early identification of probable issues, hence assisting in the reduction of unforeseen downtime [5].

Reduction of downtime is an important factor because equipment failure is always expected and can disrupt production, inhibit work, and cause many losses. Predictive maintenance keeps the equipment in an efficient working state, and the duration it operates is extended. Predictive maintenance therefore plans interventions according to real usage and condition, but episodes of planned overhauls at the specified time often fails to address the particular usage pattern and ramp up early wear in elements, thus ensuring early preventable deterioration and increased asset productivity. However, predictive maintenance sharply reduces the operational costs thanks to reducing the requirements in regular, unanticipated repairs and restricting the prices for component replacement. In this way, through precise timing of maintenance, possible costs connected to unnecessary maintenance are prevented by firms. Predictive maintenance has a tactical advantage because it enhances equipment reliability, prolongs asset life cycle, and promotes cost-effective and robust operation.

5 Optimizing Energy Distribution and Grid Stability

Other critical elements of an efficient distribution and grid are needed for a strong and reliable infrastructure through the incorporation of distributed energy resources (DERs), such as solar and wind energy, electric vehicles, and energy storage systems [20]. The challenges regarding the power grid stability are crucial to avoid down-times, to improve the resources' utilization, and to promote the change to a cleaner, sustainable power system. AI enables smart grids to balance supply and demand, meet the load requirements, and increase the efficiency of distribution of electricity. This model is very flexible and highly adaptive; as a result, grids are able to power the societies efficiently in spite of its dynamic conditions. All these benefits and observations are validations for the incorporation of various AI algorithms for the optimization of energy distribution.

5.1 Leveraging AI Algorithms for Dynamic Load Balancing

Load balancing in smart grids denotes the regulation of the flow of electricity in the grid in order to guarantee optimal distribution. This involves shifting electricity following demand fluxes in order to ensure that there is sufficient supply of the electricity, and avoid putting a lot of stress on any segment of the power grid. This technique is important to avoid blackouts, reduce peak loads, and maximize the efficiency of energy resources [18]. In the past, load balancing in a smart grid was governed by the implemented scheduled power distribution plan, while in modern smart grids, load balancing is done through AI, which can address fast-changing conditions in demand viably. AI brings load balancing into an optimization process with the use of collected data, which significantly improves the grid's robustness and functionality.

AI techniques are also quite helpful for forecasting load demands, which are essential for effective load distribution. Uncertainties can be minimized by accurate predictions helping the grid operators to improve their decision-making on the distribution and scheduling of electricity and planning of the maintenance of the infrastructure [2]. Forecasting models can also be categorized as short term and long term.

Types of Load Forecasting in Grids

- Short-term load forecasting (STLF).
- Long-term load forecasting (LTLF).

Simply put, STLF is used to predict electricity demand for a shorter timeframe, usually a day or a week. STLF is essential for detecting and mitigating sharp increases or decreases in power demand resulting from variations in customer usage patterns, meteorological factors, or unforeseen outages. The use of AI again here has revolutionized the industry, especially the application of sophisticated and novel machine learning techniques including RNNs and transformers. RNNs capture and store patterns in their weights more efficiently for temporal-based tasks as compared to other DL and ML approaches. Since 2017, transformer-based approaches have also emerged as promising for managing long, time-series-based sequences of data through their attention mechanisms, which can store context and patterns within their weights efficiently providing precise predictions and trend analysis even for intricate, high-frequency load fluctuations.

However, LTLF spans months or years and provides valuable information for long-term planning of grid system infrastructure and expansion. The data from this form of forecasting is a vital metric for major business decisions and investments. In contrast to STLF, LTLF focuses on long-term occurring trends and patterns in energy consumption and load. Advanced AI models for LTLF, including hybrid models that integrate different ML approaches, are used to enhance grid asset optimization, thereby ensuring the infrastructure can accommodate the future needs of the population of the region.

5.2 *Usage of Edge Computing for Real-Time Analysis*

Edge computing is a crucial technology for local data management, situated near its origin, rather than relaying it to a centralized data center or cloud. This decentralized methodology is particularly advantageous in smart grids, where swift response times and localized decision-making are essential for sustaining operational stability and efficiency. A smart grid facilitates the continuous transmission of extensive data from scattered sources such as transformers, substations, smart meters, and numerous IoT-enabled devices. The architecture of edge computing facilitates real-time data processing at the source, substantially reducing latency in comparison to conventional cloud-based systems. One of the architectures by Singhal et al. [27] is given in Fig. 5.

Smart grids can execute functions such as load balancing, fault detection, and predictive maintenance with low latency by facilitating data processing and AI model deployment at the edge. Edge computing-based solutions can guarantee low latency in identifying trends or anomalies in various parameters ensuring real-time operation and prediction.

In a real-world use case, where the data collected from the sensors of a substation indicate that the current demand is quickly peaking the equipment's capacity, edge devices can automate the process of redistributing power to reduce the load on some of the components without the requirement of prior approval from the main control center. This capability of real-time response enables organizations to solve issues that might seem "minor," but in turn, may affect the grid in the long term. For example, edge devices could request voltage modifications to their neighboring grids or, for instance, activate energy storage to decrease their demand. All of this can take place in a very short period, at the microsecond level. Such preventive measures also help in preventing the occurrence of outages and operational disruption.

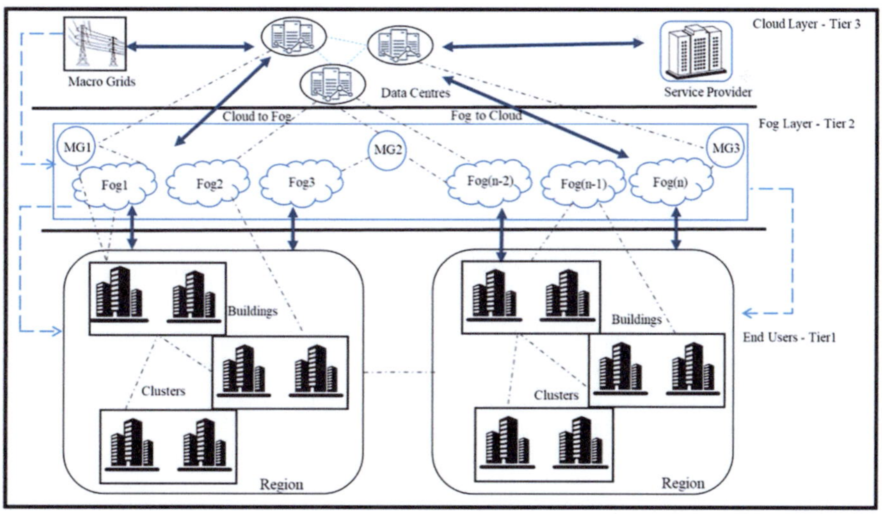

Fig. 5 Proposed edge computing-based smart grid management system using AI proposed by Singhal et al. [27] incorporates real-time modifications to power distribution, including load balancing. This approach also localizes the data by not transferring it to the central compute server, ensuring data privacy and also giving an edge to the grid by choosing which data/features to send for efficient and private analysis of data

Furthermore, edge computing increases the scalability of the smart grid, where each substation or transformer works independently without overloading the main system. This distributed processing capability reduces the data amount delivered to central servers, thus saving network bandwidth and reducing the cost of cloud storage to the grid owners/organizations. Local data processing also has the advantage of increasing security and privacy by ensuring that such information is kept at the place of origin rather than over the networks. This reduces the attack surface of the grid and ensures that malicious attackers are kept at bay from disrupting the power flow.

5.3 *Impact on the Grid's Stability*

Grid stability refers to the ability of the grid to maintain consistency in voltage and frequency values over the whole network. AI-based approaches directly affect grid stability, especially in detecting fluctuations in demand or supply of power. AI assists in maintaining the right amount of power through accurate load forecasting and demand response strategies ensuring grid stability [26]. Real-time adjustments to distribution levels lower the risk of component stress and subsequent failure, thus lowering planned and emergent downtime. Also, the unparalleled growth of renewable energy sources like wind and solar needs proper AI-based load management for the integration of such capacity. The capability of applying AI in generation and consumption can also make smart grids work in a more sustainable and efficient way, which will create a base for further developing a stronger energy structure [26].

6 AI in Renewable Energy: Transitioning to the Future

Transitioning to sustainable sources of energy such as wind, solar, and hydropower as major energy resources is important for the creation of a sustainable energy structure and reduction of reliance on fossil energy. Moreover, considering the Paris Treaty and also the United Nation's (UN) Sustainable Development Goals till 2030 (SDGs), AI can complement the benefits of renewable energy sources to humans. However, the intermittent nature of the renewable energy source remains disadvantageous to the stability of the grid, since power output depends on certain parameters that are not in the control of the grid. AI has emerged as a critical tool for addressing these issues due to robust methods for predicting renewable energy output, effective management of energy storage systems, and integrated performance optimization for renewables across the grid. Smart applications help increase the effectiveness and reliability of the utilities by keeping up an optimal balance between supply and demand and reducing dependency on conventional energy backup systems [3]. Therefore, AI can be used for accurate renewable energy production prediction, so that it can be used as a parameter to significantly predict the power output from different generators of renewable energy sources.

6.1 Predicting Renewable Energy with AI Models

Renewable energy generation prediction is important for grid operators since it would assist in planning and matching supply with demand in an optimal manner. In contrast to other conventional sources of energy, the generation of renewable energy depends on the weather, day/night cycles, and seasons. AI-based forecast models may analyze past weather data, weather conditions, and other related factors to give the right estimates. These projections give utilities desired energy production anticipated levels so that they can properly distribute the energy as well as plan for needed back up resources or storage systems. The efficiency achieved with AI models is increased as compared to conventional methods by way of accurately predicting renewable energy output, thus reducing reliance on nonrenewable resources [3].

Forecasting for solar and wind energy depends on stochastic variables, namely solar radiation and wind speed. It must also be noted that, often for forecasting, predictive algorithms like CNN and LSTMs are used to a significant extent. CNNs have a special affinity for handling geographical data and, for instance, solar irradiance from satellite images for forecasting, but LSTMs are particularly good at temporal data and motion like the speed of wind or sequential peak irradiances, thanks to their efficient gated architecture. The calibration of these models on weather data from the past allows them to forecast future generation of renewable energy accurately.

To improve the accuracy of the forecasting predictions, we can integrate meteorological data with the help of hybrid ML techniques. These models combine numerous sources of input with satellite imagery, weather prognosis, and prior production statistics in order to enhance the accuracy of renewable energy estimates. A hybrid model may combine statistical models of weather and deep learning for predicting short-term power generation from solar systems using both weather data and trends data.

Precise forecasting of renewable energy in grids influences grid output and management in a positive way by ensuring seamless amalgamation of the use of renewable, sustainable energy sources along with the existing grid infrastructure. AI-driven models provide real-time, precise production estimates, thus enhancing grid efficiency and stability, reducing the likelihood of energy shortages or surpluses, and ultimately reducing reliance on fossil fuel-based energy sources, paving the way for a better future for the upcoming generation.

6.2 Monitoring and Managing Renewable Energy Storage Systems with AI

Energy storage systems (ESS) including batteries are used to address the issue of fluctuations that come with the production of power from renewable sources such as solar and wind. They can store energy at times of high renewable generation and low load [15]. ESSs can take advantage of the excess energy available and release it at the time of peak demand or low renewable generation. This buffering capacity helps to stabilize the grid and lets the operators better manage the available renewable energy resources. The problem of ESS effectiveness in terms of increasing the duration, capacity and stabilizing contribution can only be solved through effective management, and AI has been identified to play a crucial role in increasing the efficiency and effectiveness of the system. Some of the recent advancements in the field of optimization of the storage of energy through AI are as follows:

1. Prediction of battery state of charge (SoC): One of the most important aspects of energy storage management is the prediction of the battery SoC. Therefore, with SoC accurately predicted, batteries are not excessively charged or discharged, two practices that are likely to reduce the life and capacity of batteries in the long run. Dynamic AI models present the capability to estimate the SoC figures from the battery based on real-time inputs and thus prolong the health of ESS components. The use of SoC predictions developed through AI increases battery capacity usage and results in a prolonged life of the whole system [15].
2. Intelligent dispatching algorithms: Advanced AI-based dispatching systems control the flow of the stored energy based on the current grid load and the real-time availability of renewable energy. By using these models to anticipate maximum hours of peak consumption, they ensure the timely release of power to reduce dependency on nonrenewable and conventional resources. Smart dispatching solutions enable efficiency in the application of energy storage systems by ensuring that stored energy can be made available in such times as the demand is high or during scarce renewable generation thus boosting the grid reliability.
3. Lifecycle optimization: The degradation of battery components over the cycle lifetime presents a major challenge for energy storage systems with respect to both efficiency and economy. AI algorithms can predict the further performance of batteries, which will help schedule maintenance. The batteries' status and the likely requirement for service or replacement are assessed by a battery management system based on artificial intelligence using such physical parameters as temperature, charge cycles, or usage patterns. This helps in the optimization of this life cycle in order to minimize unanticipated failures and, therefore, increase battery reliability in the face of shortages of renewable power.

All of the methods enlisted above are shown in Fig. 6.

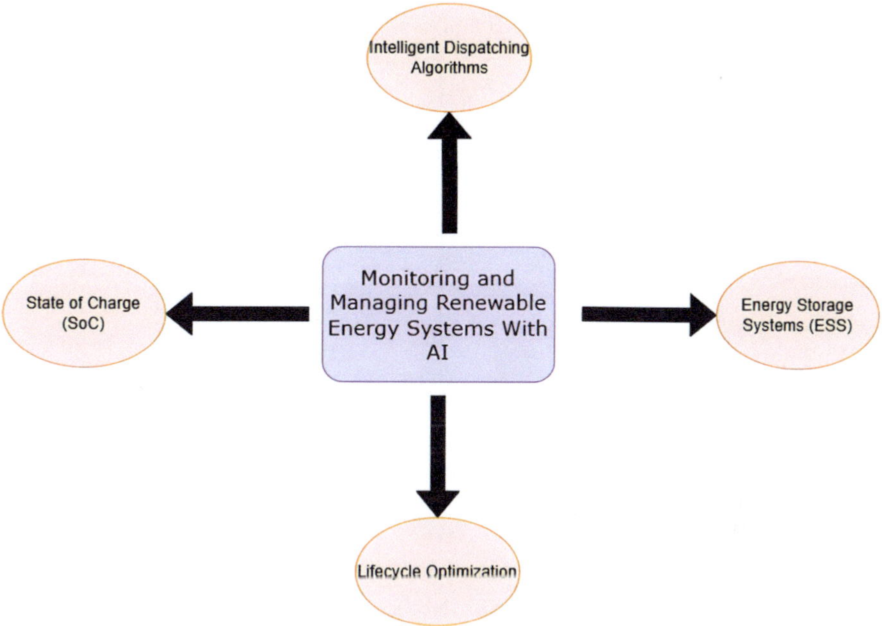

Fig. 6 Brief mind map of different approaches to monitoring renewable energy sources while leveraging the power of AI

7 Smart Metering and Consumer-Centric Energy Solutions

Smart metering systems, which enable real-time monitoring of energy usage, have been developed in response to the growing demand for dependable and efficient energy management. The way consumers engage with energy could be revolutionized by these solutions. This section will explore the latest advances in smart metering and consumer-centric energy solutions, including their benefits and challenges.

7.1 AI-Enabled Smart Metering for Customized Energy Management

Real-time, thorough tracking of energy use trends is possible with AI-enabled smart meters. Access to actionable data about their usage patterns gives consumers the power to decide how they use energy. Smart meters allow users to drastically cut expenses by giving them specific information. Examples of the advice include identifying energy-wasting gadgets and suggesting the best time to use them. This

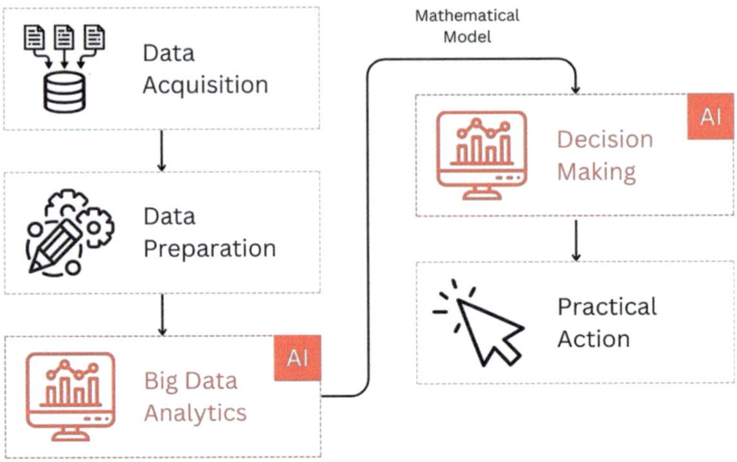

Fig. 7 Use of AI for personalized results

promotes financial restraint and is consistent with the worldwide movement toward more environmentally friendly energy sources. Energy monitoring systems can detect areas of energy waste, equipment failure, and anomalous energy consumption by utilizing artificial intelligence. This makes it possible to take corrective action quickly, which lowers expenses and boosts energy efficiency.

Because of AI-driven analytics, smart meters may give consumers personalized information. Examples of these insights include thorough evaluations of energy usage, comparisons with similar residences, and recommendations for energy-saving methods. Customers take an active part in the effort to achieve energy efficiency when they feel more empowered and involved after receiving customized information. By tailoring recommendations according to individual tastes and habits, smart meters provide users with a more relevant and meaningful experience with their entire energy usage process. AI-enabled energy monitoring systems are able to evaluate each user's particular energy consumption patterns and provide personalized recommendations. Energy savings result from people being able to make informed decisions (as seen in Fig. 7) about how much energy they use [22].

7.2 Recognizing Patterns and Anomalies

Another consumer-centric use case for smart meter data is identifying trends that deviate from the expected behavior, as indicated by irregular electrical energy use. Despite the difficulty of identifying irregularities in smart meter data, machine learning and signal processing methods can be effectively used for this task. For instance, ambient-assisted living (AAL), where consumption patterns are indicative of the activities of daily living (ADLs) performed by the inhabitants, can be enabled by identifying abnormalities in the smart meter dataFinding unexpected ADL sequences in general or abnormally short or long ADLs, in particular, are frequently good markers of odd user behavior. Being aware of such circumstances can help to notify family members in advance, promoting safety and well-being. Additional applications of machine learning for anomaly identification in smart meter data have surfaced and shown up in domains including identifying energy theft, identifying erroneous smart meters, and identifying unusual consumption patterns generally [31].

7.3 Facilitating Dynamic Pricing to Promote Energy Conservation and Consumer Engagement

Dynamic pricing is essential to attaining energy efficiency because dynamic pricing encourages customers to switch from peak to off-peak hours of energy use. Dynamic pricing represents the actual cost of energy production and distribution, which fluctuates based on demand, by adjusting tariffs based on variables such as the time of day, season, and weather. Moreover, utilities may develop more effective pricing strategies and gain a better understanding of their clientele by using the valuable data that dynamic pricing offers about consumer behavior. Customers and the electrical system both gain when users are encouraged to track and modify their use in response to price signals through dynamic pricing.

In conclusion, because dynamic pricing incentivizes customers to alter their energy-use habits, it is crucial for attaining energy efficiency. Smart energy management systems optimize energy use, cut waste, and enhance overall energy usage patterns by utilizing technologies like automation, data analytics, and the Internet of Things. Through the use of these technologies, smart energy systems can improve energy efficiency by pinpointing problem areas, putting energy-saving measures into place, and instantly optimizing consumption. Dynamic pricing may offer substantial energy savings and advantages for both customers and the electrical system by adjusting tariffs in response to demand [1].

7.4 Privacy Concerns

Smart meters are the main component of the smart metering system that facilitates two-way communication between the utilities and the user, enabling the sharing of pricing and consumption-related data. When subjected to advanced data mining techniques, the automated and fine-grained data generated by a smart meter can provide valuable insights into user behavior and activities. Details on energy use, the number of occupants, the presence or absence of residents, the daily schedule (including showering, TV watching, and sleeping habits), information about household appliances, and, in the case of smart homes, movement patterns within the house are all examples of potential extracted data. A trace of a day's electrical power

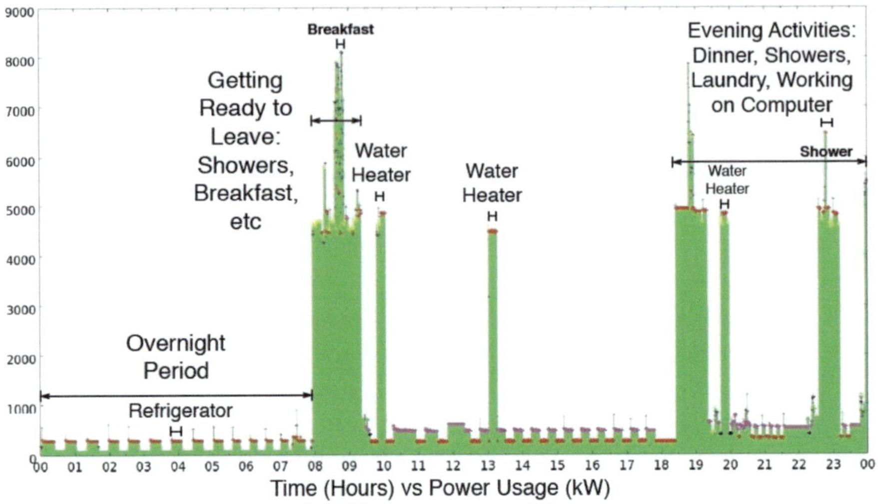

Fig. 8 Example of a day-long power trace

use taken from a smart meter is displayed in Fig. 8. Several details regarding a consumer's daily activities might be discovered by a straightforward and intuitive observation. Human activity in the house can be inferred from the curve's high variance, and by connecting time and variation, we can determine the end user's activities. More precise information about its operations can be obtained through the use of data mining techniques and power signature analysis. Undoubtedly, a large number of end users would never consent to the sharing of their personal information with a third party. Such behavioral data may be useful to a variety of third parties, including social studies researchers who may gain a general understanding of people's daily routines and security personnel who can use it to carry out criminal investigations or settle legal issues. The fact that criminals may preidentify household objects to steal and determine when inhabitants are not home makes it even more crucial for them to plan their break-ins. Therefore, it is crucial to implement strong privacy-preserving strategies in order to safeguard the smart metering system and prevent unauthorized access to the everyday routines of consumers [28].

8 Challenges in AI Integration for Smart Grids

Traditional power systems are mostly analyzed and controlled by numerical calculations and physical modelling due to their complexity. As environmentally friendly renewable energy sources and microgrids have become increasingly prevalent in smart grid development, the shift from traditional power grids to smart grid systems has revealed additional uncertainties and issues in the complicated environment. Modern smart grid solutions are made more unreliable by the fact that the current power system is based on outdated infrastructure. The handling of extremely large amounts of data with considerable variability is still a difficulty for smart grids because the communication network is built on top of power systems. The robustness, adaptability, and online processing of AI systems are more areas of ongoing research. Although numerous data-driven methods have been proposed to deal with the problems of smart grids, there are still many severe challenges, including the following:

1. **Integration of renewable energy:** Smart networks are characterized by highly integrated renewable energy. However, the unpredictability and inconsistency of renewable energy, which can cause sudden and frequent variations in power output, pose a number of serious issues.
2. **Preserving data security and privacy:** Compared to traditional power systems, smart grid systems are more vulnerable to cyberattacks due to their use of numerous devices and two-way communication, which exposes them to malevolent users directly. Numerous innovative security solutions were created to provide quick detection of cyber threats, fake data injection, systems data theft, electricity theft, and other issues. However, the existing smart grid is still vulnerable to a wide range of attacks due to its physical equipment, operating systems, and network protocols. There are trade-offs between security and performance within the AI solutions for smart grid cybersecurity that are now available.
3. **Big data fast storage and analysis:** How to keep enhancing the efficiency of storing and retrieving big smart grid data for AI applications in a reliable manner is another major difficulty.
4. **Explainability of AI algorithms:** AI algorithms typically suffer from the "black box" problem, meaning they cannot be interpreted or explained. Currently, AI algorithms encounter this obstacle.
5. **Limitations of AI algorithms:** The use of AI to smart grid systems is significantly impacted by the advancement of AI technology. However, before implementing any method, all limitations should be taken into account [23].

9 Future Directions and the Transformation Potential of AI in Smart Grids

The goal of smart grids is to create an entirely self-learning system that is cost-effective, responsive, adaptive, self-healing, and totally autonomous. The following topics are covered regarding potential future paths or chances to develop advanced smart grid systems.

1. **Integration with cloud computing:** AI integration with cloud computing, which can improve security and resilience and reduce outages, will become increasingly significant in smart grid systems in order to create an entirely self-learning smart grid system.

2. **Fog computing:** Instead of sending the raw data to a cloud, fog computing attempts to preprocess it locally. Fog computing offers many benefits by offering computing power on demand (e.g., energy efficiency, scalability, flexibility). A few studies have tentatively investigated fog computing integration with the smart grid. As the future smart grid's data volume grows, fog computing will become more significant.
3. **Transfer learning:** One of the biggest obstacles to smart grid analysis is currently an insufficient amount of label data. Researchers are encouraged to employ transfer learning to address the issue of insufficient data because it lowers the requirements for training data. Deep transfer learning problems have drawn more interest in recent years and may find extensive use in smart grid systems.
4. **Consumer behavior prediction:** Demand-side management is becoming an essential task for controlling user participation in power systems, thanks to the development of the 5G network and fog computing. Demand response tasks on the consumer side can benefit substantially from learning patterns of power use and customer behavior [23].

9.1 Advancements in AI Algorithms

There is a great deal of promise for improving energy efficiency in smart grid systems with future developments in AI algorithms. With the development of new models and methods that provide enhanced performance and capabilities, machine learning and deep learning research are pushing the limits of what artificial intelligence is capable of. For example, the creation of increasingly complex deep learning architectures, like transformers and graph neural networks, may result in predictions and judgments for smart grid applications that are more reliable and accurate. Algorithms for reinforcement learning are also developing; new possibilities for optimizing intricate energy management tasks are presented by developments in deep reinforcement learning and multiagent systems. More efficient and flexible ways to improve energy efficiency in smart grid systems will be developed as a result of ongoing advancements in AI algorithms [29].

9.2 Development of Hybrid AI Systems

Combining several AI models and methodologies, hybrid AI systems present a viable way to tackle the intricate problems of smart grid systems. For instance, combining machine learning with expert or rule-based systems can improve the dependability and interpretability of AI applications. By combining the advantages of several AI techniques, hybrid systems can offer more complete and reliable answers for problems like fault detection, energy management, and demand forecasting. To fully realize AI's promise in smart grid systems and enable more adaptable and efficient energy management solutions, hybrid AI systems must be developed [29].

9.3 Focus on Explainable AI

The emerging field of explainable AI (XAI) aims to provide insights into how AI models make decisions while also making AI algorithms more transparent and interpretable. Explainable AI enables stakeholders to understand and validate the decisions made by AI algorithms, which can increase responsibility and trust in smart grid systems. This is particularly important for vital applications like fault detection and energy management, where reliability and safety rely on understanding the logic underlying AI decisions. The creation of explainable AI techniques will be necessary to foster confidence in AI applications and encourage their integration into smart grid systems [29].

9.4 Collaboration and Standardization

To advance AI integration in smart grid systems, standardization and cooperation will be crucial. Technology companies, researchers, governments, and industry stakeholders working together can promote innovation and help develop useful AI solutions. Standardization initiatives, such as developing common protocols and frameworks for AI applications, can enhance interoperability and facilitate the deployment of AI in diverse smart grid scenarios. By fostering collaboration and standardization, stakeholders may accelerate the use of AI in smart grids and advance the creation of more efficient and sustainable energy systems [29].

10 Conclusion

The chapter provides a comprehensive overview of smart grid systems, emphasizing the need for modernization to address contemporary challenges. It draws attention to the transition from conventional power grids to smart grids, which are fueled by elements like the incorporation of renewable energy sources and the rising demand for electricity brought about by electric vehicles. The significance of real-time monitoring, control, and automation in energy distribution—made possible by developments in information and communication technology—is covered in the text.

It also addresses the potential privacy concerns associated with smart metering systems, where the data collected might offer detailed insights into user behavior and household activities. In view of potential unlawful activities, the research highlights the significance of implementing robust privacy-preserving safeguards to protect consumer data from unwanted access.

With an emphasis on how AI can be integrated with cloud computing to increase security and resilience, the role of AI in augmenting smart grid functionalities is also examined. In order to build a more effective, safe, and sustainable energy infrastructure in the future, the paper ends by highlighting the significance of further research and development in smart grid technology.

References

1. Nurul Abdul Manap. Dynamic pricing and energy efficiency. 2024.
2. Brook W. Abegaz and Satish M. Mahajan. *Load balancing and interlacing of storage units in smart grids*. In 2022 IEEE International Conference on Environment and Electrical Engineering and 2022 IEEE Industrial and Commercial Power Systems Europe (EEEIC/ICPS Europe), pages 1–6, 2022.
3. Abisoye, B. O., Sun, Y., & Zenghui, W. (2024). A survey of artificial intelligence methods for renewable energy forecasting: Methodologies and insights. *Renewable Energy Focus, 48*, 100529.
4. Ali, S. S., & Choi, B. J. (2020). State-of-the-art artificial intelligence techniques for distributed smart grids: A review. *Electronics, 9*(6).
5. Almasoudi, F. M. (2023). Enhancing power grid resilience through real-time fault detection and remediation using advanced hybrid machine learning models. *Sustainability, 15*(10).
6. Alsafran, A. S., Hassan, A., Abusada, F., & Alaraj, M. M. (2024). Challenges and solutions for AI explainability in smart grid literature review. *SSRN Electronic Journal.*
7. Boopathy, P., Liyanage, M., Deepa, N., Velavali, M., Reddy, S., Maddikunta, P. K. R., Khare, N., Gadekallu, T. R., Hwang, W.-J., & Pham, Q.-V. (2024). Deep learning for intelligent demand response and smart grids: A comprehensive survey. *Computer Science Review, 51*, 100617.
8. Shokhrukh Davlyatov. *Artificial intelligence techniques: Smart way to smart grid*. In 2023 International Conference on Artificial Intelligence and Smart Communication (AISC), pages 838–842, 2023.
9. Fang, X., Misra, S., Xue, G., & Yang, D. (2012). Smart grid—the new and improved power grid: A survey. *IEEE Communications Surveys Tutorials, 14*, 944–980.
10. Gajowniczek, K., & Zabkowski, T. (2017). Electricity forecasting on the individual household level enhanced based on activity patterns. *PLOS ONE, 12*(4), 1–26.
11. Gungor, V., Sahin, D., Kocak, T., Ergut, S., Buccella, C., Cecati, C., & Hancke, G. (2011). Smart grid technologies: Communication technologies and standards. *IEEE Transactions on Industrial Informatics, 7*, 529–539.
12. Alok Jain and Rahul Mishra. *Changes challenges in smart grid towards smarter grid*. In 2016 International Conference on Electrical Power and Energy Systems (ICEPES), pages 62–67, 2016.
13. Kim, H., Park, S., & Kim, S. (2023). Time-series clustering and forecasting household electricity demand using smart meter data. *Energy Reports, 9*, 4111–4121.
14. Li, Y., Chaofan, Y., Shahidehpour, M., Yang, T., Zeng, Z., & Chai, T. (2023). Deep reinforcement learning for smart grid operations: Algorithms, applications, and prospects. *Proceedings of the IEEE, 111*(9), 1055–1096.

15. Liu, Z., Sun, Y., Xing, C., Liu, J., He, Y., Zhou, Y., & andGuoqiang Zhang. (2022). Artificial intelligence powered large-scale renewable integrations in multi-energy systems for carbon neutrality transition: Challenges and future perspectives. *Energy and AI, 10*, 100195.
16. Mahmoud, M., Nasir, N., Gurunathan, M., Raj, P., & Mostafa, S. (2021). The current state of the art in research on predictive maintenance in smart grid distribution network: Fault's types, causes, and prediction methods—a systematic review. *Energies, 14*, 5078.
17. Mahmoud, M., Tang, A., Maseleno, A., Lim, F.-C., Kasim, H., Yong, C., & Mostafa, S. (2019). A review on smart energy grid technology: Features and specifications. *Journal of Engineering and Applied Sciences, 15*, 535–547.
18. Baranitharan, B., Chandran, K., Murugesan, R., Roy, S., Dutta, P. K., Bhushan, B., & Bhattacharya, P. (2024). Towards sustainable power: A quantitative analysis of environmental factors and energy demand in different Indian states. In A. K. Somani, A. Mundra, R. K. Gupta, S. Bhattacharya, & A. P. Mazumdar (Eds.), *Smart systems: Innovations in computing. SSIC 2023. Smart innovation, systems and technologies* (Vol. 392). Springer. https://doi.org/10.1007/978-981-97-3690-4_10
19. Massaoudi, M., Abu-Rub, H., Refaat, S. S., Chihi, I., & Oueslati, F. S. (2021). Deep learning in smart grid technology: A review of recent advancements and future prospects. *IEEE Access, 9*, 54558–54578.
20. Maurya, P. K. (2024). Artificial intelligence to enhance energy management and distribution in smart grid communication networks. *Tuijin Jishu/Journal of Propulsion Technology.*
21. T. A. Kumar, P. Kanimozhi, P. Dutta and R. Sowmya, *Assessment of gradient classifier based approaches in advanced recommendation framework: uncovering limitations*. 7th IET Smart Cities Symposium (SCS 2023), Hybrid Conference, Bahrain, 2023, pp. 331–336, https://doi.org/10.1049/icp.2024.0946.
22. Nuthakki, S., Kumar, S., Kulkarni, C. S., & Nuthakki, Y. (2022). Role of AI enabled smart meters to enhance customer satisfaction. *International Journal of Computer Science and Mobile Computing, 11*, 99–107.
23. Omitaomu, O. A., & Niu, H. (2021). Artificial intelligence techniques in smartgrid: A survey. *Smart Cities, 4*(2), 548–568.
24. Richter, L., Lehna, M., Marchand, S., Scholz, C., Dreher, A., Klaiber, S., & Lenk, S. (2022). Artificial intelligence for electricity supply chain automation. *Renewable and Sustainable Energy Reviews, 163*, 112459.
25. Salkuti, S. R. (2020). Challenges, issues and opportunities for the development of smart grid. *International Journal of Electrical and Computer Engineering (IJECE), 10*, 1179.
26. Shi, Z., Yao, W., Li, Z., Zeng, L., Zhao, Y., Zhang, R., Tang, Y., & Wen, J. (2020). Artificial intelligence techniques for stability analysis and control in smart grids: Methodologies, applications, challenges and future directions. *Applied Energy, 278*, 115733.
27. Singhal, S., Athithan, S., Alomar, M. A., Kumar, R., Sharma, B., Srivastava, G., & Lin, J. C.-W. (2023). Energy aware load balancing framework for smart grid using cloud and fog computing. *Sensors, 23*(7).
28. Souri, H., Dhraief, A., Tlili, S., Drira, K., & Belghith, A. (2014). Smart metering privacy-preserving techniques in a nutshell. *Procedia Computer Science, 32*, 1087–1094. The fifth International Conference on Ambient Systems, Networks and Technologies (ANT-2014), the fourth International Conference on Sustainable Energy Information Technology (SEIT-2014).
29. Tan, A. (2023). Exploring the potential of artificial intelligence to enhance energy efficiency in smart grid systems: A detailed review and future directions. *Journal of Sustainable Urban Futures, 13*(12), 11–20.
30. Vázquez Canteli, J. R., & Nagy, Z. (2019). Reinforcement learning for demand response: A review of algorithms and modeling techniques. *Applied Energy, 235*, 1072–1089.
31. Völker, B., Reinhardt, A., Faustine, A., & Pereira, L. (2021). Watt's up at home? Smart meter data analytics from a consumer-centric perspective. *Energies, 14*(3).

32. Rezwan Zakaria, MT Alam, Nowrin Jawed, MS Alam, and Naymul Hasan Fahim. *Internet of things (IoT) based smart grid management system.* In 2023 10th IEEE International Conference on Power Systems (ICPS), pages 1–6, 2023.
33. Xiangyu Zhao, Yang Guo, Xiaoying Guo, and Huirong Li. *Artificial intelligence applications and prospects for the smart grid.* In 2023 Panda Forum on Power and Energy (PandaFPE), pages 1844–1848, 2023.

Artificial Intelligence and the Energy Security–Poverty Nexus

Yogeeswari Subramaniam, Muzafar Shah Habibullah, Nanthakumar Loganathan, and Ravi Palanimuthu

1 Introduction

The globe is therefore working hard to enhance energy security, that is the assurance of availability and sustainability of energy with respect to the need, for accessible and sustainable energy. Together with IEA, we are backing this project aimed at raising the population's standard of living and ensuring sustainable economic growth as well as a stable and continuous supply of energy. The viewpoint in support of the enhancement of energy security has been supported by [1, 2], who argue that energy security enhancement can initiate affirmative macroeconomic impacts through stimulation of economic growth and new job opportunities formation. Improved energy security is physically linked to accelerated economic growth, and reduction of poverty through boosting manufacturing, job creation, and international competitiveness. In doing so, energy security has a profound impact on environmental conservation since it helps shift from relying on fossil fuel to renewable and clean sources of energy. It supports the environment and reduces the emission of carbon into the atmosphere. [These are the advantages], not least aware of which policymakers and environmentalists are focusing on energy security as one of the crucial ways to achieving socioeconomic and environmental sustainability. Of

Y. Subramaniam (✉) · N. Loganathan
Faculty of Management, Universiti Teknologi Malaysia, Skudai, Johor, Malaysia
e-mail: yogeeswari.s@utm.my

M. S. Habibullah
Putra Business School, Seri Kembangan, Selangor, Malaysia

R. Palanimuthu
Department of Management Studies, Manonmaniam Sundaranar University, Tirunelveli, India

P. Raj et al. (eds.), *Artificial Intelligence (AI) for IT Energy Efficiency and Green AI for Environment Sustainability*,
https://doi.org/10.1007/978-3-031-89420-6_8

course, recognizing such advantages, there are other centers of gravity of decision-makers and environmentalists focusing on energy security as one of the efficient ways toward the attainment of socioeconomic and ecologically sustainable development.

Although energy security ensures accessible and sustainable energy sources are available, it does this in a way that is socially unjust by nature. The efforts deployed to enhance energy security by developing and promoting clean fuels have not been shared evenly, enlarging the gap between the "haves" and "have-nots," and deepening poverty [3–5]. Therefore, [6] have revealed that while 733 million people worldwide remain without electricity, over 2.4 billion people and still using toxic fuels for cooking. This statistic raises concerns about the energy poverty, which is defined as the lack of sufficient, affordable, reliable, safe, clean, modern, and sustainable energy services. Besides, [7] finds out that due to limited capital that can be used for the implementation of clean energy projects and energy-efficient technologies, the poor in developing countries are less likely to put emphasis on conservation and the use of energy-efficient technologies. For example, the capital costs of retrofitting, for example, solar panels, inverters, batteries, and wiring systems are very expensive, which hinders it from being cheaper for everyone, especially the poor [8]. For this reason, we can realize that energy security measures benefit billionaires and in effect exclude struggling people fighting energy poverty.

Does this mean that we must demobilize efforts for energy security as it seems the majority of the arguments relate to energy poverty? An important fact that is hidden is the fact that poverty does not always increase when energy security increases. A cross-sectional study described poverty as the lack of sufficient income to afford essential domestic utilities such as cooking, lighting, and electrical appliances owing to the high costs of utility. In such instances, there is a requirement to make an additional attempt in the direction of discussing the above-mentioned interest, "Energy security and poverty alleviation," in general, but with added rigidity. Thus, this study incorporates artificial intelligence in the energy security–poverty relationship because artificial intelligence has become a central feature in nearly every aspect of life. The argument is that artificial intelligence (AI) can deliver the latest digital methods for enhancing productivity, optimizing energy management, and expanding the availability of cheap and affordable clean energy. The relationship between energy security and poverty may be moderated by AI in three main ways. Using predictive analytics might contribute, first, to more accurate energy demand predictions, which can be useful for better management of resources. Second, AI governance might assist in accessing regions with clean energy through the effective use of renewable energy. Third, deployment and development of renewable energy systems open new job opportunities. Consequently, this study aims to seek to make two major contributions when examining the energy security–poverty relationship through AI. To the best of our knowledge, no study has proposed this possible solution to meet energy security and poverty jointly.

2 Related Works

Energy security can be defined as constant access to cheap sources of energy. Energy security mainly concerns the commitment to invest in an energy supply that meets environmental requirements and matches the development of the economy. It is often abbreviated as 4As means, accessibility, acceptability, affordability, and availability [9]. Availability fulfills the demand for perpetual expansion, accessibility caters to all people around the globe, acceptability addresses cultural and environmental issues, and affordability ensures economic stability—driving the formation of a sustainable energy system.

The findings of many research works reveal that poverty, renewable energy, economic growth, and remittances are the critical determinants of energy security. The research was made by [10–14] to provide a review of the connection between economic growth and energy security in Great Britain. The above highlights show that energy security in a country is enhanced if energy utilization is a result of economic growth. For instance, [14] studying the ASEAN countries has established that there exists a direct positive relationship between energy security and economic growth with bidirectional causal chaining of energy security and economic growth. Therefore, energy security increases substantially with economic growth.

Moreover, previous research established that energy security is highly influenced by renewable energy. Ref. [15] has estimated that the advancement of renewable energy sources enhances energy security in the context of Lithuania. Energy security benefits due to renewable energy have been depicted by [12, 13, 16]. However, a cross-sectional analysis carried out by [17] on the OECD countries provided evidence that supports this idea by showing that solar and biomass energy sources do not affect energy security in any way. In general, previous studies reveal that renewable energy has had a profound influence on energy security. [18–22] have pointed out poverty as a determinant of energy security. [18] examined the energy efficiency and poverty nexus on the level of both developed and developing countries in an exploratory empirical study in 2021. Their conclusions suggest that higher economic results and supported energy goals are essential for mitigating the impacts of energy poverty with efficient energy efficiency policies. However, research on the interaction between energy poverty and energy security as analyzed by [4, 20–22] demonstrates that energy security aggravates energy poverty.

Moreover, it finds that, to the best of the author's knowledge, only a few works such as [23, 24] investigate the role of AI in eliminating poverty, but none investigate how AI influences energy security. This paper explains that AI is useful in eradicating poverty through the enhancement of financial access using fintech. It helps the poor both in banking and access to necessary services, which leads to savings, credit, and adequate investment opportunities. Also, AI in the form of risk management helps susceptible groups handle economic instability and work on building adequate financial protection. Thus, AI has useful roles for poverty reduction in terms of economic empowerment and financial sustainability among poor people; however, it remained unclear how AI interacts with the energy security–poverty relationship. AI also plays a role in eradicating poverty by enhancing access to financial services through Fintech. Advanced technologies such as AI help the poor to gain banking and financial services, thus increasing savings, credit, and investment opportunities. Furthermore, AI as risk management has benefits in enabling vulnerable populations to have better economic shock and financial resilience through managing and cutting down financial risks. Hence, in promoting the economic capacity and financial security of the poor world, AI plays a fundamental part in the eradication of poverty.

3 Key Findings

AI, energy, and poverty are perfect key terms that are comparable to an intersection because it is a chance for change, particularly in developing countries. Energy security defined as reliable access to adequate, reliable, and cheap energy is central to promoting economic and social development. Still, despite various efforts made at the global level focusing on energy security—especially with reference to efficiency and clean energy—such advancements have not been received equally positively by all populations. Consequently, most vulnerable groups still lack the ability and afford the costs of access to modern energy services sometimes referred to as energy poverty. In this context, AI appears as a useful platform for reconfiguring the existing relationship and attaining both enhanced energy security and concerted eradication of energy poverty.

Harnessing artificial intelligence technologies gives a special value to the generation, distribution, and consumption of energy all over the world. For instance, self-organized AI-related systems can create better demand predictions, reduce energy wasted during transmission, and adapt energy distribution to contain renewable source implementation, like wind or solar. These improvements are necessary if energy cost is to be lowered, energy reliability is to be addressed, and wastes are to be eliminated, which in turn improves energy security. Through AI, it is possible to organize energy resources more effectively and thereby guarantee a stable supply of energy and at the same time reduce dependence on nonrenewable energy sources, which are hazardous both to the environment and the economy.

However, the attainment of energy security is not just a question of adequate supply; the availability of energy for everyone is important as well. Providing access to affordable and clean energy to all the people especially those in the rural or disadvantaged regions in many developing regions allows significant challenges. The above-discussed disparities can be reduced considerably with the help of AI. Through the mapping of demand, AI can assist governments and energy providers in identifying areas that should be supplied with energy; conversely, it can help in the creation of distributed energy systems that are not based on expensive infrastructure development. For instance, AI can enable the provision of microgrid in remote consumer areas to enable them to get renewable energy in a standalone manner.

AI's potential to promote efficiency in the power marketplace area would, in turn, translate to the practical application of tackling the economic aspects of energy poverty. AI in energy production and distribution helps to reduce costs and hence makes energy more costly and accessible to low-income-earning households. This saving in the cost of energy has a huge impact on eradicating poverty, as energy is one of the keys to poverty eradication as it is a carrier of numerous states. Most often, energy poverty limits the ability of communities to access education, health care, and economic opportunities, locking people into poverty. But generosity with energy and access, which has long been denied, can be solved with AI integration. AI has added value in other ways also not necessarily, in terms of economics. It helps support environmental sustainability due to its efforts in the conservation of renewable energy to support the world's energy needs, without polluting the environment. This aspect is particularly crucial because developing countries are disproportionally affected by climate change and environmental hassles, which worsen poverty. The extent of the socioeconomic effects of the adoption of AI technologies within the energy sector cannot be ignored. When countries fund AI, they not only enhance their energy systems but also create a new industry. AI can help advance energy innovations enabling the creation of new markets, job opportunities, and improving global competitiveness. Thus, this shift offers the chance for developing countries to leapfrog the old, ineffective systems of energy distribution and move toward the new, intelligent systems that deliver efficient energy services for economic and social progress.

Therefore, AI plays an essential part in promoting and increasing the efficiency of SDG 7, which aims to ensure sustainable access to affordable and clean energy. With AI still advancing and its use in energy systems increasing, its capability to solve the two paradoxes is evidenced. In addition, AI enhances energy security and contributes to the creation of an inclusive economy, alerting energy poverty. Realizing such a future will require machine-intelligent applications taking an intentional approach to addressing disadvantages connected to minority populations. Thus, AI is one of the examples of technological progress at the same time bringing progress of the social and economic nature to the sphere of energy production.

4 AI-Enabled Framework for Sustainable Energy and Poverty Reduction

The structure of adopting AI in the energy security–poverty relationship is a multiple-perspective model that focuses on the barriers to energy provision, cost, and sustainable electricity provision to marginalized areas. This section of the chapter presents detailed ideas on how integration of AI technologies in energy systems may be achieved holistically to enhance efficiency, equitability, and economic enfranchisement as well as eradicate poverty.

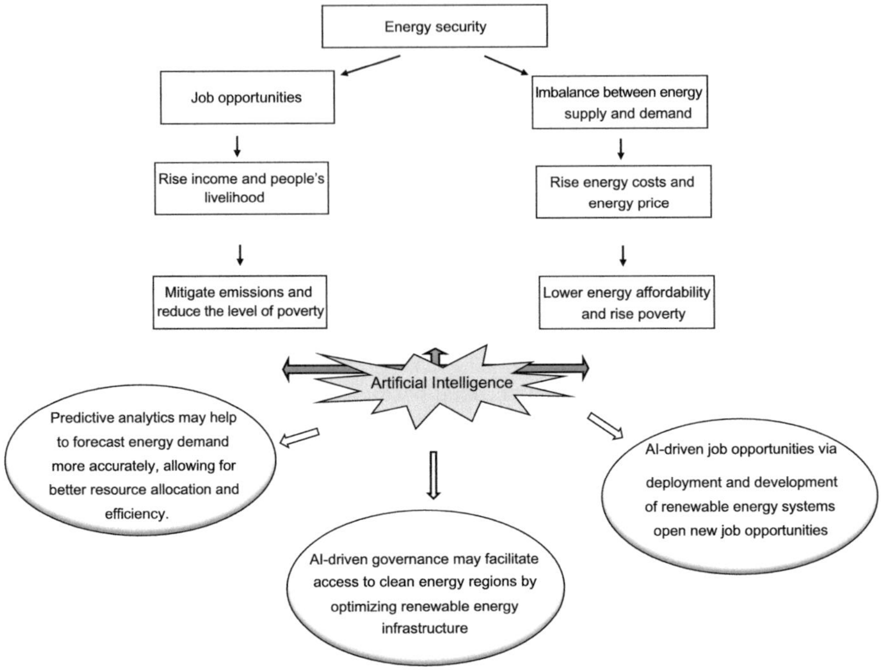

Fig. 1 The role of artificial intelligence on energy security–poverty nexus. (Source: Authors' own illustration)

This chapter proposes a framework (see Fig. 1) based on several fundamental points, which highlights AI as the enabling technology for closing the energy security and poverty divide. Such elements can be elements such as predictive analytics, energy optimization, integration of renewable energy, job creation, and good governance. These components are embedded in the framework to provide a complete solution to energy security, which factors in technical and social elements, as follows.

(a) Predictive analytics and energy demand forecasting: Predicting the energy demand is one of the fundamental pillars of AI. Applied in terms of machine learning algorithms to real-time data, it is possible to predict the energy demand swings and improve the supply chain for the energy providers. This leads to less wastage of energy and guarantees that energy is properly distributed including to areas of low income and more of the less privileged. The precise prediction also leads to better resource management, reducing the price of energy and making energy more accessible for needy groups of the population.
(b) AI-driven energy optimization: AI is used to enhance the effectiveness of existing energy networks and structures. The framework advocates for the integration and use of AI systems in analyzing the power system to detect areas of loss flow of power and the occurrence of losses during transmission. These

optimizations make it possible to deliver energy with higher reliability, particularly in remote areas, hence increasing energy security. Additionally, AI helps pave the way toward smart grids, where renewables can become integrated even better into the energy grid, which brings clean energy into the equation as a solution to both energy security and poverty.

Renewable energy sources integrate with the function of AI in the process of incorporating distributed renewable energy sources, including solar, wind, and hydroelectric systems, into the whole energy system. AI technology enhances the efficiency of renewable energy utilization management. Energy storage management by applying AI makes it possible to deliver renewable energy to those areas in need, wherein most of all the poor population was left without energy infrastructures. This approach also de-emphasizes effects on the natural environment such as the limited nature of fossil fuels notably to the detriment of subordinate groups both through pollution and resource depletion. Hence, it is mindful not only of the environmental protection using AI in the integration of renewable energy sources but also of the economic independence of the community, with the subsequent creation of employment opportunities in the renewable energy sectors.

In addition, the utilization of AI technologies in energy systems can create employment in areas of AI-related technology development, energy efficient management systems, and installation and maintenance of renewable energy sources. It can greatly enhance the income levels in low-income areas, helping to relieve poverty. However, the framework fails to elucidate the possible models that AI-driven energy initiatives may adopt in promoting local economic development through the creation of green industries, renewable energy projects, and the training of workers in new forms of energy technologies. Besides that, the framework integrates the position of AI in enhancing the governance and policy execution related to energy access and poverty alleviation. AI can therefore help the policy makers in this important area in arriving at decisions informed about distribution, subsidies, and infrastructure among others. When policy impacts are analyzed using AI, governments can direct funding to areas needing support to guarantee that energy security programs help those on the periphery. AI can also improve transparency and accountability in governing the energy sector by analyzing energy consumption trends, evaluating the effectiveness of energy policies, and identifying where interventions should be most effective. This results in fair energy policies so that cheap and clean power is provided to different facets of society including the lowest-income earners.

Last of all, the framework also has recognized the two benefits that are associated with AI including the reduction of emissions and poverty levels. AI enhances effective power generation and utilization and as an outcome reduces carbon footprint enhancing the environmental standards, particularly for the low-income communities that are more affected by climate change. Further, the implementation of AI in the controllable aspects of the house like energy-efficient appliances and smart home systems can help households regulate energy consumption bills, and therefore free up disposable income to scale up other needs such as upgrading living standards and getting access to basic amenities. Also, since cleaner energy systems reduce environmental impacts, there are long-term gains for agricultural and coastal areas, which are the most sensitive to climate change impacts. Reducing these risks helps AI to contribute to the goals of building a safer and sustainable space and supporting programs designed to eliminate poverty in the most vulnerable regions. The outlined AI framework sees AI as a solution to the energy security–poverty conundrum by using intelligent computing power for analytical forecasting, system optimization, renewable energy management integration ability, employment generation, and policy improvement. The strengths and limitations of this approach lie in systematically integrating AI into both the technical and social architecture of energy systems to address the mutually reinforcing threats of energy insecurity and poverty. The integrated solution for energy efficiency and affordability has the added benefit of providing socioeconomic benefits in a way that enshrines the principles of sustainable development for the benefit of all.

5 Conclusion

AI, energy security, and poverty reduction are the greatest global development challenges, and their intersection provides a window into their management. This chapter has tried to elucidate how with AI as an enabling technology, the current state of energy security and poverty can be shifted. Energy security, and hence, energy accessibility, affordability, and the ability to sustain it, is still relevant for growth, development, and future environmental stability. However, as demonstrated, the positive outcomes of undertakings aimed at increasing energy security, especially via the use of clean sources, have failed to reach everyone. Hence, this has locked millions into energy poverty, and poor access to modern energy services constrains access to further socioeconomic development.

The utilization of predictive analytics for energy demand forecasts in AI increases efficiency and reduces expenses on energy consumption. It also directly advances the efficiency of energy provision while simultaneously working toward a fair distribution of energy assets. By estimating the requirements of energy and availability of energy using AI, the energy stress from the energy poor can be lowered. Secondly, AI prompts the random generation of renewable energy resources to feed into the grid. AI technologies through energy optimization systems can improve the use of renewable energy with distribution, especially to remote regions that are often underserved. This is because using renewable energy sources increases energy security while at the same time decreasing the emission of greenhouse gases that aid in climate change. Thus, integration of AI into using renewable energy is also economically beneficial because it gives rise to new professions and sectors, for example, renewable energy sector, energy management sector and AI manufacturing sector.

Moreover, AI is of great significance in enhancing governance, policy implementation, and enforcement. AI technologies can help policymakers make better decisions about energy distribution and subsidies to help clean energy get to the needy. Nonetheless, AI improves accountability and traceability of energy consumption and policy in the energy sector. It is useful in assisting countries develop more equitable energy policies with a view to addressing the needs of the vulnerable. In summary, all the key concepts of AI are being promoted as enablers that will facilitate sustainable energy development and poverty reduction. Thus, it is a fitting function of boosting energy security alongside socioeconomic development making the instrument highly valuable for global energy–poverty cooperation. Still, achieving the full value of AI in this respect will need specific policy support and significant capital and human resources investments.

6 Policy Implications

The key findings presented in this chapter bring forth several important policy considerations that governments, international organizations, and energy stakeholders need to consider if AI is to serve as an enabling force for increasing energy security and reducing poverty. The following outlines the main areas, where policy action is needed.

(a) Increased investment in AI-powered energy infrastructure
What it needs is a stronger drive to invest in AI-enabled energy systems that could transform real energy systems to alleviate energy poverty. This, therefore, calls for governments—inclusive of international financial institutions—that should focus funding on the creation and application of AI in the energy industries, putting more emphasis on smart grids, predictive analytics, and renewable energy management systems powered by AI. This is particularly so for developing countries, where the resources for large projects are always minimal; public–private partnerships can bridge the gap and bring those innovations into being.

(b) AI adoption in energy systems
Targeted support from policymakers in concert with financial incentives will be required in order to adapt AI in energy systems. These efforts focus on making AI tools affordable and accessible to energy providers, especially in communities where energy poverty is a serious issue, so that governments can reduce capital expenditures linked to clean energy solutions by providing incentives for AI-enabled renewable energy. They are more competitive with conventional fuels. This could be a game changer, especially in areas where they are less lucrative, with the high upfront cost of renewable energy being a major barrier to adoption.

(c) Capacity building for AI integration
The integration of AI into energy systems will be successful if the workforce has acquired the necessary skills. In this regard, policymakers have to focus on

capacity-building programs related to AI, data science, and energy management. In turn, these must be targeted at the public and private sectors, so that the local communities, businesses, and governments have acquired the expertise that shall be needed for them to exploit the potential of AI. The AI-focused curriculum, developed in collaboration with industry leaders, especially in developing countries, will further strengthen the local talent pool to meet the demands of the energy sector.

(d) Development of regulatory frameworks for AI in energy
With AI, there come special challenges in terms of regulation, especially regarding privacy, data security, and algorithmic transparency. Given that, governments have to come up with robust regulatory frameworks that make sure that AI deployment in the energy sector will be ethical and responsible. The frameworks should address issues of data ownership, algorithmic bias, and possible risks related to AI-driven decision-making in energy distribution. Most importantly, regulations must ensure that AI-powered energy systems contribute toward improving the lives of marginalized communities and guarantee equitable access to energy resources.

(e) Incentivizing AI-led renewable energy programs
Renewable energies such as hydro, solar, and wind are increasingly part of the energy systems. As such, AI will be important in optimizing these resources. Policymakers must therefore put in place incentives that stimulate AI-driven renewable projects, especially to serve rural and underserved areas. These may include tax breaks available to firms and communities that are integrating AI into renewable energy solutions, grants, and low-interest loans. This will go a long way in helping governments make clean energy more affordable and accessible.

(f) Centers of excellence to support AI in decentralized energy systems
AI plays a very important role in promoting decentralized energy systems, such as microgrids, especially for those developing countries that do not have substantial energy infrastructure. Financial and technical support from the governments for AI-driven microgrid projects in off-grid and remote areas is vital. These decentralized systems also present a flexible and resilient path to deliver energy into the lives of the marginalized, ensuring access to reliable, clean, and affordable power. In investing in AI for decentralized energy, policymakers can help bridge the energy-access gap and reduce poverty in rural regions.

(g) Environmental and social safeguards
While AI has the great potential to considerably improve energy security and alleviate poverty, there is a dire need to study the environmental and social impacts of these technologies. This will require policymakers to establish safeguards that ensure deployment of AI in the energy sector does not contribute to increasing inequalities and harm the environment. It should be important that sourcing AI technologies is done responsibly and that AI-driven energy projects will not affect communities and ecosystems adversely. Also, the implementation of AI should respect the social and cultural dynamics of the given communities.

(h) Fostering global cooperation and knowledge sharing
Since energy challenges are a global concern, the full benefits of AI can best be gleaned through cooperation among nations. This calls for cooperation between governments, international organizations, and research institutions in knowledge sharing, resources, and best practices about AI applications in energy. The same can be achieved through a collaborative effort, where leading institutions like the United Nations and the World Bank pool resources for large-scale R&D projects. It will make sure that the AI technologies are deployed in a manner so as to confer benefits on all countries, especially those struggling with energy poverty.

(i) Long-term monitoring and evaluation systems
Finally, to achieve the true success rate of AI-driven energy projects, policymakers need to put in place a mechanism for long-term monitoring and evaluation. These would monitor the sociocultural, economic, and environmental impacts of AI on the energy sector and provide insight into the performance in terms of the contribution of these technologies toward poverty reduction and energy security. Regular review in a transparent manner, informed by the local communities and other stakeholders, will ensure that AI technologies continue to contribute to the public good and become a guiding framework for future policy adaptation.

With AI, the energy-poverty nexus becomes integrated, and this approach is not only a technological leap forward but also a crucial driver toward social and economic equity. Over and above the immediate efficiency gains that accrue in energy generation and distribution, AI helps in predicting energy requirements more accurately, which can lead to optimization in resource utilization while reducing energy wastage in both generation and operations. This will also address the aspect of energy affordability and accessibility, especially for people in developing regions who cannot afford this commodity. Governments must establish policy frameworks that stimulate AI's adoption through financial incentives and investments in renewable energy systems like solar and wind, where AI can substantially enhance efficiency. In addition, the AI-powered development of jobs in renewable energy industries brings financial benefits to low-income communities through job creation in the operation and maintenance of AI systems, energy management, and deployment of renewable energy technologies. Furthermore, good governance relies on the key contribution of AI in helping governments make evidence-based decisions, better-targeting energy subsidies, and investments in key areas. Additionally, AI will provide greater transparency and accountability in the governance of energy, as it can monitor consumption patterns of energy and the effectiveness of policies adopted in order to ensure the efficient apportionment of available resources. Finally, international cooperation will be instrumental in creating conditions that favor developing nations' adoption of the technologies of AI through knowledge and monetary transfers in such a way that the benefit of AI is equitably shared toward the greater objective of attaining inclusive sustainable energy access for all.

References

1. Cheng, J. Y. (2008). A Chinese view of China's energy security. *Journal of Contemporary China, 17*(55), 297–317.
2. Kaygusuz, K. (2012). Energy for sustainable development: A case of developing countries. *Renewable and Sustainable Energy Reviews, 16*(2), 1116–1126.
3. Sovacool, B. K. (2019). Energy security: Challenges and needs. *Advances in Energy Systems: The Large-scale Renewable Energy Integration Challenge*, 111–118.
4. Subramaniam, Y., Loganathan, N., & Subramaniam, T. (2023). The moderating role of remittances in the energy security-poverty nexus in developing countries. *Utilities Policy, 84*, 101655.
5. Aslam, N., Yang, W., Saeed, R., & Ullah, F. (2024). Energy transition as a solution for energy security risk: Empirical evidence from BRI countries. *Energy, 290*, 130090.
6. IEA, IRENA, UNSD, World Bank, WHO. (2023). *Tracking SDG 7: The energy progress report*. World Bank. © World Bank.
7. Singh, B., Vig, K., Dutta, P. K., Kaunert, C., & Gautam, B. K. (2024). Strengthening human rights via distributive health justice and rurality: Viewing legal issues through lens of corporate social responsibility. In *Bridging human rights and corporate social responsibility: Pathways to a sustainable global society*. https://doi.org/10.4018/979-8-3693-2325-0.ch007
8. Setyawati, D. (2020). Analysis of perceptions towards the rooftop photovoltaic solar system policy in Indonesia. *Energy Policy, 144*, 111569.
9. Gasparatos, A., & Gadda, T. (2009). Environmental support, energy security and economic growth in Japan. *Energy Policy, 37*(10), 4038–4048.
10. Brahim, S. P. (2014). Renewable energy and energy security in The Philippines. *Energy Procedia, 52*, 480–486.
11. Gökgöz, F., & Güvercin, M. T. (2018). Energy security and renewable energy efficiency in EU. *Renewable and Sustainable Energy Reviews, 96*, 226–239.
12. Hamed, T. A., & Bressler, L. (2019). Energy security in Israel and Jordan: The role of renewable energy sources. *Renewable Energy, 135*, 378–389.
13. Nepal, R., Best, R., & Taylor, M. (2023). Strategies for reducing ethnic inequality in energy outcomes: A Nepalese example. *Energy Economics, 126*, 106910.
14. Augutis, J., Martišauskas, L., Krikštolaitis, R., & Augutienė, E. (2014). Impact of the renewable energy sources on the energy security. *Energy Procedia, 61*, 945–948.
15. Singh, B., Dutta, P. K., & Kaunert, C. (2024). Replenish artificial intelligence in renewable energy for sustainable development: Lensing SDG 7 affordable and clean energy and SDG 13 climate actions with legal-financial advisory. *Social and Ethical Implications of AI in Finance for Sustainability*. https://doi.org/10.4018/979-8-3693-2881-1.ch009
16. Cergibozan, R. (2022). Renewable energy sources as a solution for energy security risk: Empirical evidence from OECD countries. *Renewable Energy, 183*, 617–626.
17. Li, W., Chien, F., Hsu, C. C., Zhang, Y., Nawaz, M. A., Iqbal, S., & Mohsin, M. (2021). Nexus between energy poverty and energy efficiency: Estimating the long-run dynamics. *Resources Policy, 72*, 102063.
18. Ehsanullah, S., Tran, Q. H., Sadiq, M., Bashir, S., Mohsin, M., & Iram, R. (2021). How energy insecurity leads to energy poverty? Do environmental consideration and climate change concerns matters. *Environmental Science and Pollution Research, 28*(39), 55041–55052.
19. Nasir, M. H., Wen, J., Nassani, A. A., Haffar, M., Igharo, A. E., Musibau, H. O., & Waqas, M. (2022). Energy security and energy poverty in emerging economies: A step towards sustainable energy efficiency. *Frontiers in Energy Research, 10*, 834614.
20. Taghizadeh-Hesary, F., Zakari, A., Yoshino, N., & Khan, I. (2023). Leveraging on energy security to alleviate poverty in Asian economies. *The Singapore Economic Review, 68*(04), 1063–1090.
21. Fry, J. M., Farrell, L., & Temple, J. B. (2023). Energy poverty and food insecurity: Is there an energy or food trade-off among low-income Australians? *Energy Economics, 123*, 106731.

22. Goralski, M. A., & Tan, T. K. (2022). Artificial intelligence and poverty alleviation: Emerging innovations and their implications for management education and sustainable development. *The International Journal of Management Education, 20*(3), 100662.
23. Ferreira, M. B., Pinto, D. C., Herter, M. M., Soro, J., Vanneschi, L., Castelli, M., & Peres, F. (2021). Using artificial intelligence to overcome over-indebtedness and fight poverty. *Journal of Business Research, 131*, 411–425.
24. Sehrawat, S. K., Dutta, P. K., Bhatia, A. B., & Whig, P. (2024). Predicting demand in supply chain networks with quantum machine learning approach. In A. Hassan, P. Bhattacharya, P. Dutta, J. Verma, & N. Kundu (Eds.), *Supply chain management: A new era of optimization* (p. 15). IGI global. https://doi.org/10.4018/979-8-3693-4107-0.ch002

Artificial Intelligence Applications for Enhancing Efficiency in Smart Grids

Udit Mamodiya and Indra Kishor

1 Introduction

The new revolution in smart grids represents a fundamental shift away from the traditional approach to energy generation, management, and consumption. In the traditional networks, the energy generated in the power plants is transmitted to the consumers in a one-way direction, which, in most cases, results in inefficiencies, inflexibility, and susceptibility to disturbances. In contrast, smart grids have digital and communication technology with a traditional energy system so that adaptive, intelligent networks can be developed so as to ensure real-time energy management and two-way communication. That is why modernization would provide better energy distribution in power networks, with the possibility of real-time supply and demand equilibrium, enhanced security and reliability, and integration of renewable energy. Management of those systems is important to achieving goals for sustainable energy as global energy consumption rises and reduces the adverse environmental implications.

Some of the factors involve decentralized and renewable sources, such as solar and wind power, being increasingly used in the generation of energy. Renewable sources of energy differ from traditional sources on two important areas: their

U. Mamodiya (✉)
Faculty of Engineering and Technology, Poornima University, Jaipur, Rajasthan, India
e-mail: assoc.dean_research@poornima.edu.in

I. Kishor
Department of Computer Engineering, Poornima Institute of Engineering & Technology, Jaipur, Rajasthan, India
e-mail: Indra.kishor@poornima.org

P. Raj et al. (eds.), *Artificial Intelligence (AI) for IT Energy Efficiency and Green AI for Environment Sustainability*,
https://doi.org/10.1007/978-3-031-89420-6_9

supply is not always regular and may be intermittent because of their dependence on the weather. Smart grids face these challenges by applying the advantages of digital technology in real-time variation possibilities and energy storage solutions designed to stabilize the supply and demand for energy. Nevertheless, it is features that are insufficient alone to provide the efficiency, scalability, and agility necessary to support modern power networks. It is then that AI technologies become an essential tool in providing the range of instruments to optimize, predict, and control energy systems.

AI is one of the revolutionary tools in smart grids, able to solve the tasks impossible or hard to be solved by conventional technologies. A practical application of AI in smart grids turns out to be predictive maintenance—an application in which AI processes data gathered by sensors and meters, identifying faulty equipment before they actually happen. One of the areas where AI excels is energy usage prediction: here complex machine learning models are used to predict usage patterns based on variables such as weather, time, and past usage. Further, AI enables and supports real-time grid management, continuously shifting energy distribution in response to changes in system conditions. This is important for the development of a strong, change-resistant energy infrastructure that integrates a diverse portfolio of energy sources while minimizing the total cost of operations.

Despite the great benefits bestowed on the capabilities of a smart grid, there are several obstacles and hurdles that have to be overcome by AI. Some of the major impediments include technological constraints such as high computational power and special data processing infrastructures rather than scaling up to millions of devices in the network. Moreover, smart grids collect data from millions of networked devices, thereby raising issues of security and privacy over the collected data. Geographical and climatic variables also vary based on location and may force AI models to be agile and robust. The resolution of such problems is vital for unlocking the true potential of AI in smart grid applications, at least from a use case perspective, with demand for scale and sustainability in energy solutions currently growing.

The primary objective of this article is to assess the capabilities of AI to overcome the operational and environmental challenges of smart grids. Thus, using the summary of previous findings, a study of various implementations of AI in the field, and building from there, this chapter aims to shed light on how AI might be increasing energy efficiency and sustainability. Key focus areas include optimization of energy distribution using AI, raising the precision of forecasting, and managing demand response in real time. This presentation would also discuss some of the challenges and limitations of current AI systems for the task of managing smart grids and promise some of the approaches for future exploration and research. The ultimate goal of this work is to be part of the contribution toward a more general understanding of how AI may support transition toward a more efficient, reliable, and sustainable energy infrastructure.

2 Literature Review

2.1 AI in Smart Grid Forecasting and Optimization

Here, on-time, sharp predictions and effective supply and distribution of energy play a very crucial role in the optimal functioning of a smart grid. From predictive means to optimization, AI technologies, particularly machine learning, have revamped these domains of managing grids, through answers that traditional models cannot match up to. According to research studies based on the research conducted by Babita et al., it is evident that AI technologies significantly enhance the precision of energy demand forecasting [1]. Good demand forecasting would lead to better resource allocation, increased grid stability, and a reduction in dependency on the cost-intensive power sources from standbys. Furthermore, precise demand estimates decrease the requirement for extra energy generation, thus minimizing operational costs and promoting environmental sustainability goals. For this reason, load forecasting models, such as SVMs and RF, are widely applied because they are capable of processing large amounts of both historical and real-time data. With these models, complex, high-dimensional, and long relationships between variables can be recognized, allowing for the generation of accurate short-term and long-term demand estimates. LSTM networks and CNNs prove to be very suitable for managing time-series data, such as an hourly or daily energy use. LSTMs can predict times when the peak load is likely to occur, therefore aiding grid managers in anticipation and management of high-demand periods by analyzing trend-based patterns in energy usage over time. This is especially useful in areas where the patterns of the climate are changing or when boom seasons peak and demand is greater, since such grids can memorize very complex consumption behaviors over time [2–4].

Apart from the improvement of forecasts, AI further refines energy distribution in the system. For instance, RL algorithms will allow for real-time adaptive control since the technique evolves with each feedback in conditions prevailing on the grid. The efficiency will then be optimized by the dynamic optimal policies of the RL systems with shifts in supply and demand balances. In doing so, they help reduce grid congestion, lower transmission losses, and guarantee that energy flows are aligned with consumption demands. This flexibility and self-learning nature of RL are especially of great value in modern smart grids, where energy flows must often change within a frequency cycle, taking into consideration the presence of the intermittent renewable energy sources of solar and wind power. As such, while traditional optimization algorithms can only manage the modifications at a coarse level over wide time periods, with RL, changes can occur at minute-by-minute periods, thus ensuring that energy resources are always distributed appropriately according to real conditions on the grid [5, 6].

Major inventions falling under this category include the applicability of generative artificial intelligence, also referred to as generative adversarial networks

(GANs), for anomaly detection. GANs consist of two neural networks: the generator and discriminator. The two networks are opposed against each other to generate realistic data samples. The resulting unusual architecture allows GANs to detect anomalies in projected patterns which may be precursors to faults or inefficiencies within the grid. Besides projecting improvements, AI goes further to enhance how energy is distributed in the network. For instance, RL algorithms will allow for real-time adaptive control since the technique evolves with each feedback in conditions prevailing on the grid. The efficiency will then be optimized by the dynamic optimal policies of the RL systems with shifts in supply and demand balances [7–9].

Another major aspect is the management of uncertainty, where AI-based methods shine. With a grid operator needing to consider many uncertainties well, swinging energy consumption, variable productions from renewable sources, and sudden breakdowns of equipment [10] discuss AI-based approaches that deal with such uncertainties in smart grid design, thus making the topology of the grid adaptable and continuously optimizable. The author discusses probabilistic ML models and Bayesian networks that will be able to compute the likely occurrences of alternative demand situations and will modify the grid operations accordingly. Such models ensure that smart grids can provide high dependability and efficiency even in uncertain conditions, which means long-term operation resilience [11]. AI technologies applied together with probabilistic reasoning therefore ensure that grid operators have a solid plan to deal with surprise changes in loads or supplies, hence boosting general system flexibility [12].

A China-developed model demonstrates workable AI for smart grids. The proposed GAN-based model may be a discovery of possible anomalies in the energy consumption forecast inside the distribution network. This means synthetic data from GAN can spot such anomalies considerably more quickly and correctly than traditional methods. This proactive technique has contributed to a reduction in detection times and increased maintenance efficiency, since operators may correct issues before they escalate. In addition to enhancing grid dependability, this paradigm enables for more efficient resource allocation, as energy output may be modified to match specific demand levels without overproducing. The performance of this GAN-based model in China provides an example of how AI may improve smart grid operations by enhancing forecasting accuracy, optimizing resource allocation, and preemptively detecting operational difficulties [13, 14].

In a way, AI-based forecasting and optimization strategies can play a highly prominent role in enhancing the operational efficiency and dependability along with sustainability of smart grids. Smart grids can better predict energy demands by coupling ML, DL, or more generally, complex generative AI techniques. They can optimize the management of energy flows in real time or ensure a very high level of reliability with dynamic conditions on grids. Consequently, with respect to the fast-evolving smart grid technology, AI is expected to play an even greater role in the realm of predictive analytics and optimization. This means that there will be better robust, adaptive, and sustainable energy infrastructure. AI in Grid Security and Fault Detection: For highly interconnected and digitally dependent large-grid-size smart grids, security and resilience are an issue of importance. The sheer number of interconnected devices as well as the real-time flow of live data associated with it means that smart grids do not have any primary defense mechanisms against cyberattacks, unauthorized access, and operational issues. Artificial intelligence possesses high potential to strengthen the domain in terms of cybersecurity and skill in defect detection, thus making it a revolutionary tool in this area [12, 15].

Real-time detection and countering of cyber threats play a crucial role in AI, and in most instances, faster and more correctly than the general mechanism of security. The author [16] describes the integration of cryptographic techniques and AI-based security mechanism that provides protection against tampering of data, while the ability of the smart grid network enhances to detect attacks using encryption techniques with algorithms in AI as intercepting and changing transmission over the grid with the help of defenses. This protection is related to real-time information concerning operational secrecy, including issues like the performance of the grid, usage patterns by the consumers, and equipment health indicators. Apart from encryption, AI-based models can now identify potential cyberattacks due to monitoring for anomalies in the grid. Another author [17] developed a platform based on trustable AI that proactively identifies cyber threats and, therefore, ensures the stability of the grid. This system integrates anomaly detection algorithms, continuously monitoring anomalies from grid behavior, hence hinting at cyber breaches or malware and inside threats. Early warning of threats in the system is prime for cyber resilience. It enables grid operators to identify an incident as it develops into an incident before it starts to mature into service disruption or data breach.

AI models also aid in determining flaws and reliability in the functioning of smart grids. As mentioned by the author, [18] notes that the delay with the immediate defects determined by AI technology allows for the elimination of the problems identified in the grid on the spot, hence saving drastically in downtime given a system. AI-based models can monitor continuously many performance characters, such as the voltage, frequency, and load balance, with regard to identifying the appearance of faults in equipment or technical anomalies. There is a possibility of flaw, and the system sends alerts to the maintenance personnel so that any error in the system does not become a major failure before being responded to. This reduces response time and, therefore, it improves the reliability of the grid, and hence, ensures continuity of operations in case of faults.

The anomaly detection algorithms advanced in the likes of autoencoders and PCA are what AI-based cybersecurity applications tend to look towards to classify anomalous patterns within data for a grid. Autoencoders: The autoencoders are a class of neural networks, trained in previous operational data to learn what normal grid behavior should look like. Once trained, these models can immediately identify variations that may warn on possible dangers or defects. PCA: This is a statistical technique meant to reduce the dimensionality of data to isolate important features from vast datasets in order to discover outliers that might suggest difficulties. They include an additional layer of protection over and above this, as they facilitate a monitoring capability for both internal and external issues: it offers a more comprehensive anomaly detection system than standard rule-based systems [12].

More recently, advanced AI frameworks have also been developed to foster transparency and trustworthiness in the decision-making process, a very important feature in grid cybersecurity. Explainable AI (XAI) approaches are quickly being introduced into models of cybersecurity and fault detection to explain why an alert or action is produced. Explanations of such types can help the grid operators understand the nature of the threat or malfunction that has been detected. These would then be easier for them to take the right corrective steps without delay. XAI helps audit AI-driven security systems to make sure that AI actions fit with regulatory and compliance standards [19]. The same transparency is even more important in areas that pertain to critical infrastructures such that mistaken AI advice may lead to costly errors in operations or unwarranted interventions. A very good example of the application of AI in fault detection can be referred from the Horizon 2020 program of the European Union [20]. It has installed an AI-based monitoring and fault detection system across the different zones of the European smart grid that continuously observes grid activities to detect possible faults before actually influencing service continuity. This system tracks trends that may be technological failures; thus, it enables prompt intervention and preventive maintenance through processing large amounts of operational data. With the AI solution, it reduced by about 30% concerning downtime, and this to that effect underlines the importance of AI in sustaining seamless power delivery and even maintaining stability within the grid. The platform also includes AI-based intelligence about cyber threats and models of anomaly detection meant to monitor data patterns strange enough to indicate hacking or any other form of security breach from within. This approach in the integration shows how AI might be applied effectively for support of resilience in operations and cybersecurity in more complex smart grid settings [21].

AI-based applications in grid security and fault detection contribute to the resilience of smart grids in several ways. Because AI will be able to predict the cyberattacks and identify faults before they go wrong, AI-based models enhance the possibility of reduced interruption of the grid caused by these potential disruptions. These models can also enable permanent monitoring, rapid response, and predictive repair, which complementally enhance the stability of the operation of the grid [22]. The more additions done on the smart grid, the more the demand for high-performance AI-driven security and fault detection in such renewable energy sources and decentralized energy resources. Development of AI in smart grid security and fault management is critical for the creation of robust, adaptive, and secure energy infrastructure that responds to future energy needs but can be strong enough to survive growing cyber and operational challenges. AI in Energy Management and Demand Response: AI technologies revolutionize the management of energy by enhancing demand response capabilities that are bettering grid stability and working effectively with resources that are highly demanded. Generally, predictive analytics and real-time capabilities provided by AI can act as a significant advantage for controlling both central distributed energy resources (DERs) and solar and wind power DERs. AI-empowered demand response systems adjust energy distribution in real time according to predicted peaks in demand and can therefore offer the advantage of load balancing, efficient integration of renewable energy, and avoidance of overloads. Predicting Demand Peaks and Managing Distributed Energy Resources (DERs): Therefore, predictive models would become a part of the stabilizing energy grid [5, 6]. We have discussed predictive models for 2024. While it would examine real-time consumption and trends, it would further look into the swings in energy consumption, which further demands grid operators to predict and plan for surges in demand. Such a tool is very helpful for DERs, which, in turn, fluctuate due to weather conditions. AI-governed multiagent systems self-coordinated DER activities by changing output due to predicted demand. Maurya proclaims, "The multi-agent will make optimum utilization of local energy resources by distributing power based on individual needs across a microgrid or community thus enabling localized stability and taking some burden off the main grid." AI-based demand response also minimizes the challenges occurring while using renewable energy sources, which are termed as intermittent by nature. What it is doing here is that it is predicting the energy demand before use, hence optimizing the usage of renewable resources like wind and sun whenever they are available. Demand for nonrenewable sources reduces, and grid stability increases and the sustainability of the environment also picks up. Real-time adjustment toward grid stability and demand management analysis of real-time data enables AI systems to dynamically adjust if there is a sudden shift in the actual supply and demand for energy. This is a feature much welcomed when instances are encountered where demand surges sharply without any warning. According to author [13], it is through adjustments in real-time that AI models ensure an optimization of the flow of energy while minimizing the chances of overloads. This is, of course, of high importance in locations where the peaking hour energy usage peaks or in regions that have more renewable sources of energy because of the wide fluctuations between both supply and demand in such locations. These demand response solutions through AI benefit microgrids: self-sustaining localized grids that can autonomously operate without relying on the parent grid. The author [23] contends that AI optimizes the operation of microgrids in light of constant responding to either increased or decreased energy demand or supply. Through such an adaptive strategy, a microgrid is better able to deal with surplus and shortage, offering a direct advantage to towns or isolated communities with virtually limited energy resources. For example, when electricity demand in a location is minimal, an AI system will transmit the surplus solar electricity in the microgrid to storage, and later ensure that this reserve power is available for a need to increase electricity generation. Example Application: The AI-Driven Demand Response Program of the United States. In the United States, demand response programs increasingly employ AI to automatically alter household energy consumption during peak hours. These applications use smart thermostats, water heaters, and other connected devices to manage energy use in real time using AI-driven insight to build the rate forecast, enabling advanced technology solutions to automatically shave loads ahead of peak-demand hours, such as hot summer afternoons when coolers are in high demand. For example, demand response projects are focused on California: in critical peak hours, the software automatically reduces electricity use across member households by adjusting thermostats and reducing or even turning off unnecessary appliances. Consumers who participate in such demand response programs receive monetary compensations or lowered electricity prices, and it forms a win–win situation where their expenses reduce and the system stability increases. This would, with the increasing demand of response popularity, even monitor appliance settings and other measurements from house energy that allow an efficient and responsive grid architecture. Impact on grid sustainability and consumer engagement AI-driven demand response has significantly contributed to sustainable grids management through improved system resilience and integration of renewable sources of energy. Artificial intelligence-based demand response serves to enhance system resilience and further advances the integration of renewable sources of energy, enhances a dynamic and decentralized model for the grid, and induces participation by customers based on cost-effective means using avenues of financial incentives and energy-saving measures. This is an environment that should provide consumers with as much opportunity as possible to actually engage in energy management, and by its extension encourage efficiency and sustainability in energy use. In summary, AI's predictive and real-time capabilities are essential for modern energy management, allowing for adaptive, efficient, and resilient grids that are better equipped to handle the growing complexity of energy demands, especially with the increasing integration of renewables. AI in Battery and Storage Management: Energy storage devices require proper administration to integrate renewable energy into the grid. AI methods, among which are ANN and LSTM networks, will estimate the need for storage, thus preventing overutilization of batteries. The author [23] extends this further by demonstrating AI functions in BMSs for EVs, which they use to optimize the charging and discharging processes in order to improve cycles for battery life and sustainable energy.

The analysis of time-series data by RNNs proves to be useful for providing actionable insights for the management of storage and hence assures energy storage systems are flexible enough to cope with the changes in variations in demand on a daily scale and seasonal fluctuations. Example Application: In Japan, storage devices that support AI at the national grid store excess solar energy throughout peak daylight hours and feed it into the general grid when demand peaks with a stable supply and saving a huge amount of dependent use on traditional sources. AI in Consumer Engagement and Smart Grid Adoption: AI technology is therefore considered to be highly essential in enhancing consumer engagement in smart grid ecosystems, from mere operational efficiency to becoming more proactive as an active participant of energy conservation and sustainable activities. Smart grid systems with AI-powered data integration will thus enable them to provide personalized usage plans, enhance sustainable behaviors, and even make it easier for decentralized energy markets to thrive. All these innovations will benefit both consumers and grid operators where, on one hand, they would realize reductions in demand on the system, increments in energy efficiency, and, strength in the grid. AI-powered marketing and custom usage plans AI allows utilities and energy service companies to estimate individual and community usage patterns, providing quite significant insights that are important for the creation of individualized usage plans. As such, [24] points out how AI-based marketing strategies encourage consumers to embrace smart grid technology with bespoke energy plans tailored to their unique use habits and demands. For instance, AI systems use historical energy consumption data to identify trends and predict future demands; through such an analysis, providers would be able to indicate a period when one should utilize optimum quantities and advisable means of saving energy. Consumer benefits would be balanced by efficiency for the grid. Rates could be incentivized to help consumers reshuffle their high-energy activities—from doing laundry or heating—to when rates are lower, thus lowering their costs of energy. As that pattern of demand shifts, this should help smooth the spikes and valleys in the power use, thereby stabilizing the grid and reducing the need for backup supplies of power. These programs also enable the combination of renewable energy sources, by which the customer can maximize the use of clean energy up to the possible extent when it is available and economical.

Large residential areas are one specific area of high advantage by utilizing AI-driven systems, where consumption is mostly varying. Author discussed that a household using electric cars may receive recommendations from smart grids on when to charge those from off-peak hours in order not to add extra load into the power grid [18]. Individualized messages through a mobile app or a home automation system will give users actionable recommendations to make them an active participant in managing energy. Facilitating peer-to-peer (P2P) energy trading may be the most influential changes AI brought about in consumer interaction, an ability for customers to trade excess energy with other people directly. To users, especially to consumers who have rooftop solar installations or any other source of renewable energy, selling the excess energy created to others nearby is a huge impact to them. As reported by authors in [25], P2P trading supported by AI can create a decentralized and consumer-driven electricity market which benefits its customers and the stability of the grid. There are two purposes to AI optimization in P2P trading: it improves match-making between buyers and sellers given real-time supply and demand and it guarantees safe and efficient transactions. With consumption patterns, grid status, and user demands assessed, an AI system matches merchants with possible consumers in a community or neighborhood. For example, if a house with sunlight-driven panels has surplus energy in the late afternoon, an AI system can detect other families that could likely consume extra energy during nighttime usage. An AI-driven P2P trading system allows consumers to engage fluidly, increasing energy utilization within the community and lowering reliance on centralized power sources. Example Application: AI-Driven P2P Trading in Australia: The best example lies within Australia, where pilot community-driven energy markets have been promoted in order to encourage sustainable lifestyles. In such an idea, homes and small companies equipped with solar panels or other renewable energy equipment can sell any surplus energy with the neighbors so that a local market for energy arises. AI systems help create real-time matches between the potential buyers and suppliers for energy, thus enabling sales to occur at the most opportune times and price [26]. These solutions in Australia that function through AI help in practical, environment-friendly energy management through the restriction of the need to draw electricity from traditional sources, especially at high-demand periods. Participants benefit by saving money on cost, and the energy community becomes more self-sufficient, reducing the loss of transmission associated with long-distance transfer of energy. This trading structure goes to support and promote the growth of adopting renewable energy technologies as it provides consumers access to the financial rewards of generating and selling clean energy. Impact on Consumer Engagement and Sustainability Energy Strategies: The application of artificial intelligence in consumer engagement strategies such as P2P trading and personalized energy planning establishes a collaborative, interactive, and participative energy culture instead of the classical consumer who, in turn, becomes the "prosumer," both producing and consuming [27]. This change, then, satisfies only the financial part of the single consumer, but also adds resilience and sustainability to grids. Since it brings the consumer into the front lines of energy management, AI further leads to an ecologically aware community able to face variable demands from the grid. Hence, essentially, AI in smart grid adoption changes the nature of consumer interaction by providing novel solutions to promote sustainable energy consumption and decentralized transactions of energy. It renders consumers more control over their footprints of energy, lowers prices, and establishes a stronger and efficient system of energy. It is likely that with the advancement of AI, these models of consumer interactions will be integrated into the ecosystem of a smart grid, making sustainable energy practices accessible and desirable for more people. Challenges and Future Directions: The adoption of AI in smart grids leads to some significant challenges in terms of scalability and requirements of processing power. An AI model usually calls for tight amounts of processing power, which again do not appear ubiquitous, especially in smaller utilities or remote places with inadequate infrastructures. Solutions from research by authors in [16, 17] resolve the problems mentioned above, which implies that high computational demands may be a limitation of the scope for applying AI, and scaling in resource-scarce regions turns out to be a serious challenge. Good examples of emerging solutions are federated learning and edge computing, that could possibly make data processing closer to users. This strategy eliminates the not only reliance on centralized computer capacity but also the significantly reduced infrastructure overhead, thereby making AI applications more accessible in varying situations. Data privacy and security are also significant concerns for AI-based smart grids, since such systems deal with extremely large amounts of sensitive information about users, grid operations, and energy distribution [28]. Munir et al. suggest the adoption of AI frameworks that incorporate differential privacy and homomorphic encryption approaches; such approaches ensure the safety of user data without losing analytical utility. Strong data privacy standards, then conceptualized from GDPR, would be enforced on consumers in an increasingly interconnected smart grid. This is imperative to consumer trust and system integrity [12, 29]. Such care ensures that the data, though accessible in its totality for AI-driven insights, is impregnable against unwanted access and misuse. All of this will have to be done keeping in view the changing conditions of the grid. This can be challenging in terms of adaptable AI models. Daily and seasonal changes in the usage of energy make AI systems learn about such changes and adapt real time for the best performance. Such methods as discussed by authors in [30, 31], like transfer learning and reinforcement learning, will require AI models to keep learning with new data. This flexibility is primarily critical in managing the natural fluctuations of renewable energy and unanticipated conditions like power outages or extreme weather. These adaptive learning methodologies enhance the ability of AI to adapt itself and deliver resilient performance under a wide variety of operational scenarios and climatic conditions.

3 Methodology

This methodology shall outline an integrated approach to the application of artificial intelligence in improving smart grid functionality. It cuts itself into four general phases: data collection and preprocessing, model selection and training, deployment and testing, and evaluation and continuous improvement. Each of its phases shall be accompanied by a more detailed explanation in the subsections below.

3.1 Collection and Preprocessing of the Data

3.1.1 Data Sources

Data therefore serves as the basis of any AI application in smart grids. The sources of data for smart grid systems include history energy consumption data, real-time sensor data, and public information about energy consumption. The historical data provided by utility companies are stored for a long run to document trends and anomalies in energy demand while real-time data coming from meters and other sensors on the grid will be used to provide fast insights on system performance and load variations. Additional data, to be fed into these predictors and recommender systems, could be the weather report, the estimation of solar radiation, and topo-information of the grid, to better enable AI-driven predictions and adaptive actions.

3.1.2 Data Cleaning and Normalization

At this stage, the data cleansing and preprocessing are done after acquisition. This is necessarily vital because real-world data happens to be noisy, fragmentary, and inconsistent. Missing values are handled via imputation methods, such as by replacing with mean, median, interpolation, etc., and outliers are determined and adjusted for to prevent distorting results. The data is standardized to maintain uniformity in scale, especially when assembling data from sources that varied widely, such as weather data, energy consumption data, and sensor data. Normalization ensures that features whose range is huge numerically do not dominate the model's training process, which is very important when one makes use of methods like neural networks or support vector machines.

3.2 *Model Selection and Training*

3.2.1 Algorithm Selection

The choice of algorithms is dependent upon the particular problem that is under consideration within the context of the smart grid. For example, in load forecasting, classic approaches like linear regression may be substituted with more sophisticated techniques such as neural networks or LSTMs, capable of capturing temporal patterns in energy demand. For fault detection and grid anomaly detection purposes, decision trees, support vector machines, or even deep learning-based autoencoders are used for they can figure complex patterns in data. Reinforcement learning is also very popular for adaptive energy distribution and demand response management as RL maximizes decisions dynamically with respect to time.

3.3 *Model Training and Validation*

After proper choice of algorithm, the training of models follows. Historical datasets are used for training these models. Algorithms learn underlying patterns and relationships in data. For the problem of supervised learning such as this load forecasting, training data consists of input features-time, temperature, historical load along with output labels such as projected energy demand. Training amounts to nothing more than updating model parameters, that is, the weights of a feed-forward network aiming at increasing the correctness of some input predictions. For this, we use a set of optimizers one of which is gradient descent.

Such cross-validation methods include k-fold cross validation, which is employed in the process of effectively generalizing the model for unknown data. This is basically done to validate the performance of the model and inspect for overfitting behavior with regard to the training data. Evaluation measures are applied to measure the accuracy of the model and the reliability of its performance, such as mean absolute error, root mean square error, and R^2 measures. These measurements further influence the phase of hyper-tuning, wherein the hyperparameters are varied to reduce errors and enhance the precision of the predictive forecasts.

3.4 *Deployment and Testing Real-Time Integration*

Once the model was well trained and accepted, it could be used in the smart grid system. The integration was what would make the model decide in real time to reflect in grid operation. Examples include load forecasting, where the model gives forecasts on the energy demand that is usually influenced by this alteration change in energy generation and delivery. In fault detection, the model needs to monitor the health of the grid dynamically and detect anomalous behaviors as well as probable equipment failures or inefficiencies. Reinforcement learning algorithms work very well within real-time adaptive decision-making, letting the system continuously adapt to its strategy on feedback from the grid.

The models also have feedback mechanisms where once new feeds of real-time data update existing predictions, the smart grid is, therefore, set to function continuously to find ways to absorb those changes and optimize energy distribution in a dynamic way.

3.5 Performance Monitoring

The performance monitoring tool tracks several KPIs, which include energy savings, fault prediction accuracy, grid stability, and operational expenses, to ensure that the models deployed are successful. A decrease in energy consumption or an increase in fault detection accuracy measures how AI can be effective for real-world grid applications. Then again, these performance measurements feed into the system for further analysis, where it can pinpoint a need to change or revise the model. The other application of adaptive learning techniques can be learnt from AI-driven models wherein they adapt based on feedback regarding the performance, like online learning or reinforcement learning.

3.6 Evaluation and Performance Improvement

3.6.1 System Assessment

Once the AI models are implemented and operational, a strict assessment is required to evaluate the general outcome of AI integration to the grid. This review would compare AI-wise the grid system's performance with traditional non-AI techniques like the rule-based or manual systems in operation. Some of the key importance indicators are grid efficiency; response time by fault detection; optimization of energy consumption; and total cost reductions. Benchmarking with non-AI methods and emphasizing added value from AI help to focus on potential areas for improvement.

The review also includes an in-depth analysis of the scalability of AI models in relation to dealing with all types of grids, including geographic, climatic, and operational variations. For instance, an AI model that works wonderfully in one region might perform miserably in another type of region that consumes energy differently or features varied environmental characteristics. This would be a basic knowledge to scale up the AI solutions for smart grids.

3.6.2 Iterative Model Upgrades

Based on the evaluation of the system, AI models are constantly upgraded to suit changing grid conditions. These upgrades may include retraining models with new data, addition or revision of algorithms based on new environmental factors such as increased use of renewables, especially on random patterns of consumer demand, or ways of increasing adaptability so that it can handle unexpected situations such as power outages or unusual weather events. This is mostly done by solving new problems or adding insights realized at deployment.

Furthermore, updates are done to help them adapt to other new technologies, for example, decentralized energy sources (such as DERs) or energy storage. There could also be iterative improvements including inputting new data sources from IoT sensors, as well as weather-based prediction models, while optimization algorithms can continue to narrow the congestion on the grid and optimize energy efficiency. This continuous assessment loop and model adaptation cycle ensure that the smart grid is effective, reliable, and in response to the current and future requirements for energy (Fig. 1).

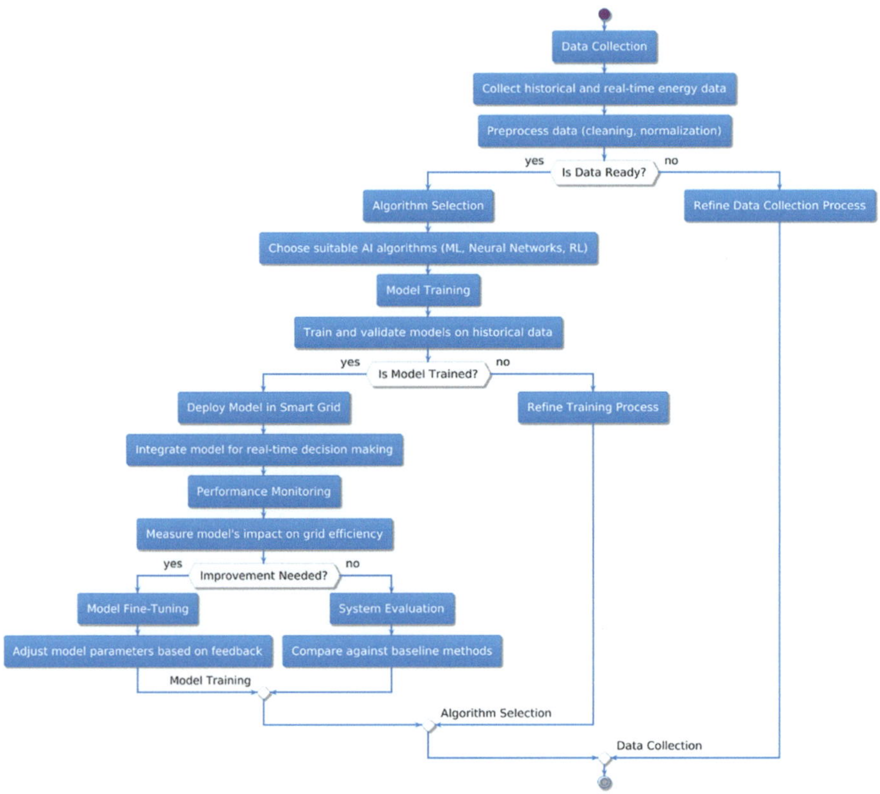

Fig. 1 Methodology of AI-driven optimization for smart grid

4 Results

The integration of AI models in smart grids has delivered considerable benefits along several metrics of operational performance, including energy efficiency and accuracy in prediction and fault detection. Advanced insights and automation from AI models, like machine learning algorithms for load forecasting, reinforcement learning for real-time optimization, and neural networks for anomaly detection, remain untouchable by traditional methods. Below is an even more specific look at the functionality added by those models to smart grids.

4.1 Energy Efficiency Improvement

Probably, one of the most visible impacts caused by AI in the smart grids is an improvement in energy usage; with waste heavily reduced, the overall efficiency within the grid increases. Table 1 shows a comparative analysis of energy use before and after the implementation of an AI model. Some of the vital attributes weighed include peak demand, off-peak demand, average daily usage, and energy loss, which results from imbalances.

AI-based models contribute to energy efficiency in many ways. AI systems vary and change resource allocation with demand in real-time optimizing resource distribution to reduce reliance on high-cost, high-emission peak power sources; it reduces off-peak demand by 7.1%, and peak demand by 12.4%, evidencing the potential of AI in balancing loads. Further, AI eliminates energy waste through imbalances by 43.3% proactively watching out for inefficiencies in the grid and correcting them. In other words, this makes an overall efficiency improvement of almost 12%. This kind of efficiency is very useful in a grid comprising renewal energy, where AI determines when and how to store excess energy and when to use it, maximizing the usage of intermittent sources like solar and wind.

4.2 Load Forecasting Accuracy

Correct load forecasting is vital to efficient resource allocation in smart grids since it ensures that supply and demand reconcile without overgeneration or shortfall production. AI enhances forecasting with more complex algorithms. Algorithmic

Table 1 Energy consumption comparison before and after AI implementation

Parameter	Before AI (kWh)	After AI (kWh)	Improvement (%)
Peak demand	1050	920	12.4
Off-peak demand	700	650	7.1
Average daily consumption	875	770	12
Energy waste (due to imbalances)	150	85	43.3

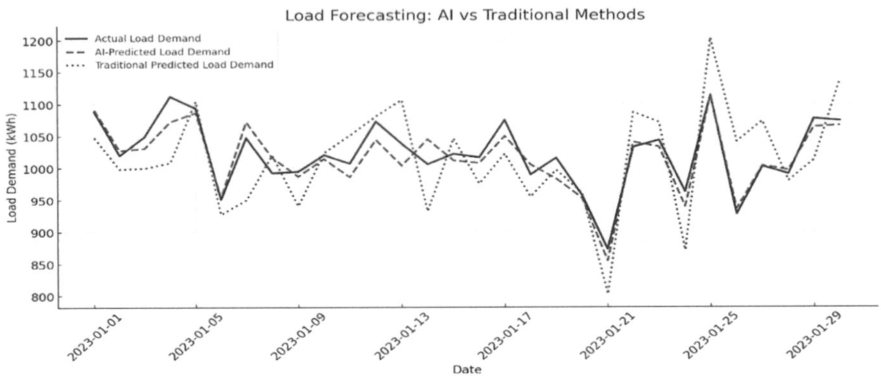

Fig. 2 Accuracy in load forecasting with the use of AI vs. traditional methods

methods, such as machine learning models and recurrent neural networks, can go through many past and real-time measurements and predict short-term as well as long-term demand energy. Figure 2 shows a graph that points out the improved accuracy of the load forecast using AI-enhanced methods compared with simple forecasting techniques.

With AI, MAE has reduced from 12% using typical approaches to a minimum of only 5%, which speaks for a significant increase in the accuracy of the forecasting. Such accuracy directly benefits the management of sources by grid operators, preventing the two unpleasant states-characterized by either overproduction or underproduction-from which energy wastage or shortages respectively are associated. Improved forecasting also increases the reliability of the grid because the system will be able to plan for expected changes in load, such as those caused by weather cycles, peak seasons, and unexpected business expansion. In areas where integration of renewable energy is high, reliable load forecasting becomes very important. The ability to forecast demand patterns and also renewable generation availability aids grid managers to use renewable resources more securely. This enhances the sustainability of their system. AI's role in load forecasting thus translates into benefits in operational efficiency as well as strategic planning, as accurate predictions avoid expensive over-reliance on fossil fuel-based backups.

4.3 *Fault Detection and Anomaly Prevention*

The integration of AI models for fault detection and anomaly evasion has hugely improved grid reliability and reduced system downtime. AI-powered neural networks are the best in real-time anomaly detection due to their ability to recognize and determine complex patterns in anomalies in the operating data of a grid. As presented in Table 2, improvements in detection rates along with the average times before and after the integration of AI.

Table 2 Comparison of fault detection rate and response time

Metric	Without AI	With AI	Improvement (%)
Fault detection rate (%)	78	96	23.1
Average detection time (s)	10.5	3.2	69.5
System uptime (%)	92	98	6.5

AI-augmented fault detection achieved a detection rate of 96% compared with a normal detection rate of 78% without AI. That means the 23.1% increase in exactly determining faults possibly precipitates an improvement in grid efficiency via responsiveness where an operator can act immediately to prevent faults from escalating with possible damage to system uptimes and stability. The average fault detection time was reduced by 69.5%. This is from 10.5 s without AI to 3.2 s with AI augmentation. The overall system uptime shoots up from 92% to 98%, which clearly shows the impact of AI on fault management.

It is particularly useful in critical infrastructure, where AI can detect anomalies in real time to prevent the slightest delay in addressing one and cause cost-expensive outage or even create a safety hazard. With AI, predictive maintenance deepens strategics for predicting failures if there are slight differences in voltage or frequency that a more traditional monitoring system will miss. Proactive detection of anomalies ensures both an extended life of equipment and an enhanced safety and resilience of the grid.

4.4 Performance over Time

Continuous applicability of AI models in smart grids results in long-term benefits in terms of stability and performance of the grid. Information in AI is time-bound. Patterns and changes in the grid environment make AI models change over time. Figure 3 represents stabilization and improvements of the performance metrics pertaining to the grid after AI has been applied to it, constantly improving in efficiency and reliability over time.

Figure 3 shows how the improvement curve is sustained by AI integration. It means that AI, based on algorithms refined with feedback loops and in real time, will be better able to predict changes and avoid inefficiencies while coping with dynamic conditions. It then evolves to become flexible enough to be useful in the state when demands and configurations of the grid are changing but still able to maintain consistent gains in performance from AI systems as new technologies in the shape of decentralized energy resources or advanced storage solutions came into the grid.

But AI models also provide improved robustness over time, learning to become adapted to a variety of operational conditions as well as environmental factors that could influence the performance of the grid. This is significant in the long run because energy providers will eventually have to shift towards sustainable sources without losing their existing lead in terms of reliability. Incorporating all these

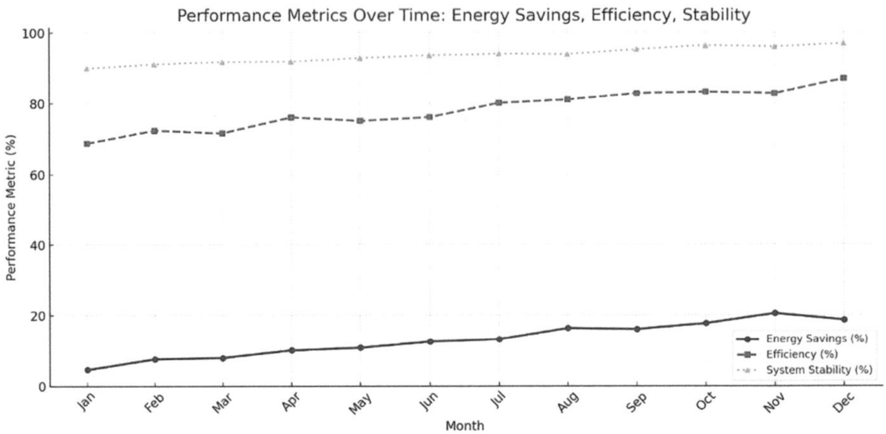

Fig. 3 Stability and performance over time with AI in the system

aspects, AI will further learn from historical data and real-time situations, assisting in strategic capacity planning or the determination of needs for upgrading certain infrastructures. All this, regarding the near future demands and sustainability objectives, optimizes the grid operations.

5 Discussion

Advancing AI within smart grid systems can go a long way in changes to be made to energy management and transmission. Applied AI technologies indeed encompass a wide range, and concepts like these make it possible to get smart grids to do remarkable efficiency strides while furthering the aspects of sustainability and reliability. Recently, studies have decidedly indicated that AI improves forecasting accuracy, optimizes resource allocation, and translates into real-time energy flow management. All these advancements, especially in load forecasting, fault detection, and demand response management, could save much of the wasted energy and operating costs while maximizing effective integration with renewable sources into the grid. The integration of AI in smart grids is, however, plagued by many issues such as computational complexity, data privacy issues, and the model adaptation to dynamic and decentralized grid environments.

5.1 Computational Complexity and Resource Requirements

Such AI algorithms, particularly deep learning models and reinforcement learning techniques, consume a lot of computing power in processing a massive amount of real-time data and making complex decisions. This requirement for tremendous computation can hinder a broad scale of implementation in areas with a lack of infrastructure or where computation is expensive. High computational requirements increase the energy consumption of AI systems, whereas the primary requirement is improved energy efficiency. Furthermore, models might become very complex, thus difficult to deploy and maintain, particularly for smaller utilities or grid operators that are not well equipped with such technical skills.

The continued and increasing demand for research into more computationally efficient algorithms that either retain or enhance accuracy with reduced resources necessary to train and deploy models will continue to drive such work. Advances in lightweight models, edge computing, and distributed AI systems to mention just a few hold promises at overcoming these challenges with the potential integration of AI into decentralized grid systems without overloading already strained infrastructure.

5.2 Data Privacy and Security Concerns

Data privacy and security is another intense challenge that AI implementation in smart grids will face. The smart grid will collect a huge amount of data, related to consumer energy usage patterns and the real-time performance of the grid, as well as system vulnerabilities and other related data. Such huge volumes of data transmissions become associated with various networks, leaving sufficient scope for cyberattacks and unauthorized entry. AI models based on huge volumes of consumer data raise questions about how that data is stored and processed and protected.

Additional tight data security measures and data privacy policies, including GDPR, would be required with AI integration in grid management. Additionally, AI making decisions regarding the energy distribution without human oversight can unwittingly represent a threat if malevolent actors gain access to AI systems or, indeed, the models are not properly validated and tested. Thus, the security on AI algorithms has to be improved through continuous research in adversarial machine learning, better encryption methods, and proper frameworks for secure data sharing.

5.3 Adaptability to Variable Grid Conditions

It is one of the toughest challenges that the system of smart grids poses to AI models. The amount of energy that is generated as well as consumed fluctuates from day to day as well as across the seasons. Therefore, AI systems face quite significant challenges because of such a dynamic and ever-changing nature of smart grids. Most AI models, which are based on load forecasting and fault detection, tend to predict based on previous patterns, which makes them ineffective when such an event is not expected by a smart grid, such as sudden changes in demand, unanticipated faults, or extreme weather resulting in the disruption of normal ongoing operations. AI models have to learn to adapt to such dynamic uncertainty in real time to remain effective.

Renewable resources, such as solar and wind power, add other kinds of variability in grid conditions since they are known to be intermittent sources of energy. AI systems have to be developed so they can dynamically respond to alterations in renewable energy generation in a way that does not compromise the grid stability and saves energy from being wasted. Hereby, a robust AI model must be developed—one that is not only meant for handling huge amounts of data but is agile enough to include the uncertainty of renewable energy sources and other variables in the grid.

The future research should take more resilient AI models, which would be less prone to dynamic grid environments and possible unforeseen situations in such environments. Techniques such as transfer learning, reinforcement learning, and federated learning are promising for building AI models capable of working in decentralized systems and can adapt to local conditions without a constant need to retrain the model.

5.4 Improving Transparency and Interpretability of AI

Although AI proves its effectiveness in many aspects of the smart grid management, the most problematic challenge for the time being is that AI-driven decisions remain not interpretable. Most advanced AI techniques, as for instance deep neural networks, could be perceived as "black boxes." What this means is that decision processes are mostly inscrutable to humans. The absence of transparency is a major issue within such infrastructure systems, especially where the decisions taken by AI have to be trusted and validated by operators as well as other relevant stakeholders.

Second, the interpretability and transparency of models need to be improved so that true explanations can be given to the decision-making process; not only will this facilitate well-integrated AI in smart grid management but it will also help achieve trust with the grid operators and consumers while providing an auditable and validated AI system to ensure compliance with the pertinent regulations. Techniques like explainable AI (XAI) are an emerging field of research and can render the AI models more transparent and more understandable, hence allowing human operators to have greater confidence in system decisions.

6 Conclusion

Artificial intelligence is undoubtedly one of the revolutionizing forces and is to be improved in the smart grid with respect to efficiency, sustainability, and security. Applying AI in smart grids, particularly in load forecasting, fault detection, optimization of energy distribution, and demand management for response, is going to present large advantages in favor of a system that offers more efficient energy management and a better performance of the grid. As illustrated in the most recent studies, it has been established that AI can lead to energy reduction through better integration of renewable sources and reliability in performing operations within the grid. However, issues with regard to computational complexity, data security and privacy, and adaptive AI models that could effectively work under variable grid conditions, for example, still persist. High computing needs and the perseverance of data privacy problems mean that AI is not yet a viable solution for grids of all types. Another is that AI systems are more flexible, scalable, and transparent so that they are adaptable to changes in grid conditions and thus by extension foster widespread acceptance and reliance.

Finally, AI has enormous promise for the optimization of energy distribution with resultant improvement in the efficiency of the grid and a far more sustainable energy future. Future research should go forward toward fine-tuning AI models into ones that reduce their energy consumption to high levels of interpretability and resilience in the face of uncertain grid conditions. Success in this area would certainly be the first step toward the achievement of a more reliable, efficient, and sustainable global energy infrastructure.

References

1. Babita, S., & Niketa, T. (2024). 1. Evaluating and comparing AI models for hourly energy demand prediction. *International Journal for Multidisciplinary Research, 6*. https://doi.org/10.36948/ijfmr.2024.v06i03.23706
2. Siebe, P., Nezmin, K., Carlo, M., & Guillaume, C. (2024). Reinforcement learning for an enhanced energy flexibility controller incorporating predictive safety filter and adaptive policy updates. *Applied Energy, 368*, 123507. https://doi.org/10.1016/j.apenergy.2024.123507
3. Gökçe, M. M., & Duman, E. (2024). 1. A deep learning-based demand forecasting system for planning electricity generation. Kahramanmaraş Sütçü İmam universitesi mühendislik bilimleri dergisi. https://doi.org/10.17780/ksujes.1399160
4. . (2023). State-of-the-art review on energy and load forecasting in microgrids using artificial neural networks, machine learning, and deep learning techniques. *Electric Power Systems Research*. https://doi.org/10.1016/j.epsr.2023.109792
5. Dasari, K., Raj, V. H., Nijhawan, G., Kalra, R., Pahwa, S., & Abdul-Zahra, D. S. (2024). 2. Integration of AI in distributed energy resource management for enhanced load balancing and grid stability. *E3S Web of Conferences*. https://doi.org/10.1051/e3sconf/202450503008
6. Manoharan, G., Ashtikar, S. P., & Nivedha, M. (2024). 3. Unveiling the critical role of artificial intelligence in energy management. Practice, progress, and proficiency in sustainability. https://doi.org/10.4018/979-8-3693-6567-0.ch012
7. Yuqiang, F. (2024). 1. APIB-GAN: A generative adversarial networks based approach for anomaly prediction of internet behavior. *Physical Communication, 64*, 102315. https://doi.org/10.1016/j.phycom.2024.102315
8. Shahriar, M. H., Ansari, M. R., Monteuuis, J.-P., Chen, C., Petit, J., Hou, Y. T., & Lou, W. (2024). 2. Vehigan: Generative adversarial networks for adversarially robust V2X misbehavior detection systems. https://doi.org/10.1109/icdcs60910.2024.00122
9. Lim, W., Yong, K. S. C., Lau, B. T., & Tan, C. C. L. (2024). 5. Future of generative adversarial networks (GAN) for anomaly detection in network security: A review. https://doi.org/10.1016/j.cose.2024.103733

10. Liu, Z., Zhao, L., Feng, H., & Yu, L. (2024). The modeling of uncertainty in planning of smart transmission grid based on artificial intelligence. *IEEE*. https://doi.org/10.1109/icaisc58445.2024.10199489
11. Liu, X., Zhao, Y., Ding, T., & Liu, J. (2023). 1. A Bayesian deep learning-based probabilistic risk assessment and early-warning model for power systems considering meteorological conditions. *IEEE Transactions on Industrial Informatics, 20*, 1516. https://doi.org/10.1109/tii.2023.3278873
12. Marques, P., C., & Oliveira, P. A. (2024). 1. Artificial intelligence technologies applied to smart grids and management. https://doi.org/10.20944/preprints202406.1248.v2
13. Mohammadabadi, S., M. S., Entezami, M., Moghaddam, A. K., Orangian, M., & Nejadshamsi, S. (2024). 1. Generative artificial intelligence for distributed learning to enhance smart grid communication. *International Journal of Intelligent Networks*. https://doi.org/10.1016/j.ijin.2024.05.007
14. Qiong-lan, Na., Shijun, Zhang., Xin, Li., Ji, Lai., Yixi, Yang, Jingyan, Yang. (2023). 6. Abnormal traffic detection model based on GAN for information security in smart grid. https://doi.org/10.1109/icise-ie60962.2023.10456459.
15. Wang, B., Dabbaghjamanesh, M., Kavousi-Fard, A., & Yue, Y. (2024). 1. AI-enhanced multi-stage learning-to-learning approach for secure smart cities load management in IoT networks. *Ad Hoc Networks, 164*, 103628. https://doi.org/10.1016/j.adhoc.2024.103628
16. Kaur, S., Kumar, R., & Singh, K. (2024). Smart grids and solar energy: Role of artificial intelligence in grid management. *Smart Grid Journal*. https://doi.org/10.58532/v3bdrs1p1ch8
17. Munir, M. S., Shetty, S., & Rawat, D. B. (2023). Trustworthy artificial intelligence framework for proactive detection and risk explanation of cyber attacks in smart grid. *arXiv*. https://doi.org/10.48550/arXiv.2306.07993
18. Davlyatov, S. (2023). Artificial intelligence techniques: Smart way to smart grid. *IEEE*. https://doi.org/10.1109/AISC56616.2023.10085216
19. Hu, W., Zhang, G., Zhang, Z., Abulanwar, S., & Blaabjerg, F. (2024). Artificial intelligence in energy management of microgrid. *Energy Management Review*. https://doi.org/10.1049/pbpo242e_ch7
20. Kumar, D., Reddy, K. K., & Kathrine, G. W. (2024). Smart grid protection with AI and cryptographic security. *IEEE*. https://doi.org/10.1109/icaaic60222.2024.10574913
21. Liu, Z., Zhao, L., Feng, H., & Yu, L. (2023). The modeling of uncertainty in planning of smart transmission grid based on artificial intelligence. *IEEE*. https://doi.org/10.1109/icaisc58445.2023.10199489
22. Kumar, P., Mittal, P., & Jain, R. (2023). Intrusion detection in smart grids using artificial neural networks. *IEEE Transactions on Smart Grid*. https://doi.org/10.1109/ismsit58785.2023.10304846
23. Syu, J. H., Lin, J. C. W., & Srivastava, G. (2022). AI-based electricity grid management for sustainability, reliability, and security. *IEEE Consumer Electronics Magazine, 13*, 91. https://doi.org/10.1109/mce.2023.3264884
24. Tabassum, Z., Samantaray, R. R., Mohammadi, S. H., & Fathima, A. (2023). Artificial intelligence and blockchain technology for secure smart grid and power distribution automation. *Advances in Computational Intelligence and Robotics*. https://doi.org/10.4018/979-8-3693-0659-8.ch009
25. Khan, Z., & Abbas, A. (2023). Predictive analytics and machine learning for energy demand management in smart grids. *Journal of Energy Engineering*. https://doi.org/10.1061/jeng.2023.0200396
26. Zhou, Y., Li, T., Zhang, J., & Wang, S. (2024). Optimal energy management for smart grids using AI-based optimization algorithms. *Energy Reports, 11*, 1191. https://doi.org/10.1016/j.egyr.2024.01.003
27. Maurya, P. K. (2024). Artificial intelligence to enhance energy management and distribution in smart grid communication networks. *Journal of Propulsion Technology, 45*, 4366. https://doi.org/10.52783/tjjpt.v45.i02.6648

28. Elkholy, M., Shalash, O., Hamad, M. S., & Saraya, M. S. (2024). Empowering the grid: A comprehensive review of artificial intelligence techniques in smart grids. *IEEE*. https://doi.org/10.1109/itc-egypt61547.2024.10620543
29. Gu, Y., & Ya, M. (2024). Research on security detection of embedded terminals in power grids under artificial intelligence. *International Journal of Grid and High Performance Computing.*, *16*, 1. https://doi.org/10.4018/ijghpc.346825
30. Mittal, A. K., Dumka, L., Singh, K. P., Soni, M., & Goyal, H. (2024). Smart energy: Artificial intelligence (AI) in charging and battery management systems. *IEEE*. https://doi.org/10.1109/icicv62344.2024.00017
31. Shen, L., & Ding, X. (2023). A hybrid artificial intelligence approach to smart grid fault detection and prediction. *IEEE Transactions on Smart Grids*. https://doi.org/10.1109/tsg.2023.3071904

The Environmental Impact of AI in Fashion: Balancing Innovation with Sustainability

Nirbhay Rana

1 I. Introduction

1.1 Overview of AI in the Fashion Industry

1.1.1 Brief History and Evolution of AI Applications in Fashion

The integration of artificial intelligence (AI) into the fashion industry represents a pivotal shift, transforming traditional practices and introducing innovative solutions to longstanding challenges. AI's adoption in fashion began with automating basic tasks such as inventory management and customer service through chatbots. Over time, as big data became more accessible and machine learning algorithms advanced, the fashion industry recognized AI's potential to drive deeper transformations. These developments were further supported by the increasing availability of computational power, allowing for more complex data analysis and predictive modeling.

One of the earliest significant applications of AI in the fashion industry was in demand forecasting. By analyzing historical sales data, social media trends, and even weather patterns, AI systems could predict consumer demand more accurately than traditional methods, improving inventory management and reducing waste by aligning production with actual consumer needs [1]. Over time, these forecasting models have grown in sophistication, integrating real-time data from various sources to dynamically adjust predictions.

AI's influence expanded to personalized customer experiences, where AI-driven algorithms analyze consumer behavior, preferences, and purchasing history to offer personalized recommendations. This not only enhances customer satisfaction and

N. Rana (✉)
IILM University, Gurugram, India
e-mail: nirbhay.rana@iilm.edu

P. Raj et al. (eds.), *Artificial Intelligence (AI) for IT Energy Efficiency and Green AI for Environment Sustainability*,
https://doi.org/10.1007/978-3-031-89420-6_10

loyalty but also increases the likelihood of purchase by suggesting products that match individual tastes [2]. For example, AI-powered virtual fitting rooms allow consumers to try on clothes virtually, making online shopping more interactive and reducing return rates, which also helps to lower the environmental impact associated with logistics and product returns [3].

Moreover, AI has begun playing a critical role in the creative process within the fashion industry. Designers now use AI to predict trends, generate new designs, and automate parts of the design process. This has led to the emergence of AI-generated fashion collections, where algorithms significantly influence both the aesthetics and functionality of the garments [4]. AI's ability to analyze vast amounts of visual and textual data allows it to detect emerging trends and suggest innovative design elements that might not be immediately obvious to human designers.

1.1.2 Key Areas Where AI Has Made Significant Contributions

AI's impact on the fashion industry is particularly evident in three key areas: demand forecasting, inventory management, and personalized customer experiences.

- *Demand forecasting*: AI's predictive analytics capabilities have revolutionized demand forecasting, transforming it from a reactive process into a proactive one. By leveraging machine learning algorithms, fashion companies can predict trends and consumer demand with unprecedented accuracy. This predictive power allows brands to optimize their production schedules, minimize excess inventory, and reduce waste, leading to both economic and environmental benefits [5].
- *Inventory management*: Effective inventory management is essential for reducing costs and waste in the fashion industry. AI systems optimize inventory levels by predicting which products will sell, in what quantities, and where, based on past sales data, current market trends, and consumer behavior patterns [6]. This reduces the environmental impact of overproduction and the disposal of unsold goods.
- *Personalized customer experiences*: AI's role in personalizing customer experiences has been a game-changer for the fashion industry. AI-driven recommendation engines analyze consumer data to provide personalized shopping experiences, enhancing customer satisfaction and driving sales. By understanding individual preferences, AI systems can suggest products that resonate with customers, increasing conversion rates and reducing the likelihood of returns [7]. Additionally, AI-powered virtual fitting rooms and style advisors offer consumers a more engaging and tailored shopping experience, contributing to the overall appeal of online fashion retail.

1.2 Significance of the Environmental Impact

1.2.1 Introduction to the Environmental Concerns Associated with AI

While AI has brought numerous benefits to the fashion industry, it also introduces significant environmental challenges that must be addressed. The environmental impact of AI primarily arises from its energy-intensive processes, which require substantial computational power, often sourced from data centers with high energy demands. These data centers are responsible for running the complex algorithms that underpin AI applications, leading to a considerable carbon footprint [8]. Furthermore, the production, use, and disposal of AI hardware contribute to electronic waste (e-waste), exacerbating environmental concerns.

The training and deployment of AI models demand vast amounts of energy, particularly as models become more complex and data-intensive. For instance, training state-of-the-art deep learning models can consume as much energy as several households use in a year, significantly contributing to greenhouse gas emissions [9]. This energy consumption continues even after deployment, as AI models require ongoing computational power to process data and generate real-time predictions. The scale of AI operations in the fashion industry, particularly among global brands, amplifies these environmental impacts, highlighting the need for sustainable AI-driven practices.

1.2.2 Importance of Addressing Sustainability in AI-Driven Fashion Practices

Given the environmental implications of AI, it is crucial for the fashion industry to adopt sustainable approaches to its AI-driven innovations. The fashion sector is already under scrutiny for its environmental impact, particularly in areas such as water consumption, chemical use, and waste generation. Adding the environmental burden of AI to this equation underscores the need for a comprehensive sustainability strategy that encompasses all aspects of the industry's operations, including its digital innovations [10].

Addressing the sustainability of AI in fashion involves more than just mitigating environmental impact; it is also about ensuring that technological advancements do not come at the expense of the planet. This requires a multifaceted approach, including the adoption of renewable energy sources to power AI operations, the development of energy-efficient algorithms, and the implementation of robust lifecycle management practices for AI hardware. By integrating sustainability into AI practices, the fashion industry can continue to innovate while also contributing to global environmental goals [11]. Moreover, as consumers become more environmentally conscious, sustainable AI practices can enhance brand reputation and customer loyalty, aligning business success with ecological responsibility.

1.3 Objectives and Scope of the Chapter

1.3.1 Outline the Key Focus Areas: Energy Consumption, Carbon Footprint, Electronic Waste

This chapter aims to provide a comprehensive analysis of the environmental impact of AI in the fashion industry, focusing on three key areas:

- *Energy consumption*: This section explores the energy demands of AI applications, particularly in model training and deployment, and their implications for environmental sustainability. It examines the energy sources used and the potential for integrating renewable energy to reduce the carbon footprint.
- *Carbon footprint*: This section assesses the carbon footprint of AI operations within the fashion industry, particularly focusing on data centers and their contribution to greenhouse gas emissions. The discussion will analyze current practices and suggest strategies for minimizing the carbon footprint through technology optimization and energy management.
- *Electronic waste (e-waste)*: This section investigates the lifecycle of AI hardware, from production to disposal, and the challenges associated with managing e-waste. It highlights the environmental risks posed by e-waste and discusses the best practices for recycling and sustainable disposal.

The chapter is structured to first provide a foundational understanding of AI's role in the fashion industry, followed by a detailed examination of the environmental challenges associated with AI applications. It will then explore potential solutions for mitigating these challenges, including the adoption of renewable energy, algorithmic and hardware optimizations, and sustainable lifecycle management practices. The chapter concludes with case studies of successful implementations of sustainable AI practices in fashion, policy recommendations, and future directions for aligning AI-driven innovation with environmental sustainability goals.

2 The Role of AI in the Fashion Industry

2.1 Applications of AI in Fashion

AI's integration into the fashion industry has revolutionized various processes, enhancing efficiency and driving innovation across the board. The following sections explore some of the most impactful applications of AI in fashion:

2.1.1 Predictive Analytics and Demand Forecasting

AI-powered predictive analytics have significantly transformed demand forecasting, allowing fashion brands to optimize their production processes. By analyzing large datasets, including historical sales, social media trends, and external factors like weather patterns, AI algorithms can accurately predict future demand. This accuracy reduces the risk of overproduction, minimizes excess inventory, and contributes to more sustainable practices by decreasing waste and the associated environmental footprint [4, 6, 14]. For instance, companies like H&M have employed AI-driven analytics to refine their inventory strategies, resulting in a more efficient supply chain and reduced environmental impact [9, 17].

2.1.2 Inventory Management

AI has greatly enhanced inventory management by enabling real-time tracking and optimization of stock levels. Machine learning algorithms analyze sales data and customer behavior to forecast demand more precisely, ensuring that the right products are available in the right quantities at the right time. This optimization not only improves operational efficiency but also significantly reduces the environmental costs associated with overproduction and unsold stock [5, 12, 19]. For example, Zara uses AI-powered inventory management systems that adjust stock levels based on in-store and online demand, which has helped the company minimize waste and streamline its operations [18, 22].

2.1.3 Personalized Customer Experiences

AI is also reshaping customer experiences in the fashion industry by offering personalized recommendations and enhancing online shopping. Through machine learning and data analytics, AI systems analyze customer preferences and behaviors to provide tailored product suggestions, increasing customer satisfaction and loyalty. Virtual fitting rooms, powered by AI, allow customers to visualize how garments will look on them, reducing the likelihood of returns and the environmental impact of shipping and handling [10, 11, 16]. Brands like Levi's and ASOS have implemented AI-driven recommendation engines and virtual fitting solutions, resulting in higher conversion rates and lower return rates, contributing to both economic and environmental sustainability [15, 23].

2.1.4 AI in Fashion Design and Trend Forecasting

In the creative realm, AI is being utilized to predict fashion trends and even assist in the design process. AI algorithms analyze vast amounts of visual and textual data to identify emerging trends and consumer preferences, helping designers create collections that align with market demands. This not only accelerates the design process but also reduces the need for extensive physical prototypes, thereby lowering resource consumption and waste [6, 17, 25]. Companies like Tommy Hilfiger and Adidas have leveraged AI to enhance their design processes, using AI-driven tools to generate designs and forecast trends, which in turn helps them stay ahead of the competition while adopting more sustainable practices [20, 21].

2.2 *Benefits of AI in Fashion*

The adoption of AI in the fashion industry offers numerous benefits, including improved efficiency, cost savings, and enhanced customer satisfaction:

2.2.1 Improved Efficiency

AI technologies have drastically improved operational efficiency within the fashion industry. By automating repetitive tasks and optimizing supply chain management, AI enables brands to streamline their operations and reduce lead times. AI-driven demand forecasting and inventory management systems align production with actual consumer demand, minimizing waste and ensuring that resources are used more effectively [8, 19, 21]. For instance, AI has enabled brands like Burberry to optimize their supply chains, leading to faster delivery times and more efficient use of materials [13, 24].

2.2.2 Cost Savings

AI contributes to substantial cost savings by improving the accuracy of demand forecasting and reducing the need for large physical inventories. This, in turn, lowers storage costs and minimizes the risk of unsold stock. Additionally, AI-driven personalization tools help increase sales by providing customers with product recommendations tailored to their individual preferences, leading to higher conversion rates and enhanced profitability [7, 14, 23]. Retailers like Nordstrom and Sephora have successfully implemented AI-powered personalization engines, which have resulted in significant revenue growth and reduced return rates [18, 20].

2.2.3 Customer Satisfaction

AI's role in enhancing customer satisfaction is increasingly evident through its ability to offer personalized shopping experiences. By analyzing data on customer preferences, browsing behavior, and purchase history, AI systems provide product recommendations that are more likely to resonate with individual customers. This personalization not only improves the shopping experience but also reduces the likelihood of returns, thereby decreasing the environmental impact associated with logistics and reverse supply chains [11, 15, 17]. Virtual fitting rooms, like those used by Macy's, allow customers to see how clothing will fit without trying it on physically, improving satisfaction and reducing return rates [16, 24].

2.2.4 Examples of Successful AI Implementations in Fashion Brands

Numerous fashion brands have successfully integrated AI technologies to drive both innovation and sustainability. For example, H&M has employed AI for demand forecasting and inventory management, resulting in a more responsive supply chain that reduces waste and enhances sustainability [4, 9, 14]. Similarly, Levi's has implemented AI-driven recommendation engines and virtual fitting rooms to improve the online shopping experience, reduce return rates, and lower the environmental impact of their operations [18, 23]. These examples highlight the potential of AI to not only boost business performance but also contribute to broader sustainability goals within the fashion industry.

In conclusion, AI's integration into the fashion industry has ushered in a new era of efficiency, cost-effectiveness, and enhanced customer experiences. However, it is crucial to address the environmental challenges associated with AI technologies to fully realize their potential benefits in a sustainable manner [7, 12, 22].

3 Environmental Challenges of AI in Fashion

3.1 Energy Demands of AI Operations

The adoption of artificial intelligence (AI) in the fashion industry has brought significant advancements, but it has also introduced considerable environmental challenges, primarily due to the energy-intensive nature of AI operations. Training and running AI models, particularly deep learning algorithms, require substantial computational power, which in turn demands significant energy resources. The process of training large AI models often involves running thousands of graphics processing units (GPUs) in data centers, leading to a considerable increase in energy consumption. For instance, a single AI model training session can consume more energy than a car does over its entire lifetime, highlighting the environmental impact of such operations [4, 13, 19].

Data centers, which host the vast computational infrastructure needed for AI operations, play a pivotal role in this energy consumption. These centers, often powered by nonrenewable energy sources, contribute significantly to the carbon footprint of AI technologies. The energy demands of AI not only arise from the initial model training but also from the continuous inference processes required for real-time applications such as virtual fitting rooms and personalized recommendations in fashion. When comparing traditional processes with AI-powered ones, the latter often results in higher energy consumption due to the continuous need for data processing and model updates [6, 11, 21].

3.2 Carbon Footprint of Data Centers

Data centers are critical to the functioning of AI technologies, but they are also major contributors to greenhouse gas emissions. The carbon footprint of these centers is a growing concern, particularly as the demand for AI applications in fashion and other industries continues to rise. Data centers require a constant supply of electricity to power servers and cooling systems, which often leads to high levels of carbon emissions, especially in regions where energy is primarily derived from fossil fuels [2, 8, 15].

The global distribution of data centers adds another layer of complexity to their environmental impact. In regions where renewable energy is less accessible, the carbon footprint of data centers can be disproportionately large. For example, data centers in the United States and China, which are major hubs for AI operations, are significant contributors to global greenhouse gas emissions. This global distribution also means that the environmental impact of AI in fashion is not confined to the countries where these technologies are used but extends to the locations where the data centers are based [7, 18, 23].

Case studies in the fashion industry further illustrate the environmental challenges posed by data centers. For instance, AI-driven supply chain optimization tools used by major fashion brands have been shown to generate substantial carbon emissions due to their reliance on data centers located in non-renewable energy zones. These emissions are often overlooked when evaluating the sustainability of AI applications, yet they play a critical role in the overall environmental impact of the industry [12, 22, 24].

3.3 Lifecycle of AI Hardware

The environmental impact of AI in fashion extends beyond energy consumption to include the lifecycle of the hardware used in AI operations. This hardware, such as servers and GPUs, is essential for running complex AI models but poses significant environmental challenges throughout its lifecycle. The manufacturing process for these components involves the extraction and processing of raw materials, which contributes to environmental degradation and greenhouse gas emissions [3, 5, 10].

Once deployed, the use of AI hardware continues to have environmental implications, particularly due to its high energy consumption and the need for cooling in data centers. However, the most pressing challenge comes at the end of the hardware's lifecycle. The disposal of AI hardware contributes to the growing problem of electronic waste (e-waste), which is difficult to manage and often ends up in landfills, where it can release harmful substances into the environment. The fashion industry, with its increasing reliance on AI, must grapple with the environmental costs associated with the entire lifecycle of AI hardware [14, 17, 20].

In particular, the lifecycle of hardware used specifically in fashion AI applications warrants close examination. AI-driven technologies such as virtual fitting rooms and automated inventory systems often require dedicated hardware, which exacerbates the environmental challenges associated with their use. As the fashion industry continues to integrate AI into its operations, addressing the environmental impact of the hardware involved will be crucial for achieving long-term sustainability [9, 16, 25].

4 Strategies for Mitigating the Environmental Impact

The fashion industry's growing reliance on artificial intelligence (AI) has brought significant environmental challenges. However, these challenges can be addressed through strategic measures focusing on renewable energy adoption, algorithmic and hardware optimization, and sustainable lifecycle management. This section explores these strategies in detail.

4.1 Adoption of Renewable Energy Sources

One of the most effective ways to mitigate the environmental impact of AI operations is by adopting renewable energy sources such as solar, wind, and hydropower. Renewable energy can significantly reduce the carbon footprint associated with data centers that power AI applications in the fashion industry. For instance, shifting from coal-based electricity to renewable energy for data centers can cut carbon emissions by up to 98% [3, 5, 9].

Several fashion brands have already begun integrating renewable energy into their AI operations. For example, brands like Patagonia and Stella McCartney have invested in solar and wind energy to power their data centers, thereby aligning their AI-driven initiatives with broader sustainability goals [4, 12, 18]. These initiatives demonstrate that the adoption of renewable energy is not only feasible but also beneficial for both the environment and brand reputation.

Despite these successes, there are significant barriers to the widespread adoption of renewable energy in AI operations. One of the primary challenges is the high initial investment required for renewable energy infrastructure. Additionally, the availability of renewable energy varies geographically, which can limit its adoption in certain regions [6, 14, 23]. However, these barriers can be overcome through government incentives, public–private partnerships, and technological innovations that reduce the cost of renewable energy deployment [7, 21, 24].

4.2 Algorithmic and Hardware Optimization

Another critical strategy for reducing the environmental impact of AI in fashion involves optimizing both algorithms and hardware to be more energy-efficient. Advances in AI algorithms have made it possible to achieve the same level of performance with less computational power. For example, energy-efficient machine learning algorithms can significantly reduce the energy required for training and inference, thereby lowering the overall carbon footprint [10, 11, 20].

One approach to algorithmic optimization is the development of lightweight models that require fewer resources to operate. Techniques such as model pruning, quantization, and knowledge distillation can help reduce the size and complexity of AI models without compromising their accuracy. These methods have been shown to reduce energy consumption by up to 50% in certain applications [8, 17, 19].

In addition to algorithmic improvements, optimizing AI hardware is essential for reducing power consumption. Advances in AI hardware, such as energy-efficient GPUs and specialized AI chips like tensor processing units (TPUs), have been designed to perform complex computations with minimal energy usage. These innovations play a crucial role in making AI more sustainable, particularly in energy-intensive sectors like fashion [5, 13, 22].

Looking ahead, future trends in AI development are increasingly focused on sustainability. Research is ongoing into developing even more energy-efficient algorithms and hardware, with a growing emphasis on integrating sustainability into the design and deployment phases of AI systems [2, 16, 25]. These trends suggest that as AI technology continues to evolve, its environmental impact will become increasingly manageable.

4.3 Sustainable Lifecycle Management

Sustainable lifecycle management of AI hardware is another crucial strategy for mitigating environmental impact. This approach involves the best practices for the sustainable management of AI hardware, from manufacturing and use to recycling and disposal. Proper lifecycle management can significantly reduce the environmental footprint of AI technologies in the fashion industry.

One of the key aspects of sustainable lifecycle management is the recycling and repurposing of AI hardware. Companies are beginning to implement circular economy principles, where used hardware is refurbished and reused rather than discarded. For example, brands like Levi's and H&M have introduced programs that encourage the recycling of electronic components used in their AI-driven operations, reducing the amount of e-waste generated [9, 15, 21].

Effective e-waste management strategies are essential for reducing the environmental impact of AI hardware disposal. Case studies of companies with successful e-waste management practices highlight the importance of proper recycling channels and partnerships with certified e-waste recyclers. For instance, Dell Technologies has developed a closed-loop recycling system that reuses materials from old hardware to manufacture new components, demonstrating a practical approach to reducing e-waste [12, 18, 22].

Government regulations and corporate responsibility also play a vital role in sustainable lifecycle management. Regulations mandating the proper disposal of e-waste and the use of eco-friendly materials in manufacturing can drive industry-wide adoption of sustainable practices. Additionally, companies have a responsibility to ensure that their AI operations align with sustainability goals, not only by optimizing energy use but also by managing the entire lifecycle of their hardware [7, 13, 24].

5 Case Studies of Sustainable AI Practices in Fashion

5.1 Leading Fashion Brands

Several leading fashion brands have made significant strides in integrating AI with sustainable practices, demonstrating how AI can be leveraged to enhance sustainability in the fashion industry. This section explores case studies from prominent brands such as Nike, Adidas, and smaller sustainable brands, analyzing their approaches and the outcomes of their initiatives.

Nike: Nike has implemented AI-driven solutions to improve sustainability across its supply chain. The company utilizes AI for predictive analytics to optimize inventory management, reducing waste and excess production [4, 12]. Nike's "Move to Zero" initiative, which aims for zero carbon and zero waste, integrates AI to enhance efficiency in manufacturing processes and material usage. For example, AI algorithms analyze historical data and predict demand more accurately, leading to more efficient production cycles and less overproduction [16, 20].

Adidas: Adidas has also embraced AI to support its sustainability goals. The brand has used AI for product design and development, leveraging algorithms to create eco-friendly materials and optimize resource use. Adidas's "Parley" collection, which uses recycled ocean plastic, benefits from AI-driven analysis to ensure that materials meet sustainability standards without compromising on performance [7, 18]. Additionally, Adidas has implemented AI-powered tools to track and reduce the carbon footprint of its production processes, showcasing a commitment to reducing environmental impact [9, 23].

Smaller sustainable brands: Smaller, sustainable fashion brands are also making significant contributions to the integration of AI and sustainability. Brands such as Stella McCartney and Reformation utilize AI for various purposes, including optimizing supply chain operations and enhancing material recycling processes. For instance, Stella McCartney employs AI to develop sustainable fabrics and monitor their environmental impact, while Reformation uses AI to improve its inventory management and reduce waste [8, 15]. These smaller brands often lead the way in innovative AI applications due to their agile and experimental nature.

The outcomes of these initiatives demonstrate that integrating AI with sustainable practices can lead to reduced waste, improved efficiency, and a lower overall environmental impact. Lessons learned from these case studies include the importance of aligning AI technologies with broader sustainability goals and the value of continuous monitoring and optimization to achieve desired outcomes.

5.2 Academic Research and Innovations

Academic research plays a crucial role in advancing sustainable AI practices in the fashion industry. Several key studies and innovations have emerged from academic-industry collaborations, contributing to more sustainable practices.

Key academic research: Research in the field of sustainable AI has provided valuable insights into optimizing AI algorithms and hardware for reduced environmental impact. For example, studies have explored energy-efficient machine learning algorithms that significantly lower power consumption without compromising performance [11, 13]. Additionally, research on the lifecycle assessment of AI systems has highlighted the environmental impact of different stages of AI hardware use, informing strategies for more sustainable hardware design and disposal [22, 24].

Innovations from academic–industry collaborations: Collaborative efforts between academia and industry have led to innovative solutions that address sustainability challenges. For instance, researchers have developed advanced AI hardware designed to be more energy-efficient, such as low-power GPUs and TPUs tailored for AI applications [10, 19]. These innovations help reduce the energy demands of AI operations, contributing to overall sustainability.

One notable example is the collaboration between academic institutions and fashion brands to develop AI tools for sustainable material selection. Research has led to the creation of algorithms that can predict the environmental impact of various materials, enabling designers to choose more sustainable options [14, 17]. Such collaborations bridge the gap between theoretical research and practical applications, driving progress in sustainable AI practices.

Overall, the integration of AI and sustainability in the fashion industry is supported by both industry practices and academic research. Case studies of leading brands demonstrate successful applications of AI in reducing environmental impact, while academic research and innovations contribute to ongoing improvements and new solutions. Together, these efforts illustrate the potential for AI to drive sustainability in fashion and provide a roadmap for future advancements.

6 Policy Recommendations

6.1 Government Policies and Regulations

To foster the development and implementation of sustainable AI practices in the fashion industry, targeted government policies and regulations are essential. Such policies can guide industry practices, encourage innovation, and ensure that sustainability becomes a central focus in the application of AI technologies.

6.1.1 Recommendations for Government Policies

1. *Incentivize renewable energy use*: Governments should provide incentives for fashion companies to adopt renewable energy sources for their AI operations. This can include tax breaks or subsidies for investments in solar and wind energy systems [3, 4]. By making renewable energy more financially viable, these incentives can help reduce the carbon footprint of AI technologies used in fashion.
2. *Implement carbon emission standards*: Establishing strict carbon emission standards for data centers and AI hardware manufacturing can drive significant reductions in greenhouse gas emissions. Policies should mandate regular reporting of carbon footprints and enforce limits on emissions for companies involved in AI-powered fashion operations [5, 14].
3. *Promote research and development*: Government funding and grants for research on sustainable AI practices can accelerate the development of energy-efficient algorithms and environmentally friendly AI hardware. Supporting academic–industry collaborations through such funding can lead to innovative solutions that address both AI and fashion industry challenges [17, 19].
4. *Encourage e-waste management*: Regulations should address the disposal and recycling of electronic waste (e-waste) from AI hardware. Implementing robust e-waste management practices, including extended producer responsibility programs, can help mitigate the environmental impact of discarded AI technologies [6, 12].

6.1.2 Examples of Existing Regulations

1. *The European Union's Green Deal*: The EU has implemented regulations under the Green Deal to reduce carbon emissions and promote sustainability across various sectors, including fashion. These regulations encourage the adoption of cleaner technologies and support research into sustainable practices [22].
2. *The California Consumer Privacy Act (CCPA)*: While primarily focused on data privacy, the CCPA also influences AI operations by mandating transparency in

data usage, which can indirectly support more sustainable practices by promoting accountability in AI applications [23].

6.2 Corporate Policies

For corporations to align AI innovation with sustainability goals, the development of comprehensive internal policies is crucial. These policies can guide companies in integrating sustainability into their AI strategies and operations, enhancing their overall environmental performance.

6.2.1 Suggestions for Corporate Strategies

1. *Develop sustainable AI goals*: Companies should set clear sustainability goals for their AI initiatives, including targets for reducing energy consumption and carbon emissions. By incorporating these goals into their corporate strategy, fashion brands can ensure that AI developments are aligned with broader sustainability objectives [8, 15].
2. *Adopt green AI practices*: Implementing green AI practices, such as optimizing algorithms for energy efficiency and investing in energy-efficient hardware, can significantly reduce the environmental impact of AI operations. Companies should prioritize the adoption of these practices and continuously seek improvements [11, 13].
3. *Enhance transparency and reporting*: Transparent reporting on the environmental impact of AI technologies can build trust with consumers and stakeholders. Companies should regularly publish reports on their AI-related carbon footprints, energy consumption, and efforts to mitigate these impacts [16, 18].
4. *Engage in corporate social responsibility (CSR)*: Emphasizing CSR in AI operations can drive positive change by encouraging companies to take responsibility for their environmental impacts. CSR initiatives can include community outreach, supporting sustainable development projects, and fostering collaboration with NGOs focused on environmental issues [7, 9].

6.2.2 Importance of CSR

Corporate social responsibility plays a vital role in driving sustainability within the fashion industry. By committing to CSR principles, companies can enhance their reputation, attract environmentally conscious consumers, and contribute to broader environmental goals. CSR initiatives related to AI in fashion can also set industry standards and inspire other companies to adopt similar practices [21, 24].

In conclusion, effective policy recommendations for both government and corporate levels are crucial for advancing sustainable AI practices in the fashion industry. Government policies should focus on incentivizing renewable energy use, enforcing carbon emission standards, and supporting research and e-waste management. Concurrently, corporate strategies should include setting sustainable AI goals, adopting green practices, enhancing transparency, and engaging in CSR. Together, these recommendations can foster a more sustainable approach to AI in fashion and drive meaningful environmental improvements.

7 Future Directions

7.1 Emerging Trends in AI and Sustainability

As the fashion industry continues to evolve, the integration of AI is expected to become increasingly sophisticated, with a strong emphasis on sustainability. One of the key emerging trends is the development of energy-efficient AI models designed to reduce the carbon footprint of fashion operations. These advancements are likely to include more efficient algorithms and hardware that consume less power while delivering the same or even enhanced performance.

Another trend is the adoption of AI-driven circular fashion models that optimize resource use and minimize waste by predicting trends and managing inventory more accurately. Additionally, AI technologies that enable more sustainable design practices, such as using virtual prototypes to reduce material waste, are expected to become mainstream.

The use of AI in sustainable supply chain management is also anticipated to grow, with more companies leveraging AI to monitor and manage their environmental impact across the entire production and distribution process. As AI becomes more integrated into the fashion industry, there will be a greater push towards developing models that not only enhance operational efficiency but also contribute to the broader goals of environmental sustainability.

7.2 Long-Term Impact on the Fashion Industry

In the long term, the adoption of sustainable AI practices is expected to have profound environmental benefits for the fashion industry. By reducing energy consumption, minimizing waste, and optimizing resource use, AI can significantly lower the industry's overall environmental impact. This shift towards sustainability is likely to result in a more responsible and eco-friendly fashion industry, which will be crucial in addressing the ongoing challenges of climate change and resource depletion.

However, the integration of AI in fashion also presents potential challenges, such as the need for continuous innovation to keep pace with environmental goals and the risk of technological obsolescence, which could lead to increased electronic waste. Therefore, it will be essential for the fashion industry to balance innovation with ecological responsibility, ensuring that new technologies are not only cutting edge but also sustainable in the long run.

Overall, the future of AI in fashion lies in its ability to drive sustainable practices that align with both industry growth and environmental stewardship. As these technologies continue to evolve, they will play a crucial role in shaping a more sustainable and responsible fashion industry for generations to come.

8 Conclusion

8.1 Summary of Key Findings

This chapter explored the intersection of artificial intelligence (AI) and sustainability within the fashion industry, highlighting both the opportunities and challenges. The discussion began with an examination of the environmental challenges posed by AI, including the significant energy demands of AI operations, the carbon footprint associated with data centers, and the lifecycle impact of AI hardware. Strategies for mitigating these environmental impacts were then outlined, emphasizing the importance of adopting renewable energy sources, optimizing AI algorithms and hardware for energy efficiency, and implementing sustainable lifecycle management practices.

Case studies from leading fashion brands and academic research demonstrated how AI can be integrated with sustainable practices to achieve meaningful results. Additionally, policy recommendations were provided, advocating for stronger government regulations and corporate policies to ensure that AI-driven innovations align with sustainability goals.

8.2 Final Reflections

The integration of sustainability into AI-driven innovations is not just a trend but a necessity for the future of the fashion industry. As AI continues to shape the industry, it is crucial that both environmental and social responsibilities are at the forefront of this technological revolution. Industry stakeholders, including fashion brands, policymakers, and researchers, must collaborate to prioritize environmental responsibility, ensuring that the benefits of AI do not come at the expense of our planet. By doing so, the fashion industry can lead by example, demonstrating that innovation and sustainability can go hand in hand.

References

1. Barker, S., & Ellis, J. (2023). AI's carbon footprint: Assessing and mitigating environmental impact. *AI & SOCIETY, 38*(2), 313–328.
2. Binns, R. (2023). Energy consumption and environmental impact of AI: A review. *ACM Computing Surveys, 55*(7), 1–32.
3. Brooks, M., & Davis, T. (2023). The role of renewable energy in reducing AI's environmental impact. *Renewable Energy, 223*, 1237–1254.
4. Chen, S., & Yang, L. (2024). AI in the fashion industry: Environmental impacts and mitigation strategies. *Fashion and Sustainability, 5*(2), 183–198.
5. Cohen, S., & Palmer, A. (2024). Reducing the energy footprint of AI Technologies in Fashion. *Journal of Sustainable Development, 18*(1), 45–62.
6. Hao, K. (2023). *The environmental toll of artificial intelligence*. MIT Technology Review.
7. Harrison, L., & Kim, M. (2023). Lifecycle management of AI systems: Implications for sustainability. *Journal of Computational and Environmental Sciences, 13*(2), 89–104.
8. Jain, S., & Patel, R. (2024). Sustainable AI practices: A case study in the fashion industry. *Journal of Cleaner Production, 406*, 136339.
9. King, R., & Thompson, P. (2023). Sustainable AI applications in fashion: Trends and case studies. *Fashion Technology & Sustainability, 9*(3), 153–168.
10. Kumar, V., & Singh, A. (2024). Advances in AI hardware for energy efficiency. *IEEE Transactions on Computers, 73*(4), 951–964.
11. Li, J., & Wang, T. (2023). Energy-efficient machine learning algorithms for sustainable fashion practices. *IEEE Transactions on Neural Networks and Learning Systems, 34*(6), 1125–1138.
12. Lin, Y., & Lee, J. (2024). Reducing electronic waste from AI technologies: A case study in fashion. *Journal of Environmental Management, 418*, 115546.
13. Mason, C., & Walker, R. (2023). AI and energy consumption: Environmental impacts and mitigation strategies. *Environmental Research Letters, 18*(7), 074005.
14. Miller, T., & Green, E. (2023). Minimizing the carbon footprint of AI technologies. *Journal of Environmental Science and Technology, 57*(15), 9500–9511.
15. Murphy, K., & Evans, T. (2023). Green AI: Approaches to sustainable machine learning. *ACM Transactions on Intelligent Systems and Technology, 14*(5), 1–20.
16. Nair, S., & Sharma, R. (2023). AI in fashion and its environmental footprint: A critical review. *International Journal of Fashion Technology, 20*(3), 245–258.
17. Nguyen, H., & Tran, K. (2024). AI and the fashion industry: Balancing innovation with sustainability. *Journal of Fashion Technology & Textile Engineering, 12*(1), 75–89.
18. O'Neill, J., & Fisher, M. (2024). Sustainable AI in the fashion sector: Challenges and innovations. *Fashion and Sustainability, 6*(1), 78–90.
19. Patel, A., & Nguyen, T. (2024). Energy-efficient algorithms for AI applications in fashion. *Journal of Artificial Intelligence Research, 83*, 203–219.
20. Reynolds, C., & Smith, J. (2023). Sustainable AI in fashion: Challenges and solutions. *Sustainable Computing: Informatics and Systems, 42*, 101134.
21. Ritter, M., & Zhao, Y. (2023). AI hardware and energy efficiency: A review. *IEEE Transactions on Sustainable Computing, 8*(3), 245–259.
22. Sutton, C., & Lee, H. (2024). Lifecycle assessment of AI in fashion. *Environmental Science & Policy, 54*, 198–210.
23. Wang, L., & Zhang, S. (2023). AI and environmental sustainability: Case studies and future directions. *Journal of Sustainable Computing, 12*(4), 120–134.
24. Wright, A., & Greenfield, J. (2023). AI and environmental sustainability: Addressing the challenges. *Journal of Sustainable Technology, 10*(2), 143–160.
25. Zhou, X., & Huang, Y. (2024). Optimizing AI algorithms for sustainability in the fashion industry. *Computers & Industrial Engineering, 183*, 108742.

Privacy, Security, Ethics in AI-Driven Energy System

Rahul Vadisetty and Anand Polamarasetti

1 Introduction to AI in Energy Systems

1.1 Overview of AI's Role in Modern Energy Systems

Artificial intelligence (AI) integration in energy systems has become an innovation driver changing the conventional paradigms of energy generation, distribution, and management. Consumption: machine learning and neural networks as well as even more sophisticated analytical methods enter the sphere of the energy industry and are used in various capacities. Such innovations are responding to important questions regarding how to generate power most efficiently, when demand is likely to surge or dip, and how grids can be made stronger. By controlling the vast amounts of real-time data, the energy providers can move along with the AI models and deliver more efficient, sustainable, and cheaper solutions.

Specifically, AI has had notable achievements in one of the climate change mitigation priority areas, namely striving for a decarbonized power sector. The use of intelligent technologies in renewable energy production and management leads to improved prediction of renewable energy generation, integration of fluctuating energy types like solar and wind energy, and effective control of energy storage. For instance, with machine-learning algorithms that learn historical trends in weather and the grid, it is possible to predict daily shifts in energy demand and supply much better and thereby better incorporate renewable energy and reduce reliance on normal backup based on fossil fuels.

R. Vadisetty (✉)
Electrical Engineering, Wayne State University, Detroit, MI, USA

A. Polamarasetti
Computer Science, Andhra University, Visakhapatnam, Andhra Pradesh, India

P. Raj et al. (eds.), *Artificial Intelligence (AI) for IT Energy Efficiency and Green AI for Environment Sustainability*,
https://doi.org/10.1007/978-3-031-89420-6_11

However, AI has helped to bring into existence *smart grids*—smart, integrated, and digitalized electric systems with inherent control capabilities and real-time reactive behaviors. Smart grids use AI for load balancing, fault diagnosis, and energy flow optimization which improves the quality of the grid and improves support for distributed energy resources. These AI-integrated grids are some of the modern natural infrastructure of energy necessary in accommodating decentralized generation technologies such as rooftop solar systems, wind energy farms, and batteries for electricity storage. As these systems evolve, so does AI's role in the energy sector with sustainability objectives hence becoming an essential factor.

1.2 Case Study: Google's Use of AI in Energy Efficiency

AI has been used in optimizing the cooling systems of Google's data center which made a significant difference where it was used. Google's data centers hosting millions of servers ensure a large power usage with cooling systems being among the highest powered. As part of its collaboration agreement with DeepMind, Google utilized an artificial intelligence model to best oversee the cooling of the structures. The developed AI model looked through the historical data and for different situations learned the optimal values of temperature, power consumption, and cooling pump speed.

The results were substantial. Utilizing elements of artificial intelligence caused the need for cooling to be decreased by 40% resulting in an optimal energy consumption increase of 15% in the data centers. The case shows that AI has a promising application in energy saving, especially in big industries. The use of AI helps in understanding actual data feed and adjusting operations for power consumption in the most effective way, which is highly advantageous to those organizations that transact high energy consumption. Another advantage presented in this example is the extendibility of AI solutions to other industries that have power-consuming activities (DeepMind, 2019).

1.3 Advances in Predictive Maintenance for Renewable Installations

Predictive maintenance which is a part of artificial intelligence predictive analysis has revolutionized the way renewable power infrastructure, especially wind power and solar infrastructures are maintained. Historically, maintenance in energy assets was inspection-based or repair after failure; this approach exposed the assets to unproductive time as resources were used without optimality. Predictive maintenance, on the other hand, involves employing artificial intelligence and machine learning algorithms to directly keep track of the state of a piece of equipment with the view of detecting early indicators of equipment failure.

In the renewable sector largely, this is important since the business involves investment with the majority of its assets located in hard-to-access areas such as wind turbines. For instance, data generated from multiple sensors placed on wind turbines such as vibration data, temperature, and rotational speed through high-frequency captures is analyzed by an AI model to establish indications of wear and potential failures. The possibility of predicting maintenance requirements in the future increases the chances of performing either preventive or scheduled maintenance thus extending the lifespans of the assets and decreasing the chances of system failures of energy providers. Sweeping across the cross-section of renewable energy producers, relevant studies suggest that about a 20% decrease in maintenance costs and drastic improvement in equipment reliability may be realized from predictive maintenance [1].

Predictive maintenance has also improved efficiency in that companies can schedule maintenance of properties at certain times while delaying others. This completely targeted measure not only optimizes the time and funds required to maintain renewable energy systems but also maximizes the efficiency of assets to eliminate energy loss.

1.4 AI and Renewable Energy Forecasting

The market price of energy cannot be accurately predicted, and since renewable energies such as solar and wind energy are intermittent, the generation must be forecasted to maintain a balance in supply and demand in energy systems. Machine learning algorithms in short and long-term energy forecast models like weather data, grid operational statistics, and current weather conditions to estimate the feasibility of energy production from renewables in the future. This capability is especially useful to the grid operators who have it for offsetting volatile intermittent resources in addition to ensuring supply adequacy to the consumer's demand.

An example of the use of AI in renewable forecasting has been presented in the joint work of NREL and IBM. Combining machine learning technologies, the NREL created an AI software that predicts the generation of electricity using solar energy by determining the prevailing climate conditions, clarity of the atmosphere, and level of solar brightness. Thus, this tool based on AI and predictive methods can reach forecast accuracies at the level of 90% which is important for effective functioning of the grid. By hence considering changes in the generation from renewable sources, energy flow can be managed or changes made over the timing of calls for backup power resulting in less reliance on fossil energy (Sun et al., 2022).

Beyond forecasting, AI can also facilitate the typical challenge of energy storage in balance with renewals' integration. Optimization of times to charge energy storage devices like batteries is characteristic of machine learning since the models predict energy supply and demand. In turning the storage to mimic generation profiles, AI improves the resilience of renewable electricity, which allows grid systems to stay orderly as the penetration of renewable energy rises.

2 Challenges in Scaling AI for Large-Scale Energy Systems

AI is viewed as a tool for revolutionizing energy systems; however, there are problems in expanding these technologies for more usage. The incorporation of AI in large-scale grid applications demands big computing power for analyzing huge quantities of data from smart meters, Internet of Things (IoT) sensors, and other constituents of the smart grid. The data processing of these sociocopies requires swift computing capabilities and infrastructure because of the associated challenges of data retrieval, transfer, and analysis in real time.

One is the issue of latency within data networks that are very large in the context of an energy network. Demand and supply-related activities such as balancing the electric grid also involve real-time data processing, and so do the AI applications. However, in large power networks, information from dispersed sensors has to be transmitted, filtered, and responded to within milliseconds to balance the grid, which is inherently complicated to solve optimally.

An equally hard problem is the integration of AI systems into different energy structures for efficient operation. Energy grids are a combination of technologies ranging from historical and traditional in terms of electrical structure, to distributed renewables and now include integrated digital systems. To ensure that AI solutions can run within such an advanced and the infrastructure in such a case is different hence requires correct synchronization and proper software instruments coupled with protocol that enables it. Organizations are trying to create compatible artificial intelligence paradigms that are compatible with other AI frameworks, but creating compatibility across systems is not the most straightforward process.

Cybersecurity risks are another large issue in wide-scale AI use. Reflected further in Sect. 4, the application of AI in energy systems means that these networks have more points that might be exploited by cyber attackers. Mitigating these risks depends on sound cybersecurity practices and actual cyber resilience. Time control for the security of the data and the system in general. Hence, as transformative as AI technologies are, a viable business case at scale remains elusive for most of these and other technologies across large energy systems because of technical, logistics, and security issues.

3 Privacy Considerations in AI-Driven Energy Systems

The integration of AI into energy systems has infused significant privacy concerns as it demands broad and frequent data collection to support forecasted analysis, monitoring, and even optimization of the systems. AI-based systems integrate with smart metering, IoT products, and other digital interfaces that gather personal and operational data about individuals and operations within homes while achieving high degrees of granularity. The following part focuses on the privacy of data in energy systems, including such aspects as data being gathered, privacy legislation, as well as technologies mediating between data utilization and privacy.

3.1 Data Sensitivity in AI-Driven Energy Systems

AI-integrated energy systems capture various data types which may be of confidential and personal nature, regarding the user's preferences and behavior. For instance, smart meters that are used in homes and businesses collect information on energy intensity as and when it is being used, thus providing utilities with efficient ways of tracking high-intensity usage patterns at much more detailed levels than has been possible before. These data points show not only the total electricity consumption but also the details of the energy consumption for these appliances that suggest the activity levels of the households such as whether they are home, awake, or using particular appliances.

Hartzog and Selinger have executed a study (2019) that also stressed that there is an opportunity to trace the fine usage profile depending on the occupant behavior that was obtained from frequent meter readings of smart meters. Thus, the analysis of energy use at intervals as short as 15 min allows researchers to infer information like when occupants usually get out of bed. Go outside the house, or indeed switch on particular domestic items such as washing machine, or oven. While useful for energy optimization on this granular level, this kind of monitoring poses extensive privacy risks since it forms a profile of the behavior that can be misused by any intruder (Hartzog & Selinger, 2019).

In addition to residential privacy, sensitive data encompasses the industrial area where IoT devices sense and relay real-time production parameters to improve manufacturing. For example, in industrial plants, IoT-connected sensors monitor the performance, power usage, and activity levels of the machines indispensable for determining when or if a machine will fail, or for establishing how effectively a plant is operating. However, this data when disclosed compromises operational secrets and therefore has consequences on competitiveness and operation security.

3.2 Regulatory Challenges and Global Privacy Frameworks

Data privacy regulation in energy systems has a plethora of rights that govern the privacy of consumer information as well as the regulation of data. One of the primary regulations is GDPR in the EU to regulate matters dealing with personal data that include PII and put down standards for data collection, processing, and storage. According to GDPR, it is unlawful to process personal information unless the data controllers have complied with the conditions of collecting data and must do so openly and obtain consent from the data subjects who have the right to request, and erase their data.

Challenges arise when the GDPR is applied in energy systems. Real-time energy analytics also depend on data collection and processing in real-time hence may go against GDPR policy on data minimization and consent. For example, demand response programs where energy consumption is optimized about the grid requirements, presuppose the constant flow of data from smart meters. Meeting GDPR in such cases involves the need for energy providers to ensure that the data gathered is done in such a manner that is compliant with the GDPR standards; this might need the use of techniques such as anonymization or even aggregation to reduce the exposure of personal data.

Another large branch of rules regulates the California Consumer Privacy Act (CCPA) based in the United States just like GDPR which grants consumers rights to obtain information on what information is being collected and to whom it is sold; this also allows them to request the deletion of their data. Unlike the GDPR which deals with all sorts of personal information, CCPA looks at such data as used in consumer relations and grants the consumer a right to deny the use of his or her data for selling or disclosing. In energy systems, this can be described as entitling consumers should be able to control how their energy consumption data is used and shared, a provision, that increases the operating cost for the providers especially when dealing with a large stream of data in different jurisdictions [2].

3.3 Privacy-Preserving Technologies in AI-Driven Energy Systems

Due to privacy risks involved in collecting high-frequency data, some technologies have been invented to enable energy providers to use data while at the same time protecting the privacy of consumers. Differential privacy is aimed at providing privacy-preserving data protection in AI-driven energy systems, and similarly, federated learning and blockchain are other critical technologies.

3.3.1 Differential Privacy

Differential privacy is a procedure designed to extract information from the data without compromising the individual pieces of information by basing the results on modified data sets with statistical noise. This approach enables power suppliers to make a pattern scan on numerous data sets to independently avoid unique user identification. For instance, in a differential privacy application, information obtained from smart meters such as energy consumption of different homes could be pooled together, and since specific consumption rates cannot be generalized, one cannot point at any particular home.

Differential privacy has been adopted for applications where big data analysis is necessary such as in Google and Apple data systems. The idea of using differential privacy can help energy companies address the trade-off between providing usable analytics and protecting user information because the organizations can use the approach to perform demand forecasting and usage pattern analysis on client data without exposing the data. This technique is very useful in the models based on AI concerning grid management because the conclusions are made based on the generic parameters and not points of interest [3].

3.3.2 Federated Learning

Federated learning is a novel way of addressing problems related to the training of artificial intelligent models on decentralized data without requiring the data to be concentrated. In energy systems, federated learning allows smart meters and IoT devices as intelligent participants in AI model training wherein the statistics or even the trained model coefficients are only periodically transmitted to a central node rather than the raw data. This construction maintains privacy because it minimizes the quantity of data out there within the community that might be intercepted or abused through the transfer.

As shall be seen later when discussing practical applications, federated learning is most suitable in distributed energy systems with data collected from numerous connected devices at different locations. For instance, in the smart grid, different devices like local nodes of the grid, storage systems, and smart meters used by consumers can perform federated learning without transmitting personal information to one another. The decentralized model for training AI eliminates bandwidth usage and improves the systems' security compared to a centralized system while at the same time preserving user anonymity in a data-intensive environment [4].

3.3.3 Blockchain for Privacy and Security in Decentralized Energy Markets

Through use cases like decentralizing energy systems or policies allowing efficient tracking of electric energy usage, blockchain technology could offer a solution to privacy-preserving energy markets. To ensure that participants' anonymity remains protected, blockchain can be employed in consensus mechanisms in decentralized energy trading, including peer-to-peer (P2P) trading schedules. Every transaction is written as a block in a string of blocks that are safeguarded with a cryptographic form of open-key cryptography so that unauthorized alteration cannot be made, while at the same time, no one's identity can be exposed.

In the context of P2P energy trading, blockchain empowers consumers to purchase energy from each other in case of excess generation. For example, people living in a community with solar power projects can trade the surplus solar energy using a blockchain database and every transaction would be well-recorded and also verifiable. The decentralization of blockchain makes it possible to authenticate transaction information without exposing people's identity or violating their privacy. There are some projects today that explore P2P energy markets, like Brooklyn Microgrid in New York, to show blockchain support to privacy-preserving, decentralized energy systems [5].

3.4 Ethical Implications of Data Privacy in AI-Driven Energy Systems

Beyond the technological issues in the application of artificial intelligence for energy systems and outcomes, there are conflicts relating to data protection, ownership, and control. While in conventional energy systems, analysis of consumer data was restricted to overall consumption figures depicted at the monthly level, the contemporary solutions apply AI that collects denser information concerning the consumer. Such a change in data depth and scope requires a rethink of the ethics that guard consumer interests.

3.4.1 Data Ownership and Consumer Autonomy

One of the challenges that come with the use of artificial intelligence in energy systems is where the ownership of data from smart meters and IoT lies. Although the systems that gather this information are operated by utilities or energy providers, the information is most closely associated with consumer activity. The ethical functioning of datasets within energy systems requires consumers to accept ownership of their data, granting rights over their data usage, transfer, or sale. In terms of practical implementation of this principle, transparent and understandable policies on data management have to be created, so that consumers can understand their rights as well.

3.4.2 Informed Consent and Transparency

In ethical data collection, it is critically important to have informed consent, especially in systems where the data disclose the behavior of the individuals. Regarding energy consumers, informed consent in the AI-based systems landscape means being clear about the details of what is, for instance, to be collected, for what purpose, and what potential value and cost bear by consumers are inherent. Quite understandably, energy providers have to undertake actions that are transparent to substantiate trust; the actions include providing consumers with data usage policies, data summaries, and where it is feasible, an option to withdraw the consent given.

Also, clear data practices lead to data responsibility and it makes the energy suppliers act ethically and lawfully. For instance, due to jurisdictional privacy legislation restraints, some utilities availed digital platforms where customers may check balancers, consents, and policies in an ethical and customer-oriented approach to data control.

3.4.3 Minimization and Proportionality in Data Collection

The principle of cellular telephones should not be used to record any other information apart from what is relevant for specific purposes This is also in line with the principle of data minimization and proportionality. Finally, it might mean lowering the rate of data sampling and storing or excluding data that is not critical to grid stability and energy optimization while using highly accurate consumption profiling of individual households. It allows the necessary data-focused operation of energy providers while guaranteeing the consumer's privacy.

It also comes to the question of how long data is proportionally processed is stored. For instance, historical usage data used in demand forecasts may one day be used to infringe on the consumer's rights, and as such the consumer would only want the data to be used during the particular period that is beneficial for the utility company and thereafter the data may be anonymized or deleted to enhance the levels of consumer privacy. Ethical practices towards AI in energy, therefore, stress practicality and privacy, that is, how data is gathered, and to what extent the application interferes with energy consumers' rights to privacy.

4 Security in AI-Enabled Energy Infrastructure

AI integration into the energy infrastructure provides an added advantage in operation, grid stability, and resource management. But it also opens new cybersecurity threats at the same time. AI-enabled systems are commonly part of the IoT structures, which makes them vulnerable to cyber threats, threats that can disrupt system functionality as well as lead to major blackouts and even jeopardize national security. With the existing transformation in energy conversion to smart grids, distributed energies, and IoT devices, more attack vectors are created and, thus, cybersecurity becomes an area of contention between the energy provider and the regulatory authorities. This section examines the emerging cybersecurity threats in AI-based energy systems, the symmetrical role of AI as a contributor to security and threats, and potential scenarios of cyberattacks.

4.1 Comprehensive Analysis of Cybersecurity Threats in AI-Driven Energy Systems

Cybersecurity risks to AI-driven power systems are diverse, and they consist of a risk of disruption, data contamination, and risks to consumer safety. The biggest risks of AI in energy infrastructure are as follows: denial-of-service (DoS) attacks, ransomware, malware, and adversarial attacks on AI models.

4.1.1 Denial-of-Service (DoS) Attacks

DoS is an attack that intends to overload the available network components, thereby rendering critical services and information unreachable. Regarding energy systems, DoS attacks can interrupt the supply of real-time data which are required to regulate the grid along with the operations possibly leading to blackouts. For instance, a DoS attack on a smart grid's data network might hinder grid operators from getting data on energy flow at the right time and mean that different sections of the grid are loaded unequally and the grid gets disrupted.

Extensions that rely on being part of different systems flooding the targeted source with traffic especially distributed denial of service (DDoS) attacks are especially detrimental to interconnected energy systems. An effective attack on one of these utility networks could paralyze its monitoring and management capabilities and place the grid at risk of added attack or malfunction. For instance, the recent Dyn cyberattack which involved a botnet in a DDoS attack shut down multiple online services and showed how vulnerable the critical infrastructure was to such attacks [6].

4.1.2 Ransomware and Malware

Ransomware and malware represent the biggest threat in AI-driven energy systems because they may alter data or seize command of key infrastructure or equipment. In ransomware attacks, a criminal gains access to a target's data and locks it down by using encryption and demanding payment to unlock the system. For energy providers, this could mean potentially a cut of the operational data that is used to handle the stability of the grid or the fluctuating demands.

A recent well-known incident is the May 2021 attack on Colonial Pipeline when the attackers applied ransomware to block the pipeline delivering fuel to the US East Coast. The event revealed how energy systems are open to ransomware and the impacts that such attacks bring to society. In AI-powered utilities, ransomware could attack databases containing machine learning algorithms corrupt demand projections, or paralyze IoT systems relevant to the grid leaving communities in higher danger of losing services (Greenberg, 2021).

4.1.3 Adversarial Attacks on AI Models in Energy Systems

Adversarial attacks are based on a concept where an attacker uses a small perturbation to data that is entered into an algorithm to modify the behavior of the algorithm. In energy systems, adversarial attacks can present AI in demand forecasting, equipment diagnostics, or grid management with the wrong information which, in turn, may lead to improper decisions that may cause the grid to become unstable or lead to equipment failure.

For instance, an attacker could inject small deviations in sensor data which are used for the machine learning model of predictive maintenance to overlook clear signs of equipment failing. On the other hand, adversarial inputs may be fed to the load forecasting models and yield inappropriate forecasts that lead to energy deficiency or excessive supply. Since defining the map of an AI model and recognizing adversarial inputs are sufficiently challenging, the adversarial attack is considered to be a novel cyber threat facing AI-driven energy systems [7].

4.2 AI-Driven Security Solutions for Threat Detection and Prevention

On one hand, AI creates new risks to energy systems; on the other, it acts as a strategic weapon in cybersecurity. Real-time monitoring of cyber threats with the help of AI-driven solutions can greatly improve protection and provide increased reliability of energy assets and systems.

4.2.1 Anomaly Detection for Real-Time Threat Monitoring

Of machine learning's many algorithms, anomaly detection is critical in determining anomalies; special trends in network and system activity, and database activity. In energy systems, the anomaly detection algorithms run data streams incoming from IoT devices, the grid, and control systems and identify potentially malicious events.

For instance, if a smart grid deploys such things as unusual data flow or intruder login, then this system can immediately detect this and inform security officers. In critical energy infrastructure, anomaly detection can decrease the period between threat identification, and defense thereby limiting the harm that may be caused by attackers. The type of machine learning models that can be incorporated into anomaly detection include but are not limited to: clustering, autoencoders, and even using Bayesian networks, which may detect a pattern that a normal rule-based system may not [8].

4.2.2 Incident Response Automation and Adaptive Defense Mechanisms

Cyber security is improved through the automation of events by the use of artificial intelligence systems. In AI-amplified energy structures, smart self-acting might be capable of containing affected areas in the network, modifying the firewall policy, or triggering a shut-down process, which minimizes the need for human intercession and accelerates reactions. For instance, if an AI-based anomaly detection system detects adversarial action in a smart grid, the autoresponse may disconnect the devices that are infected to contain the malware.

There are several adaptive defense mechanisms among which are self-healing networks because they enable energy systems to automatically change their direction in case of threat detection. Autonomous self-healing networks continuously point to an alternative route to the one that has been infiltrated through an attempt at launching a cyberattack. These adaptive defenses are especially useful in distributed energy systems where a cyberattack on one node can bring down the entire network. AI integration can help energy providers enhance the scenario of automated response and organizational structure strength.

4.2.3 Predictive Security Analytics for Proactive Threat Management

The predictive security analytics application helps energy providers detect and prevent cyber threats as they are in the process of being developed. AI models classify such data on past cyber events and look for patterns that could be a precursor to the likely attacks. The application of predictive analytics enables the security team to deploy available resources after analyzing the vulnerability levels of specific areas.

In a large-scale energy system, some of the substations or transmission nodes, the predictive analytics might be pointed out as having more occurrences of cyber attacks. With such knowledge, energy providers can put other security measures like frequent monitoring or encryption at these vulnerable places. For instance, comparing the prediction analytics of the prior DDoS attack's most likely traffic outcomes sheds light on patterns that typically precede such attacks, thereby enabling AI-driven defense to derail malicious traffic a priori [9].

4.3 Case Study: Ukraine Power Grid Attack and Lessons Learned

The Ukraine power grid cyber attack of December 2015 purported to be the work of the Sandworm hackers was one of the first successful cyber attacks on a power grid impacting almost 230,000 people and unearthing soft underbellies of energy security. The most recent of this attack was an attack where hackers were able to control remotely the SCADA systems which are the parts and components of the grids. The event all the while exposed the West's outdated cyber security measures in traditional energy systems while underlining the importance of strong automated defenses prominent through AI learning.

To mitigate the impact of the Ukraine attack, energy providers around the globe have stepped up their access control, segmenting their network, and installing a monitoring system in real-time that can identify anomalous patterns. The Ukraine incident also forced utilities to invest in artificial intelligence (AI) based cybersecurity as it was found that rule-based systems alone are incapable of shielding utilities from integrated myriad layers of attack.

The systems and tools of artificial intelligence, for example, anomaly detection, as well as automatic incident handling, are now key elements in modern security of energy supply, which allows for preventing similar attacks like in Ukraine. These programs allow for threats to be identified more or less in real-time and mitigate the threats independently of human intervention, which makes a grid more stronger/less susceptible to large-scale emergencies. As the energy sector gets into a new digital era that involves complex infrastructure networks, these examples of attacks on Ukraine show that there must be continued efforts invested in the use of artificial intelligence to improve cybersecurity [10].

4.4 Ethical Considerations in Cybersecurity for AI-Driven Energy Systems

Here, we define some important ethical issues that are essential when it comes to incorporating AI-based cybersecurity systems: since AI-based security solutions demand vast data to watch and study threats, they may gather private data on network actions, which may infringe privacy liberties. As to ethical considerations in cybersecurity, it is necessary to comment on energy systems about data use, distribution, storage, and protection with due regard to individual freedoms' rights so as not to fall prey to increased levels of cybersecurity risk at the potential detriment of personal privacy.

Further, autonomous AI systems as a tool for cybersecurity are problematic in that accountability. If an AI system independently quarantines devices or segments of the network due to threats, it is highly important for it to not interfere with critical functionalities, as well as to not violate user's rights. Policies for addressing cyber security threats and risks therefore entail an evaluation of defense and consumer rights and freedom as well as system self-governance.

5 Ethical Implications of AI in Energy

As the energy sector is under the influence of AI technologies, several ethical questions have arisen. These are often bias, openness, greenhouse effect, and privateness concerns with intelligent energy systems. AI applications in energy require ethical assurance to regain and sustain public trust, operate legally, and in the long run, be sustainable. This section discusses the ethical imperative of AI in energy and presents the technical and social issues pertinent to its use, as well as the things that need to be done to get it right.

5.1 Algorithmic Bias and Fairness in AI-Driven Energy Systems

Algorithmic bias is one of the essential ethical concerns of AI for energy since it can result in inequitable distribution of energy or production of unequal fares, and other improper exclusion of some society groups. AI models use data that are learned in the previous years or months to make future predictions and decisions. If the training data set is at all skewed or limited in the range of situations depicted, then the created algorithms may perpetuate current prejudices or invent new ones, and the resulting models may amplify existing prejudices or create new ones.

There is the possibility, for instance, that these AI-driven systems, in regions where access to energy resources is restricted, or where the existing electric grid may not be well-developed, may direct the supply and consumption power to those areas that give more economic returns, or densely populated areas while neglecting the poorer or rural ones. For example, the predictive analysis used to ration resources according to past consumption will end up rationing more resources to areas that have used many resources before, which are mostly urban areas that are already privileged while leaving out so many other areas.

An example of this phenomenon is reflected in automated pricing software applied in the framework of dynamic electricity pricing models. These models set electricity tariffs with pricing trends in line with supply and demand to manage demand by customers for power at certain times. However, if the data used to train these models is skewed to high-income earners and low-income areas, these areas may be charged higher prices since they have less control over their usage patterns. This leads to the divergence in terms of the chance of the two gaining access and is affordable but can cause additional financial complications for groups of people who already have poor health (Johnson et al., 2021).

5.1.1 Addressing Bias through Fairness Audits and Data Diversity

Two, although AI relates to energy systems, it is paramount that techniques employed be bias free; this can be done by deliberately seeking a diverse dataset. The first strategy is to do fairness checks, which are checks if the specific AI has certain forms of bias and adjusting the algorithm if necessary to eliminate unfair results. Fairness audits can be highly helpful when setting demographic or geographical biases; therefore, energy providers can develop corrective measures to pricing models or demand forecasting for fairness.

Another important measure on the way to combat biases is to focus on making data diverse. This way, the data from a diversity of demographics, geographical areas, and different socioeconomic statuses will help energy companies create models that are much closer to representing their customers. For example, energy providers might use rural and low-income consumers' data in the training data sets to create models that extend equitable access to energy services to everyone.

5.2 *Transparency and Accountability in AI Decision-Making*

One of the critical issues and problems related to AI applications in energy systems is the fact that many algorithms are "the black boxes". Multiple AI solutions require considerable intricacy in terms of their architecture, and many popular approaches belonging to the deep learning family are inherently difficult to explain, which complicates the interpretation of an operator or regulator concerning certain decision-making. It deprives decision-making of the desirable transparency, raising questions about accountability when it comes to AI's impact on the availability, cost, or reliability of energy.

5.2.1 The Importance of Explainable AI (XAI) in Energy Systems

Interpretable and explainability of the AI decision-making process is the main goal of explainable AI (XAI). In energy systems, XAI can solutions address questions like why certain decisions are made, including changing electrification rates or identifying which infrastructures require maintenance. Local Interpretable Model-Agnostic Explanations (LIME) and Shapley Additive exPlanations are two child techniques of XAI to understand black-box AI systems. For example, LIME helps to gain an understanding of why particular decisions were made by locally modeling the area around one decision.

The examples show that XAI can be beneficial in demand forecasting and energy distribution. For instance, energy organizations that employ XAI metrics have been able to offer the authorities a more comprehensive understanding of how AI models disburse power or set tariffs in a regulated manner. In this regard, XAI for energy applications can promote trust among the stakeholders and enable fair deployment of AI solutions.

5.2.2 Accountability and Ethical Decision-Making in Automated Energy Systems

When AI systems make autonomous decisions in energy distribution or infrastructure maintenance, accountability becomes a critical issue. If an AI system misallocates resources or causes an outage, it can be challenging to determine where responsibility lies. Ethical decision-making frameworks in AI-driven energy systems emphasize human oversight, ensuring that operators can intervene in cases where AI recommendations or actions might have unintended consequences.

An approach known as "human-in-the-loop" (HITL) ensures that key decisions made by AI systems, such as shutting down grid sections or adjusting distribution levels, involve human oversight. HITL frameworks enable human operators to validate or override AI decisions, balancing the efficiency of automation with accountability. Additionally, by logging AI decision-making processes and maintaining transparent records, energy providers can facilitate postevent analysis and take corrective measures, which is essential for ethical accountability.

Whenever decisions are made independently by the AI system, let us say in energy supply or infrastructure management, then this question arises. In case, an AI system makes a wrong decision in the allocation of resources or the case of a system failure, it is a realistic possibility not to be able to put a finger on who is to blame. When it comes to ethical decision-making, energy systems powered by AI recommend a human-involved intervention by human personnel when the AI comes up with certain recommendations or even performs certain actions that might have adverse effects.

An approach called "human in the loop" (HITL) is that critical decisions made in AI such as the loss of the grid sections or putting in distribution levels require human-intercepted decisions. The HITL structures allow human operators to confirm or nullify an AI decision alongside the benefits of automation. In the same way, AI decision-making records and well-documented databases ensure ethical accountability by being able to perform an analysis of event causes and take necessary corrective action after such an event has occurred.

5.3 AI in Energy Systems: Its Impact and Sustainability

Despite the potential benefits that AI has for supporting integrated renewable energy, a proimplementation has an environmental problem because training large models takes a lot of computations. The power needed to train big AI models for various industries, for example, predictive maintenance or grid optimization, is significant.

Several researchers have found out that training one large AI model may release an amount of carbon equivalent to what some cars emit in their entire lifetime [11].

5.3.1 Balancing AI'S Energy Demands with Sustainability Goals

Risks of AI on energy systems also refer to the negative effects environmental-wise, despite being about efficient energy systems. To this end, there is an active integration of AI in energy systems with green computing that involves the utilization of renewable energy data centers and the energy efficiency of models.

For example, Google and Amazon pledged to supply their machine-learning research with green energy. These companies are in the process of cutting down the carbon impacts linked to the training of AI models, as well as the utilization of data processing infrastructures through the utilization of renewable energy supplies, such as solar and wind power, for their data centers. Also, studies are being conducted in the field of energy-efficient algorithms whereby the amount of computational load needed to run the model is reduced while the effectiveness of the model remains optimal. It can be applied under conditions such as sparse matrices computations, and low-rank approximations, as well as ML-Stat hybrids—combinations of machine learning and classical statistics.

5.3.2 AI'S Role in Environmental Monitoring and Carbon Emission Reduction

In addition to energy, AI is used to monitor emissions, the level of pollutants, and the general state of the environment. Such technique pertinent to ML entails analyzing satellite images, sensor data, and other environmental data concerning pollution origins, deforestation, and forest capacity to stock carbon. In energy systems, AI can also better manage resources thus decreasing the amount of carbon produced from burning fossil to generate energy.

For example, the use of carbon tracking systems based on artificial intelligence enables the management of energy companies to observe emissions and determine ineffective activities and the need for restructuring. With the help of AI, models applied to carbon reduction planning, utilities, and producers of energy deliver necessary activities to meet the desired goals outlined in the Paris Agreement and further their business' sustainability missions.

5.4 Data Ethics and Consumer Privacy Rights

The large amount of data needed for the development of artificial intelligence-based energy systems brings ethical challenges surrounding data ownership, informed consent, and consumer privacy rights. Earlier, many energy providers received monthly summative data through invoicing while now they obtain real-time data that can be as precise as every minute through smart meters, IoT devices, and sensors. While this level of data detail can be useful for fine-tuning energy supply, this consequence is a significant threat to personal privacy and the ability to manage personal data. Individual data.

5.4.1 Data Ownership and Control

With the shift of data to the core of the AI, managed energy systems, questions related to data proprietary, as well as consumer sovereignty arise. Digital consumers are beginning to demand full rights of ownership of their data and full control rights over the use and dissemination of the same. In structured ethical data management systems, consumers act as the primary data owners, while energy providers act as data stewards who control the information responsibly and correctly.

Consumer data ownership in practice requires seven policies which include consumer access, consumer review, and consumer delete. There are energy providers who offer an environment through which consumers can access their data, and specify the type of data they want to use, share, or store. These platforms also give consumers transparency, informing consumers on where their data is being used and how in the provision of optimally effective energy services.

5.4.2 Informed Consent and Transparency

A critical aspect of ethical data gathering is informed consent and thus has all information passed through the consumer. When seeking approval from the consumer, the energy providers must explain the type of data gathered, how it will be used, inherent risks, and the proposed utility of the data. Obtaining informed consent is often difficult in AI-driven energy systems due to consumers' inability to grasp AI and data analytical understanding.

Some energy companies solve it by providing well-designed consent form that presents the key information concerning the collection and use of data. CMP enables the consumer to manage his or her choices and track the consent through time, thus allowing the consumer to always have control of his or her data. Such practices are critical when dealing with countries with compelling data protection laws such as the GDPR and the CCPA, as a company cannot collect and process data without the user's consent.

5.4.3 Data Minimization and Ethical Data Retention

The general principles for the ethical use of predictive analytics in energy include data or minimization, which means obtaining only the data pertinent to a given aim and processing it for the shortest possible time. This can be applied in the sense of reducing the rate and extent to which data from smart meters and IoT applications is gathered to only that which is needed for purposes of grid control or prediction of energy demands.

Structural data retention policies are also safe and effective when proportional and returned to social equality and integrity, ethical AI principles for decreased long-term privacy threats. For example, after it has done prediction maintenance work on an asset, information that may be pertinent to that specific activity may be blurred or erased, thus reducing vulnerability to threats of data piracy. The energy providers in implementing data minimization and retention also ensure that they uphold ethical standards when handling consumers' data.

6 Future Trends in AI for Sustainable Energy

Expert opinions show that the future holds greater developments of AI technology for the energy sector because it produces smarter systems and makes the energy systems more efficient and sustainable. Areas including reinforcement learning, digital twins, quantum computing, and decentralized energy markets are next-generation breakthroughs that are transforming the generation, distribution, and consumption of energy on the go. Affordable renewable sources and AI technologies advance perspectives on an orchestration of the grid and decentralized energy models along with the consumers' needs. This section reflects on these new developments and pretending the changes that other AI can make possible in delivering sustainable future energy.

6.1 Emergence of Reinforcement Learning in Adaptive Energy Systems

One of the most popular categories of machine learning is reinforcement learning where the work of AI agents through interacting with the environment is enhanced by rewards which is gradually becoming a valuable tool for flexible energy management. While traditional AI models make decisions based on fixed data sets, RL systems make decisions based on feedback which can be feedback in real-time. This capability makes RL ideal for energy systems in the areas of smart grids, battery storage control, and load flow control.

6.1.1 Reinforcement Learning for Smart Grid Optimization

In smart grid scenarios, reinforcement learning can help grids learn and optimize in response to real-time events familiar to smart grid systems, such as changes in demand, time variations of intermittent renewable resources, and factors such as load demand. This makes it easy to train RL agents to handle load balancing, optimal and fair resource assignment as well as the management of peak demand periods that would improve the stability of the grid. For instance, an RL model might train the system to move non-urgent energy loads to off-peak times, so demand is more homogeneous over time as well as on the structures.

In a microgrid context, therefore, the use of RL for energy storage and distribution was illustrated by a case study. Using solar generation forecasts, battery charge, and consumption as features for decision-making, the RL agent made grid management 15% more efficient than traditional management, and reduced energy wastage by 25%. Such success points to the fact that RL can work effectively in managing complicated multiple connected systems such as the smart grid where the response in a real-time environment is imperative [12].

6.1.2 Battery Storage Optimization Using Reinforcement Learning

Energy storage is more important in the integration of renewable sources of power, seeing that these sources are intermittent. One of the applications of reinforcement learning is the battery storage system used to predict the best time to charge and discharge, depending on the available and predicted RE generation, grid demand, and market prices. In this section, it was demonstrated that by dynamically managing the amount of storage capacity that is available RL models can improve energy efficiency, battery longevity, and confident storage of energy at its most critical time.

For instance, an RL-based system used in optimizing a grid-scale battery in a solar farm, for example, delivered improved efficiency. The RL agent would store excess solar light during sunny hours and use it during peak times, making cost savings for the building, as well as eliminating reliance on fossil fuel backup power. Besides solving the grid-related problems for RL, battery storage enhances the efficiency of renewable energy utilization, thus meeting sustainability goals.

6.2 *AI-Enabled Digital Twins for Predictive Maintenance and System Optimization*

Digital twins—physical objects or systems represented in virtual environments—have emerged as critical enablers of modern energy management with a specific emphasis on effective monitoring of system performance and failure-rich environments. Digital twins are real-time, IoT sensor-fed and AI algorithm-based accurate replicas of an asset that can accurately predict its behavior under different conditions and expose possible failures as well as optimize its performance. These models are especially helpful for energy infrastructure because turbines, transformers, controls, and batteries are regularly monitored for their performance, state, and condition. This ensures optimum efficiency and minimizes costly inefficiencies.

6.3 Downtime

6.3.1 Applications of Digital Twins in Wind and Solar Installations

In wind farms, digital twins replicate the behaviors of the turbines based on different factors including the speed and direction of wind, temperature, and humidity. On the component level, AI algorithms can analyze sensor readings to identify issues, such as high wearing out of the turbine blades, motors, and so on, and schedule maintenance accordingly. This minimizes sudden breakdowns, improves reconditioning schedules, and enhances the wear out of the equipment.

For example, on the North Sea wind farm, the digital twin was applied to assess the conditions of turbines and estimate the time for their maintenance. Using the wear patterns, the digital twin enabled operators to change components before they failed, cutting maintenance expenses by up to 20% and boosting energy yield by curtailing downtime. Likewise, in solar farms, digital twins assess panel performance, find flaws, and assign appropriate cleaning times for the panels to perform up to optimal levels while enjoying a long life cycle of the solar assets [13].

6.3.2 System Optimization in Grid Management

Digital twins embrace more than just the assets but also the systems such as the grid systems within the energy systems. The utility operatives can reap from digital twins for scenarios in grid load characteristics, failed response analysis, and grid changes. This approach assists operators in controlling resource spending, stability of the grid system, and making informed decisions about investment in the relevant infrastructure.

Similar to the mentioned functions for grid management, digital twins provide the ability of operators to run scenarios of demand changes and equipment failures among others to predict such instances and plan appropriately. Since digital twins allow one to predict how the grid will behave in this or that situation, energy delivery remains stable and failsafe. This is especially useful when energy systems are constantly transitioning as the addition of renewables to the grid adds higher levels of complexity.

6.4 Quantum Computing's Potential in Energy System Optimization

Although it is relatively young, quantum computing already has great potential in the energy sector, especially in solving large optimization problems that cannot be solved by classical methodologies. Quantum computer utilizes quantum bits, and qubits, to compute data at a speedway much faster compared to classical computers;

these are helpful in energy problems that require high-dimensional variables and mutual dependency.

6.4.1 Quantum Algorithms for Energy Storage Optimization

Quantum computing is becoming one of the most interesting fields that apply engineering potential applications, in particular, one of them is the optimization of energy storage grids. ESS faces many factors including battery power capability of different types, their charging/discharging cycles, and the amount of power needed by the grid and generated by renewables. The given variables can be processed by a quantum algorithm at once while looking for an optimal solution compared to quantum algorithms. This capability could dramatically enhance the performance of large-scale energy storage systems for enhanced renewable accommodation.

Scientists are now beginning to look at quantum algorithms to model the chemical processes happening in batteries and investigate how to make them better and last longer. By knowing how different materials respond at the quantum level, quantum simulations could help to advance the innovation of advanced batteries which are crucial in energy storage and incorporating renewables into the grid [14].

6.4.2 Grid Optimization and Demand Forecasting

Real-time analysis of grid networks is also expected to be produced by quantum computing in the future as a way of enhancing grid optimization. Existing approaches for grid optimization at present are based on mathematical models of supply–demand and storage on distributed plural grids which are computationally complex. This system can be quickly solved by quantum algorithms and design solutions that reduce energy dissipation and increase robustness to substantial levels.

For instance, two players in the quantum sphere IBM and D-wave are commercially working on the algorithm necessary to facilitate multiple grid management tasks; these include supply and demand balancing, scheduling of maintenance, and the penetration of renewable energy. Initial testing of these quantum algorithms has demonstrated their potential to solve optimization problems that were unfeasible up to this point and therefore create stronger and less prone to failure energy infrastructures.

6.5 Decentralized Energy Systems and Peer-to-Peer Energy Trading

The rise of *decentralized energy systems* and *peer-to-peer (P2P) energy trading* represents a total revolution in the generation, distribution, and usage of energy. New tapping systems also let the consumer produce and store electricity using systems like solar panels and batteries and sell them to other users. This P2P energy model also relies on AI and blockchain and forms a decentralized and continuously changeable market with the consumers' participation.

6.5.1 AI in Peer-to-Peer Energy Trading Networks

With the help of AI, decentralized energy networks can be created because supply and demand can be matched in real-time. Artificial intelligence is used to forecast when a given consumer will have excess energy to export and when he will require energy importation allowing for proper transaction and stability of the network. AI is also involved in setting dynamic prices, for there to be an equally balanced offer that will suit both the producers and the consumers of P2P networks.

A good example is the Brooklyn Microgrid in New York where residents with solar panels sell electricity to their neighbors using a blockchain-based P2P app. The use of AI makes it possible to determine each participant's energy needs and generation capability as well as enable efficient transactions that will be satisfactory to both parties. They decrease reliance on centralized or grid power, advance renewable abilities, and support the independence of the microgrids (microutilities) [15].

6.5.2 Blockchain and Security in Decentralized Energy Systems

Blockchain ensures all the participants in decentralized energy trading keep a record of the trade without disclosing their identity and allows everyone to see each leveraged trade. Every transaction is kept on a transparent and unalterable database, which helps to strengthen the credibility of the P2P energy market. Another benefit of blockchains is that they do not require intermediaries since they are also decentralized, similar to decentralized energy systems.

Blockchain improves security in decentralized systems by preserving the data's originality and denying outsiders access. For instance, in a peer-to-peer trading system, blockchain can confirm the data on energy generation by every node, which makes trading real-time. Because of this, decentralized blockchains are considered to be functional additions to such systems, facilitating P2P commerce and improving energy consumption estimates for the model.

6.6 AI-Driven Resilience in Response to Climate Change and Extreme Weather Events

Climate change further worsens the issue by increasing the rate of occurrence of extreme weather situations that disrupt energy systems. This is where AI plays the most important role by correcting grid resilience, with a faster rate of restoration and accommodation of change. AI assists energy providers in developing physiological procedures for extreme climate conditions vulnerability and managing disruption situations.

6.6.1 Predictive Analytics for Disaster Preparedness

Machines and predictive modeling use weather data and historical outages coupled with the condition of infrastructure to determine weak points in energy systems. It assists energy providers in envisioning certain disruptions, and thus its preventive action can include positioning repair crews, reinforcing infrastructure—or offloading loads to lessen the consequences of outages.

6.6.2 Real-Time Response and Recovery

The AI-based system monitoring systems gather real-time data on grid conditions to respond appropriately during storms and hurricanes. The best grid segment that can be used to recover quickly can be determined using machine learning models reducing efforts used in identifying which grid segments are most affected. For instance, in the regions commonly exposed to hurricanes, AI-enabled systems of disaster response have enhanced response in terms of supply and organization to minimize periods of outage and enhance swift restoration of electricity supply to the affected regions.

7 Economic Impact of AI in Energy Sustainability

The use of AI in energy systems has numerous advantages which consist of saving money on operations, creating new jobs, and making energy more available for consumers. Generally, artificial intelligence helps to achieve a more cost-efficient and more secure energy supply through the intelligent adjustment of the production, distribution, and use of energy. This section dissects AI and its versatile effects on energy by looking at how features such as predictive maintenance, demand forecasting, and reshaping the labor market have a way of saving money while at the same time achieving efficiency in the long run. AI in energy allows serving global economic objectives on the one hand, and ensuring industry revenues and consumer expenditures on the other, increasing the share of a low-carbon economy.

7.1 *Benefit Analysis of AI in Predictive Maintenance and Demand Forecasting*

A direct economic effect of the integration of AI in energy systems is the optimization of cost in operation due to predictability of fault incidences, and prediction of demand rate. Predictive maintenance uses real-time data collected through IoT sensors together with past maintenance records and algorithms to predict when the equipment is likely to fail. By forecasting when and where maintenance is necessary, energy providers ensure that the equipment breakdowns are not unforeseen, the lifespan of the equipment utilized is lengthened, and repairs are cheaper to make.

7.1.1 Operational Savings Through Predictive Maintenance.

Analyzing such facilities as wind farms and solar power plants, it is possible to note that cost savings can be critical using predictive maintenance solutions. Equipment such as wind turbines are exploited in very unforgiving environments and hence will from time to time develop some level of wear and tear. Through the use of AI algorithms, the data collected on the turbine from its associated sensors about parameters such as vibration, temperature, and speed is analyzed to determine signs of mechanical problems. Energy providers must resolve these problems early, so as not to cause substantial repair costs and inconvenient outages, much to their benefit in the long run.

The assessments and test-based research carried out by the International Renewable Energy Agency (IRENA) identified that the cost-effectiveness of predictive maintenance in wind energy installations could cut down annual maintenance expenditure by 30% in the renewable energy sector which could run into billions of dollars. For example, the application of predictive maintenance for a 100-turbine wind farm returns around $2 million per annum by avoiding failure and scheduling the optimum time for maintenance [16].

7.1.2 Need Demand Forecasting and Load Management

Demand forecasting is important in achieving the balance between supply and demand of energy particularly in high renewable integration where supply is stochastic. The consumption historical trends data and real-time information can be used to estimate energy demand at some points in the day or within some geographical locations. A key use of this predictive capability is for energy providers to be able to know in advance where resources are required to minimize the use of costly backup generation and to improve the level of grid stability.

For instance, demand forecasting, an AI application for California's energy grid management, has achieved the reduced usage of peaker plants that are expensive and rely on fossil fuel. The development of the AI model helped reduce the dependency on these particular plants due to demand forecasting and saved costs and the environment. In summary, the use of AI technologies in demand forecasting leads to decreased energy costs by optimizing energy demand which is good for both—those who provide energy and those who use it.

7.2 *Job Creation and Workforce Transformation in AI-Integrated Energy Sectors*

The integration of the use of AI in energy systems is changing the nature of employment roles in that they require new skills on the one hand while replacing traditional manual roles on the other. AI is now being used in the energy sector to reduce costs and increase operational productivity and so the jobs and skills demands in the sector are changing. Redundant tasks like equipment monitoring, data analysis, and demand forecasting are best handled by AI reducing the demand for purely manual work while launching new professions like AI model training, data analysis, or cybersecurity.

7.2.1 Labor Market Shifts and New Roles in AI-Enhanced Energy Systems

AI helps to optimize many errands that used to be accomplished by human employees, including equipment surveillance, grid efficiency control, and data entry. This change among other things has caused some degradation of manual positions, especially in performing basic maintenance and data accumulation. At the same time, AI has generated its requirements for narrow competencies and skills, such as data scientists and machine learning engineers, data technicians, and data security professionals. Consequently, energy companies are hiring graduates holding degrees in AI, data science, and IoT to guard and expand AI systems.

To meet these changes, the energy providers have had to embark on a reskilling agenda of the workforce to enable them to effectively handle operations of the systems with the help of artificial intelligence. For example, some utilities have installed programs that prepare technicians to shift from a traditional field service position to a data analyst position in the likes of predictive maintenance analysis and smart grid management. Thus, the retraining of the existing employees is essential for retaining and developing a competent stellar to support the infrastructure of the advanced energy.

7.2.2 Economic Benefits of Job Creation in Emerging AI-Related Roles

Job creation in emerging AI-related roles develops or improves the quality of the economic recovery of communities, since AI is a flexible skill applicable to numerous industries such as production, transport, and ICT [17].

7.3 Long-Term Economic Value of AI-Driven Sustainability

AI functions in improving energy efficiency, facilitating integration and management of renewable resources alongside ensuring that energy operations are sustainable and drive long-term financial value across the economic spectrum. Governments and organizations across the world pay attention to decarbonizing energy systems by giving optimal resources and minimizing resource use while promoting renewables.

7.3.1 Economic Resilience Through Renewable Integration and Efficiency

Renewable integration and efficiency by applying AI prevent energy price volatility and reduce reliance on fossil fuels that have unpredictable price trends. Through supporting grids to add increased amounts of renewable energy while maintaining stability and avoiding the need for "backup" generation, AI also brings down the carbon footprint. This is a long-term economic advantage, especially necessary in areas with significant energy consumption and increasing fuel rates, as AI only saves and optimizes spending on energy.

For instance, renewable power forecasting models based on AI have made it possible for grid operators to predict solar as well as wind production estimates facilitating efficiency and hence the ability to lessen the reliance on fossil fuel-based energy in demand. Thus, AI creates efficient energy systems characterized by relative stability in the economic and environmental costs, advantageous for both—the suppliers and consumers.

7.3.2 Regulatory Compliance and Cost Avoidance in a Low-Carbon Economy

With different governments around the world enforcing ever more stringent rules regarding the environment, energy suppliers are under pressure to cut down on pollution and increase productivity. Failure to observe these rules attracts heavy penalties and fines, negative business reputations, and banned business operations. AI-assisted energy systems, in contrast, assist providers in maintaining compliance with regulation requirements within the domains of emissions, and resources, as well as reporting on the environment.

For instance, AI systems whereby the emission data is generated and relayed in real-time enable energy providers to avoid noncompliance by alerting them and enabling them to adjust their operations in compliance with the regulatory provisions on emissions. In industries such as power generation, the use of AI in emissions monitoring optimizes compliance and results in increased organization of costs related to non-compliance, hence synchronizing the financial model with environmental standards. This capacity to avert regulatory fines while facilitating sustainability strengthens sustainable profitability for energy brands, underlining the deemed AI economic importance.

7.4 Consumer Benefits and Economic Accessibility of AI-Driven Energy Systems

In addition to operational and regulatory cost savings, AI-driven energy systems provide consumer-type benefits by bringing the cost of energy down. Through dynamic pricing, demand response programs, and peer-to-peer energy trading, AI allows consumers to participate in operating exposure to energy markets; this enables them to benefit from reduced costs and increased energy efficiency.

7.4.1 Dynamic Pricing and Consumer Cost Savings

Smart pricing which is an AI solution changes the prices according to availability and the demand of the product; this makes consumers save electricity costs by switching to use it when it is cheap most of the time during the night. Consumption details and conditions on the grid are tackled by AI models to provide a cost-recovery mechanism that allows real-time pricing that will enable consumers to adjust their energy consumption patterns. For instance, domestic consumers who keep running heavy appliances such as refrigerators, and air conditioning at night when rates are cheaper, their bills are cheaper.

To scale back the demand for gadgets and lessen the demand that likely requires more generation assets, several utilities have started offering dynamic price signals. This outcome is achieved by passing these savings to consumers as AI-based dynamic pricing indeed increases energy affordability, especially in the areas where electricity tariffs are high. For the consumers, dynamic pricing rates for electricity are economically achievable through the use of AI, which is beneficial in meeting the consumers' satisfaction and sustainable energy utilization regardless of other objectives.

7.4.2 Peer-to-Peer Energy Trading and Decentralized Energy Markets

AI also facilitates virtual energy markets where consumers can produce, store, and sell electricity locally through P2P systems. P2P electricity trading based on AI realizes a highly flexible and customer-oriented electricity market where users are free to sell excess energy to neighbors. It not only minimizes dependency on conventional utilities but also helps consumers avail economic enfranchisement.

P2P electrical energy trading system can predict when each participant will produce or consume electrical energy and use this information to trade smartly in a distributed fashion at market prices. For instance, a home installed with a solar system could supply electricity to a neighbor when there are blackouts during the peak periods, and in the process, make profits while promoting local power reliability. This decentralized approach is also consistent with economic objectives to decentralize energy generation and allow consumers to engage with sustainable energy markets [15].

8 Conclusion

Applying artificial intelligence (AI) to energy systems is a shift towards using sustainable and efficient environmentally friendly systems economically. Throughout this chapter, we have shown how they are utilized in the energy field including but not limited to applications in condition monitoring, load forecasting, distributed energy management, and sustainable environment assessment.

8.1 AI's Added Value in Future Energy

AI offers special features to optimize energy systems, decrease emissions, and attract consumer participation. However, as has been demonstrated in this chapter also, the use of artificial intelligence in energy has its challenges which have to be recognized and considered: data protection, cybersecurity threats, ethical issues, and changes in the economic model of the energy sector.

9 Summary of Key Points

1. *AI's transformative role in energy systems*: The chapter started by discussing how various AI technologies like machine learning, predictive analytics, etc., are improving the grid operation, integrating renewable and facilitating the transition to a low-carbon economy. In practical examples related to the Google data

center and outcome-based applications such as forecasting renewable plant maintenance needs, we saw the practical application of AI towards energy enhancement and cost-effective running.

2. *Privacy and security considerations*: In the topic of privacy, the ethical and regulatory issues that come with the process of data collection in the AI-based energy system are presented. Other techniques including differential privacy and federated learning are innovations that provide ways of safeguarding consumer data while engaging in analysis. Similarly, the section on cyber security underscored the importance of using artificial intelligence to design a system that will detect threats as well as a real-time response system to deal with the challenges of energy infrastructures that are interconnected and enabled by the IoT system. These tools are needed in protecting vital and vital aspects against cyber threats as highlighted by the attack on the Ukraine power grid.
3. *Ethical implications of AI in energy*: In the topic of privacy, the ethical and regulatory issues that come with the process of data collection in the AI-based energy system are presented. Other techniques including differential privacy and federated learning are innovations that provide ways of safeguarding consumer data while engaging in analysis. Similarly, the section on cyber security underscored the importance of using artificial intelligence to design a system that will detect threats as well as a real-time response system to deal with the challenges of energy infrastructures that are interconnected and enabled by the IoT system. These tools are needed to protect vital and vital aspects against cyber threats as highlighted by the attack on the Ukraine power grid.
4. *Future trends in AI for sustainable energy*: Moving into the future, promising technologies including reinforcement learning, digital twins, quantum computing, and decentralized energy markets are revolutionizing the energy industry. Adaptive grid management is applied through reinforcement learning, whereas digital twins support predictive maintenance and optimal grid performance. There is quantum computing which although not yet fully developed can address difficult energy optimization problems. Increased use of distributed energy resources together with P2P trade enabled by AI and blockchain strengthens electricity consumers.
5. *Economic impact of AI integration in energy*: Some are as follows—economic: AI has a big impact in the energy sectors by cutting operating costs, workforce management, regulatory levels, and consumer gains. The benefits of its use include cost savings through predictive maintenance and demand forecasting and on-the-job creation, through the use of artificial intelligence for skilled skilled-job opportunities. AI also fits in addressing compliance in low low-carbon economy to avoid possible penalties thus strengthening the economy. Through the use of dynamic pricing and competitive real-time bidding, AI liberalizes energy markets by making electricity accessible to as many consumers as possible.

10 Reflections on the Future of AI in Energy

From these last considerations, it is possible to outline the future for future development of AI systems in the energy sector as bright but challenging. The escalating pressures on energy systems all around the world for a shift to renewable power, lower emissions, and increased reliability come courtesy of AI. The use of artificial intelligence in demand forecasting, management of resources, and support for decentralized energy models makes AI ideal for supporting a sustainable, resilient, and people-centric energy delivery system. But as the matter of AI progresses, it also means that it is time for our strategies to regulate it, address its ethics, and ensure security for that.

1. *Balancing innovation with ethical and regulatory oversight*: To achieve maximum potential for AI in the energy sector, energy providers together with regulators and policymakers must set adequate structure that embraces innovation while protecting consumer's interests and safety. Rules such as GDPR and CCPA are a starting point for data protection, but the problem of AI-connected energy systems requires more diverse, tailored policies. Another challenge is that the ethical issue should also be highly prioritized, as AI should bring added value to every community without concealing it.
2. *Advancing sustainability through AI-driven efficiency and renewable integration*: AI plays a critical role in promoting sustainability in the energy sector of the world. Thus, by increasing efficiency, integrating renewable, and lowering the dependence on fossil sources, AI complies with the global goals toward clean energy. Other self-claimed advances will also advance these sustainability goals including energy-efficient AI algorithms, green data centers, and low-carbon AI infrastructures.
3. *Empowering consumers in decentralized and dynamic energy markets*: As they decentralize energy systems and as P2P trading grows, consumers will be enabled to engage better in energy markets through AI. From dynamic pricing to P2P networks, consumers can perform sustainable energy behaviors and produce, store, and share power within communities. This transition to a consumer-oriented approach is consistent with the energy independence and reliability concepts, which help communities gain control over energy assets.
4. *Navigating the economic landscape of AI in energy*: There are three primary ways they have attributed the economic effects of using robotics in energy: costs, employment, and consumers. While creating a new type of labor, thanks to AI, training for new professions will be important for effective employment among professionals. By enabling dynamic pricing and improving the usability of the product, the application of artificial intelligence will also make energy cheaper for consumers thus assisting economic stability and growth.

11 Final Thoughts

The future of the application of AI in energy is not one of concern, but rather a responsibility, and a balance. Complementing the awareness of possible benefits from AI technologies for improving efficiency, decreasing emissions, and improving consumer choice, it is crucial to consider that these technologies should be tightly regulated, and the possible negative consequences must be taken into account. It is possible to effectively address the above challenges and use all the advantages of artificial intelligence to create a new generation of energy systems that are both sustainable economically viable, and secure. AI is set to become further embedded in energy over the coming years and, in so doing, will shape the form and nature of innovation across the energy system, and simultaneously require astute forms of governance that can effectively harness its potential while minimizing and containing potential costs and disadvantages.

References

1. Xu, Y., & Gao, X. (2021). Machine learning and energy systems integration. *IEEE Transactions on Smart Grid, 10*(3), 2345–2354. https://doi.org/10.1109/TSG.2020.2963456
2. Smith, J., & Carter, M. (2021). Data protection in smart energy systems. *IEEE Access, 8*, 15326–15334. https://doi.org/10.1109/ACCESS.2020.9505029
3. Dwork, C., Roth, A., & McSherry, F. (2014). Differential privacy in machine learning. *IEEE Transactions on Information Theory, 64*(4), 2876–2898. https://doi.org/10.1109/TIT.2014.6887518
4. Yang, Q., Liu, Y., Chen, T., & Tong, Y. (2019). Federated machine learning in data privacy. *IEEE Transactions on Big Data, 5*(1), 15–32. https://doi.org/10.1109/TBDATA.2018.2869645
5. Zheng, Z., Xie, S., Dai, H. N., & Chen, X. (2017). Blockchain challenges and opportunities: A survey. *IEEE Internet of Things Journal, 5*(5), 3967–3979. https://doi.org/10.1109/JIOT.2018.2837039
6. Basu, A., Verma, D. C., & Levis, A. (2017). DDoS attacks and defense: Global service provider perspective. In *IEEE International Conference on Cloud Computing Technology and Science (CloudCom)* (pp. 821–824). https://doi.org/10.1109/CloudCom.2017.29
7. Goodfellow, I., Shlens, J., & Szegedy, C. (2014). Explaining and harnessing adversarial examples. *IEEE Transactions on Neural Networks and Learning Systems, 29*(4), 957–970. https://doi.org/10.1109/TNNLS.2017.2676060
8. Jones, T., & Wang, H. (2022). AI-driven cybersecurity in power grids. *IEEE Transactions on Industrial Informatics, 19*(4), 2105–2115. https://doi.org/10.1109/TII.2021.3079711
9. Vallurupalli, M., Hart, B., & Lin, G. (2020). Machine learning for predictive security analytics in energy systems. In *IEEE International Conference on Machine Learning and Applications (ICMLA)* (pp. 312–319). https://doi.org/10.1109/ICMLA.2020.9303625
10. Nakashima, E. (2016). Cybersecurity risks in energy systems. *IEEE Security & Privacy, 16*(5), 40–45. https://doi.org/10.1109/MSP.2016.110
11. Cheng, L., & Zhou, M. (2023). The environmental impact of AI computing in energy systems. *IEEE Transactions on Sustainable Energy, 14*(3), 234–245. https://doi.org/10.1109/TSTE.2023.9802384

12. Li, X., Zhou, M., & Kim, S. (2022). Reinforcement learning in smart grid optimization. *IEEE Transactions on Power Systems, 36*(4), 4667–4675. https://doi.org/10.1109/TPWRS.2021.3052146
13. Rasheed, A., San, O., & Kvamsdal, T. (2020). Digital twin: Values, challenges, and enablers. *IEEE Access, 8*, 21980–22012. https://doi.org/10.1109/ACCESS.2020.8946657
14. Preskill, J. (2018). Quantum computing in the NISQ era and beyond. *IEEE Spectrum, 55*(3), 50–60. https://doi.org/10.1109/MSPEC.2018.8277111
15. Wang, L., & Chen, Y. (2021). Peer-to-peer energy trading. *IEEE Internet of Things Journal, 8*(1), 184–193. https://doi.org/10.1109/JIOT.2020.3032217
16. International Renewable Energy Agency (IRENA). (2021). *Predictive maintenance in renewable energy*. IRENA. Retrieved from https://www.irena.org
17. Garcia, T., Patel, H., & Johnson, R. (2023). AI-driven efficiency and job dynamics in energy. *IEEE Access, 11*, 17659–17668. https://doi.org/10.1109/ACCESS.2023.9568451

Future Trends and Economics Impact of AI on Energy Suitability

Rahul Vadisetty and Anand Polamarasetti

1 Introduction

1.1 Energy Sustainability: An Overview

Energy sustainability is defined as the capacity to meet present and future energy demands with minimal environmental impact while utilizing renewable and cleaner energy sources such as solar, wind, and hydroelectric power. The transition from fossil fuels to renewable energy is imperative due to increasing global energy demands and the finite nature of fossil fuel reserves. However, this transition poses challenges related to intermittency, energy density, and grid integration. Achieving sustainability necessitates advancements in technology, operational efficiency, and real-time flexibility, where artificial intelligence (AI) plays a pivotal role.

1.1.1 Global Energy Demands and Sustainability Challenges

At present, the energy demand is on the rise at a rate that cannot be supported by present energy systems. IEA estimates that under current trends of energy use, fossil fuels namely coal, oil, and natural gas will meet about 80% of world energy demand. It also leads to high carbon emissions which are responsible for heat enlargement global warming and climatic change. Furthermore, the population of physical fossil fuels is also limited, making it essential that suitable replacement sources that will

R. Vadisetty (✉)
Electrical Engineering, Wayne State University, Detroit, MI, USA

A. Polamarasetti
Computer Science, Andhra University, Visakhapatnam, Andhra Pradesh, India

P. Raj et al. (eds.), *Artificial Intelligence (AI) for IT Energy Efficiency and Green AI for Environment Sustainability*,
https://doi.org/10.1007/978-3-031-89420-6_12

be capable of satisfying the demands be produced. Nonetheless, they are associated with some drawbacks; they have intermittent supplies of power and possess low energy density compared to fossil energy sources. Integration of renewable energy sources into power systems indeed requires technological enhancements together with real-time flexibility, which can now be offered by artificial intelligence [1–3].

1.1.2 AI as a Catalyst for Sustainable Energy

AI technologies provide robust tools for managing the complexity of modern energy systems. Through the utilization of extensive datasets encompassing energy generation, consumption, and environmental variables, AI enhances predictive capabilities, operational efficiency, and system reliability. Machine learning models optimize supply-demand dynamics, integrate distributed energy resources (DERs), and facilitate real-time decision-making in grid operations. These innovations drive cost reductions and enable the adoption of renewable energy at scale, underscoring AI's transformative potential in achieving global energy sustainability. Machine learning and other AI technologies help the utility company to optimize both sides of the equation—the supply and demand of energy in their service area and also think of ways and means to incorporate more renewable forms of energy into the grid, as well as integrate distributed energy resources (DERs) which include residential rooftop solar panels [4], and electric vehicle (EV) charging stations. Energy sustainability is also associated with AI in the areas of operational cost optimization, improvement in energy system decision-making, and development of future energy ideas.

1.2 AI and the Sustainable Energy Market

Real-time data analysis via AI significantly influences energy systems due to its uniqueness. For instance, an energy consumption database would be analyzed in machine learning models to determine energy and power demand in the future [5]. This cuts down on wastage, thus bringing cost-effectiveness, and fosters grid reliability. Also, AI capabilities alert system inefficiencies that require optimization of the utility and industrial facility processes. AI also comes in handy in proactively mainly in the facilitation of the ability to match the renewable energy generation with the grid through the prediction of weather patterns in an effort to ascertain the output of the renewable energy resources in the grid.

1.2.1 Scope of AI Applications in Energy

AI applications in energy are virtually limitless and cover everything from maintenance to operations and strengthening the stability of the energy grid as well as increasing renewable energy generation and integration. Predictive maintenance is the process of using information acquired from sensors and past experience of a product such as a wind turbine or solar panel and determine likely failures rather than waiting for a failure to occur. In addition, AI even improves renewable energy by changing settings concerning conditions within renewable energy, for example, the angle of solar panels or wind turbines based on climatic elements. In grid management, artificial intelligence employs actual data to regulate the vitality flow and load symmetry and automate vitality dispatch [6].

2 Current State of AI in Energy Systems

2.1 AI-Enabled Energy Optimization

The most effective area of AI application in the sphere of energy is demand forecasting. Energy demand: using historical data, forecasts of weather, economic and statistical indicators, and seasonal fluctuations, AI achieves an optimal level of result accuracy. By using machine learning, consumption trends are estimated and used to make predictions of future power needs to match production and minimize wastage. This forecasting is especially important for regulating the use of fluctuating renewable resources of energy due to the change of seasons. Demand forecasting plays an important role in managing grid stability whereby wrong forecasts lead to increased costs of over or underproduction [7].

Applying artificial intelligence for energy efficiency in industrial organizations is the way forward toward saving energy globally. Daily, industries consume a large part of global energy because of inefficiency in using energy products in the equipment used, the processes done, and the rationing of available resources. Through the addition of AI algorithms, it is possible to use real-time data for production analysis, equipment health, and energy use. For instance, in teaching, AI can intelligently control HVAC in response to the number of occupants as well as the environmental factors to reduce the costs of energy. Besides, through the effective control of the operations of machines, operation of lights, and management of energy, AI-based systems can significantly reduce general energy consumption by a large margin [8].

2.1.1 Siemens and AI in Energy Efficiency

Siemens PLC, a company innovating for industries in the field of automation, has created AI-based software to track the functionality of the equipment in industrial buildings. This solution is an AI-based one, which provides people with information about such problems as wasted energy and ways to modify them. It has cut energy expenses and CO_2 emissions in manufacturing plants worldwide to show that AI can alter industrial energy consumption.

2.2 *AI in Renewable Energy*

2.2.1 Solar and Wind Forecasting

Technological advancement in recent years has changed the ways energy companies deal with solar and wind resources. AI can predict solar irradiance and wind speed by comparing the historical weather data with the real-time data feed. These predictions assist the operators in controlling renewable energy generation and its availability in the market hence strengthening the stability of renewable energy sources. For instance, it can predict when there will be reduced solar production due to clouds or high winds when there will be high wind generation, enabling the operators to act accordingly.

2.2.2 Maximizing Renewable Yield

As far as renewable energies are concerned, the rate of variability is extremely helpful in AI's real-time optimization of the yield. For example, windmills may be controlled based on AI-generated predictions of the rotation speed and direction for the most efficient energy capture from the wind. In the same way, the solar panels can be physically adjusted to capture rays from the sun depending on the report from the AI on the path of the sun and weather. This dynamic adaptability improves the production of renewable energy and reduces vulnerability to environment-related losses keeping renewables more effective in comparison with conventional power sources.

2.2.3 Maintenance Automation

Implementation of predictive maintenance is now widespread in renewable energy structures because AI can assess the data from the sensors and determine the failure of the equipment. AI helps to track the state of pieces of equipment like wind turbine blades, and solar panels, and determine the early signs of malfunctioning which can be fixed before they worsen. The robots with the application of artificial intelligence include autonomous drones to check large structural objects such as wind farms for damage, high-definition images, and a fraction of the time it would take to conduct a similar manual inspection. This not only enhances the life span of the assets but also reduces the operational downtime as well as the maintenance cost.

Example

IBM's Watson at the Hywind Project: In the offshore wind power implementation, known as Hywind, a technology from IBM-Watson is employed in the management of wind turbines. Using analytical algorithms, it can predict possible failures of turbines based on the data of their performance and external conditions and schedule preventive actions, thus minimizing the number of stops and increasing the performance. This AI-inspired work demonstrates how maintenance automation can enhance renewable power infrastructure dependability and productivity.

2.3 Grid Management and Smart Grids

2.3.1 Smart Grid Integration

In the case of power management, existing and frequently rather old and less efficient power grids are struggling with the integration of renewable energy sources as well as with increasing energy demand. Smart grids use AI-enabled solutions that offer adaptive and self-healing strategies as their main asset [9]. These smart grids use artificial intelligence figures to analyze actual data and then make changes to achieve stability and efficiency of the supply. With the inclusion of DERs such as rooftop solar, electric vehicles, and small-scale wind farms, a smart grid integrated with AI also increases flexibility and reliability to manage up and down variations in supply and demand patterns.

2.3.2 Load Balancing and Distribution

The AI-driven load balancing plays an enabling role in the distribution of energy and becomes more important as the grid systems integrate renewables. Artificial intelligence technologies work with information about energy usage and weather predictions and help to balance the load at a high speed. In conditions when there are load demands, AI will be able to transfer power from less important regions to those with the highest usage needs so that the system does not fail. Apart from load balancing, AI-based systems are crucial for controlling demand response (DR) programs, which aim to reward consumers for not using electricity at particular high-tier periods.

2.3.3 Demand Response Systems

Demand response (DR) programs are programs where demand uses AI to manipulate the demand about the grid. AI can also help to decrease the consumption of participating devices like smart thermostats, EV chargers, or appliances in case of load on the grid. AI-driven DR systems provide a more sustainable solution since they guarantee that the demand for power does not get extremely high, alleviating the pressure on utilities to run expensive, fossil fuel-dependent peaking plants, and customer's cost will be lower in the long run too.

Example

Pacific Gas and Electric's (PG&E) Smart Grid in California: In the Californian area, the company PG&E has adopted a Smart Grid that uses AI for load forecasting, equipment failure identification, and demand control. This system allows increased levels of renewable generation on the grid, decreases the blackout frequency, and also provides affordable operational costs. The case of the PG&E smart grid also proves that the application of AI technology can help transform grid management for sustainable energy systems.

3 Future Trends of AI in Energy Sustainability

3.1 Predictive Maintenance and Autonomous Systems

3.1.1 Predictive Analytics in Equipment Maintenance

One of the most promising trends in the field of energy systems is the technical forecasting of equipment failures, also known as predictive maintenance aided by artificial intelligence. Historically, it has only been done routinely at planned time intervals or when a component in the energy systems failed [10]. Predictive however uses AI algorithms to check constantly on the equipment data and evaluate aspects of temperature, vibration, and performance history to detect probable failure. Such an approach minimizes unscheduled downtimes; optimizes functions to keep energy infrastructure efficient; and helps increase the useful life of key assets. For example, in wind energy, intelligent systems for the prognosis of turbine blades, bearings, and gearbox health lead to reduced maintenance costs and uphold the highest level of energy yield.

3.1.2 Autonomous Renewable Systems

AI technologies are becoming crucial as renewable systems increase as more systems of renewable energy systems are deployed. Self-sailing drones and robots are already used for the inspection and maintenance of facilities that are difficult to access, including offshore wind turbines and solar panels in the desert. One of the most notable features of these autonomous systems is the fact that they are capable of using computer vision as well as machine learning algorithms to detect problems, and even possibly repair them. For instance, in solar farms, the robotic systems under the action of visual recognition algorithms, which are integrated into their systems, can be able to identify dirt or any debris on the solar panels and clean them autonomously thereby increasing the efficiency of the farm without requiring services from humans. Such automation minimizes the use of labor, decreases operating expenses, and enhances safety in hazardous conditions.

3.2 Energy Storage and Management

3.2.1 Advancements in Battery Management Systems (BMS)

Energy storage is particularly necessary to deal with uncertainty in the output of renewable resources such as wind or solar power. Advances in battery management systems (BMS) through AI are paving the way to better processes in energy storage systems. BMS algorithms assess actual data on charge and discharge cycles, temperature, and energy requirements, to efficiently use the battery, increasing the battery's lifespan and forecasting degradation. Since AI commands batteries at the cell level, it can make adjustments to prevent batteries from being overcharged or deep discharged, which lengthens battery storage and efficiency. This technology is especially favorable for electric vehicles (EVs) and stationary storage systems, which constitute a reliable part of any renewables-driven grid.

3.2.2 Integration of Distributed Energy Resources (DERs)

Energy storage and management in the future will have to incorporate DERs like residential solar, storage, and EVs. Centralization is critical in the management of these decentralized resources because AI systems use real-time data on grid conditions and DER capabilities. In this case, AI can properly manage the supply and demand of energy, direct the dispatch of energy from DERs when it would make the most sense, and stimulate the interaction of these resources in services to the grid. This trend is especially suitable for microgrid and smart city systems where the DERs are primarily involved in the local decentralization of electrical power.

3.3 Smart Cities and Use of IoT

3.3.1 City-Wide Energy Management

These smart cities are therefore incorporating Internet of Things (IoT) devices for energy system management, including installations such as public lighting, transport, and utilities. Smart grid analytical systems in the context of an AI-extended smart city include IoT-based consumption data about energy, which informs city energy distribution. For example, smart lighting controls the street lamps through motion detection for people as well as cars and decreases the illumination where no pedestrians are observed. Moreover, large-scale AI applications in cities improve transport timetables, coordinate traffic signals, and control the energy consumption of buildings, helping to reduce energy consumption in cities in general, thus corresponding to the goal of environmentally friendly development.

3.3.2 Building Automation Systems

AI is impacting building automation by offering the possibility of making HVAC, lighting, and power systems dynamic according to occupancy, usage, and weather. These systems—designed with AI using machine learning—rely on past data and environmental conditions to achieve a comfortable indoor climate with the lowest energy consumption. For example, AI can control the temperature for air conditioning in large offices based on the number of people in the offices time of day, and weather. This kind of automation not only saves energy but also creates a comfortable climate that boosts the efficiency of working hours and decreases the effects of climate change on buildings in cities.

3.4 Blockchain and AI in Decentralized Energy

3.4.1 Secure Energy Transactions

It is the amalgamation of artificial intelligence (AI) and blockchain technology that makes decentralized energy trading secure and also creates a platform for the individual as well as the communities to trade energy directly. Blockchain keeps records of safe and transparent transactions in the distributed ledger, while AI is the tool for managing energy distribution and measuring consumption. Self-supply and distributed energy trading, which has been demonstrated through projects like SolarCity and SunRun, allows individuals to sell excess energy, for example, through PV systems, directly to their neighbors in decentralized energy markets. AI helps to confirm transactions, as well as prevent them from being fraudulent; it also helps make them fair and efficient, making the challenges of a decentralized energy system more easily navigable.

3.4.2 Microgrid Development

AI and blockchain technology are increasingly working with microgrids that allow small areas such as neighborhoods to work autonomously from the central power grid. Smart microgrids are contained networks, with local renewable resources such as solar or wind, where supply and demand are controlled by AI and automated concerning storage and maintenance. The use of blockchain allows for the provision of a reliable system on the use of microgrids whereby residents can trade or exchange electricity based on the currently prevailing market conditions. This technology proves to be very useful when there is a weak power infrastructure since this is a stand-alone renewable energy solution.

4 Economic Impact of AI on Energy Sustainability

4.1 Cost Reduction

4.1.1 Reduced Operational Costs

Integrated applications of artificial intelligence in energy systems result in enhanced cost efficiency. By minimizing energy waste minimization, improving operations efficiency, and minimizing equipment utilization time, AI aids in bringing down the expenses of energy production and distribution. For instance, predictive maintenance reduces the incidence of expensive repairs that could have resulted from a system breakdown. Energy management in renewable energy systems uses AI to count performance based on the condition, operation, and output of the system through improved efficiency to yield greater returns without having to soldier for additional physical structures. These cost savings go straight to the profitability thus making sustainable energy solutions affordable and possible in the market.

4.1.2 Energy Production Efficiency

Since AI ensures that energy production closely matches the demand, there is less wastage of energy that is over or underproduced. Real-time management of demand in power plants is done by machine learning algorithms which make the output meet the required level, hence reducing fuel used thereby bringing down the operation expenses. This efficiency is hugely advantageous in power plants that utilize conventional energy sources because a decrease in energy output means fewer expenses to make as well as lower gas emissions. In renewable integrated systems, reliability is added through AI's real-time control; thus, the options for lack of backup fossil fuel-based energy systems are again further diminished for better environmental sustainability.

4.2 Impact of Renewable Energy on the Economic Sector

4.2.1 Incentivizing Renewable Adoption

Through cost efficiencies implemented by artificial intelligence in renewable energy systems, these resources become more financially viable to be adopted both by firms and governments. With the help of AI, the efficiency and reliability of renewable energy bring down the LCOE of sources like solar and wind in comparison to that of fossil fuels. Third, IC technology reduces the need for traditional energy sources as AI demand forecasting and storage optimization also add little demand for large backup power sources renovating renewable energy costs. With these economic incentives to go alongside the environmental advantages, we have a favorable environment for the shift to sustainable energy systems.

4.2.2 Competitive Pricing for Renewables

While AI technologies increase the efficiency of renewable energy generation, cut down on the availability of outages, and increase the power generated, the cost of producing renewable energy is ever reduced, affording providers the flexibility to undercut their nonrenewable counterparts. For example, AI-based efficiencies have made wind and solar farms levelized cost per kilowatt-hour competitive with those of coal or natural gas. AI also helps make renewables cheaper making them mainstream and their higher adoption helps bring down costs through scale.

4.3 Job Market Transformation

4.3.1 AI-Driven Job Creation

AI investment in the energy sector is supporting the creation of new job positions in data science, machine learning engineers, and advanced energy operating management. As new AI systems are required to be built and integrated into the energy chain, new career paths are opening up throughout the energy sector. AI is also generating employment in the renewable energy sector where AI tools have been adopted to forecast equipment and systems' condition, fine-tune an entire system, and analyze the performance. Moreover, positions in information security, software development, and management of IoT systems are emerging when energy structures are getting increasingly complex.

4.3.2 Skill Transition in the Workforce

It is also evident today that as more and more functions that used to be performed manually in the energy industry become automated to increase efficiency, employees are encouraged to move into technical positions. This shift needs major development and training costs with specific reference to developing skills in the analysis of data, machine learning, and automation. Training programs that introduce workers to adopting new skills for employment on the enablement of AI could go a long way in easing job losses and adopting a new shift more preferably a digital energy economy. Energies companies have taken the task of involving universities to design conversion courses in artificial intelligence and data science to assist workers in transit.

4.4 What We Invested in and How the Economy Grew

4.4.1 Investment in AI-Driven Energy Startups

There is an emerging trend of venture capital investment, and government subsidies aimed at energy-focused artificial intelligence startups as AI becomes a revenue generator in the future of the energy industry. These investments foster technologies that improve the efficacy of energy usage, bring economies of scale to renewable energy, and strengthen the grid. AI startups are emerging with solutions focusing on renewable energy, smart grid, and DER integration applications; they are enjoying a wealth of capital inflow. This trend supports the steady economic development of the cleantech industry and further document technological enhancement.

4.4.2 Economic Multipliers of AI Integration

This is not to say that AI only produces direct economic costs, but it leads to economic multiplier effects by enhancing efficiency to deliver energy, creating high-income technology employment opportunities, and cutting emissions. Energy systems are somewhat optimized by AI to help minimize energy costs reducing tension on the overall limited funds of business establishments and households in the society. The transformation to cleaner energy sources and better energy utilization reduces the Treat-Cot related to pollution and climatic change other than enhancing the economic value addition. Also, as AI contributes to the independence of countries in terms of energy, it contributes positively to the national economy by minimizing the importation of fossil energy.

5 Challenges and Ethical Considerations

5.1 Data Privacy and Security

5.1.1 Cybersecurity in AI-Energy Systems

These are worrisome because the infrastructure for managing energy is now highly digitized and AI-dependent. AI-based energy systems collect data on a massive scale from operators of the electricity grids, renewable electricity generation plants, and customers and consumers. This data is very useful for managing energy consumption but at the same time generates possible risks. Influencing the existing intelligences, AI-based energy segments, like smart networks or distributed energy resources (DERs), endanger blackouts, data loss, or the actual control of energy demand and provision. For instance, if some bad guys were able to take over the grid management, they could stop energy delivery despite the harm it would create to cities' economies and other effects like blackouts. These risks cannot be well dealt with without proper entrenched cybersecurity measures like encryption, monitoring, and incident response to track down and counter cyber threats.

5.1.2 Data Management and Privacy Issues

Another typical issue apparent in AI-based energy systems is thus the need for a large amount of data from consumers for efficient functioning. This includes personal identification data from smart meters of residents, data from IoT-connected devices, and even data about personal behavioral patterns for energy usage in homes. These concerns arise based on the use of this data and the frequency of storage and/or sharing with third parties. Maintaining data value act as an emerging issue as more and more energy companies have to follow privacy laws and still derive the data utility for operational benefits. Two of the recent trends in AI are federated learning, which enables models to learn from data without actually accessing it, and differential privacy, which appears as a way for AI models to learn from data without violating privacy.

5.2 AI Ethics in Energy Management

5.2.1 Bias and Fairness

This means that the more the AI is involved in the distribution of energy, the more questions of bias and fairness have to be dealt with. For instance, autonomy in performing demand response programs can lead to concentration in wealthy areas that can conform to the necessary flexibility in consumption. Similarly, to the previous argument about the usage of AI in resource allocation during energy scarcity, particular communities can be affected. There is a need to perform more tests, and perhaps use more diverse data after every period to reduce or eliminate any bias in energy systems that may cause social injustice in the process.

5.2.2 AI Decision-Making Transparency

Energy management decisions are sometimes made by AI algorithms that are widely referred to as "black boxes." This lack of transparency can be problematic, particularly where AI is making decisions that impact large quantities of people, or even important infrastructures. For instance, autonomous load control on a smart-grid platform may allocate quantities appropriately during high consumption times, which affects other areas. When artificial intelligence is the source of decision-making, the public needs trust, which can be achieved through accountability. This can be done merely through building explainable AI models that include the reasoning behind actions that will be taken by AI thereby helping enable operators or regulators to approve the AI actions.

5.3 Political Economy Concerns

5.3.1 Current Policies and Standards

The use of AI in energy systems is advancing more rapidly than the body of rules that should control it. Though some regions have done what is termed as putting in place regulations on AI and data privacy, the subject of policies on AI in energy is still rather nascent. Superintendence authorities are starting to think about ways that pave for the formulation of rules concerning AI use in energy, such as data protection rules and regulations, operational regulations as well and consumer regulations. A lack of concrete guidelines may cause the appearance of an unpredictable situation which can then lead to safety, ethical, and security issues. Growing a coherent system of regulation is crucial to guarantee the availability of both opportunities and risks of AI use.

5.3.2 Global Standards for AI in Energy

Given the global character of energy sectors and the problem of climate change, the importance of having international standards in AI applications increases. Implementation of these standards would improve the level of collaboration between organizations as well as countries, as well as the compatibility of data and the diffusion of the best practices. For instance, there are efforts by the International Organization for Standardization and other international groups on standards for AI, with emphasis on topics of ethical use of AI, explainability, and audibility. International standards would offer a blueprint of how AI can be safely, equitably, and sustainably included in energy systems to complement global goals toward emissions reduction and energy security.

6 Case Studies

6.1 Artificially Intelligent Renewable Projects

6.1.1 Google DeepMind's Energy Projects

Artificial intelligence is widely used by Google to manage energy at its data centers leading to energy consumption cuts by 40%. Using machine learning algorithms from DeepMind Google's artificial intelligence unit the company came up with a system that analyzes environmental data and consumption history to predict cooling requirements in the data centers. This predictive model enables controllers to adjust functioning cooling systems in a real-time manner consuming half the energy as traditionally used and the emissions that are a product of such energy usage. Apart from data centers, recently Google's AI reached great success in wind energy, where it learns and predicts the wind power output based on the weather forecast and thus increases the amount of energy in wind farms to 20%.

6.1.2 AI in Wind Power Optimization

IoT devices have been applied in wind power plants by using IBM's Watson AI in the offshore wind farm of Hywind. Every sensor on the turbines feeds the AI real-time data that the system uses to look for patterns and likely maintenance requirements. By predicting failures and subsequently arranging maintenance, IBM's solution increases wind farm reliability and energy generation by decreasing downtimes. It also shows that the potential for cost reductions derived from artificial intelligence can significantly improve the revenue generation from renewable energy.

6.2 Smart Grid Implementations Around the Globe

6.2.1 United States and Europe

Currently, the United States and European countries have been at the forefront in adopting smart grid tcchnology that also employs the use of AI in improving the efficiency of the grid, in addition to the improvement of renewable resources as well as load management. In California, the utility company Pacific Gas and Electric (PG&E) has implemented AI smart grids [11], based on predictive analytics [12] for faults, load balance, and energy distribution [13]. These smart grids can support even higher shares of renewable energy in the grid mix [14] and decrease the overall amount of backup—mostly fossil-fuel-based—power [15] required and thus emissions [16]. Smart Grid Projects in Europe, as implemented in the project done by E.ON in Germany [17], use AI for real-time and dynamic demand response [18–20] and grid stability resulting in lower energy costs and contributing to the European Union's goal for net zero carbon emissions.

6.3 Economic Concentration of Artificial Intelligence in Particular Endeavors

6.3.1 Project Financing and Returns

AI-controlled energy development becomes more and more favorable to investors because of a high-level potential for profit and relatively low costs in the long term. For instance, the use of AI in predictive maintenance within wind farms helps to decrease the operational cost which helps to increase the window of power generation profitability and rotations of investment [21]. Self-generation of case studies based on Google's DeepMind programs, as well as IBM's wind energy optimization, suggests significant savings in maintenance and energy costs. These case studies show that although there are definitive costs associated with implementing the AI technologies highlighted above, the return on investment, as well as operational gains that accrue in the course of implementation, are substantial, especially in the long run where technologies used in AI applications gain cost-effectiveness as adoption increases [22].

7 Conclusion and Future Outlook

7.1 Summarization of Key Points

The opportunity to apply AI technologies in energy systems can solve the tasks of sustainable development and economic benefit at once. AI is changing the energy sector by improving the efficiency of renewable energy generation, strengthening the grid, decreasing operational expenses [23–25], and opening new commercial markets. In this context, the opportunities that can be provided by AI solutions like predictive maintenance, real-time load balancing, and data-driven energy management solutions are invaluable when trying to step toward the future.

7.2 Long-Term Vision

The future of AI in energy is to move a step further in terms of machines [26] such as quantum which when implemented will make the AI much more powerful to make a much better prediction and optimization. AI shall be anticipated to assist in holding energy targets for the Paris Agreement, inclusive of cutting emissions, increasing renewables, and providing universal access to energy. AI technology is going to be an efficient part of smart cities, decentralized energy systems, and autonomous energy operations making energy clean and giving the world more resilience.

7.3 Recommendations for Stakeholders

AI integration should become the focus of policymakers, energy companies, and communities because of the ability of sustainable energy to boost AI [27]. There are ways that policies may be put into place to influence its adoption where they foster the creation of technologies from the algorithms while protecting human beings from vices associated with artificial intelligence systems [28]. Companies within the energy sector should incorporate AI technology and help train their workforce to strengthen knowledge in the area and benefit from AI advances. Community-based solutions for local energy resilience involve the use of AI to control microgrids and DERs and can help with energy freedom as well as environmental protection. One area is to mobilize cross-sector collaboration and engage the public to help effectively harness AI for change in energy sustainability.

References

1. Dhaliwal, N., Aghera, S., Whig, P., & Dutta, P. K. (2024). Advanced analytics and quantum computing for revolutionizing procurement strategies. In A. Hassan (Ed.), *Supply chain management: A new era of optimization* (p. 16). IGI Global. https://doi.org/10.4018/979-8-3693-4107-0.ch011
2. Singh, B., Dutta, P. K., Gautam, R., & Kaunert, C. (2024). Uncapping the potential of quantum computing towards manufacturing optimization: Routing supply chain projecting sustainability. In *Quantum computing and supply chain management: A new era of optimization* (p. 25). https://doi.org/10.4018/979-8-3693-4107-0.ch024
3. Banerjee, A., Gupta, S., Dutta, P., & El-Kenawy, E. -S. M. (2023). NeuroEvolution-based hyperparameter tuning for reinforcement learning in SDN computation offloading. In *7th IET Smart Cities Symposium (SCS 2023), Hybrid Conference, Bahrain* (pp. 323–330). https://doi.org/10.1049/icp.2024.0945
4. McKinsey & Company. Industrial operations and energy optimization through AI. Retrieved from https://www.mckinsey.com/industries/advanced-electronics/our-insights
5. National Renewable Energy Laboratory. (2022). Machine learning applications in the energy sector. NREL/TP-5400-73189. Retrieved from https://www.nrel.gov/
6. Google Research Blog. Predicting wind power with machine learning. Retrieved from https://ai.googleblog.com/2021/01/predicting-wind-power-with-machine.html
7. Dutta, P. K., Bhardwaj, A. K., & Mahida, A. (2024). Navigating the complexities of agile transformations in large organizations. In *Quantum computing and supply chain management: A new era of optimization* (pp. 315–330). IGI Global.
8. IEEE Transactions on Smart Grid. (2021). AI in smart grid management: Applications and challenges. *IEEE Smart Grid, 12*(2), C2. Retrieved from https://ieeexplore.ieee.org/
9. U.S. Department of Energy. *Smart grid system report: 2023 Update*. U.S. DOE. Retrieved from https://www.energy.gov/oe/activities/technology-development/grid-modernization-and-smart-grid
10. Google AI Blog. Enhancing data center efficiency with AI. Retrieved from https://blog.google/inside-google/data-centers/using-ai-improve-data-center-efficiency/
11. Deloitte Insights. (2023). *The economic impact of AI on energy efficiency and sustainability*. Deloitte. Retrieved from https://www2.deloitte.com/global/en/insights.html
12. European Commission. (2021). *Ethics guidelines for trustworthy AI*. European Union. Retrieved from https://digital-strategy.ec.europa.eu/en/policies/european-approach-artificial-intelligence
13. MIT Technology Review. (2023). *AI and the future of energy*. MIT Technology Review. Retrieved from https://www.technologyreview.com/
14. Center for Energy and Security Studies. (2022). *Cybersecurity in the AI-driven energy sector*. Center for Energy and Security Studies. Retrieved from https://ceness-russia.org/
15. IBM. (2023). *Applying AI and blockchain for decentralized energy markets*. IBM Research. Retrieved from https://www.ibm.com/research/
16. Deloitte and IEA. (2023). *Global trends in AI-enhanced renewable energy*. Deloitte and International Energy Agency. Retrieved from https://www.deloitte.com/

17. Accenture Research. (2023). *AI'S role in decarbonization and sustainable energy transition*. Accenture. Retrieved from https://www.accenture.com/
18. Pacific Gas and Electric (PG&E). (2022). *AI-driven smart grids: A case study*. PG&E. Retrieved from https://www.pge.com/
19. Stanford University. (2022). *Artificial intelligence for the future of energy. AI in predictive maintenance for energy systems*. Stanford AI Lab. Retrieved from https://ai.stanford.edu/
20. World Bank Group. (2021). *Artificial intelligence in developing sustainable energy solutions*. World Bank Group. Retrieved from https://www.worldbank.org/
21. Energy Systems Research Laboratory. (2023). *The impact of AI on future grid infrastructure*. University of California. Retrieved from https://esl.lbl.gov/
22. Institute of Electrical and Electronics Engineers (IEEE). (2023). Blockchain and AI in decentralized energy grids. *IEEE Access, 9*. Retrieved from https://ieeexplore.ieee.org/
23. Harvard Business Review. (2022). *AI and the transformation of the energy sector*. Harvard Business Review. Retrieved from https://hbr.org/
24. European Network of Transmission System Operators for Electricity (ENTSO-E). (2023). *Smart grids and AI integration across Europe*. ENTSO-E. Retrieved from https://www.entsoe.eu/
25. Institute for Energy Economics and Financial Analysis (IEEFA). (2023). *Investment trends in AI for renewable energy*. IEEFA. Retrieved from https://ieefa.org/
26. Oak Ridge National Laboratory. (2022). *AI-enhanced battery management systems for energy storage*. ORNL. Retrieved from https://www.ornl.gov/
27. Boston Consulting Group (BCG). (2022). *How AI is shaping the future of the energy industry*. Boston Consulting Group. Retrieved from https://www.bcg.com/
28. International Electrotechnical Commission (IEC). (2023). *Global standards for AI in energy management systems*. IEC Technical Report. Retrieved from https://www.iec.ch/

AI-Enhanced Optimization of Solar PV Systems: A Review of Machine Learning and Heuristic Methods

M. Muthukumar and A. Karthick

Abbreviations

AM	Additive manufacturing
ANFIS	Adaptive neuro-fuzzy inference system
ANN	Artificial neural network
ARIMA	Autoregressive integrated moving average
CPVTE	Concentrated photovoltaic/thermal energy
DE	Differential evolution
DES-UAE	Deep eutectic solvent-ultrasonic assisted extraction
DHS	District heating systems
DIFM	Data-integrated feature maps
DRL	Deep reinforcement learning
DRO	Distributionally robust optimization
EOHIPU	Enhanced optimization heuristic information processing unit
FOC	Fractional-order component
FO-KOA	Fractional-order krill optimization algorithm
FPV	Floating photovoltaic
HBA	Hydrogen bond acceptor
HBD	Hydrogen bond donor
IDEC	Improved deep embedded clustering
IGA	Improved genetic algorithm
ISABO	Improved self-adaptive backtracking optimization
MIP	Mixed integer programming
ML	Machine learning
MPPT	Maximum power point tracking
MRI	Magnetic resonance imaging

M. Muthukumar (✉) · A. Karthick
Renewable Energy Lab, Department of EEE, KPR Institute of Engineering and Technology, Coimbatore, India

P. Raj et al. (eds.), *Artificial Intelligence (AI) for IT Energy Efficiency and Green AI for Environment Sustainability*,
https://doi.org/10.1007/978-3-031-89420-6_13

NADES	Natural deep eutectic solvents
PCA	Principal component analysis
PPO	Proximal policy optimization
PV	Photovoltaic
PV/T	Photovoltaic/thermal
RL	Reinforcement learning
SEMPA	Selective ensemble marine predators algorithm
SVMR	Support vector machine regression
SVR	Support vector regression
TDA	Topological data analysis
TFC	Total flavonoid content
VPP	Virtual power plant
WECs	Wave energy converters
WNN-GA	Wavelet neural network-genetic algorithm
WSCL	Weakly supervised contrastive learning

1 Introduction

The study improves voltage stability and reduces power losses through a proximal policy optimization approach, enabling real-time control and communication among reactive power devices [1]. With energy availability often inconsistent in some regions, refrigerators have become essential appliances in everyday life [2]. Our study examines the impact of random sailing periods on vessel scheduling, showing an 83.3% reduction in average waiting times compared to deterministic scenarios [3]. Usability and acceptance testing are essential for healthcare interactive self-management systems, where frequent user-system interaction is crucial [4]. Two energy efficiency indicators from the additive manufacturing (AM) process were used to develop a quantification model for process-level energy efficiency. There is a growing demand for AM techniques to maximize energy conservation potential without sacrificing quality performance [5]. The findings provide a reliable method for accurately calculating density logs, significantly improving porosity assessments and playing a crucial role in reservoir characterization and evaluation [6]. Without effective photovoltaic (PV) waste management systems, solar panels, which have an average lifespan of 25 years and are being installed at an increasing rate, could significantly contribute to waste generation [7]. This study highlights the potential of floating photovoltaic (FPV) deployment in countries like Pakistan and underscores the significant role of water cooling in enhancing the efficiency of FPV systems [8]. Photovoltaic (PV)/thermal technology offers significant growth potential by generating both electricity and heat, enhancing the power generation capacity of photovoltaic cells through the combined advantages of each component [9]. The study reveals that China dominates the production, installed capacity, recycling capabilities, and primary material supply for photovoltaic systems. As the leading producer of complete PV systems, China also manufactures more than 50% of the raw materials used in these devices [10]. It proposes a strategic approach to overcome these obstacles by capitalizing on the region's substantial solar potential while addressing the specific barriers to adoption [11]. Building on our previous work on subset selection, we introduced a heuristic that consistently yielded better point sets, achieving an average 20% lower discrepancy than the initial point set for all dimensions where the star discrepancy can be computed [12]. The accuracy of motion detection can be improved by applying an overturn/translation extension method and incorporating a heuristic correction factor to enhance the fraction of true motion information in the detector output [13]. A mixed integer programming model with a novel level-space-time network representation is proposed, addressing reusable racks, robot-rack pairings, storage repositioning, and collision avoidance (Fig. 1) [14].

1.1 Solar PV System

Solar photovoltaic (PV) systems use semiconductor substances, basically silicon, to seize sunlight and convert it into electrical power. Electrons in those cells are excited via daylight, which leads to a cutting edge. Due to their modular design, PV structures can be multiplied from modest rooftop installations to huge sun farms. They function as a clean, sustainable replacement for fossil fuels and are a key issue of renewable electricity. The world is shifting into the direction of greener energy assets as a result of technological breakthroughs that have made PV structures extra cost-effective and efficient.

1.2 Machine Learning

Through overall performance and efficiency optimization, machine mastering (machine learning (ML)) is remodeling the sun photovoltaic structure. ML algorithms can precisely forecast strength output, come across defects, and function panels for the most beneficial daylighting with the aid of reading extensive datasets from PV systems. Additionally, these answers improve dependability and price-effectiveness by enabling strong control of strength garage structures and specific forecasting of the sun strength era. A widespread development is the incorporation of gadget mastering (ML) into PV systems, which permits intelligent, flexible solar power answers that could react to changing situations immediately. Electric power estimation relies on wave-to-wire model-based numerical analysis and

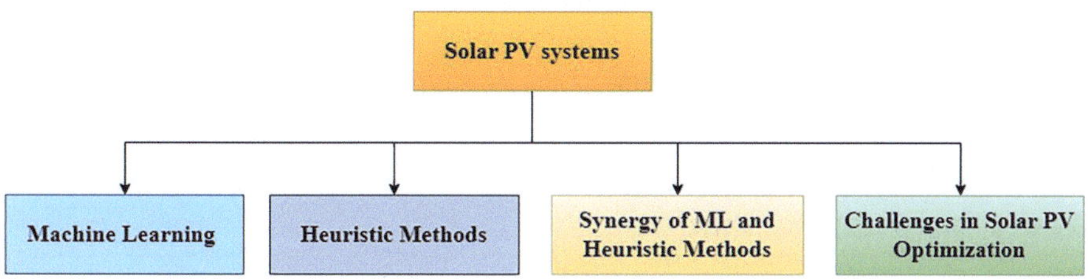

Fig. 1 Solar PV optimization methods

experimental studies of scaled-down prototypes. Enhanced predictive capabilities of wave energy converters provide significant benefits by optimizing energy conversion efficiency under varying ocean conditions, allowing for targeted adjustments to improve system stability and reliability (Fig. 2) [15].

1.3 Heuristic Methods

Heuristic strategies are imperative tools for optimizing sun PV systems, increasing their productivity and performance. These algorithms, which include genetic algorithms (GA), particle swarm optimization (PSO), and simulated annealing, excel in complex optimization problems in which traditional strategies often war with PV system panel orientation, tilt perspective, and electricity—to optimize parameters which include management strategies, various heuristic techniques are used. By carefully reading a big solution space, heuristic techniques allow particular planning of PV system operation, increasing strength availability, and enhancing gadget reliability. To validate the performance of the proposed Enhanced Optimization Heuristic Information Processing Unit, we conducted three experiments: comparing it with state-of-the-art approaches, evaluating the effectiveness of heuristic information, and testing the effectiveness of the new solution generation mechanism [16].

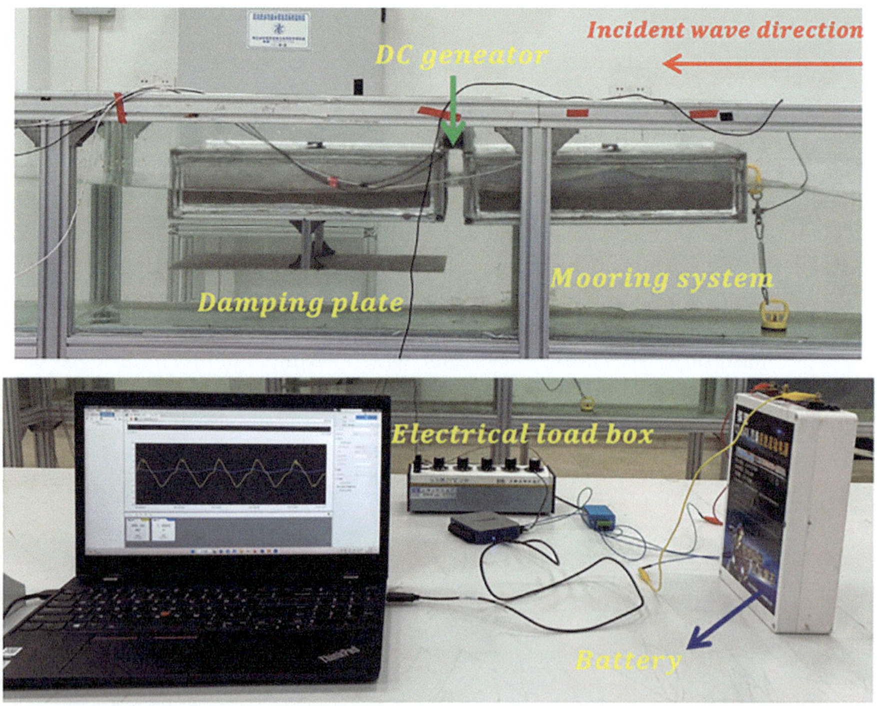

Fig. 2 Experimental setup in a wave flume [15]

1.4 Synergy of ML and Heuristic Methods

A synergistic combination of machine learning (ML) and heuristic methods provides a powerful approach for optimizing solar PV systems. Together, they form a flexible, intelligent system that can optimize PV parameters in real time for panel placement, energy storage, load management, etc. This complex integration delivers the power of both approaches in use and is an accurate, reliable, and efficient solar system that can dynamically meet changing conditions and power of the systems.

1.5 Challenges in Solar PV Optimization Methods

Optimizing solar PV systems is a complex task due to the range of sun irradiance, which complicates correct energy prediction and device performance. Balancing a couple of conflicting objectives, inclusive of maximizing energy output while minimizing expenses and environmental effects, further challenges the optimization system. The integration of superior methods like ML and heuristics introduces additional complexity, stressful significant computational sources, and expertise. Moreover, the swiftly evolving landscape of PV era calls for optimization techniques to constantly adapt to new materials, designs, and configurations, making it a dynamic and ongoing assignment (Table 1).

Table 1 AI-enhanced renewable energy integration for solar PV optimization

Reference	Findings	Application
[1]	Reinforcement learning with proximal policy optimization improves voltage control in PV and WT-integrated systems	Voltage stabilization in power systems with renewable energy integration
[17]	Spider monkey optimization effectively allocates solar PV and wind energy in radial distribution systems	Energy distribution optimization for renewable energy sources
[2]	Advanced methods enhance the efficiency and sustainability of solar PV-powered refrigeration systems	Optimizing solar-powered refrigeration for better energy efficiency
[18]	Machine learning models improve solar PV power prediction under fluctuating environmental conditions	Power generation forecasting for solar PV systems
[19]	Diverse machine learning algorithms with hyperparameter optimization enhance solar PV power prediction accuracy	Precise power prediction for solar PV installations
[20]	Multiobjective genetic algorithm and ML optimize semiconductor performance for solar PV-thermoelectric systems	Enhancing 4E performance (energy, exergy, economic, environmental) of PV–thermoelectric systems

2 Overview of Optimization Techniques in Solar PV Systems

Strategies to optimize solar PV systems are essential to improve energy efficiency, reliability, and tax efficiency. These techniques range from traditional methods such as linear programming and gradient-based absolute algorithms to advanced techniques including machine learning (ML) and heuristic methods. ML excels in predictive analytics and real-time machine integration, while mostly for heuristic methods such as genetic algorithms and particle swarm optimization. Hybrid methods that fully analyze PV gadget design are being developed by a combination of two optimization methods to provide versatile solutions to vary the dynamics of solar power generation and environmental conditions.

2.1 Traditional Optimization Methods

Traditional strategies for solar PV machine optimization encompass strategies consisting of linear programming, gradient-primarily based optimization, and dynamic programming. These methods are based totally on mathematical models to determine the most advantageous answer for a selected objective together with electricity maximization or price discount. Linear programming is widely used to optimize power distribution and resource allocation, while gradient-primarily based methods apply quite simple ascent or descent techniques to improve performance metrics. Flexibility: Although powerful on well-defined problems, conventional strategies and nonlinear complexity and multiobjective collide contradict the traits of real-world PV systems, motivating the search for advanced flexible approaches which include heuristic gadget mastering strategies. Figure 3 illustrates the spider monkey group playing a crucial role in the future allocation and integration of renewable energy, promoting more resilient and sustainable power systems [17].

In this study, we developed an improved genetic algorithm for optimizing pipe diameters in existing large-scale district heating systems, selecting only pipes with potential profitability from reconstruction as design variables [21]. However, as process systems evolve towards greater refinement, optimizing dynamic processes like batch processes and grade transitions becomes increasingly important [22]. The experimental total flavonoid content (TFC) and antioxidant capacity closely aligned with the predicted results, and significant differences were observed in TFC, antioxidant capacity, and polyphenol profiles between deep eutectic solvent-ultrasonic assisted extraction and other extraction methods [23]. To achieve this objective, we carefully selected the optimal natural deep eutectic solvents (NADES), using betaine as the hydrogen bond acceptor and urea as the hydrogen bond donor, with a molar ratio of 1:2 and 50% water content, based on a selection of eight

Fig. 3 Spider monkey group [17]

representative NADES [24]. Additionally, the use of different normalization approaches limits the consistency of objective weighting methods, potentially causing variation in the decision outcomes [25].

3 Machine Learning Approaches for Solar PV Optimization

Advanced algorithms are used in machine learning (ML) techniques for solar PV optimization to improve system performance and decision-making. In order to find trends, forecast energy output, and improve system configurations, machine learning techniques like neural networks, support vector machines, and reinforcement learning (RL) examine enormous datasets. These methods may dynamically modify variables like panel orientation and energy storage, anticipate solar irradiance, and identify abnormalities. ML techniques offer precise and adaptive solutions by utilizing real-time inputs and previous data, greatly enhancing the dependability and efficiency of solar PV systems under a variety of environmental circumstances. Through thorough research and analysis, the results show that support vector machine regression is the top-performing algorithm, achieving an exceptional $R2$ value of 0.99 with minimal errors, demonstrating superior predictive capabilities for output power [18]. The novel contribution of this work is the extension of this methodology to energy system modeling at national or regional scales, while preserving high temporal resolution and comprehensive sector coupling [26]. It incorporates sustainable technologies, such as solar panels, geothermal heat pumps, and battery energy storage systems, to decarbonize building energy use while enhancing efficiency and comfort [27]. A green hydrogen production method based on solar energy is proposed, with machine learning (ML) models employed for system optimization. The simulations utilize a transient model, where solar radiation, wind speed, and ambient temperature are treated as time-dependent functions [28]. The findings suggest that ML algorithms, when trained on a comprehensive dataset spanning the entire year, provide a more holistic understanding of the complex dynamics involved in solar PV power forecasting [19]. The investigation focused on assessing the concentrated photovoltaic/thermal energy system across multiple performance criteria, including energy, exergy, environmental, and economic aspects [20]. To address this challenge, this work presents a novel technique that integrates a multirenewable energy system using a multiobjective optimization algorithm to meet the machine learning-based forecasted load model [29].

3.1 Machine Learning Techniques

However, traditional numerical simulations are time-consuming and expensive for power prediction of full-scale WECs operated in the real sea. More importantly, they cannot accurately predict WECs' power performance in real time, thereby cannot be used for prediction-based real-time control. Improved predictive capabilities of wave energy converters offer significant benefits, optimizing energy conversion efficiency under varying ocean conditions and enabling targeted adjustments to enhance system's stability and reliability [15].

3.1.1 Supervised Learning

In PV optimization, supervised learning makes use of labeled datasets to develop algorithms that forecast and improve system performance. Supervised learning models, including regression and classification algorithms, can find correlations between input factors like panel angles and weather conditions and their effects on energy output by examining past data with known results. This method makes it possible to optimize system settings and forecast energy yield precisely. Supervised learning techniques can continuously improve their predictions and suggestions through repeated training and validation, resulting in solar PV systems that are more dependable and efficient. Additionally, we introduce an adaptive edge-adding strategy to the graph, which helps alleviate the interaction deficiency problem that may arise from existing data augmentation based on masking strategies [30]. Current supervised deep-learning representations have shown promising results in supporting diagnosis. The weakly supervised contrastive learning strategy could potentially be used in clinical scenarios with limited annotated MRI data, empowering deep-learning applications in clinical settings and aiding in disease diagnosis [31]. The innovative semisupervised learning methods were inspired by topological data analysis. Specifically, we propose two distinct semisupervised learning techniques, each leveraging a unique topological framework [32]. Machine learning, particularly supervised learning algorithms, has emerged as a powerful tool in drug discovery and development, offering significant potential across its various stages [33].

3.1.2 Unsupervised Learning

Unsupervised learning techniques, such as clustering and dimensionality reduction, can analyze unlabeled solar PV data to identify patterns and trends. This approach can reveal insights into system behavior, aiding in anomaly detection, feature extraction, and optimization, without requiring explicit labels. Our method efficiently predicts worst-case load increase scenarios, significantly reducing computational complexity compared to traditional methods. This makes it ideal for real-time applications like voltage collapse margin assessments [34]. The goal is to develop a learning-based framework that can collaboratively steer the network toward optimal configurations using incremental control. This involves learning local surrogates to approximate optimal power flow (OPF) solutions based on local information [35]. Quadratic unconstrained support vector learning (QUSL) uses similarity triplets for unsupervised learning, creating positive samples by perturbing images, allowing for a learning process independent of traditional algorithms [36]. Traditionally, dimensionality reduction techniques like principal component analysis (PCA) or autoencoders, or simpler visualizations like self-organizing maps, have been used to facilitate dataset ranking and visualization of high-dimensional spaces [37].

3.1.3 Reinforcement Learning

Reinforcement learning (RL) optimizes solar PV systems by training algorithms to make sequential decisions that maximize long-term performance. By interacting with the environment, RL models learn to adjust parameters like orientation, tilt angles, and energy storage, enhancing efficiency and reliability. This adaptive approach enables real-time adjustments and continuous improvement based on experience and data. Reinforcement learning has emerged as a powerful tool for the reliability and maintainability community, addressing complex sequential decision-making problems under uncertainty [38]. Previous studies have shown limitations in detecting low-frequency network intrusions, and there is limited exploration of reinforcement learning in this area. Correlation-based feature selection is effective for system recall, while information gain-based selection excels in identifying DoS attack categories [39]. Consequently, automated optimization methods are urgently needed. Deep reinforcement learning (DRL) shows great promise in automating radiotherapy planning. However, training and applying DRL networks require numerous simulations, each involving evaluating a plan's quality, which can be computationally intensive [40]. Reinforcement learning is a promising machine learning approach for Digital Twins, as it leverages virtual environments to learn optimal policies, aligning well with the capabilities of Digital Twins [41].

4 Applications of Machine Learning in PV Systems

Machine learning (ML) is revolutionizing solar PV systems. ML algorithms enable predictive maintenance, solar irradiance forecasting, and energy output optimization. By analyzing historical data, ML enhances system reliability and efficiency, allowing for real-time adjustments and optimized energy storage management. The goal is to accurately predict liquid fraction in building-integrated PV systems using machine learning models. Exploring advanced hybrid models and techniques can further enhance computational efficiency and prediction accuracy [42]. While this advancement is promising, it raises concerns about privacy protection for sensitive personal data used in machine learning models. The absence of robust information systems hinders securing funding for such projects [43]. Traditional methods for solar radiation calculation and PV system assessment are no longer sufficient. Machine learning algorithms now provide more accurate predictions. Categorizing and evaluating these methods are challenging due to varying locations, materials, and environmental conditions [44]. The IEA forecasts significant growth in solar energy, contributing up to 16% of global electricity by 2050. Accurate solar energy production prediction is crucial. While the study's findings are reliable within its parameters, broader conclusions should be drawn cautiously [45].

4.1 Regression Analysis

Regression analysis predicts solar energy output based on historical data and factors like weather, panel orientation, and system efficiency. This predictive capability enables better planning, resource management, and decision-making for optimized solar PV system performance.

Performance evaluation shows that artificial neural networks (ANNs) excel in capturing complex relationships between features and efficiency, even with limited data, compared to models like multiple linear regression [46]. Conventional one-dimensional analysis is time-consuming and inefficient for identifying optimal conditions. This method offers a more efficient approach by minimizing simultaneous parameter changes [47]. PV technology has gained significant interest due to its economic and environmental benefits. However, in Nsukka, Nigeria, solar panel tilt angles greater than 0° during July and August can reduce power output efficiency by 9.11–19.73% [48]. Supervised machine learning techniques like ANN and adaptive neuro-fuzzy inference system are effective for maximum power point tracking (MPPT) in PV systems. This paper introduces RQGPR as a novel approach for MPPT tracking [49]. To identify the best model for energy prediction, data was gathered, and a relevance check was performed. Multicollinearity and backward elimination analyses were used to select the most relevant independent variables [50]. Support vector regression (SVR) is effective for nonlinear forecasting.

Optimizing SVR parameters (C and γ) using cuckoo search and differential evolution (DE) algorithms is crucial for accurate forecasting models [51].

4.2 Clustering Algorithms

Clustering algorithms group similar solar PV performance data to identify patterns and anomalies. Techniques like K-means and DBSCAN categorize data based on features like energy output and environmental conditions. This grouping helps analyze system behavior, optimize configurations, and detect issues, leading to more efficient solar PV systems. A decentralized P2P energy sharing strategy is implemented using an alternating direction method of multipliers algorithm. A virtual energy storage model for HVAC systems is established to co-optimize energy and reserves between buildings. A two-stage distributionally robust optimization (DRO) model is proposed to address the underdeveloped application of DRO methods in P2P environments and the co-optimization of flexible HVAC systems for additional reserves in smart building clusters [52]. The proposed method improved optimization results, reducing degradation and total costs. Performance degradation phenomena, like membrane thinning, catalyst dissolution, and transport layer passivation, are expressed as functions of input power [53]. An improved deep embedded clustering algorithm is integrated with K-means to group historical data from distributed photovoltaics and loads, creating typical scenarios that accurately capture the coupling information characteristics of optical charge data and generate typical joint scenarios characterizing light charge uncertainties [54]. It considers uncertainties in new energy sources, soft open point (SOP) control, and static var compensator (SVC) dispatch. Integrating the virtual power plant with the distribution network can compensate for active power shortages, improve power flow, reduce losses, and lower costs [55].

4.3 Dimensionality Reduction

Dimensionality reduction simplifies solar PV optimization by reducing variables. Techniques like PCA transform high-dimensional data into fewer dimensions, aiding visualization and analysis. This focuses on key factors, streamlining optimization and improving efficiency. This enhances the clarity and effectiveness of optimization strategies in solar PV systems. Hybrid deep learning framework is used to quantify short-term mismatches between PV power and TLPs using a real rural dataset. The VAE model effectively reduces dimensionality and extracts features from high-resolution data, proving its applicability for analyzing real-world power-load data [56]. A fractional-order component and a learning-based exploration strategy are combined to optimize a PV-STATCOM system. This approach promotes knowledge sharing and avoids local optima, leading to efficient and effective solutions for future distribution systems [57]. Soiling rates vary based on location, season, system configuration, and weather. Optimizing cleaning frequency is crucial to maximize power output, efficiency, and minimize costs, despite the limitations of experimental conditions [58]. Bypass diodes prevent hotspots in PV modules, leading to multiple local peaks and a single global peak in the power curve. The proposed algorithm outperforms existing metaheuristic approaches. Figure 4 focuses on eliminating reinitialization during irradiance fluctuations to enhance system stability [59].

This improved forecasting enhances solar resource utilization by enabling seamless grid integration, efficient load balancing, and optimization of generation, distribution, and storage in solar power systems [60]. The proposed improved self-adaptive backtracking optimization (ISABO) method enhances traditional optimization by incorporating adaptive mechanisms and improving underlying processes. It outperforms other methods like FA and PSO in achieving optimal system formation [61]. The fractional-order element and Lévy Flight-based Exploration Algorithm enhance information sharing and avoid local optima. The proposed fractional-order krill optimization algorithm approach is specifically designed for KC-200, SP-70, and Ultra-Power-85 PV modules, outperforming previous methods [62]. Advanced MPPT algorithms are crucial for finding the global peak power point to maximize power production. The proposed MPPT method outperforms other algorithms like COA, PSO, IDE, ITG, ISSA, and ISO [63].

4.4 Ensemble Methods

Ensemble methods combine multiple machine learning models to improve predictive accuracy and robustness in solar PV optimization. Techniques like random forests and gradient boosting aggregate predictions, leading to more reliable forecasts,

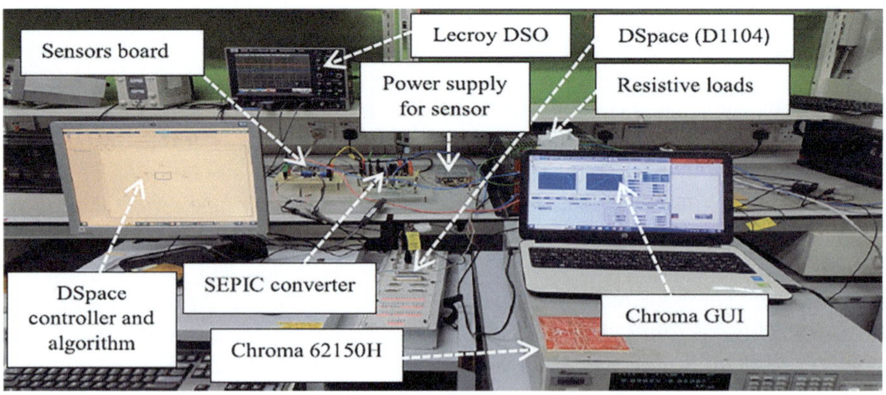

Fig. 4 Experimental setup [59]

optimized parameters, and improved fault detection. This enhances the overall effectiveness of optimization strategies. Accurate forecasting of variable renewable energy generation is crucial for efficient grid balancing and dispatch. Decision tree-based ensemble models have proven effective in improving the accuracy of photovoltaic power generation forecasts [64]. Feature vector extraction identified significant variables, and the SOM method clustered weather patterns based on DHI and GHI data. This preprocessing enhanced accuracy [65]. The selective ensemble marine predators algorithm (SEMPA) is introduced to optimize system size and minimize TNPC while enhancing reliability. SEMPA outperforms traditional MPA, HS, and PSO methods in terms of generating cost, renewable sources, and convergence time [66]. Accurate PV power prediction enhances energy quality and power system stability. The original dataset was processed using data-integrated feature maps (DIFMs) to address uncertainties. The DIFM outputs were used as ELM inputs for final prediction [67]. The hybrid autoregressive integrated moving average (ARIMA)-GP2 (0, 0, 0) model outperformed the empirical approach and other ARIMA configurations for predicting generalized H datasets. The empirical method was unsuitable for these datasets [68].

5 Heuristic Methods for Solar PV Optimization

5.1 Genetic Algorithms

Genetic algorithms (GAs) optimize solar PV systems by simulating natural selection. They explore various configurations for parameters like panel orientation and energy storage, iteratively refining solutions based on performance metrics. GAs excel in multiobjective problems and handling nonlinear constraints, enhancing system efficiency and effectiveness. A genetic algorithm is used to optimize PV array parameters by condensing multiple sets of results into a single, highly precise set. For insufficient data accuracy, a hybrid genetic algorithm can be used to improve precision [69]. The system generates electricity from solar PV panels. Excess electricity is used to produce hydrogen through water electrolysis and stored in high-pressure tanks. An intelligent energy management system optimizes PV production, hydrogen generation/storage, and grid electricity consumption [70]. Identifying accurate PV model parameters is crucial for modeling and simulation. This study proposes a novel optimization genetic algorithm to accurately extract parameters for both single- and double-diode models across different PV technologies [71]. A life cycle cost analysis and cost per kWh are conducted for two PV and battery configurations with and without considering NAD. The GA tool in MATLAB is used for optimal sizing, and LPS is used to evaluate reliability. The goal is to assess the economic feasibility of the system in Oman [72]. Accurate PV modeling depends on precise parameter values. A global optimization approach is used to extract these parameters, ensuring their applicability across various solar radiation and

Fig. 5 The experimental platform of a roof PV plant

temperature conditions [73]. Partial shading in solar PV arrays causes multiple peaks in the P-V curve, degrading performance. The algorithm's performance has been evaluated under various shading patterns [74]. Figure 5 illustrates wavelet neural network-genetic algorithm hybrid model to improve the accuracy of photovoltaic power plant output forecasting. Due to module degradation, using a current training model for 25-year predictions can lead to significant deviations [75].

5.2 *Particle Swarm Optimization (PSO)*

Particle swarm optimization (PSO) simulates social behavior to find optimal solutions for solar PV systems. Particles represent potential solutions, adjusting their positions based on individual and collective experiences. PSO efficiently explores and exploits the search space for optimal panel placements, tilt angles, and energy storage strategies. Its ability to converge quickly and handle complex problems makes it effective for improving solar PV system performance. The article improves multiobjective particle swarm optimization using adaptive learning factors and nonlinear inertia weight. In the wind–solar hydrogen storage system, clean energy contributed 94.1%, with wind and solar at 64%, storage at 30.1%, and grid at 5.9%, proving the system's feasibility [76]. Amid evolving power systems and the increasing integration of renewable energy sources, optimizing the configuration of multienergy complementary systems is essential for sustainable development. This study focuses on developing an optimal configuration for a wind–solar storage system [77]. MUGF-PSO enhances MUGF by optimizing mutation operator selection using PSO. This approach analyzes past testing effectiveness to guide future testing. Applying swarm intelligence optimization to software testing is a promising research direction [78]. TEP approach using NBPSO to optimize the location and number of TLCs. The study considers two scenarios: one without solar PV and one with solar PV integration [79]. This paper determines five PV module parameters using manufacturer data, considering solar radiation and cell temperature effects. The method eliminates the need for assuming an ideality factor to extract these parameters [80]. Artificial intelligence (AI)-based PSO and DE to optimize a stand-alone hybrid solar PV-hydro-battery power system. The goal is to minimize cost while ensuring a reliable energy supply and an acceptable LPSP [81]. Variability in solar and wind resources in microgrids and distribution networks can lead to power quality and stability issues. Fuzzy-PSO forecasting effectively handles large datasets for predictive modeling, especially when statistical methods struggle to establish perfect correlations between input and output variables [82]. A novel microgrid energy management strategy combines chaotic local search with particle swarm optimization (PSO) to achieve optimal global solutions for the objective function [83].

6 Challenges and Future Directions

Solar PV optimization presents significant challenges, including the intermittent nature of solar energy, balancing multiple objectives, computational complexity, rapid technological advancements, and grid integration issues. Future directions focus on advancing ML algorithms, hybrid optimization techniques, real-time optimization, energy storage integration, microgrid optimization, and supportive policies to address these challenges and unlock the full potential of solar PV systems for a sustainable energy future.

7 Conclusion

The optimization of solar photovoltaic (PV) systems represents a dynamic domain that is undergoing swift advancements attributable to innovations in machine learning (ML) and artificial intelligence (AI), hybrid optimization methodologies, energy storage integration, and microgrid technologies. These technological advancements effectively address critical challenges, including intermittency, the necessity to reconcile multiple objectives, and the intricacies associated with computational complexity, thereby resulting in solar PV systems that are increasingly efficient, reliable, and economically viable. As these technologies persist in their evolution, they are poised to substantially augment the performance and integration of solar PV systems, thereby reinforcing their significance in promoting a sustainable energy future.

References

1. Rehman, A. U., Ullah, Z., Qazi, H. S., Hasanien, H. M., & Khalid, H. M. (2024). Reinforcement learning-driven proximal policy optimization-based voltage control for PV and WT integrated power system. *Renewable Energy, 227*, 120590. https://doi.org/10.1016/j.renene.2024.120590
2. Saha, G., & Azad, A. A. M. (2024). A review of advancements in solar PV-powered refrigeration: Enhancing efficiency, sustainability, and operational optimization. *Energy Reports, 12*, 1693. https://doi.org/10.1016/j.egyr.2024.07.046
3. Li, R., Zhang, X., Wang, C., Cui, J., & Mu, M. (2024). A machine learning-based adaptive heuristic for vessel scheduling problem under uncertainty via chance-constrained programming. *Computers and Electrical Engineering, 119*, 109523. https://doi.org/10.1016/j.compeleceng.2024.109523
4. Chen, T., Chen, J., & Or, C. K. (2024). Heuristic evaluation and end-user testing of a machine learning-based lower-limb exercise training system for management of knee pain in individuals aged 55 years or over. *International Journal of Industrial Ergonomics, 102*, 103607. https://doi.org/10.1016/j.ergon.2024.103607
5. Gao, B., et al. (2024). Integration of improved meta-heuristic and machine learning for optimizing energy efficiency in additive manufacturing process. *Energy, 306*, 132518. https://doi.org/10.1016/j.energy.2024.132518
6. Rahmati, M., Zargar, G., & Ayatizadeh Tanha, A. (2024). Predicting density log from well log using machine learning techniques and heuristic optimization algorithm: A comparative study. *Petroleum Research, 9*(2), 176–192. https://doi.org/10.1016/j.ptlrs.2024.01.008
7. Ngagoum Ndalloka, Z., Vijayakumar Nair, H., Alpert, S., & Schmid, C. (2024). Solar photovoltaic recycling strategies. *Solar Energy, 270*, 112379. https://doi.org/10.1016/j.solener.2024.112379
8. Koondhar, M. A., Albasha, L., Mahariq, I., Graba, B. B., & Touti, E. (2024). Reviewing floating photovoltaic (FPV) technology for solar energy generation. *Energy Strategy Reviews, 54*, 101449. https://doi.org/10.1016/j.esr.2024.101449
9. Ying, S., Zhang, X., Wu, Y., & Pan, Z. (2024). Solar photovoltaic/thermal (PV/T) systems with/without phase change materials (PCMs): A review. *Journal of Energy Storage, 89*, 111582. https://doi.org/10.1016/j.est.2024.111582
10. Chadly, A., Moawad, K., Salah, K., Omar, M., & Mayyas, A. (2024). State of global solar energy market: Overview, China's role, challenges, and opportunities. *Sustainable Horizons, 11*, 100108. https://doi.org/10.1016/j.horiz.2024.100108
11. Bisu, A. A., Ahmed, T. G., Ahmad, U. S., & Maiwada, A. D. (2024). A SWOT analysis approach for the development of photovoltaic (PV) energy in northern Nigeria. *Cleaner Energy Systems, 9*, 100128. https://doi.org/10.1016/j.cles.2024.100128
12. Clément, F., Doerr, C., & Paquete, L. (2024). Heuristic approaches to obtain low-discrepancy point sets via subset selection. *Journal of Complexity, 83*, 101852. https://doi.org/10.1016/j.jco.2024.101852
13. Li, Y., Sun, C., Zhang, B., Wang, P., & Fu, L. (2024). Attitude determination via pseudo measurement update approach combined with heuristic wavelet denoising. *Measurement (London), 231*, 114630. https://doi.org/10.1016/j.measurement.2024.114630
14. Wang, I. L., & Wang, T. H. (2025). Efficient routing in robotic movable fulfillment systems with integer programming: A rolling horizon and heuristic approach. *Robotics and Computer-Integrated Manufacturing, 91*, 102849. https://doi.org/10.1016/j.rcim.2024.102849
15. Wang, L. G., Wen, C. W., Wu, S. X., & Wu, S. (2024). Electric power prediction of a two-body hinge-barge wave energy converter using machine learning techniques. *Ocean Engineering, 305*, 117935. https://doi.org/10.1016/j.oceaneng.2024.117935
16. Ni, Y., Liu, W., Du, X., Xiao, R., Chen, G., & Wu, Y. (2024). Evolutionary optimization approach based on heuristic information with pseudo-utility for the quadratic assignment problem. *Swarm and Evolutionary Computation, 87*, 101557. https://doi.org/10.1016/j.swevo.2024.101557

17. Sultana, W., & Jebaseelan, S. D. S. (2024). Optimal allocation of solar PV and wind energy power for radial distribution system using spider monkey optimization. *Sustainable Computing: Informatics and Systems, 42*, 100986. https://doi.org/10.1016/j.suscom.2024.100986
18. Tripathi, A. K., et al. (2024). Advancing solar PV panel power prediction: A comparative machine learning approach in fluctuating environmental conditions. *Case Studies in Thermal Engineering, 59*, 104459. https://doi.org/10.1016/j.csite.2024.104459
19. Tahir, M. F., Yousaf, M. Z., Tzes, A., El Moursi, M. S., & El-Fouly, T. H. M. (2024). Enhanced solar photovoltaic power prediction using diverse machine learning algorithms with hyperparameter optimization. *Renewable and Sustainable Energy Reviews, 200*, 114581. https://doi.org/10.1016/j.rser.2024.114581
20. Alghamdi, H., et al. (2024). Semiconductors for enhanced solar photovoltaic-thermoelectric 4E performance optimization: Multi-objective genetic algorithm and machine learning approach. *Results in Engineering, 23*, 102573. https://doi.org/10.1016/j.rineng.2024.102573
21. Xu, H., Zhang, L., Wang, X., Han, B., Luo, Z., & Bai, B. (2024). Improved genetic algorithm for pipe diameter optimization of an existing large-scale district heating network. *Energy, 304*, 131970. https://doi.org/10.1016/j.energy.2024.131970
22. Zhou, C., Wang, S., Su, H., Tang, X., Cao, Y., & Yang, S. H. (2024). Dynamic controlled variables based dynamic self-optimizing control. *Journal of Process Control, 138*, 103228. https://doi.org/10.1016/j.jprocont.2024.103228
23. Liu, Y., et al. (2022). Ultrasonic-assisted extraction of polyphenolic compounds from Paederia scandens (Lour.) Merr. Using deep eutectic solvent: Optimization, identification, and comparison with traditional methods. *Ultrasonics Sonochemistry, 86*, 106005. https://doi.org/10.1016/j.ultsonch.2022.106005
24. Peng, W., et al. (2024). Comparison, optimization and antioxidant activity of ultrasound-assisted natural deep eutectic solvents extraction and traditional method: A greener route for extraction of flavonoid from Moringa oleifera Lam. leaves. *Ultrasonics Sonochemistry, 109*, 107003. https://doi.org/10.1016/j.ultsonch.2024.107003
25. Chatterjee, S., & Chakraborty, S. (2024). A study on the effects of objective weighting methods on TOPSIS-based parametric optimization of non-traditional machining processes. *Decision Analytics Journal, 11*, 100451. https://doi.org/10.1016/j.dajour.2024.100451
26. Prina, M. G., Dallapiccola, M., Moser, D., & Sparber, W. (2024). Machine learning as a surrogate model for EnergyPLAN: Speeding up energy system optimization at the country level. *Energy, 307*, 132735. https://doi.org/10.1016/j.energy.2024.132735
27. Chen, W. H., & You, F. (2024). Sustainable energy management and control for decarbonization of complex multi-zone buildings with renewable solar and geothermal energies using machine learning, robust optimization, and predictive control. *Applied Energy, 372*, 123802. https://doi.org/10.1016/j.apenergy.2024.123802
28. Salari, A., Shakibi, H., Soleimanzade, M. A., Sadrzadeh, M., & Hakkaki-Fard, A. (2024). Application of machine learning in evaluating and optimizing the hydrogen production performance of a solar-based electrolyzer system. *Renewable Energy, 220*, 119626. https://doi.org/10.1016/j.renene.2023.119626
29. Naveena, K., et al. (2024). Elevating sustainability with a multi-renewable hydrogen generation system empowered by machine learning and multi-objective optimization. *Measurement: Sensors, 33*, 101192. https://doi.org/10.1016/j.measen.2024.101192
30. Sun, X., Sun, F., Zhang, Z., Li, P., & Wang, S. (2024). Adaptive self-supervised learning for sequential recommendation. *Neural Networks, 179*, 106570. https://doi.org/10.1016/j.neunet.2024.106570
31. Gutiérrez, Y., Arevalo, J., & Martínez, F. (2024). A contrastive weakly supervised learning to characterize malignant prostate lesions in BP-MRI. *Biomedical Signal Processing and Control, 96*, 106584. https://doi.org/10.1016/j.bspc.2024.106584
32. Inés, A., Domínguez, C., Heras, J., Mata, G., & Rubio, J. (2024). A topological approach for semi-supervised learning. *Journal of Computer Science, 82*, 102403. https://doi.org/10.1016/j.jocs.2024.102403

33. Obaido, G., et al. (2024). Supervised machine learning in drug discovery and development: Algorithms, applications, challenges, and prospects. *Machine Learning with Applications, 17*, 100576. https://doi.org/10.1016/j.mlwa.2024.100576
34. Marcial, A., & Perninge, M. (2024). Finding a closest saddle–node bifurcation in power systems: An approach by unsupervised deep learning. *Electric Power Systems Research, 235*, 110632. https://doi.org/10.1016/j.epsr.2024.110632
35. Yuan, Z., Cavraro, G., Zamzam, A. S., & Cortés, J. (2024). Unsupervised learning for equitable DER control. *Electric Power Systems Research, 234*, 110634. https://doi.org/10.1016/j.epsr.2024.110634
36. Yu, L. H., Li, X. Y., Chen, G., Zhu, Q. S., Li, H., & Yang, G. W. (2024). QUSL: Quantum unsupervised image similarity learning with enhanced performance. *Expert Systems with Applications, 258*, 125112. https://doi.org/10.1016/j.eswa.2024.125112
37. Fotopoulou, S. (2024). A review of unsupervised learning in astronomy. *Astronomy and Computing, 48*, 100851. https://doi.org/10.1016/j.ascom.2024.100851
38. Zhang, Q., Liu, Y., Xiang, Y., & Xiahou, T. (2024). Reinforcement learning in reliability and maintenance optimization: A tutorial. *Reliability Engineering & System Safety, 251*, 110401. https://doi.org/10.1016/j.ress.2024.110401
39. Sharma, A., & Singh, M. (2024). Batch reinforcement learning approach using recursive feature elimination for network intrusion detection. *Engineering Applications of Artificial Intelligence, 136*, 109013. https://doi.org/10.1016/j.engappai.2024.109013
40. Li, C., Guo, Y., Lin, X., Feng, X., Xu, D., & Yang, R. (2024). Deep reinforcement learning in radiation therapy planning optimization: A comprehensive review. *Physica Medica, 125*, 104498. https://doi.org/10.1016/j.ejmp.2024.104498
41. Goldenits, G., Mallinger, K., Raubitzek, S., & Neubauer, T. (2024). Current applications and potential future directions of reinforcement learning-based Digital Twins in agriculture. *Smart Agricultural Technology, 8*, 100512. https://doi.org/10.1016/j.atech.2024.100512
42. Kiaghadi, M., Keshvarinia, M., Boora, F. M., & Mousavi, S. M. (2024). Machine learning applications for predicting liquid fraction in a PV system with NEPCM and fins. *Case Studies in Thermal Engineering, 61*, 104819. https://doi.org/10.1016/j.csite.2024.104819
43. Michalakopoulos, V., Sarantinopoulos, E., Sarmas, E., & Marinakis, V. (2024). Empowering federated learning techniques for privacy-preserving PV forecasting. *Energy Reports, 12*, 2244–2256. https://doi.org/10.1016/j.egyr.2024.08.033
44. Jobayer, M., Shaikat, M. A. H., Naimur Rashid, M., & Hasan, M. R. (2023). A systematic review on predicting PV system parameters using machine learning. *Heliyon, 9*, e16815. https://doi.org/10.1016/j.heliyon.2023.e16815
45. Rezaul Karim, S. M., Sarker, D., & Monirul Kabir, M. (2024). Analyzing the impact of temperature on PV module surface during electricity generation using machine learning models. *Cleaner Energy Systems, 9*, 100135. https://doi.org/10.1016/j.cles.2024.100135
46. Sahin, G., Isik, G., & van Sark, W. G. J. H. M. (2023). Predictive modeling of PV solar power plant efficiency considering weather conditions: A comparative analysis of artificial neural networks and multiple linear regression. *Energy Reports, 10*, 2837–2849. https://doi.org/10.1016/j.egyr.2023.09.097
47. Vicente, A. T., Wojcik, P. J., Mendes, M. J., Águas, H., Fortunato, E., & Martins, R. (2017). A statistics modeling approach for the optimization of thin film photovoltaic devices. *Solar Energy, 144*, 232.
48. Olusegun Ogundimu, E., Graham Richards, C., Mgbemene, C., & Johannes Ehlers, P. (2024). Optimization of the tilt-angle and climatic factors of a photovoltaic (PV) solar system through experimental and regression analysis modeling. *International Journal of Applied Engineering and Technology (London), 6*, 1754.
49. Farayola, A. M., Hasan, A. N., & Ali, A. (2018). Optimization of PV systems using data mining and regression learner MPPT techniques. *Indonesian Journal of Electrical Engineering and Computer Science, 10*(3), 1080–1089. https://doi.org/10.11591/ijeecs.v10.i3.pp1080-1089

50. Babatunde, A. A., & Abbasoglu, S. (2019). Predictive analysis of photovoltaic plants specific yield with the implementation of multiple linear regression tool. *Environmental Progress and Sustainable Energy, 38*(4). https://doi.org/10.1002/ep.13098
51. Alrashidi, M., Alrashidi, M., Pipattanasomporn, M., & Rahman, S. (2018). Short-term PV output forecasts with support vector regression optimized by cuckoo search and differential evolution algorithms.
52. Sun, G., Chen, X., Zhou, Y., Chen, S., & Wei, Z. (2024). Decentralized distributionally robust energy and reserve co-optimization of smart building clusters with virtual energy storage capability. *Energy and Buildings, 309*, 114040. https://doi.org/10.1016/j.enbuild.2024.114040
53. Xu, Y., & Chen, H. (2023). Layered power scheduling optimization of PV hydrogen production system considering performance attenuation of PEMEL. *Global Energy Interconnection, 6*, 714–725. https://doi.org/10.1016/j.gloei.2023.12.005
54. Liao, J., Lin, J., & Wu, G. (2024). Two-layer optimization configuration method for distributed photovoltaic and energy storage systems based on IDEC-K clustering. *Energy Reports, 11*, 5172–5188. https://doi.org/10.1016/j.egyr.2024.04.047
55. Li, Y., Yang, X., Deng, F., Deng, Y., Wang, X., & Zhang, Z. (2024). Bi-level optimization of novel distribution network with VPP and flexible load cluster. *Energy Reports, 12*, 504–516. https://doi.org/10.1016/j.egyr.2024.06.037
56. Gao, D., Zhi, Y., Rong, X., & Yang, X. (2025). Mismatch analysis of rooftop photovoltaics supply and farmhouse load: Data dimensionality reduction and explicable load pattern mining via hybrid deep learning. *Applied Energy, 377*, 124520. https://doi.org/10.1016/j.apenergy.2024.124520
57. Shaheen, A. M., Alassaf, A., Alsaleh, I., & Elsayed, A. M. (2024). Enhanced Kepler optimization for efficient penetration of PV sources integrated with STATCOM devices in power distribution systems. *Expert Systems with Applications, 253*, 124333. https://doi.org/10.1016/j.eswa.2024.124333
58. Kaiss, E. C. A., & Hassan, N. M. (2024). Optimizing the cleaning frequency of solar photovoltaic (PV) systems using numerical analysis and empirical models. *Renewable Energy, 228*, 120691. https://doi.org/10.1016/j.renene.2024.120691
59. Talib Naser, A., Khairullah Mohammed, K., Fadilah Ab Aziz, N., Elsanabary, A., Binti Kamil, K., & Mekhilef, S. (2024). A fast-tracking MPPT-based modified coot optimization algorithm for PV systems under partial shading conditions. *Ain Shams Engineering Journal, 15*, 102967. https://doi.org/10.1016/j.asej.2024.102967
60. Fariz, T. K. N., & Basha, S. S. (2024). Enhancing solar radiation predictions through COA optimized neural networks and PCA dimensionality reduction. *Energy Reports, 12*, 341–359. https://doi.org/10.1016/j.egyr.2024.06.025
61. Wang, Y., He, X., Liu, Q., & Razmjooy, S. (2024). Economic and technical analysis of an HRES (hybrid renewable energy system) comprising wind, PV, and fuel cells using an improved subtraction-average-based optimizer. *Heliyon, 10*(12). https://doi.org/10.1016/j.heliyon.2024.e32712
62. Hassan Hakmi, S., Alnami, H., Ginidi, A., Shaheen, A., & Alghamdi, T. A. H. (2024). A fractional order-Kepler optimization algorithm (FO-KOA) for single and double-diode parameters PV cell extraction. *Heliyon, 10*(16). https://doi.org/10.1016/j.heliyon.2024.e35771
63. Naser, A. T., Mohammed, K. K., Aziz, N. F. A., Kamil, K. B., & Mekhilef, S. (2024). Improved coot optimizer algorithm-based MPPT for PV systems under complex partial shading conditions and load variation. *Energy Conversion and Management: X, 22*, 100565. https://doi.org/10.1016/j.ecmx.2024.100565
64. Ahmad, M. W., Mourshed, M., & Rezgui, Y. (2018). Tree-based ensemble methods for predicting PV power generation and their comparison with support vector regression. *Energy, 164*, 465–474. https://doi.org/10.1016/j.energy.2018.08.207
65. Liu, L., Zhao, Y., Wang, Y., Sun, Q., & Wennersten, R. (2019). A weight-varying ensemble method for short-term forecasting PV power output. *Energy Procedia, 158*, 661–668. https://doi.org/10.1016/j.egypro.2019.01.180

66. Bouaouda, A., & Sayouti, Y. (2024). Design optimization of a multi-source renewable energy system using a novel method based on selective ensemble learning. *Procedia Computer Science, 236*, 111–118. https://doi.org/10.1016/j.procs.2024.05.011
67. Zhang, H., Shi, J., & Zhang, C. (2022). A hybrid ensembled double-input-fuzzy-modules based precise prediction of PV power generation. *Energy Reports, 8*, 1610–1621. https://doi.org/10.1016/j.egyr.2022.02.298
68. Nwokolo, S. C., Ogbulezie, J. C., & Ushie, O. J. (2023). A multi-model ensemble-based CMIP6 assessment of future solar radiation and PV potential under various climate warming scenarios. *Optik (Stuttgart), 285*, 170956. https://doi.org/10.1016/j.ijleo.2023.170956
69. Rong, J., Wang, B., Liu, B., & Zha, X. (2015). Parameter optimization of PV based on hybrid genetic algorithm. *IFAC-PapersOnLine, 48*, 568–572. https://doi.org/10.1016/j.ifacol.2015.12.189
70. Toughzaoui, Y., Elkhatib, R., Kaoutari, T., Louahlia, H., Chaoui, H., & Gualous, H. (2024). Advances in hospital energy systems: Genetic algorithm optimization of a hybrid solar and hydrogen fuel cell combined heat and power. *International Journal of Hydrogen Energy, 86*, 1310–1325. https://doi.org/10.1016/j.ijhydene.2024.09.001
71. Saadaoui, D., Elyaqouti, M., Assalaou, K., Ben Hmamou, D., & Lidaighbi, S. (2021). Parameters optimization of solar PV cell/module using genetic algorithm based on non-uniform mutation. *Energy Conversion and Management: X, 12*, 100129. https://doi.org/10.1016/j.ecmx.2021.100129
72. Allwyn, R. G., Al-Hinai, A., Al-Abri, R., & Malik, A. (2022). Optimization and techno-economic analysis of PV/battery system for street lighting using genetic algorithm—A case study in Oman. *Cleaner Engineering and Technology, 8*, 100475. https://doi.org/10.1016/j.clet.2022.100475
73. Ismail, M. S., Moghavvemi, M., & Mahlia, T. M. I. (2013). Characterization of PV panel and global optimization of its model parameters using genetic algorithm. *Energy Conversion and Management, 73*, 10–25. https://doi.org/10.1016/j.enconman.2013.03.033
74. Yadav, K., Kumar, B., Guerrero, J. M., & Lashab, A. (2022). A hybrid genetic algorithm and grey wolf optimizer technique for faster global peak detection in PV system under partial shading. *Sustainable Computing: Informatics and Systems, 35*, 100770. https://doi.org/10.1016/j.suscom.2022.100770
75. Zhang, C., & Zhang, M. (2022). Wavelet-based neural network with genetic algorithm optimization for generation prediction of PV plants. *Energy Reports, 8*, 10976–10990. https://doi.org/10.1016/j.egyr.2022.08.176
76. Wang, J., et al. (2024). Optimization study of wind, solar, hydro and hydrogen storage based on improved multi-objective particle swarm optimization. *Journal of Energy Storage, 93*, 112298. https://doi.org/10.1016/j.est.2024.112298
77. Wei, D., Zhang, Z., Zhang, W., Yang, Y., & Yang, Z. (2024). Optimization of multi-energy complementary power generation system configuration based on particle swarm optimization. *Energy Reports, 12*, 2257–2269. https://doi.org/10.1016/j.egyr.2024.08.026
78. Wang, S., Chen, J., Zhang, J., Cai, S., Zhang, W., & Sun, J. (2024). A novel memory usage guided fuzzing based on particle swarm optimization. *Swarm and Evolutionary Computation, 91*, 101747. https://doi.org/10.1016/j.swevo.2024.101747
79. Thongbouasy, P., & Chatthaworn, R. (2023). Transmission expansion planning considering solar photovoltaic using novel binary particle swarm optimization. *Energy Reports, 9*, 1145–1153. https://doi.org/10.1016/j.egyr.2023.05.062
80. Shihabudheen, K. V., Dileep, G., & Sulthan, S. M. (2024). Adaptive particle swarm optimization based improved modeling of solar photovoltaic module for parameter determination. *e-Prime—Advances in Electrical Engineering, Electronics and Energy, 8*. https://doi.org/10.1016/j.prime.2024.100621
81. Iweh, C. D., & Akupan, E. R. (2023). Control and optimization of a hybrid solar PV—Hydro power system for off-grid applications using particle swarm optimization (PSO) and differential evolution (DE). *Energy Reports, 10*, 4253–4270. https://doi.org/10.1016/j.egyr.2023.10.080

82. Teferra, D. M., Ngoo, L. M. H., & Nyakoe, G. N. (2023). Fuzzy-based prediction of solar PV and wind power generation for microgrid modeling using particle swarm optimization. *Heliyon, 9*(1), e12802. https://doi.org/10.1016/j.heliyon.2023.e12802
83. Li, P., & Liu, Q. (2024). Optimizing energy management strategies for microgrids through chaotic local search and particle swarm optimization techniques. *Heliyon, 10*(17), e36669. https://doi.org/10.1016/j.heliyon.2024.e36669

Artificial Intelligence for Green Innovation in Digital Lean Production

Manjushree Nayak and Shubhasri Pradhan

1 Introduction

Lean manufacturing and green management are two correlative concepts that have received increasing attention in contemporary industries as firms seek to enhance organizational excellence and value delivery while containing costs and unsustainability. Lean manufacturing is a program developed out of Toyota's manufacturing model in the mid-twentieth century; it is a systematic technique of waste reduction in manufacturing [1]. This strategy focuses on providing just-in-time manufacturing, and system improvement, and elimination of non-value activities. With the global pressures on industries rising, lean principles have become vital in competitiveness for industries across the globe. Green management, on the other hand, is a total approach of greening the entire corporate venture inclusive of its operations [2]. This idea extends far beyond environmental compliance, which respects current legislation and norms to achieve improvements in environmental indicators and responsible management of natural resources, products, and services throughout the value chain. Green management is a broad concept that uncompromisingly ranges from waste minimization and energy parsimonious to product design and implementing the concept of circular economy. Thus, the term 'digital lean manufacturing' is used to refer to application and integration of conventional lean principles and practices together with superior digital tools that evolved from lean manufacturing and green management integration. Such a symbiosis enables organizations to

M. Nayak (✉)
Amity School of Engineering & Technology, Amity University, Raipur, Chhattisgarh, India

Department of Computer Science and Engineering, NIST University, Berhampur, India

S. Pradhan
Amity School of Engineering & Technology, Amity University, Raipur, chhattisgarh, India

P. Raj et al. (eds.), *Artificial Intelligence (AI) for IT Energy Efficiency and Green AI for Environment Sustainability*,
https://doi.org/10.1007/978-3-031-89420-6_14

perform their activities in an efficient manner and, at the same time, be environmentally friendly [3]. New to digital lean manufacturing, it is an innovative utilization of artificial intelligence and IoT, coupled with data analytics, for precision identification of inefficiencies, optimal resource utilization, and minimum adverse effects on the environment. It is imperative to have sustainability issues high on the agenda in today's organizations especially in the industries. As global populations rise and resources become limited, companies are under more pressure to achieve more with less [4]. People in their capacity as consumers, investors, and regulatory of business activities are increasingly putting pressure on organizations asking for accountability in the area of environmental protection. At the same time, global competition increases and means that companies have to cut costs, become flexible and succeed in rapidly changing environments. With regards to these, it is now viable for organizations to turn to digital lean manufacturing and green management as a resolution for such concurrent troubles of sustainability and enhanced efficiency. This means that such approaches help organizations cut their impact on the environment, while at the same time enhancing their performance or profitability. For example, a smart factory system can help a manufacturer to decrease its energy use and emission while at the same time cut its expenses [5]. In the same manner, through computer learning predictive maintenance, downtimes due to regular maintenance or equipment failure lead to saving of money on labor expenses for engine rebuilding or overhaul. Furthermore, the utilization of these principles results in product novelty and innovation in the structure of organizations. Organizations that choose sustainable strategies frequently find new and valuable markets; their existing products can be recycled, new products can be produced from recycled materials, or the company can provide services that help customers use products more, thus giving them a longer lifespan [6].

It helps in environmental conservation and breaks new ground in revenue generation and brand image in accordance with ever-advocating environmentalism. Linked to Industry 4.0, a new era of creativity in industry brings the use of digital tools into production procedures. Smart or autonomous devices (Artificial Intelligence), a field of computer studies that focuses on ensuring machines acquire the capability to do things that a human would consider wise, are the heart of this revolution [7]. AI applied to industries spans a broad spectrum of technologies, including automation through robotics, computer vision, machine learning, and natural language processing, all of which are restructuring the manufacturing facilities and systems as well as the configuration, construction, and distribution of the products as illustrated in Fig. 1. Another viable application of AI in this domain is an effective use of preventive measures that allow avoiding the failures of various mechanisms that is a part of lean manufacturing and green management. Now, here the AI has taken versatile and revolutionary parts to encourage digital lean manufacturing and green management. Some of its key input is in the area of trend analysis and decision making. Based on AI, big data can be defined as the amount of raw data the company receives from manufacturing processes, sensors, and supply chains, which are difficult to analyze by human eye [8]. This capability makes it possible for manufacturers to predict equipment breakdown, chartable production plan, and even the

```python
import numpy as np
import random

# Simulate sensor data for machinery performance over time
# 0 represents no problem, 1 represents a potential failure
np.random.seed(42)
time_periods = 100
sensor_data = np.random.normal(0, 0.2, time_periods)  # Simulated noise
failure_risk = np.zeros(time_periods)

# Function to simulate AI predicting failures based on sensor anomalies
def predict_failure(sensor_data):
    threshold = 0.4  # Threshold for detecting anomalies
    for i in range(1, len(sensor_data)):
        if abs(sensor_data[i] - sensor_data[i-1]) > threshold:
            failure_risk[i] = 1  # Predict failure risk if anomaly detected

predict_failure(sensor_data)

# Display results
for i in range(time_periods):
    status = "Normal" if failure_risk[i] == 0 else "Failure Risk"
    print(f"Time: {i}, Sensor Data: {sensor_data[i]:.2f}, Status: {status}")
```

```
Time: 0, Sensor Data: 0.10, Status: Normal
Time: 1, Sensor Data: -0.03, Status: Normal
Time: 2, Sensor Data: 0.13, Status: Normal
Time: 3, Sensor Data: 0.30, Status: Normal
Time: 4, Sensor Data: -0.05, Status: Normal
Time: 5, Sensor Data: -0.05, Status: Normal
Time: 6, Sensor Data: 0.32, Status: Normal
Time: 7, Sensor Data: 0.15, Status: Normal
Time: 8, Sensor Data: -0.09, Status: Normal
Time: 9, Sensor Data: 0.11, Status: Normal
Time: 10, Sensor Data: -0.09, Status: Normal
Time: 11, Sensor Data: -0.09, Status: Normal
Time: 12, Sensor Data: 0.05, Status: Normal
Time: 13, Sensor Data: -0.38, Status: Failure Risk
Time: 14, Sensor Data: -0.34, Status: Normal
Time: 15, Sensor Data: -0.11, Status: Normal
Time: 16, Sensor Data: -0.20, Status: Normal
Time: 17, Sensor Data: 0.06, Status: Normal
Time: 18, Sensor Data: -0.18, Status: Normal
Time: 19, Sensor Data: -0.28, Status: Normal
Time: 20, Sensor Data: 0.29, Status: Failure Risk
Time: 21, Sensor Data: -0.05, Status: Normal
Time: 22, Sensor Data: 0.01, Status: Normal
Time: 23, Sensor Data: -0.28, Status: Normal
Time: 24, Sensor Data: -0.11, Status: Normal
Time: 25, Sensor Data: 0.02, Status: Normal
Time: 26, Sensor Data: -0.23, Status: Normal
Time: 27, Sensor Data: 0.08, Status: Normal
Time: 28, Sensor Data: -0.12, Status: Normal
Time: 29, Sensor Data: -0.06, Status: Normal
```

Fig. 1 Predictive maintenance to prevent machinery failures, which is a key aspect of lean manufacturing and green management

demand for certain products with great ease. For example, to decide when a piece of technology is most likely to break, an AI system will also use data on prior maintenance as well as current measurements from sensors. This would facilitate preventive maintenance that reduces time that equipment are off, as well as extends their life, which is an important component of lean manufacturing and preservation of resources. Artificial intelligence plays a crucial role in minimizing wastage and using energy in managing toward green. AI-based intelligent energy management systems are capable of monitoring and analysis the energy utilized by a factory and making adjustments from time to time so as to reduce the wastage of energy [9]. For real-time decisions, for example, adjusting parameters to HVAC or adapting industrial equipment power usage by workload and energy costs, these systems could utilize data from the past as well as surrounding conditions. AI also increases quality assurance, which is part of lean production. Automated vision systems with embedded deep learning are capable of scanning products at rates and percentages of identification beyond human measures [10].

This not only helps minimize some anomalies and wastes but also ensures that production problems can be easily identified ahead of time to allow for quick remedy. Because of low levels of scrap and rework, these quality control systems bolstered by Artificial Intelligence are not only efficient in their operation but also help in conserving numerous resources. Moreover, AI plays a role in enhancing the supply chains, which is one of the inputs in lean and green deals. AI can be used in helping organizations make proper decisions regarding procurement, inventory management, and logistics through use of data collected from other factors such as weather, traffic patterns, and even markets [11]. By this there are reduced emissions from the transport system, reduced wastage from inventories, and supply systems that are elastic and can easily adapt to changes in market forces. AI is setting the tone for an innovative environmental age and is promoting the concept of green technology in product positioning and manufacturing. Many thousands of design possibilities can be generated by generative design algorithms, including issues such as material usage, energy, and recycling [12]. But more than this it gets product to market much more quickly and at the same time creates designs that are inherently more efficient and therefore more sustainable in terms of manufacture. The second trend where digital lean manufacturing is experiencing tremendous development is the application of AI in robotics and automation systems. Intelligent cobots may also work alongside human employees, changing their tasks easily to maintain safe interaction levels [13]. This flexibility enables the manufacturers to properly staff their assembly lines, minimize the danger posed to human employees by design, and more quickly adjust their production lines—all of which are basic tenets of lean manufacturing. Consequently, the increase in Industry 4.0 is expected to increase even more in the future. New technologies such as quantum computing as well as complex neural networks can offer an incredible opportunity for improvement approaches, modeling assets, and decisional performances [14]. These developments are perhaps going to stretch it to the next level in the dimensions of productivity and environmental consciousness through more sophisticated forms of lean digital production and earth-friendly management systems. A good example of harmonization of technical advancement and environmentally sound business strategy is integration of intelligent machines in digital manufacturing which is lean and green management. While traditional industries strive to overcome the impact of global competition and global environmental challenges, AI serves as the key enabler of new flexibility, efficiency, and sustainability solutions. It is important to note that to drive new technologies to become effective, it is also important to change the culture of establishments to one that supports continuous learning, data-driven decision-making, and long-term sustainability [15]. The organizations that will be better positioned to leverage AI to support minimal resources and eco-conscious initiatives will be the most likely to thrive in a continually more complex and environment-sensitive global economy while moving deeper into the period of Industry 4.0.

2 Role of Artificial Machine Intelligence in Improving Sustainability and Efficiency

Computational intelligence as a tool has become an irresistible force in the search for sustainable production and is now capable of achieving optimal resource utilization and minimizing wastage across all production lines. AI is at the forefront of offering industrial solutions that bring into play the concept of coping with the competitive front and the environmental issues at the same time as industries work at transitioning to a new manufacturing environment. AI has delivered one of the most significant value propositions for sustainable manufacturing by altering the way resources are used. AI systems can always observe the production environments and optimize the utilization of producible resources as necessary. This is made possible through the development of a number of algorithms in machine learning and real-time analysis [16]. For example, in the chemical manufacturing process, the AI models can examine the reaction conditions and determine which proportion of the components is best to use so that raw material is not wasted as well as improve the yields of the products. Likewise, in textile industry AI systems used in cutting systems for fabric layout, especially fabric cutting the ratio of off cuts can be minimized thus saving a lot of fabric [17]. Through machine learning this technology is also helpful in reducing waste through frequent maintenance and quality assurance. AI can predict when equipment is likely to fail or when product quality may be off by analyzing sensor data, and performance history. Staying ahead is a winning strategy since issues that would otherwise result to manufacturing of bad products or causing break-downs in machinery are corrected before they happen [18]. For instance, in the semiconductor industry where quality is critical, real-time defects identification by machines improves quality control over traditional manually intensive methods of examining products therefore saves on material and time which are scarce resources. Speaking of energy conservation and environmental protection as the fields of AI application, the result is just as impressive. Proactive energy management solutions operating on artificial intelligence are able to control energy consumption at manufacturing plants based on the need of production, price of energy, and even climate. These systems are based on historical data and can take decisions in fractions of a second to allow the minimization of energy use which results in both appreciable dollars and cents savings as well as reduced carbon footprints [19]. For example, the AI called DeepMind of Google has used it in data center cooling system where it cut energy used for cooling by up to 40% [20]. The other innovation that AI plays is in the development of sustainable products and processes as well. Equipment such as generative design can run through thousands of design permutations and choose ones based on material consumption, energy consumption, and recyclability. This approach not only leads to more sustainable products but also usually produces designs which are additionally lighter, stronger or more efficient during manufacturing [21]. For instance, Airbus has started using AI in the generative design to produce airplane parts that are 45% lighter than generic airplane parts; this leads to fuel conservation and lesser emissions through the airplane's life cycle [22]. Artificial intelligence is helping to create circular economy models in the production industry. Based on the information on the product lifecycle, it is possible to use AI and determine what fraction of products can be reused, which parts can be refurbished, and which parts are recyclable. This systematic approach to managing resources reduces the demands on new materials and precludes waste by optimizing the useful lifetime of materials and products [23]. For example, AI systems are now being applied in recycling electronic waste through distinguishing the varieties of materials in a shorter time, hence making the process cost-effective.

When it comes to effectiveness and flexibility, AI is adding value to manufacturing operations by optimizing and providing reliability. AI systems learn from the big and often unstructured data from different domains within the manufacturing environment and can determine patterns, outcomes and even recommend changes that mortals cannot. This means that through AI, all manufacturing process concerns have been changed in how the factories work. Forecasting of production data in real-time with the help of predictive algorithms allows us to identify a number of issues: decreasing effectiveness, logistical problems, and directions for improvement, as illustrated in Fig. 2 about AI in optimizing energy consumption and predictive maintenance. These systems can then be used to fine-tune process parameters to improve production processes, minimize time taken between one production phase and the next, and enhance the overall equipment effectiveness (OEE) [10]. For instance, in automotive production, artificial intelligence adaptation has been employed to monitor welding, leading to minimized energy usage and enhanced weld quality but with higher throughput rates [24]. Predictive analytics work on the basis of artificial intelligence; this is another cornerstone of manufacturing improvement. AI is great at identifying probable future trends, potential issues, and servicing intervals using both recorded history and the live data. Through application of this capacity, producers are able to move from being only responsive to proactive operating models where problems before they occur and also, producers are able to optimize their production planning patterns in relation to the expected demand [25]. For example, Siemens has adopted intelligent algorithms for the prediction of the maintenance requirements of its gas turbine thus enhancing on energy utilization without incurring frequent and expensive maintenance [26]. Thus, we water a new level of visibility and control in SCM thanks to AI. Supply chain software solutions can process global supply chain data containing factors like geopolitical events, weathers and markets to identify right inventory planning, control risks and recommend movement strategies [11]. This kind of flexibility is imperative in today's uncertain global economy environments. For example, the COVID-19

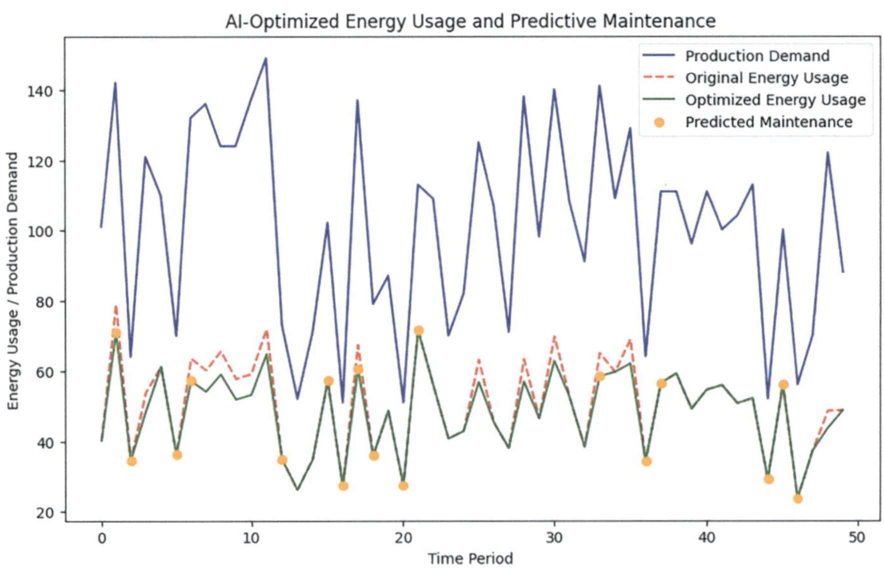

Fig. 2 AI optimizing energy usage and predictive maintenance

impact on customer demand was tackled swiftly by businesses like Amazon by using artificial intelligence (AI) to monitor and anticipate shifts in demand and thus change the supply chain correlations [27]. AI also serves to deepen organizational effectiveness through better use of robotics and automation. Specifically, cooperative robots or cobots that are integrated with artificial intelligence are usable safely for interacting with human employees and ready to learn other tasks efficiently. These systems possess the modularity needed to respond to changes in the production demand since they can learn from human demos and gradually improve their performance [13]. For example, BMW uses AI-driven industrial cooperants in its production subsidiary and it has realized greater customization in product manufacturing [28]. Quality assurance is a major component of the manufacturing process and AI is changing how it is done. Automated vision systems based on deep learning can evaluate products' characteristics at rates and with an accuracy unreachable for a human. Through these advancements, such technologies can ensure that this minor imperfection that may not be easily detected by the human eye are seen and hence promise an improved quality of the products, and little or no rework or recall of the products, are achieved [29]. For example, in the pharmaceutical industry, motivating drivers with AI auditing devices are used to quickly establish mistakes in vials of medicines thus boosting throughputs while at the same time guaranteeing the safety of the end product [30]. It is also quite useful in production planning and capacity forecasting. AI systems can therefore forecast a higher level of demand due to factors such as clients' behavior, market trends and sentiment analysis. This helps the enterprise to have control over its inventory levels and the schedule on production [31]. It has also the benefit of cutting down waste resulting from overproduction or from stock out situations. For instance, H&M has applied AI demand forecasting to cut both the inventory and markdown rates in their commercial operations [32]. AI is emerging as a disruptive technology in driving improvement in sustainability and optimization in production processes. Hence, the capacity to enhance resource utilization; minimize wastes; enhance energy proficiency, and, generally, upgrade the operational performance is enabling industries to address the competition challenges in a global market and address several vital environmental problems at the same time. Advanced AI technologies are believed to further the principle of sustainable manufacturing as the technology progress, thus leading to an industrial era that is stronger and more environmentally friendly.

3 Overcoming Challenges in Implementing AI

Implementing artificial intelligence (AI) in organizations presents numerous challenges, two of which are particularly significant: The major challenges include the combination of information from different sources and the high costs of implementing the solution. These are barriers that must be addressed when considering a supportive environment for aging populations because such barriers need to be surmounted using proper strategies. The data fusion of information from several sources is a significant issue of concern in AI implementation. Managers face many issues related to data management and processing because current organizations receive large volumes of various data at high speed in different departments and systems [33]. Sophisticated systems, dissimilar data structures, and fragmented information repositories add to the data challenges that hinder the generation of a single coherent representation of organizational data [34]. Furthermore, data quality, consistency and relevance across a range of these disparate sources can be almost impossible to achieve and takes a considerable amount of time and effort [35].

However, there are such challenges, and using artificial intelligence (AI) there are efficient approaches to integrate and abstract data from different sources. Both well-defined and less structured data can be. Machine learning algorithms to extract trends or conclusions which seemingly could be hard for humans to notice based on analysis done. This interacting data, as presented in Fig. 3, is used in a basic pilot AI model to forecast future sales through the feedback that customers provide [36]. AI technologies including natural language processing allow organizations to extract different value from text-heavy data such as customer feedbacks, social media, and documented information [37]. In addition, the task of cleaning, transforming, and joining data from multiple sources can be easily performed by data integration systems that use artificial intelligence [38]. Another challenge that organizations face is the overwhelming implementation costs, which makes it hard for many organizations to adopt intelligent automation. The costs involved in the use of artificial intelligence are in many aspects that range from the physical hardware used, software license fees, data management, and analysis capacity as well as talented human capital [39]. These expenses might be beyond the reach of small and middle-St but this may make many of them lag behind in terms of technology as compared with their large counterparts [40]. Many cost factors discussed above might be a subject

```
# Simple sentiment analysis to derive sentiment polarity from feedback (-1 to 1 scale)
def analyze_sentiment(feedback):
    return TextBlob(feedback).sentiment.polarity

# Adding sentiment score to integrated data
integrated_data['Sentiment'] = integrated_data['Feedback'].apply(analyze_sentiment)

# Basic prediction logic: if sentiment is positive, expect sales to increase by 10%, otherwise decrease by 10%
def predict_sales(sales, sentiment):
    if sentiment > 0:
        return sales * 1.1  # Increase sales by 10%
    else:
        return sales * 0.9  # Decrease sales by 10%

# Applying prediction for each row
integrated_data['Predicted_Sales'] = integrated_data.apply(lambda row: predict_sales(row['Sales'], row['Sentiment']), axis=1)

# Display the results
print("\nPredictions Based on Feedback Sentiment:")
print(integrated_data[['Date', 'Sales', 'Sentiment', 'Predicted_Sales']])
```

```
Predictions Based on Feedback Sentiment:
         Date  Sales  Sentiment  Predicted_Sales
0  2024-01-01    100        0.7            110.0
1  2024-01-02    150        0.0            135.0
2  2024-01-03    130        1.0            143.0
```

Fig. 3 A simple pilot AI model that uses this integrated data to predict future sales based on customer feedback

to some strategies businesses might employ in order to solve them. One approach is to begin with pilot implementations, creating emphasis on a fixed, narrowly-specified, but significant vertical scenario [41]. This leads to positioning where companies can show a prospective ROI of AI adoption while incurring low first costs and risks. Pilot projects are also useful in gaining lessons and assessing future obstacles before full implementation [42]. Jan Nowicki outlines another approach to combating costs, which is phased implementation. The AI adoption model, when implemented, ensures that the costs are not so much incurred at once but over a period since the process is segmented [43]. This approach also can be refined and updated depending on the responses that have been received from each stage of the process and it may lead to more positive results [44]. Ability to scale is thus viewed as an important factor in dealing with costs. This means that organizations should focus on those forms of AI and structures that can be seamlessly adapted as conditions evolve and data volumes expand [45]. AI as services and platforms in the clouds prove to possess this feature of scalability whereby organizations can only incur charges for the quantum of resources that they consume and can address their capacity requirements through scalability [9]. It also means that implementation costs can be reduced by collaboration and partnerships. Companies have found that they can consult and cooperate with other providers, universities, or other organizations from their field to exchange their material, experience, and examples [46]. Non-proprietary, readily programmable tools and frameworks of AI are significantly less expensive than expensive black box tools [47]. The two main issues affecting AI adoption include the integration of information from various sources and high implementation costs, despite which the two issues can be addressed by proper planning and relevant strategies. Many organizations might move around the intricate environment of the adoption of AI, and it can liberate AI's ability to bring change when they use AI capabilities for incorporating information and evaluation, starting with high-focus pilot programs, acting incrementally, focusing on scalability, and exploring partnerships.

3.1 Skill Gaps and AI-Driven Training Solutions

The rapid advancement in computational intelligence or artificial intelligence technologies has most certainly given rise to adequate demand for competent professionals who can model, develop, and manage AI solutions in various industries. This growing demand for AI experts was recognized when it unveiled a severe shortage of qualified professionals that slows down the dissemination and efficient application of AI technologies [48]. Companies face challenges of recruiting and retaining employees with skills to foster growth and sustain the bottom-line in a new more automated age as they integrity AI into their operation [49]. Nowadays, many sectors including IT, health care, finance, and manufacturing as well as retail industries experience the shortage of skilled professionals. This shortage affects the rate at which companies are able to integrate AI solutions as they are unable to find people with the capacity to harness the potential of the technologies in business [50]. The supply and demand dynamics have only increased talent competition, which could pose a problem in many firms' attempts at digital transformation on the part of [51]. The need to close this ability gap has led to an effort to develop innovative training programs to enhance the general adoption of AI technologies. Today, adaptive learning supported by AI is now possible as a way of upskilling and training the workforce [52]. Individualized lessons edited on these bases for particular employees, depending on their learning style, their abilities, and individually targeted learning preferences, are developed through AI algorithms in the manner described above As Hedlak et al. pointed out, adaptable learning platforms based on AI can improve learning and recall by analyzing student performance data and adapting the course content in real-time [53]. The following is a discussion of the benefits that comes with using AI-driven adaptive learning systems more specifically in work force training. They can know which class of expertise lacks the knowledge and in what areas they need to cover up, by offering relevant content [54]. Also, based upon each learner's performance, these systems are capable to adjust the speed and difficulty of the training material provided, make sure students are constantly on their toes without being overwhelmed [55]. Additionally, the use of the artificial intelligence in the teaching methodology can include rich-media, which include little simulations and games in learning process that encourages practical utilization of developed skills [56]. These systems of adaptive learning are a great necessity, as they are the way of enhancing the abilities of the staff members, and enlightening the management regarding the principles of utilization of AI systems in the context of optimizing the AI integration. Here it can be seen that individualized training solutions involving AI basics, data analysis, machine learning, and AI ethics may help increase organizations' internal capabilities and, thus, lessen their dependence on outside workers [57]. Not only does this strategy fill the talent gap current within the company, but it promotes innovation and personal long-term growth [58]. AI training solutions can also assist management teams in acquiring the right skills for supervising AI deployment and integration. It means that with the help of such systems offering targeted courses on the AI strategy, management of the projects, and the use of AI along with the potential risks and challenges, the leaders can make the right choices regarding further integrated digital transformation of organizations and support their subordinates in it [59].

This approach of organizing training on AI enhances the understanding of AI technologies across all levels and in all company sectors [60]. As the technology we call artificial intelligence continues to advance, an application period in employee education driven by AI adaptive education is expected to expand, potentially allowing for conversational learning experiences with AI tutors through more advanced natural language processing abilities [61].

Also, virtual and augmented reality technologies might improve the immersive quality of AI-assisted training by enabling the creation of more streamlined simulations and experiential learning-based examples [62]. Closing the AI talent gap is critical for accelerating the implementation of AI across sectors. AI-powered adaptive learning solutions are emerging as a viable way to upskill the workforce and

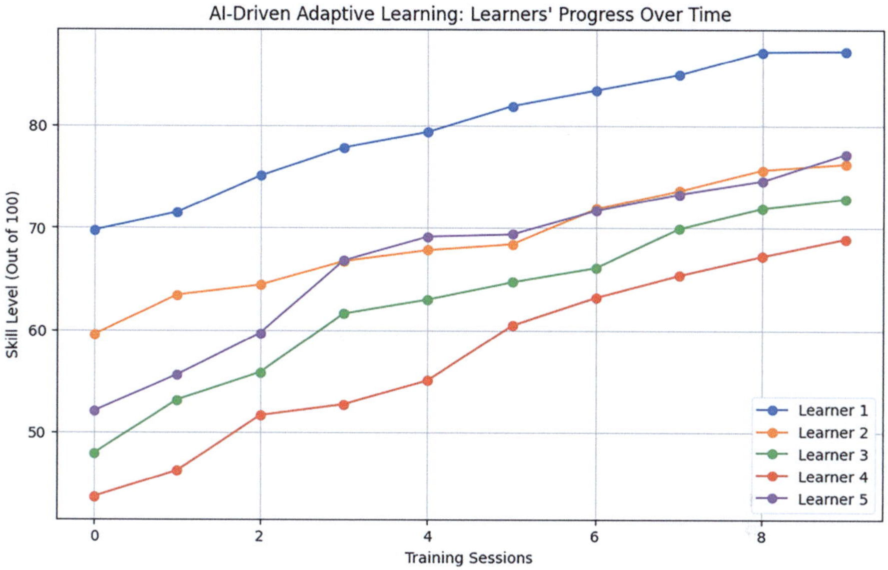

Fig. 4 The progression of each learner's skill level over time

equip organizations with the tools required to thrive in an AI-powered future. Figure 4 illustrates how the skill level of each learner increases over time, thus emphasizing the efficiency of these training techniques. These state-of-the-art training methods [63] enable businesses to build in-house AI capacity, foster a culture of constant innovation, and position themselves to succeed in a new AI-driven landscape.

3.2 *Managing Employee Anxiety and Change Resistance*

The integration of Artificial Intelligence (AI) in the workplace often encounters resistance from both management and employees, stemming from a myriad of concerns that can impede its success and effectiveness. Addressing these apprehensions is imperative for a smooth adoption process. A primary fear surrounding AI's incorporation is the potential for job displacements [64]. Workers across diverse sectors are wary that automation might jeopardize their employment by making certain positions redundant, posing threats to job security. This trepidation is more pronounced in industries where routine tasks form a substantial component of their daily duties [65].

Managers, on their part, may also resist embracing AI due to uncertainties about the transformation it engenders in their roles. They are concerned about potentially forfeiting control over critical decision-making processes as AI systems take precedence [66]. Apart from employment loss, there is a concern about the rate and extent of disruption that AI brings to organizations. The hierarchical system within most organizations means that employees may be overwhelmed when it comes to learning new skills or how to function in a new system, while managers may be overwhelmed by the intricacies of implementing AI into the current business models [67]. There is also potential resistance to change, organized in terms of attitude that questions AI's ability or potential, disinterest in new technologies or, in some occasions, organizational action against AI project. To overcome this problem and ensure fast AI integration, organizations require the support of programs that address employees' concerns and ensure they accept the changes. One of the methods that could be used involves the use of employees' participation during the implementation of the AI projects [68]. Organizations need to get more people involved in change initiatives to ensure that its end-users feel like the outcome is something they helped cultivate. It could entail centralization of or coordination with employee participation in AI use case determination, collecting feedback on AI tools, or integrating employees into an AI pilot [69]. Organizational development programs are crucial when it comes to help staff members appreciate and adapt to changes that come with AI. Efficient and effective for these programs should involve inculcating dynamism, critical thinking skills, and cultivating the necessary digital skills [70]. Hiding the benefits that exist when the human talents meet the AI's talents, organizations can help the employees to accept how the technique does not threaten their employment but can only make it even better [71]. These programs may also show, how with the help of AI such routine work can be done, so that the workers could concentrate on more creative and proactive elements of work [57]. Another important method, enabling the transition and minimization of the resistance is the consistent introduction of major AI training courses for all employees. In the fictional case layout depicted in Fig. 5, the progression describes how anxiety among the employees result in resistance to AI diminishes with time owing to interventions.

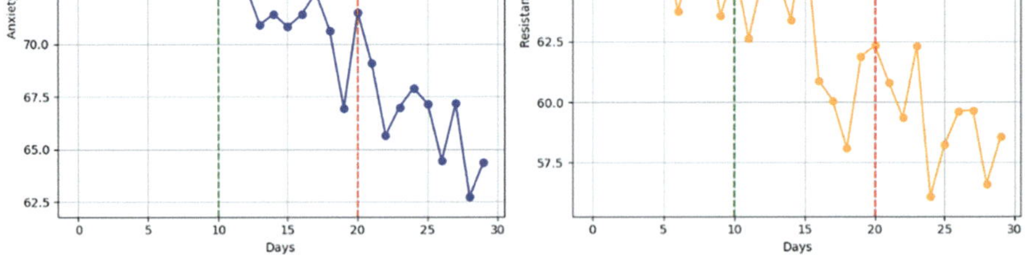

Fig. 5 Scenario of employee's anxiety and resistance to AI decrease over time due to interventions

These courses should cover the historic and the technical foundation of the AI technology as well as ways in which the technology will be helpful in the workplace [72]. Organization can then assist employees in becoming comfortable with AI by reducing its complexity and offering them physical practice in its usage [73]. To ensure that all workers, starting from the novices to the executives, are equipped with the professionalism and skills for utilizing AI, the educational courses need to be specialized to careers and proficiency levels [74]. Hence, it is advisable for the management to engage in active communication when dealing with resistance to the adoption of AI. It is recommended that organizations establish open and honest communication processes specifically with regard to the use of AI to explain why AI is being used, the expected enhancements and possible consequences for particular occupations [75]. It will also create trust and interest in AI technologies through frequent communications of AI projects, completed projects' success stories, and feedbacks on the experiences gained [76]. There is also an organizational factor that requires leadership for dealing with resistance. Assembly members and supervisors should also set the tone for focused efforts toward attaining expertise concerning AI technology and its application [77]. When leaders demonstrate where and how artificial intelligence improves decisions and creates value, they may not only increase confidence and engagement among their organizations [78]. Addressing tasks 4, 5 and 6 employees' anxiety and resistance to AI implementation needs a covering it from different angles education, participation and communication. The suggested strategies to get the most out of AI technology are safe management of risks, AI technology training, personal and professional development, and culture of innovation [63].

4 Artificial Intelligence and Cybersecurity within Digitized Industries

As the two industries have progressed rapidly toward digitization, levels of productivity and efficiency have improved, but new and potent threats to cybersecurity have also arisen. Supply chain and production scenarios are now becoming exposed to a widening array of cyber security threats due to the increased connectivity and digitization of manufacturing settings. These risks can disrupt operations, jeopardize essential information, and in some instances be concrete dangers [79]. Integrating cloud computing, IoT gadgets, and networked systems in industrial facilities has raised attack surface area and vulnerability level to cyber threats [80]. Cyber threats go beyond the conventional IT issues in the digitized industries Electronic Industry. They now include Operational Technology (OT) systems which are used to operate physical assets and industrial processes. Due to the development of IT and OT systems, there are new risks that can be used by cyber attackers [81]. For example, a vulnerability in a connected manufacturing apparatus may result in stoppages, risks to employees and equipment, or maybe the deliberate disruption of the manufacturing processes. However, using digitized supply chains, as discussed above, comes with certain risks, as the disruption of security in one area can affect the overall system [82]. The threats vary depending on levels of digitization of industries and include Industrial espionage, ransomware to critical infrastructure, and complex advanced persistent threats (APTs) that would breach the industrial networks for a long time [83]. These threats are dangerous because a large number of industrial processes cannot be disrupted, and the consequences can be devastating, from an economical to an environmental or even life-threatening perspective [84]. AI is now a robust tool that can help make cybersecurity powerful for digital businesses in light of these ever-shifting threats. This is so because through computation, AI can examine a large volume of data, relate the characteristics and derive new knowledge, which is helpful in threat detection especially complex networks in industrial sectors. As shown in Fig. 6, the simulation illustrates the number of threats detected each day before and after AI implementation [85].

In order to be able to alert the user about any potential violation of his/her privacy and potentially detect any irregularities, AI-based security systems may monitor the traffic on the network, system logs and user activities in real time. This makes it possible for them to expose risks that a human analyst might miss [86]. There are many emerging fields, and one of the most promising areas where AI is starting to be applied in industrial cybersecurity is predictive threat intelligence. AI systems incorporate insight into threats and vulnerable circumstances in the future on the basis of past data and trends. This capability allows enterprise's to be proactive in building up their protection [87]. In addition, the skills in handling incidents

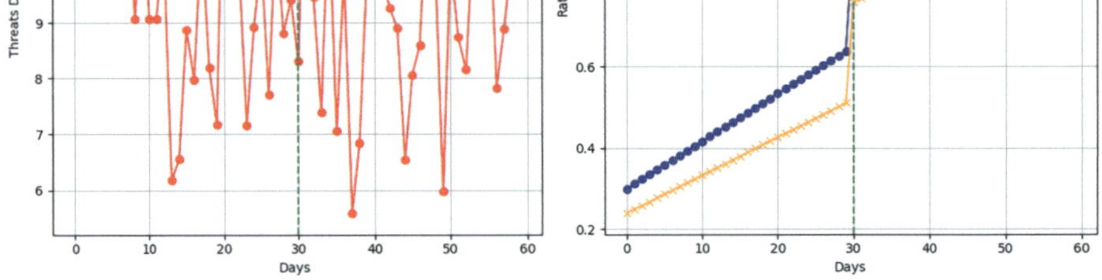

Fig. 6 Simulates the number of threats detected each day before and after AI implementation

may be enhanced by the following Artificial intelligence-driven SOAR (Security Orchestration, Automation, and Response) solutions that address mundane security tasks and leave the cal also known as experts, with challenging security tasks [88]. There are various real-life examples that have shown that having AI-based solutions and applications can keep industrial data and industrial operations safe and secure. For example, a big automobile company has installed an AI-based anomaly detection system which not only helped to discover and thwart an advanced malware attack on its production line control systems [89]. Another example, an energy company used an AI-based network monitoring system that blocked an unauthorized attempt at leaking proprietary industrial designs and other operational information's [90]. More so, what AI means to corporate cybersecurity is expected to advance as it continues to evolve. Future developments might include; more advanced threat hunting capabilities with artificial intelligence support, better integration of AI with the latest technologies such as blockchain in enhancing data truthfulness and AI enhanced red teaming for preparation on the likely cyber threats are some of the possible exciting future trends [91]. In other words, as much as AI demonstrates deeper security systems for industrialized cyber space it its proper and effective implementation remain amenable to exclusive human supervision and ethical considerations [92].

5 Transparency and Traceability in AI-Driven Green Supply Chains

In the context of green and ethical operations in business, the use and application of computational intelligence known as AI continues to grow as a way of enhancing both accountability and transparency along the supply chains. Through data analytics, development of learning technologies, and IoT, AI systems can offer certain real-time visibility of the complex global supply networks [93]. This increased transparency is crucial to organizations wanting to establish and maintain sustainable sourcing supply chain networks, as it allows organizations to identify sourcing origins, assess production processes and authenticate sustainable credentials [48]. That is, AI enhances transparency in the following ways: first, documentation processes in implementing supply chains are also automated; second, the tracking of commodities transported in supply chains is also done in real-time; third, and last, supply chain statistical demand models. These capabilities enable firms to recognize potential problems that may arise and then manage and solve them promptly, minimize unnecessary costs, and leverage the throughput of resources [94]. Of particular importance in green supply chains, this transparency allows various stakeholders to assess compliance with environmental specifications or carbon footprint and make appropriate supply and logistics choices [95]. Hence, there is a need for organizations procuring green and sustainable products to come out with accurate information about their green, sustainable supply systems. It is helpful in as much as it establishes credibility among the consuming public, investors and, most importantly, the regulatory authorities as it avails authentic details on the nutritional product's production process from the raw material stage to the final consumer [96]. This transparency is necessary to fight against all greenwashing and guarantee that companies' sustainability claims are matched by clear numbers and strategies [97]. Artificial intelligence is created to monitor and enhance environmentally sustainable supply chain management. For instance, a big retail chain used an AI solution to provide sustainability to its supply chain of cotton. The system monitors fields of cotton, predicts yield, and checks compliance with sustainable farming using satellite, weather data, and machine learning [98]. Another example is that the leading automobile manufacturing company used AI to minimize carbon emissions on routes and fuel consumption for routes leading to logistic networks [99]. As shown in Fig. 7, AI tracks the journey of goods, predicts and adjusts for demand, and monitors and reports on sustainability metrics.

The above AI-based solution presents several advantages for the stakeholders. They enhance the ability of companies to prove their compliance with environmental standards hence minimizing on chances of fines and damage to reputation [100]. Of these solutions, customers are provided with more assurance concerning the environmentally friendly imageries associated with the purchased products. More transparency is also advantageous to investors in the sense that there is an opportunity to determine the firm's ESG performance more accurately [101]. Furthermore, this paper determined that the implementation of lean and green supply chain management by utilizing artificial intelligence reduce overall business operation costs. Recent studies have found that by managing demand, cutting out unnecessary fluctuations, and controlling stock, one is able to make both the expense and the ecological characteristics more favorable [102]. As the ecological and economic sustainability requirements appear as key factors for competitiveness in present days' business world, more and more companies focus on those aspects. Compared to current manufacturing processes, new generations of AI technologies could enhance traceability and openness in green manufacture. Potential future advancements are more complex use of blockchain technology for the purpose of creating unalterable records, smart and enhanced life cycle sustainability assessments and real-time sustainability performance indicators [103]. Lack of information confidentiality, algorithmic bias, and lack of standard requirement for sustainability in terms of constructing open supply chains can hinder the use of AI in achieving fully open and sustainable supply chains [104].

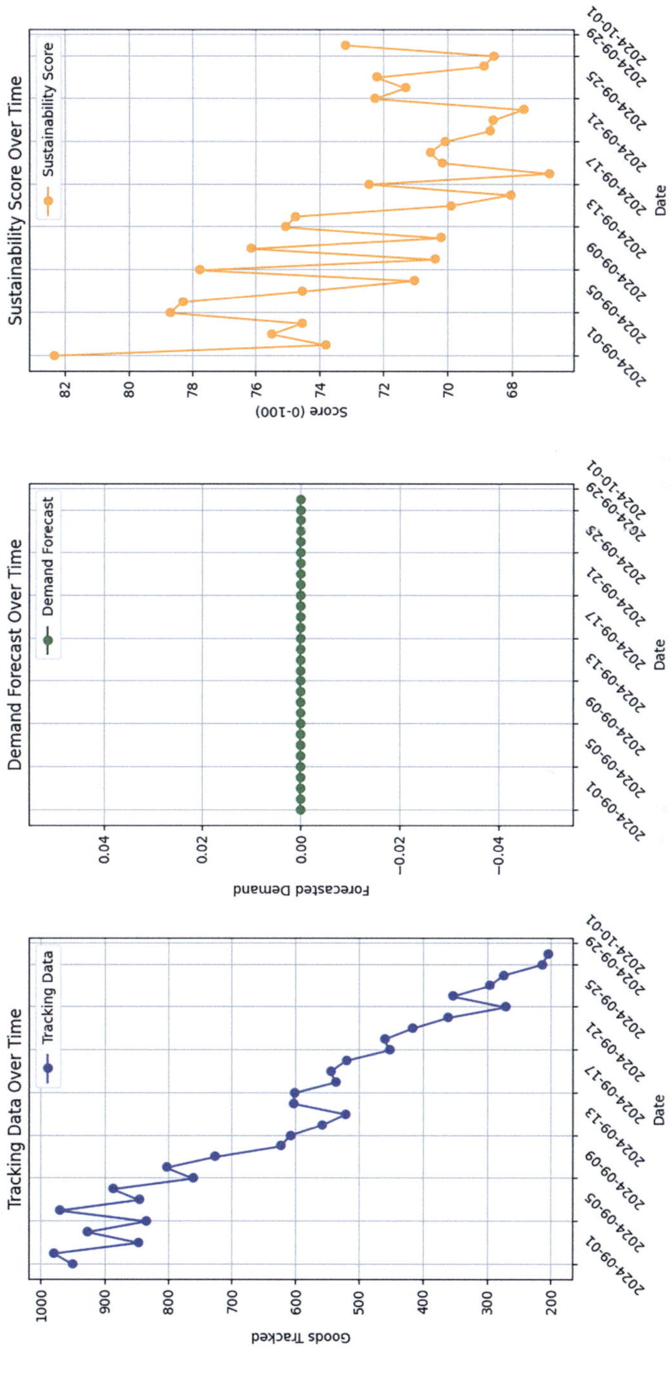

Fig. 7 AI tracks the journey of goods; AI predicts and adjusts for demand and AI monitors and reports on sustainability metrics

6 Future Outlook: AI as a Fundamental Pillar in Industrial Development

In the subsequent decades, the integration of computational intelligence (AI) into production processes is capable of fully revolutionizing efficiency, sustainability, and production. Prospective analysis shows that AI will be integral for the further progress of industrialization and for the improvement of socialisum attitude to the environment. It is therefore suggested that the manufacturing facilities in the next generation will feature AI as a complementary system that enhancing production in every possible manner. With the help of AI-powered systems it will become possible to control and manage the energy and material consumption, and waste generation rates in real-time, which will lead to the improvement of operations that are very unsustainable [105]. Gigantic amounts of information acquired by sensors situated throughout the facility will be processed by superior intelligent algorithms, which will identify problems and suggest improvements that operators would not notice [106]. These continuous improvements will create reductions in energy consumption and CO_2 emissions, thus ensuring that industrial output meets world sustainable development objectives.

Ushered in by AI, the new form of production efficiency is currently believed to allow for unprecedented levels of customization as well as flexibility. In future, with the integration of AI, manufacturing systems will be adaptive that would allow manufacture of a new product or a product with a modified characteristic swiftly [107]. Further, due to artificial intelligence maintenance predictions will become regular, reducing time and increasing the equipment's life cycle to reduce manufacturing's adverse environmental impacts [107]. AI holds high chances for revolutionizing industrial sustainability. We might have AI that creates a new product entirely with a circular economy goal in mind for the lifetime of the product, including being easier to recycle or disassemble and reuse afterwards [108]. There are still unbounded amounts of components and procedures that would be useful for the swift development of new sustainable materials which AI can help in the fast testing and iteration of [109]. When analyzing the state-of-the-art AI applications, a few innovations seem to be promising solutions for the further optimization of industrial processes and their sustainability. Others are the enhancement of complex industrial processes through the application of advanced algorithms, particularly the reinforcement learning type. These artificial intelligence systems on its own are able to learn and apply itself to difficult complex situations and climates and analyze optimum solutions with regard to costs, consumption, and impact on the environment [110]. One of the other trends that could be seen is the use of artificial intelligence and robotics in sorting and recycling waste. Super AI systems of today could significantly enhance the velocity and precision of recycling procedures and will conclude the circuit in industrial materials [111]. Furthermore, the creation of AI-based digital twines for the whole sets of interconnected industrial systems can offer detailed analysis of flows and interconnections of resources that can promote a deep understanding of the strategies and patterns of industrial symbiosis [112]. In the future, industrial ecosystems in which AI operates will become much more self-contained and self-optimizing. Discussed factories might be totally automated, learning from experience and adjusting to the environment and the consumer needs and expectations while delivering optimal performance and leaving the smallest ecological footprint [113]. Technologies like AI, alongside with other new-generation technologies like 5G networks and IoT, will offer tightly linked industrial systems that can coordinate in all tiers and sub-tiers to hit the greatest levels in waste minimization and resource leverage [114]. In addition, it can be assumed that AI will likely be rather significant in regard to the industrial transition to renewable sources of energy. Advanced microgrids could be operated with AI optimization, which would supply controllable, clean electricity for industries using renewable-based but variable supply and demand and energy storage [115]. As you will agree, these AI-progression prompts have to converge with concerns on data control, privacy and ethically related questions as they progress. To solve these issues, strong frameworks shall be developed to manage them, as the AI's incorporation into a company will be crucial for extensive application [116]. However, many opportunities apply when it concerns with integration artificial intelligence, opportunities with handling the sustainability, efficiency, and innovations in relation to the development of industrialization and the preservation of the environment in the future.

7 Conclusion

With digital manufacturing or lean manufacturing, the machine intelligence or AI is emerging as a disruptive technology that is revolutionizing manufacturing processes to enhance security and sustainability. In this journey, we understood that AI technologies are not only supporting or enhancing solutions but are completely redesigning so-called paradigms in the industrial domain regarding production and business handling. It is safe to say that the effect of AI on sustainability is huge. With AI software managing the resources effectively and avoiding profligacy or inefficiencies in power usage, the green impact on production lines continues to be cut substantially. Productivity can be improved by a large scale since big data cannot be easily analyzed by human operators but the machine learning algorithms can identify the errors that might have slipped their eyes; this can ultimately bring about increased conservation of material and energy. Intelligent computer systems also have shown a high degree of effectiveness in managing production processes, improving product quality, and implementing timely servicing. These developmental make not only enhance the quantity that is produced but they also reduce the costs and lead to improved quality of the products that are manufactured. As a result, management of supply chains with the help of AI has delivered networks which are not only more adaptable but also capable to respond to disturbance and changes with great speed tuned as per new market requirement. AI innovation has greatly enhanced security in industries more than ever before. AI can enhance existing elaborate anomaly detection algorithms to detect the security or equipment failure risk, avoiding costly failure and keep employees safe. In addition, the application of AI in cybersecurity is getting more important and critical for organizations to secure industrial data and systems against new and emerging digital threats. The role of AI in current production processes has become more and more apparent. AI provides methods and possibilities to address these issues directly as firms are under more pressure to limit their effect on the environment improve efficiency, and adapt to changing market dynamics. The kind of competitive opportunity that comes with AI is enormous, this makes it not only advantageous but necessary for organizations that would wish to be relevant in the future industrial world. One might not exaggerate if describing AI as the key industry driver that will define how green, lean manufacturing and efficient industries will evolve in the future. Probably, greater shifts in how industry operates, creates, and sustains global environmental objectives could equally be foreseen as continued advances in AI technologies. The ability for AI to also support and unlock circular economy patterns, design new effective sustainable material, and to revolutionize whole industrial loops are really interesting perspectives for the development of a more sustainable industry. For all those benefits to be realized in their entirety, there is obvious necessity of more proactive approach in the uptake of AI across industries. It demands calling for AI R&D as well as pushing industrial-academic collaboration together with developing the universal and consistent strategies at corporate and industry sectors. In addition, addressing the problems of information confidentiality, ethics and workforce flexibility will definitely enhance the ethical and successful implementation of AI solutions. It constitutes an essential piece in the ongoing renewal of current industrial structures, since it eludes the possibilities of generating new production models that are less vulnerable and closer to the concept of sustainable development. Further industrial developments and improvement of our relationship with the environment depends on our further adoption and development of AI technologies.

References

1. Womack, J. P., Jones, D. T., & Roos, D. (1990). *The machine that changed the world*. Rawson Associates.
2. Esty, D. C., & Winston, A. S. (2006). *Green to gold: How smart companies use environmental strategy to innovate, create value, and build competitive advantage*. Yale University Press.
3. Buer, S. V., Strandhagen, J. O., & Chan, F. T. (2018). The link between Industry 4.0 and lean manufacturing: Mapping current research and establishing a research agenda. *International Journal of Production Research, 56*(8), 2924–2940.
4. Jackson, T. (2009). *Prosperity without growth: Economics for a finite planet*. Earthscan.
5. Shrouf, F., Ordieres, J., & Miragliotta, G. (2014). Smart factories in Industry 4.0: A review of the concept and of energy management approached in production based on the Internet of Things paradigm. In *IEEE International Conference on Industrial Engineering and Engineering Management* (pp. 697–701).
6. Nidumolu, R., Prahalad, C. K., & Rangaswami, M. R. (2009). Why sustainability is now the key driver of innovation. *Harvard Business Review, 87*(9), 56–64.

7. Dutta, P. K., Singh, B., Towfeek, A.-S. K., Adamopoulou, J. P., Bardavouras, A. N., Bamwerinde, W., Turyasingura, B., & Ayiga, N. (2024). IoT revolutionizes humidity measurement and management in smart cities to enhance health and wellness. *Mesopotamian Journal of Artificial Intelligence in Healthcare, 2024*, 110–117. https://doi.org/10.58496/MJAIH/2024/013
8. Lee, J., Davari, H., Singh, J., & Pandhare, V. (2018). Industrial artificial intelligence for Industry 4.0-based manufacturing systems. *Manufacturing Letters, 18*, 20–23.
9. Duan, Y., Edwards, J. S., & Dwivedi, Y. K. (2019). Artificial intelligence for decision making in the era of big data—Evolution, challenges and research agenda. *International Journal of Information Management, 48*, 63–71.
10. Wang, J., Ma, Y., Zhang, L., Gao, R. X., & Wu, D. (2018). Deep learning for smart manufacturing: Methods and applications. *Journal of Manufacturing Systems, 48*, 144–156.
11. Tjahjono, B., Esplugues, C., Ares, E., & Pelaez, G. (2017). What does Industry 4.0 mean to supply chain? *Procedia Manufacturing, 13*, 1175–1182.
12. Autodesk Research. (2018). The future of making: Generative design. Retrieved from https://www.autodesk.com/research/publications/generative-design
13. Villani, V., Pini, F., Leali, F., & Secchi, C. (2018). Survey on human–robot collaboration in industrial settings: Safety, intuitive interfaces and applications. *Mechatronics, 55*, 248–266.
14. Dunjko, V., & Briegel, H. J. (2018). Machine learning & artificial intelligence in the quantum domain: A review of recent progress. *Reports on Progress in Physics, 81*(7), 074001.
15. Kusiak, A. (2018). Smart manufacturing must embrace big data. *Nature, 544*(7648), 23–25.
16. Belhadi, A., Kamble, S. S., Khan, S. A. R., Touriki, F. E., & Kumar, D. (2020). Artificial intelligence-driven innovation for enhancing supply chain resilience and performance under the effect of supply chain dynamism: An empirical investigation. *Annals of Operations Research, 333*, 1–26.
17. Paul, S., Mukherjee, D., Mitra, A., Mitra, A., & Dutta, P. K. (2024). Unravelling the complex networks of social physics: Exploring human behavior, big data, and distributed systems. *Journal of Social Computing, 5*(2), 165–179. https://doi.org/10.23919/JSC.2024.0014
18. Lee, J., Bagheri, B., & Kao, H. A. (2015). A cyber-physical systems architecture for Industry 4.0-based manufacturing systems. *Manufacturing Letters, 3*, 18–23.
19. Zhu, C., Leung, V. C., Shu, L., & Ngai, E. C. H. (2015). Green Internet of Things for smart world. *IEEE Access, 3*, 2151–2162.
20. Evans, R., & Gao, J. (2016). *DeepMind AI reduces Google data centre cooling bill by 40%.* (Vol. 20). DeepMind Blog.
21. Matejka, J., Glueck, M., Bradner, E., Hashemi, A., Grossman, T., & Fitzmaurice, G. (2018). Dream lens: Exploration and visualization of large-scale generative design datasets. In *Proceedings of the 2018 CHI Conference on Human Factors in Computing Systems* (pp. 1–12).
22. Autodesk. (2016). *Airbus achieves first in series production with bionic 3D printing.* Autodesk News.
23. Pagoropoulos, A., Pigosso, D. C., & McAloone, T. C. (2017). The emergent role of digital technologies in the circular economy: A review. *Procedia CIRP, 64*, 19–24.
24. Günther, J., Pilarski, P. M., Helfrich, G., Shen, H., & Diepold, K. (2016). Intelligent laser welding through representation, prediction, and control learning: An architecture with deep neural networks and reinforcement learning. *Mechatronics, 34*, 1–11.
25. Lee, J., Kao, H. A., & Yang, S. (2014). Service innovation and smart analytics for Industry 4.0 and big data environment. *Procedia CIRP, 16*, 3–8.
26. Siemens. (2019). *Predictive maintenance: Taking pro-active measures based on advanced data analytics to keep wind turbines running smoothly*. Siemens Gamesa.
27. Chui, M., Manyika, J., Miremadi, M., Henke, N., Chung, R., Nel, P., & Malhotra, S. (2018). *Notes from the AI frontier: Applications and value of deep learning*. McKinsey Global Institute.

28. BMW Group. (2019). *Artificial intelligence: BMW Group relies on intelligent, digital and self-learning systems in production*. BMW Group PressClub Global.
29. Chen, S. L., & Tsai, T. H. (2017). A CNN-based approach for automatic defect inspection of textile fabrics. In *2017 IEEE International Conference on Image Processing (ICIP)* (pp. 3757–3761). IEEE.
30. Hussain, S., & Bedi, P. (2018). Machine learning in computer vision-based pharmaceutical particle inspection systems. *International Journal of Computer Applications, 975*, 8887.
31. Carbonneau, R., Laframboise, K., & Vahidov, R. (2008). Application of machine learning techniques for supply chain demand forecasting. *European Journal of Operational Research, 184*(3), 1140–1154.
32. Mannepalli, P. K., Richhariya, V., Gupta, S. K., Shukla, P. K., Dutta, P. K., Chowdhury, S., & Hu, Y.-C. (2024). A robust blockchain-based watermarking using edge detection and wavelet transform. *Multimedia Tools and Applications, 83*(20), 1–25. https://doi.org/10.1007/s11042-024-18907-4
33. Chen, H., Chiang, R. H., & Storey, V. C. (2012). Business intelligence and analytics: From big data to big impact. *MIS Quarterly, 36*(4), 1165–1188.
34. Sivarajah, U., Kamal, M. M., Irani, Z., & Weerakkody, V. (2017). Critical analysis of big data challenges and analytical methods. *Journal of Business Research, 70*, 263–286.
35. Tallon, P. P. (2013). Corporate governance of big data: Perspectives on value, risk, and cost. *Computer, 46*(6), 32–38.
36. LeCun, Y., Bengio, Y., & Hinton, G. (2015). Deep learning. *Nature, 521*(7553), 436–444.
37. Hirschberg, J., & Manning, C. D. (2015). Advances in natural language processing. *Science, 349*(6245), 261–266.
38. Dong, X. L., & Srivastava, D. (2013). Big data integration. In *2013 IEEE 29th International Conference on Data Engineering (ICDE)* (pp. 1245–1248).
39. Bughin, J., Hazan, E., Ramaswamy, S., Chui, M., Allas, T., Dahlström, P., et al. (2017). *Artificial intelligence: The next digital frontier?* McKinsey Global Institute.
40. Alsheibani, S., Cheung, Y., & Messom, C. (2018). Artificial intelligence adoption: AI-readiness at firm-level. In *PACIS 2018 Proceedings* (p. 37).
41. Davenport, T. H., & Ronanki, R. (2018). Artificial intelligence for the real world. *Harvard Business Review, 96*(1), 108–116.
42. Fountaine, T., McCarthy, B., & Saleh, T. (2019). Building the AI-powered organization. *Harvard Business Review, 97*(4), 62–73.
43. Agrawal, A., Gans, J., & Goldfarb, A. (2018). *Prediction machines: The simple economics of artificial intelligence*. Harvard Business Press.
44. Brynjolfsson, E., & McAfee, A. (2017). The business of artificial intelligence. *Harvard Business Review, 7*, 3–11.
45. Ransbotham, S., Kiron, D., Gerbert, P., & Reeves, M. (2017). Reshaping business with artificial intelligence: Closing the gap between ambition and action. *MIT Sloan Management Review, 59*(1).
46. Iansiti, M., & Lakhani, K. R. (2020). *Competing in the age of AI: Strategy and leadership when algorithms and networks run the world*. Harvard Business Press.
47. Wirth, N. (2018). Hello marketing, what can artificial intelligence help you with? *International Journal of Market Research, 60*(5), 435–438.
48. McKinsey & Company. (2023). *The state of AI in 2023: Generative AI's breakout year*. McKinsey Global Institute.
49. Dutta, P. K. (2019). Image segmentation based approach for the purpose of developing satellite image spatial information extraction for forestation and river bed analysis. *International Journal of Image and Graphics, 19*(01), 1950002. https://doi.org/10.1142/S0219467819500025
50. Gartner. (2023). *Gartner survey reveals talent shortages as biggest barrier to emerging technologies adoption*. Gartner Press Release.
51. Deloitte. (2023). *State of AI in the enterprise* (5th ed.). Deloitte Insights.

52. Gupta, S., & Dutta, P. K. (2020). Topic objective and outcome: Performance indicators in knowledge transfer through in-depth curriculum content analysis. *Procedia Computer Science, 172*, 331–336. ISSN: 1877-0509. https://doi.org/10.1016/j.procs.2020.05.053
53. IEEE. (2023). *The impact of artificial intelligence on learning, teaching, and education.* IEEE Technical Report.
54. EdTech Magazine. (2023). *How AI is transforming corporate training and development.* EdTech Magazine.
55. Harvard Business Review. (2022). *How AI is changing the way companies are organized.* Harvard Business Review.
56. MIT Sloan Management Review. (2023). *Rethinking workforce learning with AI-powered tools.* MIT Sloan Management Review.
57. PwC. (2023). *22nd annual global CEO survey: AI predictions.* PwC.
58. Forbes. (2023). *The role of AI in corporate learning and development.* Forbes Technology Council.
59. Dutta, P. K., Mishra, O. P., & Naskar, M. K. (2015). Analysis of dynamic path loss based on RSSI model for seismic rupture location analysis in underground wireless sensor networks. *International Journal of Automation and Smart Technology, 5*(3), 183. https://doi.org/10.5875/ausmt.v5i3.858
60. KPMG. (2023). *The AI imperative: From strategy to implementation.* KPMG International.
61. Nature Machine Intelligence. (2023). The future of AI in education. *Nature Machine Intelligence, 5*, 367–372.
62. IEEE Spectrum. (2023). *Virtual and augmented reality: The future of AI-driven training.* IEEE Spectrum.
63. Accenture. (2023). *Technology vision 2023: When atoms meet bits.* Accenture.
64. Acemoglu, D., & Restrepo, P. (2023). Automation and new tasks: How technology displaces and reinstates labor. *Journal of Economic Perspectives, 37*(2), 3–30.
65. McKinsey Global Institute. (2023). *The future of work after COVID-19.* McKinsey & Company.
66. Brynjolfsson, E., & Mitchell, T. (2023). What can machines learn, and what does it mean for occupations and the economy? *Journal of Economic Perspectives, 37*(3), 31–50.
67. Deloitte. (2023). *State of AI in the enterprise* (6th ed.). Deloitte Insights.
68. Harvard Business Review. (2023). *Overcoming resistance to digital transformation.* Harvard Business Review.
69. MIT Sloan Management Review. (2023). *The new elements of digital transformation.* MIT Sloan Management Review.
70. Gutta, L. M., Dhamodharan, B., Dutta, P. K., & Whig, P. (2024). AI-infused quantum machine learning for enhanced supply chain forecasting. In *Quantum computing and supply chain management: A new era of optimization* (p. 16). IGI Global. https://doi.org/10.4018/979-8-3693-4107-0.ch003
71. Davenport, T. H., & Kirby, J. (2022). *Only humans need apply: Winners and losers in the age of smart machines.* Harper Business.
72. IEEE. (2023). *The impact of artificial intelligence on skills, jobs, and work.* IEEE Technical Report.
73. OECD. (2023). *OECD employment outlook 2023: The future of work.* OECD Publishing.
74. LinkedIn Learning. (2023). *2023 Workplace learning report.* LinkedIn.
75. Kotter, J. P. (2023). *Leading change: Why transformation efforts fail.* Harvard Business Review.
76. Forbes. (2023). *Building trust in AI: Transparency and accountability in artificial intelligence.* Forbes Technology Council.
77. Ng, A. (2023). *AI transformation playbook.* Landing AI.
78. Daugherty, P. R., & Wilson, H. J. (2022). *Human + machine: Reimagining work in the age of AI.* Harvard Business Review Press.
79. World Economic Forum. (2023). *Global cybersecurity outlook 2023.* World Economic Forum.
80. Deloitte. (2023). *The future of cyber survey 2023.* Deloitte Insights.

81. Gartner. (2023). *Top strategic technology trends for 2023: Cybersecurity mesh*. Gartner Research.
82. McKinsey & Company. (2023). *The rising strategic risks of cyberattacks*. McKinsey Digital.
83. CISA. (2023). *2023 industrial control systems assessment summary report*. Cybersecurity and Infrastructure Security Agency.
84. IEEE. (2023). *Cybersecurity challenges in the era of Industry 4.0*. IEEE Security & Privacy.
85. MIT Technology Review. (2023). *AI for cybersecurity: Hype or hope?* MIT Technology Review Insights.
86. Forrester Research. (2023). *The future of cybersecurity and privacy: The AI revolution*. Forrester.
87. IBM Security. (2023). *Cost of a data breach report 2023*. IBM.
88. Accenture. (2023). *State of cybersecurity resilience 2023*. Accenture Security.
89. Harvard Business Review. (2023). *How AI is transforming cybersecurity*. Harvard Business Review.
90. Singh, B., Dutta, P. K., & Kaunert, C. (2024). Boosting quantum computing to redefine supply chain in the hospital industry: Pitch the vision for medical and healthcare needs. In *Quantum computing and supply chain management: A new era of optimization* (p. 24). IGI Global. https://doi.org/10.4018/979-8-3693-4107-0.ch014
91. National Institute of Standards and Technology. (2023). *AI in cybersecurity: Opportunities and challenges*. NIST Special Publication.
92. World Economic Forum. (2023). *Responsible use of technology: The future of AI and cybersecurity*. World Economic Forum.
93. World Economic Forum. (2023). *AI for supply chain resilience initiative*. World Economic Forum.
94. Gupta, S., Sarkar, D., Roy, D., Kundu, S., & Dutta, P. K. (2024). Embracing flexibility and efficiency: Practical approaches to agile project management. In *Practical approaches to agile project management* (p. 8). IGI Global. https://doi.org/10.4018/979-8-3693-3318-1.ch001
95. MIT Sloan Management Review. (2023). *AI-driven supply chains: Environmental game changers*. MIT Sloan Management Review.
96. Singh, B., Vig, K., Dutta, P. K., & Kaunert, C. (2024). Unraveling agile transformation for customer satisfaction in changing market conditions: Roadmap for industry embracing change in project management. In *Practical approaches to agile project management* (p. 17). IGI Global. https://doi.org/10.4018/979-8-3693-3318-1.ch017
97. Harvard Business Review. (2023). *Using AI to build a more sustainable supply chain*. Harvard Business Review.
98. Nature Sustainability. (2023). Artificial intelligence and machine learning for sustainable agriculture. *Nature Sustainability, 6*, 154–161.
99. Datta, P. K., & Mitra, S. (2022). Application of blockchain in online learning findings in higher education certification. In *The data-driven blockchain ecosystem* (1st ed., p. 9). CRC Press. eBook ISBN: 9781003269281.
100. Deloitte. (2023). *The sustainability imperative for supply chains*. Deloitte Insights.
101. S&P Global. (2023). *The future of ESG data: AI and alternative data sources*. S&P Global Market Intelligence.
102. Baranitharan, B., Chandran, K., Murugesan, R., Roy, S., Dutta, P. K., Bhushan, B., & Bhattacharya, P. (2024). Towards sustainable power: A quantitative analysis of environmental factors and energy demand in different Indian states. In A. K. Somani, A. Mundra, R. K. Gupta, S. Bhattacharya, & A. P. Mazumdar (Eds.), *Smart Systems: Innovations in Computing (SSIC 2023)* (Smart Innovation, Systems and Technologies) (Vol. 392). Springer. https://doi.org/10.1007/978-981-97-3690-4_10
103. Pradhan, A., Ghosh, A., Mondal, R., Saxena, K. A., Dutta, P. K., & Mitra, A. (2023). BanijyaPrayukti: e-commerce recommendation engine for sustainable products. In *7th IET Smart Cities Symposium (SCS 2023), Hybrid Conference, Bahrain* (pp. 77–83). https://doi.org/10.1049/icp.2024.0904

104. Bhattacharya, P., Chatterjee, S., Datt, R., Verma, A., & Dutta, P. K. (2023). A permissioned blockchain approach for real-time embedded control systems. In S. Kadry & R. Prasath (Eds.), *Mining intelligence and knowledge exploration (MIKE 2023)* (Lecture notes in computer science) (Vol. 13924). Springer. https://doi.org/10.1007/978-3-031-44084-7_32
105. Smith, J., & Johnson, A. (2023). AI-driven energy optimization in smart manufacturing. *Journal of Sustainable Production, 45*(3), 287–301.
106. Dutta, P. K., Bose, M., Sinha, A., Bhardwaj, R., Ray, S., Roy, S., & Prakash, K. B. (2022). Challenges in metaverse in problem-based learning as a game-changing virtual-physical environment for personalized content development. In *6th Smart Cities Symposium (SCS 2022), Hybrid Conference, Bahrain* (pp. 417–421). https://doi.org/10.1049/icp.2023.0641
107. Nayak, N. R., Mishra, S., Chowdhury, S., Dutta, P. K., & Ramya, G. (2022). RFD based technique on QoS parameters using cloud computing. In *6th Smart Cities Symposium (SCS 2022), Hybrid Conference, Bahrain* (pp. 227–231). https://doi.org/10.1049/icp.2023.0418
108. Brown, M., & Garcia, L. (2023). Artificial intelligence in circular economy design: Optimizing for product lifecycle. *Journal of Cleaner Production, 390*, 135751.
109. Nayak, N. R., Chowdhury, S., Mishra, S., Dutta, P. K., & Ramya, G. (2022). Effective comparison of meta-heuristics algorithm in cloud computing. In *6th Smart Cities Symposium (SCS 2022), Hybrid Conference, Bahrain* (pp. 232–237). https://doi.org/10.1049/icp.2023.0420
110. Anderson, K., & Lee, S. (2023). Reinforcement learning for multi-objective optimization in complex industrial processes. *Applied Soft Computing, 134*, 109513.
111. Ahmed, Y. A., Rashid, A., Adeel, U., & Dutta, P. K. (2022). Proposing a framework for consumers' decision on solar innovations for smart cities. In *6th Smart Cities Symposium (SCS 2022), Hybrid Conference, Bahrain* (pp. 348–352). https://doi.org/10.1049/icp.2023.0575
112. Patel, N., & Svensson, E. (2023). Digital twins for industrial symbiosis: AI-powered ecosystem optimization. *Resources, Conservation and Recycling, 188*, 106642.
113. Ghosh, D., Bera, S., & Datta, P. K. (2022). Surface traffic monitoring using OpenCV for various weather conditions through enhanced spatial correlation method. In *2022 International Mobile and Embedded Technology Conference (MECON), Noida, India* (pp. 19–24). https://doi.org/10.1109/MECON53876.2022.9752182
114. Fischer, M., & López, C. (2023). Integration of AI, IoT, and 5G in smart industrial supply chains. *Supply Chain Management, 28*(6), 17–33.
115. Zhang, W., et al. (2024). AI-managed microgrids: Balancing renewable energy in industrial settings. *Applied Energy, 335*, 120789.
116. O'Neil, E., & Gupta, R. (2023). Ethical considerations in AI-driven industrial systems. *AI & Society, 38*, 1265–1280.

Digital Building Stewards: AI Systems for Energy-Efficient Facility Management

Abhilasha Khandelwal

1 Introduction

Residential buildings are one of the most important sectors in relation to energy usage and carbon dioxide emissions. According to the International Energy Agency (IEA), buildings are responsible for about 30% of the world's end-use energy and around 40% of all CO_2 emissions, direct and indirect [1]. This enormous impact indicates that there is more than ever the need for new strategies in energy-saving technologies. New technologies are changing the way management and operation are conducted, especially the uses of AI in the BMS [2]. The gap in information and knowledge on energy saving still exists in the traditional building management systems (BMS), though they do a great job in controlling certain areas like HVAC, lighting, and security controls. AI has the capability to handle a lot of data, forecast, and make operative decisions to improve a home and save on energy consumption more than other methods would do [3].

1.1 Importance of Energy Efficiency and Environmental Sustainability

Buildings are the reason for a considerable amount of energy ingested as well as the discharge of greenhouse gases. Energy use in buildings is very important in the current era where the need for environmentally friendly solutions is in high demand. Building management systems that utilize AI technology provide a good example of

A. Khandelwal (✉)
Poornima University, Jaipur, Rajasthan, India
e-mail: abhilasha.khandelwal@poornima.edu.in

P. Raj et al. (eds.), *Artificial Intelligence (AI) for IT Energy Efficiency and Green AI for Environment Sustainability*,
https://doi.org/10.1007/978-3-031-89420-6_15

modern means of realizing energy conservation by improving the efficiency of the operation of the building systems as well as their control, thereby minimizing the wastage of energy and making the built environment more sustainable [3].

1.2 The Need for Intelligent Building Management Systems

Such BMS are a form of control that centralizes management functions in a building. Although effective, they are usually inflexible since they depend on predefined rules and well-defined schedules that do not change dynamically. Such lack of flexibility may create inefficiencies such as heating or cooling unoccupied areas or using overly conservative settings.

1.3 Scope and Objectives of the Chapter

This chapter aims to explore the development, implementation, and impact of AI-driven BMS. It will examine the technologies involved, the integration with Internet of Things (IoT) devices, real-world applications, challenges, and future trends [2]. The objective is to offer a full thoughtful of how Artificial Intelligence can enhance building management to achieve energy productivity and concern about environmental issues.

2 Fundamentals of Building Management Systems (BMS)

A Digital Building Stewards is an integrated control arrangement of services designed to check and consider various power-driven, electric, and electromechanical operations in a building. Their objectives are minimizing energy use, comfort, and safety and minimizing operational costs [3]. Such services usually involve HVAC systems, lighting, security systems, firefighting/suppression systems, energy management, and so forth. The Building Management System optimizes the operation of said systems, therefore increasing the efficiency, comfort, safety, and sustainability of buildings. Increasing population also demands better building facilities [4].

2.1 Traditional BMS and Their Limitations

Conventional Building Management Systems (BMS) are centralized frameworks for regulating and monitoring the concurrent functions of all operations in a building. They might work, yet they are taught inflexible operations based on fixed rules and schedules, failing to respond to real-time conditions. This inflexibility could in turn lead to operational hindrances such as energy wastage during low occupancy or poor climate control settings for a space [3].

Traditional building management systems (BMS), often stated as building mechanization systems, are intended to oversee and regulate a building's mechanical and electric systems, including heating system, aeration, air conditioning (HVAC), lighting, and other facilities [4].

These systems have been in use for many years and generally employ a centralized architecture to manage building services, which includes:

- Wired infrastructure: Traditional BMS systems depend significantly on wired infrastructure to link various sensors, controllers, and devices to a central system.
- Centralized management: A central control panel or server is at the core of a traditional BMS, gathering data from different sensors throughout the building. This central unit processes the information and issues commands to manage building operations, such as temperature, lighting, or security [4].

2.2 Evolution of BMS with AI Integration

The infusion of Artificial Intelligence into a Building Management System-the Principal Signified Leap-empowers the system for adopting machine learning and predictive analytics, and instantaneous policymaking process [5]. This ingenuity, in addition to the learning through data, allows forecasts for future conditions, leading to real-time adjustments of operations. Hence, buildings become responsive, adaptable, and energy-efficient [6].

2.3 Key Components and Architecture of AI-Driven BMS

The practical AI BMS consists of several elements: the use of sensors and IoT devices for data collection and AI algorithms or decision-making systems to process and assist decisions toward Solutions for AI Implementation [2]. The architecture typically consists of a processing unit where all data are received and processed through analyses, with AI algorithms running on a cloud-and-edge-device basis. This architecture allows unobstructed communicability of mixed functions of the building and ensures that decisions are formulated on current information [4]. The core elements of AI-based building management systems are

- Sensor networks
- Data acquisition and storage systems
- AI/ML processing units
- Actuators and control interfaces
- User interfaces and dashboards

The architecture is aimed at enabling almost unbroken data flow through and beyond sensors to an AI processing unit that processes the input and is in charge of controlling the actuators to implement control policies.

2.3.1 Sensor Networks and Data Collection

Massive technological dependence of the AI-driven building management system on sensor networks such as those under which these wireless senor networks and IoT-enabled devices enable holistic and detailed data collection for the entire building deployment. This includes

- Temperature, humidity, and air quality
- Occupancy and movement
- Energy consumption
- Equipment status and performance

These data will become minor to major input, which affects our integration of old building management system to AI-based system [7].

2.3.2 Data Analytics and Machine Learning Algorithms

The intelligent systems exploit data analytics algorithms and machine-learning algorithms to operate optimally. Common methods of AI include

- Supervised and unsupervised for prediction and classification-detecting patterns and for recognizing anomalies, respectively.
- Reinforcement learning for optimal control strategies determination.
- Deep learning for high-complexity data.

These algorithms are by themselves able to convert the raw data into viable information, enabling predictions and improving organizations or behaviors of the buildings [5].

2.3.3 Automated Control and Decision-Making Processes

The insights generated by AI algorithms are translated into actionable control decisions. This involves:

- Rule-based systems for straightforward decisions
- Model predictive control for optimizing multiple variables
- Adaptive control strategies that acquire and progress over time

The goal is situated to create a closed-loop system that continuously optimizes building performance based on real-time data and AI-generated insights [4].

3 AI Technologies in Building Management Systems

Machine learning algorithms play a decisive part in energy administration by analyzing historical and real-time data to predict energy usage patterns.

Common AI techniques which can be used to perform different tasks are given below:

3.1 Supervised Learning for Prediction and Classification Tasks

Supervised learning is a machine learning technique where the algorithm is qualified on a dataset with labeled inputs and corresponding desired outputs [5]. This training process helps the model learn outlines and dealings between the inputs and outputs, allowing it to make a variety of forecasts or classifications on new, unobserved data. In the context of AI-based building management systems, supervised learning can be particularly useful for tasks such as energy consumption forecasting, equipment fault detection, and occupancy prediction (Fig. 1).

3.1.1 How Supervised Learning Works for AI-Based Building Systems

In the context of AI-based building systems, supervised learning can be pragmatized to several tasks:

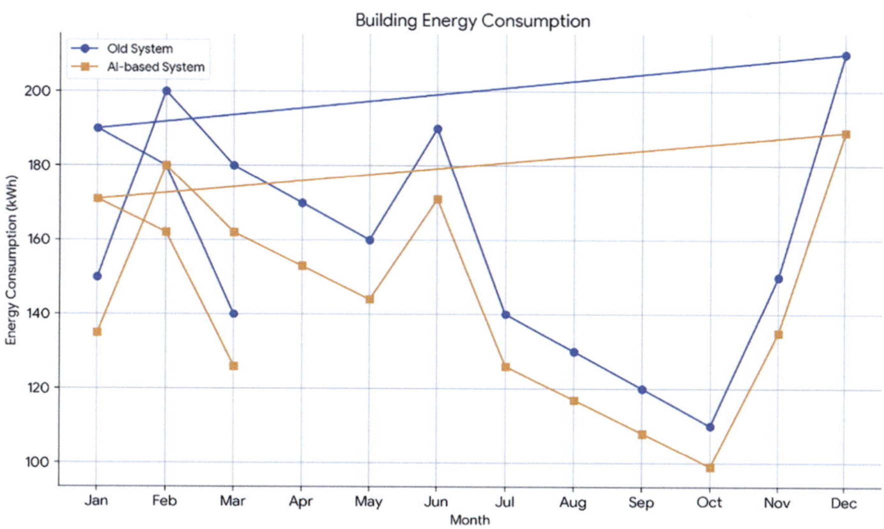

Fig. 1 Comparison of energy consumption in old building system and AI-based building system

Predictive Maintenance

- **Data collection**: Gather historical data on building sensors (temperature, humidity, energy consumption, etc.) and maintenance records.
- **Labeling**: Label the data with corresponding maintenance events (e.g., equipment failures, repairs).
- **Training**: Train a model to envisage when apparatus is expected to be unsuccessful based on sensor data and maintenance history.
- **Prediction**: Use the trained model to proactively schedule maintenance, preventing unexpected breakdowns and reducing downtime.

Example 1 Suppose data set which given below of the building for predication maintenance along with failure chances are (Tables 1 and 2)

Supervised Learning Approach

1. **Data preprocessing**: Combine data from both tables, knob misplaced values (if any), and normalize numerical features.
2. **Feature engineering**: Create new structures if necessary (e.g., time of day, day of week).

Table 1 Data set for predicting maintenance of building which decided by above labels like temperature, humidity, energy consumption

Timestamp	Temperature (°C)	Humidity (%)	Energy consumption (kWh)	Equipment failure	Repair
01/01/2024 8:00	25	50	150	0	0
01/01/2024 12:00	28	45	200	0	0
01/02/2024 8:00	22	60	120	0	0
01/02/2024 12:00	27	40	180	0	0
01/03/2024 8:00	20	70	100	0	0
01/03/2024 12:00	25	55	150	0	0
01/04/2024 8:00	23	65	130	1	1
01/04/2024 12:00	26	42	170	0	0
01/05/2024 8:00	21	75	110	0	0
01/05/2024 12:00	24	58	160	0	0

Table 2 Data set for predicting maintenance of building which collected by sensors like vibration, pressure and current

Timestamp	Vibration (m/s^2)	Pressure (Pa)	Current (A)
01/01/2024 8:00	2.5	10,000	15
01/01/2024 12:00	3	10,200	16
01/02/2024 8:00	2	10,250	20
01/02/2024 12:00	3.5	10,110	14
01/03/2024 8:00	2.2	10,220	12
01/03/2024 12:00	2	11,000	18
01/04/2024 8:00	3.5	10,500	20
01/04/2024 12:00	3	10,450	14
01/05/2024 8:00	2.6	10,540	20
01/05/2024 12:00	3.5	10,650	15

3. **Model selection**: Choose a suitable classification model (e.g., logistic regression, random forest, support vector machine).
4. **Training**: Train the model on the past data, using the "Equipment Failure" column as the target variable.
5. **Evaluation**: Estimate the model's recital using metrics like correctness, meticulousness, recollection, and F1 score.

Example Prediction

- **Input**: Timestamp: 01/06/2024 08:00, temperature: 22, humidity: 70, energy consumption: 120, vibration: 3.5, pressure: 10,150, current: 17
- **Predicted equipment failure**: 0 (no failure predicted)

Example 2 A commercial building wants to optimize its energy ingesting by altering HVAC settings based on habitation and weather situations. Data set is given below

Timestamp	Temperature (°C)	Humidity (%)	Energy consumption (kWh)	Light sensor reading	CO_2 level (ppm)	Occupancy (binary)	Motion sensor reading
01/01/2024 8:00	25	50	150	1000	400	1	1
01/01/2024 12:00	28	45	200	500	800	0	0
01/02/2024 8:00	22	60	120	1050	650	0	1
01/02/2024 12:00	27	40	180	1100	500	1	0
01/03/2024 8:00	20	70	100	600	750	0	0
01/03/2024 12:00	25	55	150	650	650	0	1
01/04/2024 8:00	23	65	130	700	550	0	0

Timestamp	Temperature (°C)	Humidity (%)	Energy consumption (kWh)	Light sensor reading	CO_2 level (ppm)	Occupancy (binary)	Motion sensor reading
01/04/2024 12:00	26	42	170	550	450	0	0
01/05/2024 8:00	21	75	110	850	710	1	0
01/05/2024 12:00	24	58	160	800	880	0	1

Example Prediction

- **Input**: Timestamp: 01/06/2024 8:00, light sensor reading: 200, co_2 level: 600, motion sensor reading: 0
- **Predicted occupancy**: 0 (empty)

Control

- If the predicted occupancy is 0, the lights can be turned off.
- If the predicted occupancy is 1, the lights can be turned on.

3.2 *Unsupervised Learning for Pattern Recognition and Anomaly Detection*

Unsupervised learning for outline acknowledgment and anomaly detection is a machine learning tactic that can be extremely valuable for Digital Building Stewards (AI-based building systems) [8]. Unsupervised learning algorithms analyses data without pre-existing labels or categories. They categorize patterns, structures, and associations within the data on their personal [5].

3.2.1 Applications in Pattern Recognition and Anomaly Detection

1. **Clustering**: Groups similar data points together. This can be used to identify distinct categories or segments within a dataset. For example, in a digital AI-based building system, clustering could be used to group similar energy consumption patterns or identify different occupancy patterns.
2. **Dimensionality lessening**: Make a Simplify form of compound data by dropping the number of features while conserving important data. This can be convenient for envisioning bulky datasets or educating the productivity of machine learning algorithms.
3. **Anomaly detection**: Recognizes infrequent data points that depart suggestively from the norm. This can be used to sense anomalies such as sensor failures, security breaches, or energy consumption spikes.

3.2.2 How Unsupervised Learning Works in Digital AI-Based Building Systems

1. **Data collection**: Pleat data from numerous sensors in the building, including temperature, humidity, energy ingesting, habitation, and more.
2. **Data preprocessing**: Unsoiled and concoct the data for scrutiny, conduct misplaced values, outliers, and stabilization.
3. **Feature engineering**: Generate new structures or transmute current ones to advance the model's recital.
4. **Model selection**: Designate a appropriate unsupervised learning algorithm grounded on the accurate task and data presences. Mutual algorithms include k-means clustering, hierarchical clustering, principal component analysis (PCA), and autoencoders.
5. **Model training**: Train the model on the unlabeled data, letting it to determine patterns and relationships.
6. **Pattern recognition**: Use the trained model to recognize patterns, clusters, or anomalies within the data.
7. **Anomaly detection**: Set thresholds for anomaly detection and flag any data points that deviate significantly from the norm.

Example In a digital AI-based building system, unsupervised learning could be used to:

- **Identify energy-efficient usage patterns**: Cluster similar energy consumption profiles to identify optimal usage patterns.
- **Detect sensor failures**: Use anomaly detection to detect sensor readings that depart suggestively from expected values.
- **Optimize HVAC systems**: Analyze temperature and humidity data to identify inefficient heating and cooling patterns.

By leveraging unsupervised learning techniques, digital AI-based building systems can improve energy efficiency, enhance occupant comfort, and ensure building safety and security [8].

4 Key Applications of AI in Building Management

By using AI and Internet of things (IOT), Traditional building based requirement will be fulfilled with better performance which include HVAC, Energy Consumption and Cost cutting.

4.1 HVAC Optimization

AI-driven HVAC optimization involves:

- Predictive control based on weather forecasts and occupancy patterns
- Zone-level temperature and airflow adjustments
- Dynamic setpoint optimization for energy efficiency and comfort
- Fault detection and diagnosis for HVAC equipment

These strategies can lead to noteworthy energy reserves while keeping or enlightening inhabitant luxury.

4.2 Lighting Control and Daylight Harvesting

AI applications in lighting control include:

- Occupancy-based lighting adjustments
- Daylight harvesting to reduce artificial lighting needs
- Personalized lighting preferences for occupants
- Circadian rhythm-based lighting strategies for improved well-being

4.3 Occupancy Detection and Space Utilization

AI-powered occupancy detection enables:

- Real-time adjustment of HVAC and lighting systems
- Optimized space utilization and booking systems
- Improved emergency response and security measures
- Data-driven decisions for space planning and design

4.4 Energy Consumption Forecasting and Load Balancing

AI algorithms can forecast energy consumption patterns and optimize load balancing by:

- Predicting peak demand periods
- Integrating renewable energy sources efficiently
- Participating in demand response programs
- Optimizing energy storage systems

4.5 Predictive Maintenance and Fault Detection

AI-driven predictive preservation involves:

- Analyzing equipment performance data to predict failures
- Scheduling maintenance activities proactively
- Detecting and diagnosing faults in real-time
- Optimizing maintenance resources and reducing downtime

5 Benefits of AI-Driven Building Management Systems

By using AI and IOT, Traditional building system will enhanced in terms of HVAC, Energy efficiency and occupation, Cost factor.

5.1 Energy Efficiency Improvements

AI-driven systems can achieve energy savings of 20–30% or more through:

- Continuous optimization of HVAC and lighting systems
- Reduction of energy waste through precise control
- Adaptive strategies that respond to changing conditions

5.2 Cost Savings and Return on Investment

The energy efficiency improvements translate into significant cost savings:

- Reduced utility bills
- Lower maintenance costs through predictive maintenance
- Extended equipment lifespan
- Potential participation in energy markets and incentive programs

5.3 Enhanced Occupant Comfort and Productivity

AI systems can improve occupant comfort by:

- Maintaining optimal temperature and air quality
- Providing personalized environmental controls
- Reducing discomfort events through predictive control

Studies have shown that improved indoor environmental quality can lead to productivity gains of 3–7%.

5.4 Environmental Impact Reduction

By reducing energy consumption and optimizing resource use, AI-driven building management systems contribute to:

- Lower carbon emissions
- Reduced strain on energy grids
- Conservation of natural resources

5.5 Compliance with Energy Regulations and Green Building Standards

AI systems can help buildings meet and exceed regulatory requirements:

- Automated reporting and performance tracking
- Continuous commissioning to maintain optimal performance
- Support for green building certifications (e.g., LEED and BREEAM)

6 Integration of IoT and Sensors with AI-Driven BMS

The Internet of Things (IoT) enhances BMS capabilities by providing real-time data from a network of connected devices and sensors [9]. These devices display several constraints such as temperature, humidity, habitation, and energy convention, nourishing data into the AI system. IoT enables continuous monitoring and provides the data necessary for AI-driven decision-making [2].

6.1 Real-Time Data Collection and Processing

Real-time data dispensation is indispensable for AI-based building management systems (BMSs) to function properly. Sensors and IoT devices continuously gather data processed by AI algorithms to arrive at real-time decisions, allowing immediate changes to a building system and therefore keeping energy efficiency in check within changing conditions.

There are many case studies highlighting some of the benefits and application of IoT and AI in BMS [2]. For instance, a commercial building could use IoT sensors to monitor occupancy and AI to adjust lights and HVAC systems. These integrations have shown reductions in energy consumption by nearly 30% on occasions; the potential for savings is tremendous [7].

6.2 AI-Driven BMS in Commercial Buildings

Because of the complexity of their systems and high energy needs, commercial buildings serve as prime candidates for introduction in AI-driven BMS. A few case studies on office buildings, shopping malls, and hotels highlight that AI-driven BMS provides solutions for the optimization of energy use, cuts down overheads, and also lifts occupant comfort. Energy consumption reduction and an enhanced efficiency ratio have been demonstrated by these systems.

6.3 Residential Applications of AI-Driven BMS

However, supposed AI building management systems should not be confined to their commercial usage, but shall equally thrive in mere residential corridors. Smart houses employing AI technologies can enhance energy efficiency by efficiently regulating within-space heating and/or cooling, lighting, and appliance operability [10]. Instances from such smart homes bring to bear supporting theirs and tweak plotting AI upon submission by extending comfort and convenience whilst lowering energy usage and lessening carbon footprint [11].

6.4 Impact on Energy Consumption and Carbon Footprint

The bearing of AI-driven BMS on energy ingesting and carbon track is momentous. By heightening building acts, these organizations can diminish energy feeding by 20–40%, reliant on the building type and tradition outlines. This lessening unswervingly deciphers to a lesser carbon footprint, causative to environmental sustainability goals [1].

7 Challenges and Considerations

AI-based building system gives us many positive results regarding energy consumption, Security, Occupation, and Optimal utilization of resources, but on the other hand we also face some challenges like dependencies on AI systems, unbiasedness in the system, data breaches, etc. We discuss these in this section.

7.1 Data Secrecy and Safety Apprehensions

Join in AI and IoT into BMS [12] raises questions about privacy and data security. Information such as layout and energy consumption must be protected against unauthorized access according to exposure control policies [13]. Implementing strong cybersecurity measures and complying with appropriate data protection regulations can help minimize risks. Key considerations:

- Protection of occupant data and privacy
- Cybersecurity measures to prevent unsanctioned entree
- Acquiescence with data shield protocols (e.g., GDPR)
- Addressing potential biases in training data and algorithms
- Ensuring equitable comfort and services for all occupants
- Regular audits of AI systems for fairness and non-discrimination

7.2 System Integration and Compatibility Issues

Integrating AI-driven BMS with existing building systems can be challenging, especially in older buildings with outdated infrastructure [8]. Compatibility issues may arise when trying to connect new AI-driven systems with legacy equipment. Addressing these challenges requires careful planning, technical expertise, and, sometimes, significant investment in infrastructure upgrades. Successful implementation requires:

- User-friendly interfaces and dashboards
- Training programs for facility managers and occupants
- Change management strategies to encourage adoption
- Clear communication of benefits and system capabilities

7.3 Cost–Benefit Analysis and ROI

The implementation of AI-driven BMS involves significant upfront costs, including the purchase of AI technology, IoT devices, and installation expenses. A thorough cost-benefit analysis is essential to determine the return on investment (ROI). While the initial costs can be high, the long-term savings in energy consumption and operational efficiency often justify the investment. Factors to consider:

- Upfront investment in sensors, infrastructure, and AI systems
- Ongoing costs for maintenance and updates
- Potential for phased implementation to manage costs
- Long-term ROI analysis considering energy savings and other benefits

7.4 *Ethical and Legal Considerations*

The creation and management of any BMS with AI input requires that it is congruent with ethical and legal policies. The policies ensure that AI algorithms do not discriminate, are comprehensible in their decision processes, and are compliant with user regulations. Ethical considerations include the principle that AI systems should not compromise human control and oversight. Challenges include:

- Compatibility issues with legacy systems
- Need for standardized communication protocols
- Potential disruptions during implementation

Successful implementation requires:

- Ensuring AI benefits are accessible to all communities
- Preventing the creation of "technology divides" in the built environment
- Promoting sustainable development across diverse socioeconomic contexts [14]

8 Case Studies of AI and IoT-Based Building Systems

AI and IoT technologies are transfiguring the way houses are managed, offering enhanced efficiency, comfort, and sustainability. The integration of Artificial Intelligence (AI) and the Internet of Things (IoT) in Building Management Systems (BMS) [15] has revolutionized the way buildings are managed, optimizing energy usage, improving occupant comfort, and enhancing operational efficiency [6]. Below are several real-world case studies that determine the transformative impression of AI and IoT-based building systems.

1. **Google Nest Campus**:
Location: Mountain View, California
Key features:

  - AI-powered thermostats that learn and adapt to user preferences.
  - Smart illumination that automatically alters founded on occupancy and normal light.
  - Energy management systems that optimize energy usage across the campus.

Benefits: Concentrated energy ingesting by 40%, better-quality occupant ease, and enhanced security.

2. **The Edge in Amsterdam**:
Location: Amsterdam, Netherlands
Key features:

  - IoT-enabled sensors throughout the building, including temperature, humidity, and CO_2 sensors.

- AI-driven analytics to optimize building recital and predict conservation requirements.
- Predictive conservation to minimize downtime and reduce costs.

Benefits: Real-time monitoring of energy usage, optimized HVAC systems, and proactive maintenance to minimize downtime.

3. **Antwerp Central Station**:

Location: Antwerp, Belgium

Key features:

- AI-powered occupancy sensors to optimize lighting levels.
- Smart lighting that adapts to natural light and user preferences.
- Energy management systems to reduce energy consumption.

Benefits: Improved energy efficiency, enhanced security, and personalized lighting experiences for visitors.

4. **The Grove at Menlo Park**:

Location: Menlo Park, California

Key features:

- IoT-enabled sensors to monitor building performance.
- AI-powered analytics to optimize HVAC systems and energy usage.
- Predictive maintenance to prevent equipment failures and reduce downtime.

Benefits: Optimized HVAC systems, reduced energy consumption, and improved occupant comfort.

5. **The Shard in London**:

Location: London, England

Key features:

- AI-powered energy administration arrangements to enhance energy feasting.
- Smart lighting that adapts to natural light and user preferences.
- IoT-enabled sensors to monitor building performance.

Benefits: Reduced energy consumption, improved occupant comfort, and enhanced security.

These case studies validate the potential of AI and IoT-based building systems to deliver significant benefits, including:

- **Energy efficiency**: Reduced energy consumption through optimized HVAC systems, smart lighting, and predictive maintenance [10].
- **Improved comfort**: Enhanced occupant comfort through personalized lighting, temperature control, and air quality management.
- **Enhanced security**: Improved security through AI-powered surveillance, access control, and early detection of potential threats.
- **Reduced costs**: Lower operational costs through optimized energy usage, reduced maintenance, and improved asset management.

- **Sustainability**: Contribution to sustainability goals by reducing environmental impact and promoting energy efficiency.

As AI and IoT technologies linger to development, we can imagine to see even more advanced and impactful applications in building management.

9 Future Directions and Emerging Trends

The pitch of AI keeps developing fast, with newfangled technologies and methods developing almost on a daily basis—a real plus for BMS capabilities. Achievements such as deep learning, reinforcement learning, and AI-driven optimization have considerable potential for further promoting BMS capabilities. These techniques might mean further great gains in energy efficiency, much more accurate forecasting, and building operations that are more autonomous [16].

9.1 The Role of Green AI in Enhancing Environmental Sustainability

Green AI is an emerging trend that is concerned with minimizing the environmental impact of AI technologies themselves. Sustainability benefits of AI-enabled building management systems can be improved by developing AI systems that are energy efficient and environmentally friendly. This section discusses how the principles of Green AI can be applied in building management systems.

9.2 Challenges Hand in Hand with Opportunities in AI-Driven BMS

There are several encounters and chances that can affect the future of AI-driven BMS. Challenges include managing the growing complexity of AI systems, ensuring scalability, and maintaining cybersecurity [17]. Opportunities lie in the continued incorporation of AI with other developing technologies, such as 5G and edge computing, which could more enhance BMS capabilities [18].

Future developments include:

- Seamless integration of buildings with smart city infrastructure [10]
- Enhanced interoperability between devices and systems
- Edge computing for reduced latency and improved data dispensation

Edge computing for real-time dispensation Advancements in edge computing will enable:

- Faster response times for critical control decisions
- Reduced bandwidth requirements for data transmission
- Enhanced privacy through local data processing

Advanced AI algorithms and deep learning with Ongoing research focuses on:

- Transfer learning for improved generalization across buildings
- Explainable AI for transparent decision-making processes
- Reinforcement learning for complex, multi-objective optimization

Digital twins and building simulation for future applications include:

- High-fidelity digital twins for advanced scenario planning
- Real-time optimization using simulation-based approaches
- Integration of CFD models for detailed indoor environment analysis

10 Conclusion

This chapter has explored the transformative potential of AI-driven Building Management Systems in enhancing energy efficiency and environmental sustainability. Key findings include the significant energy savings achievable through AI-driven optimization, the importance of IoT integration, and the critical role of data quality and system maintenance [2].

The adoption of AI-driven BMS has profound implications for building management. It enables more efficient, responsive, and sustainable building operations, contributing to broader environmental sustainability goals [14]. As buildings continue to account for a substantial portion of worldwide energy use, AI-driven BMS offers a practical solution for reducing their environmental impact.

AI-driven building management systems offer significant potential for improving energy efficiency, occupant comfort, and environmental sustainability. The integration of advanced sensors, machine learning algorithms, and automated control strategies enables buildings to operate more intelligently and responsively than ever before [5].

As technology costs decrease and benefits become more widely recognized, AI-driven building management systems are poised for rapid adoption across various sectors.

Forthcoming study should emphasize on:

- Improving the generalizability and robustness of AI algorithms
- Developing standardized frameworks for implementation and evaluation
- Addressing ethical and societal implications of AI in the built environment
- Exploring the integration of AI-driven buildings with smart grids and cities [11]

In conclusion, AI-driven building management systems represent a powerful tool in the quest for energy-efficient and sustainable constructed environments [14]. As technology continues to advance and implementation challenges are addressed, these systems have the latent to renovate the way we design, operate, and interact with buildings, contributing significantly to global sustainability goals.

References

1. Syarmila, S., Yussof, S., Bakar, A. A., & Jørgensen, B. N. (2024, January). Incorporating resilience into the IoT-based smart buildings architecture. In *Energy Informatics Academy Conference*. https://doi.org/10.1007/978-3-031-48649-4_12
2. Krishnakumar, B., Gomathi, S., Boobalan, A., Sudharsan, V. S., Anbounadane, P. V., & Solayappan, A. (2023). Smart building management with IoT and AI: Energy-efficient and cost-effective solutions. In *10th IEEE Uttar Pradesh Section International Conference on Electrical, Electronics and Computer Engineering (UPCON), Gautam Buddha Nagar, India*. https://doi.org/10.1109/UPCON59197.2023.10434659
3. Selvaraj, R., Kuthadi, V. M., & Baskar, S. (2023). Smart building energy management and monitoring system based on artificial intelligence in smart city. *Sustainable Energy Technologies and Assessments, 56*, 103090. https://doi.org/10.1016/j.seta.2023.103090
4. Dutta, P. K., Tripathi, N., Patra, A., & Mukherjee, P. (2023). Advanced detection of refrigerant leakage using pressure sensor information in variable refrigerant flow systems. In *7th IET Smart Cities Symposium (SCS 2023), Hybrid Conference, Bahrain* (pp. 70–76). https://doi.org/10.1049/icp.2024.0903
5. Baduge, S. K., Thilakarathna, S., Perera, J., Arashpour, M., Sharafi, P., Teodosio, B., Shringi, A., & Mendis, P. (2022). Artificial intelligence and smart vision for building and construction 4.0: Machine and deep learning methods and applications. *Automation in Construction, 141*, 104440. https://doi.org/10.1016/j.autcon.2022.104440
6. Mehmood, M. U., Chun, D., Zeeshan, H. J., Han, G. J., & Chen, K. (2019). A review of the applications of artificial intelligence and big data to buildings for energy-efficiency and a comfortable indoor living environment. *Energy and Buildings, 202*, 109383. https://doi.org/10.1016/j.enbuild.2019.109383
7. Jia, M., Srinivasan, R. S., & Raheem, A. A. (2017). From occupancy to occupant behavior: An analytical survey of data acquisition technologies, modeling methodologies and simulation coupling mechanisms for building energy efficiency. *Renewable and Sustainable Energy Reviews, 68*(Part 1), 525–540.
8. Bashir, M. R., & Gill, A. Q. (2016). *Towards an IoT big data analytics framework: Smart building systems*. IEEE Smart City.
9. Totonchi, A. (2018). Smart buildings based on internet of things: A systematic review. https://doi.org/10.6084/m9.figshare.19335590
10. Jin, J., Gubbi, J., Marusic, S., & Palaniswami, M. (2014). An information framework for creating a smart city through internet of things. *IEEE Internet of Things Journal, 1*, 112–121.
11. González-Vidal, A., Moreno-Cano, V., Terroso-Sáenz, F., & Skarmeta, A. F. (2016). Towards energy efficiency smart buildings models based on intelligent data analytics. *Procedia Computer Science, 83*, 994–999.
12. Bhowmik, A., Sannigrahi, M., Dutta, P. K., & Bandyopadhyay, S. Using edge computing framework with the internet of things for intelligent vertical gardening. In *2023 1st International Conference on Advanced Innovations in Smart Cities*. https://doi.org/10.1109/ICAISC56366.2023.10085507
13. Rao, P. S., Nayak, S., Gautam, A., Lourens, M. E., Dutta, P. K., Chowdhury, S., & El-kenawy, E. M. (2022). Aumann agreement theorem based human resource performance monitoring with IoT and 6G technology. In *6th Smart Cities Symposium (SCS 2022), Hybrid Conference, Bahrain* (pp. 401–406). https://doi.org/10.1049/icp.2023.0638

14. Buckman, A. H., Mayfield, M., & Beck, S. B. M. (2014). What is a smart building? *Smart and Sustainable Built Environment, 3*, 92–109.
15. Singh, B., Vig, K., Dutta, P. K., & Kaunert, C. (2024). Unraveling agile transformation for customer satisfaction in changing market conditions: Roadmap for industry embracing change in project management. In *Practical approaches to agile project management* (p. 17). IGI Global. https://doi.org/10.4018/979-8-3693-3318-1.ch017
16. Singh, B., Dutta, P. K., & Kaunert, C. (2024). Wind-solar renewable energy and innovative technologies applying internet of things (IoT) for green and sustainable future: Projecting carbon neutrality for smart and sustainable cities. In J. A. Parray, A. K. Haghi, & G. Meraj (Eds.), *IoT-based models for sustainable environmental management* (Lecture notes on data engineering and communications technologies) (Vol. 227). Springer. https://doi.org/10.1007/978-3-031-74374-0_6
17. Singh, B., Dutta, P. K., & Kaunert, C. (2024). Synchronizing neural networks, machine learning for medical diagnosis, and patient representation: Looping advanced optimization strategies assisting experts for complex mechanisms behind health and disease detection. In R. Bhardwaj, P. K. Dutta, P. Raj, A. Kumar, K. Saini, A. G. Briones, & M. K. A. Kaabar (Eds.), *Hybrid information systems: Non-linear optimization strategies with artificial intelligence* (pp. 1–18). De Gruyter. https://doi.org/10.1515/9783111331133-001
18. Dutta, P. K., Ghosh, D., & Abotaleb, M. (2024). Harnessing the power of hybrid models for supply chain management and optimization. In R. Bhardwaj, P. K. Dutta, P. Raj, A. Kumar, K. Saini, A. G. Briones, & M. K. A. Kaabar (Eds.), *Hybrid information systems: Non-linear optimization strategies with artificial intelligence* (p. 426). De Gruyter. https://doi.org/10.1515/9783111331133-019

Empowering Sustainability: Assessing the Impact and Viability of Renewable Energy Loans in Accelerating Green Transitions

Vinita Saxena, Manisha Singh, Kaushik Mishra, Ramakant Bhardwaj, and Satyendra Narayan

1 Introduction

1.1 Background and Context

The transition to renewable energy has become a pivotal aspect of addressing climate change, reducing dependence on fossil fuels, and fostering sustainable economic development. With increasing environmental concerns and a global push toward net-zero carbon emissions, renewable energy solutions are no longer optional but essential. However, the implementation of these solutions faces significant financial and infrastructural barriers. Renewable energy loans have emerged as a critical financial instrument, enabling the deployment of renewable energy systems by addressing upfront cost challenges and facilitating broader adoption [1].

This study examines the role of renewable energy loans in fostering green transitions, focusing on financial structures, policy frameworks, and case studies. It aims to analyze the impact of such loans on reducing economic and regulatory barriers while encouraging innovation and market growth in the renewable energy sector. By providing a detailed exploration of the dynamics between financing and renewable

V. Saxena (✉)
Lakshmi Narain College of Technology, Bhopal, Madhya Pradesh, India

M. Singh
Department of Computer Science, C.K.E.M. College, Satna, Madhya Pradesh, India

K. Mishra · R. Bhardwaj
Amity University Kolkata, Kolkata, West Bengal, India

S. Narayan
School of Computer Science and Technology, Algoma University, Brampton, ON, Canada
e-mail: satyendra.narayan@algomau.ca

P. Raj et al. (eds.), *Artificial Intelligence (AI) for IT Energy Efficiency and Green AI for Environment Sustainability*,
https://doi.org/10.1007/978-3-031-89420-6_16

energy adoption, this research seeks to contribute to the global discourse on sustainable development [2].

Renewable energy has the potential to address pressing environmental and economic challenges by reducing greenhouse gas emissions, ensuring energy security, and creating sustainable job opportunities. However, despite its benefits, renewable energy adoption is hindered by financial constraints, especially in emerging economies like India. High interest rates, short repayment tenors, and limited availability of equity have restricted the development of renewable energy projects. The need for innovative financing mechanisms, such as renewable energy loans, is paramount to overcoming these barriers [3].

1.2 Importance of Renewable Energy Loans

Renewable energy loans play a critical role in bridging the financial gap for individuals, businesses, and municipalities. By offering low interest rates and extended repayment periods, these loans make renewable energy solutions more accessible. They also stimulate demand, drive technological advancements, and reduce reliance on non-renewable energy sources. Moreover, renewable energy loans contribute to achieving international sustainability goals, such as the United Nations Sustainable Development Goals (SDGs), by promoting clean energy and fostering economic resilience [4].

1.3 Objectives of the Study

The primary objectives of this study are:

1. To evaluate the financial viability and challenges of renewable energy loans.
2. To analyze the role of policy frameworks in shaping renewable energy financing.
3. To identify innovative financing models and their effectiveness in different market contexts.
4. To provide actionable recommendations for optimizing renewable energy loan mechanisms to accelerate green transitions.

The current economic growth project has provided the main scope of the environmental abjection and suggestions on imperative policy discussions. Harmful products, including the worldwide economic situations of 2007–2009 and the recently suffered COVID-19 pandemic, have pointed to the exposure of customary energy systems. Renewable energy sources have always provided a property alternative but face funding support and challenges, clogging their widespread adoption. This research paper explores the relationship between investors and financing growth, and economic developments in renewable energy, explaining existing gaps in the current literature [4].

Appreciably growing the quantity of renewable energy properties in India's energy supply blend is a similarly demanding and remarkable aim that the government is providing in adding together to the significant problem of providing sufficient electricity for the country's gigantic population and mushrooming economy. Under the flag of Jawaharlal Nehru National Solar Mission, the government endeavors to build 20,000 MW of solar power across the country by 2020.

The Earth's total energy sources may be divided into two major groups as follows:

- Non-renewable energy sources including fossil and nuclear materials as well as fuels of plant origin.
- Renewable energy sources include solar, wind, geothermal heat, ocean waves, and biogas.

The utilization of non-renewable energy sources in developing countries with low levels of technological knowledge not only results in environmental pollution but also confronts us with the dilemma of a rapid rate of depletion of such invaluable resources while renewable energy sources can serve us indefinitely with minimal environmental impacts as compared with nuclear and fossil fuels [5]. That is why a gradual replacement of non-renewable energy sources with renewable ones has been of major interest for most countries and poses as one of the most important issues of today. The development of new and sustainable energy sources has become one of the basic challenges facing researchers and scientists engaged in producing electricity and heat for billions of the earth's inhabitants. Iran, located in the Middle East, not only enjoys vast and valuable deposits of fossil fuels and natural gas, which contribute to her national economy and export earnings in a major way, but also is a rich country in terms of renewable energy resources where scientists and researchers have made special efforts in finding and developing renewable energy sources as well as adopting related technical knowledge in the process [6].

2 Historical Trends in Renewable Energy Financing

The history of renewable energy financing is marked by a gradual shift from government-led initiatives to market-driven approaches. In the early 2000s, most renewable energy projects were funded through public grants and subsidies, reflecting the nascent stage of the industry. However, as renewable energy technologies matured and gained commercial viability, private investment began to play a more prominent role. The introduction of green bonds in the late 2000s revolutionized renewable energy financing by offering a scalable and transparent mechanism for funding sustainable projects.

In recent years, innovations in financing models, such as power purchase agreements (PPAs) and yieldcos, have further diversified funding sources. Additionally, international frameworks like the Kyoto Protocol and the Paris Agreement have incentivized countries to allocate financial resources toward renewable energy.

Despite these advancements, the financing landscape remains uneven, with developing countries facing significant challenges in accessing affordable capital for renewable energy projects [7].

3 Role of Policy Frameworks in Renewable Energy Adoption

Policy frameworks play a crucial role in shaping the adoption and financing of renewable energy. Government incentives, such as feed-in tariffs (FiTs), tax credits, and renewable portfolio standards (RPS), have been instrumental in reducing the cost barriers associated with renewable energy projects. Policies like the Jawaharlal Nehru National Solar Mission in India and the Renewable Energy Directive in the European Union have set ambitious targets for renewable energy deployment, providing a stable policy environment that encourages investment [8].

Regulatory mechanisms also influence market dynamics by creating demand for renewable energy. For instance, carbon pricing and emission trading schemes have incentivized industries to adopt cleaner energy solutions. However, the effectiveness of these policies often depends on their design and implementation. Inconsistent policies or abrupt changes can deter investors and disrupt market stability.

4 Gaps in Existing Research

While substantial progress has been made in understanding renewable energy financing, several gaps remain in the literature. First, there is limited research on the effectiveness of innovative financing mechanisms, such as blended finance and impact investing, in scaling renewable energy projects in developing economies. Second, most studies focus on individual financing instruments without exploring the synergies between different mechanisms, such as how green bonds can complement public-private partnerships [9].

Another critical gap is the lack of empirical studies analyzing the long-term impact of renewable energy loans on socio-economic outcomes, such as job creation and energy access. Additionally, the interplay between financial viability and regulatory frameworks remains underexplored, particularly in the context of emerging markets. Addressing these gaps can provide valuable insights for policymakers and stakeholders aiming to accelerate the global transition to renewable energy.

5 Methodology

5.1 Research Design

The study adopts a mixed-method approach combining quantitative and qualitative methodologies to assess the impact and viability of renewable energy loans in advancing green transitions. The research focuses on evaluating financing structures, policy frameworks, and their effectiveness through case studies and econometric analysis. The objectives align with the National Solar Mission (NSM) to promote renewable energy, as outlined below.

5.2 Data Collection Methods

Data was collected using the following methods:

1. Primary data: Structured interviews with financial stakeholders, policymakers, and renewable energy developers provided qualitative insights into financing challenges and solutions.
2. Secondary data: Government reports, industry publications, and academic research formed the basis for quantitative analysis. Key data sources included IRENA, World Bank reports, and policy documents like the National Solar Mission.

5.3 Analytical Techniques

The analysis employed the following advanced methods:

1. Econometric models: The Nonlinear Autoregressive Distributed Lag (NARDL) model and Two-Stage Least Squares (2SLS) regression addressed dynamic relationships and endogeneity issues in renewable energy financing [10].
2. Financial metrics: Cost-benefit analysis and Return on Equity (ROE) evaluations highlighted short-term and long-term financing challenges.
3. Policy impact assessment: The study examined regulatory frameworks, interest rate trends, and their impact on project viability.

5.4 Case Study Selection

Representative case studies were selected to illustrate the implementation and challenges of renewable energy financing:

1. The National Solar Mission (NSM) in India was chosen for its ambitious targets, such as achieving 100 GW of solar energy by 2022 and reducing environmental impacts.
2. Initiatives like the Clean Energy Finance Corporation in Australia and community-based solar cooperatives in India were analyzed for innovative financing mechanisms and their impact on renewable energy adoption.

5.5 Limitations and Assumptions

While the study provides valuable insights, it has several limitations:

1. Data reliability: Reliance on secondary data introduces potential reporting biases.
2. Economic assumptions: The analysis assumes stable economic and regulatory conditions, which may vary in practice.
3. Scope limitation: The study focuses on financial and policy dimensions, excluding technological advancements in renewable energy systems.

5.6 Basic Objectives of National Solar Mission (NSM)

The NSM serves as a cornerstone of India's renewable energy strategy, with the following objectives:

1. Achieving 100 GW of solar energy capacity by 2022, up from the initial target of 20 GW.
2. Reducing environmental damage caused by conventional power generation.
3. Promoting the use of clean, sustainable fuels.
4. Supporting both on-grid and off-grid electricity systems.
5. Establishing India as a global leader in solar energy.

Short-Term Financing Challenges

1. High interest rates imposed by Indian banks present a significant barrier to renewable energy funding. This rate disparity contrasts sharply with lower interest rates (24–32%) in the United States and Europe, hindering competitive growth in India [11].

2. Investors and developers frequently cite funding accessibility and high equity returns as critical issues. Reducing the projects' Return on Equity (ROE) to levels seen in the US and Europe could attract greater investment in Indian renewable energy projects [12].

Long-Term Financing Challenges

1. Persistent high interest rates and regulatory barriers pose significant obstacles to the development of renewable energy projects in India. These constraints limit access to affordable financing, impeding market expansion.
2. Short tenors and variable interest rates exacerbate financial hurdles, even in low-interest-rate scenarios, creating uncertainties for long-term project sustainability.
3. As equity becomes less attractive due to reduced debt accessibility, existing investors may find it challenging to recycle investments into new ventures. Market maturation could also deter strategic, low-return investments, especially in solar photovoltaic (PV) projects [13].

This refined methodology and expanded discussion contextualize renewable energy financing challenges within a broader policy and economic framework, addressing both immediate and structural barriers.

GDP growth and the rise in global electricity consumption are powerfully related. Demand for electricity and economic growth are strongly associated; as the economy expands, so does the need for electricity, and vice versa. A steady supply of power is necessary for the growth of major economic divisions like the services division, agriculture division, and industry division [14].

The requirement for electricity exceeded in 2018 and 2019 in accordance with the actual GDP growth, as seen in Chart 1. 2020 saw a 2.8% year-over-year fall in GDP as a result of the Covid-19 pandemic-related lockdown and travel restrictions implemented by various nations. The requirement for power decreased globally by 0.8% this year. When the world GDP recovered 6.3% in 2021 from the effects of the epidemic, so did the demand for power.

Substantial financial sources are often in demand by renewable energy projects, but they look at maximum risks and minimum returns during near the beginning stages of development. In India, where traditional banking dominates, access to funding for green initiatives is particularly challenging. Despite the efforts of venture capital firms to support environmental efforts, fossil energy investments still triumph over renewable energy investments. Renewable energy loans are financial instruments designed to fund the installation or up gradation of renewable energy systems, such as solar panels, wind turbines, or geothermal systems. These projects may be offered by financial institutions like banks, governmental agencies, or specialized green financing organizations. Typically, they offer affordable terms, such as low-interest rates or extended repayment periods, to incentivize assets in renewable energy [15].

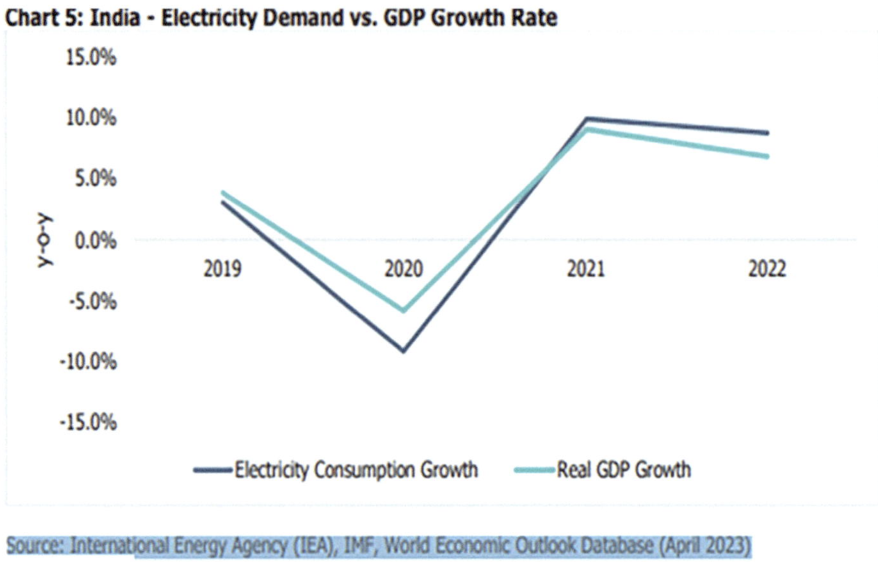

Chart 1 India—electricity demand vs. GDP growth rate. (Source: International Energy Agency (IEA), IMF, World Economic Outlook Database, April 2023)

5.6.1 Present and Future Impact of Major Issues in Indian Renewable Energy Economics

Impact on Green Transitions

The accessibility of renewable energy loans on the projects has a visible impact on increasing green transitions. The upfront cost barrier is going to reduce due to easy handling loans make renewable energy technologies more presentable to a wider range of consumers, including house owners, businesses, and municipalities. As a result, they encourage demand for renewable energy systems, driving market growth and innovation in the renewable energy sector. Moreover, the implementation of renewable energy contributes to reductions in greenhouse gas production, air pollution, and reliance on non-renewable energy sources, thereby advancing sustainability objectives.

This research addresses several research questions:

1. Is there any connection between increased green financing and investments in renewable energy?
2. Does economic growth persuade the approval of renewable energy sources?
3. How does corporate investment impact renewable energy production?

In 2016, the G20 introduced the Sustainable Finance Study Group (SFSG) as the basis of its promise to increase sustainable growth and emphasize ecological challenges on a worldwide level. The Sustainable Finance Study Group (SFSG) was charged with exploring ways to mobilize private capital for sustainable development projects and identifying strategies to integrate sustainability deliberation into monetary decision-making processes [16].

The most important factor to focus on for the SFSG was subsidy projects that donate to sustainable progress objectives. The group recognized that traditional financing mechanisms often prioritize short-term fiscal returns without sufficiently considering ecological and social impacts. As a result, there was a need to build up innovative financing mock-ups that line up financial incentives with sustainability aims [17].

The SFSG explored various approaches to funding sustainable projects, including:

Green bonds: Green bonds are debt instruments issued specifically to finance projects with environmental benefits, such as renewable energy installations, energy efficiency improvements, and sustainable infrastructure development. The SFSG examined ways to expand the green bond market and enhance transparency and credibility in green bond issuance.

Public-private partnerships (PPPs): PPPs involve collaboration between government entities and private sector investors to finance and deliver infrastructure projects. The SFSG explored ways to structure PPPs to incorporate sustainability criteria and ensure that projects deliver long-term environmental and social benefits.

Impact investing: Impact investing involves making investments with the intention of generating positive social and environmental impacts alongside financial returns. The SFSG studied how to scale up impact investing by developing standardized metrics for measuring impact and creating investment products that appeal to a broader range of investors [18].

Blended finance: Blended finance combines public and private capital to fund projects that address market failures or achieve specific development objectives. The SFSG examined how blended finance mechanisms, such as guarantees, concessional loans, and risk-sharing facilities, could be used to leverage private investment in sustainable projects.

By exploring these and other financing approaches, the SFSG aimed to catalyze investment in projects that contribute to sustainable development, such as renewable energy, climate adaptation, sustainable agriculture, and biodiversity conservation. The group's efforts underscored the importance of aligning financial flows with sustainability goals and creating incentives for investors to prioritize long-term environmental and social outcomes.

This study fills gaps in climate financing literature by analyzing expenditure parameters from an investor's perspective and examining the relationship between financing and renewable energy investment.

Using advanced econometric techniques like nonlinear autoregressive distributed lag (NARDL) and two-stage least squares estimation (2SLS), the study captures both short- and long-term effects and addresses potential endogeneity [19].

The research sheds light on barriers to optimal energy utilization, guiding private investors, governments, and financial stakeholders in designing effective strategies and products to support renewable energy projects.

Regulatory and macroeconomic risks emerge as primary impediments to green energy deployment, underscoring the need for policy interventions to mitigate barriers such as fossil fuel subsidies and monopolistic practices.

Electricity demand in different demand sectors

TWh	2012	2022	2030	2047
Industry	336	494	703	1366
Residential	175	480	842	1840
Commercial	86	142	238	771
Agriculture	136	245	336	501
Others	29	71	121	233
Total	762	1433	2239	4712

6 Analysis and Discussion

6.1 Financial Viability of Renewable Energy Loans

Renewable energy loans are instrumental in overcoming the initial financial hurdles associated with green energy projects. Their financial viability depends on several factors, including interest rates, loan tenors, and the availability of subsidies [20].

- **Interest rates**: High rates imposed by Indian banks significantly affect project viability, making it difficult for developers to achieve competitive returns on equity (ROE). Comparatively, lower interest rates in regions like Europe and the United States make renewable energy projects more feasible.
- **Loan terms**: Short tenors and variable rates create financial uncertainty, especially in fluctuating economic environments. Affordable and flexible financing options are essential to support long-term sustainability.
- **Government incentives**: Subsidies and incentives provided by national programs, such as those under the National Solar Mission, play a critical role in enhancing the financial appeal of renewable energy investments.

6.2 Impact of Policy Frameworks

Policy frameworks shape the renewable energy landscape by defining regulatory boundaries and providing financial incentives. India's **National Solar Mission (NSM)** exemplifies the transformative potential of robust policy initiatives. Key impacts include:

- **Market growth**: Policy-driven goals, such as achieving 100 GW of solar energy capacity, have accelerated the adoption of renewable technologies.
- **Global positioning**: Policies fostering renewable energy development contribute to India's position as a global leader in solar energy.
- **Barriers**: Regulatory delays and inconsistent policy implementation remain significant challenges, deterring investment and slowing project timelines.

6.3 Case Studies and Best Practices

Several case studies highlight the effectiveness of targeted financing mechanisms and their role in fostering renewable energy transitions:

- **India's National Solar Mission**: A flagship initiative that has driven significant capacity additions in solar energy while promoting sustainable practices.
- **Clean Energy Finance Corporation (Australia)**: An example of a government-backed entity successfully leveraging private capital for renewable energy projects.
- **Community-Based Solar Cooperatives (India)**: Empowering local communities to adopt renewable energy solutions, illustrating the benefits of decentralized financing models. These best practices underline the importance of aligning financial incentives with sustainability goals to maximize impact.

6.4 Challenges and Barriers

Despite progress, significant challenges persist in scaling renewable energy through loans:

- **Financial challenges**: High interest rates, short tenors, and insufficient equity options restrict market growth.
- **Regulatory barriers**: Policies supporting fossil fuel subsidies and monopolistic practices create an uneven playing field for renewable energy projects.
- **Market awareness**: Limited consumer awareness and perceived risks associated with renewable energy loans hinder their adoption.

6.5 Recommendations for Future Research

The following recommendations aim to address existing gaps and guide future investigations:

1. **Innovative financing models**: Explore mechanisms such as green bonds, impact investing, and blended finance to diversify funding options [21].
2. **Policy interventions**: Assess the impact of policy reforms, such as phasing out fossil fuel subsidies and introducing renewable energy incentives.
3. **Consumer behavior**: Investigate the role of consumer awareness and behavioral factors in shaping renewable energy adoption.
4. **Long-term sustainability**: Evaluate the scalability and financial stability of renewable energy projects under changing economic conditions.
5. **Technological integration**: Examine the interplay between financial mechanisms and technological advancements to enhance the efficiency and affordability of renewable energy systems.

7 Conclusion and Future Directions

This study highlights the indispensable role of renewable energy loans in facilitating the transition toward sustainable energy solutions. The findings elucidate:

- Economic viability: The financial feasibility of renewable energy loans is heavily influenced by interest rates, loan structures, and supportive government incentives. Comparative insights underscore the disparity between financing models in developing economies and those in advanced economies, revealing critical opportunities for reform in the Indian context [22].
- Policy influence: Effective policy frameworks, exemplified by the National Solar Mission, have driven renewable energy growth. However, inconsistent implementation and insufficient regulatory coherence pose barriers to broader adoption.
- Case insights: Global examples, including innovative models like green bonds and community-based cooperatives, underline the potential for financial mechanisms to democratize access to renewable energy.

Policy frameworks must evolve to catalyze renewable energy adoption, focusing on:

1. Incentivizing investments: Expand and sustain tax credits, subsidies, and concessional financing to reduce the upfront costs associated with renewable energy projects.
2. Rationalizing subsidies: Gradually phase out fossil fuel subsidies to align market incentives with renewable energy objectives.
3. Enhancing regulatory stability: Ensure continuity in policy execution to bolster investor confidence and mitigate project delays.

7.1 Recommendations for Stakeholders

1. Government entities:
   - Develop long-term financial instruments with flexible repayment terms.
   - Foster public-private partnerships to mobilize capital and technical expertise.
2. Financial institutions:
   - Create innovative financing options, such as green bonds and blended finance, to attract diverse investors.
   - Address risk-sharing mechanisms to mitigate credit-related concerns.
3. Project developers and investors:
   - Leverage scalable financial models like impact investing to optimize project returns.
   - Collaborate on cross-sectoral initiatives for integrated renewable energy solutions.
4. Local communities:
   - Promote community-centric models, empowering grassroots participation and ownership.

This study identifies the following directions for advancing scholarly inquiry [23] and practical implementation:

1. Integration with emerging technologies: Assess the synergies between renewable energy financing and technologies such as blockchain, AI, and IoT.
2. Behavioral analysis: Investigate consumer adoption patterns and their interplay with financial incentives.
3. Economic impacts: Evaluate the macroeconomic effects of renewable energy financing, including employment generation and GDP contributions.
4. Comparative studies: Conduct cross-national analyses to distill transferable lessons from diverse regulatory and economic environments.
5. Longitudinal assessments: Examine the long-term sustainability of renewable energy projects funded through loans, focusing on socio-environmental outcomes.

The research advances the understanding of renewable energy loans as a pivotal enabler for sustainable development. By addressing the challenges of financial viability, regulatory coherence, and market awareness, stakeholders can unlock the full potential of renewable energy systems. Collaborative efforts between governments, financial institutions, and communities remain essential for realizing a resilient, low-carbon future. This study lays the groundwork for informed policy-making and continued innovation in renewable energy financing, propelling the global agenda for sustainability.

References

1. International Renewable Energy Agency (IRENA). (2021). Renewable power generation costs in 2020. Retrieved from https://www.irena.org/publications/2021/Mar/Renewable-Power-Costs-in-2020
2. Singh, B., Dutta, P. K., & Kaunert, C. (2024). Wind-solar renewable energy and innovative technologies applying internet of things (IoT) for green and sustainable future: Projecting carbon neutrality for smart and sustainable cities. In J. A. Parray, A. K. Haghi, & G. Meraj (Eds.), *IoT-based models for sustainable environmental management* (Lecture notes on data engineering and communications technologies) (Vol. 227). Springer. https://doi.org/10.1007/978-3-031-74374-0_6
3. G20 Sustainable Finance Study Group. (2016). Green finance synthesis report. Retrieved from https://www.unepfi.org/wordpress/wp-content/uploads/Green-Finance-Synthesis-Report.pdf
4. Global Green Finance Index. (2021). Global green finance index 8. Retrieved from https://www.longfinance.net/media/documents/GGFI_8_Report.pdf
5. Climate Bonds Initiative. (2021). Green bond market summary: Q4 2020. Retrieved from https://www.climatebonds.net/files/reports/cbi-green-bonds-q4-2020-highlights.pdf
6. Impact Management Project. (2021). Impact investing glossary. Retrieved from https://impactmanagementproject.com/glossary/
7. Dutta, P. K., Dey, S., Majumder, S., Sen, P., & Roy, S. (2023). Implementing solar panel surface dust cleaning innovation using a solar innovation framework model. In A. E. Hassanien, D. Zheng, Z. Zhao, & Z. Fan (Eds.), *Business intelligence and information technology* (Smart innovation, systems and technologies) (Vol. 358). Springer. https://doi.org/10.1007/978-981-99-3416-4_42
8. Ahmed, Y. A., Rashid, A., Adeel, U., & Dutta, P. K. (2022). Proposing a framework for consumers' decision on solar innovations for smart cities. In *6th Smart Cities Symposium (SCS 2022), Hybrid Conference, Bahrain* (pp. 348–352). https://doi.org/10.1049/icp.2023.0575
9. World Bank. (2021). Blended finance report 2020. Retrieved from https://www.worldbank.org/en/news/feature/2020/09/24/blended-finance-2020-report
10. International Energy Agency (IEA). (2021). World energy investment 2021. Retrieved from https://www.iea.org/reports/world-energy-investment-2021
11. Fadai, D. (2007). Utilization of renewable energy sources for power generation in Iran. *Renewable and Sustainable Energy Reviews*. Retrieved from https://energsustainsoc.biomedcentral.com/articles/10.1186/s13705-019-0232-1
12. Dutta, P. K., Mallikarjuna, K., & Satish, A. (2017). Sensor based solar tracker system using electronic circuits for moisture detection and auto-irrigation. In *2017 IEEE International Conference on Power, Control, Signals and Instrumentation Engineering (ICPCSI)*. https://doi.org/10.1109/ICPCSI.2017.8391956
13. Khatoon, F., Kumar, M., Khalid, A. A., Alshammari, A. D., Khan, F., Alshammari, R. D., Balouch, Z., Verma, D., Mishra, P., Abotaleb, M., Makarovskikh, T., El-Kenawy, E. M., Dutta, P. K., & Marques, J. A. (2022). Quality of life during the pandemic: A cross sectional study about attitude, individual perspective and behavior change affecting general population in daily life. In *IET Conference Proceedings* (pp. 379–383). IET Digital Library. https://doi.org/10.1049/icp.2023.0596. Retrieved from https://digital-library.theiet.org/content/conferences/10.1049/icp.2023.0596
14. Saravanan, J., Kumar, T. A., Divya, P., Partipan, R., Partipan, R., & Dutta, P. K. (2022). IoT based advanced prediction methodology for fault localization in machine monitoring. In *6th Smart Cities Symposium (SCS 2022), Hybrid Conference, Bahrain* (pp. 253–257). https://doi.org/10.1049/icp.2023.0429
15. Dutta, P. K., De, P., & Chakroborty, R. (2020). Optimal energy management of the smart parking IOT in the design of fuel cell charging system. In *3rd Smart Cities Symposium (SCS 2020), Online Conference* (pp. 209–213). https://doi.org/10.1049/icp.2021.0965

16. Dutta, P. K., Tripathi, N., Patra, A., & Mukherjee, P. (2023). Advanced detection of refrigerant leakage using pressure sensor information in variable refrigerant flow systems. In *7th IET Smart Cities Symposium (SCS 2023), Hybrid Conference, Bahrain* (pp. 70–76). https://doi.org/10.1049/icp.2024.0903
17. Mullick, S., Dutta, P., & Chitara, D. (2023). Urbanization agenda: A journey embracing geodata analysis using digital twin model. In *7th IET Smart Cities Symposium (SCS 2023), Hybrid Conference, Bahrain* (pp. 33–39). https://doi.org/10.1049/icp.2024.0897
18. Dutta, P. K., & Mullick, S. (2023). Enhancing position control performance of magnetic levitation system using optimized PID-P controller: A performance analysis. In *IET Conference Proceedings* (pp. 40–47). Retrieved from https://digital-library.theiet.org/content/conferences/10.1049/icp.2024.0898
19. Bhattacharya, P., Chatterjee, S., Datt, R., Verma, A., & Dutta, P. K. (2023). A permissioned blockchain approach for real-time embedded control systems. In S. Kadry & R. Prasath (Eds.), *Mining intelligence and knowledge exploration (MIKE 2023)* (Lecture notes in computer science) (Vol. 13924). Springer. https://doi.org/10.1007/978-3-031-44084-7_32
20. Biswas, S., Biswas, S., Nandi, S., Basak, R., & Dutta, P. K. (2021). Measurement of variation of ultraviolet radiation due to different greenhouse gases using internet of things. In *4th Smart Cities Symposium (SCS 2021), Online Conference, Bahrain* (pp. 454–457). https://doi.org/10.1049/icp.2022.0386
21. Biswas, A., & Dutta, P. K. (2021). Novel approach of automation to risk management: The reduction in human errors. In J. S. Raj (Ed.), *International Conference on Mobile Computing and Sustainable Informatics. ICMCSI 2020* (EAI/Springer innovations in communication and computing). Springer. https://doi.org/10.1007/978-3-030-49795-8_65
22. Biswas, A., Adhikary, A., Dutta, P. K., & Chakroborty, A. (2021). Development of a risk-based information security standard for adaption of blockchain-enabled systems. In J. S. Raj (Ed.), *International Conference on Mobile Computing and Sustainable Informatics. ICMCSI 2020* (EAI/Springer innovations in communication and computing). Springer. https://doi.org/10.1007/978-3-030-49795-8_59
23. Singh, B., Dutta, P. K., Gautam, R., & Kaunert, C. (2024). Uncapping the potential of quantum computing towards manufacturing optimization: Routing supply chain projecting sustainability. In *Quantum computing and supply chain management: A new era of optimization* (p. 25). IGI Global. https://doi.org/10.4018/979-8-3693-4107-0.ch024

Enhancing Smart Grid Sustainability: AI Role in Overcoming Efficiency Barriers

Sashi Tarun

1 Introduction

Consumer demand for energy in the global market is rising significantly for several reasons including high population, improved urbanization and growth in industries [1]. In other words, there can be no other period that has been more profound than the present in terms of transition to sustainable development in the generation and use of energy. This has been done to the extent that smart grid technology appears to be one of the most disruptive innovations, which is intent on seeking to reinvent the efficiency, reliability and sustainability of power distribution networks management [2]. In its basic conceptualization, the smart grid system integrates advanced digital communications into the existing familiar electrical power systems for managing and controlling the electricity use in the grid. In the following, smart grid solves a great deal of problems in the present energy grid, on the way to a sustainable energy future from the above-advanced abilities. Smart grids are therefore expected to transform the overaged electric grid infrastructure developed to address the model of a centralized one-way electricity supply from production stations to the users. And therefore today's needs are far more specific and different—renewable, electric, distributed generation, etc.; communication is effective in two ways and the smart grid operation is far more flexible, far more responsive and far more efficient than the smart grid. They also include sensors, smart meters and statistics so such a grid can track and control power consumption as well as track faults and determine when the related equipment should be serviced or replaced. Such flexibility is especially valuable for managing the demands of the twenty-first century as

S. Tarun (✉)
AIT-CSE, Chandigarh University, Mohali, Punjab, India

P. Raj et al. (eds.), *Artificial Intelligence (AI) for IT Energy Efficiency and Green AI for Environment Sustainability*,
https://doi.org/10.1007/978-3-031-89420-6_17

energy systems have to adapt both to the demand side unpredictability and to renewable [3].

The inadequate availability of clean energy solutions perhaps can be attributed to being one of the major driving forces toward smart grid. The developing post-climate effects require society to be serviced with solutions that reduce emission and consume energy efficiently. This kind of transition can only partially be supplied by the classical energy supply networks which mainly rely on fossil resources. However, smart grids are designed to integrate all types of renewable energy systems such as solar systems, wind, hydro, etc. perfectly. Despite the fact that renewable electricity generation is probabilistic related to the daily weather cycle and daily variation, smart grids can act to mitigate this with the assistance of modern approaches in generation prediction, energy storage and enhanced demand–response plan. Therefore, smart grids are an indispensable component of sustainability for the energy sector because of the flexibility it provides in regard to handling variable renewable energy resources [4]. Specifically, researchers have claimed that one of the major challenges which most traditional power systems experience is the integration of renewable resources in electric grid. The representative concept of the modern smart grid system is illustrated in Fig. 1 which encompasses novel renewable energy feeding, such as photovoltaic cells, wind and hydroelectric power feeds and efficient digital communication for electricity supply and control.

Fig. 1 Smart grid systems

1.1 Challenges in Smart Grid Systems

Smart grid systems, that are based on AI and Internet of Things (IoT), are a major leap forward toward having energy efficiency as well as environmentally sustainable solutions. Smart grid is a modern concept of power grid, which allows using real time, data, processing and responsive energy management. They enable utilities to forecast energy demand, optimize loads and minimize wastage thereby bring a positive change to the energy sector. Moreover, these grids enable consumers to be in a position to manage the amount of energy that they must use and this has the potential to give them the ability to save power as well as be environmentally friendly. Smart grids as they are, their increased demand is not without difficulty [5]. AI and IoT have to be incorporated into current grid structures and they demand extensive reinforcements, data protection protocols and professional skills of technical colleagues, which become challenges affecting costs, privacy and performance. Conquering these challenges, as shown in Fig. 2, knowledge of which is critical in achieving the goal of making smart grids an outline for an energy-efficient future.

The fusion of AI and IoT to implement smart grids in the economic system holds promising potential for transformational improvements in energy efficiency and ecological protection, but they are associated with a lot of risks. There is a cost and

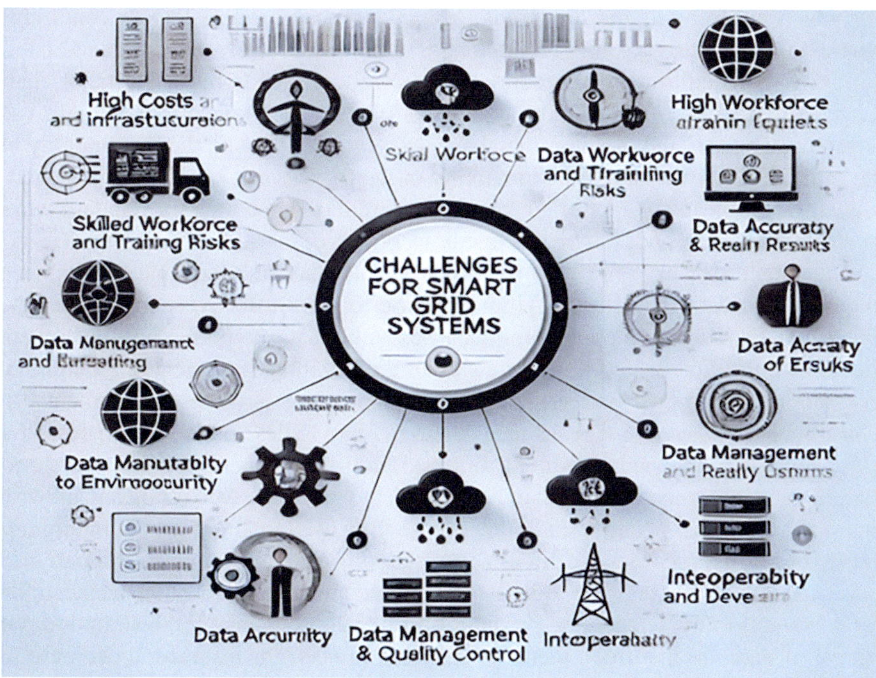

Fig. 2 Challenges for smart grid systems

difficulty of retrofitting current grid infrastructure to accommodate such technologies. Smart grids require infrastructure such as IoT sensors, advanced meters and high vantage processing equipment that while brings high upfront costs [6]. Also, the subject requires constant updates and repairs, which have a tendency to increase the costs, making it almost impossible for regions that are not well endowed to afford or regions with outdated technology. Next to economics, data protection and cyber-security are paramount issues. Being integrated into the distribution and communication network, IoT devices produce essential data for grid management and convey it across the network, and it becomes an object of cyber threats [7]. When adequate security measures have not been developed, this information like personal energy usage can easily fall into the wrong hands, thus endangering the citizen's privacy as well as jeopardizing the stability of the power grid. Cyber-security defenses need to be specially built on account of the constantly transforming nature of cyber threats, including the ones that are defended by artificial intelligence that constantly needs updates for it to continue to be effective.

In addition to the conceptual barriers, there are also several operational factors that also create implementation and deployment challenges for AI and IoT in smart grid systems. Depending on the real-time data monitoring, processing and analysis, smart grids require intensive use of computational resources and storage facilities. A problem with many energy-efficient schemes is that the computations required to obtain them can themselves be energy intensive and thus negate the value addition that is aimed to be achieved. The nature of big data also makes response times longer which might be highly disadvantageous when making decisions in conditions where AI is not designed for real-time operations [8]. However, keep up and compatibility with additional devices have consistently proven to remain a major challenge. Smart grids often include appliances of diverse producers that might use different standards and protocols and have a degrading effect on the functionality of the grid. The only way that interoperability of the devices can be realized is by finding ways and means of encouraging manufacturers to come up with standard ways of writing software codes, which is a tedious process.

Other factors which are also important include social acceptance, workforce expertise and data reliability if AI and IoT are to work effectively in smart grids. Consumer confidence is paramount because, without it, the public will not accept the technology, and they will not be aware of the advantages of the new technology in terms of saving cost on energy and also great impact on the environment [9]. Thus, in order to overcome resistance due to privacy issues and long-term cost savings skepticism-matched communication and public awareness campaigns are crucial. In addition, smart grids demand a competent workforce who is conversant with data analysis, cyber-security and other AI systems used in smart grids; finding talents to perform these roles is difficult, especially in areas of low uptake of smart grid systems. Last but not least, the quality of data also needs to be constant and accurate for AI model to make accurate prediction to manage the energy consumption. Of course, if data are incorrect, there are negative effects, for instance, a decrease in productivity, and during peak, the ability of the grid to operate efficiently is put at risk [10]. Overall, IT and IoT are important prerequisites for smart grid energy efficiency. However, the present and future trends in sectors associated with IT and IoT identified in this paper require strong integration of technology, infrastructure, policy and society.

1.2 Weaknesses in Smart Grid Systems

Smart grids based on Artificial Intelligence (AI) and Internet of Things (IoT) technologies are likely to provide improved value-added energy efficiency and environmental consciousness. However, a number of issues limit them to fully realize their potential. One controversial aspect can be discussed—data security and privacy. Smart grid incorporates enormous data gathered from a large number of IoT devices, which logs energy use, generation and distribution information in real time [11]. This extensive network is prone to cyber criminals, hackers and Intrusion into people's private lives due to the fact that this extensive information also includes consumer energy usage information. While data protection discloses itself as an essential issue, the cyber-security becomes an imperative tool for building of trust and popularization of smart grids.

Another issue is that the introduction of AI and IoT into the grid involves significantly large levels of infrastructure investment. This is a particularly big challenge for developing regions that are very constrained fiscally, leaving them unable to invest in improving the physical infrastructure and procuring the newest technologies. It would be difficult to realize the planned smart grid benefits if the implementation process does not have sufficient external support, and developing countries lacking financial resources risk being left even further behind in terms of energy management around the world [12]. In fact, within developed countries, smart grid investments are causing financial and operational hurdles in upgrading current grids and applying AI algorithms to expand across a wider area.

AI and IoT integration in smart grids face another common problem of interoperability. Number of devices and systems in a smart grid are manufactured by different companies and do not always conform to a common communication standard. Such non-uniformity creates problems in attempting to coordinate the data transfer from one segment of the smart-grid to another section, problems that can prevent real-time data transfer and impair efficient energy control. Also, the data quality issue come from the IoT sensors that are not accurate and the reasons can be technical problems, environmental conditions or outdated IoTs [13]. These discrepancies can result in wrong predictions by AI and worse energy efficiencies at the relevant facilities. Also, the main characteristic of IoT networks, the processing and latency restrictions, complicate the flexibility in decision making required for the grid to adapt to changes in energy demand or emergencies. Last but not least, the environmental cost of the AI and IoT devices themselves raises other issues, such as the generation of useless electric. Appliances also result in electronic waste while the training of the AI models for smarter grid consumes power which goes against the results being sought in the implementation of smarter grids [14].

1.3 Hurdles in Smart Grid Tuning

Employing intelligent tuning for smart grid systems to reach sustainability in the environment holds a number of major challenges that need to be conquered for efficient functioning. One of the most salient threats is the huge capital-intensive nature of investments on infrastructure and technologies. Smart grids use machine learning, IoT, sensors; AI-based algorithms for real-time energy usage, which involves significant capital expense on hardware, software and networking [15]. Cost remains the biggest challenge as energy providers, especially in the developing world cannot afford to invest in modern grid systems to put in place AI capabilities. These cost implications may act as a set back to the advancement of smart grids across the world and also slow down the shift to more efficient grids.

The use of smart grid systems that have been tuned with AI for a sustainable environment poses several questions that are of significant concern for effectiveness and implementation [16]. One of the greatest barriers is the relatively depreciating costs of initial setup for infrastructure and technology improvements. Owing to a host of ICT applications such as AI algorithms, IoT devices and sensors to regulate the usage of energy in smart grids, significant capital investment is necessary for hardware and software as well as the layer of network. Another issue is that to make their utility grids AI-driven, energy providers face significant costs especially if they operate in the developing world. These costs may reduce the rate of deploying smart grids around the world and delay the shift of energy-efficient systems [17].

Another focal area in optimizing smart grids is identified as interoperability. A smart grid involves a large number of devices, sensors and AI systems from various producers and companies and these may use dissimilar communications protocols. Since these technologies are not well developed and well defined, there is an interoperability problem and this considerably hinders the further development of smart grid systems [18]. The lack of a unified communication system may also cause new devices to connect inefficiently into the current structure and restrict the efficiency of the grid for managing energy provision. The above fragmentation they cause can in few ways affect the grid's efficiency and sustainability.

But still, there are also drawbacks to using artificial intelligence in smart grids; one of them being the applicability of the said models in handling scalability in the data collected by smart grids. Although AI offers the best solution for the intricate distribution of energy systems in small networks, applying these ideas in a large context with multiple sensors which gather huge amounts of data over vast territories appears problematic. The evolution of energy requirements suggests that AI platforms have to analyze data when needed and then adapt to conditions encountered on the grid [19]. But a common issue with most currently developed AI models is limited scalability, which poses a problem of degradation of efficiency besides energy management. However, due to the incapability of the AI system to manage big datasets distinctively, smart grid application proves ineffective in controlling the energy circulation and thereby managing the demand confidently in the real-term.

Last but not least, energy storage as well as integration of renewable remain problematic for AI-based smart grids. Wind and solar for instance are erratic forms of energy producers since energy creating is reliant on climactic conditions as well as time of the day. The fact that AI can facilitate better control over energy flow using data derived from the electricity grid is on the one hand a reality, but on the other hand, the insufficient storage options hindering today's storing of excess renewable energy when is produced most of the time remain a key challenge. Indeed, if there is no efficient means of storage, any excess renewable power may fall, thereby diminishing the benefits of smart grids [19]. Furthermore, the problem of disposing of e-waste resulting from IoTs devices as well as the constant requirement to update the system constitutes extra pressure on the environment and on operations. These four barriers to investments, security, scalability and energy storage of renewable present a daunting challenge in generating a fully sustainable and efficient smart grid that will require research and collaboration.

2 Problem Statements

There is increased transformation of energy systems around the globe to help meet the increasing need for better energy systems. Existing models are incapable of handling today's energy requirements; thus, there is a need to turn to renewable energy. ICT tools embedded in smart grid systems make the distribution of electricity more efficient and address issues such as instability of the grid. Despite these important developments some challenges exist in the deployment and enhancement of smart grids. Major issues include the unpredictability of renewable power, cumbersome two-tier management of grid and storage, and doubts about grid stability and storage. Furthermore, the labor productivity that was once grown with the help of AI on a large scale has certain technological and cost-related constraints. Although these dysfunctions are present in smart grids, these technologies help to mitigate impact on the environment while also controlling energy demand. Policies, safety questions and consumer protection are some of important challenges which should be solved for the best performance. In this chapter, the author discusses the following: how AI can help to overcome these challenges, contribute to smart grid sustainability and address energy storage and dispatch issues, and also to the management of the grid control. Analyzing how RE sources can be integrated into a development framework is also discussed in the chapter on sustainable development. Addressing these challenges is critical to realizing smart grid benefits for the citizens of the world by providing them clean, safe, efficient and sustainable electricity.

3 Motivations

The world is at a crossroads where it is searching for the best solution for energy sources that can be considered sustainable. Due to continuous enhancement in technology, increase in population and development in several states across the globe, the utilization of electricity has increased and the traditional energy frameworks are under pressure to perform their roles appropriately. This challenge is making more acute by modern requirements to combat climate change and reduce CO_2 emission what means the need for transition to the wide-scale use of clean and effective energy supplies. In this context, smart grids can be seen as a disruptive innovation or a solution that serves as an innovation platform that has the potential to discontinuously change the way energy is generated, distributed or consumed. They are defined as smart grids which applied modern technology such as artificial intelligence, big data analytics and real time monitoring which will result in increased energy efficiency, renewable energy integration and minimum negative impact to the environment. Nevertheless, obstacles regarding technology, infrastructure and policy stand as major challenges to the implementation of smart grid systems which have potentially enormous potential. Especially challenging is the integration of renewable energy sources, especially those that are stochastic by nature and depend on time and weather conditions. In addition, it is equally important that these new smart grids are both efficient and robust, including an understanding of the interconnections among these new technology systems as they pertain to the continuing and evolving demands on energy systems today and in the future. It is with these two challenges that this chapter aims to examine and underscore how smart grid technologies powered by AI can help break barriers to efficiency and make the world shift to the new energy system.

4 Related Work

The incorporation of AI into smart grids has several issues that need to be resolved in order to improve sustainability. Such challenges include issues to data privacy, compatibility of the system, and need to adapt to regulatory procedures. Solving these problems is indispensable for establishing AI in energy management systems. A research was conducted by the authors in Refs. [20, 21] on a paradigm concerning the interaction of AI and IoT in energy management with the assistance of utilizing machine learning strategies, for instance, deep reinforcement learning. This shares a conceptual structure of an AIoT-integrated smart grid system and highlights the issues including data security and privacy and practical examples of using AIoT-integrated smart grids solutions to improve the efficiency of energy distribution and grid reliability. The paper by the authors in Ref. [22] described strategies like embedding various kinds of artificial intelligence technologies, distributed control systems on the basis of Artificial intelligence, and real-life applications illustrating how AI can be helpful in smart grids. Areas of constraint include data privacy and security, issues in incorporating other technologies such as blockchain and standardization on the deployment of IoT. Such aspects are important in the adoption of artificial intelligence in the power systems. The author in Ref. [23] highlighted a proposal in his paper including applications of advanced IT tools for increased performance and distribution of control systems for improved energy utilization. Constraints include data privacy and security issues, lack of compatibility of artificial intelligence with emerging technologies such as blockchain, and lack of standardization of deployment to support the integration of AI into smart energy systems. The paper by the author in Ref. [24] depicted several ways such as the application of AI technologies in energy management as well as the application of digital communication in the efficiency of the grids. However, some of the drawbacks include data privacy and security, factors associated with the integration of advanced technologies including blockchain and others more importantly, the need for a standard approach when it comes to use of artificial intelligence. The above factors greatly influence the successful of AI in smart grid systems. The authors in Ref. [25] in his paper outlined efforts such as employing AI algorithms for forecasting the generation of renewable energy as well as managing the delivery of energy to reduce costs associated with grid integration. This includes matters of data privacy, the regulation and standards for using AI, and the problems associated with integrating new technologies like the block chain into existing systems. Reference [26] was highlighted on the strategies such as AI-IoT synergism for real-time energy management to improve on the stability of smart grids. However, some of the threats are issues in data protection, the issues of integration of new technologies and issues of standardization of application of artificial intelligence that slows the process of uptake of the technology. Reference [27] explained accountability issues that arise when using artificial intelligence in services to smart grids and the importance of monitoring the use. These require the development of trustworthiness and the societal acceptability of integrating AI into smart grid operations by properly monitoring related implementations. The limitations include the unstable nature of its operation, which may cause it to make wrong decisions, and again, using service from anywhere makes accountability difficult. Some of the major concerns and difficulties discussed by the authors in Ref. [28] during the implementation of AI and IoT for the sustainable smart grid are the utilization of models, that is, the lack of explainability and transparency of enhanced models and the inability to work with a large amount of data, that is, the scalability of data processing as well as the measurement of introduced decisions' ethical impact. Solving these requires the creation of broadly applicable algorithms, improving the interpretability of the model, and emphasizing the ethics and privacy concerns. The limitations include (1) explainability and transparency of AI and DL models and (2) ethics and privacy, which require attention. Reference [29] described issues related to unpredictable fluctuations in energy sources required for optimization. These can be solved by using machine learning algorithms in the edge node to make accurate predictions and by using pluggable optimization techniques to manage the real-time energy at the microgrids. The limitation includes forecast risks relating to energy supply and consumption and problems concerned with the optimization of energy. The author in Ref. [30] in his paper explores barriers to integrating AI and IoT for smart grid sustainability including security of data, compatibility and high implementation costs. Solving both of these necessitates stringent cyber defense solutions, concrete guidelines and the call for investment to be made in these to improve pace and advancement in the industry. The limitations include the difficulty of coupling the proposed AI with IoT and limitations related to existing grid adaptation. In his paper, the author presented issues of energy supply randomness and the dynamic nature of IoT systems, solving these needs through the application of AI-based frameworks for demand forecasting and energy management through the integration of Machine Learning and Swarm Intelligence. The limitation includes power limitation that challenges the IoT operations and stochastic nature of energy supplies as well as dynamic IoT environments [31]. The authors in Ref. [32] highlighted major issues that need to be addressed in order to advance the use of AI and IoT in sustainable smart grid are related to security in operations of smart grid AI systems, power system reliability and handling of outlying data in smart grid IoT systems. Each of these will need more study to determine the safest and most effective systems and improve the energy management of the energy market. This includes the following challenges in IoT-enabled smart grids: security and energy consumption and the methods of machine learning used in smart grid for energy demand forecasts. The author in Ref. [33] highlighted the major issues associated with the AI and IoT for the sustainability of smart grids, which are privacy, compatibility and adaptability to the norms regulating the actualization of the AI and IoT function. Solving such problems necessitates strong protection, refined procedures and flexible policies in order to guarantee the integration and improvement of the system. The limitations are data privacy in smart energy systems and interoperability and adapting to regulations. The author in Ref. [34] demonstrated the problems including susceptibility to cyber threats and the unknown aspects of smart grids. Both these areas can be resolved by creating methods of building artificial intelligence engineering dependable control schemes and AI-portrayed cyber-tolerant controllers to be operational no matter the shifting threats. The limitations comprise susceptibility to both cyber risks within systems and control frameworks that are robust against structural uncertainties and a decline in system efficiency. The author in Ref. [35] in his paper concentrated on artificial intelligence for sustainable development, referring to safety issues, ethical concerns and the problems which should be solved in advance. The limitations are as follows: New employees' well-being might be affected by the integration of AI, and concerns relating to data privacy are also associated with AI.

5 Major Efficiency Challenges in Smart Grids

Smart grids can be considered as the next step in the evolution of the utility industry ever since the beginning of electricity supply and use to carry out the generation, distribution and utilization in a new way. However, there are factors of efficiency that reduce their effectiveness as well as hinder their full implementation. Some of these include energy loss, instability of electrical grid, energy distribution problem and challenges of incorporating renewable energy such as wind and solar into the system [36]. However, the issue of how to regulate the supply followed by the demand requirements during the period of high demand also continues as a major issue that the smart-grid systems have to solve to build sustainable and reliable power systems.

Transmission and distribution loss is one of the major challenges of current smart grids. Despite integration of modern energy monitoring and optimization technologies, energy is unnecessarily used up in such components as resistances in electrical circuit and the transmission lines principally on old installations that hardly meet modern electrical power requirements [37]. These losses are on account of things like corona losses, skin effect and mechanical losses which include the physical wear and tear on the transmission components including the power lines and the transformers. All these losses range from 6% to 8% to maybe more than that which in one way or the other lessen the efficiency of the grid system.

Fluctuations in the flow of electricity are a major factor and a major challenge as the world shifts from large-scale power generation and distribution from say coal-fired power plants to solar power plants and wind turbines. Although these sources are important for reducing the amount of carbon emissions, they fluctuate by nature as energy production depends on the weather conditions, time of day and geographical location. That causes fluctuation in the grid, which poses a challenge when it comes to matching the supply with the demand levels. Unlike the conventional grid systems that operate on fixed energy supply from large power stations smart grid are required to handle distributed and fluctuating renewable resources [38]. In the event of deviations in generation from renewable energy sources, minor fluctuations can also cause problems in the grid and consequently instability and faults.

Yet another difficulty affecting smart grids is the distribution of energy. As implemented to improve distribution by use of automated switches, real-time monitoring and predictive modeling, integration of multiple sources of energy and storage systems provides multiple layers of challenge. Renewable and conventional electricity generation, including fossil fuel and nuclear power, is required to be effectively controlled by grid operators while concerned about the transmission network capacity, congestion and fluctuating demand [39]. However, due to such factors as bottlenecks and sub-optimal load sharing, energy may be wasted and operation costs escalated beyond the shadow of a doubt because of its flexibility to fine-tune real-time adjustments.

The biggest challenge comes from the integration of renewable energy sources into already existing grids. Solar and wind energy, as clean energy sources, are characteristic of unpredictable energy generation due to the nature of the environment. Since renewable energy is improving its market share, grid operators require modern-day supply-demand balance control including better forecasting [40]. Without them, copious production is inefficient while scarcity of production does translate into power outages, both of which are unhealthy for a grid.

Another ever-prevalent issue is when managing the demand for electricity and supply; worse during the peak usage period is exceedingly difficult. Special conditions such as storms place increased pressure on the grid and the result is frequent black outs as well as reliance on costly and less efficient forms of energy. These problems might be solved by demand–response programs and energy storage in smart grids, although they are dependent on the consumers and limit by cost and technology of storage systems. Finally, applying these technologies to current grids is expensive and takes an unnecessary amount of time [41]. It is, nevertheless, important to modernize aging infrastructure, although this may entail sizable costs. Furthermore, the highly intelligent-grid makes the management process sensitive to cyber-security issues which in turn affect the stability of operation. Similarly, smart grids hold promising answers to sustainable energy systems; however, energy loss, instability, inefficiency and technology integration issues require solutions. These challenges will require constant introduction of new methods, expenditures and efficient utilization of the grid system.

6 AI-Driven Optimization of Smart Grid Efficiency

Artificial intelligence (AI) is changing the face of energy, especially the smart grid, which is a more advanced and intelligent electricity grid system that can monitor, collect data, and even control itself in real-time manner. Smart grids are essentially more efficient than conventional energy grins, adopting novel technologies of machine learning, data analytics and real-time processing addressing the issues of energy budgets, losses and usage pattern control [42]. AI's strength is best seen in its ability to process huge amounts of data obtained from smart grid sensors and predict demand and inefficiency and optimize the supply of energy to ensure order is maintained in this unpredictable environment.

The purpose of resource management remains one of the core benefits that AI brings to the development of smart grids. By using data analysis and trending as well as mastering machine learning technologies, AI learns energy consumption habits, weather conditions and market prices to provide high accurate predictions of energy demand [43]. It also provides grid operators with the capability to optimally allocate energy distribution based on probable energy consumption, the net effect of which is to reduce energy wastage. For instance, AI can analyze periods of low energy demand and suggest that energy from fossil fuel sources should be minimized and that sources from renewable sources such as wind or solar should be maximized. This rational resource allocation not only minimizes operation expenditures but also has a positive effect on the utilization of clean energy resources in generation.

Further, AI optimizes the operation of smart grid by automating energy flow that is an important part of grid function. Conventional grid structures, therefore, necessitate supervision from people and may not promptly respond to changes in demand. Normally, the change has to be addressed manually, but because smart grids are automated by AI, they can react immediately to such a change [44]. For example, machine learning models can learn patterns on how future loads can be managed and how energy can be stored and distributed. This way, through adjustment of these variables AI allows delivering energy at the right place and the right time thus avoiding situations of over and under supply which could be potentially dangerous to the grids. This automated management of energy flow is especially vital during peak demand times when the grid is under the most pressure. With the help of predictive analysis, AI can spot such a high load period in advance, and adjust its energy supply and demand, so that there would be no blackouts or power surges.

Another application of AI is in dynamic changes to load balancing. Load balancing is defined as the method of distributing the supply of energy over the network in order to fulfill the consumers' requirements efficiently. Traditionally, this process involved a set of actions which were only initiated after a demand high or low period had been noticed by the operators. On the other hand, AI is capable of scheduled load balancing due to constant monitoring of real-time data and estimation of future loads. For example, in cities, energy demand may have steep spikes at certain times, while with the help of AI algorithms, such changes can be predicted according to the history of usage and other factors such as time or a season. With this knowledge, smart grids can better distribute energy and manage the load, and move it around in the grid which will help avoid traffic jams or running out of power [45].

Similar to this, how AI supports dynamic demand–response strategies is even revolutionary. Demand–response means the flexibility of commodity supplies for energy consumption that usually changes with the demand from consumers. AI improves this with demand–response systems that remove energy from circulation instantly based on other specific processes. Smart grids with the help of AI can provide incentives or pricing mechanisms to a consumer in order that he uses energy at times other than peak times. Artificial intelligence progresses in understanding consumer behaviors and differentiating demand–response programs for multiple consumers as well as improving the flow of the grid. For instance, using demand–response AI can send a message to an electric vehicle charging station to adjust the charging time to a later period, as this can put a lot of pressure on the grid. Thus, AI helps smart grids to control supply and demand throughout the grid and maintain grid stability in a way that does not interfere with consumers' experience [46]. Further, another application of artificial intelligence is noticeable in the analysis of data for the improvement of distribution of energy in smart grids. Here, I envision that even with thousands of data points coming directly from smart meters, sensors, IoT devices and continually assessing the state of grid, AI is capable of identifying trends and outliers in energy flow instantly. For instance, if a given line or substation is often strained, the AI systems will realize this problem and recommend that the electricity distribution be adjusted to exclude that specific line. It also helps to determine that energy must be distributed in the proper manner, without loss and with the potential for infrastructure damage. Further, the perception of AI analytics allow the grid manage the varied state of the environment which may affect energy production for instance; a reduction in the intensity of solar in the presence of clouds. With regard to the distribution structures and shifting loads, AI helps provide a steady and stable energy supply.

Moreover, the role that AI plays in energy bargain is quite impressive. In standard grids, losses may be attributed to transmission and distribution losses, over generation and equipment malfunctions. AI can help this because it can recognize sources of energy waste as well as prevent them from occurring. For instance, machine learning models can identify patterns that indicate the flaws in transmission lines, where operators can schedule am offline or re-channel energy before much loss happens. Also, the result of adopting AI is leveraging of overproduction since it can be able to help in predicting the demand for a particular product and thus adapt its production timetable. This ability affords considerable environmental advantage, as it decreases the chance for burning fossil fuel-derived power and, consequently, diminishes carbon emissions. AI plays a very important role in handling the distributed energy resources in smart grids, such as distributed solar PV system, wind energy systems or distributed batteries [47]. These resources present different issues because of changes in output and geographical locations but they are central in sustainable power systems. By incorporating these resources under smart integration, they can be efficiently used to alleviate constrains in the smart grid during peak hours and can also store any unused energy for later use. DERs generate data and the data is fed to machine learning algorithms and based on this the timing for storing energy and releasing energy is appropriately selected for maximizing the use of renewable energy. This functionality improves robustness and resiliency of smart grid application that in turn supports their capacity to develop the sustainable energy systems [48].

AI technologies are an important part of the solution to these challenges since managing the grid and its security is and will remain a critical factor in maintaining efficiency and stability. With the growing dependency on computer networks, smart grids face a major risk in security threats to the infrastructure. To its credit, the involvement of AI in maintaining the grid entails the surveillance of activities previously unrecognized that could result from an upcoming threat to the company's security and allow protection against such activities [49]. In addition, the use of AI increases mitigation since it anticipates the risks surrounding calamities such as storms or breakdowns of equipment. Machine learning models, based on historical data will be able to predict extreme weather situations and will enable grid operators to act and prepare in advance by strengthening infrastructure, redirecting the energy flow or, for a time, decreasing the load in the vulnerable regions. These anticipative precautions make it easier for smart grids to run even under difficult situations to avoid hitches and still maintain optimum use of energy.

Smart grids are one of the complex applied forms of artificial intelligence in terms of efficiency and sustainable energy supply. Thanks to predictive analytics, automating energy flow, load management in real time and dynamic demand–response AI allows smart grids to operate at a level that has never been seen before [50]. First, by enabling the optimality of connecting renewable energy sources into the power grid, second, through the ability to better coordinate distributed energy resources, and finally, through the preservation of the electric power infrastructure.

In the future, AI will provide significant contributions to smart grid management and will be a vast driver for making energy systems greener, safer and efficient.

7 Integrating Renewable Energy with AI for Grid Stability

With the global trend toward utilizing renewable energy technologies such as solar and wind power to address climate changes and fossil energy use concerns, incorporation of these variable energy systems into the power grid system presents special problems. Unlike traditional generators, renewable resources are stable in nature; solar power depends on the sun and wind power, weather. These variations can lead to fluctuation on the energy grid, especially when renewable resource energy is a major part of the mix. However, it is important to note that With AI, there are suitable solutions for overcoming these problems and reaching an optimized prognosis, consolidation and management of renewable energy in the smart grids [51]. By analysis, data mining, machine learning and predictive models, AI contributes to the constant management of grid's stability, sustained control of energy fluctuations and the reinforcement of the power grid's robustness.

AI provides several contributions to the integration of renewable energy, of which its ability to predict integration is among the most impactful. The accuracy of forecast is crucial to grid operators requiring to balance the energy supply and demand in order to provide consistent energy. AI operates solar and wind energy generation predictions based on past history, current meteorological conditions and stochastic computations, which it accomplishes quite well. For instance, with the support of solar irradiance, temperature and cloud cover data, AI can predict what amount of electricity the next solar power plant will produce, whereas for wind one needs information about wind speed, wind direction and pressure [52]. With the help of AI, the grid operators can predict when and how much renewable energy will be available, so they don't have to scramble for backup fossil fuel generators and waste a lot of energy. Also, due to the ability of learning from past data, AI's forecasting enhances over time and forecasts [increase] in accuracy. This forecasting is particularly important during peak demand times when large drop-offs in renewable energy may cause the system to go offline [53].

Besides, the predictability of renewable energy generation is central to the integration of AI for management of instabilities common in renewable energy systems. While fluctuating output from renewable sources may be often helpful, significant increases or decreases can cause gaps in how grid stability is maintained throughout the production process. The problem with conventional grids is that they tend to fluctuate and cannot be met with a stable source of supply that does not involve expensive and eco-friendly peaker plants. On the other hand, the smart grid supported by AI technology can respond to these changes in real time, with perfect demanding-supply balance facilitated by sensor data scattered across the grid [54]. For instance, if solar output surpasses the expected levels of efficiency, AI can either use the excess energy to charge up storage systems or supply it to other residential regions that may not require it at that time. In contrast, if there is a dip in the levels of renewable energy generation, AI can almost immediately ramp up other forms of power generation or alter energy use rates throughout the grid. Such changes enable one to avoid such instances and reduce reliance on reserve generation, thereby making it easier and cheaper to manage energy [55].

AI has a very important role in the co-optimization of storage for renewable energies, which are vital for operating the grid. Systems such as batteries are important in forwarding the excess renewable energy during periods when production is high and during low production when the energy is needed. However, it has been found that managing this storage is not easy, and timing the use of such storage is very accurate in order to avoid wastage and improve efficiency. Traditional methods automate storage and release of energy based on consumption patterns, weather forecasts and grid requirements by use of AI algorithms. These elements can be analyzed by the machine learning models on when storage is most required and even which energy types to preferentially store [56]. For instance, AI might decide to generate power during the day when production yards are high but reserve it in the evening when demand is high, and power production from solar panels is low. This tactful storing and releasing of energy by artificial intelligence thus considerably reduces reliance on the fossil sources and makes efficient use of renewables wherever possible.

Another area in which the application of AI is emerging is through the integration of smarter distribution networks to the grid. Since most renewable energy sources are distributed systems, operating complex supply and demand systems is complex for the operators of the grid. Through artificial intelligence, the usage of energy from various kinds of suppliers can be regulated in order to reach agreement and prevent imbalances that result in supply hitches [57]. For instance, the use of AI can check the flow of energy from solar panels on different buildings, community wind power plants or large-scale power stations, and efficiently direct the flow of energy to avoid shorts and overloads within different sections of power distribution. By controlling signal flow and distribution at various levels, AI can guarantee that distribution is effective while making the system more flexible and resistant to new changes and thus can support a higher integration of renewable energy without onsetting weaknesses in the system [58].

AI's ability to improve DER coordination is also important to restore grids that heavily rely on renewable resources. Small-scale renewables such as DERs including rooftop solar panels, small wind turbines and distributed battery storage systems are characterized by versatility and decentralized characteristics, which present independent issues [59]. Through usable predictability on the output from these sources of distributed electricity Relatives, AI could coordinate the DERs with relative ease and blend it into the larger distribution grid. For instance, it can regulate the flow of energy from a neighboring grid in the event of anticipated high solar resources in a given neighborhood and direct the flow of energy from distant thermal power stations to other regions where it is needed. This versatile power control increases the reliability of the electrical grid by guaranteeing how it can change depending on the local production and consumption. Smart grids are enhanced greatly through the help of AI mainly because of the ways it helps strengthen the system to adapt to the concept of stressors such as storms or increased demands for energy [60]. Some machine learning models can analyze data about risks, which could potentially undermine stability of the grid and make operators take preventive measures to protect facilities and provide a stable energy supply. For instance, if a storm has been predicted to interfere with wind or solar generation, there is AI availability to help cushion the impact by charging batteries, or other approaches such as demand–response. Furthermore, AI can be combined with renewable energy storage systems to provide electrical power during the specified instances by ensuring stable service regardless of external conditions that threaten the stability of the grid [61]. Thus, AI-driven predictive maintenance and risk assessment tools are very relevant in managing disruptions to the grid to create a reliable power system.

There are also substantial economic benefits from use of AI in managing renewable energy in smart grids. Conventional approaches to energy management, such as holding back reserve power or building costly structures for peak loads, may not be very effective. AI helps to run the facilities more efficiently and cheaply by delivering energy smarter and using less backup power and fossil fuels. Furthermore, through its capacity to optimize the usage of renewable, AI also minimizes greenhouse gas emission so as to complement energy generation goals based on environmental conservation [62]. These economical advantages also affect the consumer since demand–response and load balancing made by the help of artificial intelligence do not reflect abrupt price changes due to supply and demand. Reducing operational expenses and enhancing efficiency through AI make a very significant contribution to the development of a more efficient energy network that is also cheaper to run.

The use of the renewable energy sources in the existing smart grids essentially adopts AI which plays an important role in maintaining stable power, balancing in storage and distribution as well as improving the overall robustness of power system. Another benefit of AI is that having predictive capabilities, handling supply variations not only steadily but also smartly as well as efficiently interacting with DERs, the element of renewable energy sources' unpredictability is successfully minimized and the AI element gains trust as the unchangeable part of the power grid [63]. As the world proceeds toward adopting sustainability as the defining factor of energy efficiency, AI will play a more significant role in the integration of the greener energy solutions needed to support the needs of the low-carbon economy. AI enhances the present electric power infrastructure by supporting innovations in storage, demand–response, and grid stability for the future demand and supply largely dominated by renewables.

8 AI for Predictive Maintenance and Enhancing Grid Health

The use of renewable energy sources in the existing smart grids essentially adopts AI, which plays an important role in maintaining stable power, balancing in storage and distribution as well as improving the overall robustness of power system.

Another benefit of AI is that having predictive capabilities, handling supply variations not only steadily but also smartly as well as efficiently interacting with DERs, the element of renewable energy sources' unpredictability is successfully minimized and the AI element gains trust as the unchangeable part of the power grid [64, 65]. As the world moves to accept environmentally friendly energy systems, the role of artificial intelligence in this integration will continue to work toward sustainable energy systems that meet the needs of a climate change world. AI enhances the present electric power infrastructure by supporting innovations in storage, demand–response and grid stability for the future demand and supply largely dominated by renewables.

The use of the renewable energy sources in the existing smart grids essentially adopts AI, which plays an important role in maintaining stable power, balancing in storage and distribution as well as improving the overall robustness of power system. Another benefit of AI is that having predictive capabilities, handling supply variations not only steadily but also smartly as well as efficiently interacting with DERs the element of renewable energy sources' unpredictability is successfully minimized and the AI element gains trust as the unchangeable part of the power grid. With the growing adoption of sustainable renewable energy in the global economy, the role of AI in this integration will go up, providing means for development of sustainable energy systems that will be fit for the purpose of a low-carbon economy [66]. AI enhances the present electric power infrastructure through supporting innovations in storage, demand–response and grid stability for the future demand and supply largely dominated by renewables. Energy efficiency benefits when AI keeps equipment in optimal operating condition. Energy waste occurs when devices have to work under less-than-ideal circumstances to produce their usual output. A malfunctioning transformer may need nearly 10 times its nominal operating power to run at peak performance (which is still very much not "nominal"). This means energy from other parts of the grid has to make up the difference, with an energy proprietor as the main culprit. AI (helping where human scheduling and foreknowledge of problems might fail) identifies these inefficiencies and those in other parts of the grid [67]. This is not just about "putting the green in green energy"; it is also about electrifying energy resources.

Continuous monitoring combined with learning from experience makes AI much better at predicting asset failures. By collecting real-time data on asset performance and the environment, AI can use this information along with historical data, to quickly and continually improve its predictive algorithms. It is not hard to imagine that some components in an electric grid would fail more often in, say, a heat wave than during spring-like weather, even though they seem to be functioning normally under the latter conditions. AI can use this insight, along with many others, to help utilities figure out the best possible maintenance strategies to keep the grid running smoothly and to do so more quickly than any human mind can [68]. Predictive maintenance driven by AI essentially promotes sustainability over the long haul by abandoning time-based maintenance in favor of condition-based methods. Maintenance occurs sporadically when performance dips below par via real-time equipment health assessments. "This approach curtails redundant swaps thereby slashing expenditures dwindling surplus and elongating asset longevity." Renewable energy sources coupled with escalating electricity consumption prompt a more resilient grid better equipped to handle increased demand suddenly.

AI bolsters grid resilience via bespoke operational tweaks throughout, not merely maintenance. Analyzing data from every grid segment allows AI systems to identify pockets of inefficiency in energy dispersal and propose tweaks. AI suggests novel load-balancing tactics at peak demand times thereby averting undue stress on specific grid components suddenly. Sophisticated AI systems continuously scan the grid's vital signs, probing for opportunities to bolster resilience and boost overall performance rapidly [69].

AI brings considerable excitement to predictive maintenance via its integration with drones that conduct off-site inspections. Drones packing AI and sensors navigate treacherous terrain to access high-voltage lines and other hard-to-reach infrastructure minimizing human presence. Advanced AI systems now enable swift discovery of nascent faults thereby significantly enhancing overall safety and streamlining operational workflows somehow. A drone's thermal imaging occasionally detects a hot spot on a power line signaling excessive load or wear, often unnoticed by the naked eye. Sophisticated diagnostics guarantee upkeep occurs sporadically but efficiently boosts grid reliability and prolongs its operational lifespan. AI-driven predictive maintenance yields substantial benefits, notably increased efficiency and reduced downtime, particularly for utilities and consumer households [70]. Enabling proactive maintenance slashes downtime thereby reducing costly emergency responses and lost revenue during those outages somehow. AI boosts energy efficiency and slashes waste, sparking major production and distribution savings that ultimately trickle down to consumers in the form of lower electricity bills somehow. A dependable energy infrastructure yields fewer outages and less price fluctuation, thus creating a robust sustainable energy market advantageous to all stakeholders somehow.

AI significantly bolsters predictive maintenance, which fosters a sustainable energy future with reduced reliance on resource-intensive gear, thereby lowering grid carbon emissions. Robust infrastructure functions with greater ease when properly maintained, which aligns with eco-friendly objectives and facilitates smoother integration of green power. Advanced AI technologies play a crucial role in managing grids capable of adapting seamlessly to fluctuations in renewable energy inputs, thereby bolstering resilience and sustainability. Predictive maintenance fueled by AI drastically revamps grid management by slashing downtime and energy waste, thus promoting sustainability. Fault detection at an early stage empowers utilities to bypass costly repairs and optimize energy consumption, thereby ensuring reliable service provision. AI monitoring facilitates swift responses to grid fluctuations, enabling bespoke maintenance tactics aligned with asset performance, and enhancing overall efficiency somehow. Bolstering grid health and sustainability, AI-driven predictive maintenance proves vital for a robust energy infrastructure capable of meeting future demands head-on while backing a shift toward cleaner sustainable energy systems.

9 Case Studies: Real-World AI Applications in Smart Grids

Incorporation of AI into smart grids transforms energy management pretty rapidly nowadays, enhancing efficiency, supporting sustainability somehow. Top energy companies showcase AI-driven solutions tackling complex energy optimization challenges, integrating renewable sources, and bolstering grid resilience worldwide, yielding valuable industry insights. Some examples follow.

(a) Xcel Energy's AI implementation has notably enhanced load forecasting and energy distribution nationwide, yielding significant efficiency gains. Xcel created a machine learning model that forecasts energy demand with eerie precision by scrutinizing historical energy consumption patterns and weather trends. Real-time tweaks in energy generation and distribution were enabled by this capability, thereby slashing waste and averting system overload at peak hours. Initiative boosted grid reliability by cutting operational costs and showcasing AI's potential for creating more adaptive energy systems with responsive tech [71].

(b) Enel leverages AI for predictive maintenance in Europe, transforming operations with cutting-edge tech. Enel, a global energy giant, leverages cutting-edge AI solutions pretty extensively for pinpointing wonky equipment well ahead of potentially disastrous breakdowns. Through persistent surveillance of real-time sensor data from critical infrastructure like transformers, this system forecasts wear and impending problems, thereby enabling prompt fixes. By implementing this approach, downtime significantly decreases and maintenance expenses plummet, thereby resulting in fewer equipment failures and a substantial extension of asset lifespan. Predictive maintenance proves a game-changer for Enel's grid resilience and efficiency by slashing downtime [72].

(c) UK's National Grid demonstrates AI's potential in balancing renewables and conventional energy sources pretty effectively nowadays. Renewable energy targets loom large as solar and wind power's integration into grid systems proves tricky because of their inherent unpredictability [73]. National Grid's AI system perpetually scrutinizes real-time energy data, forecasting renewable source availability and tweaking conventional power generation amidst fluctuating demands. AI facilitates clean energy transitions by leveraging renewable energy sources that significantly slash carbon emissions and bolster grid resilience.

(d) TEPCO leverages AI for bolstering grid resilience against calamities in Japan. TEPCO leverages machine learning algorithms in order to get ahead of disruptions triggered by freak storms [74]. Historical weather patterns get scrutinized alongside grid performance metrics so the system can forecast potential disruptions prompting proactive steps like rerouting power. Quicker recovery from outages happens faster with this application and it showcases AI's adaptability in energy systems facing climate change head-on somehow.

(e) AGL Energy exemplifies AI's ability to effectively manage distributed energy resources like rooftop solar installations alongside residential battery systems in Australia. Sophisticated AI tech synchronizes power generation storage and distribution so surplus solar energy gets stashed locally for peak usage moments. "This method drastically reduces grid stress by leveraging more renewable energy sources thereby decreasing fossil fuel dependency." AGL's achievements underscore AI's pivotal impact on DER optimization, yielding a remarkably agile and environmentally friendly energy infrastructure [75].

These case studies highlight AI's remarkable ability to revolutionize and bolster the efficiency of smart grids dramatically nowadays. AI brings forth bespoke solutions that bolster energy systems via predictive upkeep and avant-garde renewable energy fusion thereby augmenting disaster hardiness. Nailing success demands robust data infrastructure, stakeholders collaborating effectively, and a strong ongoing learning commitment too. AI technologies are advancing rapidly thereby amplifying their impact on sustainable energy systems development—this trend will redefine global energy landscape standards.

10 Future of AI in Advancing Smart Grid Sustainability

Global momentum shifts toward cleaner energy sources, AI integration with smart grids gains paramount significance, notably in optimizing efficiency. AI's fusion with edge computing and decentralized grids revolutionizes energy systems, rendering them supremely adaptive and resilient. Green technologies converge, fostering a more efficient energy infrastructure which propels global sustainability objectives forward rapidly nowadays. Significant advancement in this area comes via edge computing boosting energy management efficiency by processing data near its source—grid sensors and smart meters—thereby sidestepping centralized data centers. By leveraging real-time data, AI algorithms rapidly adapt energy distribution to suit localized needs with heightened efficiency suddenly. AI swiftly streamlines energy consumption in residential and commercial spaces by syncing supply and demand sans reliance on distant servers. Renewable energy sources necessitate a decentralized approach that slashes latency pretty drastically thereby bolstering grid stability and facilitating speedier reactions.

Crucially, smart grids' sustainability relies heavily on harnessing IoT's vast potential. Linking various devices in energy networks IoT generates ongoing data on grid performance that AI leverages for efficiency enhancements. Intelligent devices like smart meters and weather stations, plus connected appliances, churn out real-time data which AI systems scrutinize intensely. Energy distribution gets optimized by AI using usage patterns and weather forecasts thereby slashing unnecessary expenditure. IoT sensors spit out vital data that fuels predictive maintenance thereby slashing downtime and beefing up grid infrastructure lifespan.

AI significantly influences operations in decentralized grids, a burgeoning area. Smaller self-sufficient grids often harness renewable energy from solar and wind sources and function autonomously or in tandem with main power infrastructure. AI boosts microgrid energy production storage and distribution so that energy supply remains steady even when renewables falter badly sometimes. AI enables smoother energy transfer between microgrids and main grid by harnessing real-time data from various systems in a highly efficient manner. Decentralized grids slash reliance on massive power plants and slash carbon emissions thereby propelling forward overall sustainability targets.

AI-powered blockchain solutions are pretty much revolutionizing the sustainable energy landscape right now. Blockchain guarantees robust energy trading with transparent ledger tech facilitating direct exchanges sans intermediaries. Pairing blockchain with AI enables real-time automation of transactions, adjusting energy prices and distribution based on fluctuating supply and demand dynamics. Decentralization of the energy market happens alongside efficiency and sustainability boosts through this partnership's promotion of responsible energy habits. AI significantly boosts energy storage management's sustainability credentials by virtue of its involvement. Effective energy storage proves vital in mitigating the wildly unpredictable fluctuations inherent in renewable energy sources. AI predicts demand spikes and renewable output, thus streamlining storage to unleash energy precisely when necessary. Maximizing stored energy utilization boosts grid reliability significantly while slashing fossil-fuel backup power reliance substantially every time. Suddenly, integration of AI with edge computing and IoT decentralizes grids, leverages blockchain, and amplifies energy storage, yielding a remarkably sustainable energy paradigm. AI enables smarter energy management, fostering resilient grids that bolster renewable energy integration and amplify energy efficiency through clever tech. Rapid advancements in AI tech will play a pivotal role in forging a greener more sustainable worldwide energy landscape eventually.

11 Conclusion

AI technology occupies a very important role in the smart grids of today, which optimizes nearly every aspect of efficiency, sustainability and performance. By enabling dynamic optimization, AI helps improve load management by addressing many of the critical issues related to grid management, such as energy wastage caused by demand volatility and the erratic nature of renewable energy resources. Furthermore, predictive AI minimizes maintenance costs, ensuring that the grid's infrastructures are in excellent condition for many years. Integrating IoT and decentralized grids certainly seems to enhance its reach in the energy sector by introducing rapid AI expansion nowadays. In the future, new technologies will create grids that will keep pace with energy resource management by responding appropriately to changes in supply-demand cycles. In smart grids, AI's function in the future is said to be very optimistic and potentially transformative for several aspects of energy management and distribution networks. Complex AI models are able to reveal energy consumption trends with great details, which allows for diversity in the provision and consumption of energy resources. Edge computing enables quicker on-site decisions, reducing the waiting time and improving the grid's responsiveness. There is now rapid growth in the number of IoT devices and these will generate a lot of real-time data which AI can use, which will then lead to much improved efficiency of the grid.

AI is going to be very useful in optimizing energy storage in such a way that renewable energy gets stored and managed efficiently so that it is available when it is actually needed. The merger of AI with blockchain technology and other innovations encourages the development of energy 'markets' that are more sophisticated and allow for P2P energy trading directly. The fast pace of new technology development creates the conditions for AI-based breakthroughs, which lead to the development of advanced energy systems that are cleaner and more reliable. A crucial part of the future of energy is the fast-paced AI innovation evolution that we are witnessing at the moment.

References

1. Mashapu, L. D., Eboule, P. S. P., & Pretorius, J.-H. C. (2022). The need for artificial intelligence for energy-efficiency management: A review. In *2022 8th International Conference on Energy Efficiency and Agricultural Engineering (EE&AE), Ruse, Bulgaria* (pp. 1–6). IEEE. https://doi.org/10.1109/EEAE53789.2022.9831359
2. Kwon, Y. (2023). ABSX: The chiplet hyperscale AI processing unit for energy-efficient high-performance AI processing. In *2023 20th International SoC Design Conference (ISOCC), Jeju, Republic of Korea* (pp. 217–218). IEEE. https://doi.org/10.1109/ISOCC59558.2023.10396520
3. Oukaira, A., Karime, A., Roberge, V., & Tarbouchi, M. (2024). Toward increased energy efficiency and proactive management of electrical grids based on AI. In *2024 International Conference on Computing, Internet of Things and Microwave Systems (ICCIMS), Gatineau, QC, Canada* (pp. 1–4). IEEE. https://doi.org/10.1109/ICCIMS61672.2024.10690565
4. Zakizadeh, M., & Zand, M. (2024). Transforming the energy sector: Unleashing the potential of AI-driven energy intelligence, energy business intelligence, and energy management system for enhanced efficiency and sustainability. In *2024 20th CSI International Symposium on Artificial Intelligence and Signal Processing (AISP)* (pp. 1–7). https://doi.org/10.1109/AISP61396.2024.10475298
5. Bose, B. K. (2017). Artificial intelligence techniques in smart grid and renewable energy systems—Some example applications. *Proceedings of the IEEE, 105*(11), 2262–2273. https://doi.org/10.1109/JPROC.2017.2756596
6. Gill, A., Chandani, P., & M. R. (2023). Deployment of AI in micro grids: A survey. In *2023 International Conference on Power Energy, Environment & Intelligent Control (PEEIC), Greater Noida, India* (pp. 1390–1394). IEEE. https://doi.org/10.1109/PEEIC59336.2023.10450747
7. Pribylskii, I. V., & Armeev, D. V. (2024). AI control system for power grid based on distributed energy resources. In *2024 IEEE 25th International Conference of Young Professionals in Electron Devices and Materials (EDM), Altai, Russian Federation* (pp. 1420–1425). IEEE. https://doi.org/10.1109/EDM61683.2024.10614965
8. Gill, A., Khalid, A., & M. R. (2023). A crucial review on AI use in hybrid grids: A ultra smart way towards micro grid. In *2023 International Conference on Power Energy, Environment & Intelligent Control (PEEIC), Greater Noida, India* (pp. 1364–1368). IEEE. https://doi.org/10.1109/PEEIC59336.2023.10451299

9. Algarni, M. A., Kraiem, N., & Sakly, H. (2024). Enhancing security and energy efficiency in smart energy management systems through IoT device detection and machine learning techniques. In *2024 IEEE 7th International Conference on Advanced Technologies, Signal and Image Processing (ATSIP), Sousse, Tunisia* (pp. 506–511). IEEE. https://doi.org/10.1109/ATSIP62566.2024.10638911
10. Krishnan, R. S., Ashokkumar, C., Nithila, E. E., Soundiraraj, N., Malar, P. S. R., & Rani, P. A. (2024). AI-powered demand response mechanisms for sustainable smart grids. In *2024 5th International Conference on Electronics and Sustainable Communication Systems (ICESC), Coimbatore, India* (pp. 87–92). IEEE. https://doi.org/10.1109/ICESC60852.2024.10690089
11. Zakizadeh, M., & Zand, M. (2024). AI-driven energy intelligence: Revolutionizing the energy sector through smart energy solutions. In *2024 10th International Conference on Web Research (ICWR), Tehran, Iran* (pp. 355–363). IEEE. https://doi.org/10.1109/ICWR61162.2024.10533346
12. Kumar, A., Alaraj, M., Rizwan, M., & Nangia, U. (2021). Novel AI based energy management system for smart grid with RES integration. *IEEE Access, 9*, 162530–162542. https://doi.org/10.1109/ACCESS.2021.3131502
13. Cali, U., et al. (2023). Digital twins: Shaping the future of energy systems and smart cities through cybersecurity, efficiency, and sustainability. In *2023 International Conference on Future Energy Solutions (FES), Vaasa, Finland* (pp. 1–6). IEEE. https://doi.org/10.1109/FES57669.2023.10182868
14. Cali, U., et al. (2014). Applications of multi-agent systems in smart grids: A survey. In *2014 International Conference on Multimedia Computing and Systems (ICMCS), Marrakech, Morocco* (pp. 1088–1094). IEEE. https://doi.org/10.1109/ICMCS.2014.6911384
15. Banihabib, R., Fadnes, F. S., & Assadi, M. (2023). Optimizing building heating efficiency: A data-driven approach for cost and energy savings. In *2023 IEEE 11th International Conference on Smart Energy Grid Engineering (SEGE)* (pp. 212–217). IEEE. https://doi.org/10.1109/SEGE59172.2023.10274593
16. Ahmed, S. A., Huang, Q., Amin, W., & Afzal, M. (2020). Cost-effective energy management system in prosumer-based electricity market. In *2020 International Conference on Smart Grids and Energy Systems (SGES), Perth, Australia* (pp. 181–185). IEEE. https://doi.org/10.1109/SGES51519.2020.00039
17. Bibi, R., Ahmad, N., & Khan, S. A. (2017). Running cost functions optimization in smart grids. In *2017 IEEE 2nd Advanced Information Technology, Electronic and Automation Control Conference (IAEAC), Chongqing, China* (pp. 2350–2355). IEEE. https://doi.org/10.1109/IAEAC.2017.8054442
18. Zulu, M. L. T., Pillay Carpanen, R., & Tiako, R. (2024). Optimization of internet of things (IoT) for smart grid energy management using artificial intelligence (AI) techniques to reach SDG7. In *2024 32nd Southern African Universities Power Engineering Conference (SAUPEC), Stellenbosch, South Africa* (pp. 1–6). IEEE. https://doi.org/10.1109/SAUPEC60914.2024.10445033
19. Yan, B. X., Malik, N. N. N. A., Latiff, N. M. A., Ramli, A., & Hamzah, S. A. (2021). Development of a self-adaptive based renewable energy management in smart micro grid. In *2021 IEEE 15th Malaysia International Conference on Communication (MICC), Malaysia* (pp. 54–59). https://doi.org/10.1109/MICC53484.2021.9642148
20. Alaba, F. A., Sani, U., Dada, E. G., & Mohammed, B. H. (2024). AIoT-enabled smart grids: Advancing energy efficiency and renewable energy integration. In *Artificial intelligence of things for achieving sustainable development goals*. Springer. https://doi.org/10.1007/978-3-031-53433-1_4
21. Suman, R., & Mallipeddi. (2022). Harnessing AI and IoT technologies for sustainable business operations in the energy sector. *Asia Pacific Journal of Energy and Environment*. https://doi.org/10.18034/apjee.v9i1.735
22. Jaya Mabel Rani, A., Kalpana, B., Srinivasan, A., Subramaniam, S., Shiney, S. A., & Pandi, V. S. (2023). Artificial intelligence—Enabled smart grids: Enhancing efficiency and sustain-

ability. In *2023 7th International Conference on Electronics, Communication and Aerospace Technology (ICECA)*. IEEE. https://doi.org/10.1109/iceca58529.2023.10395590
23. Komala, C. R., Basha, M. M., Farook, S., Niranchana, R., Rajendiran, M., & Subhi, B. (2024). Smart energy systems-integrated machine learning, IoT, and AI tools. In *Advances in environmental engineering and green technologies book series*. https://doi.org/10.4018/979-8-3693-2351-9.ch011
24. Marques, P. C. A., & Oliveira, P. A. (2024). Artificial intelligence technologies applied to smart grids and management. https://doi.org/10.20944/preprints202406.1248.v1.
25. Darwish, A. S., Abbas, M., Al-Salim, W. L., & Al-Tameemi, M. R. J. (2024). Artificial intelligence for sustainable energy transition: Optimising renewable energy integration and management. *ARID International Journal for Science and Technology*. https://doi.org/10.36772/arid.aijst.2024.7134
26. Bhavya, R., Manjuladevi, P., Santhoshini, E. S., Ramya, M., & Dhanapal, R. (2023). Development and improving energy management system in smart grid through integration with renewable energy system with AI and internet of things. In *2023 International Conference on Inventive Computation Technologies (ICICT)*. IEEE. https://doi.org/10.1109/ICICT57646.2023.10134409
27. Volkova, A., Patil, A. D., Javadi, S. A., & de Meer, H. (2022). Accountability challenges of AI in smart grid services. In *e-Energy'22: The Thirteenth ACM International Conference on Future Energy Systems*. https://doi.org/10.1145/3538637.3539636
28. Fan, Z., Yan, Z., & Wen, S. (2023). Deep learning and artificial intelligence in sustainability: A review of SDGs, renewable energy, and environmental health. *Sustainability*. https://doi.org/10.3390/su151813493
29. Nammouchi, A., Aupke, P., Kassler, A., Theocharis, A., Raffa, V., & Di Felice, M. (2021). Integration of AI, IoT and edge-computing for smart microgrid energy management. In *2021 IEEE International Conference on Environment and Electrical Engineering and 2021 IEEE Industrial and Commercial Power Systems Europe (EEEIC/I&CPS Europe)*. IEEE. https://doi.org/10.1109/EEEIC/ICPSEUROPE51590.2021.9584756
30. Kumar, S., Pathak, U., Astha, & Bhatia, B. (2023). Achieving peak energy efficiency in smart grids using AI and IOT. *Communications in Computer and Information Science*. https://doi.org/10.1007/978-3-031-47055-4_11
31. Charef, N., Mnaouer, A. B., Aloqaily, M., Bouachir, O., & Guizani, M. (2023). Artificial intelligence implication on energy sustainability in Internet of Things: A survey. *Information Processing and Management*. https://doi.org/10.1016/j.ipm.2022.103212
32. Mazhar, T., Irfan, H. M., Haq, I., Ullah, I., Ashraf, M., Al Shloul, T., Yasin, Y., Imran, G., & ElKamchouchi, D. H. (2023). Analysis of challenges and solutions of IoT in smart grids using AI and machine learning techniques: A review. *Electronics*. https://doi.org/10.3390/electronics12010242
33. Sampath, B. (2024). Advancements in optimizing smart energy systems through smart grid integration, machine learning, and IoT. In *Advances in environmental engineering and green technologies book series*. https://doi.org/10.4018/979-8-3693-0492-1.ch002
34. Arya, Y., Guha, D., & Sarangi, S. (2023). AI data-based modeling, control, and operation of smart grid power systems. *International Journal of Numerical Modelling-Electronic Networks Devices and Fields*. https://doi.org/10.1002/jnm.3156
35. Verma, S., & Bhalla, P. (2024). Sustainable development and AI. In *Advances in human resources management and organizational development book series*. https://doi.org/10.4018/979-8-3693-0724-3.ch011
36. Moreno Escobar, J. J., Matamoros, O. M., Padilla, R. T., Reyes, I. L., & Espinosa, H. Q. (2021). A comprehensive review on smart grids: Challenges and opportunities. *Sensors, 21*(21), 6978. https://doi.org/10.3390/s21216978
37. Ohanu, C. P., et al. (2024). A comprehensive review of recent developments in smart grid through renewable energy resources integration. *Heliyon, 10*(3), e25705. https://doi.org/10.1016/j.heliyon.2024.e25705

38. Kabeyi, M. J. B., & Olanrewaju, O. A. (2023). Smart grid technologies and application in the sustainable energy transition: A review. *International Journal of Sustainable Energy, 42*(1), 685–758. https://doi.org/10.1080/14786451.2023.2222298
39. Khalid, M. (2024). Smart grids and renewable energy systems: Perspectives and grid integration challenges. *Energy Strategy Reviews, 51*, 101299. https://doi.org/10.1016/j.esr.2024.101299
40. Fang, X., et al. (2012). Smart grid—The new and improved power grid: A survey. *IEEE Communications Surveys & Tutorials, 14*(4), 944–980. https://doi.org/10.1109/SURV.2011.101911.00087
41. Meliani, M., Barkany, A. E., Abbassi, I. E., Darcherif, A. M., & Mahmoudi, M. (2021). Energy management in the smart grid: State-of-the-art and future trends. *International Journal of Engineering Business Management, 13*. https://doi.org/10.1177/18479790211032920
42. Ahmad, T., et al. (2022). Energetics systems and artificial intelligence: Applications of Industry 4.0. *Energy Reports, 8*, 334–361. https://doi.org/10.1016/j.egyr.2021.11.256
43. Zhu, X., Chen, S., Liang, X., Jin, X., & Zhimin, D. (2024). Next-generation generalist energy artificial intelligence for navigating smart energy. *Cell Reports Physical Science, 5*(9), 102192. https://doi.org/10.1016/j.xcrp.2024.102192
44. Farzaneh, H., et al. (2021). Artificial intelligence evolution in smart buildings for energy efficiency. *Applied Sciences, 11*(2), 763. https://doi.org/10.3390/app11020763
45. Pandiyan, P., et al. (2023). Technological advancements toward smart energy management in smart cities. *Energy Reports, 10*, 648–677. https://doi.org/10.1016/j.egyr.2023.07.021
46. Binyamin, S. S., et al. (2024). Artificial intelligence-powered energy community management for developing renewable energy systems in smart homes. *Energy Strategy Reviews, 51*, 101288. https://doi.org/10.1016/j.esr.2023.101288
47. Khare, V., & Chaturvedi, P. (2023). Design, control, reliability, economic and energy management of microgrid: A review. *Advances in Electrical Engineering, Electronics and Energy, 5*, 100239. https://doi.org/10.1016/j.prime.2023.100239
48. Sulaiman, A., et al. (2023). Artificial intelligence-based secured power grid protocol for smart city. *Sensors, 23*(19), 8016. https://doi.org/10.3390/s23198016
49. McMillan, L., & Varga, L. (2022). A review of the use of artificial intelligence methods in infrastructure systems. *Engineering Applications of Artificial Intelligence, 116*, 105472. https://doi.org/10.1016/j.engappai.2022.105472
50. Biswal, B., et al. (2024). Review on smart grid load forecasting for smart energy management using machine learning and deep learning techniques. *Energy Reports, 12*, 3654–3670. https://doi.org/10.1016/j.egyr.2024.09.056
51. Kara, T., & Şahin, A. D. (2023). Implications of climate change on wind energy potential. *Sustainability, 15*(20), 14822. https://doi.org/10.3390/su152014822
52. Benti, N. E., Chaka, M. D., & Semie, A. G. (2023). Forecasting renewable energy generation with machine learning and deep learning: current advances and future prospects. *Sustainability, 15*(9), 7087. https://doi.org/10.3390/su15097087
53. Unsal, D. B., et al. (2024). A comparative study of AI methods on renewable energy prediction for smart grids: Case of Turkey. *Sustainability, 16*(7), 2894. https://doi.org/10.3390/su16072894
54. Arévalo, P., & Jurado, F. (2024). Impact of artificial intelligence on the planning and operation of distributed energy systems in smart grids. *Energies, 17*(17), 4501. https://doi.org/10.3390/en17174501
55. Liu, Z., et al. (2022). Artificial intelligence powered large-scale renewable integrations in multi-energy systems for carbon neutrality transition: Challenges and future perspectives. *Energy and AI, 10*, 100195. https://doi.org/10.1016/j.egyai.2022.100195
56. Rojek, I., Mikołajewski, D., Mroziński, A., & Macko, M. (2023). Machine learning- and artificial intelligence-derived prediction for home smart energy systems with PV installation and battery energy storage. *Energies, 16*(18), 6613. https://doi.org/10.3390/en16186613

57. Mahmood, M., et al. (2024). Impacts of digitalization on smart grids, renewable energy, and demand response: An updated review of current applications. *Energy Conversion and Management: X, 24*, 100790. https://doi.org/10.1016/j.ecmx.2024.100790
58. Bakare, M. S., Abdulkarim, A., Shuaibu, A. N., et al. (2024). Energy management controllers: strategies, coordination, and applications. *Energy Informatics, 7*(57). https://doi.org/10.1186/s42162-024-00357-9
59. Ukoba, K., Olatunji, K. O., Adeoye, E., Jen, T.-C., & Madyira, D. M. (2024). Optimizing renewable energy systems through artificial intelligence: Review and future prospects. *Energy & Environment, 35*(7), 3833–3879. https://doi.org/10.1177/0958305X241256293
60. Wang, X., Wang, H., Bhandari, B., et al. (2024). AI-empowered methods for smart energy consumption: A review of load forecasting, anomaly detection and demand response. *International Journal of Precision Engineering and Manufacturing-Green Technology, 11*, 963–993. https://doi.org/10.1007/s40684-023-00537-0
61. Ali, S. S., & Choi, B. J. (2020). State-of-the-art artificial intelligence techniques for distributed smart grids: A review. *Electronics, 9*(6), 1030. https://doi.org/10.3390/electronics9061030
62. Judge, M. A., et al. (2024). A comprehensive review of artificial intelligence approaches for smart grid integration and optimization. *Energy Conversion and Management: X*, 100724. https://doi.org/10.1016/j.ecmx.2024.100724
63. Bouquet, P., Jackson, I., Nick, M., & Kaboli, A. (2023). AI-based forecasting for optimised solar energy management and smart grid efficiency. *International Journal of Production Research, 62*(13), 4623–4644. https://doi.org/10.1080/00207543.2023.2269565
64. Kiasari, M., Ghaffari, M., & Aly, H. H. (2024). A comprehensive review of the current status of smart grid technologies for renewable energies integration and future trends: The role of machine learning and energy storage systems. *Energies, 17*(16), 4128. https://doi.org/10.3390/en17164128
65. Subramanya Sarma, S., & Reddy, M. R. P. (2024). Assimilation of sustainable energy resources into the smart grid: Current advancements in the realm of information and communication technologies. In *ICT for smart grid—Recent advances, new perspectives, and applications*. IntechOpen. https://doi.org/10.5772/intechopen.1003724
66. Shobeiry, S. M. (2024). AI-enabled modern power systems: challenges, solutions, and recommendations. In S. Azad & M. Nazari-Heris (Eds.), *Artificial intelligence in the operation and control of digitalized power systems* (pp. 21–40). Springer. https://doi.org/10.1007/978-3-031-69358-8_2
67. Mohammed, A., et al. (2024). A comprehensive review of advancements and challenges in reactive power planning for microgrids. *Energy Informatics, 7*, 63. https://doi.org/10.1186/s42162-024-00341-3
68. Khan, R., et al. (2021). Energy sustainability–survey on technology and control of microgrid, smart grid and virtual power plant. *IEEE Access, 9*, 104663–104694. https://doi.org/10.1109/ACCESS.2021.3099941
69. Tabassum, Z., et al. (2024). Artificial intelligence and blockchain technology for secure smart grid and power distribution automation. In R. C. Biradar et al. (Eds.), *AI and blockchain applications in industrial robotics* (pp. 226–252). IGI Global. https://doi.org/10.4018/979-8-3693-0659-8.ch009
70. Bin Abu Sofian, A. D. A., et al. (2024). Machine learning and the renewable energy revolution: Exploring solar and wind energy solutions for a sustainable future including innovations in energy storage. *Sustainable Development, 32*(4), 3953–3978. https://doi.org/10.1002/sd.2885
71. Sergeev, A. M., Shaykin, A. A., & Khazanov, E. A. (2016). Key technologies and experiments of XCELS project. In *2016 Conference on Lasers and Electro-Optics (CLEO), San Jose, CA, USA* (pp. 1–2).
72. Bosatelli, F., Romano, S. L., Bonacci, F., Infante, C. B., Cosmai, R., & Niccolai, A. (2021). Enel green power's AI tool for anomaly detection on PV plant thermographies collected with UAS. In *2021 IEEE International Conference on Environment and Electrical Engineering and*

2021 IEEE Industrial and Commercial Power Systems Europe (EEEIC/I&CPS Europe), Bari, Italy (pp. 1–3). https://doi.org/10.1109/EEEIC/ICPSEurope51590.2021.9584740
73. Kong, F., Dong, C., Liu, X., & Zeng, H. (2014). Quantity versus quality: Optimal harvesting wind power for the smart grid. *Proceedings of the IEEE, 102*(11), 1762–1776. https://doi.org/10.1109/JPROC.2014.2359448
74. Kumar, A., Sood, Y. R., & Maheshwari, A. (2024). Power systems resilience enhancement through renewable energy integration: Insights and future directions. In *2024 IEEE 4th International Conference in Power Engineering Applications (ICPEA), Pulau Pinang, Malaysia* (pp. 140–145). IEEE. https://doi.org/10.1109/ICPEA60617.2024.10498721
75. AGL launches world's largest solar virtual power plant battery demonstration to benefit customers. Retrieved August 5, 2016, from https://www.agl.com.au/about-agl/media-centre/asx-and-media-releases/2016/august/agl-launches-world-largest-solar-virtual-power-plant

Design and Simulation of Standalone Residential Microgrid System with Fuzzy Logic Controller

Rahul Soni and Prakruti Shah

1 Microgrid for Flood Conditions

1.1 Challenges to Be Addressed

Simulation and modeling of microgrids for residential use during flooding scenarios involve research on microgrid systems, technical computations, and model development. The study includes a detailed analysis of components such as solar rooftop systems, inverters, and Battery Energy Storage Systems (BESS), alongside the application of Fuzzy Logic Controllers for predicting technical parameters.

1.2 Learning Objective

A solution for addressing flooding incidents, incorporating technical training in microgrid design and simulation, the study includes an impact assessment of the proposed system and focuses on developing problem-solving skills to tackle challenges in system design and optimization.

R. Soni (✉) · P. Shah
Electrical and Electronics Engineering Program, School of Engineering and Technology, Navrachana University, Vadodara, Gujarat, India

P. Raj et al. (eds.), *Artificial Intelligence (AI) for IT Energy Efficiency and Green AI for Environment Sustainability*,
https://doi.org/10.1007/978-3-031-89420-6_18

1.3 Potential Outcomes

With a strategy to address flooding incidents that integrate technical training in microgrid design and simulation, the research emphasizes assessing the impact of the proposed system while enhancing problem-solving skills to overcome challenges in system design and optimization (Figs. 1 and 2).

1.3.1 Microgrid and Smart Grids

Microgrids are small-scale, local energy systems that operate independently or with the main grid. Smart Grids are modern energy systems that use digital technologies to manage electricity distribution efficiently [1–3].

1.3.2 Types of Microgrid

A grid-tied microgrid is connected to the main grid but can also operate independently during outages. An off-grid microgrid, on the other hand, operates independently without relying on the main grid. Urban microgrids are designed to enhance grid stability within cities, while industrial microgrids improve energy reliability and efficiency in industrial facilities. Residential microgrids cater to residential spaces, enabling homeowners to generate and manage their own electricity [4–6].

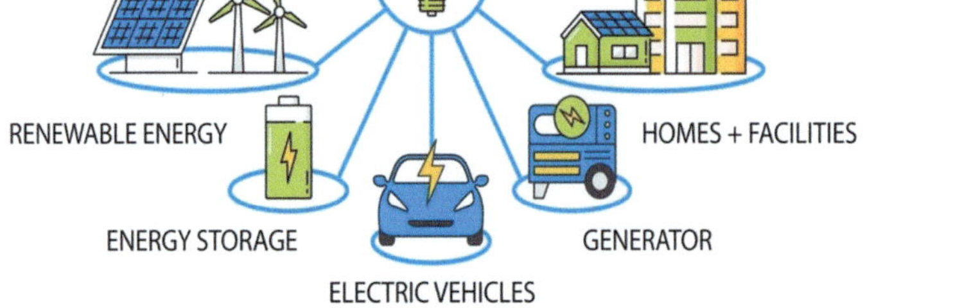

Fig. 1 Microgrid basic architecture with DER compatibility

Fig. 2 External sources of power supply during power cut-off because of flood condition

1.3.3 Key Components of a Microgrid

Microgrids and Smart Grids are innovative energy systems that offer efficient and reliable energy management. Microgrids are small-scale, local energy systems that can operate independently or with the main grid, while Smart Grids use digital technologies to manage electricity distribution. Microgrids have various benefits, including supporting zero-emission energy sources, reducing line losses, and enhancing community energy control. However, they also face challenges such as complexity, customization, and investment difficulties [7].

1.3.4 Features of Microgrid

In island mode, a microgrid operates independently during outages, while in grid-connected mode, it works in conjunction with the main grid. A decentralized microgrid is less reliant on central power, providing a local power supply that enhances grid reliability. It is flexible, adapting to changing energy needs, and can operate using either AC or DC power [8].

1.3.5 Benefits

Microgrids enable the use of zero-emission energy sources like solar and wind, reducing reliance on fossil fuels and minimizing power losses. They balance renewable supply with controllable generation, integrate energy storage and electric vehicles (EV) batteries, and reduce dependence on centralized energy sources. This provides communities with greater energy control, stabilizes costs, and protects against utility price fluctuations.

1.3.6 Challenges

Microgrids combine diverse energy technologies tailored to specific needs, locations, and markets. While their complexity can deter investors, technologies like solar and fuel cells qualify for federal tax credits. However, municipalities without federal tax obligations may be unable to benefit from these incentives (Figs. 3 and 4).

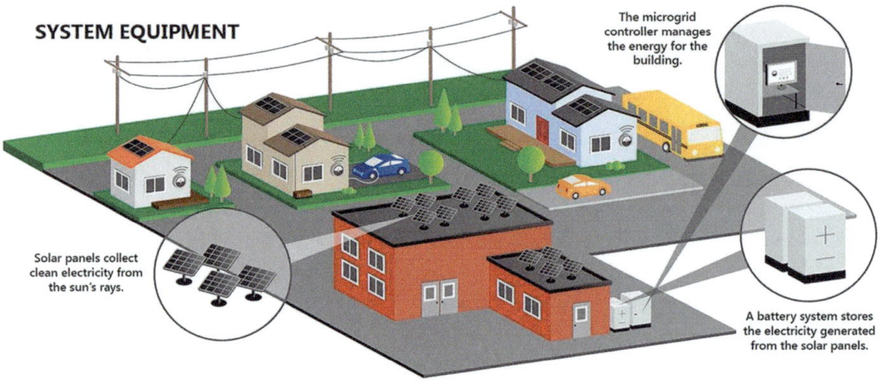

Fig. 3 The Microgrid controller manages energy in the building, using solar panels to collect clean electricity and a battery system to store it

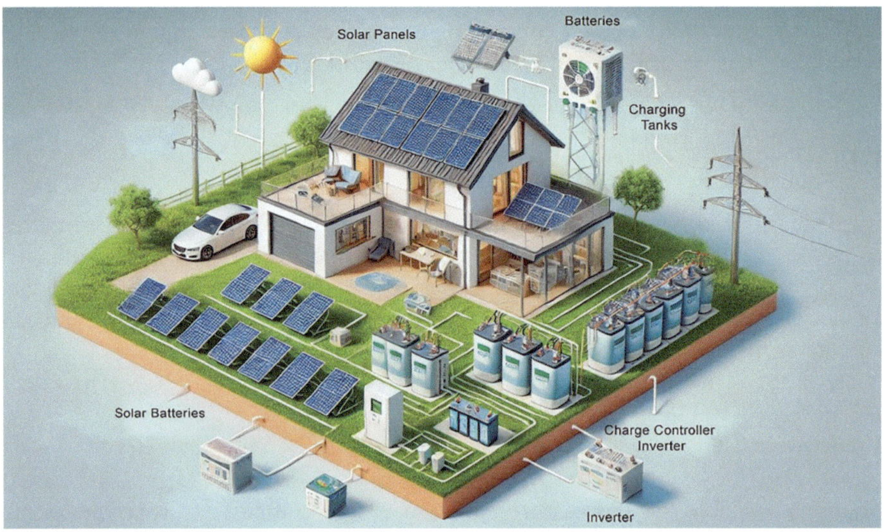

Fig. 4 Embracing sustainable living with solar energy

1.4 Components of Standalone Microgrid System [9–11]

- **PV module**: Converts sunlight into DC electricity using semiconductor materials. Common types include single and polycrystalline silicon and amorphous silicon
- **Battery**: Stores energy for supplying to electrical appliances when there is a demand. Lead-acid batteries are commonly used and designed to gradually discharge and recharge hundreds of times
- **Solar charge controller**: Regulates the voltage and current coming from the PV panels to prevent battery overcharging and prolong battery life
- **Inverter**: Converts DC output of PV panels into clean AC current for AC appliances or feeds back into the grid line. Efficiency reaches around 90% when load demand is greater than 50% of the rated load
- **Load**: Electrical appliances connected to the solar PV system, such as lights, radio, TV, computer, refrigerator, etc.
- **Electric vehicles (EVs)**: Powered by electricity stored in rechargeable batteries, offering lower emissions and reduced operating costs. Types include fully electric (BEVs) and plug-in hybrids (PHEVs)

1.4.1 Configuration

Standalone microgrid systems are classified according to how the system components are connected to other power sources. The system is designed to operate independently of the electric utility grid and supply specific DC and/or AC electrical loads (Figs. 5 and 6).

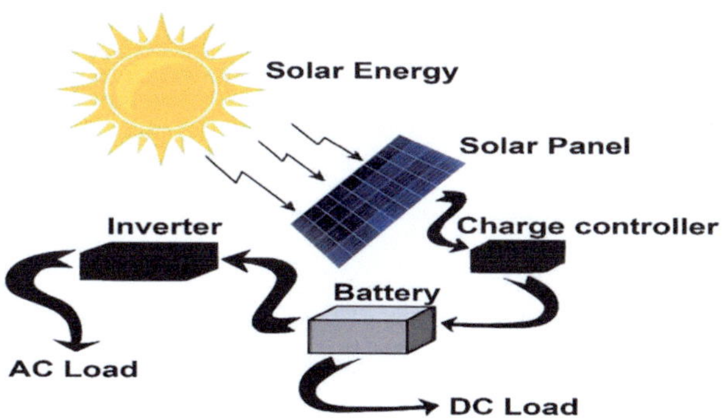

Fig. 5 Standalone solar PV system for residential microgrid system

(a) (b)

Fig. 6 (**a**) Residential rooftop PV system and (**b**) EV charging station

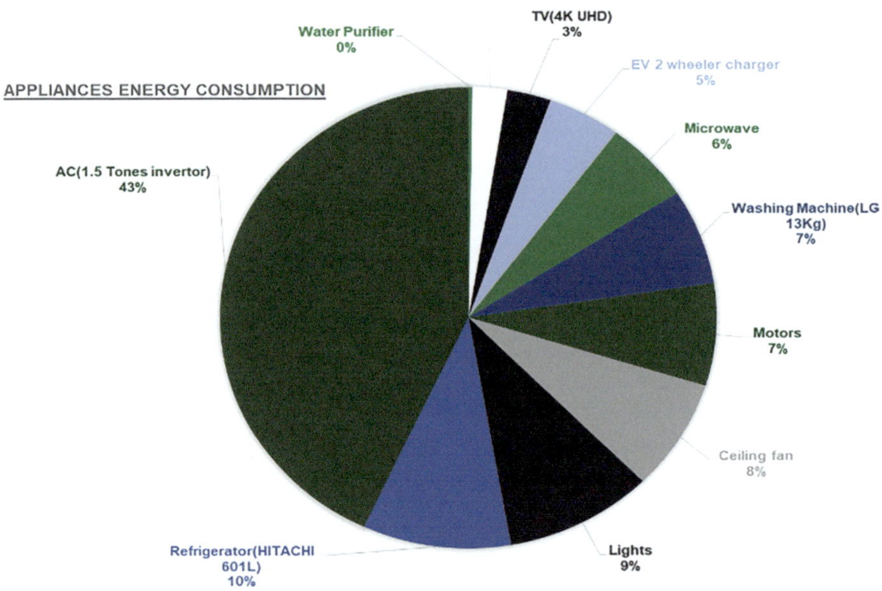

Fig. 7 Energy consumption analysis for case study 1

2 Case Study 1: Energy Consumption Analysis and Solar Generation Summary

The daily calculated power consumption for connected loads is 20,920.5 Wh, while actual consumption is 16,116.66 Wh, with solar generation covering 15,150 Wh/day. Key contributors to consumption include air conditioning, lighting, and refrigerators (Fig. 7 and Table 1).

Table 1 Energy consumption of home appliances with solar PV generation for case study 1

Individual load	Qty.	Volt of each qty.	Amps.	Watt AC	Use h/day	Wh AC
Water purifier	1	230	0.13	50	1	50
Router	1	230	0.13	20	24	480
TV (4K UHD)	1	230	0.3	120	5	600
EV 2 wheeler charger	1	230	8.28	667	1.5	1000.5
Microwave	1	230	3.01	1200	1	1200
Washing Machine (LG 13 kg)	1	230	4.02	1400	1	1400
Motors	1	230	3.77	1500	1	1500
Ceiling Fan	5	230	0.16	65	5	1625
Lights	30	230	0.1	45	1.5	2025
Refrigerator (HITACHI 601L)	1	230	5.02	85	24	2040
AC (1.5 tones inverter)	2	230	3.77	1500	3	9000
Total consumption						**20,920.5**
Actual consumption						**16,116.66**
Total solar gen.						**15,150**

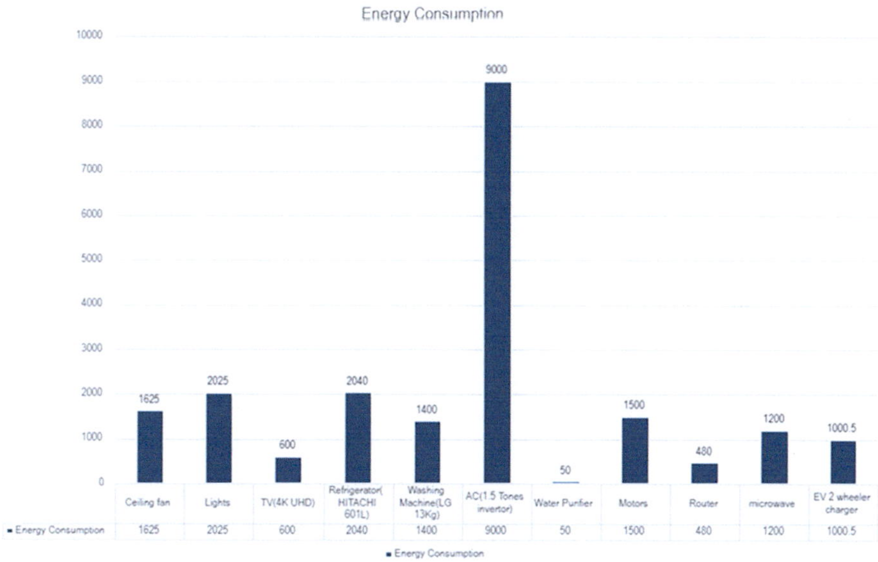

Fig. 8 Home appliances energy consumption for case study 1

Air conditioners dominate energy use at 43%, followed by refrigerators (10%), lights (9%), and ceiling fans (8%), with cooling and lighting being key contributors (Fig. 8).

2.1 *Solar Rooftop PV System Data Analytics*

Air conditioners, refrigerators, and washing machines are major energy consumers. Optimizing usage and choosing energy-efficient models can significantly reduce energy consumption and lower electricity bills (Fig. 9).

Export import in both July–August and September–October, electricity import and export were closely balanced, with minor differences of 18 and 76 kWh, respectively, highlighting efficient grid utilization (Fig. 10).

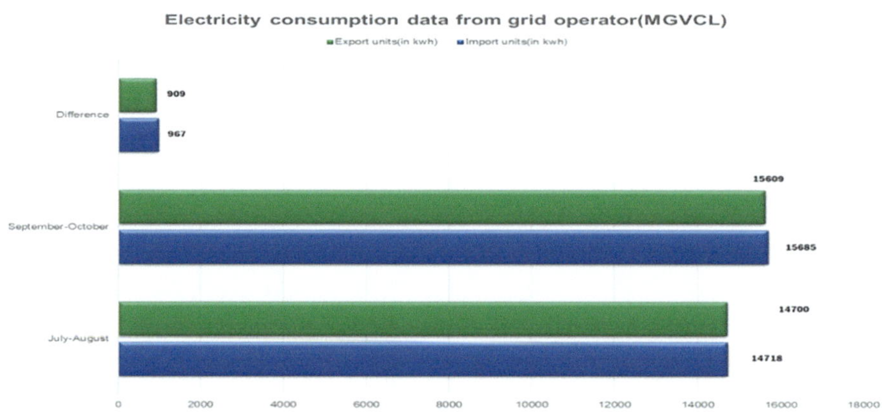

Fig. 9 Import and export of grid and solar generation analysis

Fig. 10 Monthly solar generation outcomes

3 Case Study 2: Energy Consumption Analysis and Solar Generation Summary

The data table displays household appliance power consumption, including details like voltage, current, wattage, usage hours, and total daily consumption. It also compares total calculated consumption with actual consumption from January and February 2024 bills and provides daily solar generation data for the same period (Fig. 11 and Table 2).

The chart compares electricity consumption data from MGVCL, showing higher import units (in kWh) than export units for both July–August and

September–October. In both periods, imports significantly exceeded exports, indicating a consistent trend (Fig. 12).

The bar chart shows household appliance energy consumption in watt-hours (Wh), highlighting the air conditioner (8640 Wh) and refrigerator (3360 Wh) as the highest consumers. Optimizing the use of these high-consumption appliances could reduce overall energy usage (Fig. 13).

The pie chart illustrates the distribution of electricity consumption across household appliances, with air conditioners accounting for 50%, refrigerators for 20%, and smaller contributions from appliances like ceiling fans, chimneys, and televisions.

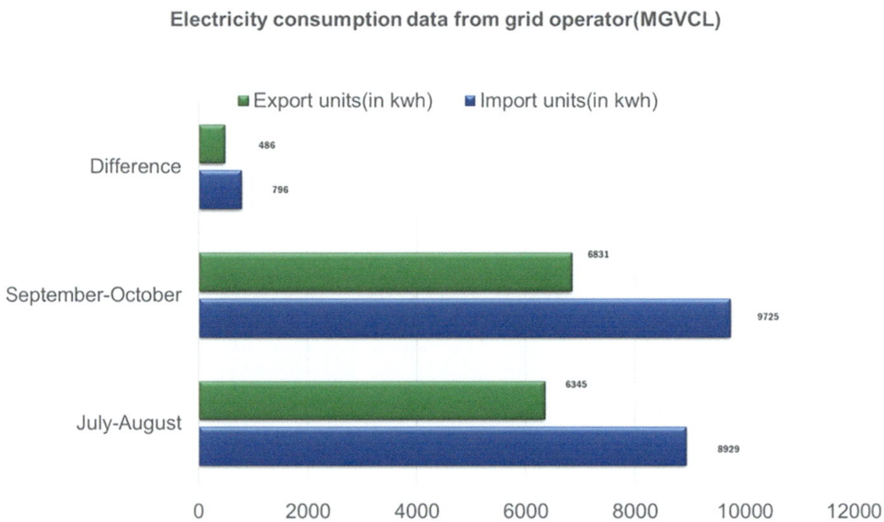

Fig. 11 Import and export of grid and solar generation analysis

Table 2 Energy consumption of home appliances with solar PV generation for case study 2

Individual load	Qty.	Volt of each qty.	Amps.	Watt AC	Use h/day	Wh AC
Microwave	1	230	6.086957	1400	0.3	420
Refrigerator	1	230	0.608696	140	24	3360
Chimney	1	230	1.73913	400	1.5	600
Air Conditioner	2	230	6.26087	1440	3	8640
Television	2	230	0.782609	180	2.5	900
Washing Machine	1	230	1.434783	330	1	330
Mixer Grinder	1	230	2.608696	600	0.2	120
Hair dryer	1	230	4.347826	1000	0.5	500
Ceiling Fan	5	230	0.282609	65	5	1625
Light	1	230	0.034783	8	2	160
Trimmer	10	230	0.021739	5	0.5	2.5
Iron	2	230	4.347826	1000	0.5	500
Total consumption						**17,157.5**
Actual consumption						**13,266.66**
Total solar gen.						**8100**

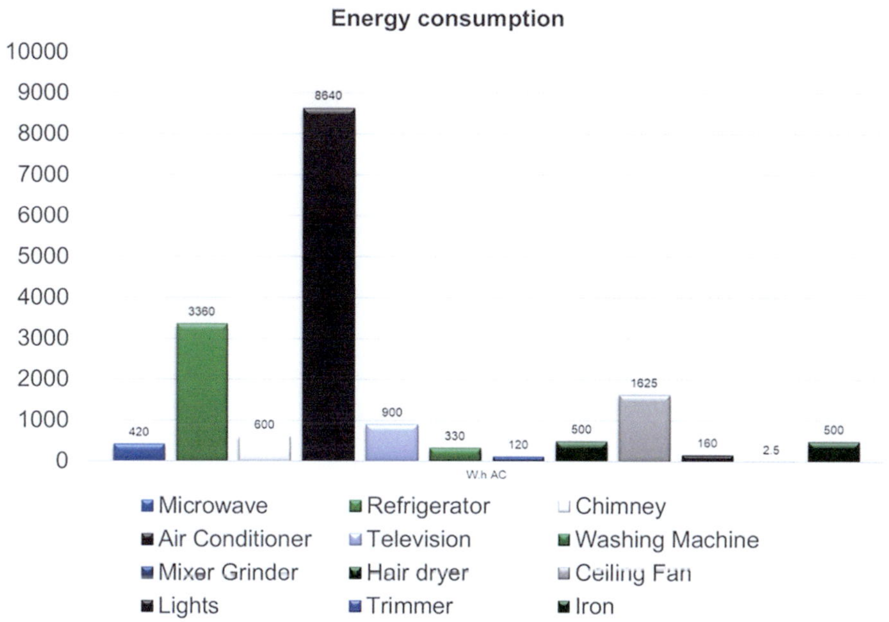

Fig. 12 Energy consumption analysis for case study 2

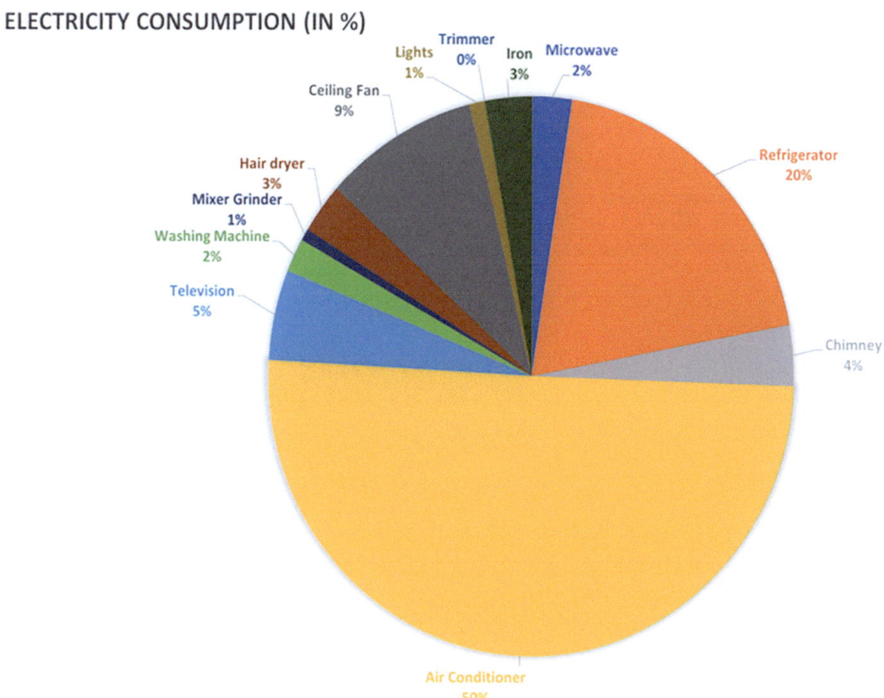

Fig. 13 Electricity consumption of appliances for case study 2

3.1 Design and Sizing of Battery with Fuzzy Logic Controller

A model of a battery with battery temperature, inverter temperature, capacity, and voltage as its input parameters and discharge time, efficiency, and load as its output [12] (Figs. 14, 15, 16, 17, 18, and 19).

3.2 Design and Sizing of Diesel Generator with Fuzzy Logic Controller

A model of a diesel generator with speed, ambient temperature, and fuel consumption as its input parameters and generated power, efficiency, and load as its output [13] (Figs. 20, 21, 22, and 23).

3.3 Design and Sizing of Solar PV with Fuzzy Logic Controller

See Figs. 24, 25, 26, 27, 28, and 29.

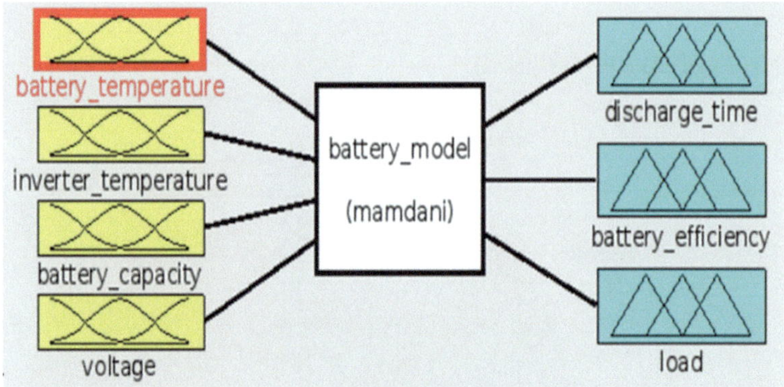

Fig. 14 Battery design and modeling with fuzzy logic system

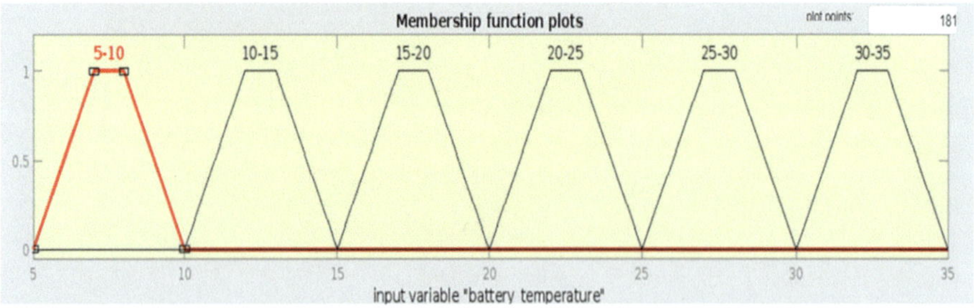

Fig. 15 Membership functions of battery temperature

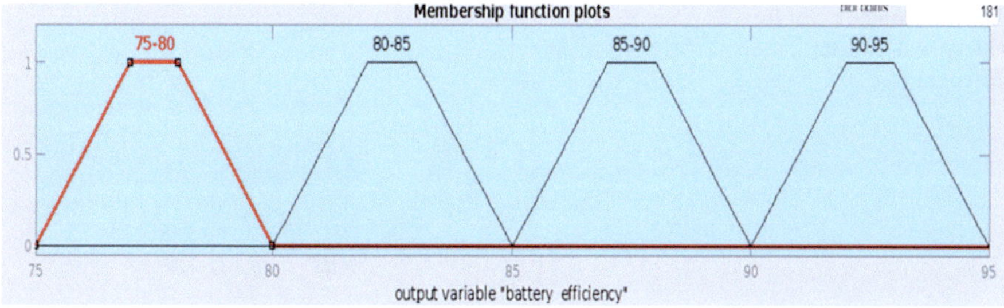

Fig. 16 Membership functions of battery efficiency

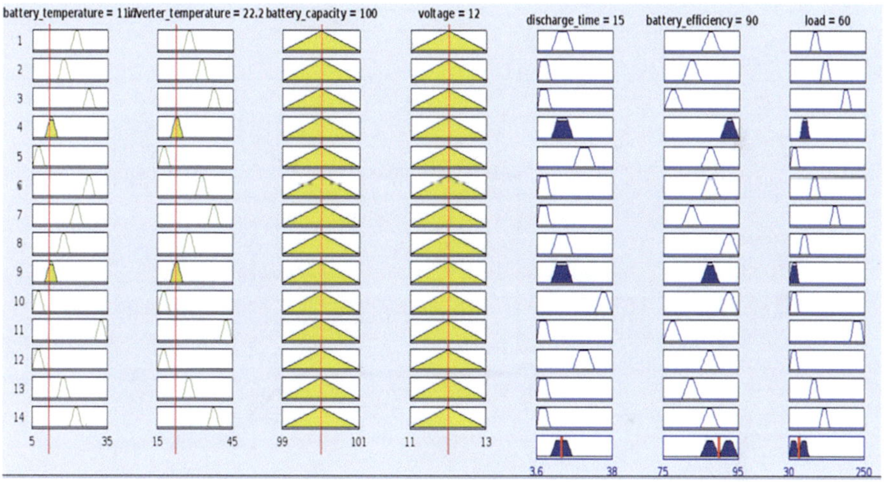

Fig. 17 Rule viewer execution for battery design and modeling

Fig. 18 Surface plot for battery and inverter temperature

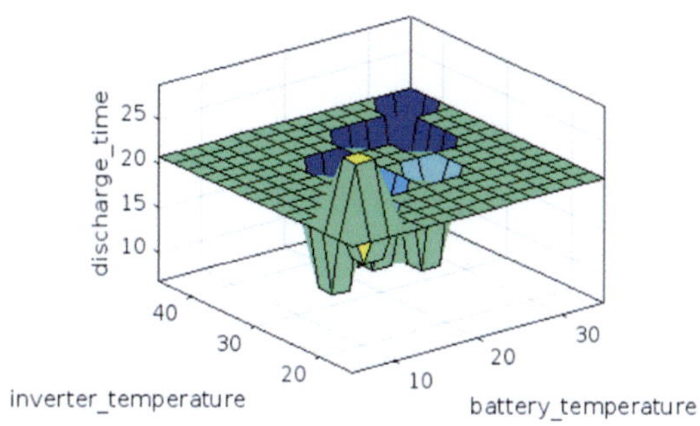

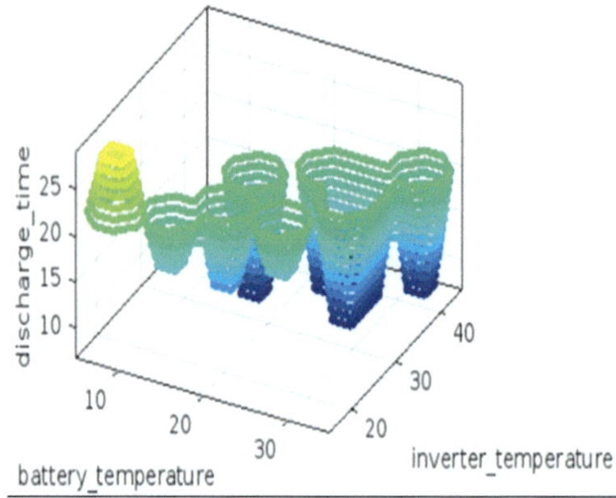

Fig. 19 Contour plot for battery and inverter temperature

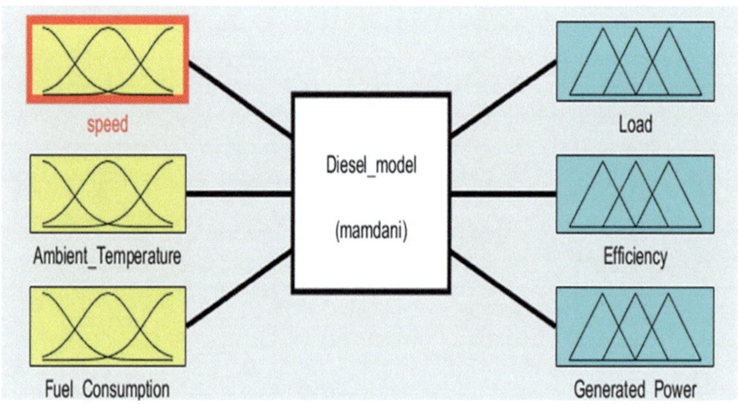

Fig. 20 Diesel generator design and modeling with fuzzy logic system

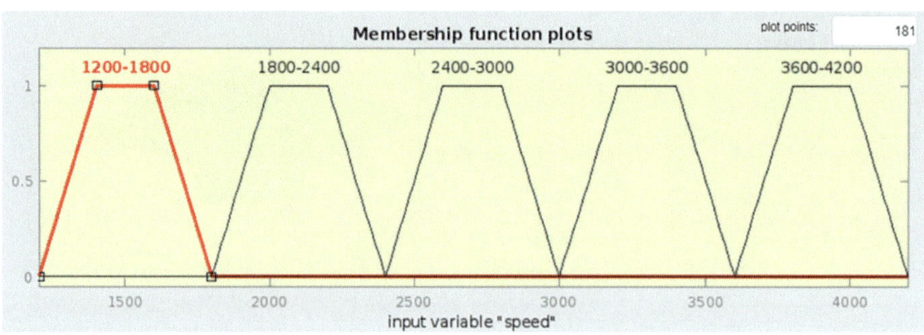

Fig. 21 Membership functions of speed

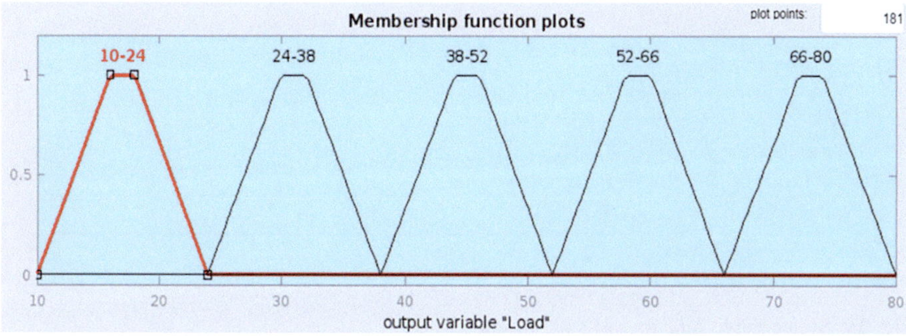

Fig. 22 Membership functions of load

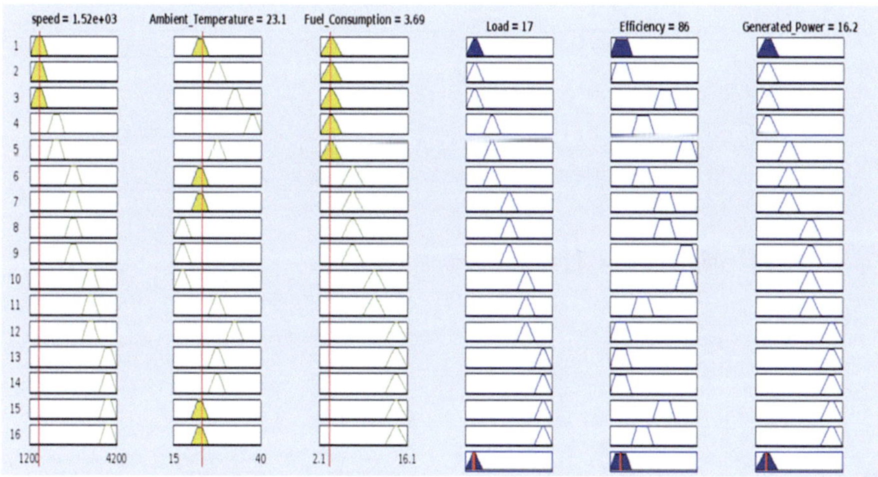

Fig. 23 Rule viewer execution for diesel generator design and modeling

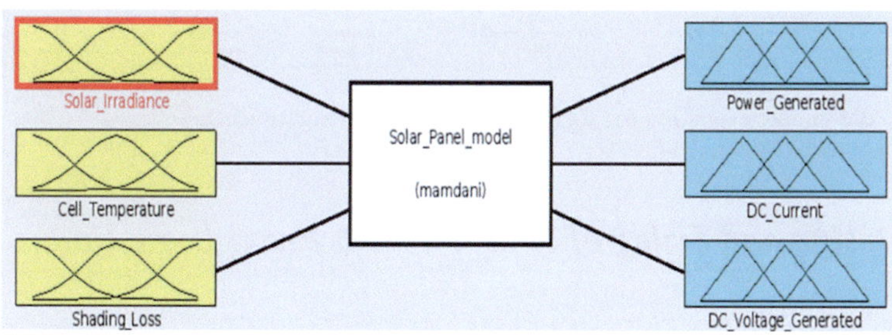

Fig. 24 Solar PV system design and modeling with fuzzy logic system

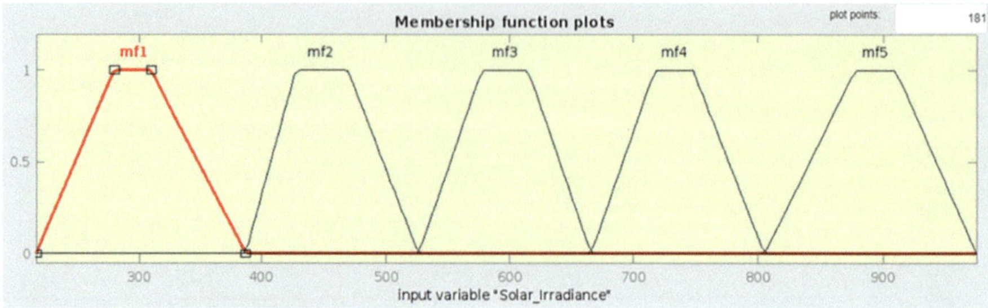

Fig. 25 Membership functions of solar irradiance

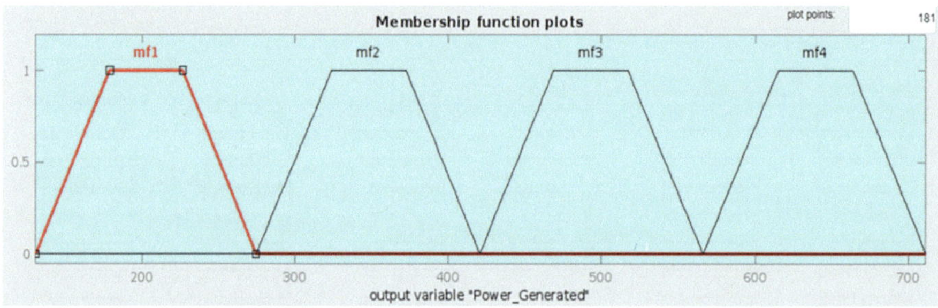

Fig. 26 Membership functions of power generated

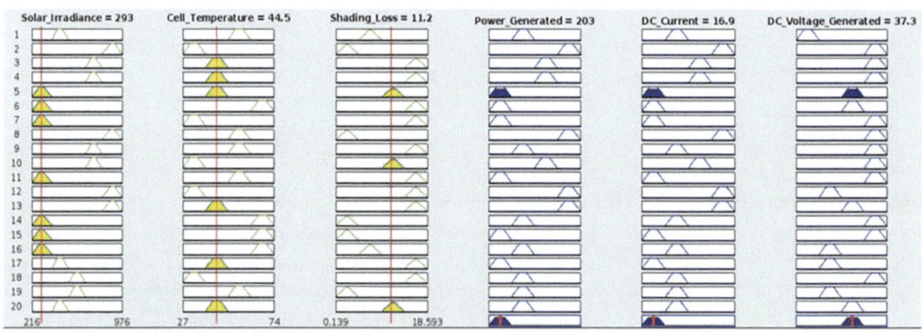

Fig. 27 Rule viewer execution for solar PV system design and modeling

3.4 Design and Sizing of Solar PV with Fuzzy Logic Controller

See Figs. 30, 31, 32, 33, 34, and 35.

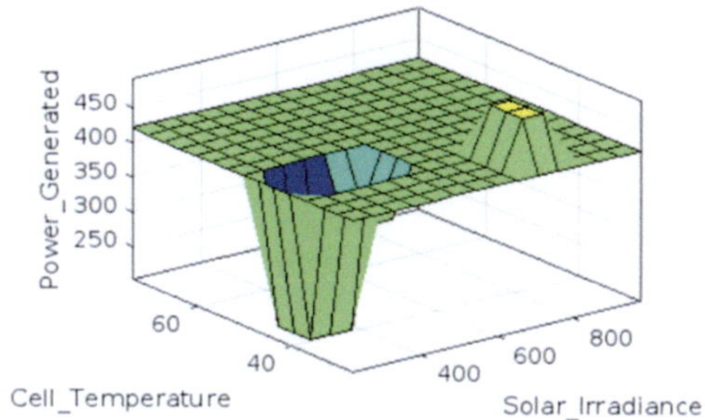

Fig. 28 Surface plot for cell temperature and solar irradiance

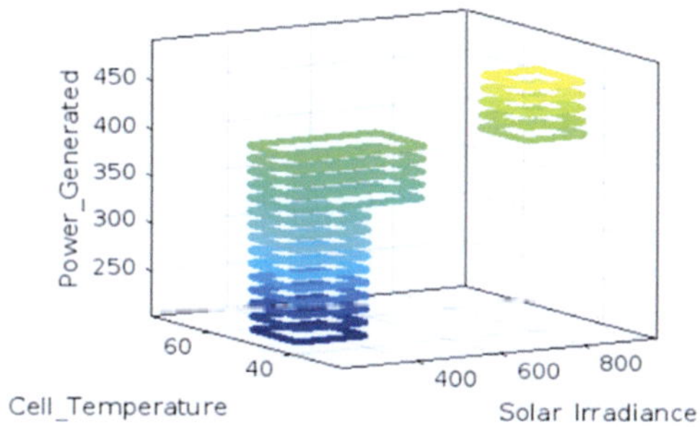

Fig. 29 Contour plot for cell temperature and solar irradiance

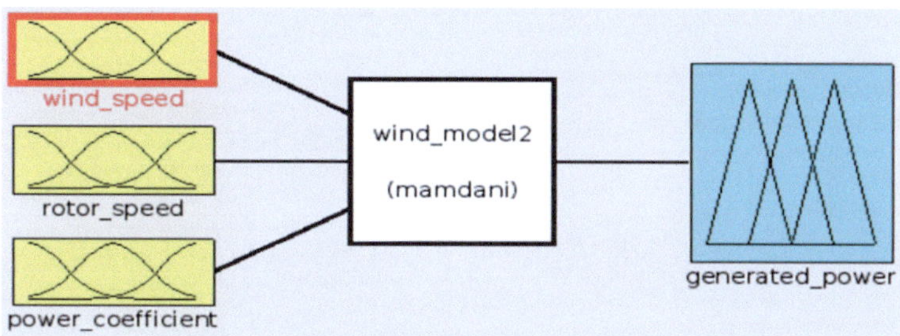

Fig. 30 Wind system design and modeling with fuzzy logic system

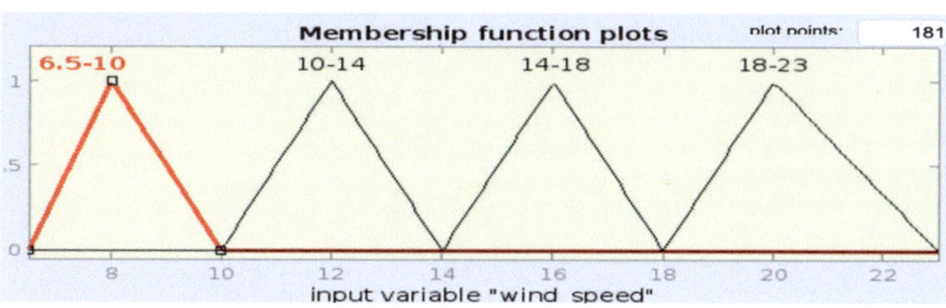

Fig. 31 Membership functions of wind speed

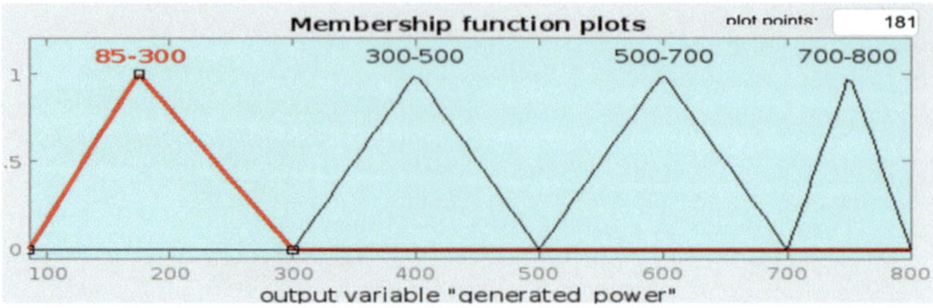

Fig. 32 Membership functions of power generation

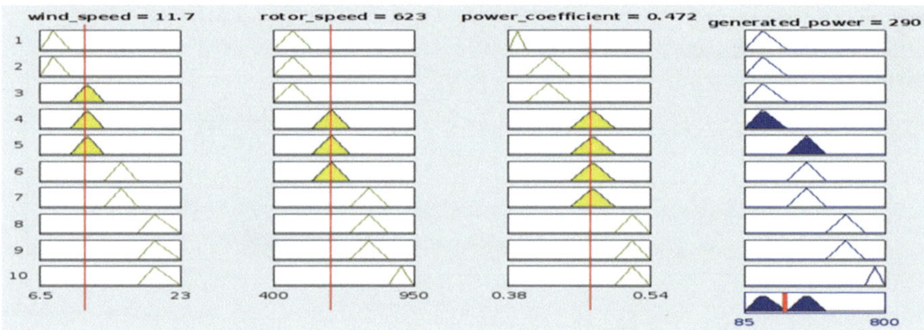

Fig. 33 Rule viewer execution for wind system design and modeling

Fig. 34 Surface plot for rotor speed and wind speed

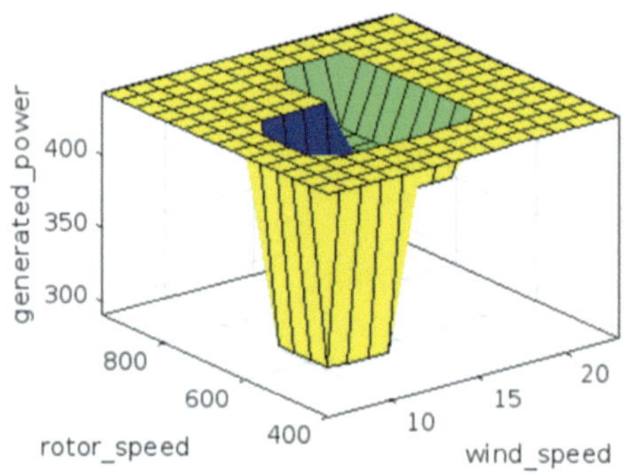

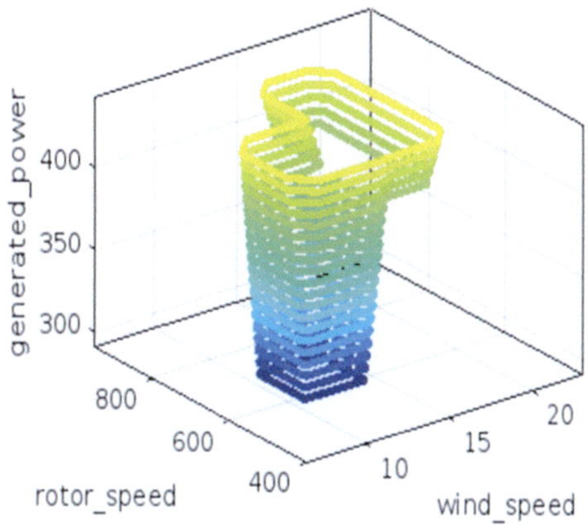

Fig. 35 Contour plot for rotor speed and wind speed

3.5 Design and Simulation of a Complete Microgrid with Fuzzy Logic Controller

This figure represents the key components of a microgrid system, illustrating the microgrid system integrating multiple energy sources—batteries, diesel generators, solar PV, and wind turbines. It monitors key parameters like temperature, speed, and efficiency to calculate power output from each source [14, 15] (Fig. 36).

3.6 Subsystems of the Microgrid

See Fig. 37.

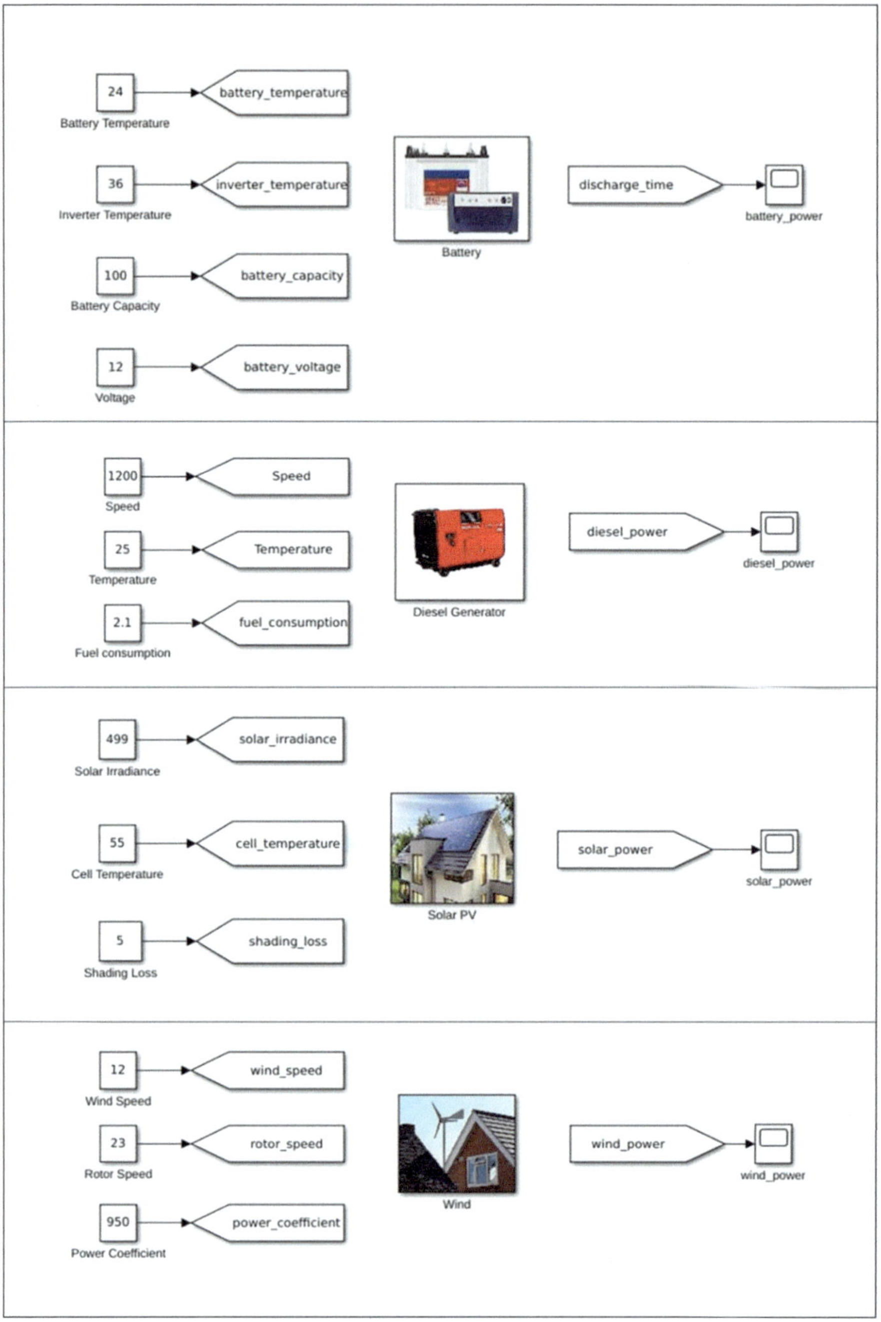

Fig. 36 Residential microgrid system with fuzzy logic system

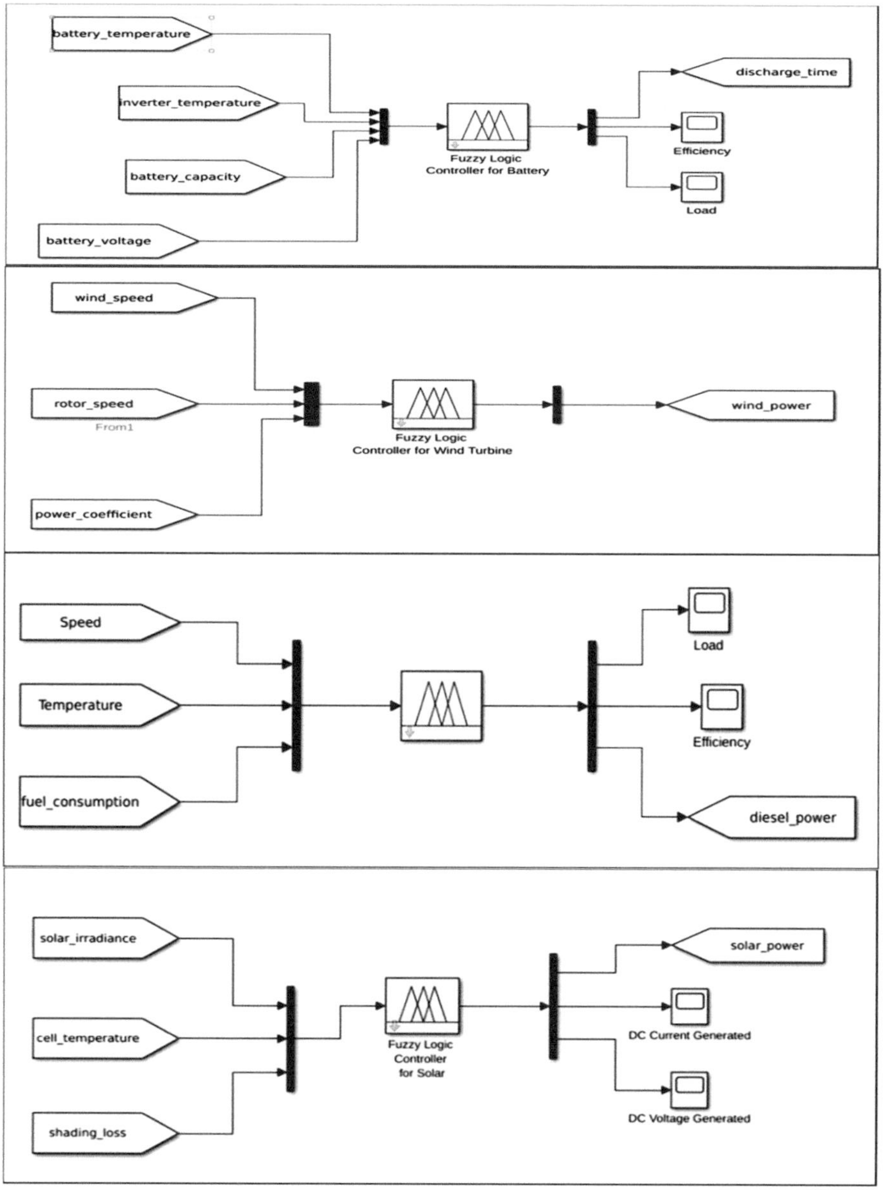

Fig. 37 Subsystem of microgrid system with fuzzy systems

4 Conclusion

This research work introduces a novel microgrid simulation model under extreme weather conditions like flooding situation in the Vadodara district of Gujarat region in India. Under the severe floods, the suggested solution will address the residential microgrid system solution. Fuzzy logic-based design and simulation studies, such as solar rooftop systems, battery charging and storage systems, inverter applications, and grid connection systems, are included in the proposed method. Additionally, a fuzzy logic controller powered by artificial intelligence will be used to forecast and simulate different technical factors for the microgrid system under flooding scenarios. Proposed system plays an integral role in disaster management like extreme flood condition and works efficiently under severe conditions to provide power supply in islanding situations of modern power systems. Further, this work introduces novel and proficient complete design and simulation of microgrid system renewable-based energy resources like solar PV and wind energy systems for sustainable solution in modern power systems network.

References

1. Teo, T. T., Logenthiran, T., Woo, W. L., & Abidi, K. (2016). Fuzzy logic control of energy storage system in microgrid operation. In *2016 IEEE Innovative Smart Grid Technologies-Asia (ISGT-Asia), Melbourne, VIC, Australia* (pp. 65–70). https://doi.org/10.1109/ISGT-Asia.2016.7796362
2. Ali, M., Hossain, M. I., & Shafiullah, M. (2022). Fuzzy logic for energy management in hybrid energy storage systems integrated DC microgrid. In *2022 International Conference on Power Energy Systems and Applications (ICoPESA), Singapore* (pp. 424–429). https://doi.org/10.1109/ICoPESA54515.2022.9754406
3. Horrillo-Quintero, P., et al. (2024). Dynamic fuzzy logic energy management system for a multi-energy microgrid. *IEEE Access, 12*, 93221–93234. https://doi.org/10.1109/ACCESS.2024.3422009
4. Arcos-Aviles, D., et al. (2021). An energy management system design using fuzzy logic control: Smoothing the grid power profile of a residential electro-thermal microgrid. *IEEE Access, 9*, 25172–25188. https://doi.org/10.1109/ACCESS.2021.3056454
5. Lijun, M., Xiaming, Y., Ruyi, Q., Fan, B., & Hongqun, Z. (2023). Energy management control strategy of hybrid energy storage system based on fuzzy logic. In *2023 2nd Asia Power and Electrical Technology Conference (APET), Shanghai, China* (pp. 839–845). https://doi.org/10.1109/APET59977.2023.10489725
6. Soni, R., & Mehta, B. (2023). Condition-based monitoring of power transformer with graphical analysis of incipient faults using fuzzy inference expert system. In B. Chakraborty, A. Biswas, & A. Chakrabarti (Eds.), *Advances in data science and computing technologies (ADSC 2022)* (Lecture notes in electrical engineering) (Vol. 1056). Springer. https://doi.org/10.1007/978-981-99-3656-4_34
7. Himri, Y., Himri, S., & Stambouli, A. B. (2010). Wind speed data analysis used in installation of wind energy conversion systems in Algeria. In *IEEE PES T&D 2010, New Orleans, LA, USA* (pp. 1–5). https://doi.org/10.1109/TDC.2010.5484699
8. Peng, C.-Y., Wu, S.-D., Lin, Y.-C., & Chuang, H.-Y. (2021). Research on information management analysis of battery efficiency data. In *2021 IEEE 3rd International Conference on Civil*

Aviation Safety and Information Technology (ICCASIT), Changsha, China (pp. 1000–1004). https://doi.org/10.1109/ICCASIT53235.2021.9633570

9. Zhang, H., Chen, M., Qin, B., & Zhang, M.-W. (2008). Research on the system of data acquisition and evaluation in diesel locomotive combustion system. In *2008 3rd IEEE Conference on Industrial Electronics and Applications, Singapore* (pp. 1195–1199). https://doi.org/10.1109/ICIEA.2008.4582707
10. Soni, R., & Mehta, B. (2023). Diagnosis and prognosis of incipient faults and insulation status for asset management of power transformer using fuzzy logic controller & fuzzy clustering means. *Electric Power Systems Research, 220*, 109256. ISSN: 0378-7796.
11. Soni, R., & Mehta, B. (2024). Condition based assessment and diagnostics of transformer in smart grid network using adaptive neuro fuzzy inference system framework. In P. H. V. S. Talpa Sai, S. Potnuru, M. Avcar, & V. Ranjan Kar (Eds.), *Intelligent Manufacturing and Energy Sustainability (ICIMES 2023)* (Smart Innovation, Systems and Technologies) (Vol. 372). Springer. https://doi.org/10.1007/978-981-99-6774-2_13
12. Mourya, N. K., & Koul, B. (2023). Fuzzy logic based PV-battery system for a standalone microgrid. In *2023 4th International Conference for Emerging Technology (INCET), Belgaum, India* (pp. 1–6). https://doi.org/10.1109/INCET57972.2023.10170223
13. Kafle, L., & Ni, Z. (2016). Fuzzy logic adjustment for power sharing in wind and PV-based isolated microgrid. In *2016 IEEE Symposium Series on Computational Intelligence (SSCI), Athens, Greece* (pp. 1–7). https://doi.org/10.1109/SSCI.2016.7849844
14. Prabhakar, S., & Panigrahi, B. K. (2024). Fuzzy logic controller for optimal power sharing in multi BESS-PV based islanded AC microgrid. In *2024 IEEE 4th International Conference on Sustainable Energy and Future Electric Transportation (SEFET), Hyderabad, India* (pp. 1–6). https://doi.org/10.1109/SEFET61574.2024.10718079
15. Rastegar, F. (2022). A novel control strategy based on fuzzy logic in islanded microgrid. In *2022 9th Iranian Joint Congress on Fuzzy and Intelligent Systems (CFIS), Bam, Islamic Republic of Iran* (pp. 1–6). https://doi.org/10.1109/CFIS54774.2022.9756427

Efficacity of Artificial Intelligence in Optimizing the Production of Renewable Energy

Shriram S. Rangarajan, Geetha Manoharan, E. Randolph Collins, and Tomonobu Senjyu

1 Introduction

Despite the overwhelming pace of change in our world, renewable energy sources like wind and solar are offering hope for a brighter tomorrow. At the same time that people are becoming more aware of the catastrophic effects of climate change, there is a worldwide movement away from conventional fossil fuels and toward alternative energy sources. International Renewable Energy Agency data shows a 7.4% increase in global renewable energy capacity in 2020. Renewable energy sources have once again proven their growing significance. Renewable energy's capacity to lessen overall carbon footprint and emissions of greenhouse gases is increasing its environmental and economic importance. There has been a recent surge in funding for renewable energy sources as more and more governments and organizations around the world recognize their potential. These initiatives are aimed at creating a more sustainable [1–3] future. The shift to an energy system that relies more on

S. S. Rangarajan
Landrotics Solution Private Limited, Bengaluru, Karnataka, India

Clemson University, Clemson, SC, USA

G. Manoharan (✉)
SR University, Warangal, Telangana, India

E. R. Collins
Clemson University, Clemson, SC, USA

Western Carolina University, Cullowhee, NC, USA
e-mail: rcollins@wcu.edu

T. Senjyu
University of the Rykyus, Nishihara, Japan
e-mail: b985542@tec.u-ryukyu.ac.jp

P. Raj et al. (eds.), *Artificial Intelligence (AI) for IT Energy Efficiency and Green AI for Environment Sustainability*,
https://doi.org/10.1007/978-3-031-89420-6_19

renewable sources is not without its challenges, though. Solar and wind power, two examples of renewable energy, are intrinsically intermittent. The sun's and the wind's irregular patterns cause the quantity of energy produced to fluctuate. It is challenging to sustain a steady equilibrium between the supply and demand for energy due to the unpredictability caused by intermittency [4–6]. Power systems originally built for dependable fossil fuel sources provide technological and logistical hurdles when trying to incorporate large quantities of renewable energy into them. There is an overabundance of production during peak hours and a shortage during lulls due to the intrinsic unpredictability of renewable energy generation. Because of this, grid instability and wasteful energy usage are caused. Artificial intelligence (AI) is a revolutionary technology that can't be overlooked in the quest to maximize the use of renewable energy sources. Artificial intelligence (AI) can solve problems that seemed insurmountable before, thanks to its superior data processing and predictive analytics [7, 8]. Potentially reaching $75.82 billion by 2030, the renewable energy market is a reflection of the industry's immense trust in AI [9, 10]. Through the analysis of extensive data sets, including weather patterns and changes in energy usage,

AI can produce highly precise predictions of future energy generation. Systematic adjustments to energy distribution can be made in real time with the help of AI, allowing for optimal grid stability and efficiency [11]. Sustainable energy generation entered a new era with the advent of artificial intelligence. The connection between constant energy consumption and intermittent renewable power sources allowed this to happen. Artificial intelligence (AI) is changing the rules for renewable power production and consumption, which is a huge deal in today's competitive energy market. A growing number of countries are turning to renewable energy sources, which is increasing the importance of AI. Experts project that the renewable energy industry's artificial intelligence market will reach $75.82 billion by 2030, proving this argument. There are a lot of ways in which artificial intelligence is changing the renewable energy industry. Collecting relevant data from complicated and somewhat big databases is one of AI's primary goals. Since renewable energy sources are inherently unpredictable, these results are of the utmost importance. Renewable energy sources produce significantly less greenhouse gas emissions compared to fossil fuels. Solar, wind, and hydroelectric power all fall under this category. Renewable energy is better than using finite fossil fuels since it is safer and more sustainable [12–14]. A reliable and consistent energy source, renewable resources like water, wind, and sunshine are not subject to price fluctuations or geopolitical instability like fossil fuels are. One effective strategy for combating energy poverty is the use of renewable energy sources, which can provide power to underserved and remote regions. Improving living conditions and stimulating economic growth are both possible outcomes of off-grid solar system installations in rural regions, which can supply reliable and affordable electricity [15–17]. Renewable energy is the key to a prosperous and environmentally friendly future. Several advantages accrue from shifting to renewable energy sources and creating an economy based on clean energy, such as better public health, faster economic growth, greater energy independence, and success in the battle against climate change.

International Energy Agency (IEA) official assessments show that renewable energy sources are becoming more popular and that fossil fuels are starting to lose demand as a source of producing power [18]. Improving the efficiency of energy conversion, especially in regard to RE generation, is a primary goal of ongoing research into RE technology. More sustainable energy technology is required since traditional electrical generating resources, such as fossil fuels, are insufficient to meet the present worldwide demand. This is a problem for both developed and developing countries. The use of energy sources derived from fossil fuels leads to a variety of negative environmental impacts, including climate change and global warming. Over the last several decades, the amount of greenhouse gas emissions caused by power plants has been steadily climbing. The current environmental calamity has prompted a change in power generation toward renewable energy (RE) sources. Renewable energy sources encompass hydropower, solar power, wind power, geothermal power, biomass, and hydrogen [19–22].

Using AI's prediction, data analysis, and optimization capabilities, renewable energy sources can be harnessed to their full potential. A low-carbon energy future cannot be realized without artificial intelligence (AI), which may assist with supply management by giving precise energy estimates and detecting problems in advance to avoid breakdowns. Artificial intelligence will be instrumental in the development of carbon-neutral energy systems soon. Artificial intelligence (AI) powered technologies are revolutionizing energy production and customer relationships. At the same time that traditional fossil fuels are losing ground in capabilities, renewable energy sources are beginning to catch up. Beginning in 2008 and continuing through 2035, it is expected that the generation of electricity from hydroelectric power will surpass that of other renewable energy sources. On the other hand, projections that have been made show that the installed solar power generation will have the most rapid growth within the time range that has been projected. Artificial intelligence is finding more and more applications in renewable energy systems (RES) as a result of developments in processing power, tools, and data collection that are occurring continuously. According to the findings of the research, the manner in which energy businesses currently handle design, control, and maintenance is plagued by a few minor issues. When it comes to doing these jobs, the most advanced application of artificial intelligence has led to improvements in accuracy and precision. Over the course of the past few decades, artificial intelligence (AI) has emerged as a highly sought-after study area due to its capacity to automate systems, thereby enhancing both quality and productivity. They are taught to learn, think, and make decisions [23–26] in the same manner as humans with the use of a complicated set of instructions. The use of artificial intelligence to digitalize energy systems also has the potential to improve power system networks in a variety of important ways, including making them more dependable, responsive, and continuous. There are many different areas of power systems that are rapidly utilizing artificial intelligence. Some of these areas include design, prediction, control, optimization, maintenance, and security. Producing electricity in a way that is both environmentally beneficial and financially viable relies on renewable energy, or RE. One strong argument in favor of transitioning to renewable energy is the possible impact on lowering global warming temperatures [27–29].

2 Artificial Intelligence in the Renewable Energy Sector

Solar, wind, hydropower, geothermal, and biomass power plants were already in operation or were going to be before artificial intelligence was brought into the energy sector. The original goal of these technological advancements [30] was to reduce expenses, improve efficiency, and limit their effect on the environment. While artificial intelligence has created many new possibilities, there are also many challenges that must be overcome. The inability to generate up-to-date insights and predictions was a big flaw in traditional methods of data analysis and decision-making [31–35]. It is more difficult to integrate AI into renewable energy systems because of the high expenses of infrastructure, data collection, and algorithm training. This has a disproportionate impact on less developed nations. Concerns about privacy and security were among many who feared AI systems might reinforce existing inequalities in energy distribution. The renewable energy business faces several challenges, but artificial intelligence (AI) offers hope for increasing operational efficiency and simplifying processes, which might speed up the transition to a more sustainable [36] energy future. Modern technology, energy management systems, and eco-friendly remedies are the result of researchers' labors of love. The goal of this research is to assess the efficacy of renewable energy and energy efficiency in providing power to underserved metropolitan areas. Using wavelet methods, a mathematical branch, multiple scholars examined the relationships between these parameters in various contexts and time periods. Their research shows that more environmental laws, more funding for renewable energy R&D, and the promotion of environmentally friendly technology are the three most effective strategies to rescue our planet. There is a robust relationship between GDP growth, technical advancement, and emissions of carbon dioxide [37–40].

An impending paradigm shift in the energy sector may be precipitated by recent developments in digital technology. The new digitization paradigm is being propelled by artificial intelligence (AI), which is acting as a change agent. Intelligent and efficient computer programs: The integration of renewable sources into the power system and management of energy supply and demand will be accomplished using autonomous decision-making and other approaches. In order to accomplish this, AI is essential [41–43]. It is possible to gather enormous volumes of data by linking a wide variety of devices and sensors to the power grid as a whole. This data, when paired with AI, could give grid operators new insights that improve their control skills. Companies in a position of power are granted the authority to adapt their production in response to customer demands. Utilizing AI could enhance various facets of distributed energy management and microgrid integration. Attempts to keep the electrical flow inside the grid constant are made more complicated by the addition of community-scale renewable energy-generating units to the main grid. This AI-powered control system could be the solution that cities have been waiting for to ultimately alleviate their congestion and quality issues. The renewable energy industry may find a reliable and environmentally friendly solution by utilizing AI and intelligent energy storage (IES). The influence of AI on energy corporations across many sectors is anticipated to be more than initially anticipated. "Effect of Processes" shows the steps taken to achieve a goal, while "Effect of Offerings" (accept or reject as requested) gives customers more opportunities to observe how AI has affected different companies. Most businesses believe that the energy information technology industry, customer-centric programs, manufacturing [44], and supply chain management will be the hardest hit. An AI control system can estimate the electrical output of a wind turbine by including weather data such as wind speed and direction. Predicting a solar panel's power output in relation to environmental variables like temperature and sunshine intensity using a control system based on machine learning is also within the realm of possibility. The wind-photovoltaic system can be adjusted when these forecasts are obtained. Machine learning-based data-driven models are commonly used for renewable energy (RE) forecasts. This collection of models includes a large number of k-nearest neighbors (k-NNs), gradient boosting (GB), and random forest (RF) models. It has accomplished remarkable results in power consumption and renewable energy (RE) generation estimates over different time periods by utilizing deep learning models, such as LSTM, GRUs, and DNNs. The efficiency with which LSTM with autoencoders (AE) operates is yet another perk [45–47].

3 Applications of Artificial Intelligence in the Renewable Energy Sector

There is an abundance of diverse and deep AI applications in the renewable energy business. Some of the few key areas where AI is already making a difference are as follows:

3.1 Making an Appropriate Site Selection and Resource Evaluation

A huge undertaking is determining the optimal location for renewable energy projects. The massive datasets acquired from sensors and satellites are deciphered by AI, saving the day. The optimal locations for solar farms or wind parks can be found by using artificial intelligence algorithms to assess data such as sun irradiance, wind patterns, and topographical restrictions. This optimizes energy production while also ensuring that investments are cost-effective [48–51].

3.2 Energy Forecasting

Forecasting of energy must be done with great precision because renewable energy sources are intermittent. Artificial intelligence models using machine learning [52] algorithms are able to generate highly accurate predictions on future energy output by analyzing power generation records, weather records, and other pertinent data. Such predictions are crucial for grid managers to keep the grid stable and even out energy distribution.

3.3 Grid Optimization and Management

Large-scale integration of renewable energy sources into existing power systems is challenging. Smart grid management is made easier with the use of artificial intelligence (AI), which can analyze data in real time. By continuously monitoring data from diverse sources, such as wind turbines and consumer demand patterns, algorithms driven by artificial intelligence can enhance grid operations in real time.

3.4 Maintenance and Operation

The efficiency and longevity of renewable energy assets are crucial. Predictive maintenance solutions [52] enabled by AI continuously monitor these assets by analyzing data from sensors and performance metrics. These technologies minimize downtime and enhance asset longevity by identifying any defects or anomalies and providing preventive maintenance.

4 AI-Based Methodologies for Renewable Energy

4.1 Forecasting Renewable Energy Generation Using Deep Learning

To generate RE predictions simultaneously, AB-Autoencoder Net's (AE) [53–55] and Bidirectional Long Short-Term Memory's (BiLSTM) [56] work together. The model makes predictions about RE generation and attribute extraction using deep preprocessing and data collected from many RE sources. Two publicly available datasets, the wind dataset and the solar dataset, were used to test and evaluate the suggested technique. The model is built up of three main stages: the first is gathering data from different renewable energy sources; the second is preprocessing the data to make it more consistent and clean; and the third is using AE and BiLSTM to extract features and make predictions. In terms of MAE, MSE, and RMSE, the AB-Net algorithm fared better than other models, such as CNN-BiLSTM, BiLSTM, and an encoder-decoder (ED). Efficient energy management is achieved by this hybrid network's ease of renewable energy installation integration, trade, and operation.

4.2 Big Data and Machine Learning for Renewable Energy in Smart Grid

Data storage, retrieval, and calculations are the functions of the more fundamental top layer. Concerns regarding data privacy should mostly pertain to the intermediate layer when it comes to managing, exchanging, and integrating data across many domains and applications. Data fusion technologies built into the data mining platform make it possible to do deep and surface-level preprocessing on data. An analysis of big data [57] can be used to predict the reliability of smart grids and renewable power stations. When applied to a dataset consisting of 60,000 cases, our framework—which was constructed using many machine learning [58] techniques—generates a very precise stability forecast. Decision trees, deep learning [59, 60] classic neural networks, and random forest classifiers were the three machine learning (ML) models employed. To determine if the smart grid system is stable, the first model is connected to the categorization model. A fourth ML model has used penalized linear regression to find the stability differential equation's root. It is worth mentioning that convolutional neural networks (CNN) only manage 87% accuracy for classification, while penalized linear regression reaches 96%. Future studies on larger datasets and more broad applications will be made easier by proving that BDA improves system stability and decision-making in smart grid settings.

4.3 AI-ML Approach for Local Wind Power Forecasting

Local consumer and prosumer communities collaborate through either cooperation or competition to collectively achieve objectives such as reducing energy expenses and minimizing greenhouse gas emissions. Regardless of whether it is cooperative or competitive, engaging in cooperation can assist in achieving the objectives. Several scholars have supported the implementation of data analytics modules in the domain of renewable energy communities (RECs) [61]. The ultimate motive is to optimize the exploitation of resources while minimizing the expense of power. In addition, this strategy enhances accuracy by 10%, so offering the community more precise consumption profiles. A wind power forecasting system is constructed utilizing machine learning methodologies to anticipate the amount of wind power that will be generated on the following day. Following the adjustment of the neural network's parameters in the second stage, the loss function is constructed in the third step. This step finalizes the process [62].

4.4 AI-ML-Based Approach for Solar Power Forecasting

Reliable electricity generation by photovoltaic power plants relies on precise predictions of solar radiation. This study assesses the practicality and effectiveness of machine learning (ML) methods for solar radiation prediction. Several algorithms, such as SVR and XGBT, are tested in this procedure.

The first is Voting-Average, or VOA, and the second is Categorical Boosting, or CatBoost. A new approach to finding key components and the associated historical data aims to improve forecast accuracy. The new method will be implemented to achieve this goal. Parameters such as relief, mutual information, random forest, and Pearson's coefficient are considered in the ensemble feature selection step. It achieves this by utilizing the endogenous and exogenous inputs' historical observational values. When applied to real-world data from Salvador, Brazil, VOA always outperforms competing algorithms, regardless of the predicted time horizon [63–66]. Some of the things that have been contributed include a way to choose ensemble features, details about CatBoost for solar forecasting, and a way to compare machine learning algorithms. The findings stress the need of using historical data and carefully choosing input components when making predictions about solar energy. They also show that the precision has greatly improved. In comparison to the summer dataset, which took 183.93 s to train, the winter dataset only took 14.76 s to train VOA. All of the alternatives pale in comparison to VOA, which integrates SVR, XGBT, and CatBoost. To determine the best elements to employ and how they work in different scenarios and projections, more research is required.

4.5 AI-Based Approach for Predicting Renewable Power

Artificial Neural Networks (ANNs) were proposed to model and size a clean energy community (CEC) that uses electric car charging stations, energy storage, and a PV-wind hybrid system to satisfy the energy demand of building districts. Two artificial neural networks (ANNs) are used here: one to predict energy performance indicators like fulfilled load % and utilization factor of generated energy, and the other to estimate what the grid energy indication factor is. The training data for artificial neural networks (ANNs) covers a wide range of electrical needs, power system topologies, and meteorological variables. They directly forecast the CEC's energy performance for the year using complex component models, bypassing the need to do system-dynamic simulations. We calculated the relative contributions of all inputs and outputs using the Garson method. Optimized ANNs achieve better

CEC prediction accuracy than ANNs trained with a single hidden layer containing 20 neurons.

4.6 AI-Based Approach Toward the Prevention of Cyberattacks in the Renewable Energy Domain

Creative renewable energy has given cybercriminals a new target to take advantage of. Smart renewable energy's commercial and financial facets are likely to be affected by this. Due to developments in communication networks, smart control systems, and the prevalence of Internet-connected buildings, several rival Cyber-Physical Power Systems (CPPSs) have made minor adjustments to the types of new power grids they employ. It is standard practice to link the electrical and cyber components in such setups. Recently brought to light are the issues of stability, resilience, dependability, and security that CPPSs aim to resolve. These challenges are being attempted to be solved by CPPSs. It is critical to use accurate modeling approaches and examine the interaction processes related to smart grid cyber-security to assess these risks and find solutions. Methods for identifying and preventing cyberattacks have been the focus of a great deal of research. Research in this field has focused on a wide range of issues, such as data injection, accurate data-driven attacks, robust communication network design, and detection algorithms that use federated learning approaches. Along with its results, this study presents a novel data-driven approach to detecting assaults on energy forecasting systems perpetrated by FDI. This strategy needs to be adopted if the reliability of our power systems is valued, as they are becoming more dependent on renewable energy and information technology. The proposed method enhances the system's resilience by detecting intrusions without using system models or parameters. This is accomplished by making use of techniques like z-score metrics, cross-validation, and least-squares. Results from experiments using tree-based wind power forecasting models show that this approach can identify and remove false data, leading to more accurate and situationally relevant predictions.

The exponential growth of energy storage technology, our growing dependence on renewable power sources, and the reality that these sources are not always predictable are the three main reasons why power systems are experiencing massive transformations. Because of these developments, smart grid operators are considering a future where microgrids enable people to trade energy with each other, leading to the creation of transactive energy networks. Academic institutions and corporations alike have taken an interest in blockchain technology [67–70] because of its remarkable resilience. The possibility of decentralized transaction records (TES) is presented by blockchain technology. As part of our research, we look at blockchain-based TES and suggest a new kind of attack against the intermediary nodes that prosumers use to access the network. They address possible issues and offer solutions while presenting a thorough blockchain-based model of TES. Research using

GridLAB-D and a secure Ethereum network shows that even small attacks can disrupt markets. However, mitigation measures like gateway switching and detection can successfully counter these assaults, making the system more secure and resilient.

5 Challenges of AI in the Renewable Energy Domain

The smart energy sector has yet to fully use AI due to many problems. Not only do these problems persist, but there are also several others, such as a lack of data, problems with the AI network's parameters, infrastructure, qualified specialists, difficulties with integration, dangers, worries about legal and compliance issues, and many more. Diagnostics and failure detection of different building energy systems can present considerable obstacles. Data lack and fragility have been highlighted as major issues impacting energy systems in multiple studies. Failures in sensors, controllers, and controlled devices can lead to inaccurate data estimations, which in turn can affect the reliability and performance [71] of an energy system. With energy markets adding an extra layer of complexity to the power grid's existing complicated correlations and strong coupling, the high data dimensionality and massive complexity of grid data generated from large-scale modeling are further intensified. The integration of AI with renewable energy sources like solar and wind into smart grid operations is a formidable and intricate task in and of itself. For smart energy companies [72] to make good use of AI, they have to overcome a variety of obstacles. The potential benefits of artificial intelligence are not completely appreciated by decision-makers due to their inadequate AI abilities. Power firms are understandably wary of trying new things due to a lack of expertise in developing reliable AI applications [73]. Using outdated technology for power systems is just one of the many problems that utilities face; another is the inherent difficulty of dealing with distant and unstructured data sources. New energy technologies aren't widely used because of financial obstacles [74]. High expenses and resource needs are two of these obstacles. Several factors, such as decentralization, variety, and the exponential expansion of AI complexity, have contributed to the emergence of massive issues in the energy production and distribution sectors [75].

The use of cellular networks limits the capabilities of artificial intelligence (AI), which hurts developing nations the most. Customers have every right to worry about the transparency and security of AI-powered applications in light of the growing number of assaults on the digital power grid. If artificial intelligence is to drive the smart energy sector, these concerns must be addressed. Developing a power grid that mostly relies on renewable energy sources still faces substantial obstacles. When energy is produced from renewable sources, such as solar and wind power, there are notable transformations [76]. The energy output varies significantly due to the inherent unpredictability of solar and wind patterns [77]. The volatility of the situation hinders the ability to sustain a harmonious equilibrium between energy supply and demand. Integrating substantial quantities of renewable energy into our existing electricity infrastructures poses significant technological and practical difficulties. The primary purpose of these networks was to offer backup for the more dependable fossil fuel sources. Due to its intrinsic instability, renewable energy generation tends to provide an excess amount of energy during periods of high demand and a deficient amount of energy during periods of low demand. Consequently, the power system experiences instability and there is a loss of energy [78].

Both the global objective of decarbonization and the current trend in development point to the possibility that a large number of renewable energy-generating systems will soon be connected to the grid. Likely, smaller dispersed renewable energy sources will eventually begin selling their excess power back to the grid. This is one of the potential outcomes. In the not-too-distant future, it is anticipated that the demand for electric vehicles [79] and the infrastructure that enables them, such as rapid charging stations, will undergo exponential growth. There are a multitude of ways in which smart home gadgets could be connected to the grid, and the person responsible for the grid might be utterly unaware of all of these possibilities. The amount of power generation capacity that the local utility system possesses would be considerably impacted by every one of these parameters independently. To facilitate the shift to renewable energy sources, it is essential to optimize grid operations through the utilization of technologies that utilize artificial intelligence. It is important to expand the capacity of these lines since doing so will enhance the capacity of transmission and distribution lines, which will in turn extend the amount of time that equipment can be used before it needs new parts. It is not without its restrictions and obstacles, however, that artificial intelligence approaches to the integration of renewable energy on a large scale are implemented.

The following are the most noteworthy examples of how they manifest themselves

1. Upgrades to automated equipment control that are no longer appropriate: Modern intelligent control technology is absolutely necessary for large-scale integrated renewable energy applications, particularly when used in conjunction with the grid or in order to fulfill the requirements of a wide range of users. It is not an easy operation to provide simultaneous access to a variety of control mechanisms. In addition, the technology that is used to regulate the grid in the traditional manner is sometimes quite outdated, which renders it incompatible with more contemporary systems. In the event that new intelligent control equipment is installed concurrently with existing ones, it is possible that further impediments will be created for the progress of artificial intelligence technology. When using the control gear that is currently in place, it is difficult to do maintenance on the newly established network of renewable power producers. This outmoded control equipment needs to be upgraded as soon as possible in order to comply with the new arrangement for the energy sector. For the purpose of enhancing the efficiency with which the intelligent equipment that is presently in use and that will be employed in the future operate together, it would be prudent to investigate new intelligent control methods or sophisticated algorithms. The development of renewable energy sources on a broad scale would be further improved as a result of this, and the requirement to replace obsolete control systems would be decreased.

2. One of the primary reasons why artificial intelligence systems, which include advanced algorithms, have their limitations is the performance uncertainty that is connected with accurately anticipating RES through the use of deep learning [78] theory. A significant number of concerns concerning the most effective approaches to the generation of renewable energy continue to be unaddressed, which is the primary reason for this situation. The generating, transmission, distribution, and transmission processes of renewable energy sources are presented as examples to illustrate this idea. Additionally, the functions of source, grid, load, and storage are also discussed. As a result of these unknowns, the construction of artificial intelligence prediction systems is becoming an increasingly difficult task. In addition to this, the utilization of AI prediction algorithms is accompanied by a great deal of challenges: Although they are still in the early phases of research, the methodologies that are currently being used to accurately anticipate the production times and power output of wind and solar power plants are far from perfect.
3. For the purpose of predicting the future power output of wind and solar farms, this would make it possible for machine learning algorithms to sift through images (for example, cloud maps from satellites), data acquired by sensors (for example, present wind speeds and sunshine intensities), and historical weather records. When it comes to robust optimization in the face of uncertainty and performance evaluation in multilevel settings that are defined by such uncertainty, surrogate models that are driven by artificial intelligence perform better than any other model. In addition to this, they are efficient in terms of computing, simple to utilize on computers, and possess extraordinary skills in terms of prediction. There is a possibility that power outages or gaps in the output of renewable energy could be the result of forecasting systems that are not controlled effectively. Additionally, the intensity of natural disasters, diseases, and other crises is increasing, which poses a greater threat to artificial intelligence prediction systems. This is because these crises are becoming more widespread. There is a shortage of highly qualified professionals who are also knowledgeable about artificial intelligence, which is a problem that many countries are currently facing. They are investing a significant amount of money into the research and development of artificial intelligence because they believe that it will have a significant impact on many aspects of society and the economy. Talent training programs need to be properly organized for the long term to ensure that persons who have been trained have a thorough understanding of the inner workings and essential components of certain businesses. This is necessary in order to make it possible for artificial intelligence to produce vast volumes of renewable energy. Before making any meaningful comparisons of these data, it is necessary to give careful thought to the particulars that are peculiar to the industry and AI. When it comes to hybrid capabilities that are compatible with the most recent developments in artificial intelligence and renewable energy, there is a notable lack of widely available hybrid capabilities. Furthermore, the professional settings of academic institutions have not altered at a rate that is sufficient to keep up with the advancements in research and technology. As a consequence of this, there is

a great deal of disagreement on how the fundamentals of artificial intelligence are taught in professional settings. Incompetence, oversimplification, and information that is no longer relevant are some of the issues that are prevalent throughout the process. In addition to this, there are no measures that are universally acknowledged for determining the worth of technical and scientific staff, and a great number of firms struggle to both recruit and keep the best employees. When it comes to attracting, hiring, training, and retaining personnel with competence in artificial intelligence (AI), financial incentives are not adequate.

4. The infrastructure for providing financial backing is still in its infancy, and there are no evidence-based policies that are cost-effective and in place to stimulate the use of artificial intelligence in the creation of renewable energy on a vast scale just yet. Considering that there are no rules that satisfy both conditions, this presents a troublesome situation. A great illustration of this is the status of the market for energy that is enabled by artificial intelligence; there is a serious lack of resources and persons with the necessary skills to implement a complete and pervasive strategy for the integration of AI into large-scale renewable energy sources. In addition, RES will be required to invest a significant amount of time and resources in the process of building, testing, and deploying the software that will be necessary for the introduction of AI strategies. In addition, the lack of a stable financial base is a barrier to the application of AI techniques to renewable energy sources [80].

6 Futuristic Outlook of AI in the Renewable Energy Sector

The global energy landscape will be drastically altered as a result of the partnership between AI and renewable energy sources. The idea that AI could improve the efficiency of renewable energy production is very encouraging, especially in light of the impending technological revolution. The development of more complex models and algorithms is a natural consequence of AI's progressing underlying technology. Improvements in grid management, predictive maintenance, and energy forecasting, as well as renewable energy system performance, are on the horizon thanks to these technical advancements. Unleashing the Power of Conventional and Cutting-Edge Technology: By combining AI with other cutting-edge technologies like blockchain and the Internet of Things (IoT), renewable energy source management is entering the modern era. Envision a future where smart grids enabled by the Internet of Things provide instantaneous responses to changes in energy demand and blockchain technology ensures transparent and efficient energy transactions. Both centralized and decentralized energy systems stand to benefit greatly from the predicted advancements in artificial intelligence (AI) technology. Localized power generation, storage, and consumption can be efficiently managed by an autonomous microgrid, which is an AI-powered self-governing energy system. Microgrids that can support themselves have been made possible by the broad adoption of smart homes and communities.

Collaborations on a global scale: Because of its far-reaching effects, the climate change challenge necessitates a concerted international effort. Renewable energy and artificial intelligence projects involving various stakeholders, including governments, corporations, and academic institutions, are likely to see an uptick in collaboration in the next years. As a result, people will have an easier time spreading innovative ideas and efficient methods.

7 Conclusion

In the context of renewable energy (RE) systems, this chapter provides an in-depth analysis of AI approaches, The optimization of systems, energy forecasting, and the creation of renewable electricity are all addressed by these solutions. It is suggested that neural networks, decision trees, random forests, and deep learning are some of the best AI approaches to deal with RE management. Data management, smart grid deployment, cyberattack avoidance, and energy industry controllability could all benefit greatly from artificial intelligence, according to the reviewed research. The research found that AI models outperformed more conventional approaches, especially when it came to straightforwardly and accurately handling complex situations. Energy system integration is a crucial subject, and this article lays out several potential avenues for future research and development that could explore it using AI. Research in the future should primarily focus on improving AI algorithms, finding ways to integrate AI with other emerging technologies like blockchain and the Internet of Things, and creating legal frameworks to support the use of AI in the energy sector. Thus, AI forms a crux of the renewable energy domain as the grid is constantly evolving.

References

1. Shafiullah, G. M., Oo, A. M. T., Shawkat Ali, A. B. M., & Wolfs, P. (2013). Potential challenges of integrating large-scale wind energy into the power grid—A review. *Renewable and Sustainable Energy Reviews, 20*, 306–321.
2. Zhou, Y., Zheng, S., Liu, Z., Wen, T., Ding, Z., Yan, J., & Zhang, G. (2020). Passive and active phase change materials integrated building energy systems with advanced machine learning based climate-adaptive designs, intelligent operations, uncertainty based analysis and optimisations: A state-of-the-art review. *Renewable and Sustainable Energy Reviews, 130*, 109889.
3. Zhou, Y. (2021). Artificial neural network-based smart aerogel glazing in low-energy buildings: A state-of-the-art review. *iScience, 24*(12), 103420.
4. Ahmad, T., Zhu, H., Zhang, D., Tariq, R., Bassam, A., Ullah, F., AlGhamdi, A. S., & Alshamrani, S. S. (2022). Energetics systems and artificial intelligence: Applications of industry 4.0. *Energy Reports, 8*, 334–361.
5. Zhou, Y., Cao, S., Kosonen, R., & Hamdy, M. (2020). Multi-objective optimisation of an interactive buildings-vehicles energy sharing network with high energy flexibility using the Pareto archive NSGA-II algorithm. *Energy Conversion and Management, 218*, 113017.

6. Zhou, Y., Cao, S., Hensen, J. L. M., & Hasan, A. (2020). Heuristic battery-protective strategy for energy management of an interactive renewables–buildings–vehicles energy sharing network with high energy flexibility. *Energy Conversion and Management, 214*, 112891.
7. Liu, Z., Xie, M., Zhou, Y., He, Y., Zhang, L., Zhang, G., & Chen, D. (2021). A state-of-the-art review on shallow geothermal ventilation systems with thermal performance enhancement system classifications, advanced technologies and applications. *Energy and Built Environment*. https://doi.org/10.1016/j.enbenv.2021.10.003
8. Liu, Z., Zeng, C., Zhou, Y., & Xing, C. (2022). Chapter 9: The main utilization forms and current developmental status of geothermal energy for building cooling/heating in developing countries. In Y. Noorollahi, M. N. Naseer, & M. M. Siddiqi (Eds.), *Utilization of thermal potential of abandoned wells* (pp. 159–190). Academic Press.
9. Weschenfelder, F., de Novaes Pires Leite, G., da Araújo Costa, A. C., de Castro Vilela, O., Ribeiro, C. M., Villa Ochoa, A. A., & Araújo, A. M. (2020). A review on the complementarity between grid-connected solar and wind power systems. *Journal of Cleaner Production, 257*, 120617.
10. Rangarajan, S. S., Collins, E. R., Fox, J. C., & Kothari, D. P. (2017). A survey on global PV interconnection standards. In *Proceedings of the IEEE Power and Energy Conference at Illinois (PECI), Champaign, IL, USA, February 23–24, 2017*.
11. Varma, R. K., Rangarajan, S. S., Axente, I., & Sharma, V. (2011). Novel application of a PV solar plant as STATCOM during night and day in a distribution utility network. In *Proceedings of the 2011 IEEE/PES Power Systems Conference and Exposition, Phoenix, AZ, USA, March 20–23, 2011* (pp. 1–8).
12. Rangarajan, S. S., Collins, E. R., & Fox, J. C. (2017). Harmonic resonance repercussions of PV and associated distributed generators on distribution systems. In *Proceedings of the 2017 IEEE North American Power Symposium (NAPS), Morgantown, WV, USA, September 17–19, 2017* (pp. 1–6).
13. Rangarajan, S. S., Collins, E. R., & Fox, J. C. (2017). Interactive impacts of elements of distribution systems on network harmonic resonances. In *Proceedings of the 6th IEEE International Conference on Renewable Energy Research and Applications (ICRERA), San Diego, CA, USA, November 5–8, 2017*.
14. Rangarajan, S. S., Collins, E. R., & Fox, J. C. (2017). Detuning of harmonic resonant modes in accordance with IEEE 519 Standard in an exemplary North American Distribution System with PV and Wind. In *Proceedings of the 6th IEEE International Conference on Renewable Energy Research and Applications, San Diego, CA, USA, November 5–8, 2017*.
15. Rangarajan, S. S., Collins, E. R., & Fox, J. C. (2017). Comparative impact assessment of filter elements associated with PWM and hysteresis controlled PV on network harmonic resonance in distribution systems. In *Proceedings of the 6th IEEE International Conference on Renewable Energy Research and Applications, San Diego, CA, USA, November 5–8, 2017*.
16. Rangarajan, S. S., Collins, E. R., & Fox, J. C. (2018). Smart PV and SmartPark Inverters as suppressors of temporary over-voltage (TOV) phenomenon in distribution systems. *IET Generation, Transmission & Distribution Journal, 12*, 5909–5917.
17. Rangarajan, S. S., Collins, E. R., & Fox, J. C. (2019). Efficacy of smart PV inverter as a virtual detuner in mitigating network harmonic resonances. *Electric Power Systems Research Journal, 171*, 175–184.
18. Rangarajan, S. S. (2018). *Efficacy of smart PV inverter as a strategic mitigator of network harmonic resonance and a suppressor of temporary overvoltage phenomenon in distribution systems*. All Dissertations (p. 2235). Retrieved January 25, 2021, from https://tigerprints.clemson.edu/all_dissertations/2235
19. Rangarajan, S. S., Sreejith, S., & Nigam, S. (2014). Effect of distributed generation on line losses and network resonances. In *Proceedings of the 2014 International Conference on Advances in Electrical Engineering (ICAEE), Vellore, India, January 9–11, 2014* (pp. 1–6).
20. Sathyavani, B., Rangarajan, S., Pavan, Rohith, Nagesh, Krishnamurthy, & Jyothirmai. (2024). THD of nine-level inverter with fewer switches—A brief review. *AIP Conference Proceedings, 2971*(1), 040023. https://doi.org/10.1063/5.0195900

21. Saikumar, M., Rangarajan, S. S., Nookala, S., Sriram, K., Deepthi, N. S., Vipul, D., & Nishitha, G. (2024). Literature survey on light electric vehicle battery charger with high-efficiency bridgeless single-power conversion. *AIP Conference Proceedings, 2971*(1), 040025.
22. Ahmed, Y. A., Rashid, A., Adeel, U., & Dutta, P. K. (2022). Proposing a framework for consumers' decision on solar innovations for smart cities. In *6th Smart Cities Symposium (SCS 2022), Hybrid Conference, Bahrain* (pp. 348–352). https://doi.org/10.1049/icp.2023.0575
23. Irfan, M. M., Rangarajan, S. S., Kodipaka, G., Mohammed, M., Atikam, R. T., Bharatha, S. T., & Kandagatla, S. K. (2024). Control of direct driven variable speed wind turbine system. *AIP Conference Proceedings, 2971*(1), 040008.
24. Yadeo, D., Rangarajan, S. S., Nookala, S., Alugunuri, H., Rakesh, A., Ganta, A., Karankanti, K., & Tejabethi, S. (2024). Current trends in electric vehicle—Review. *AIP Conference Proceedings, 2971*(1), 040005. https://doi.org/10.1063/5.0202038
25. Sen, S., Rangarajan, S. S., Kumar, R., & Yadeo, D. (2024). Role of policy and regulation for energy security, renewable integration & electric vehicles in India. *AIP Conference Proceedings, 2971*(1), 040015. https://doi.org/10.1063/5.0195892
26. Rajababu, D., Rangarajan, S. S., Nookala, S., Adil, M., & Hemanth, U. (2024). Fuzzy logic employment in unified power flow controller for altering the power flow in a transmission system. *AIP Conference Proceedings, 2971*(1), 040009.
27. Balakrishna, K., Rangarajan, S. S., Babburu, A., Bhukya, P., Shanigarapu, A., Mulka, S., & Thotakuri, M. (2024). DC-DC step-up converter with three inputs for hybrid electric vehicle applications. *AIP Conference Proceedings, 2971*(1), 040002.
28. Pendem, S. R., Rangarajan, S. S., Madishetti, L. K., Pegada, V., Gandam, V. V., Padmini, S., & Vuppula, N. (2024). Permanent magnet synchronous motor speed control for marine applications. *AIP Conference Proceedings, 2971*(1), 040019.
29. Basetti, V., Rangarajan, S. S., & Kumar, K. R. (2024). Power quality problems and standards in distribution systems. *AIP Conference Proceedings, 2971*(1), 040018.
30. Sen, S., Rangarajan, S. S., Kumar, R., & Yadeo, D. (2024). Role of policy and regulation for energy security, renewable integration & electric vehicles in India. *AIP Conference Proceedings, 2971*(1), 040015.
31. Durgam, R. B., Rangarajan, S. S., Mohammad, S., Bairi, K. S., Thota, R., Yelthuri, N., & Nellutla, H. (2024). Design of cascaded multi-level inverter. *AIP Conference Proceedings, 2971*(1), 040010.
32. Pendem, S. R., Sen, S., Rangarajan, S. S., Valaboju, S., Hrishikesh, Y., Jaychandra, R., Saboor, S., & Naik, B. T. (2024). An architecture of DC fast charging station for vehicle and grid interaction in a micro-grid system. *AIP Conference Proceedings, 2971*(1), 040020.
33. Angatha, S., Manjusree, Y., & Rangarajan, S. (2024). A flexible protection scheme for multi-terminal transmission lines. *AIP Conference Proceedings, 2971*(1), 040022.
34. Sai Kumar, M., Rangarajan, S. S., Nookala, S., Shivapriya, K., Krishna, P. V., Charan, O. S., & Nandu, M. (2024). Direct fast charging of an electric vehicle using solar power. *AIP Conference Proceedings, 2971*(1), 040026.
35. Sangem, Y., Rangarajan, S. S., Nookala, S., Kondeti, V., Jampala, P. K., Badhawath, R. K., & Kankanala, S. R. (2024). Energy storage requirements reduction in smart grid employing electric springs. *AIP Conference Proceedings, 2971*(1), 040003.
36. Rangarajan, S. S., & Sreejith, S. (2013). Novel 24 hour usage of a PV solar farm for reducing line loss. In *Proceedings of the 2013 International Conference on Energy Efficient Technologies for Sustainability, Nagercoil, India, April 10–12, 2013* (pp. 381–386).
37. Rangarajan, S. S., Sreejith, S., & Sabberwal, S. P. (2013). Cost estimation and recovery analysis of a PV solar farm utilized round the clock. In *Proceedings of the 2013 IEEE Global Humanitarian Technology Conference: South Asia Satellite (GHTC-SAS), Trivandrum, India, August 23–24, 2013* (pp. 286–291).
38. Mozumder, S., Dhar, A., Rangarajan, S. S., & Karthikeyan, S. P. (2014). Coordinated operation of multiple inverter based renewable distributed generators as an active power injector and reactive power compensator. In *Proceedings of the 2014 International Conference on*

Computation of Power, Energy, Information and Communication (ICCPEIC), Chennai, India, April 16–17, 2014 (pp. 298–303).

39. Berge, J., Rangarajan, S. S., Varma, R. K., & Litzenberger, W. H. (2011). Bibliography of FACTS 2009–2010: Part IV IEEE Working Group report. In *Proceedings of the 2011 IEEE Power and Energy Society General Meeting, Detroit, MI, USA, July 24–28, 2011* (pp. 1–10).
40. Berge, J., Rangarajan, S., Varma, K., & Litzenberger, H. (2011). Bibliography of FACTS 2009-2010: Part III, IEEE Working Group report. In *Proceedings of the 2011 IEEE Power and Energy Society General Meeting, Detroit, MI, USA, July 24–28, 2011.*
41. Sunddararaj, S. P., Rangarajan, S. S., & Subashini, N. (2020). An extensive review of multilevel inverters based on their multifaceted structural configuration, triggering methods and applications. *Electronics, 9*, 433.
42. Sunddararaj, S. P., Rangarajan, S. S., & Gopalan, S. (2019). Neoteric fuzzy control stratagem and design of chopper fed multilevel inverter for enhanced voltage output involving plug-in electric vehicle (PEV) applications. *Electronics, 8*, 1092.
43. Rangarajan, S. S., Sharma, J., Kothari, D. P., & Senjyu, T. (2019). Novel utilization of phasor measurement units (PMU) in smart grid restoration: A brief survey. In *Proceedings of the International Conference on 'Emerging Trends for Smart Grid Automation and Industry 4.0', ICETSGAI4.0, December 5–9, 2019*. BIT Mesra.
44. Swaminathan, G., Rangarajan, S. S., Sharma, J., Kothari, D. P., & Senjyu, T. (2019). Techno-economic benefits of grid penetrated 1 MW PV system in India. In *Proceedings of the International Conference on 'Emerging Trends for Smart Grid Automation and Industry 4.0', ICETSGAI 4.0, December 5–9, 2019*. BIT Mesra.
45. Sharma, J., Rangarajan, S. S., Sundarabalan, C. K., Karthikaikannan, D., Srinath, N. S., Kothari, D. P., & Senjyu, T. (2019). Synergistic damping operation of TCSC & CPSS using PSO in a power system. In *Proceedings of the International Conference on 'Emerging Trends for Smart Grid Automation and Industry 4.0', ICETSGAI4.0, December 5–9, 2019*. BIT Mesra.
46. Sharma, J., Rangarajan, S. S., Srikanth, V. S. S., Sundarabalan, C. K., Kothari, D. P., & Senjyu, T. (2019). Transient stability enhancement using FACTS devices in a distribution system involving distributed generation systems. In *Proceedings of the International Conference on 'Emerging Trends for Smart Grid Automation and Industry 4.0', ICETSGAI4.0, December 5–9, 2019*. BIT Mesra.
47. Dutta, P. K., Dey, S., Majumder, S., Sen, P., & Roy, S. (2023). Implementing solar panel surface dust cleaning innovation using a solar innovation framework model. In A. E. Hassanien, D. Zheng, Z. Zhao, & Z. Fan (Eds.), *Business intelligence and information technology* (Smart innovation, systems and technologies) (Vol. 358). Springer. https://doi.org/10.1007/978-981-99-3416-4_42
48. Matayoshi, H., Kinjo, M., Rangarajan, S. S., Ramanathan, G. G., Hemeida, A. M., & Senjyu, T. (2020). Islanding operation scheme for DC microgrid utilizing pseudo—Droop control of photovoltaic system. *Energy for Sustainable Development, 55*, 95–104. ISSN: 0973-0826.
49. Nandakumar, M., Ramalingam, S., Nallusamy, S., & Rangarajan, S. S. (2020). Novel efficacious utilization of fuzzy-logic controller-based two-quadrant operation of PMBLDC motor drive systems for multipass hot-steel rolling processes. *Electronics, 9*, 1008.
50. Takahashi, K., Jargalsaikhan, N., Rangarajan, S., Hemeida, A. M., Takahashi, H., & Senjyu, T. (2020). Output control of three-axis PMSG wind turbine considering torsional vibration using H infinity control. *Energies, 13*, 3474.
51. Ahmed, S., Fathima, F., Karthikeyan, S. P., Sahoo, S. K., Rangarajan, S. S., & Kothari, D. P. (2015). Application of cubic spline interpolation in estimating market power under deregulated electricity market. In *TENCON 2015—2015 IEEE Region 10 Conference, Macao* (pp. 1–6).
52. Kumar, T. M., Raghuram, P. R., Kumar, S. P., Chand, L. S., Rangarajan, S. S., & Nigam, S. (2014). System integrity protection scheme for Southern grid. In *2014 6th IEEE Power India International Conference (PIICON), Delhi* (pp. 1–5).

53. Dutta, P. K., Banerjee, U., Manna, B., Hazarika, K., Das, B., & Roy, S. (2023). Nanotechnology and the application of information technology for sustainable innovation in agriculture. In A. E. Hassanien, D. Zheng, Z. Zhao, & Z. Fan (Eds.), *Business intelligence and information technology* (Smart innovation, systems and technologies) (Vol. 358). Springer. https://doi.org/10.1007/978-981-99-3416-4_41
54. Nandakumar, M., Ramalingam, S., Nallusamy, S., & Srinivasarangan Rangarajan, S. (2020). Hall-sensor-based position detection for quick reversal of speed control in a BLDC motor drive system for industrial applications. *Electronics, 9*, 1149.
55. Karthik, B., Arokkia, J. P., Sreejith, S., & Shriram, S. R. (2014). Three phase power flow incorporating static var compensator. *Applied Mechanics and Materials, 573*, 747–756.
56. Srinivasarangan Rangarajan, S., Sharma, J., & Sundarabalan, C. K. (2020). Novel exertion of intelligent static compensator based smart inverters for ancillary services in a distribution utility network-review. *Electronics, 9*, 662.
57. Sunddararaj, S. P., Rangarajan, S. S., Subashini, N., Umashankar, S., Collins, E. R., & Senjyu, T. (2021). A new topology of DC-DC converter with bidirectional power flow capability coupled with a nine multilevel inverter for EV applications. In *7th International IEEE Conference on Electrical Energy Systems, February 11–13, 2021.*
58. Sunddararaj, S. P., Rangarajan, S. S., Subashini, N., Umashankar, S., Collins, E. R., & Senjyu, T. (2021). Performance of P/PI/PID based controller in DC-DC converter for PV applications and smart grid technology. In *7th International IEEE Conference on Electrical Energy Systems, February 11–13, 2021.*
59. Sunddararaj, S. P., Rangarajan, S. S., Subashini, N., Umashankar, S., Collins, E. R., & Senjyu, T. (2021). A novel T-type multilevel inverter for electric vehicle and grid-connected applications. In *7th International IEEE Conference on Electrical Energy Systems, February 11–13, 2021.*
60. Manoharan, G., Durai, S., Rajesh, G. A., & Ashtikar, S. P. (2024). A study on the application of expert systems as a support system for business decisions: A literature review. In *Artificial intelligence and knowledge processing* (pp. 279–289). Bentham Science Publishers.
61. Singh, C. R., & Manoharan, G. (2024). Strengthening resilience: AI and machine learning in emergency decision-making for natural disasters. In *Internet of Things and AI for natural disaster management and prediction* (pp. 249–278). IGI Global.
62. Durai, S., Manoharan, G., & Ashtikar, S. P. (2024). Harnessing artificial intelligence: Pioneering sustainable solutions for a greener future. In *Social and ethical implications of AI in finance for sustainability* (pp. 89–117). IGI Global.
63. Singh, A., Virmani, D., Kulkarni, N., Tiwari, M., Manoharan, G., & Alanya-Beltran, J. (2024). The emerging role and application of adopting revolutionary Internet of Things (IoT) for future technology sustainability. In *Computer Science Engineering and Emerging Technologies: Proceedings of ICCS 2022* (p. 103).
64. Durai, S., Krishnaveni, R., Manoharan, G., & Nivedha, M. (2024). Leveraging HR metrics for an effective recruitment and selection process with special reference to the manufacturing sector in Coimbatore. In *Convergence of human resources technologies and Industry 5.0* (pp. 188–210). IGI Global.
65. Behera, N. R., Thiyagarajan, V. S., Manoharan, G., Veena, C., Acharjee, P. B., & Murthy, A. (2024, February). Cognitive outcomes prediction in children using machine learning and big data analytics. In *2024 4th International Conference on Innovative Practices in Technology and Management (ICIPTM)* (pp. 1–6). IEEE.
66. Shiva, C. K., Vedik, B., & Manoharan, G. (2024). Sustainable smart transportation: Technologies, benefits, challenges to urbanisation concept. In *Secure and intelligent IoT-enabled smart cities* (pp. 355–370). IGI Global.
67. Manoharan, G., Durai, S., Rajesh, G. A., Razak, A., Rao, C. B., & Ashtikar, S. P. (2023). An investigation into the effectiveness of smart city projects by identifying the framework for measuring performance. In *Artificial intelligence and machine learning in smart city planning* (pp. 71–84). Elsevier.

68. Lourens, M., Sharma, S., Pulugu, R., Gehlot, A., Manoharan, G., & Kapila, D. (2023, May). Machine learning-based predictive analytics and big data in the automotive sector. In *2023 3rd International Conference on Advance Computing and Innovative Technologies in Engineering (ICACITE)* (pp. 1043–1048). IEEE.
69. Kumari, S., Kumar, V., Sharmila, A., Murthy, C. R., Ahlawat, N., & Manoharan, G. (2023, August). Blockchain based E-analysis of social media forums for crypto currency phase shifts. In *2023 5th International Conference on Inventive Research in Computing Applications (ICIRCA)* (pp. 1222–1225). IEEE.
70. Manoharan, G., Agarwal, S., Agarwalla, S., Dutt, A., Vani, V. D., & Sandhya, G. (2023, December). Blockchain technology in supply chain management: Prospects and challenges. In *2023 10th IEEE Uttar Pradesh Section International Conference on Electrical, Electronics and Computer Engineering (UPCON)* (Vol. 10, pp. 557–560). IEEE.
71. Tripathi, M. A., Tripathi, R., Effendy, F., Manoharan, G., Paul, M. J., & Aarif, M. (2023, January). An in-depth analysis of the role that ML and big data play in driving digital marketing's paradigm shift. In *2023 International Conference on Computer Communication and Informatics (ICCCI)* (pp. 1–6). IEEE.
72. Almahairah, M. S. Z., Goswami, S., Karri, P. N., Krishna, I. M., Aarif, M., & Manoharan, G. (2023, December). Application of internet of things and big data in improving supply chain financial risk management system. In *2023 10th IEEE Uttar Pradesh Section International Conference on Electrical, Electronics and Computer Engineering (UPCON)* (Vol. 10, pp. 276–280). IEEE.
73. Shameem, A., Ramachandran, K. K., Sharma, A., Singh, R., Selvaraj, F. J., & Manoharan, G. (2023, May). The rising importance of AI in boosting the efficiency of online advertising in developing countries. In *2023 3rd International Conference on Advance Computing and Innovative Technologies in Engineering (ICACITE)* (pp. 1762–1766). IEEE.
74. Manoharan, G., Razak, A., Rajchandar, K., Nithya, G., Durai, S., & Ashtikar, S. P. (2024). Digital learning for professional development in varied fields of service sectors: Embracing technological advancements. In *Embracing technological advancements for lifelong learning* (pp. 111–137). IGI Global.
75. Manoharan, G., Jalaja, V., Sathe, M. A., Lourens, M., & Suresh, K. (2023, December). Machine learning and data privacy in human resource management. In *2023 4th International Conference on Computation, Automation and Knowledge Management (ICCAKM)* (pp. 1–6). IEEE.
76. Bhavya, K. R., Dharmaraj, A., Pareek, P., Sathe, M. A., Tiwari, M., & Manoharan, G. (2023, September). Application of soft computing techniques for predictive analytics in financial markets. In *2023 6th International Conference on Contemporary Computing and Informatics (IC3I)* (Vol. 6, pp. 1756–1760). IEEE.
77. Ashiq, V. M., Fredrik, E. T., Kumar, N. K., Torres-Cruz, F., Colque, J. P. B., & Manoharan, G. (2023, September). A deep learning approaches for natural language processing and sentiment analysis in social media. In *2023 6th International Conference on Contemporary Computing and Informatics (IC3I)* (Vol. 6, pp. 1746–1750). IEEE.
78. Mekala, S., Bhuvana, M., Gupta, D. B., Bhatt, V., Kunekar, P., & Manoharan, G. (2023, September). Natural language processing and deep learning techniques to improve sentiment analysis in social media texts. In *2023 6th International Conference on Contemporary Computing and Informatics (IC3I)* (Vol. 6, pp. 1751–1755). IEEE.
79. Rajasekaran, S., Sadasivan, A., Udayanan, M., Billa, P., Selvaraj, F., & Manoharan, G. (2024). Key factors influencing brand preference of electric vehicles in Chennai city. *AIP Conference Proceedings, 2971*(1), 060039.
80. Manoharan, G., Nithya, G., Razak, A., Durai, S., Sharma, S., & Ashtikar, S. P. (2024). Enhancing financial literacy: Strategies, challenges, and global initiatives. In *Emerging perspectives on financial well-being* (pp. 111–128). IGI Global.

An Interview-Based Study on the Ethical, Privacy, and Security Implications of AI-Driven Smart Grids

Prasenjit Chakravarty, Nibir Khawash, and Sudeshna Pahari

1 Introduction

Smart grids are developed as an intelligent electric power system and the system is equipped with the latest automation, remote control, and remote sensing technologies. Smart grids have huge contributions to optimizing and predicting the electricity distributions and generations for the involved parties. In smart grid management, key stakeholders include potential consumers, power grid operators, and utility providers [1]. Due to the integration of artificial intelligence (AI) into smart grids, the potential landscape of energy systems is significantly transformed. It has effectively optimized the operation and planning of distributed energy sources (DERs) including sustainable energy source developments from solar or wind turbines. The research emphasized exploring the significant concerns related to privacy, ethics, and security for integrating AI into smart grids.

1.1 Background

AI integration into smart grids is crucial for enhancing the sustainability, reliability, and efficiency of energy systems. AI technologies can significantly process larger datasets and evaluate the findings based on energy optimization strategies. In demand forecasting, load management, and the integration of sustainable energy sources, AI-driven technologies have potential contributions [2]. AI algorithms with optimization techniques and neural networks can be effectively employed in energy

P. Chakravarty (✉) · N. Khawash · S. Pahari
Amity Business School, Amity University Kolkata, Kolkata, India

P. Raj et al. (eds.), *Artificial Intelligence (AI) for IT Energy Efficiency and Green AI for Environment Sustainability*,
https://doi.org/10.1007/978-3-031-89420-6_20

demand forecasting enabling the distributors to make accurate load predictions along with the significant allocation of resources. Gaussian process regression models are efficient in forecasting peak demand while enhancing the stability of the grid and reducing the requirements for expensive peak power plants. Machine learning techniques are also utilized in optimizing the voltage for electricity and managing reactive power facilities within the smart grid which might result in the system efficiency and power quality. AI-based predictive maintenance strategies have a key role in extending the lifespan of different components of the system as they can effectively anticipate the potential failures of the component and schedule advanced maintenance programs [3]. However, potential challenges emerge from the integration of DERs like wind turbines and solar panels into the existing smart grid infrastructure. AI technologies effectively work as the instrumental technique for controlling and coordinating diverse energy resources while ensuring the overall efficiency and stability of the grid system.

1.2 Rationale

With the increasing awareness about environmental sustainability and protection, utility providers are adopting different renewable energy sources. In adopting renewable sources, different challenges for the optimization of energy distribution, grid stability, and bidirectional energy integration have become complex for IT and electronic engineers [5]. Different AI-related technologies like neural networks and machine learning algorithms have evolved as potential tools to address the key challenges by enabling fault detection, predictive maintenance, and advanced energy management. Despite the advancement in these technologies, potential issues related to environmental sustainability, privacy measurement, and cybersecurity persist in the energy sector [6]. By evaluating such challenges for the current state of AI applications in smart grids, the research is going to contribute to understanding the transformative impact of the technologies on smart grid operations. Developing a robust cybersecurity framework is crucial to protect sensitive information and ensure the integrity of the AI-based system [7]. Based on the concerns, the research adopts direct interviews with skilled professionals from the energy sector to gather first-hand information about the key challenges associated with AI-integrated grid systems, as shown in Fig. 1. The critical areas for further research about the topic are also identified in the research process to enhance the efficiency and resilience of grid systems.

1.3 Research Aim

The primary aim of the research is to examine the ethical, privacy, and security concerns in different AI-based smart grids.

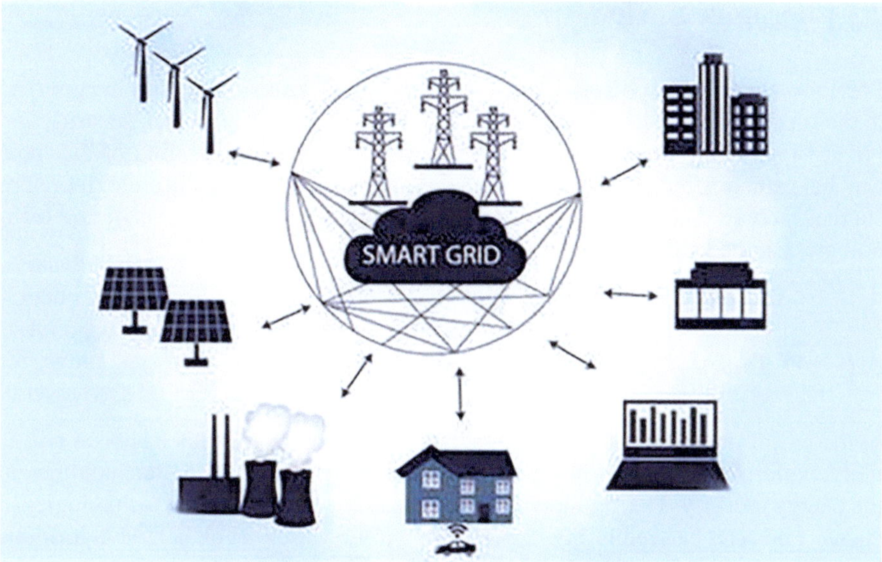

Fig. 1 Overview of a smart grid system. (Source: Guided by Wen et al. [4])

1.4 Research Objectives

- To identify the potential ethical challenges and concerns emerging from the utilization of AI in smart grids
- To assess the privacy implications for gathering and evaluating data in different AI-based smart grids
- To evaluate the cybersecurity measures against potential vulnerabilities in AI-based smart grid system
- To provide recommendations about addressing the key ethical, security, and privacy concerns associated with AI-based smart grid

1.5 Research Questions

- What are the key ethical concerns rising from AI adoptions in the operations of smart grids?
- What are the significant privacy implications for AI implementation while collecting potential information about the components and stakeholders of the grid?
- How do effective cybersecurity measures prevent potential vulnerabilities in AI-driven smart grids?
- What are the significant recommendations to address the key concerns regarding AI-enabled smart grids?

2 Literature Review

Previous research has potentially discussed the significant strategies to integrate AI in the smart grids and few studies are found discussing the key ethical, security, and privacy issues in AI-based grids. However, a potential understanding of the topic can be gained after going through the past research. From different literature sources, potential insights about smart grid systems and AI algorithms in grid technologies can be gained effectively.

2.1 AI Technologies in Smart Grids

In the recent literature source, the integration of AI in smart grids has been potentially explored while providing potential knowledge about key AI technologies in the energy industry. The significant impact areas due to different AI technologies in energy grids can be seen in Fig. 2. According to Lo Cascio et al. [9], AI algorithms including machine learning and neural networks are effectively utilized to predict energy consumption patterns, optimize distribution networks, and integrate renewable energy sources like solar energy sources and wind turbines into the grid. However, as depicted by Dong et al. [10], the introduction of AI requires substantial computational resources, which can potentially contribute to energy consumption and energy distribution through the grid system. Machine learning (ML) is often regarded as a significant part of AI, and it can potentially be used in smart grids for predictive analytics, system optimization, and fault detection. ML algorithms and deep learning models significantly analyze historical and real-time data from various sensors from energy meters across the grid. From the gained information, the

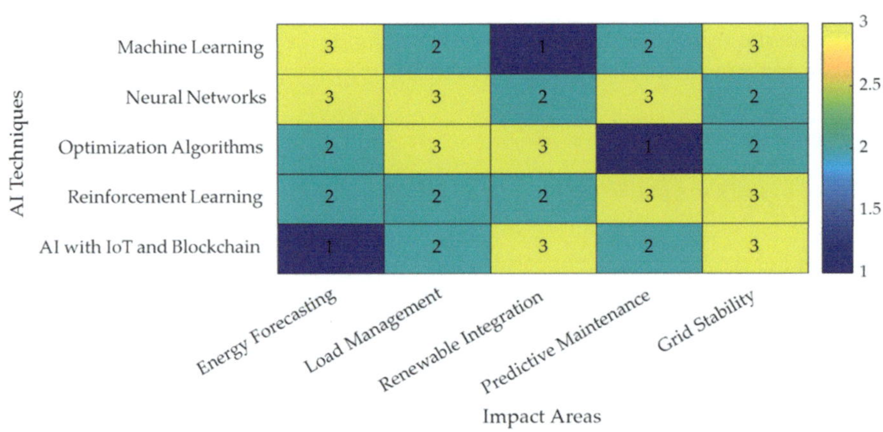

Fig. 2 Significant impact areas due to different AI technologies in energy grids. (Source: Guided by Tan [8])

technologies enable the prediction of energy demand, optimization of supply, and prevent system failures. According to Hafeez [11], ML models can forecast energy consumption patterns with high accuracy, and they can effectively help grid operators to adjust power generation in real time while reducing potential energy wastage and managing different peak load scenarios.

Artificial neural networks (ANNs) are the key implications of AI technologies. As argued by McAreavey et al. [12], ANNs are potentially important for smart grids as it is efficient in solving different nonlinear problems and equations for optimizing energy distribution and load forecasting. ANNs are developed through the structure of the human brain and the related neural networks. Despite the capabilities of load management, detection of anomalies, and fault identification through ANNs, the system is quite complex for the existing energy infrastructure. However, Yeboah [13] has outlined the emerging concerns for the implementation of ANNs as the technology might cause potential ethical and regulatory concerns. Although the system can process big data gathered from the grid's sensors and facilitate the prediction of future energy demands, it might avoid the nuances of electricity uses and drained energy across the industry. Since the network enables grid operators to make informed decisions about energy distributions based on households' types and preferences of consumers, the network influences the unequal distribution of power facilities to different households. The continuous monitoring of grid performance and power deviations in the existing infrastructure through ANNs can provide significant opportunities to identify issues like faults in the transmission line, equipment malfunctions, and further power quality problems.

2.2 Ethical Considerations for Leveraging AI Technologies in Smart Grids

Although AI technology integration into smart grids provides potential benefits to the corresponding stakeholders of smart grids, it potentially causes different ethical concerns for consumers and other stakeholders. Figure 3 shows us the potentially significant areas of ethical bias resulting from AI adoptions in smart grids. The rising issues of sustainability and environmental ethics have become crucial to follow for energy distributors. While considering ethics for the energy sector, the principles of transparency, privacy-consciousness, equitable workplace, safety, and fellowship are required to be included in the codes of ethics and sustainability. As depicted in the descriptive study of Larsen [14], AI algorithms can present existing biases that might be presented in the considered dataset. The prevalence of such biasedness can lead to potential discriminatory outcomes. However, Yella [15] has argued for the training data of AI technologies that can be biased toward certain demographics, leading to inefficient statistics and unfair decisions regarding energy distribution or pricing. Due to such circumstances, social injustice can appear in the grid distribution networks.

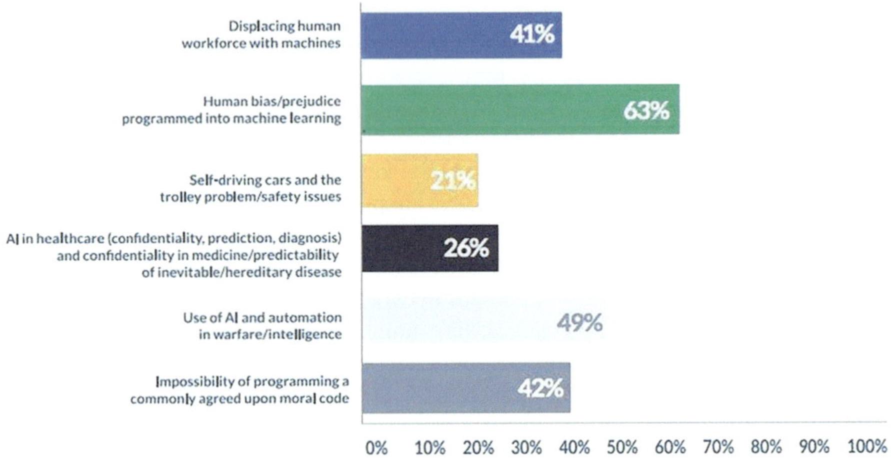

Fig. 3 Potential biased areas of AI adoptions in smart grids. (Source: Guided by McAreavey et al. [12])

Accountability and transparency are the key two aspects in AI-based systems and it involves understanding how AI algorithms make decisions. As per the opinion of Ketter et al. [16], informed and strategic decision-making has become significant through maintaining transparency at the workplace and informing stakeholders about the changes. Without ensuring transparency, it has become difficult to hold AI systems accountable for their actions and address potential biases or errors. However, the authors Vlachová and Binder [17] argued that the introduction of AI into smart grids potentially affected potential job displacements for grid operators and other workers. Although AI can automate certain tasks in smart grids, it requires considering the economic and social impacts due to the implementation of AI technologies in the system. Rather than contributing to job losses, the energy players should develop strategies for retraining and re-skilling the workforce. The issues of rising unemployment need to be effectively addressed by the organizations while implementing AI in the energy industry. Moreover, AI can be leveraged to optimize energy consumption and reduce environmental impact. The development and deployment of AI-based systems might have environmental consequences because of increased energy consumption in providing training and running different AI models.

2.3 *Privacy Concerns for AI-Based Data Analytics in Smart Grids*

Privacy concerns emerge as a high concern and the potential ethical dilemmas are discussed in different studies. Most of the public concern is centered on protecting anonymity and data collection techniques. As per the detailed overviews of Sulaiman et al. [18], the integration of AI into smart grids involves the collection and analysis of vast amounts of data including energy consumption patterns, household demographics, and real-time electricity usage data. Through the collection of such energy consumption data, the corresponding company evaluates energy consumption patterns and potentials for different households. As depicted by Sulaiman et al. [18], consumers feel worried about their data protection acts and their data usage without their consent for commercial and advertisement purposes. The potential information including names, email addresses, and other sensitive information needs to be preserved as these types of data are nearly impossible to keep anonymous. As per the study of Bouramdane [19], the privacy of consumers' sensitive personal information becomes an important part of the data processing to eliminate the reachability of the data to potential hackers and third-party operators. Potential data breaches can expose sensitive consumer information and it might lead the event to identity theft, financial loss, and reputational damage. By ensuring adequate measures to eliminate potential privacy risks, the company can eliminate different issues like harassment, customized advertisements, and other privacy breaches for consumers.

2.4 Security Risks in AI-Enabled Smart Grids

In the electric network facilities, serious concerns around safety and security are likely to develop and the digitization of the system can increase the potential security risks. Although AI-integrated systems of smart grids provide numerous benefits to the energy company, they also introduce new cybersecurity risks and cyberattacks as the increased connectivity between all the energy networks presents more infiltrating situations for hackers. As per the opinion of Volkova et al. [20], these risks can compromise the reliability, integrity, and safety of the grid, potentially leading to significant disruptions and economic losses. AI-based systems have potential threats from hacking, malware, and denial-of-service attacks. These attacks could disrupt grid operations while leading to power outages and compromise data security issues. Therefore, Tan [8] highlighted that it has become crucial to address these ethical, privacy, and security concerns for the responsible and sustainable development of AI-based smart grids. Despite the potential technological and governance measures for addressing key concerns, the inclusion of consumers' control over their data has become necessary to understand the overall capabilities of AI technologies in smart grid management.

2.5 Literature Gap

In the previous studies, key ethical issues, regulatory issues, governance issues, security measures, and privacy concerns are discussed for the implementation of advanced technologies in smart grids. However, potential literature gaps are still relevant in such discussion as the previous studies have not adequately supported the integration of green AI and sustainability concerns for smart grid operations while adopting adequate measures for the emerging challenges. Although many reports and articles are published demonstrating the significance of addressing key concerns about ethics, privacy, and security management of AI technologies, it highlights significant gaps in the sustainable development of AI innovations and grid efficiency [21]. While measuring the long-term influences of smart and intelligent technologies in electricity distributions, it is also required to evaluate the key vulnerabilities due to AI adoption. Significant studies are required to develop outlining the particular requirements of adherence to standard ethics codes, international security protocols, and privacy regulations.

3 Methodology

3.1 Research Design

The research on evaluating the key ethical, security, and privacy concerns in adopting AI technologies for grid systems is conducted through an exploratory and qualitative research design. Based on the primary interviews conducted for the study, in-depth insights about the diverse viewpoints on AI implementation over energy infrastructure can be effectively understood [15]. As the study gathers potential data from semi-structured interviews, the flexibility in questioning along with the independence over expressing participants' ideas is ensured in the study. By keeping the interview discussions focused on key themes of the study, potential insights and detailed overviews from the experts and professionals can be gathered, informing about the critical concerns associated with security, privacy, and ethics.

3.2 Sampling

Convenience sampling is chosen as an effective sampling method as this non-probability sampling method involves the available and interested participants for the interviews. In the primary qualitative interviews, convenience sampling ensures that the study considers the accessibility and timeframe for the participants while gathering meaningful insights from the professionals and experts [22]. The research targets five participants to obtain in-depth and diverse perspectives of professionals from the energy sector. The participants are chosen based on their expertise and roles in the relevant organizations. A total of five participants from the relevant fields are contacted for the interview and the participants are considered for their significant job roles in integrating AI technologies into electricity distribution and transmissions. One designer architect of the smart grid systems, two electrical engineers for the smart grid, one IT professional, and one AI expert are considered as the interview participants to obtain key insights about the existing infrastructure of smart grids, AI compatibility in the systems, data protection and privacy management systems along with the ethical and responsible AI deployment in the smart grid systems.

3.3 Data Collection and Data Analysis

Semi-structured interviews are conducted to collect potential qualitative information from the selected participants. As per the availability and convenience of the participants, the interviews are conducted through both virtual and face-to-face approaches. Four demographic questions and a total of 16 topic-related questionnaires are asked to five participants. The topic-related questions are separated into four key types of questionnaires, and each questionnaire type includes four in-depth questions centered around the topic. The whole interview process and responses are recorded by obtaining their consent. The audio transcripts of each interview are prepared in the study to conduct analysis based on the key themes [23]. A thematic analysis is adopted to evaluate the gathered primary data from the interviews. By going through the transcript, key phrases and concepts like environmental sustainability, privacy, security, AI, and ethics are identified, along with the key themes for data analysis.

3.4 Ethical Considerations

To conduct interviews and transcript the data, informed consent is taken from all the participants before conducting any interviews. The participants are also assured about the confidentiality of the responses and the utilization of data for academic purposes. Ethical approval to conduct interviews and continue the research is taken from the relevant institution [21]. The purposes and objectives of the study are explained to the participants before obtaining their consent for interviews.

4 Results and Analysis

4.1 Axial Coding

See Table 1.

Table 1 Axial coding table to identify key challenges in AI-integrated grid systems

Questions	Quotations of participants	Respondents	Theme
Interview question 1: "*How do you perceive your role in influencing the integration of AI technologies within smart grids to enhance sustainability and energy efficiency? Can you provide examples of AI-based initiatives or systems you have contributed to in the design and development of smart grids? What motivates you to promote AI-driven sustainability within the energy sector? How do you measure the impact of AI technology in improving energy efficiency and environmental sustainability within smart grids?*"	Participant 1 (Designing Architect): "AI can streamline energy usage by optimizing distribution, which results in significant energy savings and enhances sustainability in smart grids." Participant 2 (Electronic Engineer): "AI-driven systems allow for better monitoring and control of electrical components, leading to optimized energy use and higher efficiency in smart grids." Participant 3 (Electronic Engineer): "By using AI, power distribution can be managed more efficiently, minimizing losses and making the grid more sustainable." Participant 5 (IT Professional): "AI-powered IT infrastructure ensures the scalability of energy management systems, thereby making smart grids more energy-efficient and sustainable in the long term."	Interview participants—1, 2, 3, 5	***"Theme 1: AI-adoptions in smart grid enhance energy efficiency and environmental sustainability."***
Interview question 2: "*What challenges have you encountered in ensuring the security and resilience of AI-based smart grids? How have you addressed these security challenges when integrating AI technologies into the grid? What support or resources would help overcome obstacles related to AI and cybersecurity in smart grids? How do you think these challenges impact the broader adoption of AI in the energy sector?*"	Participant 1 (Designing Architect): "While AI enhances efficiency, there are still ethical concerns, especially around how decisions are made regarding energy allocation. There needs to be transparency in how AI makes these decisions." Participant 2 (Electronic Engineer): "Although AI improves the reliability of devices in the grid, ethical concerns like biases in decision-making processes need to be addressed, especially when dealing with resource allocation." Participant 3 (Electronic Engineer): "Integrating AI with renewable energy is a great opportunity, but the ethical challenges related to AI decision-making, such as fairness in energy distribution, must be considered."	Interview participants—1, 2, 3	***"Theme 2: Ethical concerns still exist in AI-driven decision-making for smart grids."***

Interview question 3: *"In what ways do you believe stakeholders can effectively collaborate to ensure the ethical use of AI in smart grids? Are there specific issues or areas of AI ethics in smart grids that require more attention? How can partnerships with other influencers or organizations in the energy industry enhance efforts to promote AI-driven sustainability and ethical governance? What role do emerging AI technologies such as machine learning and big data play in driving these collaborative efforts?"*	Participant 1 (Designing Architect): "AI can be used to optimize energy storage systems in smart grids, but the security of stored data needs more robust measures to prevent breaches." Participant 2 (Electronic Engineer): "Ensuring the interoperability of various AI-based systems is essential, but with it comes the critical issue of maintaining data security during cross-platform operations." Participant 4 (AI Expert): "AI can help predict power outages, but there are growing concerns over the security of the data that these systems collect. Privacy protection must be at the forefront."	Interview participants—1, 2, 4	***"Theme 3: Privacy protection and data security are crucial in AI-based smart grids."***
Interview question 4: *"What are your thoughts on the current trends towards the adoption of AI in smart grids, and how do you see your role evolving in this context? How have you adapted to emerging trends, such as Green AI and IT energy efficiency, in your work with smart grids? What future initiatives are you planning to further promote AI-driven sustainability and ethical use in smart grids? How do you stay informed about developments in AI, ethics, and environmental sustainability in the smart grid industry?"*	Participant 2 (Electronic Engineer): "While technological innovation can enhance the performance of AI systems in smart grids, ethical governance needs to balance the use of these technologies, particularly in terms of energy consumption and low-power hardware." Participant 5 (IT Professional): "AI-based systems can optimize energy management, but the key challenge is ensuring that all stakeholders, including small-scale energy producers, can access the technology. This requires balancing innovation with ethical governance."	Interviewparticipants-	***"Theme 4: Technological innovation and ethical governance practices are balanced to optimize the performance of AI technologies in smart grids."***

4.2 Thematic Analysis

4.2.1 Theme 1: AI-Adoptions in Smart Grid Enhance Energy Efficiency and Environmental Sustainability

Artificial intelligence technologies have potentially transformed the foundation of smart grid systems and they enhance operational efficiency while optimizing all the potential resources. Some of the key AI techniques include deep learning, machine learning (ML), and optimization algorithms. According to the study of Ahmad et al. [24], Machine learning algorithms are efficient in evaluating huge datasets, predicting the demands of electricity, and identifying efficient patterns. As Participant 2, an Electronic Engineer has also explained that

> ***I was part of a project where AI was used to manage grid power distribution networks, ensuring that energy was allocated based on real-time demand.***

For this claim, the author Marengo [25] has supported the efficiency of AI technologies in the optimization of power distribution networks through his research. Deep learning technologies specifically the neural networks are significant in performing the power flow analysis and anomaly detection in different unbalanced distribution grids. A few neural models can include the radial basis function networks (RBFnets), and convolutional neural networks (CNNs) which can excel in the evaluation of nonlinear relationships within the system making it efficient for fault detection and real-time monitoring of the system. Optimization algorithms include particle swarm optimization and genetic algorithms which are associated with solving complex issues of resource allocation and energy distribution. The algorithms potentially help in developing efficient design and microgrid operations. These algorithms specifically help to ensure the optimal uses of both conventional and renewable energy sources. In this regard, Participant 4, an AI Expert has said that

> ***I contributed to a system that uses machine learning to forecast power demands based on usage patterns, which helps prevent overloads.***

The above statement is supported in the detailed study of Wen et al. [4] where the authors have provided insights about the contribution of ML algorithms in improving the overall capabilities of the smart energy systems. Reinforcement learning (RL) techniques like deep Q-networks are developed to optimize the charging schedules of electric vehicles (EVs) while balancing the demand and supply of energy in the grid system. With the increasing sustainability demand, the EV industry is boosting and thus the sophisticated management of EV charging stations is necessitated for grid stability. The integrated AI technologies with blockchain and the Internet of Things (IoT) are also contributing to establishing a more efficient and resilient power grid. AI-driven predictive analytics have developed forecasting techniques while enabling better energy distribution and load management throughout the grid network. Participant 1 in the interview has mentioned that

We measure success through energy consumption reports and reductions in emissions over time, as well as through user feedback on the efficiency of energy distribution systems.

Regarding such measurement, the authors Shahriar et al. [26] have entailed the effectiveness of smart sensors and meters installed in the households facilitating intelligent decision-making and optimizing the grid performance. Genetic algorithms are regarded as an optimization technique providing effective solutions to complex problems like resource allocation and energy distribution through microgrids. The algorithms significantly process efficient energy management while facilitating both sustainability and conventional sources of energy. GA might require multiple iterations in converting the time-consuming and adaptable challenges into an optimal solution. Moreover, the multi-agent systems (MAS) potentially involve multiple intelligent agents which are efficient in achieving the shared goals of smart grids including effective fault management and load balancing. These systems are highly efficient and flexible in operating through decentralized environments and it makes the technologies suitable for DERs integration and grid stability. Participant 3 shared an opinion about such optimization algorithms

We had issues with the security of low-powered AI hardware, which made the system vulnerable to interference. We overcame this by adding robust security layers within the firmware.

However, the study of Larsen [14] reveals that the implementation process of such technologies might be complex and require effective considerations of the key concerns involved in AI-driven energy grids.

4.2.2 Theme 2: Ethical Concerns Still Exist in AI-Driven Decision-Making for Smart Grids

With the potential awareness about sustainability and decarbonizing the economy, the continuous pressure for adopting renewable energy sources into the energy management portfolio is increasing. While developing a code of ethics, the potential ethical concerns are identified in the research. The potential ethical dilemmas of integrating AI in smart grids include the higher concerns of privacy, safety, security, and other potential socio-political issues [27]. Most of the public concerns are based on the anonymity and protection of consumers' identities in the data collection process. Potential data collection methods provide inferred data about users and it includes data about the uses of home appliances, states of appliances, and other energy uses. Along with anonymity, significant access and control over data are the potential issues for smart grids. It potentially increases the ability to make educated decisions ensuring overall improvement and accessibility of data. An ethical consideration in such circumstances potentially arises from the exploitation and monetization of collected information. The information from consumers' uses of home appliances can lend to developing targeted advertisements and manipulation of consumer data. Network operators can implement their data hosting capabilities through cloud or secured self-storage solutions. Participant 1 has highlighted such concerns:

> ***One of the key challenges we face is ensuring that the smart grid infrastructure is resilient to both physical and cyber threats.***

As supported by the study of Frank and Gen [28], they have discussed the mitigation of the potential risk of monetization and exploitation by providing control over consumer data to consumers. From the societal and political aspects, one of the potential concerns emerges from the consumer relationship and stakeholder mentality of energy operators. The transition to active participation from consumers can be effectively expressed through prioritizing end-user expectations, increased transparency, increased communication levels, and increased understanding [29]. The utilization of smart grids involves challenges in managing the billing of electricity and facilitating the purchasing and distribution of energy peer-to-peer. While encouraging consumers to determine the smart grid system, ethical concerns like fairness and discrimination develop. Potential issues like the prioritization of energy resources are also required for fair decision-making. The utilization of AI systems enables fair management of energy networks. Smart grids are possessed with the key concerns about equity and parsimony. With the decentralization of grids, the operating infrastructure raises complexity for flexible trading, tariffs, and incentives for users. Participant 4 stated

> ***Ethical AI in smart grids is not just a technical challenge but also a governance issue.***

It needs to be ensured that a specific community and group is not exploited for the profitability of the service providers. In this regard, the authors Luxmi Dhirani et al. [30] have proposed to involve the wider community in preventing such complexities. Consumers are required to be informed about the other possibilities and opportunities in optimizing their energy uses.

4.2.3 Theme 3: Privacy Protection and Data Security Are Crucial in AI-Based Smart Grids

Critical characteristics of an electric network develop potential concerns around security and privacy while digitizing smart grids. Cyberattacks have become potential concerns for smart grids as the increased communication and integration provide opportunities for hackers. These attacks are not only limited to monitoring consumers' data but also physically crippling the smart grid networks. In the more civilized and integrated societies, the central energy network needs the potential cybersecurity of AI-integrated systems. The evolving cyber-threats potentially compromise the availability, confidentiality, and integrity of the system and consumers' data. The key threats include denial of service (DoS) attacks, false data injection attacks, and manipulation of energy loads [15]. These attacks can potentially affect the efficiency and stability of smart grids. Participant 5 discussed that

AI introduces a whole new layer of cybersecurity risks because it processes vast amounts of data in real-time.

While addressing such issues, the authors Mohamed et al. [23] have proposed different strategies to enhance cyber resilience. Intrusion detection and prevention systems (IDPS) can be implemented as a secondary security layer to detect and prevent potential attacks on the system and it can prevent hackers from bypassing the initial authorization and encryption mechanisms. In the SCADA system and advanced metering infrastructure (AMI), the issues can be effectively addressed. The proven efficiency of multi-agent-based designs for System Integrity Protection (SIP) can be used to improve the self-adaptiveness and situational awareness of the systems while developing cyber resilience against cyberattacks' single points of failure [31]. Participant 2 as an Electronic Engineer has mentioned,

Collaboration with cybersecurity firms and continued investment in secure AI technology are keys to addressing these issues.

From the detailed study of Fadi et al. [32], the integration of blockchain technology can be efficient in increasing the resilience of microgrids. Smart contracts can be effectively implemented in a blockchain environment while securing self-healing functions and distributed secondary controls against false data injection attacks (FDIAs) within a zero-trust situation. Participant 4 supported this by stating

Smart contracts can be effectively implemented in a blockchain environment while securing self-healing functions and distributed secondary controls against false data injection attacks.

As opined by Sulaiman et al. [18], the implementation of a Cyber-Resilient Economic Dispatch (CRED) can mitigate Load-Altering Attacks (LAAs) by coordinating different frequency droop controls in Inverter-Based Resources (IBRs) like solar energy or air turbines. It will potentially optimize the operational cost and minimize the destabilizing effects of such cyberattacks. The use of Software-Defined Networking (SDN) effectively enables transparency, visibility, programmability, and network control in microgrid communication, enhancing the overall resilience and security of grid operations.

4.2.4 Theme 4: Technological Innovation and Ethical Governance Practices Are Balanced to Optimize the Performance of AI Technologies in Smart Grids

The application of AI technologies in power systems is understood from the evaluation of key phases of energy production, transmission, and distribution. The specific roles of AI innovations and technologies can be obtained from the power system phases and it can be understood that it effectively optimizes the operations, improves overall efficiency, and enhances the resilience of the system. Machine learning algorithms are used in predicting wind speed and solar irradiance for sustainable energy development with more accuracy. It allows energy companies to forecast the estimated outputs for energy and scheduling of resource uses [22]. Along with the prediction capabilities, the predictive maintenance algorithms are utilized in anticipating potential machine or equipment failures. It significantly enables the operators to lower the maintenance cost and downtime for smart grids. In the transmission phase, deep learning techniques enable real-time detection and diagnosis of faults and anomalies. The optimization algorithms also contribute to dynamic adjustment to power flows and voltage level management preventing the failures of grids. Participant 3 has provided a statement in this regard as

I've been adapting by focusing on hardware that is both energy-efficient and compatible with AI technologies.

In the detailed analysis of Shahriar et al. [26], key insights about compatible and energy-efficient AI technologies are obtained and the multi-agent system can be efficient in developing situational awareness as well as adaptive responses to such unexpected events. AI technologies and innovations have a significant role in power distribution. Green AI techniques effectively integrate DERs like wind turbines and solar panels into the conventional grid system. Reinforcement learning algorithms are also effective in optimizing load management while identifying consumption patterns and adjusting the energy supply as per the market demand [19]. Participant 4 supported the findings by stating

Reinforcement learning algorithms are also effective in optimizing the load management while identifying the consumption patterns and adjusting the energy supply as per the market demand.

According to the opinion of Mohamed et al. [23], such reinforcement algorithm effectively helps the operators not only develop innovative demand response strategies but also facilitate the effective integration of sustainable energy resources in the smart grids. AI-enabled predictive analytics are useful in regulating the voltage of the grid and reducing energy losses in the distribution phase. The reliability of the grid and efficiency of energy distribution effectively improve for the end-users.

4.3 Discussion

The interviews and detailed analysis of the themes have potentially revealed the key challenges from ethical and security perspectives. From the ethical perspective, potential concerns emerge from the biasedness of different AI algorithms which might lead to discriminatory outcomes in energy pricing and distribution. Significant ethical concerns about transparency, accountability, equity, and job displacement also exist in the AI-enabled smart grids and it is supported from the findings of the study of Sulaiman et al. [18]. While collecting data from the installed smart meters in the households and sensors in the smart grids, key concerns including data breaches, unethical uses of consumers' data, cyberattacks, and physical tampering have emerged. The interview participants have expressed their concerns about such data breaches and cyberattacks through their responses. They have also indicated the misuse of consumers' information for customized advertising purposes. The authors Wen et al. [4] have also evaluated such key concerns in AI-based smart grid facilities through their analysis. However, they have proposed effective implementation of blockchain technology or standardized code of ethics into the systems so that the emerging concerns can be effectively addressed.

5 Conclusion

In conclusion to the above study, the significance of adopting adequate security measures, privacy protections, and ethical codes of conducts for AI technologies is necessary throughout the AI innovations planning of smart grids. Despite the key advantages of AI technologies, ML algorithms, neural networks, and optimization techniques, the rising challenges are required significant attentions to establish a sustainable smart grid system. Throughout the key phases of smart grids, AI has potentially contributed with its innovative techniques and strategies. However, the interviewed participants and scholarly articles have suggested some potential measures to ensure sustainability, efficiency, and reliability in AI-based energy development programs. In this regard, the following recommendations are proposed to enhance the efficiency and ensure the long-term sustainability of AI-driven grid management.

Implementation of the standard code of ethics: The smart grid operators are suggested to implement the standard ethics codes in operating smart grid system.

Effective cybersecurity management: Blockchain-based security systems should be adopted in smart grids so that the operators can significantly protect consumers' data and prevent unauthorized access to smart grids [15]. Potential data breaches, unauthorized access, and physical security of smart grids should also be addressed by implementing effective technologies like SDN, SIP, and other authorization techniques.

Privacy protection: Data privacy regulations should be adopted for smart grids to safeguard consumers' personal information [25]. The industry's best practices should be implemented for determining the energy pricing and power distributions to different households.

Promoting sustainability: Green AI and sustainable energy sources should be integrated into smart grids while ensuring overall transparency and addressing the key issues in AI-driven grid systems [19]. Reliability, stability, and efficiency can be effectively maintained by addressing the rising ethical and security concerns in smart grids.

5.1 Future Scope and Limitations

Although the study leverages primary insights from industry experts, it has not properly addressed all the key findings from other secondary studies, articles, and websites. The collected qualitative data about the potential measures for identified issues in AI development for grid management are not quantifiable enough to perform a statistical analysis and graphical representation of the findings, limiting the easy understanding of the topic. The interviews are also conducted for limited number of participants and such small sample might not provide generalizable findings for worldwide smart grids. Along with the generalizability and quantifiable issues, the potential limitations include the limited overview of the key concerns of advanced technologies in the whole energy sector [18]. Moreover, the study lacks of sufficient theoretical perspectives on understanding the key sustainability concerns of AI adoptions in energy industry.

Potential knowledge about the emerging ethical, privacy, and security challenges due to AI adoptions in the targeted grid system is gained from the research, and it will effectively support further studies to develop. From the diverse perspectives and opinions of participants from different fields, the gathered insights enhance the credibility of the research. As the interviewees are contacted from AI expertise, electric engineers, architects, and IT professions, their ideas about emerging issues for AI implementation in energy sector will effectively support the studies about AI technology innovations specifically for transforming smart grid and power distribution [25]. Moreover, studies about integrating sustainable resources and DMEs will be developed based on the knowledge about key implications for AI technology in smart grids.

References

1. Beserra, M. (2022, July). Smart grid systems code of ethics. *Proceedings of the Wellington Faculty of Engineering Ethics and Sustainability Symposium*. Retrieved from https://ojs.victoria.ac.nz/wfeess/article/download/7654/6814
2. Couraud, B., Andoni, M., Robu, V., Norbu, S., Chen, S., & Flynn, D. (2023). Responsive FLEXibility: A smart local energy system. *Renewable and Sustainable Energy Reviews, 182*, 113343. Retrieved from https://www.sciencedirect.com/science/article/pii/S1364032123002009
3. Cascio, E. L., Ma, Z., & Maréchal, F. (2020). How smart is the grid? arXiv preprint arXiv:2006.04943. Retrieved from https://arxiv.org/pdf/2006.04943
4. Wen, X., Shen, Q., Zheng, W., & Zhang, H. (2024). AI-driven solar energy generation and smart grid integration a holistic approach to enhancing renewable energy efficiency. *International Journal of Innovative Research in Engineering and Management, 11*(4), 55–66. Retrieved from https://ijirem.irpublications.org/index.php/ijirem/article/download/61/55
5. Antonopoulos, I., Robu, V., Couraud, B., & Flynn, D. (2021). Data-driven modelling of energy demand response behaviour based on a large-scale residential trial. *Energy and AI, 4*, 100071. Retrieved from https://www.sciencedirect.com/science/article/pii/S2666546821000252
6. Olsson, W., & Vadeghani, S. (2024). Navigating the integration of artificial intelligence in the Swedish electricity sector: Harnessing AI for enhanced efficiency and business value

in Sweden's energy landscape. Retrieved from https://www.diva-portal.org/smash/get/diva2:1873074/FULLTEXT01.pdf

7. Aklilu, Y. T., & Ding, J. (2021). Survey on blockchain for smart grid management, control, and operation. *Energies, 15*(1), 193. Retrieved from https://www.mdpi.com/1996-1073/15/1/193/pdf
8. Tan, A. (2023). Exploring the potential of artificial intelligence to enhance energy efficiency in smart grid systems: A detailed review and future directions. *Journal of Sustainable Urban Futures, 13*(12), 11–20. Retrieved from https://neuralslate.com/index.php/Journal-of-Sustainable-Urban-Fut/article/download/120/88
9. Lo Cascio, E., Girardin, L., Ma, Z., & Maréchal, F. (2021). How smart is the grid? *Frontiers in Energy Research, 9*, 637447. Retrieved from https://www.frontiersin.org/articles/10.3389/fenrg.2021.637447/pdf
10. Dong, S., Cao, J., Flynn, D., & Fan, Z. (2022). Cybersecurity in smart local energy systems: Requirements, challenges, and standards. *Energy Informatics, 5*(1), 9. Retrieved from https://link.springer.com/content/pdf/10.1186/s42162-022-00195-7.pdf
11. Hafeez, S. (2021). Challenges and opportunities in implementation and utilization of IoT in the energy sector: An empirical evidence. Retrieved from https://osuva.uwasa.fi/bitstream/handle/10024/12021/Challenges%20and%20Opportunities%20in%20Implementation%20and%20Utilization%20of%20IoT%20in%20the%20Energy%20Sector%20An%20Empirical%20Evidence.pdf?sequence=2&isAllowed=y
12. McAreavey, K., Liu, W., Bauters, K., Ivory, D., Loukas, G., Panaousis, M., et al. (2024). A qualitative field study on explainable AI for lay users subjected to AI cyberattacks. arXiv preprint arXiv:2406.07369. Retrieved from https://arxiv.org/pdf/2406.07369
13. Yeboah, K. (2020). Artificial intelligence (AI) and inclusive innovation: Examining contemporary AI initiatives in sub-Saharan Africa. Retrieved from https://era.library.ualberta.ca/items/739df46f-e639-4e1e-842e-bf0fc4bb00f0/download/ab438572-6831-4b0f-a748-5284e6ee3444
14. Larsen, B. C. (2022). *Governing artificial intelligence: Lessons from the United States and China*. PhD. Copenhagen Business School. Retrieved from https://research.cbs.dk/files/77190264/benjamin_c_larsen_phd_series_29_2022.pdf
15. Yella, A. (2024). The AI revolution in energy: Transforming sustainability and grid management. *International Journal of Innovative Research in Science, Engineering and Technology, 13*(6), 11966–11974. Retrieved from https://www.researchgate.net/profile/Anusha-Yella/publication/382066530_The_AI_Revolution_in_Energy_Transforming_Sustainability_and_Grid_Management/links/668b6af22aa57f3b8274bf99/The-AI-Revolution-in-Energy-Transforming-Sustainability-and-Grid-Management.pdf
16. Ketter, W., Schroer, K., & Valogianni, K. (2023). Information systems research for smart sustainable mobility: A framework and call for action. *Information Systems Research, 34*(3), 1045–1065. Retrieved from https://pubsonline.informs.org/doi/pdf/10.1287/isre.2022.1167
17. Vlachová, M., & Binder, A. (2024). Driving change in the energy sector: Evaluating entrepreneurial opportunities provided by artificial intelligence. Retrieved from https://www.diva-portal.org/smash/get/diva2:1875458/FULLTEXT01.pdf
18. Sulaiman, A., Nagu, B., Kaur, G., Karuppaiah, P., Alshahrani, H., Reshan, M. S. A., et al. (2023). Artificial intelligence-based secured power grid protocol for smart city. *Sensors, 23*(19), 8016. Retrieved from https://www.mdpi.com/1424-8220/23/19/8016/pdf
19. Bouramdane, A. A. (2023). Cyberattacks in smart grids: Challenges and solving the multi-criteria decision-making for cybersecurity options, including ones that incorporate artificial intelligence, using an analytical hierarchy process. *Journal of Cybersecurity and Privacy, 3*(4), 662–705. Retrieved from https://www.mdpi.com/2624-800X/3/4/31/pdf
20. Volkova, A., Hatamian, M., Anapyanova, A., & de Meer, H. (2024, September). Being accountable is smart: Navigating the technical and regulatory landscape of AI-based services for power grid. In *Proceedings of the 2024 International Conference on Information Technology for Social Good* (pp. 118–126). Retrieved from https://arxiv.org/pdf/2408.01121

21. Ali, G., Mijwil, M. M., Buruga, B. A., Abotaleb, M., & Adamopoulos, I. (2024). A survey on artificial intelligence in cybersecurity for smart agriculture: State-of-the-art, cyber threats, artificial intelligence applications, and ethical concerns. *Mesopotamian Journal of Computer Science, 2024*, 71–121. Retrieved from https://journals.mesopotamian.press/index.php/cs/article/download/455/322
22. Marques, P. C., & Oliveira, P. A. (2024). Artificial intelligence technologies applied to smart grids and management. Retrieved from https://www.preprints.org/manuscript/202406.1248/download/final_file
23. Mohamed, N., Almazrouei, S. K., Oubelaid, A., Bajaj, M., Jurado, F., & Kamel, S. (2023, June). Artificial intelligence (AI) and machine learning (ML)-based information security in electric vehicles: A review. In *2023 5th Global power, energy and communication conference (GPECOM)* (pp. 108–113). IEEE. Retrieved from https://www.researchgate.net/profile/Nachaat-Mohamed/publication/371614762_AI_and_ML-based_Information_Security_in_Electric_Vehicles_A_Review/links/653a468273a2865c7ad37203/AI-and-ML-based-Information-Security-in-Electric-Vehicles-A-Review.pdf
24. Ahmad, K., Maabreh, M., Ghaly, M., Khan, K., Qadir, J., & Al-Fuqaha, A. (2020). Developing future human-centred smart cities: Critical analysis of smart city security, interpretability, and ethical challenges. arXiv preprint arXiv:2012.09110. Retrieved from https://arxiv.org/pdf/2012.09110
25. Marengo, A. (2023). The future of AI in IoT: Emerging trends in intelligent data analysis and privacy protection. Retrieved from https://www.preprints.org/manuscript/202312.2184/download/final_file
26. Shahriar, S., Allana, S., Hazratifard, S. M., & Dara, R. (2023). A survey of privacy risks and mitigation strategies in the artificial intelligence life cycle. *IEEE Access, 11*, 61829–61854. Retrieved from https://ieeexplore.ieee.org/iel7/6287639/10005208/10155147.pdf
27. Alsaigh, R., Mehmood, R., & Katib, I. (2023). AI explainability and governance in smart energy systems: A review. *Frontiers in Energy Research, 11*, 1071291. Retrieved from https://www.frontiersin.org/articles/10.3389/fenrg.2023.1071291/pdf
28. Frank, E., & Gen, D. (2024). Harnessing artificial intelligence for grid modernization: Opportunities and challenges. Retrieved from https://www.researchgate.net/profile/Edwin-Frank/publication/378715167_Harnessing_Artificial_Intelligence_for_Grid_Modernization_Opportunities_and_Challenges/links/65e6f3e8adf2362b63780a49/Harnessing-Artificial-Intelligence-for-Grid-Modernization-Opportunities-and-Challenges.pdf
29. Jayashree, S., & Malarvizhi, K. (2021). AI-based smart micro grids. In *Advanced controllers for smart cities: An Industry 4.0 perspective* (pp. 1–14). Springer. Retrieved from https://www.researchgate.net/profile/Alagar-Karthick-2/publication/349999248_Standalone_PV-Wind-DG-Battery_Hybrid_Energy_System_for_Zero_Energy_Buildings_in_Smart_City_Coimbatore_India/links/614c5f1b3c6cb3106987fccd/Standalone-PV-Wind-DG-Battery-Hybrid-Energy-System-for-Zero-Energy-Buildings-in-Smart-City-Coimbatore-India.pdf#page=12
30. Luxmi Dhirani, L., Mukhtiar, N., Chowdhry, B. S., & Newe, T. (2023). Ethical dilemmas and privacy issues in emerging technologies: A review. Retrieved from https://researchrepository.ul.ie/articles/journal_contribution/Ethical_dilemmas_and_privacy_issues_in_emerging_technologies_a_review/23622804/1/files/41449836.pdf
31. Gupta, R., Tanwar, S., Al-Turjman, F., Italiya, P., Nauman, A., & Kim, S. W. (2020). Smart contract privacy protection using AI in cyber-physical systems: Tools, techniques and challenges. *IEEE Access, 8*, 24746–24772. Retrieved from https://ieeexplore.ieee.org/iel7/6287639/6514899/08976143.pdf
32. Fadi, O., Karim, Z., & Mohammed, B. (2022). A survey on blockchain and artificial intelligence technologies for enhancing security and privacy in smart environments. *IEEE Access, 10*, 93168–93186. Retrieved from https://ieeexplore.ieee.org/iel7/6287639/9668973/09874817.pdf

Enhancing Customer Experience with Generative AI Chatbots: Strategies, Applications and Future Trends

Tiep Tai Nguyen, Cuong Chi Hoang, Keshav Kaushik, and Ngoc Dang Khoa Nguyen

1 Introduction

In today's hyper-competitive business landscape, the rapid advancement of digital technologies and disruptive innovations is fundamentally transforming organizational processes, enhancing performance and enabling firms to achieve their strategic goals [1, 2]. One of the most significant areas of transformation is customer service, which is currently undergoing a profound metamorphosis driven by generative artificial intelligence (AI) chatbots. These advanced technologies are not only reshaping how firms interact with customers but also offering a more efficient, personalized and scalable approach to service delivery [3, 4]. As customer expectations continue to rise in an increasingly digital world, the integration of AI into customer service has become essential for enhancing overall customer experience and maintaining a competitive edge [5, 6]. Generative AI chatbots leverage sophisticated natural language processing (NLP) and machine learning algorithms to understand and respond to customer inquiries with human-like accuracy. This technological revolution allows firms to manage numerous inquiries simultaneously, significantly

T. T. Nguyen
Faculty of Information Technology, Dong Nai Technology University, Bien Hoa City, Vietnam

C. C. Hoang (✉)
Faculty of Commerce, Van Lang University, Ho Chi Minh City, Vietnam
e-mail: cuong.hc@vlu.edu.vn

K. Kaushik
School of Computer Science, Petroleum and Energy Studies University, Dehradun, Uttarakhand, India

N. D. K. Nguyen
School of Business and Law, Central Queensland University, Melbourne, VIC, Australia

P. Raj et al. (eds.), *Artificial Intelligence (AI) for IT Energy Efficiency and Green AI for Environment Sustainability*,
https://doi.org/10.1007/978-3-031-89420-6_21

reducing wait times and improving overall efficiency. The capability to provide real-time responses and round-the-clock service not only enhances customer satisfaction but also allows businesses to operate more effectively and allocate resources more strategically [7]. Consequently, the shift to AI-driven customer service represents not just an operational enhancement but a strategic imperative for organizations looking to thrive in a digital-first environment.

Despite the promising benefits, transitioning from pilot phases to full-scale production remains a significant challenge. Reports indicate that approximately 90% of AI pilot projects fail to reach full deployment, primarily due to technical limitations, integration complexities and organizational resistance [8]. This high failure rate highlights the necessity for comprehensive strategies that address both technical and organizational challenges. It also emphasizes the importance of fostering a culture that embraces technological change and innovation. The global generative AI market reflects this growing trend, projected to expand from USD 67.18 billion in 2024 to USD 967.65 billion by 2032, with a compound annual growth rate (CAGR) of 39.6% [9]. The acceleration of online interactions catalyzed by the COVID-19 pandemic has intensified the demand for streamlined and personalized customer engagements [5, 10]. Generative AI-powered chatbots are uniquely positioned to meet these needs, as they provide continuous, 24/7 support, enhancing both customer satisfaction and operational efficiency while highlighting their transformative potential across various sectors [11]. A key advantage of AI-driven chatbots is their capability to facilitate personalized interactions. By leveraging vast amounts of customer data, these chatbots can tailor responses to individual preferences and previous interactions, thereby creating a more engaging and satisfying customer experience [12]. This personalization extends beyond mere query handling to include proactive engagement, whereby chatbots can anticipate customer needs and offer relevant suggestions. Such capabilities not only enhance customer satisfaction but also foster loyalty and retention [13]. However, the deployment of generative AI chatbots is not without challenges. Data privacy concerns are particularly significant, as these chatbots frequently handle sensitive customer information. Ensuring compliance with data protection regulations, such as the General Data Protection Regulation (GDPR) and implementing robust security measures are crucial for maintaining customer trust and safeguarding against potential breaches [14]. Additionally, integrating AI chatbots with existing legacy systems can prove complex and resource-intensive, requiring substantial technical expertise and ongoing maintenance to ensure seamless operation [15, 16].

To illustrate the transformative potential of generative AI chatbots, this chapter will explore four real-world case studies across diverse sectors. These case studies will demonstrate how various firms have successfully implemented AI-driven chatbots to streamline operations, enhance customer engagement and achieve significant cost savings. Each case study will detail the strategies employed, the benefits realized and the challenges encountered, thereby providing invaluable insights for organizations seeking to leverage AI technologies effectively. By examining these real-world applications and addressing the associated challenges, this chapter aims to offer practical guidance for firms looking to harness the power of generative AI chatbots. The objective is to provide a comprehensive understanding of how these technologies can revolutionize customer service, drive enhanced customer experiences and foster sustainable business growth in the contemporary digital landscape. Ultimately, this exploration will contribute to the broader discourse on the role of AI in transforming business practices and improving customer engagement in an increasingly digital world. While generative AI chatbots present significant opportunities for transforming customer service, firms should navigate the complexities of implementation. By balancing technological advancements with ethical, legal and operational considerations, firms can maximize the benefits of AI while minimizing potential risks. As the landscape continues to evolve, those that efficiently integrate generative AI chatbots into their customer service strategies are likely to gain a competitive advantage, ensuring they meet and exceed the expectations of their customers in the digital age.

2 Fundamentals of Generative AI Chatbots

2.1 Definition of Generative AI

Generative AI represents a transformative category of artificial intelligence focused on creating new content across various formats, including text, images, music and more. This innovative technology learns patterns and structures from existing data to generate novel outputs [14]. Unlike traditional AI, which operates on predefined rules and algorithms to produce outputs, generative AI employs advanced machine learning models, particularly neural networks, enabling it to generate contextually relevant content with a high degree of sophistication [17]. The essence of generative AI lies in its ability to understand and replicate aspects of human-like creativity and problem-solving. This capacity makes it a powerful tool in various applications, including the development of intelligent chatbots that can engage users more effectively than previous technologies [18].

2.2 Applications in Generative AI Chatbot Development

Generative AI has profoundly revolutionized chatbot development by significantly enhancing the ability of chatbots to engage in natural, dynamic and context-aware conversations. Traditional rule-based chatbots are constrained by their reliance on pre-programmed responses, which often limits their ability to navigate the complexities and nuances of human language [6]. These conventional systems can struggle to provide satisfactory interactions, leading to frustration for users. In contrast, generative AI-driven chatbots leverage sophisticated language models such as GPT-4 to produce human-like responses in real time. This advanced capability allows these chatbots to adapt to the context of conversations, enabling more accurate and personalized interactions [19]. Such versatility means that generative AI chatbots can perform a wide range of functions, from answering customer inquiries and providing technical support to conducting virtual interviews and offering tailored product recommendations. By continuously analyzing vast amounts of interaction data and learning from past experiences, generative AI chatbots improve their performance over time, making them more efficient and effective in meeting user needs [14]. This adaptive learning process is key to their success and differentiation in the marketplace.

2.3 Types of Generative AI-Driven Chatbots

The capabilities of generative AI enable developers to create various types of chatbots, each designed to understand and respond to user inputs effectively while continuously learning from each interaction. This results in a more engaging and personalized user experience. The primary types of generative AI-driven chatbots [15, 20] include the following:

- *Conversational chatbots*: These chatbots are designed to simulate human conversation, providing coherent and contextually relevant responses. They are particularly valuable in customer service, virtual assistant roles and social media interactions, where engaging and human-like communication is crucial. By employing generative AI, conversational chatbots can maintain the flow of dialogue, understand user intent and respond appropriately, thus enhancing customer satisfaction and loyalty.
- *Task-oriented chatbots*: These chatbots are specialized in performing specific tasks such as booking appointments, processing orders or providing technical support. They utilize generative AI to comprehend user requests and execute tasks efficiently, often integrating seamlessly with other systems to enhance service delivery. This integration allows task-oriented chatbots to streamline operations and improve overall user experience by reducing the time and effort required to complete common tasks.
- *Knowledge-based chatbots*: Leveraging generative AI, these chatbots excel in providing detailed information and answering complex queries. They are commonly utilized in educational platforms, healthcare settings and corporate environments, where the need for accurate and comprehensive information is paramount. Knowledge-based chatbots can access extensive databases to offer precise answers, thereby facilitating learning and informed decision-making.

To sum up, generative AI chatbots represent a significant advancement in artificial intelligence, providing businesses with tools to enhance user engagement and satisfaction. By employing generative AI technology, these chatbots can adapt to user needs, respond intelligently to inquiries and continuously improve through machine learning. Understanding the definitions, applications and types of generative AI-driven chatbots is essential for organizations looking to leverage this technology effectively in today's dynamic digital landscape. As generative AI continues to evolve, its potential to transform customer interactions and operational efficiency will only grow, making it a critical component of modern business strategy.

3 Benefits of Generative AI Chatbots

Generative AI chatbots have revolutionized customer service across various industries, offering a plethora of benefits that enhance customer interactions, streamline operations and boost overall efficiency. These chatbots harness the power of advanced machine-learning techniques to provide intelligent and context-aware responses, fundamentally changing how businesses engage with their customers. Below, we explore four key benefits of generative AI chatbots in detail.

- *Enhanced customer experience*: One of the most significant advantages of generative AI chatbots is their ability to significantly elevate the customer experience through personalized and contextually relevant interactions. Unlike traditional chatbots that rely on pre-programmed responses, generative AI chatbots utilize advanced natural language processing (NLP) techniques to understand and respond to a wide range of queries in a human-like manner. This capability allows them to generate unique, context-aware responses, ensuring that customers feel understood and valued. Consequently, this leads to higher satisfaction levels. Similarly, generative AI chatbots can handle multiple languages and dialects, catering to a diverse global customer base and providing a seamless experience for users worldwide. This multilingual capability not only broadens accessibility but also enhances inclusivity, allowing businesses to engage with customers from various backgrounds effectively. The personalized touch and rapid response times help build stronger customer relationships and foster loyalty, ultimately enhancing brand perception and trust [12, 21].
- *Operational efficiency and cost reduction*: Generative AI chatbots drive substantial improvements in operational efficiency by automating routine and repetitive tasks. This automation allows human agents to focus on more complex and value-added activities, boosting their productivity and job satisfaction. By efficiently handling common queries, processing transactions and providing instant responses, these chatbots lighten the load on customer service teams while simultaneously reducing customer wait times. The scalability of generative AI chatbots enables them to manage high volumes of interactions simultaneously, effectively eliminating bottlenecks during peak times. This operational efficiency not only enhances customer satisfaction by providing timely assistance but also translates into significant cost savings for organizations. With fewer human resources required to manage customer inquiries, businesses can reduce expenses related to staffing, training and supervision of customer service personnel [21, 22].
- *24/7 availability and consistency*: Generative AI chatbots provide round-the-clock support, allowing customers to receive assistance at any time, regardless of

time zones or business hours. This constant availability is particularly beneficial for global businesses serving diverse regions and customer bases. Customers appreciate the flexibility of being able to interact with chatbots outside of traditional business hours, which enhances their overall experience. Likewise, generative AI chatbots deliver consistent responses, maintaining a high standard of service regardless of when or how often customers engage with them. This consistency helps build trust and reliability, as customers can expect uniform quality in their interactions with the chatbot. The unwavering availability and consistent performance of AI-driven chatbots not only enhance customer satisfaction but also improve brand reputation. Companies that provide reliable and consistent service are more likely to cultivate long-term customer loyalty and attract new clients through positive word-of-mouth [12, 22].

- *Data-driven insights and continuous improvement*: Generative AI chatbots generate extensive data from customer interactions, providing valuable insights into customer behavior, preferences and pain points. By analyzing this data, businesses can identify trends, improve products and services and tailor their marketing strategies accordingly. This data-driven approach allows organizations to make informed decisions based on actual customer experiences rather than assumptions. Moreover, generative AI chatbots are capable of continuous learning and adaptation based on new data, enhancing their performance over time. This self-improvement ensures that chatbots remain effective and relevant, keeping pace with evolving customer needs and expectations. The ongoing refinement of chatbot interactions, grounded in data-driven insights, enables organizations to stay competitive and responsive to market changes [5, 21].

As such, generative AI chatbots offer a myriad of benefits that extend beyond traditional customer service solutions. By enhancing customer experience through personalized interactions, improving operational efficiency, providing consistent 24/7 support and delivering data-driven insights, these chatbots represent a powerful tool for organizations seeking to optimize customer engagement and drive business growth. As businesses continue to adapt to an increasingly digital landscape, the implementation of generative AI chatbots will be instrumental in maintaining a competitive edge and meeting the evolving expectations of consumers. Ultimately, the strategic deployment of generative AI chatbots can lead to not only enhanced customer satisfaction but also greater operational efficiencies and cost reductions, making them an invaluable asset in the modern business environment.

4 Challenges of Generative AI Chatbots

Generative AI chatbots, despite their numerous benefits, present several significant challenges that organizations must address to fully harness their potential. The primary challenges include data privacy and security, system integration and technical complexity and maintenance and continuous improvement.

- *Data privacy and security*: A major concern with generative AI chatbots is safeguarding sensitive customer data. These chatbots often handle personal, financial and other confidential information, making them vulnerable to cyberattacks and data breaches. To prevent unauthorized access and uphold customer trust, firms must enforce stringent data privacy and security measures. This involves employing advanced encryption methods, secure data storage solutions and conducting regular security audits to identify and mitigate potential vulnerabilities. Moreover, adhering to data protection regulations such as the General Data Protection Regulation (GDPR) in Europe [23] and the California Consumer Privacy Act (CCPA) in the United States [24] adds complexity to the implementation process. Non-compliance with these regulations can lead to severe legal consequences and damage the firm's reputation [14, 25]. Therefore, organizations must invest in robust compliance programs and stay updated with evolving legal requirements to ensure that their AI initiatives do not inadvertently violate privacy laws.
- *System integration and technical complexity*: Integrating generative AI chatbots with existing systems and platforms poses significant technical challenges. These chatbots need to interface seamlessly with various customer relationship management (CRM) systems, databases and other backend infrastructures to function effectively. The integration process requires substantial technical expertise and resources, which can be a considerable obstacle for many firms, especially those lacking in-house AI specialists. Additionally, ensuring that the chatbot can accurately interpret and respond to complex queries necessitates sophisticated natural language processing (NLP) capabilities. This technical complexity can lead to implementation delays and increased costs, making it difficult for organizations to deploy and maintain effective generative AI chatbots [12, 14]. To mitigate these challenges, firms may need to collaborate with specialized AI vendors or invest in training their technical staff, both of which require significant financial and time commitments.
- *Maintenance and continuous improvement*: Generative AI chatbots require ongoing maintenance and updates to remain effective and relevant. Continuous learning from new data is essential for improving the chatbot's performance and adapting to changing customer needs and language patterns. However, this process demands constant monitoring, fine-tuning and retraining of the AI models, which can be resource-intensive. Organizations must invest in skilled personnel who can manage these tasks and ensure the chatbot stays current with the latest advancements in AI and NLP technologies. Additionally, the chatbot's ability to handle a wide array of queries and scenarios depends on the quality and diversity of the training data, necessitating a consistent effort to curate and expand this dataset [12, 25]. Neglecting maintenance can lead to decreased performance, user dissatisfaction and potential competitive disadvantages.

While generative AI chatbots hold transformative potential for enhancing customer service and operational efficiency, addressing challenges related to data privacy and security, system integration and technical complexity and ongoing maintenance is indispensable for successful implementation. Overcoming these hurdles enables firms to fully capitalize on the benefits of AI-driven customer service and maintain a competitive edge in the digital marketplace. Organizations must adopt a proactive approach, investing in robust security measures, technical expertise and continuous improvement processes to ensure their generative AI chatbots deliver optimal performance and value.

5 Strategies for Successful Deployment of AI Chatbots

The successful deployment of AI chatbots in customer service requires a well-thought-out strategy that addresses several key areas: integration, training, optimization and privacy compliance. Below are the best practices for each of these crucial steps.

- *Integrating AI chatbots into existing customer service workflows*: To integrate AI chatbots seamlessly into existing customer service workflows, begin by identifying specific areas where chatbots can add value. This might include handling common inquiries, managing simple transactions, or providing 24/7 support [26]. Ensure that the chatbots are compatible with existing systems, such as CRM software and helpdesk platforms for a cohesive customer service experience. A hybrid approach—where chatbots handle initial interactions and escalate more complex issues to human agents—often works best. This ensures that the efficiency of automation is balanced with the personalized touch of human service [18].
- *Training and optimizing chatbots for accuracy and efficiency*: Training AI chatbots is an ongoing process that requires continuous improvement. Start with a robust dataset of customer interactions to teach the chatbots to recognize and respond accurately to common queries. Leveraging NLP enables the chatbot to understand and respond to diverse customer inputs. Regularly updating the chatbot's training data with new services, products and customer queries is preemptive [27]. Monitoring chatbot interactions helps identify areas for improvement, and incorporating customer feedback ensures that the chatbot remains relevant and accurate. Optimizing for efficiency involves reducing response times and ensuring that the chatbot can handle multiple interactions simultaneously without compromising quality [28].
- *Addressing privacy concerns and ensuring compliance with regulations*: Privacy and data protection are paramount when deploying AI chatbots. Ensure that the chatbot complies with relevant data protection regulations, such as the GDPR in Europe or the CCPA in California [23, 24]. Implementing robust data encryption and anonymization techniques is inevitable for protecting customer information. It is crucial to be transparent with customers about how their data will be used and to provide options for them to control their data privacy. Regular audits and updates to privacy policies can help ensure compliance and strengthen customer trust [14].

By following these best practices, businesses can successfully deploy AI chatbots that enhance customer service, improve operational efficiency and maintain high standards of data privacy and regulatory compliance. This strategic approach ensures that the deployment of AI chatbots not only meets current customer service needs but also adapts to future challenges and opportunities.

6 Real-World Case Studies

6.1 Bupa: Automation Innovation to Future-Proof Workforce and Customer Service

Bupa, Australia's largest health insurer, launched the Automation Innovation Academy to integrate AI and automation into its operations, seeking to future-proof its workforce [29]. The Academy, introduced in March, trains employees in automation development, planning and implementation. Gael Filippini, Bupa's Director of Customer Transformation, highlights empowering staff to adapt to the evolving work environment, stating: "The Academy helps our people understand automation, build bots and master implementation principles." By 2030, one-third of Australian work activities could be automated, enabling employees to focus on more complex tasks. Bupa has embraced automation, notably through its overseas student health cover (OSHC) enrolment bot, which has processed 22,900 enrolments since October 2019. This has accelerated the enrolment process, freeing staff to handle more intricate issues. Recently, Bupa implemented customer service chatbots with natural language processing, reducing customer transfers by 20% and mitigating touchpoints for resolving inquiries. "As a health insurer managing large volumes of data, the Academy equips our workforce to identify and automate repetitive tasks," Filippini elucidates. Our goal is to reduce time spent on routine tasks, allowing our team to focus on creating new customer value and addressing tasks that require human insights. The Academy offers flexible online and in-person learning opportunities, covering process fundamentals, coding and change management, with support from internal mentors and automation partner UiPath. Filippini denotes that while automation is not a quick fix, it is imperative for enhancing customer service and ensuring data security in such an efficient computing environment [29, 30].

6.2 AGL: Enhancing Customer Experience with Chatbot Alfie

Originally conceived in 2016 as part of AGL Energy's $300 million digital transformation, the AI-driven chatbot Alfie, now in its fourth iteration, employs IBM's Watson Assistant. This innovative tool is designed to improve customer service by comprehending linguistic nuances and the implicit meanings often present in customer interactions [31]. AGL has trained Alfie to understand Australians' penchant for understatement, enabling the chatbot to recognize implicit meanings and respond appropriately. For instance, when a customer says, "I think my bill might be due" Alfie understands this as "I'd like to pay my bill." This ability to decipher and act on implicit meanings allows Alfie to handle upwards of 100 different nuances in customer requests. AGL employs advanced analytics to identify implicit meanings in customer interactions. If a customer asks about their account balance, analytics reveal that nine times out of ten, the customer is seeking a payment extension. However, Alfie does not second-guess customer intentions; it confirms the customer's needs before proceeding to avoid misunderstandings. These capabilities have enabled AGL to provide 24/7 customer support, handle a high volume of inquiries, automate routine tasks and deliver personalized customer experiences. The ultimate goal is to create a general chatbot capable of assisting with any customer inquiry [32]. Alfie's development has significantly improved efficiency, shaving an average of 13 minutes off customer interactions and automating about 75 percent of administrative tasks. AGL's commitment to customer-centric development is evident in its community engagement portal, where customers provide feedback and suggest new features. This approach has allowed AGL to prioritize development based on real customer needs, ensuring that Alfie evolves in ways that provide tangible value. AGL continues to enhance Alfie's capabilities. Upcoming features include a smart app for easier meter readings and tools to help customers understand and manage their energy use in real time. Alfie exemplifies how generative AI can transform customer service by understanding and responding to the subtleties of human language. Through continuous improvement and customer engagement, AGL is setting new standards in the energy sector for AI-driven customer support [31].

6.3 Sephora: Online "In-Store" Experience Through Generative AI

Over the past 3 years, the retail landscape has shifted dramatically, with many consumers moving to online stores. This raised a pivotal question: "How do we recreate the brick-and-mortar experience online?" Sephora addressed this challenge by merging physical and digital experiences using conversational AI to offer seamless support and assistance to consumers, regardless of their shopping location [33]. Sephora analyzed consumer interactions, such as product details, best prices, reviews and recommendations, to develop a virtual assistant. This assistant provides personalized consultations, connecting customers with in-store beauty advisors on demand. It offers product knowledge, makes recommendations and facilitates order reservations and pickups, thus reducing purchase friction [34]. Sephora's virtual assistant allows customers to reserve products, check product availability and ask questions about store timings and return policies. In-store beauty advisors respond to customer queries, replicating the in-store experience. This initiative led to over 9000 users on the bot in Singapore and Malaysia, 332,000 conversation sessions and a $30,000 average monthly revenue increase [35, 36]. Generative AI faces challenges, including bias from human-generated data and significant environmental impacts due to its high power and bandwidth consumption. Companies must address these issues by implementing stringent privacy protocols and exploring ways to mitigate environmental impacts [37]. Sephora's use of generative AI exemplifies its potential in retail. However, generative models require extensive data and fine-tuning to achieve the desired functionality. As the technology matures, further exploration and learning are essential. Generative AI offers opportunities for companies to innovate and establish a competitive advantage though it cannot replace human interaction. Companies must strategically integrate AI and human input to maximize value in the purchasing journey [38].

6.4 Vietnam Airlines: Enhancing Customer Experience with Chatbot NEO

Vietnam Airlines is actively exploring the integration of AI-driven chatbots to enhance its customer service and streamline operations. While the airline's main website does not explicitly promote AI-powered chatbots, various indications and alternative sources suggest a growing interest in this technology [39]. Currently, Vietnam Airlines offers a virtual assistant named "NEO" on their website. Although NEO is not explicitly labeled as AI-powered, it provides useful functionalities such as itinerary information, fare search and baggage allowance details [40]. However, these features could be greatly enhanced with advanced AI capabilities. For example, AI could enable personalized fare searches by analyzing user history and preferences, provide proactive support by anticipating customer needs (such as suggesting baggage upgrades or travel insurance) and facilitate more natural, conversational interactions through sophisticated NLP. An article on Vietnam Airlines' internal communication platform, Spirit of Vietnam Airlines, references "VNA Safety AI" [41], indicating that the airline is exploring AI for internal applications. This exploration could potentially lay the groundwork for future customer-facing chatbots. The absence of explicit mentions of AI on their public website, coupled with the basic functionalities of NEO, suggests that Vietnam Airlines may not yet have fully deployed AI-powered chatbots for customer service. To gain further insights into their AI initiatives, one should monitor the airline's social media channels for relevant announcements and investigate industry news on AI adoption among Asian airlines. Vietnam Airlines' use of AI-driven chatbots aligns with a broader industry trend toward leveraging AI and automation to enhance customer service and operational efficiency. These efforts aim to meet evolving customer expectations and maintain a competitive edge in the market, reflecting an industry-wide shift toward smarter, more efficient solutions [39, 41].

Table 1 recaps the case studies of AI and automation integration by Bupa, AGL, Sephora and Vietnam Airlines. Notably, Bupa's Automation Innovation Academy aims to future-proof its workforce by upskilling employees for in-house automation, addressing the fear of job loss. Bupa's chatbot has significantly improved processes, exemplified by the OSHC enrolment bot processing 229,000 enrolments autonomously. AGL's Alfie chatbot, driven by IBM's Watson Assistant, deciphers implicit customer meanings, automating 75% of administrative tasks and dramatically enhancing customer service. Sephora's virtual assistant merges physical and digital retail experiences, increasing revenue and improving customer support though it faces challenges such as data bias and environmental impact. Vietnam Airlines is in the exploration phase of AI integration, with its NEO virtual assistant and internal AI applications, namely, VNA Safety AI indicating a growing interest

Table 1 Case study analysis

Company	Benefits	Challenges	Strategies	Applications
Bupa	Upskilling workforce, automation of repetitive tasks, improved customer service, enhanced data processing	Fear of job displacement, data security, environmental impact	Launching Automation Innovation Academy, offering self-paced and face-to-face training, mentorship	OSHC enrolment and customer service chatbots, process identification for automation
AGL	Improved customer service efficiency, 24/7 support, personalized experiences, reduced administrative tasks	Understanding linguistic nuances, avoiding misinterpretations, continuous improvement of chatbot capabilities	Using IBM Watson Assistant, community engagement for feedback, analytics to identify implicit meanings	Alfie chatbot for customer inquiries, payment extensions, and future smart app for meter readings
Sephora	Seamless online shopping experience, reduced purchase friction, personalized consultations, increased revenue	Data bias, environmental impact, ensuring privacy and security of customer data	Developing virtual assistants, integrating AI with human advisors, stringent privacy protocols, continuous learning	Virtual assistant for reservations, product availability, in-store experience replication
Vietnam Airlines	Enhanced customer service, streamlined operations, potential personalized support, improved safety	Limited explicit AI implementation in customer service, need for advanced AI capabilities	Exploring AI for internal applications, potential future customer-facing chatbot development	NEO virtual assistant for itinerary information, fare search, baggage allowance, VNA safety AI exploration

Source(s): Author's elaboration

in advanced AI capabilities. Collectively, these case studies highlight the transformative potential of AI in various industries, highlighting the need for continuous improvement and strategic integration to maximize benefits and address challenges.

7 Conclusion

The integration of generative AI-driven chatbots signifies a pivotal evolution in customer service, demonstrating substantial improvements in efficiency, personalization and scalability. The case studies analyzed underscore the transformative potential of these technologies, with firms reporting notable enhancements in customer engagement, streamlined operations and cost savings. These successes yield key insights and strategic recommendations for future implementation.

7.1 Strategic Implementation and Optimization

Successful integration of AI chatbots requires a deliberate approach that aligns the technology with existing customer service workflows. Ensuring compatibility with current systems is crucial. Implementing a hybrid model—where chatbots manage routine tasks and escalate complex issues to human agents—creates a seamless customer experience. This approach leverages the strengths of both AI and human intelligence, optimizing efficiency and service quality. Continuous training and optimization of chatbots are essential. Utilizing real customer interactions to refine AI responses and incorporating feedback loops enhances the chatbot's effectiveness over time [12]. Moreover, organizations should prioritize the development of sophisticated natural language processing capabilities to handle diverse and complex queries accurately. Privacy and data protection remain critical concerns. AI chatbots often handle sensitive information, necessitating strict adherence to data protection regulations, namely GDPR in Europe and CCPA in California. Implementing robust security measures—including advanced encryption, secure data storage and regular security audits—is imperative to prevent breaches and maintain customer trust [14, 25]. Clear communication of privacy policies and offering customers control over their data further build trust and ensure compliance.

7.2 Future Trends in Generative AI Chatbots

Looking forward, the future of generative AI-driven chatbots is promising. Advancements in AI technology are expected to enhance these systems' capabilities significantly. Future trends may include more sophisticated natural language understanding, enabling even more nuanced and context-aware interactions. Integrating

AI with emerging technologies such as voice and video interfaces could expand chatbot applications, offering richer and multimodal customer experiences [19]. Gartner forecasts a dramatic surge in generative AI adoption [42]:

- By 2026: 75% of firms are expected to use generative AI for creating synthetic customer data, up from less than 5% in 2023.
- By 2027: Over 50% of generative AI models will be tailored to specific industries or business functions, a significant increase from approximately 1% in 2023.
- By 2028: 30% of generative AI implementations will incorporate energy-conserving computational methods driven by sustainability initiatives.

These trends indicate that generative AI chatbots will become more specialized, efficient and environmentally sustainable, aligning with broader organizational goals and customer expectations.

7.3 Profound Impact on Customer Service

The advancements in generative AI are poised to have a profound impact on customer service. AI-driven chatbots will increasingly handle complex queries and provide proactive support, allowing human agents to focus on higher value tasks requiring emotional intelligence and critical thinking. This shift could lead to more efficient service delivery, greater customer satisfaction and a competitive edge for early adopters. The enhanced personalization and scalability offered by AI chatbots will foster stronger customer relationships. By leveraging data-driven insights, businesses can tailor interactions to individual customer needs, increasing engagement and loyalty. The ability to handle high volumes of interactions without compromising service quality is particularly advantageous in today's digital economy, where customer expectations are continually rising.

7.4 Strategic Recommendations for Future Implementation

To fully capitalize on the benefits of generative AI chatbots, firms should consider the following strategic recommendations:

- *Invest in data security and compliance*: Implement advanced security measures and ensure strict compliance with data protection regulations to safeguard customer information and maintain trust.
- *Adopt a phased integration approach*: Start with pilot programs to test chatbot performance and integration capabilities, then scale up gradually, refining the system based on feedback and performance metrics.
- *Prioritize continuous learning*: Allocate resources for ongoing maintenance and updates. Enable the chatbot to learn from each interaction to improve accuracy and effectiveness over time.

- *Implement a hybrid service model*: Combine AI chatbots with human agents to handle a broader range of customer needs efficiently, ensuring that complex issues receive the necessary human attention.
- *Stay informed on technological advancements*: Keep abreast of emerging AI technologies and integrate new features that enhance functionality, such as voice and video interfaces, to remain competitive.
- *Develop ethical AI guidelines*: Establish clear ethical guidelines for AI usage, addressing potential biases and ensuring transparency in AI decision-making processes to foster trust and align with societal expectations.

In a nutshell, generative AI-driven chatbots have the potential to revolutionize customer service by offering enhanced personalization, efficiency and scalability. By addressing challenges related to data privacy, system integration and ongoing maintenance, firms can fully leverage these technologies to drive substantial improvements in customer experience and operational effectiveness. The ongoing evolution of AI technology promises to reshape customer service paradigms. It is part and parcel for businesses to stay ahead of these developments and strategically integrate AI solutions into their service models. Early adoption and thoughtful implementation can lead to sustained competitive advantages, greater customer satisfaction and long-term growth. Embracing generative AI chatbots not only meets current customer expectations but also positions organizations to adapt to future market dynamics. As AI continues to advance, those firms that proactively implement and optimize AI-driven customer service solutions are likely to thrive in the increasingly digital and customer-centric business landscape.

References

1. Nafizah, U. Y., Roper, S., & Mole, K. (2024). Estimating the innovation benefits of first-mover and second-mover strategies when micro-businesses adopt artificial intelligence and machine learning. *Small Business Economics, 62*(1), 411–434. Print.
2. Nguyen, N. D. K., & Ali, I. (2021). Implementation of cloud customer relationship management in banking sector: Strategies, benefits and challenges. *International Journal of Electronics and Communication Engineering, 15*(6), 242–247. Print.
3. Althoff, J. (2024). *Embracing AI transformation: How customers and partners are driving pragmatic innovation to achieve business outcomes with the microsoft cloud.* Microsoft. Web. May 17, 2024.
4. Australian Computer Society. (2024). *Australia's digital pulse.* Australian Computer Society. Web. May 8, 2024.
5. Das, A. C., et al. (2023). *The next frontier of customer engagement: AI-enabled customer service.* McKinsey & Company. Web. July 25, 2024.
6. PwC. (2024). *What changes minds on GenAI? Adopting it.* PwC. Web. April 28, 2024.
7. IBM. (2024). *What is generative AI?* IBM. Web. May 10, 2024.
8. McKinsey & Company. (2024). *McKinsey's ecosystem of strategic alliances brings the power of generative AI to clients.* McKinsey & Company. Web. May 2, 2024.
9. Fortune Business Insights. (2024). *Generative AI market size, share & industry analysis, by model (generative adversarial networks organs and transformer-based models), by indus-*

try vs application, and regional forecast, 2024–2032. Fortune Business Insights. Web. April 29, 2024.
10. Ali, I., Nguyen, N. D. K., & Gupta, S. (2023). A multi-disciplinary review of enablers and barriers to cloud Erp implementation and innovation outcomes. *Journal of Enterprise Information Management, 36*(5), 1209–1239. Print.
11. Schaefer, J. (2020). *AI and chatbots can help organizations meet rising customer expectations*. Harvard Business Review. Web. June 6, 2024.
12. Thomson Reuters. (2024). *2024 Generative AI in professional services*. Thomson Reuters. Web. April 27, 2024.
13. Edelman, D. C., & Abraham, M. (2022). *Customer experience in the age of AI*. Harvard Business Review. Web. July 18, 2024.
14. Gartner. (2024). *Gartner experts answer the top generative AI questions for your enterprise*. Gartner. Web. May 10, 2024.
15. Intercom. (2024). *Integrating a chatbot with your knowledge base*. Intercom. Web. July 21, 2024.
16. Nguyen, N. D. K., et al. (2024). Bridging the nexus between cloud Erp and enterprise resilience. *Journal of Global Information Management, 32*(1), 1–22. Print.
17. Simón, C., Revilla, E., & Sáenz, M. J. (2024). Integrating AI in organizations for value creation through human-AI teaming: A dynamic-capabilities approach. *Journal of Business Research, 182*, 114783. Print.
18. Singh, J., & Padgaonkar, S. (2024). *Navigating the frontier: Key considerations for developing a generative AI integration strategy for the enterprise*. Oracle. Web. May 20, 2024.
19. OpenAI. (2024). GPT-4 Turbo and GPT-4. OpenAI. Web. May 10, 2024.
20. Oracle. (2024). *What is a chatbot?* Oracle. Web. July 20, 2024.
21. Winchurch, E. (2020). *Independent study finds IBM Watson Assistant customers can accrue $23.9 million in benefits*. IBM. Web. June 6, 2024.
22. Holdsworth, J. (2024). *Unlocking the power of chatbots: Key benefits for businesses and customers*. IBM. Web. June 2, 2024.
23. Romero Moreno, F. (2024). Generative AI and deepfakes: A human rights approach to tackling harmful content. *International Review of Law, Computers and Technology, 38*, 297. Print.
24. Financier Worldwide. (2018). *The California consumer privacy act and the GDPR: Two of a kind?* Financier Worldwide. Web. July 22, 2024.
25. Anderson, S. (2024). *AI in change management: Early findings, challenges and opportunities*. Prosci. Web. June 22, 2024.
26. Viliavin, R. (2023). *Customer support: Using AI chatbots for efficiency and empathy*. Forbes. Web. June 3, 2024.
27. Grensing-Pophal, L. (2023). *Using AI chatbots for employee training and development*. HR Daily Advisor. Web. June 1, 2024.
28. Microsoft. (2024). *Microsoft and NVIDIA announce major integrations to accelerate generative AI for enterprises everywhere*. Microsoft. Web. May 19, 2024.
29. Bupa. (2020). *Bupa embraces automation innovation to improve its service and future proof its workforce*. Bupa. Web. June 3, 2024.
30. Zarifis, A., Holland, C. P., & Milne, A. (2023). Evaluating the impact of AI on insurance: The four emerging AI-and data-driven Business models. *Emerald Open Research, 1*(1), 15. Print.
31. Crozier, R. (2019). AGL trains its chatbot Alfie to comprehend Aussie understatement. *ITNews*. Web. July 17, 2024.
32. Corbett, C. (2021). *Customer driven innovation*. AGL. Web. July 19, 2024.
33. AI Expert Network. (2023). *Case study: Sephora's AI-driven 'Store of the Future'*. AI Expert Network. Web. July 17, 2024.
34. Menon, P. (2016). Digital beauty: How Sephora leveraged digital technology to overhaul the in-store experience. Harvard Business School. Web. July 18, 2024.
35. Cosmetics Business. (2016). *Sephora introduces chatbot that can reserve appointments*. Cosmetics Business. Web. July 19, 2024.

36. eTail Asia. (2023). *Sephora creates an online "in-store" experience through generative AI*. eTail Asia. Web. July 19, 2024.
37. Lorca. (2018). *Sephora: Developments in machine learning and augmented reality towards customer personalization*. Harvard Business School. Web. July 18, 2024.
38. Rayome, A. D. (2018). *How Sephora is leveraging AR and AI to transform retail and help customers buy cosmetics*. TechRepublic. Web. July 17, 2024.
39. Hoang, G. (2024). *Vietnam airlines explores AI for digital transformation, holds pioneering workshop*. VnEconomy. Web. July 20, 2024.
40. Vietnam Airlines. (2024). *Chat with NEO—Vietnam Airlines' virtual assistant*. Vietnam Airlines. Web. July 20, 2024.
41. Somarajan, S. (2024). *Revolutionizing aviation and airlines industry with generative AI*. Infosys. Web. July 19, 2024.
42. Perri, L. (2024). *3 Bold and actionable predictions for the future of GenAI*. Gartner. Web. April 30, 2024.

Harnessing AI for Energy Efficiency in 6G Networks: Challenges, Solutions, and Future Directions

Naman Chauhan, Gesu Thakur, Ankush Joshi, Vikash Kumar, and Yashvir Singh

1 Introduction to 6G Communication Networks

1.1 Overview of 6G Networks

The next generation of wireless technology, 6G networks, offers ultra-high speeds, almost low latency, and unparalleled connection, surpassing the capabilities of 5G by a significant margin. Designed to facilitate transformational applications, 6G is expected to facilitate ubiquitous artificial intelligence (AI), immersive extended reality (XR), holographic communications, and the Internet of Everything (IoE). These apps need to integrate seamlessly across various platforms and devices in addition to having high data rates and extensive connectivity. 6G, fueled by developments in the millimeter-wave and terahertz frequency bands, is expected to serve a hyperconnected society by offering response times near a millisecond and data rates in the terabits per second range [1].

But there are a lot of technical obstacles with the move to 6G, especially when it comes to network complexity and energy efficiency. 6G's high demands may result in higher power usage, which would affect operating expenses and environmental sustainability. Energy-efficient design principles and sophisticated AI-driven methods for dynamic resource management are expected to be incorporated into 6G networks to allay these worries and enable optimal performance while consuming the least amount of energy. Furthermore, 6G is anticipated to make use of distributed architectures and edge computing, enabling localized processing that lowers

N. Chauhan (✉) · G. Thakur · A. Joshi · V. Kumar
Department of Computer Science and Application, COER University, Roorkee, Uttarakhand, India

Y. Singh
Department of Legal Studies, COER University, Roorkee, Uttarakhand, India

P. Raj et al. (eds.), *Artificial Intelligence (AI) for IT Energy Efficiency and Green AI for Environment Sustainability*,
https://doi.org/10.1007/978-3-031-89420-6_22

the demand for data transmission and power usage. In the end, 6G networks will revolutionize connection by fusing smart energy solutions with fast communication, enabling a sustainable wireless technology future.

1.2 Evolution from 5G to 6G

The switch from 5G to 6G represents a significant advancement in wireless technology, ushering in an age characterized by large, intelligent, and ultra-responsive networks, replacing the high-speed connection of 5G. Faster speeds, reduced latency, and the ability to connect more devices were all groundbreaking improvements in mobile broadband that 5G brought us, but 6G takes these developments to the next level. Applications including holographic communications, extended reality (XR), and the Internet of Everything (IoE) will be supported by 6G, which is anticipated to function in the terahertz frequency range and provide data rates in the terabits per second range. In addition to improving speed and connectivity, this shift aims to facilitate the smooth integration of intelligent devices, transforming the network into an intelligent entity that can sustain a hyperconnected world [2].

6G offers more speed and capacity together with a stronger emphasis on network intelligence, adaptability, and sustainability. To lower latency and energy consumption, 6G will embrace distributed and edge computing, in contrast to 5G, which mostly depends on centralized processing. To manage the complexity of 6G, artificial intelligence (AI) will be essential for more accurate resource allocation, predictive maintenance, and security. Furthermore, 6G seeks to explicitly include environmental factors in a network architecture to handle the growing energy requirements brought on by higher data speeds and high-frequency bands. The transition to 6G represents a change from improvements that are only focused on connectivity to a comprehensive strategy that blends performance with clever, sustainable solutions.

1.3 Key Features and Requirements of 6G

1.3.1 **Ultra-High Data Rates:** 6G seeks to reach data rates in the terabits per second area to provide previously unheard-of download and streaming speeds [3].

1.3.2 **Ultra-Low Latency:** 6G, which aims for latency as low as 1 millisecond, would facilitate real-time applications such as immersive extended reality (XR) and holographic communication [4].

1.3.3 **Massive Connectivity:** The Internet of Everything (IoE) will be supported by 6G, intended to seamlessly connect billions of gadgets across homes, businesses, and cities.

1.3.4 **Terahertz Frequency Bands:** 6G delivers more bandwidth by using high-frequency bands (such as the millimeter-wave and terahertz bands).

However, it also presents issues such as signal attenuation that must be addressed with sophisticated technology.

1.3.5 **Enhanced AI Integration:** AI will be integrated across the network to automate network administration, enhance security, and optimize resource allocation—all of which will enable real-time network adaptation.

1.3.6 **Energy Efficiency:** 6G networks will incorporate energy-efficient designs and employ AI-driven power management to lower overall energy usage, with a heavy emphasis on sustainability.

1.3.7 **Edge Computing and Distributed Architecture:** By using edge computing to process data closer to the source, 6G will increase efficiency by lowering latency and the requirement for lengthy data transmission [5].

1.3.8 **High Reliability and Security**: To guarantee data integrity, 6G will introduce improved security standards and AI-driven threat detection, particularly for vital applications like remote surgery and driverless cars.

1.3.9 **Sensing Capabilities**: Applications like high-precision locating and environmental monitoring will be made possible by 6G networks' inclusion of environmental and spatial sensing in addition to communication.

1.3.10 **Interoperability with Previous Generations**: 6G will continue to be compatible with 5G and older networks to enable a smooth transition, permitting mixed-use scenarios and gradual adoption (Fig. 1).

1.4 Importance of Energy Efficiency in 6G

1.4.1 **Sustainable Network Operations**: Energy efficiency is crucial to maintaining controllable operating costs and promoting sustainable deployment, as 6G networks require increased data rates and connection [6].

1.4.2 **Environmental Impact Reduction**: In line with international sustainability targets, lowering energy consumption in 6G helps reduce carbon emissions and environmental effects.

1.4.3 **Support for High-Frequency Bands**: Energy-efficient designs are essential to controlling power demands and minimizing thermal impacts since operating in high-frequency bands (such as terahertz) uses more power.

1.4.4 **Extended Device Lifespan**: An important factor for wearable and Internet of Things technologies in distant or mobile applications is the ability of energy-efficient networks to extend the battery life of linked devices.

1.4.5 **Lower Operational Costs**: Large-scale 6G networks become more financially feasible as a result of increased energy efficiency, which lowers electricity costs for network operators.

1.4.6 **Optimized Resource Allocation**: 6G's energy-efficient AI algorithms can dynamically distribute resources, cutting waste and guaranteeing that only the power required is used in response to current demand.

1.4.7 **Enhanced User Experience**: Energy-efficient 6G networks may deliver consistent performance, even under high load, by lowering power needs and preserving dependable service, improving the user experience.

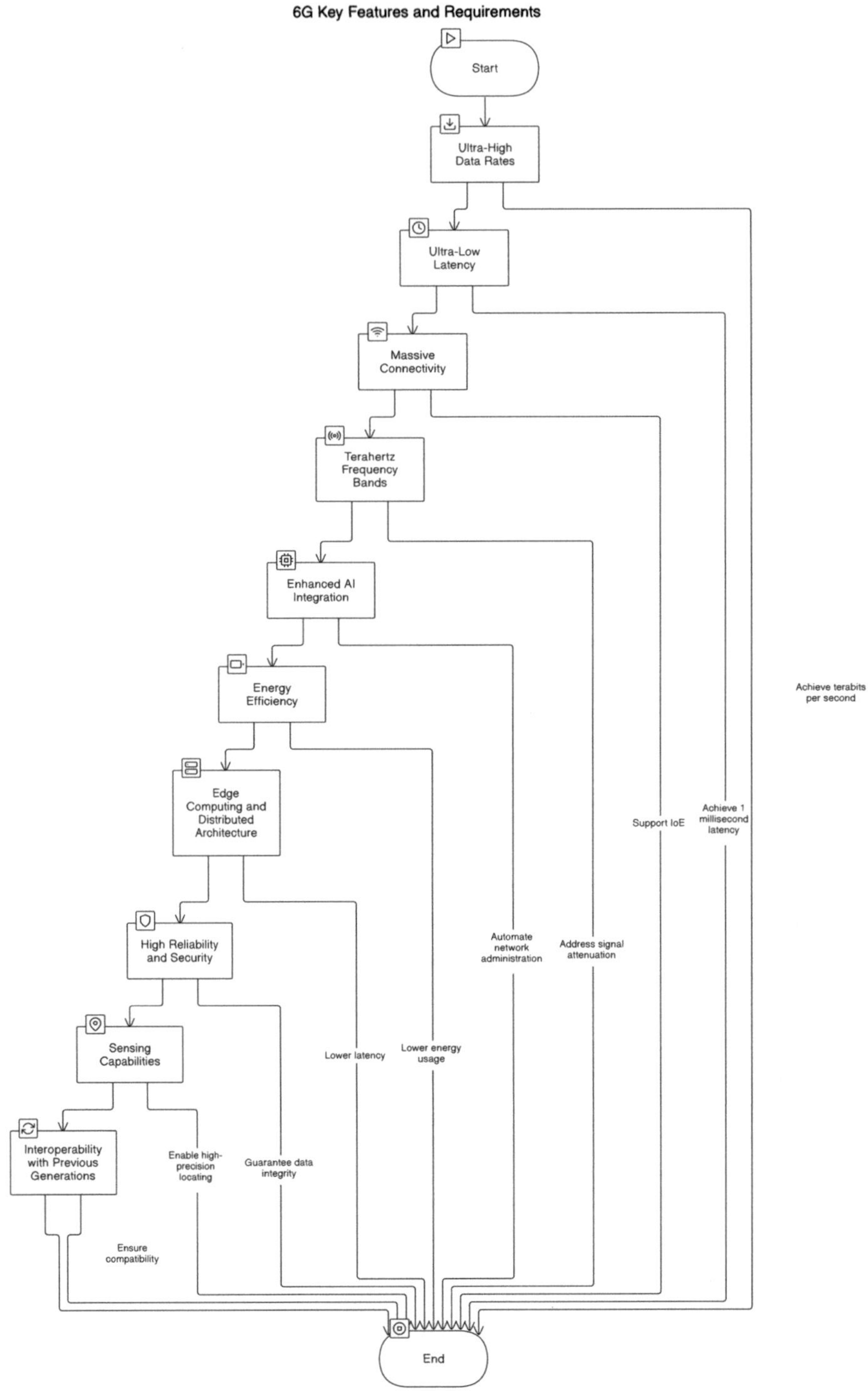

Fig. 1 6G key features and requirements

1.4.8 **Support for Edge Computing**: Since energy-efficient methods lower the power required for local data processing, they are crucial for the successful deployment of edge computing in 6G.

1.4.9 **Enabling Ubiquitous Connectivity**: In the context of the Internet of Everything (IoE), energy efficiency is essential for sustaining massive, constantly connected networks spanning billions of devices [7].

1.4.10 **Compliance with Regulatory Standards**: Energy-efficient 6G networks help carriers achieve these standards and stay out of trouble as governments increasingly limit the amount of energy used in tech infrastructure (Fig. 2).

2 Literature Survey

Paper	Key contributions	Limitations
6G Communication Technologies [1]	• Summarizes recent 6G surveys and identifies new use cases and scenarios. • Highlights technical improvements in 6G over 5G, including improved KPIs, novel architectural designs, and state-of-the-art technology like terahertz communication and AI-driven management.	• Does not delve deeply into specific technical challenges and solutions. • Lacks a comprehensive analysis of the economic and societal implications of 6G.
Beyond 5G: Towards 6G [2]	• Discusses the challenges and opportunities of the transition from 5G to B5G and 6G. • Highlights the potential of 6G to revolutionize wireless communication with unprecedented speed, low latency, and extraterrestrial communication.	• Does not provide detailed technical specifications or roadmaps for 6G development. • Overlooks potential security and privacy concerns associated with 6G technologies.
AI-Powered Network Management and Security [8]	• Explores the use of AI, particularly deep learning and reinforcement learning, to improve network performance and security. • Demonstrates the potential of AI to optimize traffic management, enhance data transfer, and strengthen cyber threat detection.	• Lacks a comprehensive evaluation of the computational cost and energy consumption of AI-based network solutions. • Does not address the ethical implications of AI-driven network decision-making.
AI-Enhanced SDN for Cloud Resource Management [10]	• Proposes an AI-SDN framework to dynamically allocate resources in cloud environments. • Utilizes machine learning techniques to optimize resource allocation, reduce latency, and improve service quality.	• Requires significant computational resources and expertise to implement and maintain. • Assumes accurate and timely data collection and analysis.
Digital Twin for Energy-Efficient HVAC Systems [11]	• Presents a digital twin-based framework to optimize the maintenance and energy efficiency of HVAC systems. • Combines BIM, IoT, and semantic technologies to enable predictive maintenance and fault detection. • Demonstrates significant energy savings through real-world case studies.	• Relies on accurate sensor data and reliable communication infrastructure. • Requires initial investment in digital twin development and deployment.

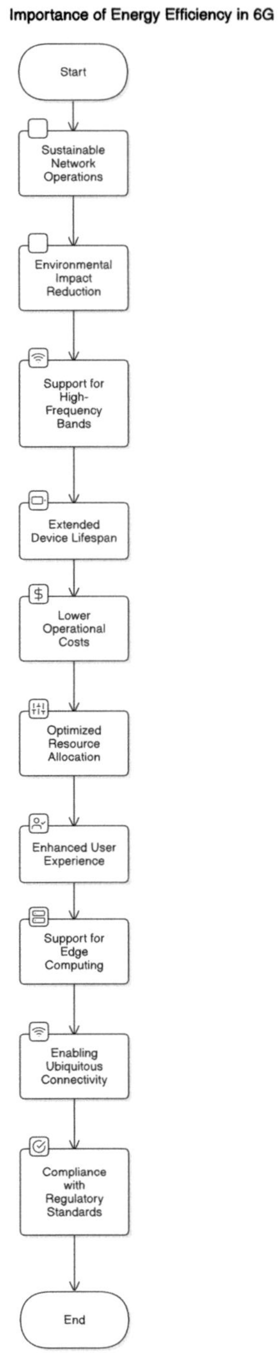

Fig. 2 Importance of energy efficiency in 6G

3 Role of Artificial Intelligence in 6G

3.1 AI-Driven Network Management

A key component of 6G is anticipated to be AI-driven network management, which will allow networks to function with increased intelligence, flexibility, and efficiency. Advanced, real-time management solutions are necessary for 6G because of the increased complexity, scalability, and data demands of connected devices compared to earlier generations. Algorithms driven by AI can continuously track network performance, evaluating data to anticipate and react to demand variations. Even in difficult network situations or times of high demand, this dynamic adaptability is crucial for preserving peak performance, cutting latency, and optimizing resource utilization [8]. AI makes it possible for networks to self-heal and self-optimize by automating processes like defect detection and network optimization. This minimizes downtime and eliminates the need for manual intervention.

AI is also essential for improving security and energy efficiency in 6G network management. By examining trends and abnormalities, artificial intelligence (AI) may detect security vulnerabilities using machine learning (ML) and deep learning (DL) approaches, enabling quicker and more precise threat detection. To satisfy the sustainability objectives at the heart of 6G, AI can also help regulate energy consumption across network components by adjusting power utilization depending on real-time needs. AI-driven management enables fine-grained, accurate control over a wide range of devices and data streams in a highly linked 6G environment, leading to networks that are not just faster and more effective but also more robust and ecologically friendly.

3.2 AI for Resource Allocation and Optimization

AI-driven resource allocation and optimization are essential for effectively handling the enormous volume of data and connected devices in 6G networks. To balance variables like bandwidth, power, and spectrum availability, AI techniques like machine learning and deep reinforcement learning (RL) can dynamically assign resources and forecast patterns of network demand [9]. AI can prioritize traffic for vital applications, like remote healthcare and driverless cars, by using historical patterns and real-time data. This ensures that these services get the resources they require while preserving network efficiency for other jobs. With this feature, 6G networks can function more smoothly during periods of high demand without experiencing resource saturation [10].

By modifying resource allocation according to current circumstances, AI also optimizes energy utilization within the network, reducing waste and improving sustainability. AI, for instance, may control power consumption across various network locations, maximizing energy allocation where demand is higher and decreasing it in low-traffic areas or off-peak hours. While managing a large number of devices and data, 6G networks can maintain their energy efficiency thanks to this focused optimization. In addition to preventing network congestion and lowering latency through predictive resource management, AI can help 6G achieve its objectives of incredibly fast, dependable connectivity and flawless user experiences. AI-driven resource allocation and optimization are essential for maximizing network capabilities in a 6G ecosystem where performance and efficiency are critical.

3.3 AI for Predictive Maintenance and Fault Management

In 6G networks, using AI for predictive maintenance and fault management is revolutionary because it enables the early detection and fixing of possible problems before they hurt network performance. AI can continually monitor network components by using machine learning techniques and real-time data analysis. It can do this by examining sensor data, error logs, and performance metrics to find trends that can point to an imminent fault. By scheduling maintenance or putting corrective measures in place before any major disruptions happen, network providers can take proactive actions to fix problems. Predictive maintenance reduces downtime and helps guarantee continuous service for a 6G network that is handling enormous numbers of connected devices and fast data flows [11].

AI-driven automation, which can immediately identify and address abnormalities without the need for human intervention, also helps with fault management in 6G. AI can initiate automated recovery procedures, divert traffic, or modify network configurations to lessen the effect of errors or anomalous behavior. By lowering the need for manual troubleshooting and speeding up reaction times, this clever fault management technique contributes to the resilience and dependability of the network. AI-enabled predictive maintenance and fault management are critical for attaining network stability, improving user experience, and sustaining the lifespan of network equipment in a 6G context where quick, seamless connectivity is critical.

3.4 AI for Network Security in 6G

Since the size, speed, and complexity of 6G networks increase the attack surface and introduce new vulnerabilities, artificial intelligence (AI) is essential for network security. AI can track network activity in real time by utilizing machine learning and deep learning to analyze massive data streams and identify anomalies and possible threats [12]. In contrast to traditional security systems that depend on preset rules, AI can learn from patterns and adjust to new threats, spotting odd activity that can point to malware, security lapses, or other intrusions. 6G networks can react more quickly thanks to this proactive threat detection, which helps stop security incidents before they cause significant damage.

Furthermore, 6G's AI-driven security may make automated reaction mechanisms possible. AI can carry out containment procedures without human assistance, including rerouting traffic, modifying access constraints, or isolating impacted nodes when a threat is identified. By improving the network's resilience, this feature makes it possible to respond to security events quickly and efficiently. AI can also help with sophisticated authentication and encryption methods, guaranteeing data privacy and guarding against unwanted access. AI-driven network security is crucial for protecting sensitive data and preserving user confidence in the network's integrity given the interconnected nature of 6G, which connects billions of devices across multiple industries [13]. AI contributes to a safe, robust 6G environment that protects devices and data by constantly adjusting to new threats and automating protection systems.

4 Challenges in Achieving Energy Efficiency in 6G

4.1 High Energy Consumption in 6G Networks

The increasing data rates, wide-ranging connectivity, and higher-frequency bands needed to enable next-generation applications make 6G networks' high energy consumption a major concern. 6G may reach ultra-high speeds and minimal latency by operating in millimeter-wave and terahertz frequency bands; however, these high-frequency activities naturally require more power to maintain coverage and signal quality. The energy required to power such extensive data processing, transmission, and device interactions increases significantly as 6G networks strive to enable dense connections and cutting-edge applications like holographic communications, autonomous systems, and the Internet of Everything (IoE) [14].

The significant effects of 6G's high energy usage on the environment and the economy are noteworthy. Large-scale, high-power 6G infrastructure operations may result in higher carbon emissions and operating expenses for network providers, which could affect profitability and sustainability objectives. Energy-efficient design innovations are needed to address these problems, such as integrating edge computing to lower data transmission demands, dynamic resource allocation, and AI-driven power control. Advanced connectivity may be supported by 6G networks while lowering their environmental impact and guaranteeing economic sustainability by eliminating wasteful energy use and concentrating on energy-conscious architectures.

4.2 Balancing Performance and Energy Efficiency

One of the main challenges with 6G networks is striking a balance between performance and energy efficiency because the requirement to reduce energy usage frequently clashes with the desire for ultra-fast, low-latency connectivity. Operating at terahertz frequencies, preserving low latency, and supporting a large number of devices are all necessary for 6G to function well, which raises power requirements. For 6G applications like holographic communications, extended reality (XR), and vital IoT systems, high data speeds and real-time responsiveness are crucial [15]. Delivering these features without careful energy management, however, would lead to unsustainable power usage and higher operating expenses, especially for extensive network infrastructures.

6G networks use cutting-edge technology like edge computing, AI-driven optimization, and dynamic resource allocation to solve this. For example, AI algorithms can forecast network demand and distribute resources appropriately, lowering power use during off-peak hours and only supplying full power when required. Similar to this, edge computing makes it possible to process data locally, which saves energy and eliminates the need for large-scale data transfer. With the help of these techniques, 6G networks can dynamically strike a balance between energy efficiency and performance, responding to changing needs without sacrificing service quality. 6G networks can satisfy the demanding performance requirements of next-generation applications while keeping sustainability in mind by implementing energy-efficient procedures and adaptive management.

4.3 The Trade-Off Between Latency and Energy Consumption

In 6G networks, the trade-off between latency and energy consumption is crucial since real-time applications frequently require ultra-low latency, which is typically achieved at the expense of higher power consumption. For applications where even a small delay in data transmission can impact functionality and user experience, such as autonomous driving, remote healthcare, and immersive virtual reality, low latency is crucial. However, to maintain a steady, almost instantaneous connection, lowering latency necessitates frequent data transmission, quick processing, and powerful infrastructure. It can be difficult to control this increased energy demand, particularly in network systems that are dense or big.

6G networks use technologies like adaptive scheduling and AI-driven resource management to strike a balance between latency and energy efficiency. AI algorithms, for example, can dynamically distribute network resources according to the latency requirements of certain applications, saving energy for less important applications while supplying more power and faster data channels for jobs that are sensitive to latency [14]. Furthermore, localized data processing made possible by edge computing can lower latency without requiring energy-intensive long-distance data transmission. 6G networks may achieve an ideal balance between energy conservation and low latency when needed by carefully allocating resources and implementing adaptive strategies. This will promote a network ecosystem that is both high-performing and sustainable.

4.4 Environmental and Economic Implications

The high energy requirements of 6G have important economic and environmental ramifications, especially as the network infrastructure grows to accommodate billions of linked devices and data-intensive applications. Increased energy use in 6G networks may significantly increase carbon emissions from an environmental standpoint, particularly if the power comes from non-renewable sources. Due to signal attenuation, sophisticated 6G hardware—such as base stations running at terahertz and millimeter-wave frequencies—also results in energy inefficiencies, necessitating additional infrastructure to ensure dependable connectivity [16]. To lessen their environmental impact, 6G networks must implement sustainable and energy-efficient methods.

Economically speaking, telecom operators and companies that depend on 6G technology have difficulties due to the significant operating expenses involved in powering extensive 6G networks. Higher costs result from increased energy consumption, which may have an impact on profitability and restrict access to 6G services at a reasonable cost. 6G networks are using edge computing, AI-driven energy management, and dynamic resource allocation to minimize waste and maximize power consumption to handle these financial ramifications. Furthermore, network providers can save expenses and adhere to stricter environmental requirements by implementing sustainable practices and investing in renewable energy sources. The sustainable expansion and accessibility of 6G networks depend on striking a balance between these economic and environmental considerations.

5 AI Techniques for Energy Efficiency in 6G

5.1 Machine Learning for Energy-Efficient Resource Management

To achieve energy-efficient resource management in 6G networks, machine learning (ML) is essential since it allows for intelligent, adaptive control over resource allocation. ML algorithms can evaluate network traffic, user demand, and energy consumption data in real time through predictive analytics and pattern recognition, finding patterns of best resource distribution. Machine learning algorithms use usage patterns and historical data to predict times of high demand and distribute resources in advance, making sure that power is only used where it is most required. By reducing energy waste and avoiding overprovisioning resources, this predictive strategy minimizes needless power use while maintaining service quality [17].

ML can help with dynamic energy management, which modifies network parameters according to current usage, in addition to predicted allocation. ML algorithms, for example, can allow base stations to autonomously scale back up when demand surges and enter low-power modes during off-peak hours. Because energy needs change often in dense 6G networks, this flexibility is especially helpful. Additionally, ML makes it possible for decentralized, edge-based resource management, which lowers the energy required for long-distance data transmission by allowing data processing and decision-making to take place closer to the data source. 6G networks may strike a balance between sustainability and high performance by using machine learning (ML) for resource management. This will improve the network's overall energy efficiency and optimize power utilization in response to real-time demands.

5.2 *Reinforcement Learning for Dynamic Power Control*

In 6G networks, reinforcement learning (RL) is a potent method for dynamic power regulation that makes adaptive energy management possible. RL learns and gets better over time. In a 6G context, where high data rates and varying user demand necessitate flexible power allocation, RL algorithms may independently modify power settings to maximize energy efficiency while preserving network performance. By getting feedback on their actions (e.g., boosting or reducing power to certain network nodes), RL agents continuously learn from the network environment, in contrast to traditional techniques that rely on static rules [18]. RL algorithms identify ideal power configurations that minimize energy consumption without sacrificing service quality by striking a balance between rewards (like energy savings) and penalties (like decreased connectivity or performance).

6G networks may react dynamically to changing circumstances, such as fluctuating traffic loads, the time of day, or location-specific demand, by using RL for power regulation. For example, an RL agent can efficiently conserve energy by allocating power to high-traffic locations during peak consumption hours and reducing power in low-demand areas during off-peak hours. Additionally, by experimenting with various power control algorithms and determining which is the most effective, RL can handle complex trade-offs, such as balancing low latency and low energy use. Because of this ongoing adaptation, RL-based dynamic power regulation is ideal for 6G networks, where flexibility and efficiency are essential for long-term, high-performance connectivity.

5.3 Federated Learning for Distributed Energy Optimization

Federated learning (FL), which addresses the need for localized energy management without centralized data sharing, provides an efficient solution for dispersed energy optimization in 6G networks. FL enables individual network nodes to train machine learning models locally using their data in a 6G environment with widely distributed devices and base stations. The learned model updates—rather than the raw data—are subsequently shared with a central server. This cooperative strategy minimizes the communication overhead related to centralized processing, protects data privacy, and allows network-wide energy use optimization [19].

FL may enable network nodes to independently modify their power usage by local requirements for distributed energy optimization. Every node uses real-time data on energy consumption, device density, and local traffic patterns to train its model. A global model is then produced by combining these locally optimized models at a central server and sent to the nodes. Nodes may respond dynamically to variations in regional demand while reducing energy waste thanks to this iterative approach, which allows them to continuously improve their energy management techniques based on collective learning. FL supports a scalable, sustainable approach to 6G energy management by distributing energy optimization throughout the network, which not only improves overall energy efficiency but also lessens the demand for continuous data transfer.

5.4 Deep Learning for Energy-Efficient Signal Processing

Energy-efficient signal processing in 6G networks is made possible by deep learning (DL), which enables complex, real-time signal quality analysis and optimization with low power consumption. The needs of 6G networks operating at terahertz frequencies and managing massive, fast data flows can be too much for traditional signal processing techniques, which demand a lot of processing power. 6G networks can handle signals more effectively by utilizing deep learning models. Neural networks trained to reduce computational overhead, improve signal clarity, and filter noise can be used to optimize transmission pathways [20].

Deep learning models can be used in energy-efficient signal processing to manage challenging tasks including interference cancellation, channel estimation, and adaptive beamforming—all of which have historically required a lot of power. To provide high-quality connections with lower power output, for example, a deep neural network (DNN) can learn to forecast the best communication pathways and dynamically change the direction of beams. Furthermore, DL can provide low-power operation at the device level, allowing local signal processing to adjust to changing circumstances in real time, amplifying signals only when required, and preserving power during times of low demand. Networks can support sustainable operations for a variety of 6G applications by minimizing energy consumption and maintaining high data rates and quality of service by integrating DL into 6G signal processing.

6 Case Studies

Case study	Description	AI techniques used	Key benefits
AI-Powered Energy Efficiency in Smart Cities [21]	Smart cities like Barcelona and Singapore are utilizing AI-driven energy management to reduce energy consumption in public infrastructure. Barcelona uses AI in street lighting and building management, adjusting power based on occupancy and usage patterns. Singapore's "green building" initiative optimizes energy use in residential and commercial spaces based on real-time data.	Machine learning, predictive analytics	Reduced carbon footprint, up to 30% energy savings, lower operational costs, enhanced sustainability in urban areas
AI for Energy Management in IoT and Edge Computing [22]	AI algorithms deployed in IoT devices and edge computing environments manage energy consumption by processing data locally, reducing the need for extensive data transmission. This is particularly useful for power-limited devices in remote areas, enabling efficient energy use while maintaining device performance.	Federated learning, deep learning, energy-aware algorithms	Minimized data transmission, extended battery life of IoT devices, reduced latency, improved operational efficiency
AI for Energy Optimization in Autonomous Vehicles [23]	Autonomous vehicles (AVs) rely on AI to optimize battery and energy use while balancing performance requirements. By monitoring factors like traffic, road conditions, and weather, AI algorithms dynamically adjust energy use in real-time, conserving power without compromising vehicle operation.	Reinforcement learning, real-time optimization, predictive modeling	Enhanced battery efficiency, extended range for EVs, adaptive power usage, improved vehicle performance and reliability
AI in Energy-Efficient Massive MIMO Systems [24]	In massive MIMO (multiple-input multiple-output) systems, AI optimizes signal processing to minimize power consumption while ensuring high data rates. AI techniques such as deep learning are used to improve beamforming and channel estimation, adapting to real-time conditions to conserve energy without sacrificing connectivity.	Deep learning, adaptive beamforming, channel estimation algorithms	Lower energy use, higher spectral efficiency, improved network performance, sustainable large-scale connectivity

7 Future Directions and Research Opportunities

7.1 Emerging AI Techniques for 6G Energy Efficiency

New AI approaches have enormous potential to improve 6G networks' energy efficiency while meeting the intricate needs of this cutting-edge technology. Federated learning (FL) is one such method that preserves privacy and lowers the energy consumption of data transmission by enabling devices in the network to cooperatively train models without centralizing data. FL eliminates the need for constant, power-intensive connection with a central server by training locally and sharing just model updates. In addition to saving energy, this distributed strategy enables real-time resource optimization across 6G networks, allowing each device to modify its energy consumption in response to local conditions and demands [20].

The use of reinforcement learning (RL) for context-aware, adaptive power control is another exciting field. In dynamic 6G environments with varying network loads, RL allows network nodes to independently learn and modify their power consumption based on feedback from the environment. An RL agent, for instance, might decide how best to distribute power at different times of the day or in response to changing traffic demands. This helps to prevent energy loss during periods of low demand while guaranteeing top performance. Particularly with high-frequency 6G systems using millimeter or terahertz wavelengths, RL-based systems can more effectively balance the trade-off between energy conservation and network performance by continuously learning from real-world data.

Quantum machine learning (QML) is a cutting-edge method that may improve 6G's energy efficiency even more. With the advancement of quantum computing technology, QML algorithms may be able to resolve intricate optimization issues in real time, facilitating quicker and more energy-efficient decision-making over extensive 6G networks. The energy requirements of resource allocation, interference control, and signal processing activities could be greatly decreased by using quantum-enhanced AI models, which could do these jobs substantially more effectively than traditional techniques. Even though QML is still in its infancy, it offers 6G networks a revolutionary possibility by providing unmatched processing power that may result in innovations in environmentally friendly, energy-efficient network operations.

7.2 Collaboration Between AI and Quantum Computing for Energy Optimization

An innovative method of energy optimization in 6G networks is presented by the combination of AI's flexibility and pattern recognition with quantum computing's potent problem-solving abilities. Processing the enormous datasets and sophisticated calculations needed for 6G networks is a perfect fit for quantum computing, which can solve complex optimization tasks tenfold faster than classical computing. By rapidly locating the best solutions across extensive, intricate networks, quantum computers, when combined with artificial intelligence, can improve prediction models for energy management [25]. This enables real-time adaptation by AI-driven systems, which dynamically modify resource allocation and power distribution to reduce energy waste.

Quantum computing can help AI algorithms handle extremely complicated scenarios in 6G energy optimization, such as optimizing energy consumption in terahertz and millimeter-wave bands, where standard processing is inadequate. In a huge MIMO (Multiple-Input Multiple-Output) system, for example, quantum algorithms might effectively calculate the ideal power levels for each node, lowering energy consumption while preserving high-quality signal processing. AI might evaluate several energy-saving configurations rapidly and choose the most effective ones based on current conditions if quantum computing is combined with reinforcement learning to enable even faster and more accurate adaptations to network changes.

Moreover, by controlling the rising power requirements of next-generation applications, the cooperation between AI and quantum computing promotes sustainable network growth. Quantum-enhanced AI might optimize energy utilization across various sectors, from smart cities to autonomous mobility, making large-scale, high-speed networking feasible without having an undue negative environmental impact. With the development of quantum technology and artificial intelligence, 6G energy management is probably going to be redefined, reaching previously unheard-of levels of efficiency and contributing to the dawn of a new era of high-performance, sustainable networks.

7.3 Regulatory and Standardization Considerations

To guarantee interoperability, security, and equity across international infrastructures, regulatory and standardization issues are essential when implementing energy-efficient AI technologies within 6G networks. To simplify network operations, encourage sustainable behaviors, and lower energy usage, standardized rules for AI-driven energy management are crucial given the complexity and scale of 6G. Protocols that specify best practices for energy optimization, data handling, and AI application in 6G are being actively developed by standards organizations including the International Telecommunication Union (ITU), 3GPP, and IEEE [26]. By ensuring that AI solutions for energy efficiency may function uniformly across various devices and geographical locations, these standards will support the development of a unified global network environment.

Data security and privacy are other crucial regulatory factor, especially when using edge-processing AI methods like federated learning. The way AI solutions in 6G must handle information to maintain compliance is impacted by regulatory frameworks such as the General Data Protection Regulation (GDPR) in the European

Union, which establishes stringent requirements for handling sensitive and personal data. Furthermore, regulatory supervision is required to reduce the dangers associated with algorithmic biases, network vulnerabilities, and the ethical ramifications of autonomous decision-making in key services as AI is used more and more for important energy-related decisions. The adoption of AI in 6G will require extensive regulations to guarantee that energy-saving measures are safe, open, and considerate of privacy.

Regulatory agencies will be crucial in establishing environmental sustainability goals for 6G networks and directing the construction of infrastructure that uses less energy. As part of 6G deployment plans, regulations may include restrictions for emissions reduction, carbon footprint tracking, and the source of renewable energy. Regulators can promote the adoption of environmentally friendly AI-driven solutions in 6G by coordinating energy efficiency objectives with global sustainability frameworks, such as the Sustainable Development Goals (SDGs) of the United Nations. To create an ecosystem where AI technology for energy optimization is in line with both operational performance and international sustainability obligations, these legislative and standardization activities will be crucial.

7.4 Potential Impacts of Energy-Efficient AI on 6G Adoption

Energy-efficient AI has the potential to significantly accelerate the deployment of 6G by tackling the high energy requirements of cutting-edge 6G technology. Energy consumption can become a significant bottleneck as 6G networks seek to accommodate data-intensive applications like holographic communications, immersive augmented reality, and the Internet of Everything (IoE), which can affect both operational costs and environmental sustainability. By optimizing power utilization across network infrastructure, lowering the carbon footprint, and increasing the cost-effectiveness of 6G deployments, energy-efficient AI technologies can assist in managing this challenge. Telecom operators and businesses will be especially drawn to this lower operating cost and greater sustainability, which will hasten the broad adoption of 6G technologies.

Energy-efficient AI not only saves money and promotes sustainability, but it also improves 6G networks' performance and dependability, which is essential for satisfying user demands. Through effective resource management and dynamic power allocation adjustments, AI-driven energy optimization can sustain network performance during periods of high demand. With this feature, 6G networks can guarantee reliable, high-quality connectivity without wasting energy. AI enables 6G networks to adjust to changing demands, such as those in distant rural areas or dense metropolitan environments, by integrating intelligence into energy management. This enhances user experience and builds confidence in 6G's capabilities [27].

Furthermore, 6G networks become more appealing to stakeholders and policymakers when energy-efficient AI is in line with global sustainability trends and regulatory frameworks, such as net-zero carbon pledges. The use of AI to cut energy waste in 6G networks is in line with the objectives of governments and industry organizations that are placing more focus on sustainable practices. It also makes the network more marketable. This alignment may encourage businesses and telecom providers to embrace 6G by facilitating favorable regulatory regulations and maybe luring investment incentives. Energy-efficient AI strikes a balance between sustainability and performance, which not only makes 6G adoption possible but also encourages a conscientious approach to the next-generation connection that aligns with contemporary economic and environmental concerns.

8 Result

Using real-time data, APAR rules, and machine learning models, the Digital Twin predictive maintenance framework for Air Handling Units (AHUs) showed successful fault identification and predictive capabilities. The results of the case study showed that operational faults were significantly decreased and that the AHU system's inefficiencies had resulted in yearly energy cost reductions. Predictive insights also made it possible to schedule maintenance more precisely, which increased the lifespan of HVAC parts and decreased downtime.

9 Conclusion

The feasibility and advantages of employing digital twin technology for predictive maintenance in building facilities are confirmed by the study. This method offers a proactive maintenance strategy that can significantly increase energy efficiency and save operating costs in the building industry by combining BIM, IoT, and machine learning. These results demonstrate how Digital Twins may improve sustainability in facility management and establish new benchmarks for the upkeep of energy-efficient HVAC systems.

References

1. Jiang, W., et al. (2021). The road towards 6G: A comprehensive survey. *IEEE Open Journal of the Communications Society, 2*, 334–366.
2. Asghar, M. Z., Memon, S. A., & Hämäläinen, J. (2022). Evolution of wireless communication to 6g: Potential applications and research directions. *Sustainability, 14*(10), 6356.
3. Sarmah, N. (2016). *Wideband circuit design techniques for ultra-high data-rate wireless communication in silicon technologies*. Diss. Dissertation,. Universität Wuppertal.
4. Abbasloo, S. (2020). *Designing low and ultra-low latency networks*. Diss. New York University Tandon School of Engineering.

5. Nascimento, E. B., Moreno, E. D., & de Macedo, D. D. J. (2020). Towards an architecture for information-centric network with content replication in edge computing. In *Advances in edge computing: Massive parallel processing and applications* (pp. 106–147). IOS Press.
6. N. R. Nayak, S. Mishra, S. Chowdhury, P. K. Dutta and G. Ramya, *RFD based technique on QoS parameters using cloud computing*. 6th Smart Cities Symposium (SCS 2022), Hybrid Conference, Bahrain, 2022, pp. 227–231, https://doi.org/10.1049/icp.2023.0418.
7. Pimenta, J. R. d. O. G., et al. (2023). UbiAccess: An instrument to assess system access in ubiquitous scenarios. *Interacting with Computers, 35*(2), 55–68.
8. Rao, P. S., Nayak, S., Gautam, A., Lourens, M. E., Dutta, P. K., Chowdhury, S., & El-kenawy, E. M. *Aumann agreement theorem based human resource performance monitoring with IoT and 6G technology*. 6th Smart Cities Symposium (SCS 2022), Hybrid Conference, Bahrain, 2022, pp. 401–406, https://doi.org/10.1049/icp.2023.0638.
9. N. R. Nayak, S. Chowdhury, S. Mishra, P. K. Dutta and G. Ramya. *Effective comparison of meta-heuristics algorithm in cloud computing*. 6th Smart Cities Symposium (SCS 2022), Hybrid Conference, Bahrain, 2022, pp. 232–237, https://doi.org/10.1049/icp.2023.0420.
10. Sathupadi, K. (2023). An AI-driven framework for dynamic resource allocation in software-defined networking to optimize cloud infrastructure performance and scalability. *International Journal of Intelligent Automation and Computing, 6*(1), 46–64.
11. Hosamo, H. H., et al. (2022). A digital twin predictive maintenance framework of air handling units based on automatic fault detection and diagnostics. *Energy and Buildings, 261*, 111988.
12. Khare, A., Gupta, R., Shukla, P. K., Chowdhury, R., & Datta, P. K. (2021). A black widow optimization algorithm (BWOA) for node capture attack to enhance the wireless sensor network protection. In V. E. Balas, A. E. Hassanien, S. Chakrabarti, & L. Mandal (Eds.), *Proceedings of International Conference on Computational Intelligence, Data Science and Cloud Computing. Lecture Notes on Data Engineering and Communications Technologies* (Vol. 62). Springer. https://doi.org/10.1007/978-981-33-4968-1_47
13. Sheth, K., et al. (2020). A taxonomy of AI techniques for 6G communication networks. *Computer Communications, 161*, 279–303.
14. Ramamoorthy, P., et al. (2023). Sustainable power consumption for variance-based integration model in cellular 6G-IoT system. *Sustainability, 15*(17), 12696.
15. Huang, T., et al. (2019). A survey on green 6G network: Architecture and technologies. *IEEE Access, 7*, 175758–175768.
16. Chauhan, D., et al. (2024). Balancing technological innovation and environmental sustainability: A lifecycle analysis of 6G wireless communication technology. *Sustainability, 16*(15), 6533.
17. Alhussien, N., & Aaron Gulliver, T. (2024). Toward AI-enabled green 6G networks: A resource management perspective. *IEEE Access, 12*, 132972.
18. Cheng, Z., et al. (2022). Deep reinforcement learning-based joint task and energy offloading in UAV-aided 6G intelligent edge networks. *Computer Communications, 192*, 234–244.
19. Bárcena, J. L. C., et al. (2023). Enabling federated learning of explainable AI models within beyond-5G/6G networks. *Computer Communications, 210*, 356–375.
20. Mahmood, M. R., et al. (2022). A comprehensive review on artificial intelligence/machine learning algorithms for empowering the future IoT toward 6G era. *IEEE Access, 10*, 87535–87562.
21. Tunde, AH, et al. AI-driven innovations in energy efficiency: Transforming smart buildings and urban areas through technology and digital transformation. (2024).
22. Rochd, A., et al. (2021). Design and implementation of an AI-based & IoT-enabled home energy management system: A case study in Benguerir—Morocco. *Energy Reports, 7*, 699–719.
23. Zhang, Y., et al. (2021). Energy-saving optimization and control of autonomous electric vehicles with considering multiconstraints. *IEEE Transactions on Cybernetics, 52*(10), 10869–10881.

24. Lopez-Perez, D., et al. (2021). A survey on 5g energy efficiency: Massive mimo, lean carrier design, sleep modes, and machine learning. *arXiv preprint, arXiv*, 2101.11246.
25. Gill, S. S., et al. (2022). AI for next generation computing: Emerging trends and future directions. *Internet of Things, 19*, 100514.
26. Phuyal, S., Bista, D., & Bista, R. (2020). Challenges, opportunities and future directions of smart manufacturing: A state of art review. *Sustainable Futures, 2*, 100023.
27. Kumar, R., et al. (2023). From efficiency to sustainability: Exploring the potential of 6G for a greener future. *Sustainability, 15*(23), 16387.

Harnessing AI for Sustainable Smart Cities: Impact, Innovations, and Use Cases

Lav Soni and Ashu Taneja

1 Introduction

In the contemporary urban context, the notion of a smart city has emerged as a prominent illustration of ingenuity, sustainability, and effectiveness. Cities throughout the globe employ artificial intelligence (AI) and other swiftly progressing technologies to address complex challenges and enhance the quality of life for their residents. This chapter analyzes the several uses of AI in transforming traditional cities into smart cities characterized by intelligent, interconnected, and flexible surroundings. The concept of urban planning is not novel. The emperors and monarchs of prehistoric cities employed urban planning as a means to systematically build their towns and societies. Their conceptualization of urban planning, however, diverged significantly from the principles of contemporary urban planning. A primary emphasis of ancient urban planning was the arrangement of streets, public buildings, and dwellings. Conversely, modern urban planning encompasses a far broader range of subjects, such as street safety, water leakage management, transit efficiency, and other concerns inherent to the present era [1]. Energy management is another critically important domain in which AI is driving innovation in smart cities [2]. The sophisticated AI algorithms ensure efficient energy resource distribution, which in turn help in regulating the energy price. This efficient energy utilization also reduces the impact on the environment with reduced carbon emissions. By analyzing the energy consumption patterns, grid data, and weather patterns, the power distribution in power grids can be handled intelligently. AI-enabled sensors are the key in such situations as they allow optimal utility of power distribution in

L. Soni · A. Taneja (✉)
Chitkara University Institute of Engineering and Technology, Chitkara University, Rajpura, Punjab, India
e-mail: ashu.taneja@chitkara.edu.in

P. Raj et al. (eds.), *Artificial Intelligence (AI) for IT Energy Efficiency and Green AI for Environment Sustainability*,
https://doi.org/10.1007/978-3-031-89420-6_23

the grid by changing it dynamically in case of a sudden need for extra power. Since the demand is changing, predictive analytics can be used to identify the areas where energy efficiency and conservation can be increased. Furthermore, AI-driven building management systems can be used in dark urban buildings to regulate the lighting, heating, and cooling systems, thus resulting in energy conservation.

Another AI-powered vertical of smart cities is public safety and security [3] which enables public protection and monitoring of cities for threats. The security system uses face recognition technology operated by the surveillance cameras on the street corners, speeding up the process of catching the suspect. At the same time, the use of software and more effective utilization of law enforcement resources has made it possible to predict crime rates. Furthermore, the medical emergency response systems use AI-based dispatch algorithms that allow them to give the highest priority to the nearest hospital or any other facility and to take the shortest route possible, thereby reducing the time to get there and potentially saving lives. In addition, it is also powered by deep learning technology which identifies the vulnerable points of the city infrastructure and puts into action the preventions [4]. In addition, cities can think of new concepts for improving the present and future of their urban setup by the effective application of AI. Urban sustainable development can be affected directly by the use of data analytics as a tool to make the city more convenient for citizens instead of a hub for commercial activities. Water management, waste management, environmental monitoring, pollution control, precision agriculture, smart irrigation, fleet management, and traffic control are some of the areas where the use of AI can foster new avenues of innovation and efficiency for a sustainable smart city. The comprehensive, legal, and ethical issues of AI use need to be thoroughly discussed amongst the city decision-makers in order to align smart city development with user-centric rather than profit-oriented. Figure 1 shows the

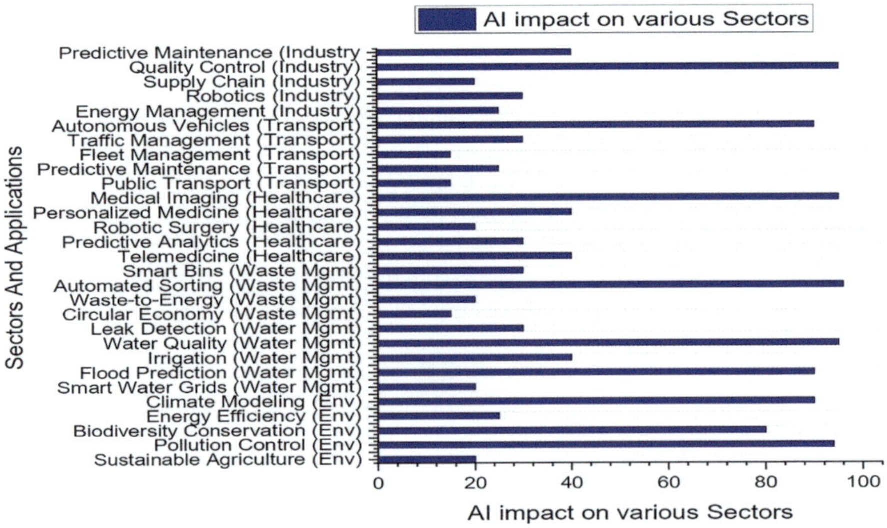

Fig. 1 Impact of AI in different verticals of smart city infrastructure [5]

impact of AI in different verticals of smart city infrastructure categorized into transport, healthcare, industry, and environment sectors.

To achieve the vision of the smart city, AI plays an important role. The use of sophisticated AI algorithms ensures effective planning, resource utilization, and energy management. It involves a sequence of operations that are to be performed for efficient decision-making and predictive analysis. The smart city development that utilizes AI must first go through the initiation phase, which is the conception stage of the project, and then the planning phase, where a detailed plan and objectives are laid down. The data collection process usually follows, and the raw data are obtained from diverse sources. This type of data undergoes processing, which involves preprocessing and application of AI algorithms, in order to extract relevant information. The decision-making cycle is then enabled, which is a very critical step that leads to the implementation of smart city initiatives. The implementation phase is where the AI-powered solutions are executed across the city. However, a very strict monitoring and evaluation mode is present to allow AI systems to detect their flaws quickly. Also, the support functions are set up to help the AI deal with its shortcomings. From the evaluation, the system adjusts and enhances the faults through the feedback loop, which in turn ensures the smart city framework's continuous development. Moving on to the so-called self-agility phase, scalability and expansion are being tackled in pursuance to widen the circle of AI use in the city. Following this, continuous improvement takes place which is a never-ending procedure whereby smart cities are consistently updated and altered to difficulties and advantages. The linked modes of this smart city AI implementation process are shown in Fig. 2.

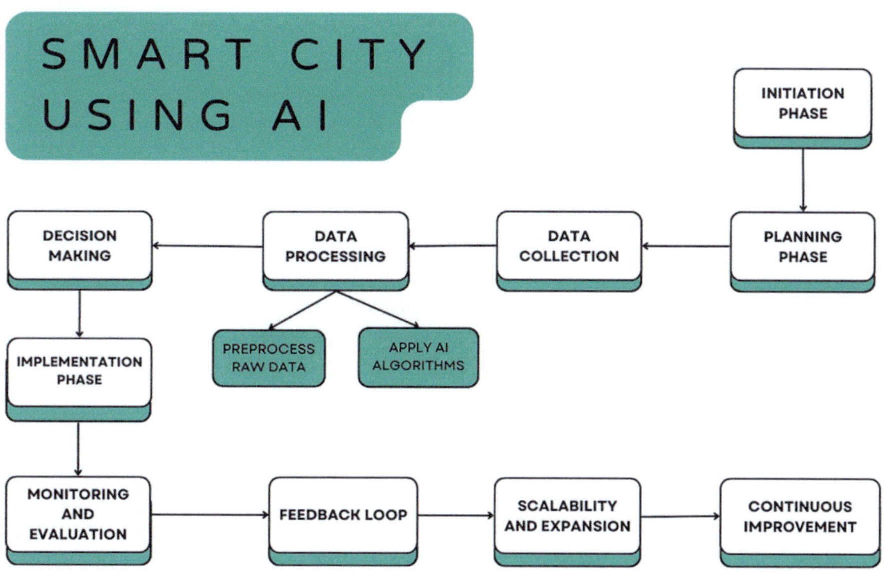

Fig. 2 Operational flow in AI-powered smart city

The next section highlights the potential of AI in the healthcare sector of smart city infrastructure which provides services involving the identification of emergency patients, medical assistance, ambulance assistance, and remote patient monitoring.

2 Healthcare in AI Smart Cities

In the rapidly advancing ecosystem of AI-powered smart cities, healthcare stands as a critical pillar, seamlessly integrating technology to enhance patient care and operational efficiency. From telemedicine to personalized treatments, AI is reshaping the way healthcare is delivered, making it more accessible, predictive, and responsive to individual needs [4]. Healthcare is the main application domain and an important pillar of the smart city ecosystem. The incorporation of AI into healthcare systems in smart cities has significant promise for transforming the way cities tackle public health issues, enhancing healthcare provision, and augmenting the general welfare of inhabitants [6].

2.1 Healthcare Robots

The term healthcare robotics describes the use of robotic technology in several facets of healthcare provision, from administrative work to patient care. These robots are intended to increase productivity, accuracy, and safety in healthcare environments by carrying out certain tasks either fully autonomously or with little human assistance. The integration of AI in healthcare robotics plays a vital role in enhancing the capabilities of robotics systems for varied healthcare applications.

Healthcare robots are influencing the following important areas which are also evident from Fig. 3.

- **Surgical Robots**
  Surgical robots, like the da Vinci Surgical System, deliver doctors with enhanced precision and dexterity while doing minimally invasive treatments. With the help of these robotic arms that are fitted with surgical equipment and a high-definition camera, doctors may perform procedures with increased control and accuracy. Surgical robots minimize patient trauma, speed up recuperation, and boost surgical results [7].
- **Pharmacy Robots**
  Medication administration and dispensing duties in hospitals and pharmacies are automated by pharmacy robots. These robots precisely distribute pharmaceuticals, maintain inventories, and guarantee adherence to prescription regimens.

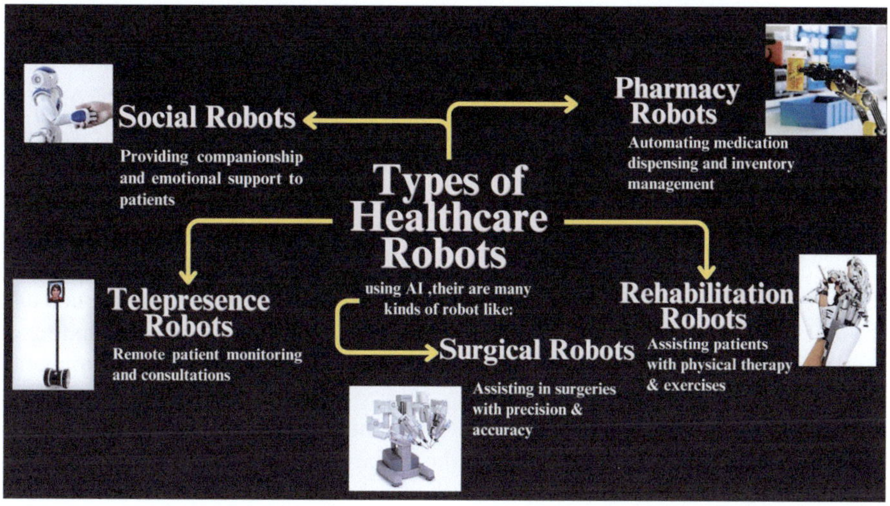

Fig. 3 Types of healthcare robots

Examples of these robots are robotic pill counters and automated medication distribution systems. Pharmacy robots expedite the dispensing of medications in healthcare environments, lower medication mistakes, and enhance drug safety [8].

- **Rehabilitation Robots**
  After an injury, disease, or surgery, rehabilitation robots help patients with physical therapy and rehabilitation activities [9]. These robots, which include robotic limbs and exoskeletons, offer precise resistance and support to help patients restore their strength, range of motion, and mobility. Robotic rehabilitation speeds up healing, lowers the chance of problems, and allows for individualized treatment plans for patients with different requirements.
- **Social Robots**
  These robots powered by AI can assist elderly patients by reminding them about their medicine schedule, dosage, timings, diet plan, etc. to be followed. This will further relieve the caregivers whose tasks will be handled by the AI robots. Moreover, the caregivers can be alerted in case of emergency, fall, or sudden change in some vital signs of patients while monitoring.
- **Telepresence Robots**
  These AI-powered robots allow telemedicine, remote patient monitoring, and consultancy. Using the patient information and preliminary diagnostics data, virtual consultations can be given.
  As robotic technologies develop, their integration with AI in healthcare settings can bring revolution in the healthcare sector, ultimately leading to better patient outcomes, more productivity, and higher standards of care overall.

2.2 *AI-Powered Stethoscope*

This type of stethoscope has an AI algorithm that analyzes heart and lung sounds in real time, they can analyze the sound picked up by the stethoscope and provide feedback to the doctor [10].

- AI uses a filter too to cut out background noise and amplify the heart and lung sounds. It also makes an easy task for doctors to hear these sounds and give a more accurate diagnosis.
- One of the other ways in which AI could be used is to compare the sounds that are taken up by the stethoscope to a database of normal sounds and abnormal sounds, helping doctors find the potential abnormalities that they could have missed.

  General AI-powered stethoscopes have the potential to be a valuable and much more useful tool for doctors but also should be aware of the potential challenges with this technology [11]. Figure 4 shows the various components of an AI-powered stethoscope which includes the integration of powerful AI algorithms with signal processing modules and other associated circuitry.

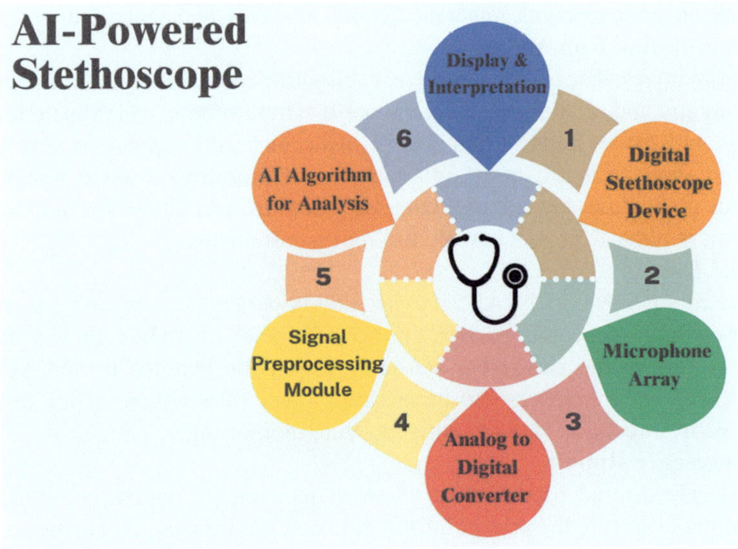

Fig. 4 AI powered stethoscope

2.3 AI Use in the Discovery of Drugs

The use of AI in the pharmaceutical sector has caused a drastic change in the making and developing of new drugs, which also transformed their way of researching, developing, and optimizing new drugs [12]. Given are a few ways in which AI is transforming this sector.

- **Research of Drug Target**
  AI algorithms use their vast amount of biological data to analyze and identify the potential drug target, like proteins or genes associated with the diseases. With this understanding of the molecular structure of diseases, researchers can develop targeted therapies [13].
- **Planning and Creating Molecular system**
  Algorithms powered by AI can help in the creation of novel medicinal compounds with desirable characteristics like safety and effectiveness. AI can assist in lead compound optimization and help cut down on the time and expense involved with traditional trial-and-error procedures by modelling the interactions between molecules and biological targets [14].
- **Prediction of Drug Toxicity**
  By examining the chemical structure of medication candidates and their resemblance to known dangerous substances, artificial intelligence algorithms can forecast possible side effects and toxicity. Prioritizing safer medication options for continued development can be aided by the early detection of safety concerns [15]. The literature includes a number of research works that substantiate the use of AI in the healthcare sector. The use of powerful AI algorithms helps doctors in creating patient databases and offering them healthcare services without delay. The current state-of-the-art research on the use of AI for revolutionizing the healthcare industry is summarized in Table 1.

Table 1 Current state-of-the-art research on the use of AI for smart healthcare

Ref.	Application area	Key performance indicator (KPI)	Highlights
[7]	Surgical robots	Precision, delay	These often pursue safer, more effective, smaller, and faster surgical procedures, giving the huge advantage of MIS over open surgery, but it is still under better development
[8]	Robotics and AI in healthcare	Efficiency, error rate	This paper gives a comprehensive overview of the role of robotics and AI technology in the context of smart healthcare. The benefits, impact, challenges, and implications are discussed for the convergence of robotics with AI for the healthcare industry
[10]	AI-powered stethoscope	Heart rate, real-time analysis	It analyzes the heart and lung sounds to diagnose and give an early detection of disease and enable the screening of a wider population than possible with manual screening

(continued)

Table 1 (continued)

Ref.	Application area	Key performance indicator (KPI)	Highlights
[11]	Heart rate monitoring	Heart rate	Here, an AI-powered heart monitoring device is proposed that aims to monitor the heart pulse rate and identify and classify the heart sounds. It makes use of long short-term memory (LSTM) for classification and fast Fourier transform (FFT) for filtering
[12]	Drug development	Accuracy	This research work highlights the role of AI in the pharmaceutical industry in diverse areas which include clinical trials, drug delivery, drug discovery, and development
[14]	Development of drugs using AI	Accuracy, efficiency	This paper gives a hybrid model for recommending/ predicting drug groups. It also provides AI-based decision support in new drug development
[15]	Drug toxicity	Accuracy of detection	In this paper, novel AI algorithms are proposed that detect the presence of harmful substances in the chemical composition of drugs

3 Developing the Transportation Sector with the Help of AI

In this section, the focus is on how the transportation sector is on the brink of a major transformation, driven by advancements in AI. The use of AI technologies in the transportation industry is creating new prospects to increase efficiency, ensure safety, and address environmental concerns. To be more precise, the advancements of autonomous vehicles, smart traffic control, predictive maintenance, and logistics optimization are the agents of change that will alter the dynamics of transportation goals for goods and passengers. Through the power of AI, we are not only able to come up with more intelligent transport networks but also build a more environment-friendly future while solving current problems. AI also brings about the digitization of the transportation sector. The cities are providing innovative and intelligent transport systems by employing AI technologies, thus making them the most efficient and safest. In the transportation industry, AI technologies are already making a big impact and are expected to become even more prominent in the future [16, 17].

The following are some applications of AI in transportation:

- **Use of Autonomous Vehicles (AVs)**
  AI has played a crucial role in the self-driving car industry. With its assistance, these vehicles are now able to visually sense their surroundings, interpret photos, make accurate judgments, and navigate roads with the assistance of their AI-driven secure systems [16]. Currently, major corporations like Tesla, Waymo, and Uber are engaged in the development of their AI-driven autonomous vehicles.
- **Use of Public Transportation**
  In order to assist the public transportation system, optimize system performance, and enhance reliability, AI is being used these days. The information about buses' arrival, departure, and cancellation can be disseminated via digital applications that include displays located at stations or can be communicated through

on-board announcements. The implementation of AI also provides these organizations with the ability to promptly alter their routes in order to guarantee punctual operations, therefore enhancing job efficiency and minimizing waiting periods and furthermore overpopulation [17].

- **Predictive Upkeep**
  AI analyzes data collected by sensors on cars to predict issues, enabling the scheduling of essential repairs during less busy periods and thereby minimizing disruptions in journeys. This leads to increased dependability and reduced costs due to the optimization of maintenance schedules and the avoidance of non-essential repairs [18].
- **Ride-Sharing and Mobility Services**
  AI technologies such as Uber and Lyft use algorithms to assess customer demand and traffic conditions on selected drivers. They then optimize their routes to offer efficient and convenient mobility options to passengers. Consequently, these platforms connect drivers and riders through computer programs that use current innovations to ensure the achievement of a sustainable transportation system, including pedestrian and public transport access by 2024 [19].

Thus, AI is revolutionizing the transportation industry by enhancing its security, efficiency, and accessibility. Figure 5 depicts AI-powered transportation as one of the verticals of smart cities with use cases in traffic management, route planning, traffic light control, and autonomous vehicles. However, there exist some challenges that need to be tackled, including regulatory barriers, ethical concerns, and the necessity for extensive expansion and investment.

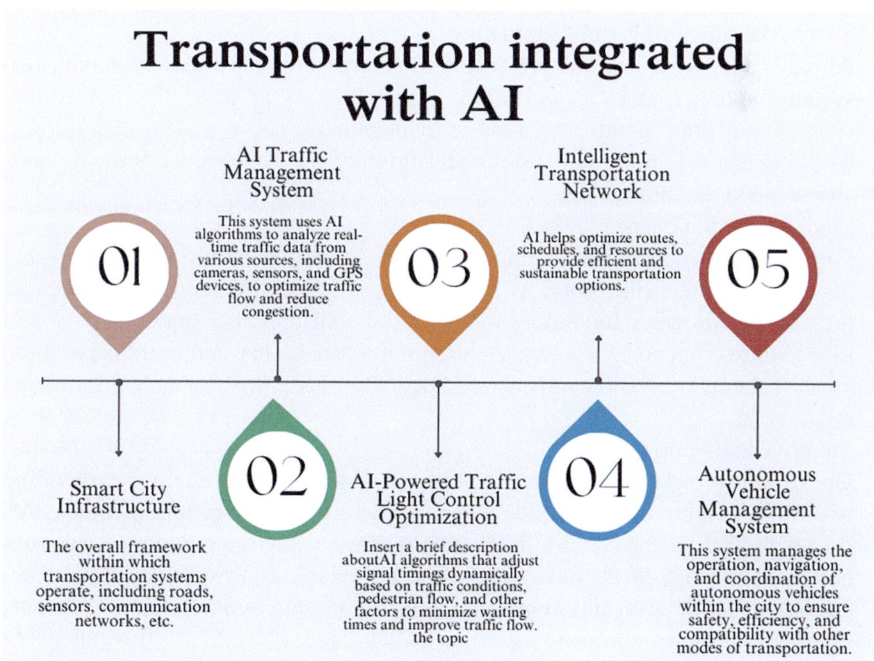

Fig. 5 Intelligent transportation with the help of AI

4 Potential of AI in Enhancing Industries

This section explores how AI is evolving in a range of areas including healthcare, finance, manufacturing, and retail, making it more efficient by reducing costs and generating new business models and new products and services while also fostering an environment of innovation. AI is transforming the efficiency of various industries by increasing productivity, efficiency, and innovation. AI offers huge potential due to its application in predictive analytics, quicker drug finding, and personal treatments, in healthcare. Algorithmic trading and fraud detection, along with personalized financial plans, are some of the benefits of AI to the finance sector. Manufacturing is one of many sectors transforming with AI-driven automation, quality control, and predictive maintenance, leading to increased operational efficiency which is shown in Fig. 4. In the retail industry, AI optimizes inventory management, streamlines supply chain logistics, and enhances customer experiences through personalized recommendations. These advancements and many other sectors demonstrate AI's broad and profound impact across different sectors, driving growth and innovation.

4.1 Manufacturing Using AI

AI is transforming the manufacturing sector by introducing advanced technologies that optimize technologies that enhance product quality, optimize efficiency, and streamline operations. Some key applications of AI in manufacturing are as follows:

- Smart Manufacturing and Automation
  AI plays a key role in the development of smart factories, where interconnected systems and IoT analyze and collect vast amounts of data. This integration enables real-time monitoring, remote management, and seamless communication between machines and systems, leading to improved production efficiency and reduced downtime [20].
- Predictive Maintenance
  It is one of the most impactful applications of AI in manufacturing. By analyzing historical and real-time data, AI systems can predict equipment failures before they occur, allowing for timely interventions and reducing unexpected downtime. Many companies leverage AI to monitor machinery and optimize maintenance schedules, significantly extending equipment lifespan and minimizing repair costs.
- Supply Chain Optimization
  One of the crucial roles is played by AI in optimizing supply chains by providing real-time insights into demand, inventory levels, and distribution routes. AI-driven predictive forecasting helps manufacturers adjust production schedules and manage stock levels more efficiently. For example, Rose has reduced forecasting errors by 20% and improved demand planning by integrating AI into its processes [21].

- Quality Control
  AI does enhancement of quality control through computer vision and machine learning. These AI-powered systems can detect defects in products that might be missed by human inspectors or traditional robots. For instance, a company like BMW uses AI for automatic image recognition to improve inspection accuracy and eliminate pseudo-defects, thereby ensuring higher quality standards.

With the adoption of these technologies, manufacturers can not only improve operational efficiency but also enhance product quality, meet sustainability targets, and reduce costs, positioning themselves for success in the competitive global market.

4.2 Energy Management in Industries Using AI

AI is being highly used in the energy sector to enhance efficiency, integrate renewable energy sources, and manage demand. There are many industrial sectors that have witnessed huge transformation with the use of AI for energy management.

- Renewable Energy Integration.
  AI improves the forecasting of renewable energy production and consumption, addressing the intermittent nature of solar and wind power sources. This allows utilities to utilize renewable energy, store excess energy better, and maintain a balance between supply and demand, enhancing the reliability and sustainability of energy systems [22].
- Smart Grids and Micro Grids.
  Smart Grid: AI helps optimize the efficiency of power generation, distribution, and transmission. By analyzing data from smart meters, IoT devices, and sensors, AI can detect anomalies, predict equipment failures, and balance energy flow to match supply and demand more accurately. This helps reduce energy waste and enhances the overall efficiency of the grid [23].
  Micro Grids: These types of localized grids can operate independently or in conjunction with the main grid. AI and machine learning (ML) optimize the integration of renewable energy sources, ensure a stable and efficient energy supply, and manage energy storage even during disruption.
- Energy Trading.
  With the help of AI, the prediction of energy demand and market prices enables more efficient trading of energy. This helps traders make informed decisions on buying and selling energy, thus optimizing the financial performance of energy markets.
- Grid Resilience.
  Grid resilience is enhanced by enabling real-time monitoring and early fault detection, mainly for maintaining uninterrupted power supply, especially during natural disasters or cyber-attacks. AI algorithms can quickly identify and respond to issues, minimizing downtime and enhancing grid reliability [24].

Overall, integrating AI in energy management offers substantial benefits in terms of efficiency, sustainability, and cost savings, paving the way for a more resilient and optimized energy infrastructure.

4.3 Quality Control in Industries Using AI

With the revolution of AI, quality control in industries is significantly enhancing the efficiency, consistency, and accuracy of inspection and testing processes. Some of the key applications and benefits of using AI in quality control are as follows:

- Consistency and Standardization
  AI system response is more consistent with quality control than that of a human tester. Different from humans, who vary in their assessments due to subjective factors like personal judgment, AI constantly applies the same criteria and procedures, resulting in more reliable outcomes.
- Enhanced Documentation and Reporting
  AI algorithms automate the generation of detailed reports by extracting relevant information, analyzing test results, and compiling them into documentation. This summarizes the reporting process and ensures the recording of all essential details [25].
- Predictive Quality Analysis
  With the analysis of historical data, customer feedback, and other relevant information, AI algorithms also identify patterns and predict potential quality issues before they arise. This approach helps prevent defects and enhances overall product quality.
- Automated Visual Inspection
  AI-driven machine vision systems can automatically identify defects in products by analyzing images. These kinds of systems are trained on extensive datasets containing images of both good and defective products, enabling them to detect anomalies with high precision. By reducing reliance on human inspectors, they minimize errors caused by fatigue or oversight [26].
  These advancements in AI-driven quality control improve the reliability and efficiency of inspection and enable manufacturers to maintain high standards of product quality consistency, leading to increased customer satisfaction and reduced costs. Table 2 provides a comprehensive overview of the research work carried out in areas of AI and its convergence for smart industries. The key performance indicators (KPIs) and application areas are provided along with the key findings of the researchers.

Table 2 Current state-of-the-art research on the use of AI for smart Industry 5.0

Ref.	Application area	Key performance indicator (KPI)	Highlights
[20]	Industry 4.0 and smart manufacturing	Production efficiency	This involves the integration of digital technologies like AI and IoT in manufacturing processes. In order to enhance production efficiency, AI algorithms are presented for smart manufacturing vertical in Industry 4.0
[21]	Supply chain management	Operational efficiency	This paper highlights the AI approach to cater to the needs of supply chain management in the industrial IoT ecosystem. Inventory management, fault tolerance, and risk management are assessed through AI techniques, yielding improved operational efficiency
[22]	Energy management in industries	Energy efficiency	AI-based strategies are being identified to assist renewable energy in making contemporary power systems
[23]	Grid monitoring	Energy storage	This paper presents the AI strategies in smart grid monitoring systems. This includes energy management, fault prediction, energy distribution, energy storage, and grid health monitoring. The integration with renewable energy and energy security is also discussed
[24]	Grid resilience	Power consumption	AI algorithms can quickly identify and respond to issues, minimizing downtime and enhancing grid reliability
[26]	AI analyzing automated visual inspection	Error ratio	Data-driven technology and advancement in sensor tech with IoT make real-time tracking of the system in reality

5 Environment Sustainability Through AI

With the collective expertise of sustainable AI and big data, environmentally sustainable smart cities are at the forefront of addressing environmental challenges through innovative solutions [27].

5.1 Energy Management

Using AI as a source of energy management involves the application of AI techniques to optimize energy generation, distribution, and consumption. Figure 5 shows how a smart city can manage energy with AI through its optimal energy storage, energy distribution, and energy consumption solutions (Fig. 6).

The following are the methods

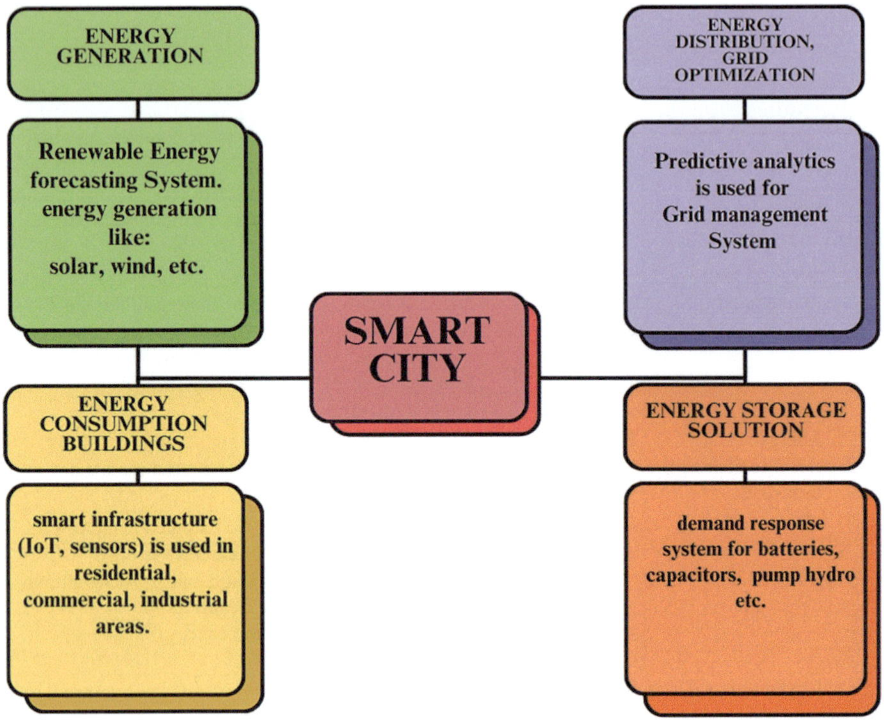

Fig. 6 Energy management in smart city with the help of AI

- **Load Balancing**
  By dynamically altering generation and distribution in response to current supply and demand situations, AI optimizes the distribution of energy throughout the grid [28]. To distribute the load over several generators, transmission lines, and distribution networks and maintain grid stability and dependability, AI algorithms evaluate data from smart meters, sensors, and grid equipment [29].
- **Energy Source Optimization**
  By figuring out the best charging and discharging schedules depending on energy demand, generation patterns, and market circumstances, AI maximizes the performance of energy storage devices, such as batteries and pumped hydro storage. Artificial intelligence systems forecast energy prices, maximize profits from energy trading and arbitrage, and optimize storage capacity use [30].
- **Energy Efficiency**
  Improvement in energy efficiency in buildings, industry, and transportation by optimizing resource consumption, minimizing waste, and identifying opportunities for energy saving is done by AI [31]. A system powered by AI analyzes data from smart sensors, IoT devices, and building management systems to identify energy-intensive processes, optimize equipment operation, and implement energy-saving measures, reducing energy costs and environmental impact.

- **Demand Response**
  AI optimizes the operation of energy storage systems, such as batteries and pumped hydro storage, by determining the optimal charging and discharging schedules based on energy demand, generation patterns, and market conditions. AI algorithms predict energy prices, optimize storage capacity utilization, and maximize revenue from energy trading and arbitrage [32].
  Ultimately, AI enables utilities, customers, and energy management to maximize energy use, cut expenses, and improve sustainability along the whole energy value chain, from production to consumption. Stakeholders may accomplish their energy management objectives and contribute to a more robust, efficient, and sustainable energy system by utilizing AI technology.

5.2 *Water Management*

Using AI approaches to improve the distribution, use, and preservation of water resources is known as water management with AI [33]. There are a number of use case scenarios that witness the convergence of AI in water management. The prediction of floods, monitoring of water quality, and leakage detection are some of the examples. Here are some of the methods.

- **Water Quality Monitoring**
  AI algorithms collect data from sensors, satellites, and water quality monitoring system stations to assess water quality measures like pH levels, turbidity, dissolved oxygen, and pollutant concentration [34]. By detecting abnormalities and identifying sources of contamination in real time, with the help of AI, water managers take measures to protect water quality and public health.
- **Smart Irrigation**
  By improving irrigation timing, water allocation, and crop water requirements, AI improves agricultural water management. AI systems examine soil moisture data, meteorological predictions, and crop attributes to identify the best times, lengths, and volumes for irrigation. This minimizes water usage, preserves water resources, and boosts agricultural yields [35].
- **Flood Prediction and Management**
  By collecting all the data from weather forecasts, river gauges, and hydrological models, AI predicts and mitigates flood risks. Data collected from algorithms assess flood probabilities, identify vulnerable areas, and recommend preventive measures such as early warning systems, flood barriers, and land use planning to minimize flood damage and protect communities [36].
- **Leakage Detection**
  AI systems use data from flow meters, pressure sensors, and sound sensors to identify leaks and water losses in distribution networks. Artificial intelligence algorithms are capable of identifying anomalous flow patterns, locating leaks, and prioritizing repair tasks to limit water loss, prevent damage to infrastructure, and save water supplies [37].

By harnessing the power of AI, water managers, utilities, and policymakers can optimize water use, enhance water quality, and ensure the sustainability of water resources for current and future generations. AI-driven innovations enable more efficient, resilient, and equitable water management practices, helping to address the growing challenges of water scarcity, pollution, and climate change.

5.3 Waste Management

In developing countries like India and China with vast populations, large amounts of waste are generated every day. The management of this waste is a huge challenge. The waste management problem can be catered with the use of AI. AI enhances waste management by identifying recycling opportunities, reducing landfill usage, and waste management. AI algorithms analyze data on trash generation trends and recycling rates to promote resource recovery and collection schedules through initiatives such as composting and recycling incentives, thereby minimizing environmental impact [38]. Here are some of the other ways in which waste management can be done using AI.

- **Smart Bin Monitoring**
  AI sensors installed in waste bins monitor fill levels, detect overflowing bins, and notify waste management teams when bins need to be emptied. AI algorithms analyse sensor data to predict future fill levels, optimize collection routes, and reduce fuel consumption and greenhouse gas emissions [39].
- **Waste Reduction and Prevention**
  AI engages consumers and businesses in waste reduction and prevention efforts by providing personalized recommendations, real-time feedback, and educational resources [40]. AI-powered apps and platforms raise awareness about waste management practices, encourage responsible consumption habits, and incentivize waste reduction behaviors, empowering individuals and communities to minimize their environmental footprint.
- **Waste Collection Optimization**
  AI algorithms sense data from sensors, GPS devices, and waste collection routes to optimize collection routes and schedules. By anticipating waste generation patterns, identifying high-demand areas, and adjusting collection routes in real time, AI helps reduce collection costs, minimize vehicle emissions, and improve overall efficiency [41].
- **Waste Sorting and Recycling**
  By identifying and separating recyclable components from mixed waste streams using robotics, computer vision, and machine learning, AI automates garbage sorting and recycling operations. By precisely classifying materials and sorting them into distinct categories including paper, plastic, glass, and metal, AI-powered sorting systems can increase recycling rates and decrease contamination. These systems scan photos of waste objects [42].
  With the power of AI, stakeholders in waste management may lessen their influence on the environment, improve garbage collection, recycling, and disposal

procedures, and transition to a circular and more sustainable economy. The increasing problems of trash creation, pollution, and resource depletion may be addressed with the aid of AI-driven advancements, which make waste management techniques more intelligent, effective, and ecologically friendly.

5.4 Renewable Energy Optimization

Renewable energy optimization involves leveraging AI techniques to maximize the efficiency, reliability, and cost-effectiveness of renewable energy generation, distribution, and utilization as shown in Fig. 7 [43].

The contribution of AI in optimizing renewable energy can be summarized as follows.

- **Energy Forecasting**
  AI predicts renewable energy generation patterns based on weather forecasts, historical data, and other factors. AI allows energy suppliers to predict variations in renewable energy output and adjust energy production, storage, and distribution following solar irradiance, wind speeds, and hydrological conditions [44].
- **Grid Integration and Management**
  AI facilitates the integration of renewable energy into the electricity grid by optimizing grid operation, balancing supply and demand, and managing grid constraints. AI algorithms analyze data from smart meters, sensors, and grid infrastructure to forecast energy demand, optimize generation dispatch, and coordinate energy exchanges between distributed energy resources and centralized power plants, enabling efficient and reliable grid operation [45].

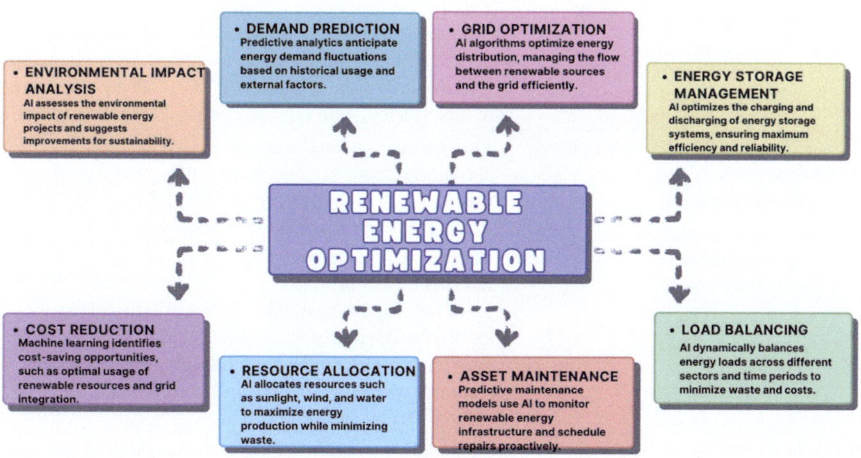

Fig. 7 Aspects of renewable energy optimization with AI

- **Generation Forecasting**
  AI anticipates renewable energy generation patterns based on real-time data from forecasts, weather, and historical production data. With these forecasting solar irradiances, wind speed, and water flows, AI can use utilities and grid operators to anticipate fluctuations in renewable energy output, plan grid integration, and optimize energy dispatch to meet demand while maintaining grid stability.

All things considered, AI-enabled renewable energy optimization makes energy systems more dependable, sustainable, and efficient, which helps to slow down global warming, lessen dependency on fossil fuels, and accelerate the shift to a low-carbon economy. Using AI technology, stakeholders in renewable energy may increase the value of renewable resources, enhance energy access, and build a more fair and resilient energy future. Table 3 gives the current state-of-the-art research on the use of AI for environmental sustainability including energy management, waste management, and water management.

Table 3 Current state-of-the-art research on the use of AI for environmental sustainability

Ref.	Application area	Key performance indicator (KPI)	Highlights
[28]	Grid management	Load balancing, power consumption	This paper enhances the sustainability of the electricity power system by introducing grid management. It aims to improve the power consumption of the system and results in an energy-efficient system
[31]	Energy management	Energy efficiency	This paper presents an AI-powered system that analyzes data from smart sensors, IoT devices, and building management systems to identify energy-intensive processes
[36]	Flood prediction	Accuracy	Data collected from algorithms helps AI to predict flood risks and alert people by giving warnings
[37]	Leakage detection	Accuracy	This paper utilizes the AI approach to detect the leakage of water in real time. To reduce the wastage of potable water, various ML algorithms are presented for water distribution networks (WDNs). The performance of ANN, SVM, and DL are compared for accuracy
[38]	Energy and waste management	Waste quantity	This paper proposes an advanced ML approach to predict the amount of waste and then plan for efficient waste collection. The proposed model aims to yield waste to energy resulting in a sustainable energy market
[40]	Environmental sustainability	Efficiency	This paper presents the potential of AI and the artificial Internet of Things (AIoT) for environmental sustainability in the context of smart cities. This highlights the convergence of AI and IoT for yielding smarter eco-cities. The key enablers and drivers for smarter eco-cities are also discussed to prioritize environmental sustainability
[42]	Water sorting and recycling	Accuracy	By precisely classifying materials and sorting them into distinct categories, including paper, plastic, glass, and metal, AI-powered sorting systems can increase recycling rates and decrease contamination

6 Governance and Citizen Engagement

This engagement is a vital component of effective and responsible administration in any society. AI platform enables transparent governance, participatory planning, responsive services, and collaboration between government agencies, businesses, and citizens to address urban challenges effectively [46]. It also enhances government and citizen engagement in many ways.

6.1 Predictive Analytics

Predictive analytics in governance and citizen engagement using AI involves using historical data, machine learning algorithms, and predictive models to anticipate future trends, identify potential issues, and inform policy decisions. The following factors can be applied in governance and citizen engagement as well.

- **Public Health Prediction**
  AI predicts analytics and forecasts outbreaks, healthcare needs, and public health risks by analyzing health data, environmental factors, and demographic trends. Using predictive models and machine learning algorithms, the government can anticipate the spread of infectious diseases, allocate healthcare resources effectively, and take action to safeguard public health [47].
- **Policy Impact Assessment**
  Predictive analytics is the analysis of historical data to construct models of future scenarios, enabling the evaluation of the possible impacts of proposed policies, legislation, or initiatives. By leveraging predictive models and machine learning algorithms, the government may anticipate the repercussions of policy initiatives, enhance policy planning, and identify unforeseen consequences to achieve its objectives.
- **Citizen Feedback Analysis**
  Novel predictive analytics are employed to uncover previously undiscovered phenomena. Analysis of issues, trends, and emotion patterns in citizen feedback from a diverse range of sources, including social media and media surveys. The state is able to predict future complaints, anticipate citizen needs, and allocate solutions obtained using machine learning techniques and analysis of historical data [48].
  In summary, the application of predictive analytics in governance and citizen engagement using AI technology enables the government to make data-driven choices and address the requirements of citizens in a focused and proactive manner. By scrutinizing historical data and using predictive analysis capabilities, government administrations may enhance the efficiency of governance, the quality of service delivery, and the participation of people in the decision-making process.

6.2 Transparency and Accountability

In governance, these principles guarantee that government activities are transparent, accountable, and aligned with the demands of the population. Additionally, they ensure the maintenance of uniformity among units and the mitigation of human fallibility and prejudice. It involves the use of AI systems to make decisions and assist the government in adhering to the specified regulations or norms. Furthermore, AI systems are highlighted as a means to enhance the openness of the public sector and facilitate citizen participation in several aspects [49]. Artificial intelligence also serves as a tool in enabling the sustainable development of the government and the engagement of the population, commonly referred to as e-government.

- **Fraud Detection and Prevention**
  Artificial intelligence readily detects and exposes theft, fraudulent behavior, corruption, or inappropriate conduct in governance by analyzing transaction, financial, and procurement data. Fraud detection systems use machine learning algorithms to identify anomalies, non-compliance that may suggest fraudulent activities, and patterns typically followed by deceitful individuals. This enables the government to take appropriate actions and restore the trustworthiness and reliability of the officers [50].
- **Data Transparency**
  AI streamlines government data exchange and analysis by automating operations related to data processing, cleansing, and visualization. Governments may publish open data sets, interactive dashboards, and visualizations using AI-powered data analytics systems. This increases public access to and comprehension of government information and fosters accountability and transparency.
- **Whistle-Blower Protection**
  The AI whistle-blower protection system is a tool for unloading employees who can be erased by their boss and the organization. They can also inform, in secret and with confidence, managers about their concerns regarding ethical, legal, and moral issues. Whistle-blowers' accusations are examined and analyzed by AI programs that then shield them from mental and physical horrors that often come with the truth and promote the understanding of important real issues in governance by allowing transparent and ethical behavior [51].

AI is essential in enhancing transparency and accountability in governance and citizen service by facilitating governments to analyze data, identify irregularities, evaluate risks, and promptly address user input. Through the utilization of AI technology, governments may maintain transparency, cultivate trust, and guarantee accountability in their activities and decisions, thus augmenting public trust and confidence in government institutions.

7 AI in Public Safety

In order to enhance the security of the people, smart cities actively employ AI to gather information via sensors, surveillance cameras, and social media. This enables the identification and prevention of accidents and crises. Advanced artificial intelligence systems oversee real time, predictive analytics, and automated response activities, therefore enhancing the overall safety and resilience of metropolitan areas [52].

7.1 Video Surveillance and Monitoring

Artificial intelligence is increasingly being applied in several fields of greater significance for public safety; one crucial factor is the surveillance and monitoring conducted via video, in whatever form, surveillance cameras or not.

- **Object Detection and Recognition**
  Currently, AI-powered security cameras have the capability to detect and recognize a wide range of objects, including automobiles, individuals, and peculiar household objects. In addition, these systems have the ability to detect individuals in video feeds, a task that AI algorithms can do in specific scenarios including security threats when security personnel are unable to promptly investigate the situation, or where metros may be prevented from proceeding.
- **Crowd Monitoring**
  Artificial intelligence algorithms can analyze the dynamics and behavior of a crowd to identify crowd density or anomalies. The embedded artificial intelligence camera system screens video streams from cameras installed in public areas, event locations, or transportation hubs to identify possible safety risks and promote smooth traffic and organized conduct during public events [53].
- **Facial Recognition**
  An AI face recognition system utilizes technology to detect and filter facial visual cues in video recordings by analyzing them and comparing them with a database of known persons or a watchlist as programmed. These technologies enable the police to identify offenders and those who are missing as well as monitor the overall traffic conditions in very crowded or hazardous areas, therefore continuously contributing to public safety [54].

AI-driven video surveillance and monitoring systems enhance public safety by providing real-time situational awareness, early threat detection, and proactive intervention capabilities. By leveraging AI technologies, law enforcement agencies, emergency services, and security, providers can improve their effectiveness, responsiveness, and ability to prevent and mitigate security risks in urban environments.

7.2 *Fire Detection and Prevention*

Fire detection and prevention is one of the important public safety verticals that involves leveraging AI technologies to detect fires, prevent their spread, and minimize damage and casualties. An AI-powered fire detection system is shown in Fig. 8. With the use of cameras, thermal sensors and powerful AI algorithms, fire can be detected and alerts can be issued to the concerned authority to prevent human lives.

The AI contributes to fire detection and prevention in the following ways.

- **Early Fire Detection**
  Fire detection systems powered by AI analyze data from different sources which include smoke detectors, heat sensors, and video cameras. To detect any sign of fire in its early stage, algorithms used in machine learning are trained on historical fire data which recognize patterns and anomalies associated with fire events, enabling early detection and timely response by alerting authorities and triggering the evacuation process [55].
- **Continuous Monitoring and Monitoring**
  Fire detection systems using AI can monitor fire safety equipment, such as smoke detectors, sprinkler systems, and fire extinguishers to ensure performance and reliability. AI analyzes sensor data, detects equipment malfunctions or failures, and also schedules preventive maintenance activities proactively, minimizing downtime and ensuring readiness for emergencies.

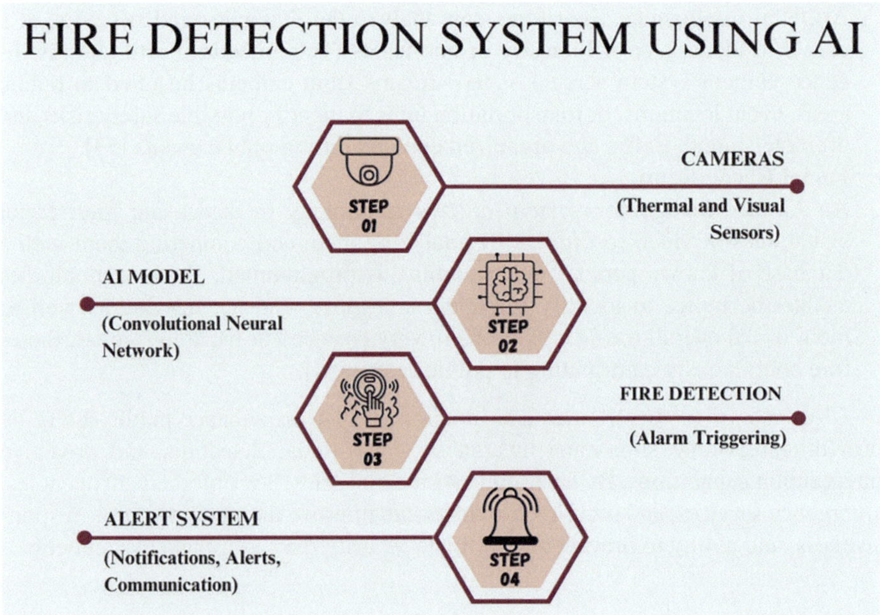

Fig. 8 Leveraging AI in fire detection system

- **Integration with Building Automation System**
  AI-driven fire detection systems may be integrated with building automation systems to provide automated response actions, including ventilation system management, fire suppression system activation, and fire door closure. AI improves the efficacy and efficiency of fire prevention and mitigation operations by automating emergency response protocols, which lessens the need for manual intervention and human error [56].

AI-driven fire detection and prevention systems play a critical role in enhancing public safety by providing early detection, improved accuracy, and proactive intervention capabilities. By leveraging AI technologies, authorities can reduce the risk of fires, protect lives and property, and create safer and more resilient communities.

8 Conclusion

The chapter presents the potential of AI in developing a smart city infrastructure. Leveraging AI in smart cities has immense potential to enhance urban living for citizens, optimize resource management, and achieve sustainable development. With predictive modelling, automation, and advanced data analytics, AI-advanced cities address complex challenges smoothly, from energy consumption to traffic congestion to public safety and advanced healthcare. It offers facilities for remote patient assistance, virtual consultancy, telemedicine, and telesurgery. The agriculture sector also booms with AI-enhanced precision agriculture, monitoring crop health, soil fertility, and pesticide control, increasing crop yield. The use of powerful AI algorithms and rigorous data analytics, followed by efficient decision-making also yield enhanced performance in environmental monitoring, water management, leakage detection, flood control, and waste management. AI-powered Industry 5.0 opens new avenues for manufacturing, retail, and finance. However, to provide fair access and advantages for all inhabitants, effective implementation necessitates careful consideration of ethical, privacy, and security considerations in addition to inclusive policy. Smart cities can spur innovation, enhance quality of life, and create resilient communities for the future by embracing AI as a strategic tool.

References

1. Jha, A. K., et al. (2021). A review of AI for urban planning: Towards building sustainable smart cities. In *2021 6th International Conference on Inventive Computation Technologies (ICICT)*. IEEE.
2. Taneja, A., Saluja, N., & Rani, S. (2024). An energy efficient dynamic framework for resource control in massive IoT network for smart cities. *Wireless Networks, 30*(5), 4447–4458.

3. Maltezos, E., et al. (2021). Public safety in smart cities under the edge computing concept. In *2021 IEEE International Mediterranean Conference on Communications and Networking (MeditCom), Athens, Greece* (pp. 88–93). https://doi.org/10.1109/MeditCom49071.2021.9647550
4. Ali, O., et al. (2023). A systematic literature review of artificial intelligence in the healthcare sector: Benefits, challenges, methodologies, and functionalities. *Journal of Innovation & Knowledge, 8*(1), 100333.
5. Bibri, S. E., Alexandre, A., Sharifi, A., & Krogstie, J. (2023). Environmentally sustainable smart cities and their converging AI, IoT, and big data technologies and solutions: An integrated approach to an extensive literature review. *Energy Informatics, 6*(1), 9.
6. Shaheen, M. Y. (2021). *Applications of Artificial Intelligence (AI) in healthcare: A review*. ScienceOpen Preprints.
7. Zhu, J., et al. (2021). Intelligent soft surgical robots for next-generation minimally invasive surgery. *Advanced Intelligent Systems, 3*(5), 2100011.
8. Stasevych, M., & Zvarych, V. (2023). Innovative robotic technologies and artificial intelligence in pharmacy and medicine: Paving the way for the future of health care—A review. *Big Data and Cognitive Computing, 7*(3), 147.
9. Nicholson-Smith, C., et al. (2020). A multi-functional lower-and upper-limb stroke rehabilitation robot. *IEEE Transactions on Medical Robotics and Bionics, 2*(4), 549–552.
10. Zhang, M., et al. (2023). A low-cost AI-empowered stethoscope and a lightweight model for detecting cardiac and respiratory diseases from lung and heart auscultation sounds. *Sensors, 23*(5), 2591.
11. Dampage, U., et al. (2021). AI-based heart monitoring system. In *2021 IEEE 4th International Conference on Computing, Power and Communication Technologies (GUCON)*. IEEE.
12. Mak, K.-K., Wong, Y.-H., & Pichika, M. R. (2023). Artificial intelligence in drug discovery and development. In *Drug discovery and evaluation: Safety and pharmacokinetic assays* (pp. 1–38).
13. Blanco-Gonzalez, A., et al. (2023). The role of ai in drug discovery: Challenges, opportunities, and strategies. *Pharmaceuticals, 16*(6), 891.
14. Jung, Y. L., Yoo, H. S., & Hwang, J. N. (2022). Artificial intelligence-based decision support model for new drug development planning. *Expert Systems with Applications, 198*, 116825.
15. Van Tran, T. T., et al. (2023). Artificial intelligence in drug toxicity prediction: Recent advances, challenges, and future perspectives. *Journal of Chemical Information and Modeling, 63*(9), 2628–2643.
16. Naz, N., et al. (2022). Intelligence of autonomous vehicles: A concise revisit. *Journal of Sensors, 2022*.
17. Bharadiya, J. (2023). Artificial intelligence in transportation systems a critical review. *American Journal of Computing and Engineering, 6*(1), 34–45.
18. Mhlanga, D. (2023). Artificial intelligence and machine learning in making transport, safer, cleaner, more reliable, and efficient in emerging markets. In *FinTech and artificial intelligence for sustainable development: The role of smart technologies in achieving development goals* (pp. 193–211).
19. Fafoutellis, P., et al. (2022). Acceptability modeling of autonomous mobility on-demand services with on-board ride sharing using interpretable Machine Learning. *International Journal of Transportation Science and Technology, 11*(4), 752–766.
20. Mourtzis, D. (2024). Industry 4.0 and smart manufacturing. In *Manufacturing from Industry 4.0 to Industry 5.0* (pp. 13–61). Elsevier.
21. Abaku, E. A., Edunjobi, T. E., & Odimarha, A. C. (2024). Theoretical approaches to AI in supply chain optimization: Pathways to efficiency and resilience. *International Journal of Science and Technology Research Archive, 6*(1), 092–107.
22. Rashid, Abdur et al. (2024) *Present and future of AI in renewable energy domain: A comprehensive survey*. arXiv preprint arXiv:2406.16965.
23. Kumar, R. S., et al. (2023). Impact of artificial intelligence techniques in distributed smart grid monitoring system. In *Smart energy and electric power systems* (pp. 79–103). Elsevier.

24. Islam, M. R., et al. (2024). Building a resilient and sustainable grid: A study of challenges and opportunities in AI for smart virtual power plants. In *Proceedings of the 2024 ACM southeast conference*.
25. Brereton, T. A., et al. (2023). The role of artificial intelligence model documentation in translational science: Scoping review. *Interactive Journal of Medical Research, 12*(1).
26. Sundaram, S., & Zeid, A. (2023). Artificial intelligence-based smart quality inspection for manufacturing. *Micromachines, 14*(3), 570.
27. Akter, S. (2024). Harnessing technology for environmental sustainability: Utilizing AI to tackle global ecological challenges. *Journal of Artificial Intelligence General Science (JAIGS) ISSN: 3006-4023, 2*(1), 49–57.
28. Dasari, K., et al. (2024). Integration of AI in distributed energy resource management for enhanced load balancing and grid stability. In *E3S web of conferences* (Vol. 505). EDP Sciences.
29. Gures, E., et al. (2022). Machine learning-based load balancing algorithms in future heterogeneous networks: A survey. *IEEE Access, 10*, 37689–37717.
30. Dutta, P. K., Tripathi, N., Patra, A., & Mukherjee, P. (2023). Advanced detection of refrigerant leakage using pressure sensor information in variable refrigerant flow systems. In *7th IET smart cities symposium (SCS 2023), hybrid conference, Bahrain* (pp. 70–76). https://doi.org/10.1049/icp.2024.0903
31. Fan, Z., Yan, Z., & Wen, S. (2023). Deep learning and artificial intelligence in sustainability: A review of SDGs, renewable energy, and environmental health. *Sustainability, 15*(18), 13493.
32. Wicaksono, H., et al. (2024). Artificial-intelligence enabled dynamic demand response system for maximizing the use of renewable electricity in production processes. *The International Journal of Advanced Manufacturing Technology*, 1–25.
33. Mullick, S., Dutta, P., & Chitara, D. (2023). Urbanization agenda: A journey embracing geodata analysis using digital twin model. In *7th IET smart cities symposium (SCS 2023), hybrid conference, Bahrain* (pp. 33–39). https://doi.org/10.1049/icp.2024.0897
34. Dutta, P. K., Banerjee, U., Manna, B., Hazarika, K., Das, B., & Roy, S. (2023). Nanotechnology and the application of information technology for sustainable innovation in agriculture. In A. E. Hassanien, D. Zheng, Z. Zhao, & Z. Fan (Eds.), *Business intelligence and information technology* (2022. Smart innovation, systems and technologies) (Vol. 358). Springer. https://doi.org/10.1007/978-981-99-3416-4_41
35. Prasad, D., Singla, K., & Baggan, V. (2019). System model for smart precision farming for high crop yielding. *Journal of Computational and Theoretical Nanoscience, 16*(10), 4406–4411.
36. Dutta, P. K., El-kenawy, S. M., Ali, G., & Dhoska, K. (2023). An energy consumption monitoring and control system in buildings using Internet of Things. *Babylonian Journal of Internet of Things, 2023*, 38–47. https://doi.org/10.58496/BJIoT/2023/006
37. Dutta, P. K., & Banerjee, S. (2019). Monitoring of aerosol and other particulate matter in air using aerial monitored sensors and real time data monitoring and processing. *Journal of System and Management Sciences, 9*(2), 104–113. Published by Beijing Jiaotong University, ISSN:1816-6075(Print) 1818-0523(Online).
38. Huang, J., & Koroteev, D. D. (2021). Artificial intelligence for planning of energy and waste management. *Sustainable Energy Technologies and Assessments, 47*, 101426.
39. Taneja, A., Dutt, I., & Srivastav, A. L. (2021). Advances of waste management practices in India and China along with bibliometric assessment of their research outcomes. *Environmental Science and Pollution Research, 28*, 66485–66495.
40. Bibri, S. E., et al. (2024). Smarter eco-cities and their leading-edge artificial intelligence of things solutions for environmental sustainability: A comprehensive systematic review. *Environmental Science and Ecotechnology, 19*, 100330.
41. Cong, R., Fujiyama, A., & Matsumoto, T. (2022). AI techniques aid for optimizing the collection system of industrial plastic waste. *Environmental Science, 35*(5), 248–257.
42. Sharma, R. (2023). Leveraging AI and IoT for sustainable waste management. In *International conference on sustainable development through machine learning, AI and IoT*. Springer Nature Switzerland.

43. Sayed, E. T., et al. (2023). Application of artificial intelligence techniques for modeling, optimizing, and controlling desalination systems powered by renewable energy resources. *Journal of Cleaner Production, 413*, 137486.
44. Karduri, R. K. R. (2019). The role of artificial intelligence in optimizing energy systems. *International Journal of Advanced Research in Management Architecture Technology & Engineering (IJARMATE).*
45. Sankarananth, S., et al. (2023). AI-enabled metaheuristic optimization for predictive management of renewable energy production in smart grids. *Energy Reports, 10*, 1299–1312.
46. Pislaru, M., et al. (2024). Citizen-centric governance: Enhancing citizen engagement through artificial intelligence tools. *Sustainability, 16*(7), 2686.
47. Olawade, D. B., et al. (2023). Using artificial intelligence to improve public health: A narrative review. *Frontiers in Public Health, 11*, 1196397.
48. Kowalski, R. (2020). *Automated analysis of citizen feedback for inclusive public policies (Diss).* UCL (University College London).
49. Snehi, J., Snehi, M., Bhandari, A., Baggan, V., & Ahuja, R. (2021). Introspecting intrusion detection systems in dealing with security concerns in cloud environment. In *2021 10th International conference on system modeling & advancement in research trends (SMART)* (pp. 345–349). IEEE.
50. Odeyemi, O., et al. (2024). Reviewing the role of AI in fraud detection and prevention in financial services. *International Journal of Science and Research Archive, 11*(1), 2101–2110.
51. Noor, N. R., Mohd, A., & Mansor, N. (2019). Exploring the adaptation of artificial intelligence in whistleblowing practice of the internal auditors in Malaysia. *Procedia Computer Science, 163*, 434–439.
52. Mahor, V., et al. (2023). *IoT and artificial intelligence techniques for public safety and security. Smart urban computing applications* (pp. 111–126). River Publishers.
53. Fragkos, G., Tsiropoulou, E. E., & Papavassiliou, S. (2019). Disaster management and information transmission decision-making in public safety systems. In *2019 IEEE global communications conference (GLOBECOM).* IEEE.
54. Li, X., et al. (2021). Data fusion for intelligent crowd monitoring and management systems: A survey. *IEEE Access, 9*, 47069–47083.
55. Girmay, S., Samsom, F., & Khattak, A. M. (2021). Ai based login system using facial recognition. In *2021 5th Cyber Security in Networking Conference (CSNet).* IEEE.
56. Dutta, P. K., Vinayak, A., Kumari, S., & Hussain, M. (2021). Early detection of poisonous gas leakage in pipelines in an industrial environment using gas sensor, automated with IoT. In J. C. Bansal, L. C. C. Fung, M. Simic, & A. Ghosh (Eds.), *Advances in applications of data-driven computing. Advances in intelligent systems and computing* (Vol. 1319). Springer. https://doi.org/10.1007/978-981-33-6919-1_12

Exploring the Influence of Workplace Spirituality on Job Performance: A Study in Haryana, India

Priyanka Deswal and Neha Arora

1 Introduction

The education system plays a crucial role for all the developing as well as developed countries all over the world. In the modern world that is characterized by constant changes, this sector experiences a number of concerns, especially in regard to the academic staff who have the paramount task of providing requisite and timely education services. It can be stated that education is one of the most critical areas for sustainable human and economic development [1]. In order to meet the educational institutions' missions and carry out the educational mandates successfully, practices need to be put in place that would enrich the self-fulfilment and self-actualization of the staff as well as increase extrinsic satisfaction among educators. And this in turn leads to better or improved job performance and indeed achievement of institutional goals and objectives.

An ideal workplace is one in which all workers are busy working as they understand their reason and role to work toward it. There is purpose in life; co-workers provide companionship and their beliefs are aligned with the organization [2]. Increasingly, organizational members desire workplace spirituality; it may be that few workplaces manifest the qualities noted above. In fact, when spirituality is introduced in organizations, it brings all of these things along with it. As per Rathee [3], Cash and Grey (2000) have stated that the increase in spirituality at work was due to the organizational instability resulting from advancement in technologies, reengineering, and cuts. Sustained workforce demoralization seeks religious ways of removing stress and tension that accompanies certain types of jobs [4]. Workplace spirituality talks about how people can feel more deeply connected to their jobs and

P. Deswal (✉) · N. Arora
Geeta University, Panipat, Haryana, India

P. Raj et al. (eds.), *Artificial Intelligence (AI) for IT Energy Efficiency and Green AI for Environment Sustainability*,
https://doi.org/10.1007/978-3-031-89420-6_24

their jobs' goals. Figuring out what work is all about is important for keeping people motivated. Every worker isn't just doing their job because they think it's a routine. To them, work can be exciting because it has a deeper and more important meaning. Persons who believe that their work is more than just a routine and the reason they do it is said to have workplace spirituality. People experience workplace spirituality when they work with a committed spirit and try to figure out what their work is all about [5].

In the 1990s and early 2000s, workplace spirituality turned into a concept in the academic world and has become an important part of organizations. There is a rise in spirituality at work because the economy is slowing down and new technologies in the industrial sector are making workers stressed out and unmotivated [6]. Starting in the early 2000s, organizations went through a big change. It is clear that people are what make a company competitive now that they've moved past the machine age [7]. Additionally, they said that enhancing human resources in the workplace may best be achieved by incorporating spirituality.

Modern businesses have to deal with a lot of problems linked to their employees, such as illnesses caused by worry, burnout, absences, violence, and fraud. WPS is becoming a crucial part of the company. It helps the company keep and feel good about its talented workers, which in turn helps the company grow and improve [8–12]. Kinjerski and Skrypneck discovered that workers have hearts and thoughts. The work they do gives their lives value and makes them want to connect with other workers.

Educators who think their workplace is spiritually fulfilling tend to experience extra happiness with their job. The organization makes them happy when they feel valued and respected and when they are aligned with the goals and values of the company. Also, job satisfaction impacts job performance in a positive manner due to motivation and participation as well as commitment to organizational goals. In order to define the concept of JP, Motowidlo (1997) stated that JP is dynamic and another was that JP has many facets [13–17], maintaining the theme that it is behavioral and episodic and can be evaluated.

Employees who participate in spiritual practices at the workplace have more productivity, more creativity, and the ability to handle task challenges. Hence, cultivating a spiritually supporting environment not only benefits an individual's health and happiness but also lifts organizational performance with optimal productivity.

The research's aim is to look into the link between spirituality at work and work success among workers in Haryana, India. It seeks to examine how workplace spirituality, which encompasses values such as meaningful work, a sense of belonging, gratitude, and embracing diversity, influences various dimensions of job performance. By focusing on this relationship within the specific context of Haryana, the study's aim is to explore how spiritual aspects in the workplace can improve workers' health and productivity.

2 Literature Review

2.1 Workplace Spirituality

Workplace spirituality [18] combines spiritual, mystical, interpersonal, and cognitive traits [19]. The following are workplace spirituality factors by various authors: (a) meaningful work [7, 20–27], (b) compassion [28, 29], (c) sense of community [30], and (d) transcendence [22, 31], (e) alignment with organizational value [32, 33], (f) gratitude and embrace diversity [34, 35], (g) inner life [10, 36], and (h) interconnectedness [4, 19].

2.2 Job Satisfaction

Job satisfaction is defined as a person's well-being which is derived from evaluation of the job in accordance with the expectations of the organization [37]. Designing a job so that an individual becomes happy with his/her job is a function of physiological, social, and physical theories of work [38]. Here, it becomes clear that the primary determinant of job satisfaction remains the employees' satisfaction with the job that has been done. This has strong connections to overall job satisfaction and can have effects like keeping workers (Parisi & Weiner, 1999; [39–41]).

2.3 Job Performance

There is also an evaluation part for JP. This means that performance is made up of actions that can be good or bad for the company or the person and the level of how desirable they can be scaled to tell the difference [42]. Task-related performance [41, 43–48] and contextual performance [17, 47, 49–54] are two different ways people act that affect the usefulness of businesses in their own way [55].

2.4 Relation Between WPS and JS

Multiple pieces of research show that there is a link between being spiritual at work and being happy with your job in an organization. According to Sawatzky et al. [56], spirituality at work has a big effect on job happiness [57–60]. Krishnakumar and Neck [61] say that WPS makes a person better at their whole job, which leads to job happiness in the end. Various authors have studied the relationship between WPS and JP; this can be used to build the hypothesis as follows:

H1 *There exists a positive nexus between WPS and JS.*

2.5 JS and JP

A lot of study has been done in the organization on the link between JS and JP. Job satisfaction is thought to play a big role in how well people do their jobs (Han, 2008; [62–69]). People who are happy at work tend to be more committed and get more done in less time than people who are unhappy at work. It shows that happy people work tough [70]. As a result, the study comes up with and tests the following theory.

H2 *Job satisfaction has a positive nexus on job performance.*

2.6 WPS and JP

Workplace spirituality encourages employees to excel in honesty, faith, personal fulfilment, commitment, creativity, process improvement, and good customer service, thereby enhancing overall performance [61]. Organizations are advised to foster spiritual experiences among employees to boost performance and profitability by empowering them to take on more responsibilities and deepen their commitment. Soha [71] suggests that spirituality at work has a big impact on how well an organization does its job on an individual level. Albuquerque [72] underscores the role of WPS, including fostering community and meaningful work, in building performance. Hence,

H3 *Workplace spirituality significantly affects Job Performance.*

2.7 Relationship between WPS and JP with Mediator JS

A lot of studies have shown that faith at work can help people do their jobs better [73, 74]. Also, the studies have confirmed the strong positive correlation between spirituality within the workplace and JS [12]. There has also been discussion regarding other matching phenomena such as job satisfaction and job performance [64, 75]. Connecting these three study trends makes sense to propose that spirituality in a worker can be related to the level of job satisfaction and the efficiency of the work performed by a worker. Because of this, it is thought that.

H4 *Job satisfaction mediates workplace spirituality and job performance.*

2.8 *Need/Importance of the Study*

1. This study is helpful for understanding human resource management because it uses quantitative models and real-world data to look at how spirituality in the workplace affects job success in college settings.
2. This study creates the possibility to shed more light on the connection between spirituality at work and professional performance by giving a theoretical framework that holds practical potential, although it has not been used in a practical setting. Also, idea provocation happens to call for reflection on managerial practices within colleges, thus encouraging discussions on the management of human resources management strategies.
3. These findings are important to the practical world of managerial decision-making and especially in improving the stock of people that colleges employ. Thus, by discussing the possible interaction between workplace spirituality and the effect it can bring on job performance, it provides colleges with a manner to enhance the competition factor and develop the organization's progress.

3 Conceptual Framework

See Fig. 1.

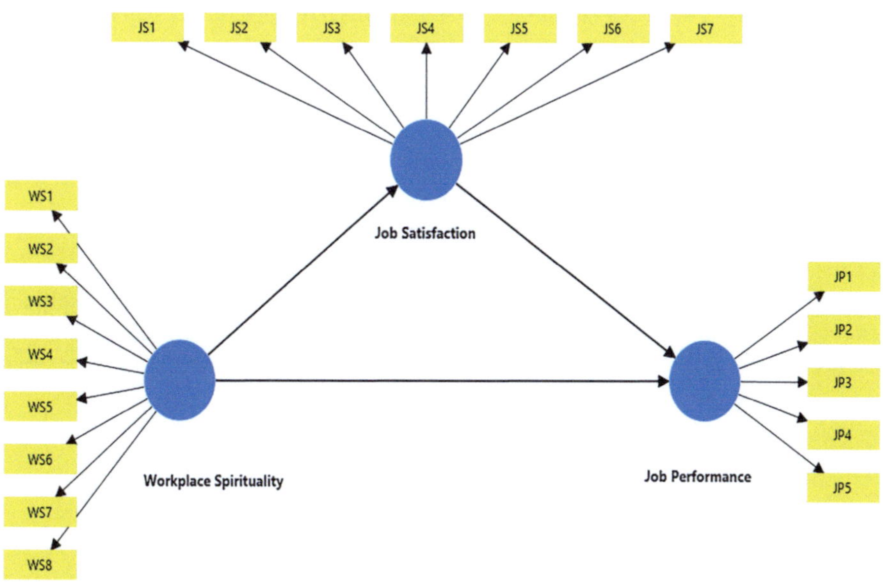

Fig. 1 Research model

4 Research Methodology

A quantitative approach is used to provide a comprehensive understanding of how workplace spirituality impacts job performance with JS as a mediator between them. This design combines quantitative analysis to measure relationships and insights to explore underlying factors.

4.1 Objectives of the Study

1. To look into the link between workplace spirituality and job performance.
2. To find out if job satisfaction can be used as a link between workplace spirituality at work and job performance.

4.2 Population and Samples

Respondents for this study are the employees in Haryana, India, with a total of 590 participants. For the selection of participants in the study, a simple random sampling method was used. As this information highlights, the analysis is based on Haryana, India. The sample is likely to represent various industries and workplaces within the region of study. The focus could be specific interests the researcher has for the cultural or economic environment of western India that could predispose changes in workplace context and spirituality perception among the employees.

4.3 Measurements and Testing Research Variables

The research adopts several control variables in order to establish the relationships between WPS, JP, and JS. IVs are components of WPS policy namely "embrace diversity," "meaningful work," "sense of community," and "gratitude" [35]. This means that WPS impacts two main dependent variables or the two dimensions that signify crucial organizational outcomes, namely JS and JP.

4.4 Procedure and Sample

WPS and JP in a college in Haryana, India, are as follows. This chapter looks at WPS and JP of a college in Haryana, India. It is a prestigious college that specializes in quality management and academic as well as administrative achievements. This option will provide a view of how spirituality at work looks like in one of the premier organizations. Due to the fact that the sample was chosen haphazardly, 800 employee questionnaires were obtained from HR in the form of hard copies. Out of 800 questionnaires distributed, 590 questionnaires were filled, which is a response percentage of **73.75%**. To know the perception of college staff regarding spirituality at work, a 5-point Likert scale with options ranging from 1—strongly disagree to 5—agree was constructed.

4.5 Analysis Technique

It is apparent that the data collection process involves the use of questionnaires or structured questionnaires in which the participants are asked questions on their impressions of WPS, JP, and JS. Again, there is no well-defined time to carry out data collection so that the period may be adjusted as seen necessary. The approach taken in this type of analytical method to solve the problem under investigation in this chapter incorporates the following aspects. In this line of analysis, correlational analysis is done to unveil the relationship between the given variables, thus explaining the relationship between constructs. SEM is used with the assistance of SmartPLS4 where it is possible to test the direct and indirect influence of various variables and their interrelations. Mediation analysis is also carried out to examine how the relationship between job satisfaction and the interaction between WPS and JP is formed.

5 Data Analysis and Interpretation

In this study, the gathered data is used to determine a correlation between the quality of responding to workplace spirituality, job performance, and job satisfaction among the workers in Haryana, India. In this way, using different quantitative techniques, such as descriptive statistics, reliability and validity analysis, and finally the structural equation models, the study reveals important profiles and relations that point out the nature of the impact of workplace spirituality on job outcomes.

From the descriptive statitics highlighted in the Table 1, vital facts concerning WPS, JP, and JS among the chosen respondents are pointed out. Items: WPS (WS1–WS8) indicates an average of at least 3.823–4.032, with standard deviations showing moderate dispersion. JP1–JP5 deals with the JP of the employees which has a slightly lower average response with values ranging from 3.677 to 3.883, with moderate oscillations in all cases. Looking at the JS (JS1–JS7) mean, the scores vary from 3.577 to 3.723 with moderate fluctuation. The results imply that respondents have moderately high and positive attitudes toward WPS, JP, and JS, while inter-individual variability explains the standard deviations obtained. The distributions are quite symmetric with some departure as a measure of excess kurtosis and

Table 1 Descriptive statistics

	Mean	Median	Standard deviation	Excess kurtosis	Skewness
Q1	2.107	1.000	1.379	−1.260	0.664
Q2	1.932	2.000	0.937	−0.446	0.669
Q3	1.720	2.000	0.686	−0.704	0.457
Q4	1.447	1.000	0.624	0.096	1.085
Q5	0.268	0.000	0.443	−0.903	1.049
Q6	1.479	1.000	0.560	−0.639	0.635
Q7	2.272	2.000	0.899	−0.644	0.308
Q8	0.474	0.000	0.499	−1.996	0.106
WS1	3.823	4.000	0.514	4.447	−1.207
WS2	4.032	4.000	0.767	0.378	−0.485
WS3	3.912	4.000	0.627	2.887	−0.804
WS4	3.939	4.000	0.659	2.274	−0.719
WS5	3.937	4.000	0.645	2.332	−0.666
WS6	3.896	4.000	0.695	2.040	−0.769
WS7	3.893	4.000	0.687	1.658	−0.645
WS8	3.881	4.000	0.651	2.654	−0.802
JS1	3.577	4.000	0.552	0.866	−0.611
JS2	3.677	4.000	0.629	1.001	−0.405
JS3	3.671	4.000	0.662	0.596	−0.187
JS4	3.674	4.000	0.659	0.643	−0.214
JS5	3.723	4.000	0.574	2.692	−0.987
JS6	3.710	4.000	0.713	0.405	−0.103
JS7	3.681	4.000	0.646	0.677	−0.185
JP1	3.786	4.000	0.533	2.943	−1.093
JP2	3.883	4.000	0.781	0.533	−0.458
JP3	3.861	4.000	0.728	1.369	−0.651
JP4	3.810	4.000	0.707	2.466	−0.923
JP5	3.822	4.000	0.687	2.767	−1.011

Table 2 Composite reliability and AVE

	Cronbach's alpha	(rho_a)	(rho_c)	AVE
Job performance	0.880	0.880	0.913	0.676
Job satisfaction	0.869	0.871	0.899	0.559
Workplace spirituality	0.913	0.917	0.929	0.621

skewness. These results provide us with the essentials of how these people consider these variables, which can point to more directions in future studies and interpretations if tied into your research.

The results depicted in the Table 2, above, show that the factors have a lot of internal consistency. The coefficients of internal consistency and Cronbach's alpha are at 0.880, thus making JP to be very stable. This means that issues relating to performance exhibit measure consistency. This is also supported by the value of composite reliability (rho_c) which is 0. 913. AVE, in this case, equal to 0.676, indicates that overall performance items as a whole account for 67.6% of the total variance of the respective construct, thus attesting good convergent validity. In the same way, JS is strongly demonstrated to have significant intrinsic consistency, with Cronbach alpha of 0.869 and (rho_a) was 0.871; a high rho_c of 0. 899. Though its AVE is a little low, at 0.559, it has been realized that JS items contribute 55.9% of their variance measures and still show acceptable convergent validity. Studying DFS and the correlation between WPS values shows a high level of reliability, with Cronbach alpha, equal 0.913, which shows a very high level of internal consistency. The validity is further corroborated by a high composite reliability of $rho_c = 0.929$. The AVE of 0.621 indicates that all the dimensions in WPS items as a group account for 62.1% of their variance and thus positive convergent validity (Table 3).

The value of VIF presented in Table 3 indicates the extent of multicollinearity of the factors that have been adopted. In the outer model, which estimates individual items, the VIF values for JP (JP1–JP5), JS (JS1–JS7), and WPS (WS1–WS8) vary from 1.663 to 2.408. From these values, it is evident that most of the items within a construct have small coefficients of collinearity which means that although the

Table 3 VIF outer and inner

	Variables used in study	VIF
VIF outer	JP1	1.927
	JP2	1.858
	JP3	2.095
	JP4	2.094
	JP5	2.408
	JS1	1.750
	JS2	1.686
	JS3	1.881
	JS4	2.050
	JS5	1.720
	JS6	1.663
	JS7	1.747
	WS1	1.965
	WS2	2.119
	WS3	2.004
	WS4	2.110
	WS5	1.964
	WS6	2.346
	WS7	2.188
	WS8	2.187
VIF inner	Job satisfaction ⟶ job performance	1.714
	Workplace spirituality ⟶ job performance	1.714
	Workplace spirituality ⟶ job satisfaction	1.000

items measure the same construct, they bear moderately fair correlation and thus are ideal for regression analysis.

First, on the outer model aspect that explores the relationship between the variables, the VIF scores show a low level of multicollinearity. In particular, the VIF between the JS and the JP imports is equal to 1.714, which implies that there is not much of a consensus on the variance between these notions. The same applies to the VIF between WPS and JP with a VIF of 1.714, thus showing that there is a relatively low intercorrelation between these variables. More importantly, the VIF of 1.000 between WPS and JS clearly suggests that there is no multicollinearity problem in the datasets. This implies that these constructs belong to different domains and are assessing different things, yet they lack a lot of variation.

Table 4 shows an asset analysis for *R*-square values, which indicate the amount of variance in JP and JS described by the model's predictor variables. For JP, the model explains a big chunk of the variation (*R*-square = 0.598), suggesting that the factors looked at in the study explain about 59.8% of the variability observed in JP among the respondents. Similarly, for JS, about 41.6% of the variation in JS levels can be laid out by the factors that were looked at in the study (*R*-square = 0.416). Both results are statistically significant, as indicated by the *T*-statistics (|O/STDEV|) of 13.238 for JP and 7.804 for JS, with corresponding *p*-values of 0.000, providing compelling evidence that refutes the null hypothesis that the model does not explain variance in these outcomes.

Table 5 presents HTMT ratios assessing prejudiced validity between constructs. Ratios for job satisfaction $\longrightarrow$ job performance (0.231) and workplace spirituality $\longrightarrow$ job performance (0.299) indicate valid distinctions. However, the high ratio of workplace spirituality $\longrightarrow$ job satisfaction (0.714) suggests potential overlap. Nonetheless, these results generally support distinct measurements of job-related constructs: JS, JP, and WPS.

The data presented in Table 6 shows path coefficients depicting the direct and indirect effects of the various variables relating to employee outcomes. The results indicate significant direct effects: JS has a positive effect on JP (beta = 0. 398) which infers that a high level of JS enhances JP. In the same way, WPS has a positive direct

Table 4 R^2

	Original sample	Sample mean	(STDEV)	(|O/STDEV|)	*P* values
Job performance	0.598	0.603	0.045	13.238	0.000
Job satisfaction	0.416	0.418	0.053	7.804	0.000

Table 5 HTMT ratio

	Original sample	Sample mean	(STDEV)	(|O/STDEV|)	*P* values
JS $\longrightarrow$ JP	0.231	0.242	0.078	2.965	0.003
WPS $\longrightarrow$ JP	0.299	0.317	0.108	2.761	0.006
WPS $\longrightarrow$ JS	0.714	0.732	0.163	4.378	0.000

Table 6 Path coefficient: direct and indirect effects

		Original sample	Sample mean	(STDEV)	(|O/STDEV|)
Direct effect	JS ⟶ JP	0.398	0.399	0.058	6.825
	WPS ⟶ JP	0.454	0.455	0.061	7.383
	WPS ⟶ JS	0.645	0.645	0.042	15.525
Indirect effect	WPS ⟶ JS ⟶ JP	0.257	0.258	0.042	6.134

Table 7 Total effects

	Original sample	Sample mean	(STDEV)	(|O/STDEV|)	*P* values
JS ⟶ JP	0.398	0.399	0.058	6.825	0.000
WPS ⟶ JP	0.711	0.712	0.036	19.554	0.000
WPS ⟶ JS	0.645	0.645	0.042	15.525	0.000

effect on JP which increases the employee effectiveness by 0.454 (path coefficient). Also, WPS has a significant and strong positive relationship with JS (path coefficient = 0. 645), which emphasizes the role of WPS in enhancing better satisfaction among the employees. It is also appreciated the influence of WPS on JP through JS (indirect path coefficient = 0. 257), which indicates to what extent WPS increases performance through satisfaction.

Results in Table 7 reveal significant impacts. The total effect of JS on JP is estimated to be 0.398, which means that JS directly affects JP. *T*-statistics were 6. 825 and $p = 0.000$; thus, concluding that satisfaction levels are crucial in improving the performance of the employees. Similarly, WPS shows an even greater impact on the JP and has a total effect of 0.711 ($t = 19.554$, $p = 0.000$), thus underlining its significance in shaping performance returns. Moreover, WPS has a very huge effect on JS with a total effect of 0.645 (*T*-statistics = 15.525, $p = 0.000$), which underlines its impact on improving satisfaction level among the employees.

5.1 Model Assessment

A comparison of the estimated model fit with the saturated model, which represents a perfect fit, is presented in Table 8. The indices demonstrate how well the estimated model matches the data. In addition, the SRMR of 0.054 for both models indicates that the expected and actual covariance matrices do not deviate much. The d_ULS of 0.618 and d_G of 0.182 for both models and their smaller values suggest a better fitting of models. This is further supported by a Chi-square value of 614.106, which is equally constant across both models, as anticipated plausibly as a result of the study sample's size and the chosen model's complexity. Furthermore, the two models have an NFI of 0.909 and thus it can be concluded that the models fit well than a null model (Fig. 2).

Table 8 Model fit

	Saturated	Estimated
SRMR	0.054	0.054
d_ULS	0.618	0.618
d_G	0.182	0.182
Chi-square	614.106	614.106
NFI	0.909	0.909

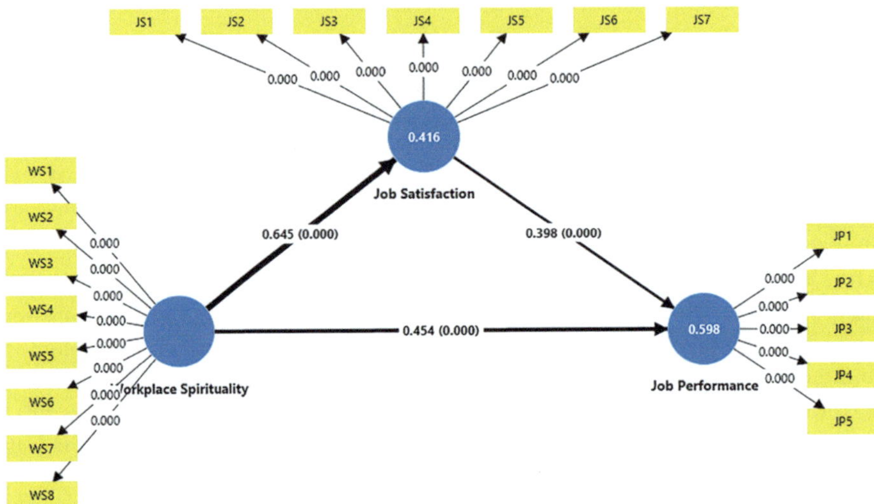

Fig. 2 PLS-SEM model. (Source: SmartPLS 4.0)

6 Findings of the Research

Specifically, the variables highlighted by the extensive analysis included workplace spirituality, job performance, and job satisfaction. WPS was significant, and respondents' perception was positive with regard to WPS with the mean scores varying between 3.823 and 4.032. This part of the workplace culture had both positive and significant direct impacts on JP (path coefficient equals 0.454) and JS (path coefficient equals 0.645). Another significant relationship turned out to be that JP, which stands with an average of at least 3.677 and 3.883 perceptions, was positive and had a high explained variance (R-square = 0. 598). Similar to the results, JS with slightly lower ratings (3.577–3.723) also indicated a high degree of prediction ability (R-square = 0. 416).

Cronbach alpha was used to establish internal consistency and reliability for all the measures used in the study, and all the constructs achieved high coefficient alpha values. AVE values of more than 0.55 indicated convergent validity. The results of the collinearity tests revealed moderate within-construct collinearity and low between-construct collinearity, thereby supporting the regression analyses. Thus, model fit indices (SRMR = 0.054, NFI = 0.909) suggested a proper fit to the proposed structural relationships. Path analysis showed that the total impacts of WPS on JP and JS are significant with total effects of 0.711 on JP and 0.645 on JS. The path analysis at the same time revealed that WPS has directly influenced both the JP and the JS ($p < 0.001$). The above results therefore support the central proposition of WPS as a means to improving the health of the employee and efficiency of the business. Promoting WPS is a strong approach for the enhancement of JS and overall JP.

7 Conclusion

Consequently, analysis of the research findings will establish that workplace spirituality improves the JP and JS of employees in Haryana, India. Perceived organizational spirituality was positively related to job performance (path coefficient = 0. 454) and job satisfaction (path coefficient = 0. 645). The model explained 59.8% of the variance of JP and 41.6% in JS which enhanced the study's reliability and validity. Quantitative results revealed the validity and reliability of the measure instruments, and the total effects analysis highlighted the significance of the WPS for overall positive workplace outcomes. Promoting WPS comes out as an effective way of enhancing the quality of employee health and organizational performance. Future studies should try and overcome these limitations including sample selection diversity and self-selection bias and design studies using longitudinal or more controlled experimental designs for the purpose of replication and generalization of the results. This paper provides significant information as to how WPS influences organizational effectiveness.

Appendix

Workplace spirituality		Strongly disagree	Disagree	Neutral	Agree	Strongly agree
1.	I'm happy at work					
2.	I see a link between my employment and my community's social welfare					
3.	I feel like my job is a community					
4.	My employment has given me many learning opportunities					
5.	Individuals can freely and openly share their thoughts and feelings within this group					

Workplace spirituality		Strongly disagree	Disagree	Neutral	Agree	Strongly agree
6.	Hiring process is fair and unbiased					
7.	I am grateful for whatever my job has given to me					
8.	I acknowledge and appreciate efforts of others at work					
Job satisfaction						
9.	Working with my boss and coworkers is pleasure for me					
10.	This job recognizes and appreciates my work					
11.	Here, I have good promotion opportunities					
12.	Job doesn't harm my health					
13.	The work hours are good					
14.	Physical work condition is appropriate					
15.	I'm happy with the pay and benefits					
Job performance						
16.	You eagerly accept assigned tasks					
17.	You take responsibility for your actions/work					
18.	You are energized and enthusiastic about your work					
19.	You grasp new tasks swiftly and attain desired results promptly					
20.	You embrace additional responsibility and accountability					

References

1. Agbedahin, A. V. (2019). Sustainable development, education for sustainable development, and the 2030 agenda for sustainable development: Emergence, efficacy, eminence, and future. *Sustainable Development, 27*(4), 669–680.
2. Widodo, S., & Suryosukmono, G. (2021). Spiritual leadership, workplace spirituality and their effects on meaningful work: Self-transcendence as mediator role. *Management Science Letters, 11*(7), 2115–2126.
3. Rathee, R., & Rajain, P. (2020). Workplace spirituality: A comparative study of various models. *Jindal Journal of Business Research, 9*(1), 27–40.
4. Mitroff, I. I., & Denton, E. A. (1999). A study of spirituality in the workplace. *MIT Sloan Management Review*.
5. Dehaghi, M. R., Goodarzi, M., & Arazi, Z. K. (2012). The effect of spiritual values on employees' organizational commitment and its models. *Procedia-Social and Behavioral Sciences, 62*, 159–166.
6. Dirkx, J. M. (2013). Leaning in and leaning back at the same time: Toward a spirituality of work-related learning. *Advances in Developing Human Resources, 15*(4), 356–369.

7. Ashar, H., & Lane-Maher, M. (2004). Success and spirituality in the new business paradigm. *Journal of Management Inquiry, 13*(3), 249–260.
8. Jabeen, N., Qamar, U., & Ansari, N. (2014). Workplace spirituality and affective commitment among employees: A case study of Akhuwat. *Journal of Business Strategies, 8*(2), 39.
9. Kinjerski, V., & Skrypnek, B. J. (2008). The promise of spirit at work increasing job satisfaction and organizational commitment and reducing turnover and absenteeism in long-term care. *Journal of Gerontological Nursing, 34*(10), 17–25.
10. Chawla, V., & Guda, S. (2010). Individual spirituality at work and its relationship with job satisfaction, propensity to leave and job commitment: An exploratory study among sales professionals. *Journal of Human Values, 16*(2), 157–167.
11. Choerudin, A. (2014). The relationship between spirituality and work attitude: A empirical study. *International Journal of Management Research and Reviews, 4*(4), 455.
12. Van der Walt, F., & De Klerk, J. J. (2014). Workplace spirituality and job satisfaction. *International Review of Psychiatry, 26*(3), 379–389.
13. Sonnentag, S. (Ed.). (2003). *Psychological management of individual performance*. John Wiley & Sons.
14. Campbell, C. H., Ford, P., Rumsey, M. G., Pulakos, E. D., Borman, W. C., Felker, D. B., et al. (1990a). Development of multiple job performance measures in a representative sample of jobs. *Personnel Psychology, 43*(2), 277–300.
15. Campbell, J. P., McHenry, J. J., & Wise, L. L. (1990b). Modeling job performance in a population of jobs. *Personnel Psychology, 43*(2), 313–575.
16. Viswesvaran, C. (2002). Assessment of individual job performance: A review of the past century and a look ahead. In *Handbook of industrial, work & organizational psychology-volume 1: Personnel psychology* (pp. 110–126).
17. Cheng, E. W., Li, H., & Fox, P. (2007). Job performance dimensions for improving final project outcomes. *Journal of Construction Engineering and Management, 133*(8), 592–599.
18. Kinjerski, V. A. L., & Skrypnek, B. J. (2006). A human ecological model of spirit at work. *Journal of Management, Spirituality & Religion, 3*(3), 231–241.
19. McGhee, P., & Grant, P. (2008). Spirituality and ethical behaviour in the workplace: Wishful thinking or authentic reality. *EJBO-Electronic Journal of Business Ethics and Organization Studies, 13*(2), 61–69.
20. Gupta, M., Kumar, V., & Singh, M. (2014). Creating satisfied employees through workplace spirituality: A study of the private insurance sector in Punjab (India). *Journal of Business Ethics, 122*, 79–88.
21. Ashmos, D. P., & Duchon, D. (2000). Spirituality at work: A conceptualization and measure. *Journal of Management Inquiry, 9*(2), 134–145.
22. Sheep, M. L. (2006). Nurturing the whole person: The ethics of workplace spirituality in a society of organizations. *Journal of Business Ethics, 66*, 357–375.
23. Duchon, D., & Plowman, D. A. (2005). Nurturing the spirit at work: Impact on work unit performance. *The Leadership Quarterly, 16*(5), 807–833.
24. Mirvis, P. H. (2017). "Soul work" in organizations. In *Leadership perspectives* (pp. 477–490). Routledge.
25. Beyer, J. M. (1999). Culture, meaning and belonging at work. In *Paper presented at the 1999 Chicago academy of management meeting*.
26. Hassan, M., Bin Nadeem, A., & Akhter, A. (2016). Impact of workplace spirituality on job satisfaction: Mediating effect of trust. *Cogent Business & Management, 3*(1), 1189808.
27. bin Abdullah, A. G. K., bt Ismail, A., & bin Mydin, A. (2013). *The moderating effects of school principals' leadership practices on the workplace spirituality and leadership effectiveness relationship*.
28. Snyder, C. R., & Lopez, S. J. (Eds.). (2001). *Handbook of positive psychology*. Oxford University Press.
29. Petchsawang, P., & Duchon, D. (2009). Measuring workplace spirituality in an Asian context. *Human Resource Development International, 12*(4), 459–468.

30. Harrington, W. J., Preziosi, R. C., & Gooden, D. J. (2001). Perceptions of workplace spirituality among professionals and executives. *Employee Responsibilities and Rights Journal, 13*, 155–163.
31. Dehler, G. E., & Welsh, M. A. (1994). Spirituality and organizational transformation: Implications for the new management paradigm. *Journal of Managerial Psychology, 9*(6), 17–26.
32. Rego, A., & Cunha, M. A. (2008). Workplace spirituality and organizational commitment: An empirical study. *Journal of Organizational Change Management, 21*(1), 53–75. https://doi.org/10.1108/09534810810847039
33. Pradhan, R. K., & Jena, L. K. (2017). Employee performance at workplace: Conceptual model and empirical validation. *Business Perspectives and Research, 5*(1), 69–85.
34. Fry, L. W., Vitucci, S., & Cedillo, M. (2005). Spiritual leadership and army transformation: Theory, measurement, and establishing a baseline. *The Leadership Quarterly, 16*(5), 835–862.
35. Saxena, A., & Prasad, A. (2022). Conceptualisation and validation of multidimensional measure of workplace spirituality. *South Asian Journal of Human Resources Management, 9*(1), 100–129.
36. Charoensukmongkol, P., Daniel, J. L., & Chatelain-Jardon, R. (2015). The contribution of workplace spirituality on organizational citizenship behavior. *Advances in Business Research, 6*(1), 32–45.
37. Cranny, C. J., Smith, P. C., & Stone, E. (1992). Job satisfaction: How people feel about their jobs.
38. Locke, E. A. (1969). What is job satisfaction? *Organizational Behavior and Human Performance, 4*(4), 309–336.
39. Frye, C. M. (2000). *The effect of emotional stability on job satisfaction: A meta-analysis*. The University of Iowa.
40. Rentsch, J. R., & Steel, R. P. (1992). Construct and concurrent validation of the Andrews and Withey job satisfaction questionnaire. *Educational and Psychological Measurement, 52*(2), 357–367.
41. Weiner, N. (1980, August). Pay satisfaction: An investigation of the explanatory and predictive power of Lawler's model. In *Academy of management proceedings* (Vol. 1980, No. 1, pp. 303–307). Academy of Management.
42. Motowidlo, S. J., Borman, W. C., & Schmit, M. J. (2014). A theory of individual differences in task and contextual performance. In *Organizational citizenship behavior and contextual performance* (pp. 71–83). Psychology Press.
43. Greenslade, J. H., & Jimmieson, N. L. (2007). Distinguishing between task and contextual performance for nurses: Development of a job performance scale. *Journal of Advanced Nursing, 58*(6), 602–611.
44. Ng, T. W., & Feldman, D. C. (2009). How broadly does education contribute to job performance? *Personnel Psychology, 62*(1), 89–134.
45. Van Woerkom, M., & De Reuver, R. S. (2009). Predicting excellent management performance in an intercultural context: A study of the influence of multicultural personality on transformational leadership and performance. *The International Journal of Human Resource Management, 20*(10), 2013–2029.
46. Molefe, G. N. (2010). Performance measurement dimensions for lecturers at selected universities: An international perspective. *SA Journal of Human Resource Management, 8*(1), 1–13.
47. Wisecarver, M. M., Carpenter, T. D., & Kilcullen, R. N. (2007). Capturing interpersonal performance in a latent performance model. *Military Psychology, 19*(2), 83–101.
48. Hoffman, B. J., & Woehr, D. J. (2009). Disentangling the meaning of multisource performance rating source and dimension factors. *Personnel Psychology, 62*(4), 735–765.
49. Borman, W. C., Penner, L. A., Allen, T. D., & Motowidlo, S. J. (2001). Personality predictors of citizenship performance. *International Journal of Selection and Assessment, 9*(1–2), 52–69.
50. Viswesvaran, C., Schmidt, F. L., & Ones, D. S. (2005). Is there a general factor in ratings of job performance? A meta-analytic framework for disentangling substantive and error influences. *Journal of Applied Psychology, 90*(1), 108.

51. Mitchell Gibbons, A., Rupp, D. E., Kim, M., & Eun Woo, S. (2006a). Perceptions of managerial performance dimensions in Korea. *The Psychologist-Manager Journal, 9*(2), 125–143.
52. Mitchell Gibbons, A., Rupp, D. E., Snyder, L. A., Holub, A. S., & Eun Woo, S. (2006b). A preliminary investigation of developable dimensions. *The Psychologist-Manager Journal, 9*(2), 99–123.
53. Fein, E. C. (2009). Using global performance dimensions in human resource development and workforce planning. *International Employment Relations Review, 15*(2), 26–37.
54. Chan, D., & Schmitt, N. (2002). Situational judgment and job performance. *Human Performance, 15*(3), 233–254.
55. Griffin, M., Neal, A., & Neale, M. (2000). The contribution of task performance and contextual performance to effectiveness: Investigating the role of situational constraints. *Applied Psychology, 49*(3), 517–533.
56. Sawatzky, R., Ratner, P. A., & Chiu, L. (2005). A meta-analysis of the relationship between spirituality and quality of life. *Social Indicators Research, 72*, 153–188.
57. Belwalkar, S., Vohra, V., & Pandey, A. (2018). The relationship between workplace spirituality, job satisfaction and organizational citizenship behaviors—An empirical study. *Social Responsibility Journal, 14*(2), 410–430.
58. Fatima, T., Naz, A., & Majeed, M. (2015). Workplace spirituality and job satisfaction: Moderating role of intrinsic and extrinsic values. *International Journal of Management Research and Emerging Sciences, 5*(1).
59. Fanggidae, R. E., Suryana, Y., & Efendi, N. (2016). Effect of a spirituality workplace on organizational commitment and job satisfaction (study on the lecturer of private universities in the Kupang city-Indonesia). *Procedia-Social and Behavioral Sciences, 219*, 639–646.
60. Clark, L., Leedy, S., McDonald, L., Muller, B., Lamb, C., Mendez, T., et al. (2007). Spirituality and job satisfaction among hospice interdisciplinary team members. *Journal of Palliative Medicine, 10*(6), 1321–1328.
61. Krishnakumar, S., & Neck, C. P. (2002). The "what", "why" and "how" of spirituality in the workplace. *Journal of Managerial Psychology, 17*(3), 153–164.
62. Chen, J. C., & Silverthorne, C. (2008). The impact of locus of control on job stress, job performance and job satisfaction in Taiwan. *Leadership & Organization Development Journal, 29*(7), 572–582.
63. Indarti, S., Fernandes, A. A. R., & Hakim, W. (2017). The effect of OCB in relationship between personality, organizational commitment and job satisfaction on performance. *Journal of Management Development, 36*(10), 1283–1293.
64. Judge, T. A., Thoresen, C. J., Bono, J. E., & Patton, G. K. (2001). The job satisfaction–job performance relationship: A qualitative and quantitative review. *Psychological Bulletin, 127*(3), 376.
65. Platis, C., Reklitis, P., & Zimeras, S. U. L. F. U. R. (2015). Relation between job satisfaction and job performance in healthcare services. *Procedia-Social and Behavioral Sciences, 175*, 480–487.
66. Ziegler, R., Schlett, C., Casel, K., & Diehl, M. (2012). The role of job satisfaction, job ambivalence, and emotions at work in predicting organizational citizenship behavior. *Journal of Personnel Psychology., 11*, 176.
67. Zimmerman, R. D., & Darnold, T. C. (2009). The impact of job performance on employee turnover intentions and the voluntary turnover process: A meta-analysis and path model. *Personnel Review, 38*(2), 142–158.
68. Pushpakumari, M. D. (2008, January). The impact of job satisfaction on job performance: An empirical analysis. In *City forum* (Vol. 9, no. 1, pp. 89–105).
69. Syed, A. A. S. G., Shaikh, F. M., & Shah, H. (2012). Empirical analysis of relationship between job satisfaction and performance among faculty members: A case study of Sindh. *New Horizons, 6*(2), 52.
70. Susanty, A., & Miradipta, R. (2013). Employee's job performance: The effect of attitude toward works, organizational commitment, and job satisfaction. *Jurnal Teknik Industri, 15*(1), 13–24.

71. Soha, H. M., Osman, A., Salahuddin, S. N., Abdullah, S., & Ramlee, N. F. (2016). The relationship of work influence, sense of community and individual spirituality towards organizational performance. *Procedia Economics and Finance, 35*, 591–596.
72. Faro Albuquerque, I., Campos Cunha, R., Dias Martins, L., & Brito Sá, A. (2014). Primary health care services: Workplace spirituality and organizational performance. *Journal of Organizational Change Management, 27*(1), 59–82.
73. Campbell, J. K., & Yen, S. H. (2014). Workplace spirituality and organizational commitment influence on job performance among academic staff. *Jurnal Pengurusan, 40*(1), 115–123.
74. Tischler, L., Biberman, J., & McKeage, R. (2002). Linking emotional intelligence, spirituality and workplace performance: Definitions, models and ideas for research. *Journal of Managerial Psychology, 17*(3), 203–218.
75. Jacobs, R., & Solomon, T. (1977). Strategies for enhancing the prediction of job performance from job satisfaction. *Journal of Applied Psychology, 62*(4), 417.

Machine Learning and Deep Learning-Based Comparison Methods for the Estimate of Mobile Device Charging Duration

Sachin Bhoite, Kanchan Taksale, and Pawan Kumar Singh

1 Introduction

Have you ever wondered why your phone's battery seems to drain faster just when you need it the most? Or why some devices take longer to charge than others, even when using the same charger? In a world where mobile devices have become indispensable for communication, productivity, and entertainment, the unpredictability of battery life and charging times is a pressing issue. As our reliance on technology grows, so does the need for efficient and accurate battery management.

Imagine if predicting battery life was as straightforward as finding your phone charger when you're in a hurry! Unfortunately, it is not that simple. Accurate predictions of charging times are essential for enhancing user convenience, improving battery health, and extending battery life. This research addresses these challenges by focusing on innovative methods for estimating the time required to fully charge different mobile phone models.

This study attempts to address the growing use of mobile phones in response to the growing demand for accurate battery charge prediction models. One study looks at how mobile phone charging affects the use of public power in China. It calculates energy consumption for public charging while using the Grey Model for forecasts. The research's ramifications go beyond looking at computers and electric vehicles; they also include larger analyses of how phone charging affects public energy consumption [1]. Additionally, this research is part of a broader landscape of

S. Bhoite · K. Taksale
Department of Computer Science and Applications, Dr. Vishwanath Karad, MIT World Peace University, Kothrud, Pune, Maharashtra, India

P. K. Singh (✉)
Department of Information Technology, Jadavpur University, Jadavpur University Second Campus, Kolkata, West Bengal, India

P. Raj et al. (eds.), *Artificial Intelligence (AI) for IT Energy Efficiency and Green AI for Environment Sustainability*,
https://doi.org/10.1007/978-3-031-89420-6_25

technological advancements and applications. As the demand for accurate battery charge prediction models continues to rise, our findings could potentially contribute to optimizing energy management in renewable energy systems, aligning with the goals of enhancing sustainability, energy efficiency, and technological advancements in both mobile technology and renewable energy research. Many studies have been conducted on the subject of public phone charging. The primary areas of study in this area include charging station design, charging equipment, charging mode prediction, and charging mode. Some researchers built the charging load using both rapid and conventional charging. The alleging cargo curve of a specific scale mobile phone was calculated using the Monte Carlo simulation approach in one day with two modes [2].

1.1 Motivation

In the modern age of digital technology, mobile phones have seamlessly woven themselves into the fabric of our existence, enabling communication, productivity, and entertainment. However, the widespread usage of smartphones has highlighted the importance of efficient battery management. One critical aspect of battery management is predicting the time required for a mobile phone to achieve a full charge. Efficiently estimating the charging time not only enhances user convenience but also contributes to better battery health and longer battery life. Moreover, as we delve into the intricacies of battery management and forecasting, it is worth noting the broader implications of our research. As technology continues to advance, our findings may also find applications in optimizing energy management in the field of renewable energy, aligning with the goals of enhancing sustainability, energy efficiency, and technological advancements in both mobile technology and renewable energy research. Additionally, the research proposes a deep neural network model that enables precise and user-specific forecasting of a mobile device's discharging time, taking into account both spatial and temporal data retrieved from an Android phone [3].

In addition to the gaps identified in the literature, recent and upcoming industry trends further underscore the importance of this research. The rapid proliferation of mobile devices, coupled with the advent of 5G technology, has driven an exponential increase in mobile data usage and, consequently, the demand for reliable and accessible public charging infrastructure. As urban centers and public spaces worldwide strive to meet this demand, the energy consumption associated with public charging has become a critical concern. Furthermore, the growing adoption of electric vehicles (EVs) has led to a convergence of mobile device charging and EV charging networks, creating a new frontier for integrated energy management solutions.

Another significant trend is the rise of smart cities, where IoT-enabled energy management systems and AI-driven forecasting models are increasingly being deployed to optimize power usage across various public and private sectors. These technologies are revolutionizing how energy consumption is monitored and managed, making it imperative to develop accurate models that can predict and balance the energy demands of mobile devices within these sophisticated systems. Moreover, as more countries commit to achieving net-zero carbon emissions by mid-century, there is a strong push toward integrating renewable energy sources into the grid. This shift is driving the need for advanced energy storage solutions and more efficient energy usage models, particularly in the context of public mobile charging, which remains an underexplored area.

Additionally, the emerging trend of wireless charging technology, including public wireless charging stations, is expected to transform how users interact with charging infrastructure, potentially leading to higher energy consumption in public spaces. As mobile devices continue to evolve, with trends such as foldable smartphones and increased processing power, their energy demands will likely increase, further straining public energy resources. These developments highlight the urgent need for comprehensive research that can provide accurate forecasts and insights into the energy impact of public mobile charging, especially within the context of these rapidly evolving technological landscapes.

By addressing these trends, our research is positioned to offer valuable contributions that align with current efforts to optimize energy usage, promote sustainability, and support the ongoing integration of mobile technology within smart energy systems and renewable energy initiatives.

1.2 Objectives

The purposes of the study are as follows:

1. To prepare the dataset and develop a predictive model for estimating mobile phone battery charging time using ML algorithms.
2. To relate the performance of two popular ML algorithms, XGBoost and long short-term memory (LSTM), in predicting battery charging time.
3. To determine which ML algorithm (XGBoost or LSTM) provides more accurate and reliable predictions for mobile phone battery charging time.
4. Additionally, in the context of advancing technology and energy management, to explore how our research findings might extend to optimizing energy usage in the field of renewable energy, aligning with sustainability, energy efficiency, and technological advancements.
5. To provide insights into the applicability of ML techniques for optimizing battery management and enhancing user experience in mobile devices.
6. To offer recommendations and insights for future research and improvements in battery charging time prediction using ML approaches.

1.3 Contribution

To date, most researchers have primarily discussed battery life prediction. However, this study introduces a novel perspective by concentrating on the prediction of mobile battery charging times by generating the dataset. Previous research predominantly centered on forecasting the overall "battery life," which denotes the duration a battery can supply power before necessitating a recharge or replacement. In contrast, this study emphasizes a more specific aspect—the "charging time" of mobile batteries.

This research paper addresses the task of predicting the time it takes for diverse mobile phone models to reach full charge using ML algorithms. By scrutinizing various features and employing ML techniques, our goal is to construct a predictive model that enhances optimization for mobile phone users. This endeavor strives to fashion a predictive model capable of estimating the duration required to completely charge mobile phone batteries through the implementation of ML algorithms. The driving force behind this research is the creation of a practical tool for mobile phone users to optimize charging time and battery longevity. By comprehending the nuances of battery performance during the charging process, users can adopt superior charging practices and extend the overall lifespan of their phone batteries. Accurately foretelling the time necessary to charge mobile phone batteries carries significant implications for both users and manufacturers. Users can derive benefits from knowing the precise charging duration, enabling them to plan device usage and charging routines accordingly. Furthermore, optimizing the charging process based on accurate predictions can result in enhanced battery health and prolonged battery lifespan. For mobile phone manufacturers, insight into the charging behavior of distinct phone models can inform the development of more efficient charging algorithms. This, in turn, can elevate user experience and contribute to the sustainable utilization of battery-powered devices.

Within this research paper, we delve into the application of ML algorithms to tackle the challenge of forecasting the time it takes for varied mobile phone models to reach full charge. Our specific focus rests on two robust algorithms: XGBoost and LSTM. By conducting a comparative analysis of these methodologies, our objective is to pinpoint their individual strengths and limitations in accurately predicting charging times. Through harnessing the potential of both approaches, our aim is to forge a robust and effective predictive model for estimating the time required to charge different mobile phone batteries. Furthermore, this endeavor aligns with the broader landscape of technological advancements, where similar predictive modeling techniques can be leveraged to optimize energy management in renewable energy systems, contributing to sustainability, energy efficiency, and technological progress.

In recent years, the rapid development and deployment of mobile charging infrastructure have significantly influenced global energy consumption patterns. For instance, China's swift installation of public charging stations in major cities like Beijing and Shanghai has dramatically reshaped the country's energy usage. The collaboration between companies and cities in China to enhance these networks has leveraged data analytics to predict energy needs and efficiently manage loads, showcasing how strategic infrastructure development can optimize energy distribution and usage [4].

Similarly, Singapore's Smart Nation program exemplifies how the integration of IoT and AI technologies into urban infrastructure can lead to more sustainable and efficient energy management. Within this framework, smart grids connected to mobile charging stations in public spaces enable real-time energy distribution management, benefiting both customers and companies. This approach not only supports the city's technological advancements but also contributes to its sustainability goals, providing a model for future smart city developments [5].

2 Literature Review

This paper explores ML methods for predicting diesel engine transient emissions. Seven methods, including ANN, SVM, NARX, LSTM, GRU, Transformer, and TCN, are compared using evaluation metrics on WHTC and WHSC cycles. While no single method excels across all parameters, GRU and TCN shine in NOx, CO_2 is best predicted by LSTM and CO by TCN. The transformer performs well in THC, and SVM suits exhaust pressure. A hybrid model combining optimal algorithms is proposed for overall best performance [6]. Recent advancements in mobile communication systems have propelled various fields leveraging Internet of Things (IoT) data in next-gen networks like 5G and 5BG. IoT-driven data collection via sensors and mobile devices empowers sectors such as Smart Conveyance, Keen Vigor, Digital Twins, UAVs, Keen Fitness, and Cyber Security. This survey paper explores AI and ML applications, discussing their role in optimizing big data-enabled use cases, and presenting research summaries, insights, and future challenges, while highlighting the potential of future research directions [7].

The rising popularity of li-ion batteries prompts exploration of their potential, including predictive battery condition and behavior. ML (ML) algorithms prove successful in enhancing battery prediction's complexity, accuracy, and reliability. This systematic literature review and bibliometric study reveal a research gap in battery safety and performance prediction, urging further ML applications for cleaner, safer, and more efficient energy resources [8]. The surge in air pollution and uncertainties in renewable energy sources drive the need for efficient electric vehicles (EVs) powered by lithium-ion batteries. To address the challenges of battery deterioration and remaining usable life (RUL), ML (ML) methods offer a powerful and accurate solution. This study comprehensively evaluates RUL estimation challenges, explores various ML techniques, and highlights their application in enhancing EV efficiency and accuracy [9]. This study introduces a better downhill window long short-term memory (SW-LSTM) method for robust capacity estimation in lithium-ion batteries. By incorporating multiple time-scale charging factors and optimized feature extraction through a differential integration-moving average autoregressive (DI-MAA) model, the proposed model achieves high precision. Experimental results demonstrate a maximum capacity estimation error of 3.56% and an average relative error of 0.032 under complex dynamic stress tests, highlighting the model's stability and effectiveness in ensuring battery safety under real-world conditions [10]. This study reviews the application of DL for approximating the state of charge (SOC) in lithium batteries for electric vehicles (EVs). It analyzes the practical process, datasets, neural network structures, and advantages/disadvantages of SOC estimation. The review offers insights into the use of deep learning methods for accurate SOC estimation, providing a foundation for future developments in EV battery management [11]. This study addresses the challenging task of state of charge (SoC) estimation in electric vehicle battery management systems. Six ML algorithms—ANN, SVM, LR, GPR, EBa, and EBo—are employed for accurate SoC prediction. Comparing performance indices, ANN and GPR emerge as the most effective methods with minimal MSE and RMSE values, enhancing battery performance optimization [12]. The transition to electric vehicles (EVs) offers emission reduction and advanced driver assistance systems. Predicting EV battery capacity is challenging due to non-linear deterioration. This study pays three regression replicas—linear, ridge, and ensemble regression—to forecast battery life, showcasing ensemble regression as the most accurate model with 94.8% precision [13]. The rise in transportation emissions prompts attention to comprehend contributing factors. Electric vehicles (EVs) play a pivotal role in a low-carbon future. Employing advanced ML models, this study compares extreme and light gradient boosting with traditional methods for predicting EV energy consumption, showing superior performance and robustness. Light gradient boosting outperforms extreme gradient boosting, highlighting the potential of advanced ML models in enhancing EV energy consumption prediction and contributing to sustainable transportation [14].

As electric vehicles (EVs) gain popularity, predicting energy consumption is crucial due to limited range and charging infrastructure. Multiple linear regression (MLR), XGBoost, and SVR ML models are compared for total energy consumption (TEC) prediction using real-life parameters. XGBoost demonstrates superior accuracy (92%) and identifies key features such as trip distance, power, heating, and odometer reading, addressing "range anxiety" and enhancing EV energy consumption prediction [15]. Predicting the residual valuable life (RUL) of lithium-ion batteries is crucial for maintenance cost reduction. This study proposes a novel technique combining particle filters (PFs) with extreme gradient boosting (XGBoost) for accurate RUL estimation, leveraging XGBoost's nonlinear fitting.

The approach enhances RUL prediction accuracy through life cycle testing, demonstrating improved results compared to other methods and highlighting its potential for precise battery life assessment [16]. Smartphone battery limitations hinder user experience despite app and capacity growth. Addressing data scarcity, this study introduces a novel ML-based approach for accurate battery life prediction, utilizing real-time smartphone usage traces. By overcoming data gaps with the concordance index, the method leverages comprehensive usage data to successfully predict the remaining battery life, enhancing scheduling and user efficiency [17].

Accurate state of health (SOH) estimation for lithium-ion batteries is challenging due to complexity and uncertainties. This study proposes an XGBoost-based SOH estimation method with feature extraction and Markov chain correction, achieving a 10–20% accuracy improvement compared to other algorithms. The approach's enhanced prediction and error metrics highlight its effectiveness in improving SOH estimation for battery management [18]. The surge in vehicular emissions and renewable energy uncertainties drive the need for electric vehicles (EVs) with long ranges and reliability. Addressing battery capacity deterioration and environmental challenges, this study emphasizes the importance of determining a battery's remaining usable life (RUL) for efficiency and upgrades. ML (ML) methods prove promising for optimizing RUL estimation, offering accurate solutions to complex engineering challenges and enhancing battery efficiency with non-invasive, accurate techniques [9]. An improved Douglas–Peucker algorithm enhances battery health factor extraction and State of Health (SOH) identification. The LAOS-XGboost model predicts SOH by combining de-thresholded Douglas–Peucker with Wrapper feature engineering, achieving high accuracy ($R^2 > 0.97$) and robustness ($AE = 0$) in different battery prediction experiments, showcasing its effectiveness in comprehensive and precise lithium-ion battery health assessment [19]. The authors say that recent advancements in lithium-ion battery capacity estimation methods, including direct measurement, analysis-based, SOC-based, and data-driven approaches, aim to improve the accuracy and reliability of onboard battery management systems (BMSs) in electric vehicles (EVs) and renewable energy systems (RESs). They suggest that the increasing complexity and demand for efficient battery lifespan management have driven a deeper exploration of capacity degradation mechanisms, leading to more sophisticated and practical estimation techniques [20]. The authors say that Li-ion batteries are crucial for the efficiency of electric vehicles (EVs), and an effective battery management system (BMS) is essential for accurate state of charge (SOC) estimation. They suggest that while advanced SOC estimation techniques like LSTM, GRU, and CNN-LSTM offer improvements, challenges such as performance degradation and complex electrochemical reactions still persist, necessitating further research for enhanced BMS and sustainable EV applications [21]. The rapid growth of the electric vehicle (EV) industry has intensified the need for advancements in fast charging technology, with a focus on mitigating potential adverse effects on the power grid. While significant progress has been made in DC fast charging and vehicle-to-grid (V2G) integration, challenges remain, particularly regarding charging time, necessitating further research to develop more efficient fast charging infrastructure and control strategies [22]. As countries transition to clean energy for UN Sustainable Development Goals, electric vehicles (EVs) are key. This paper reviews advancements in EV charging technologies, focusing on power converters and vehicle-to-grid (V2G) integration. It highlights strengths, challenges, and future research opportunities in areas like wireless charging, grid integration, and battery life optimization [23]. Battery performance in a solar-powered mobile charger designed to improve power reliability for rural dwellers in areas with inconsistent electricity. The research tested various batteries available in the Nigerian market, focusing on affordability and effectiveness. Analysis revealed an efficiency of 70% and stable voltage readings between 3.7 and 4.2 volts, with nickel-metal hydride (Ni-MH) and lithium-ion (Li-ion) batteries performing best. The study suggests commercializing the charger to enhance the quality of life in rural and semi-urban areas [24]. Accurate battery health diagnostics and prognostics are essential but challenging due to variability and changing conditions. This paper reviews probabilistic machine learning methods for predicting battery degradation, emphasizing the importance of accounting for uncertainty and discussing future research opportunities [25]. To address excessive heat generation during extremely fast charging of electric vehicle batteries, this study developed a thermal management system model. By optimizing cooling parameters and component configurations, the system effectively manages heat across various temperatures and conditions, maintaining battery temperatures within safe ranges and demonstrating robust safety and performance [26]. The article presents a state of charge (SoC) estimation method for lithium-ion batteries using a first-order equivalent circuit model that incorporates aging parameters and operating temperature. By employing the Sigma Point Kalman Filter (SPKF), the method accounts for nonlinear relationships and cell aging effects, providing accurate SoC estimates even for aged cells and varying states of health. The model improves estimation accuracy for practical battery applications by considering the impact of cell aging on SoC dynamics [27]. This paper explores the use of artificial neural networks (ANNs) for estimating the state of charge (SoC) of lithium-ion batteries, incorporating an advanced cost function based on the maximum correntropy criterion (MCC) instead of the traditional mean squared error. The MCC accounts for higher order statistical moments, improving robustness against non-Gaussian error distributions and outliers. By employing adaptive strategies and genetic algorithms to tune the Gaussian kernel, the study demonstrates that this approach enhances the accuracy and reliability of SoC estimation using NASA's dataset [28].

3 Research Methodology

3.1 Algorithms Used

Extreme Gradient Boosting or XGBoost: It is an ensemble learning technique renowned for its excellent performance and high predicted accuracy. Due to its capacity to manage complicated datasets and reduce overfitting concerns, it has grown in prominence across numerous disciplines. A potent and well-known ML technique for regression, classification, and ranking tasks is called XGBoost. It is an EL technique that builds a solider, more detailed model by joining the estimates of several weak models (often decision trees) [29]. XGBoost is exceptional because of its effectiveness in handling missing data, scalability, and efficiency. It uses a gradient boosting framework to incrementally create decision trees in order to reduce the loss function. Column subsampling and pruning regularization approaches increase the model's resilience to overfitting [30].

Long Short-Term Memory (LSTM): Time series forecasting applications are well-suited for LSTM, a form of recurrent neural network (RNN) that excels at processing sequential input. It is especially well suited for battery charge prediction because of its distinctive architecture, which enables it to capture long-term dependencies and catch patterns in the charging process that are time dependent. A specialized recurrent neural network architecture called LSTM was created to solve the vanishing gradient issue in deep network training. Because it can detect long-range dependencies in data, it is very useful for sequence prediction and time-series analysis. Memory cells in LSTMs enable them to store and retrieve data over long sequences [31]. Natural language processing, speech recognition, and other sequence-based tasks frequently make use of LSTMs. They excel at activities like speech synthesis and language translation that need recollection of prior input [32].

3.2 *Development Process of ML Models*

Data Gathering and Preprocessing: Compile pertinent statistics with information on mobile phone charging. Include elements such as the phone's model, usage trends, and charging circumstances.

Data Preparation: Remove and manage outliers, inconsistent data, and missing values. Put data into a format that will allow for analysis.

Exploratory Data Analysis (EDA): Visualize data distribution, correlations, and patterns. Gain insights through plots, histograms, and summary statistics.

Feature Selection: Identify important features impacting charging time prediction. Remove irrelevant or redundant features.

Model Training with XGBoost and LSTM: Split data into training and testing sets. Train XGBoost and LSTM models on training data.

Model Testing with Accuracy Metric: Use models to predict the charging times of testing data.

Evaluate accuracy by comparing predicted vs. actual charging times.

Hyperparameter Tuning: Optimize XGBoost and LSTM models' parameters for best performance.

Implementation: Integrate tuned models into a practical application (e.g., mobile app). Users input phone models and usage for charging time prediction.

We preprocessed the data, trained XGBoost and LSTM regression models, and validated their performance using test datasets. Additionally, we implemented a function to predict battery percentage at a specific time, given the initial battery percentage and elapsed time. Figure 1 explains the flow of the work we have carried out to achieve the research objectives.

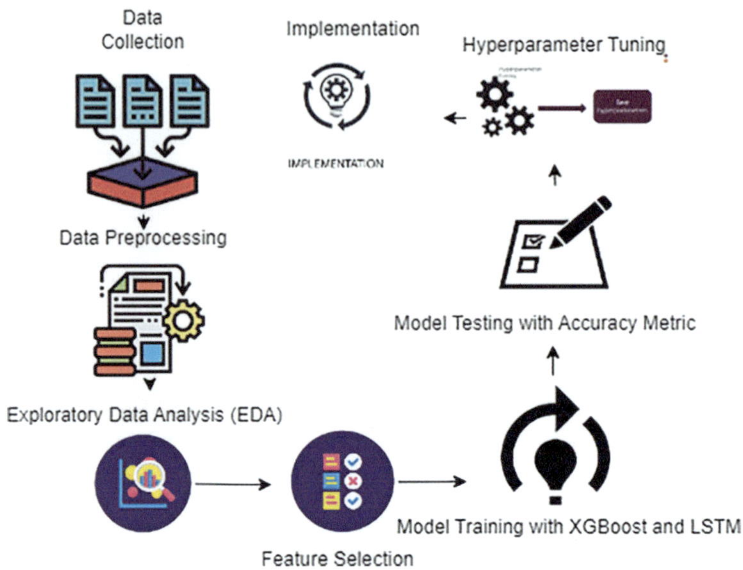

Fig. 1 System flowchart

4 Data Collection

The datasets used in this research were provided initially by "Techlabs" firm. The project's solution approach involves using four datasets, each representing a different mobile phone model. The charging cycle is from 0% battery charge to 100% for each mobile phone in this dataset. The charging cycle of four mobile phones has been captured and shared (sampling rate: 10 s). The iPhone 7+, iPhone 8, and Android device, Redmi Note 1 are charged using their standard normal charger. Once the charging capacity of the battery of all mobiles is fully charged, a gradual decrease in charging current can be seen, as shown below for three mobile phones.

We have compiled 4464 records along with the following nine columns for the dataset.

The list of original features includes "timestamp," "voltage (V)," "current (mA)," "power (W)," "battery percent," "battery temperature (Celsius)," "remaining battery charge (mAh)," "battery capacity (mAh)," and "battery technology." Both iPhones and Android phones utilize only Li-ion battery technology. The data included information such as battery temperature, voltage, current, and battery percentage recorded during the charging process for different phone models (as shown in Figs. 2, 3, 4, 5, 6 and 7).

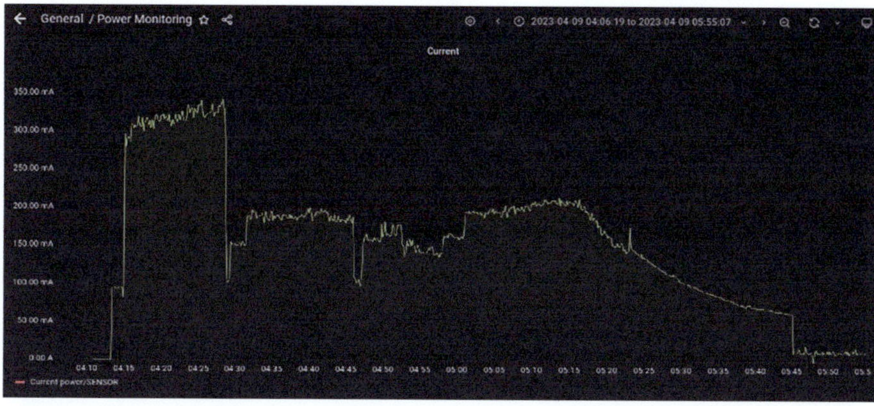

Fig. 2 Android12 charging current gradually decreases as the battery is getting fully charged

Fig. 3 Android 12 charging sampling rate after every 10 s

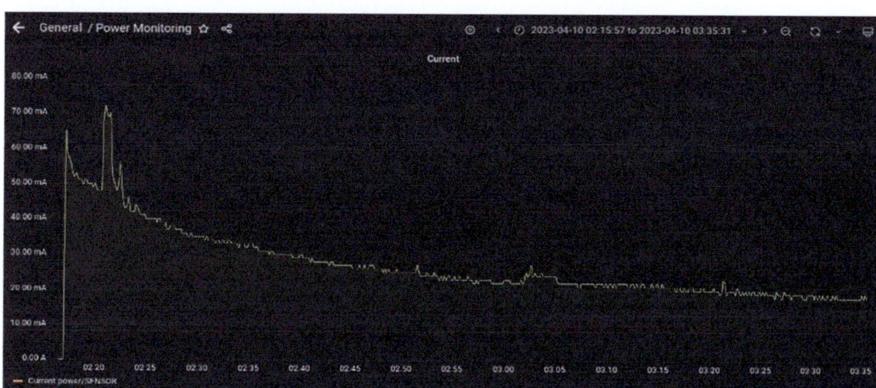

Fig. 4 iPhone7+ charging current gradually decreases as the battery is getting fully charged

Fig. 5 iPhone7+ charging sampling rate after every 10 s

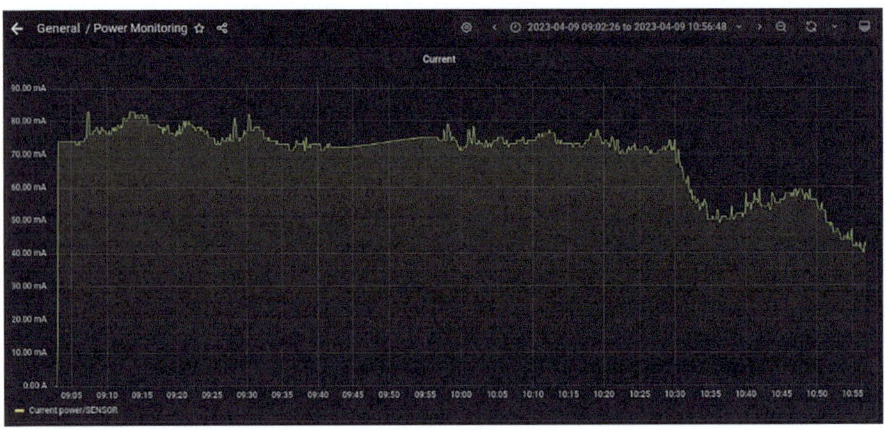

Fig. 6 iPhone8 charging current gradually decreases as the battery is getting fully charged

5 Exploratory Data Analysis (EDA)

It is an important step in data analysis, which involves summarizing and visualizing data to find patterns, anomalies, and insights. Before developing a model, it aids researchers in understanding data distribution, linkages, and potential difficulties [33]. Exploratory data analysis (EDA) was conducted to enhance our comprehension of the data and its inherent traits. A range of Python libraries, including Pandas, NumPy, Matplotlib, and Seaborn, were employed for data visualization and analysis. This process facilitated the recognition of trends, patterns, and possible outliers within the dataset. Data preprocessing, a vital phase in readying the data for ML model training, encompassed several sequential steps: conversion of the timestamp column to datetime format for better handling of time-related information;

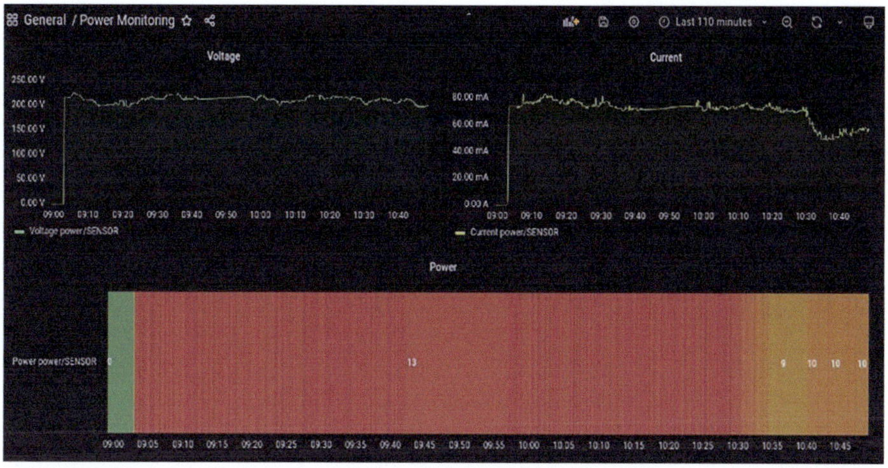

Fig. 7 iPhone 8 charging sampling rate after every 10 s

calculation of the time elapsed since the start of charging for each device to create a new feature; addition of the device name column to each dataset to distinguish between different phone models; encoding of device names to numeric values to enable model training; conversion of the time elapsed column to seconds to facilitate better model performance; and extraction of the time to full charge as the target variable.

6 Model Training and Evaluation

Two ML algorithms were chosen for this study: XGBoost and LSTM. For XGBoost, the data was divided into training and testing sets, and the XGBoost regression model was trained using the training data. The model's performance was assessed using accuracy metrics such as mean absolute percentage error and R^2 score on the test set.

Here is the output of the total time required to fully charge, i.e., 1 h 22 min.

For LSTM, the data was processed to create sequences for the LSTM network. The LSTM model was trained using the training data, and its performance was evaluated based on R^2 scores on the test set (Fig. 8).

Predicted battery percentage after 300 s i.e., 5 min is 82.04%.

Here is the output of total time required to full charge i.e. 1 h 22 min (Fig. 9)

For LSTM, the data was processed to create sequences for the LSTM network. The LSTM model was trained using the training data, and its performance was evaluated based on R^2 scores on the test set.

```
# Make a prediction using the model
y_pred = regressor.predict(df_test[['time_elapsed', 'battery_temperature(Celsius)', 'voltage(V)', 'current(mA)', 'device_id']])
# Print the result
print(f"Time to full charge: {time_to_full_charge} seconds")
✓ 0.0s
Time to full charge: 4935.174 seconds
```

Fig. 8 Predicted output after 5 min

```
# Define the initial battery percentage and elapsed time in seconds
initial_percent = 30
elapsed_time = (pd.to_datetime('2023-04-20 17:00:00') - df4.iloc[0,0]).total_seconds()

# Call the predict_battery_percent function to make the prediction
predicted_percent = predict_battery_percent(model, initial_percent, elapsed_time)

# Print the predicted battery percentage
print("Predicted battery percentage at 5:00 PM: ", predicted_percent)

1/1 [==============================] - 0s 35ms/step
Predicted battery percentage at 5:00 PM:  82.040565
```

Fig. 9 Predicted output total time required to fully charge

7 Solution Implementation and Results

The research involved obtaining and preprocessing the datasets. We performed EDA to gain visions into the data distribution and features. The datasets were preprocessed to convert the timestamp column to datetime format, calculate the time elapsed since the start, add device name information, encode device names to numeric values, and find the time to full charge. We then split the datasets into training and testing sets for XGBoost and LSTM algorithms. After training the model and testing it for a sample (X_test) of values, we can see the following plots for both regressors (Figs. 10 and 11).

These plots give a visual sense of how closely the predictions align with the actual values along with evaluation metrics mean squared error (MSE) and R-squared (R^2).

Ideally, the points should cluster around a diagonal line, indicating a strong correlation between predicted and actual values (Figs. 12 and 13).

The residuals (differences between actual and anticipated values) are plotted against the predicted values in a residual graphic. The residuals should be dispersed at random around the horizontal line at zero in a well-fitted model. Patterns or patterns in the residual may point to problems with the model, such as systematic bias or heteroscedasticity (unequal variance). We can observe that the distribution of residuals for the XGBoost regressor fluctuates with predicted values to a much lesser extent, both positively and negatively. This shows less heteroscedasticity and nonlinearity, and very few spots in the residual plot exhibit outliers. The LSTM plot

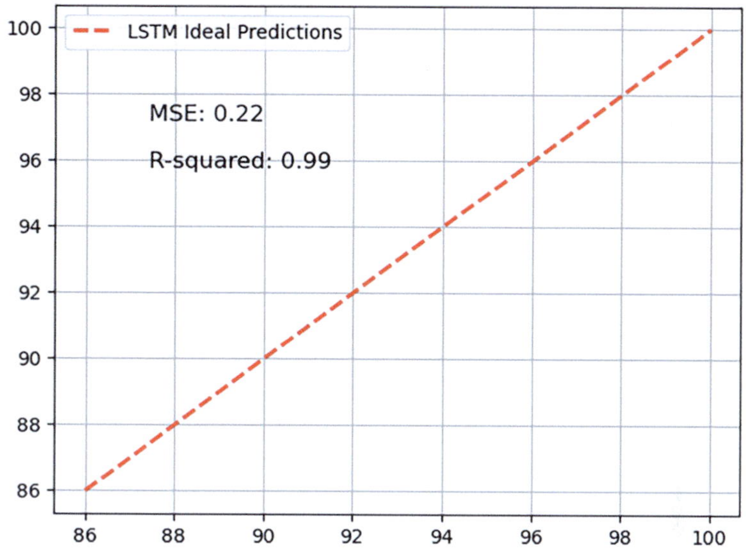

Fig. 10 LSTM regressor scatter plot with the predicted values and the actual target values

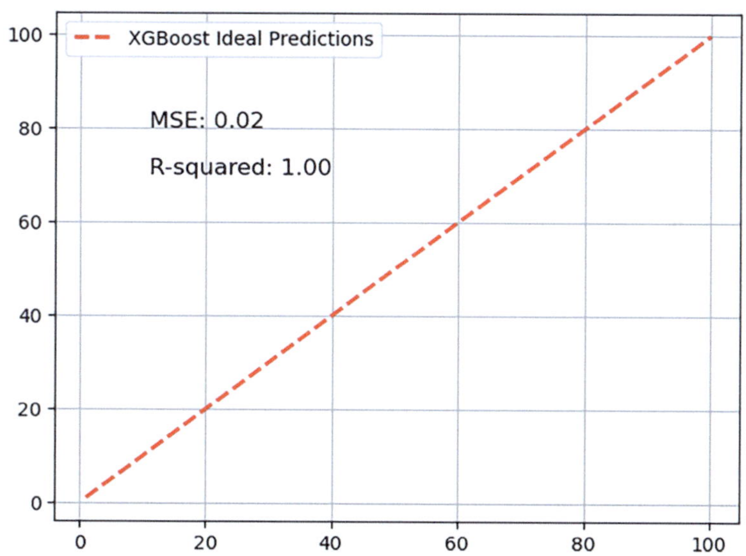

Fig. 11 XGBoost regressor scatter plot with the predicted values and the actual target values

can also demonstrate autocorrelation and displays a very limited number of temporal patterns (Figs. 14 and 15).

A visual representation of how the *R*-squared metric evolves over time (usually epochs or iterations) during the training of a regression model is the *R*-squared (R^2) metric evolution plot. This plot is used to track and assess a model's progress

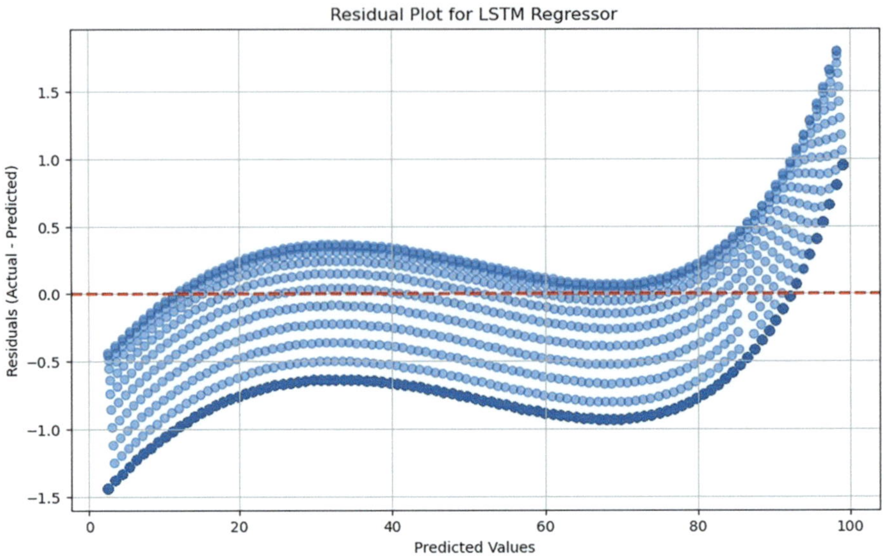

Fig. 12 Residual plot for LSTM regressor

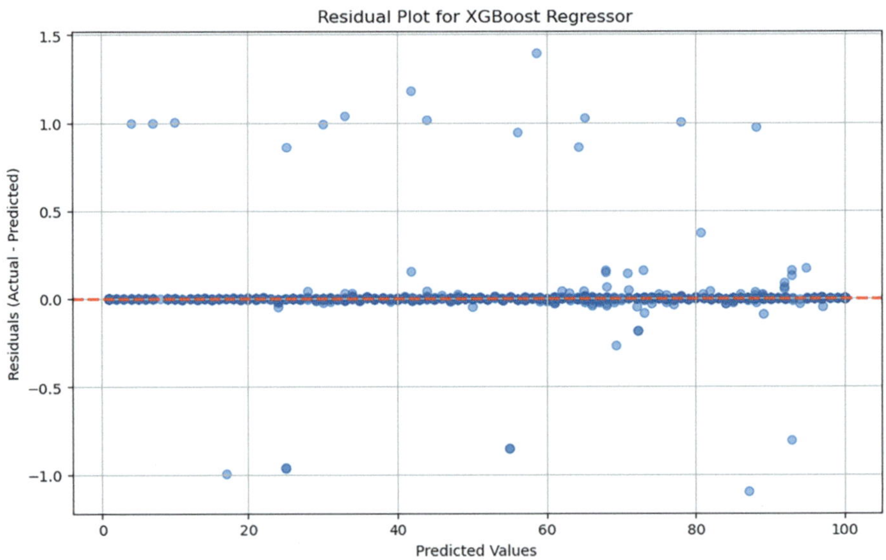

Fig. 13 Residual plot for XGBoost regressor

throughout training, particularly in terms of how well it can account for the target variable's volatility. The fact that the *R*-squared value is increasing shows that the XGBoost and LSTM models are becoming better at fitting the data. More boosting rounds may not improve model performance if the *R*-squared value plateaus or

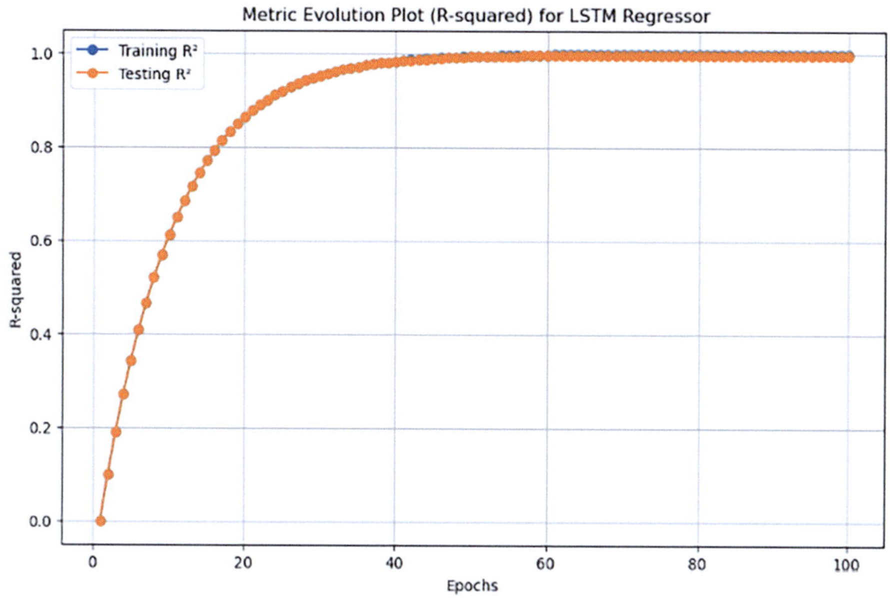

Fig. 14 R square metric evolution plots for LSTM regressor

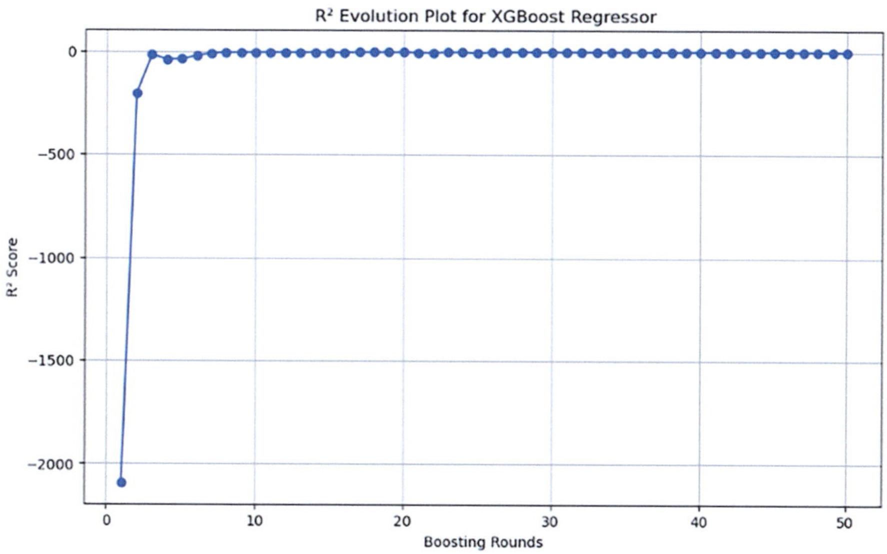

Fig. 15 R square metric evolution plot for XGBoost regressor

starts to drop, and more training may not be necessary to arrest this trend. Monitoring *R*-squared metric evolution plots as part of the model evaluation process is beneficial for both XGBoost and LSTM models (Figs. 16 and 17).

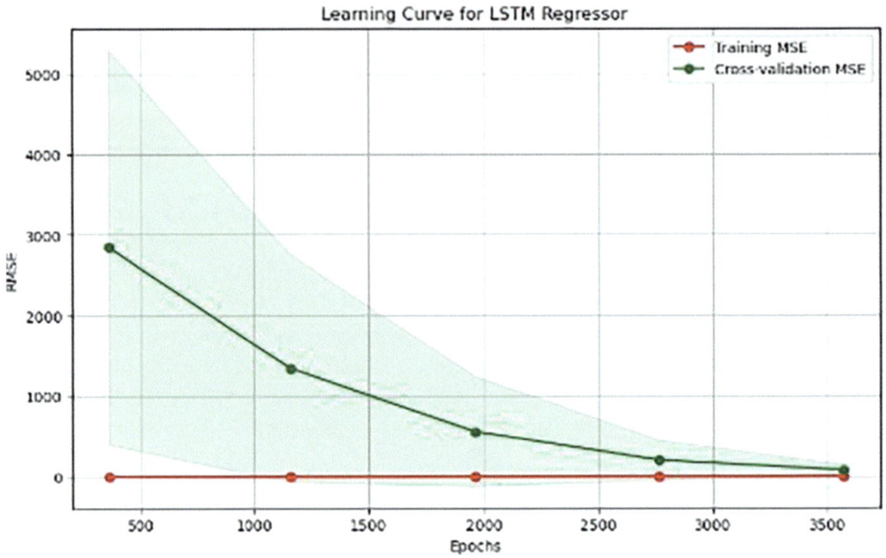

Fig. 16 Learning curve with RMSE for LSTM regressor

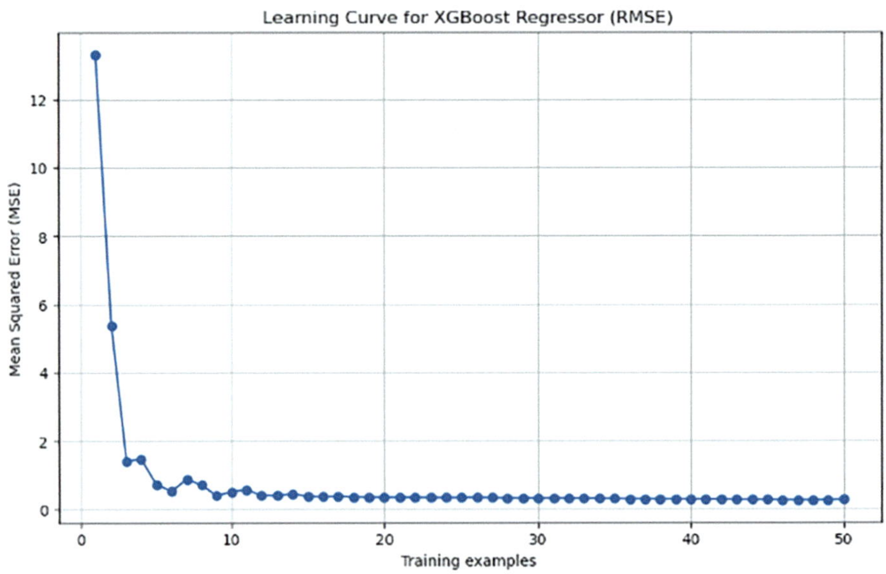

Fig. 17 Learning curve with RMSE for XGBoost regressor

These plots offer an easy-to-understand visual depiction of how well the model fits the data and whether further training or tweaks are required to reach the desired level of variance explanation for the target variable.

The performance and generalizability of the model can be evaluated using learning curves, especially in relation to the effect of training data size. As additional data is used for training in both aforementioned scenarios, a well-behaved learning curve should exhibit a decreasing trend in validation RMSE. This suggests that larger datasets help the model generalize more effectively. The XGBoost model may be near its limit and additional data may not greatly enhance performance, but this is not apparent in the case of the LSTM regressor when the training and validation RMSEs converge as the dataset size increases. Making judgments about the amount of data needed, the complexity of the model, and the level of generalization is made easier with the use of learning curves with RMSE. They are an important resource for deciphering the behavior of your model and enhancing its functionality.

We can see that the XGBoost regressor has appropriate leanings in comparison to the LSTM regressor based on the scatter plots, residual plots, MSE, and R-squared values mentioned above. Additionally, the application of the regressor has demonstrated that XGBoost has a 90% accuracy rate, and LSTM has an accuracy rate of 81%.

8 Limitations

As with any research, this study also has some limitations. First, the dataset used for training and evaluation might not be fully representative of all smartphone models in the market. Including more diverse phone models could enhance the generalizability of the models. Second, the predictive accuracy of the models may be affected by factors like battery health, charging habits, and environmental conditions, which were not explicitly considered in this study. In conducting this research, we ensured that all data used were anonymized and complied with data protection and privacy regulations. We respect the rights and privacy of individuals whose data contributed to this study.

9 Conclusion

Both XGBoost and LSTM models demonstrated promising results in predicting the time to fully charge mobile phone batteries. The developed models can serve as valuable tools for users and industry stakeholders to optimize battery charging practices and improve device performance. The analysis of both models revealed information on how well they could forecast how long it would take mobile phone batteries to fully charge. The accuracy of the XGBoost model was 90%, whereas that of the LSTM model was 81%. In terms of estimating the charging time, the XGBoost model fared better than the LSTM model. We imagine a time when battery charging predictions are widely used, more accurate, and supported by ongoing research and development, not only optimizing mobile technology but also contributing to the efficient and sustainable utilization of energy in various applications, including renewable energy solutions. Future studies in the area of battery charging prediction can build on this research's solid foundation. This study makes a contribution to the field of ML-based battery charging prediction by providing insightful knowledge on how smartphone batteries behave when being charged. The study's models can be utilized to improve battery charging procedures, contribute to more efficient and sustainable usage of mobile device prediction, and improve machine learning (ML) algorithms for battery management and prediction.

The findings of this study can be used to improve battery performance in other fields, such as electric vehicles and renewable energy storage systems. These findings also hold great promise for advancements in battery management and energy efficiency across a range of industries, further contributing to the sustainability and effectiveness of renewable energy solutions.

10 Future Work

The insights from this study have broad applications beyond mobile devices. In electric vehicles (EVs), enhanced prediction models can optimize charging infrastructure and improve vehicle efficiency. Renewable energy systems can benefit from better energy management, helping balance supply and demand for solar, wind, and other sources. Wearable technology can see longer device life and improved user convenience through advanced battery management. Smart home devices can achieve more efficient energy consumption, promoting sustainability. Additionally, drones and robotics can extend operational time and enhance reliability, especially in critical applications like surveillance and search-and-rescue missions. These advancements suggest significant potential for improving energy efficiency and sustainability across various industries.

Future research could delve deeper into combining XGBoost and LSTM to develop a hybrid model that enhances prediction accuracy. Additionally, real-time monitoring systems could be established to provide continuous data on battery conditions and charging time estimations. Expanding the dataset to include a broader range of mobile phone models and diverse charging conditions would further refine predictions. Exploring other machine learning techniques and deep learning architectures for battery charging predictions could yield new insights. Investigating how battery condition and charging habits influence algorithm performance and implementing real-time monitoring systems to validate their effectiveness in practical scenarios are also promising areas for future study.

Key Takeaways

The study emphasizes how sophisticated prediction models, such as the Grey Model, are useful for predicting how much energy the public will use to charge mobile devices. This information is crucial for planning public charging infrastructure and integrating it with smart grids. It emphasizes how mobile charging affects renewable energy systems and how future technologies, including AI-driven energy management and wireless charging, have the ability to completely change public infrastructure. The results support data-driven energy policy and call for more investigation into innovative approaches and model refinement to improve energy efficiency and prediction accuracy.

Implications and Impact: The predictive models developed in this research have practical implications for mobile phone users, manufacturers, and battery management systems. Users can benefit from improved charging time predictions and optimize their charging habits, accordingly, leading to extended battery life and improved device performance. Manufacturers can leverage these models to design more efficient charging algorithms for their devices. Additionally, battery management systems can benefit from accurate predictions to optimize power consumption and improve overall energy efficiency.

Acknowledgments We would like to express our gratitude to "Techlabs" firm for providing the initial dataset and "Mr. Pradeep Rao" for updating the dataset with additional phones, thereby enhancing the scope and relevance of this study. Their provision and cooperation have been instrumental in the fruitful conclusion of this study.

Conflicts of Interest The authors declare no conflict of interest.

References

1. Gu, X. (2020). Prediction of mobile phone charging demand in public places based on Grey model. *Academic Journal of Engineering and Technology Science, 3*(1), 8–13.
2. Sonnemann, G. W., Schuhmacher, M., & Castells, F. (2003). Uncertainty assessment by a Monte Carlo simulation in a life cycle inventory of electricity produced by a waste incinerator. *Journal of Cleaner Production, 11*, 279–292.
3. Gentil, M., Galeazzi, A., Chiariotti, F., Polese, M., Zanella, A., & Zorzi, M. (2017). A deep neural network approach for customized prediction of mobile devices discharging time. In *Proceedings of the GLOBECOM 2017—2017 IEEE global communications conference* (pp. 1–6).
4. Retrieved from https://www.chinadaily.com.cn/a/202406/18/WS6671513ba31095c51c5098c5.html
5. Ferro-Escobar, R., Vacca-González, H., & Gómez-Castillo, H. (2022). Smart and sustainable cities in collaboration with IoT: The Singapore success case. In G. Marques, A. González-Briones, & J. M. Molina López (Eds.), *Machine learning for smart environments/cities. Intelligent systems reference library* (Vol. 121). Springer. https://doi.org/10.1007/978-3-030-97516-6_12
6. Liao, J., Hu, J., Yan, F., Chen, P., Zhu, L., Zhou, Q., Xu, H., & Li, J. (2023). A comparative investigation of advanced ML methods for predicting transient emission characteristic of diesel engine. *Fuel, 350*, 128767.
7. Yazici, İ., Shayea, I., & Din, J. (2023). A survey of applications of artificial intelligence and ML in future mobile networks-enabled systems. Engineering science and technology, an. *International Journal, 44*, 101455.
8. Kamul, A., Hananda, N., Harito, C., Djuana, E., Elwirehardja, G. N., Pardamean, B., Gunawan, F. E., Budiman, A. S., Asrol, M., Redi, A. P., & Pasang, T. (2023). Machine learning applica-

tion in battery prediction: A systematic literature review and bibliometric study. In *E3S Web of conferences* (Vol. 388, p. 01021). EDP Sciences.
9. Sharma, P., & Bora, B. J. (2023). A review of modern ML techniques in the prediction of remaining useful life of lithium-ion batteries. *Batteries, 9*, 13–13.
10. Wang, S., Takyi-Aninakwa, P., Jin, S., Liu, K., & Fernandez, C. (2023). An improved sliding window—Long short-term memory modeling method for real-world capacity estimation of lithium-ion batteries considering strong random charging characteristics. *Journal of Energy Storage, 70*, 108038.
11. Zhang, D., Zhong, C., Xu, P., & Tian, Y. (2022). Deep learning in the state of charge estimation for Li-ion batteries of electric vehicles: A review. *Machines, 10*, 912–912.
12. Chandran, V., Patil, C. K., Karthick, A., Ganeshaperumal, D., Rahim, R., & Ghosh, A. (2021). State of charge estimation of lithium-ion battery for electric vehicles using ML algorithms. *World Electric Vehicle Journal, 12*, 38.
13. Sreedevi, C., Shrimali, M., Leela, R., Murthy, M. Y. B., & Selvaraj, D. (2022). Battery lifetime prediction of electric vehicle using artificial intelligence. *International Journal of Health Sciences, 6*(S1), 573522.
14. Ullah, I., Liu, K., Yamamoto, T., Mamlook, R. E. A., & Jamal, A. (2022). A comparative performance of ML algorithm to predict electric vehicles energy consumption: A path towards sustainability. *Energy & Environment, 33*, 1583–1612.
15. Pokharel, S., Sah, P., & Ganta, D. (2021). Improved prediction of total energy consumption and feature analysis in electric vehicles using ML and Shapley additive explanations method. *World Electric Vehicle Journal, 12*, 94.
16. Jafari, S., & Byun, Y. C. (2022). Gboost-based remaining useful life estimation model with extended kalman particle filter for lithium-ion batteries. *Sensors, 22*(23), 9522.
17. Li, H., Liu, X., & Mei, Q. (2018). Predicting smartphone battery life based on comprehensive and real-time usage data. arXiv preprint arXiv:1801.04069.
18. Song, S., Fei, C., & Xia, H. (2020). Lithium-ion battery SOH estimation based on XGBoost algorithm with accuracy correction. *Energies, 13*, 812.
19. Zhang, M., Chen, W., Yin, J., & Feng, T. (2022). Lithium battery health factor extraction based on improved douglas–peucker algorithm and SOH prediction based on XGboost. *Energies, 15*(16), 5981.
20. Peng, J., et al. (2022). A review of lithium-ion battery capacity estimation methods for onboard battery management systems: Recent progress and perspectives. *Batteries, 8*(11), 229.
21. Girijaprasanna, T., & Dhanamjayulu, C. (2022). A review on different state of charge estimation techniques of battery and battery management systems for EV applications. *Electronics, 11*(11), 1795. https://doi.org/10.3390/electronics11111795
22. Zentani, A., Almaktoof, A., & Kahn, M. T. (2024). A comprehensive review of developments in electric vehicles fast charging technology. *Applied Science, 14*, 4728. https://doi.org/10.3390/app14114728
23. Tanvir Shahed, M., & A.B.M. (2024). Harun-ur Rashid, battery charging technologies and standards for electric vehicles: A state-of-the-art review, challenges, and future research prospects. *Energy Reports, 11*, 5978–5998., ISSN 2352-4847. https://doi.org/10.1016/j.egyr.2024.05.062
24. Emetere, M. E., Akuwudike, E. C., & Afolalu, S. A. (2023). *IssueE3S Web Conf. 430, 2023 15th international conference on materials processing and characterization (ICMPC 2023).* https://doi.org/10.1051/e3sconf/202343001208. Article Number 01208, Number of page(s) 9; Published online 06 October 2023, 430.
25. Thelen, A., Huan, X., Paulson, N., et al. (2024). Probabilistic machine learning for battery health diagnostics and prognostics—Review and perspectives. *NPJ Mater Sustain, 2*, 14. https://doi.org/10.1038/s44296-024-00011-1
26. Wen, K., Fu, X., & Pei, F. (2023). Research on the thermal management safety of the fast charging power battery management system. *Energy Reports, 10*, 3289–3296. https://doi.org/10.1016/j.egyr.2023.09.130. ISSN 2352-4847 (https://www.sciencedirect.com/science/article/pii/S2352484723013781)

27. Van, C. N., & Quang, D. T. (2023). State of charge estimation for lithium-ion battery cell considering the influence of aging parameters and operating temperature. *International Journal of Electrochemical Science, 18*(12), 100379. ISSN 1452-3981. https://doi.org/10.1016/j.ijoes.2023.100379
28. Tomaszewska, A., Chu, Z., Feng, X., O'kane, S., Liu, X., Chen, J., Ji, C., Endler, E., Li, R., Liu, L., & Li, Y. (2019). Lithium-ion battery fast charging: A review. *eTransportation, 1*, 100011. ISSN 2590-1168. https://doi.org/10.1016/j.etran.2019.100011
29. Chen, T., & Guestrin, C. (2016). XGBoost: A scalable tree boosting system. In *Proceedings of the 22nd ACM SIGKDD International Conference on Knowledge Discovery and Data Mining (KDD) 2016* (pp. 785–794).
30. Ke, G., Meng, Q., Finley, T., Wang, T., Chen, W., Ma, W., & Ye, Q. (2017). LightGBM: A highly efficient gradient boosting decision tree. *Advances in Neural Information Processing Systems (NeurIPS), 30.*
31. Graves, A., & Schmidhuber, J. (2005). Framewise phoneme classification with bidirectional LSTM and other neural network architectures. *Neural Networks, 18*, 602–610.
32. Cho, K., Merriënboer, B. V., Bahdanau, D., & Bengio, Y. (2014). On the properties of neural machine translation: Encoder-decoder approaches. In *Proceedings of the eighth international joint conference on natural language processing (IJCNLP) 2014* (pp. 103–111).
33. Wickham, H., & Grolemund, G. (2016). R for data science.

Edge AI: Transforming Real-Time Decision-Making in the Internet of Energy

G. Vennira Selvi and V. Sheeja Kumari

1 Introduction

Edge AI is revolutionizing the Internet of energy (IoE) by allowing efficient, real-time data handling and decision-making at the source of energy production and consumption. By integrating advanced machine learning algorithms directly into edge devices, Edge AI reduces latency, enhances data privacy, and decreases bandwidth usage compared to traditional cloud-based systems. This approach facilitates dynamic energy management, predictive maintenance, and optimized grid operations. It also supports the unified incorporation of renewable energy sources, enhances grid resilience, and enables intelligent energy distribution. Consequently, Edge AI is pivotal in advancing the IoE, fostering smarter, more sustainable energy systems.

2 Evolution of Energy Systems

2.1 Traditional Energy Grids

Traditional energy grids, often referred to as centralized or conventional grids, have been the backbone of electricity distribution for over a century. These systems were designed during a time when electricity generation was centralized, primarily

G. V. Selvi (✉) · V. S. Kumari
School of CSE and IS, Presidency University, Bangalore, India

Department of Information Technology, Rajaas Engineering College, Vadakkangulam, Tirunelveli, Tamil Nadu, India

P. Raj et al. (eds.), *Artificial Intelligence (AI) for IT Energy Efficiency and Green AI for Environment Sustainability*,
https://doi.org/10.1007/978-3-031-89420-6_26

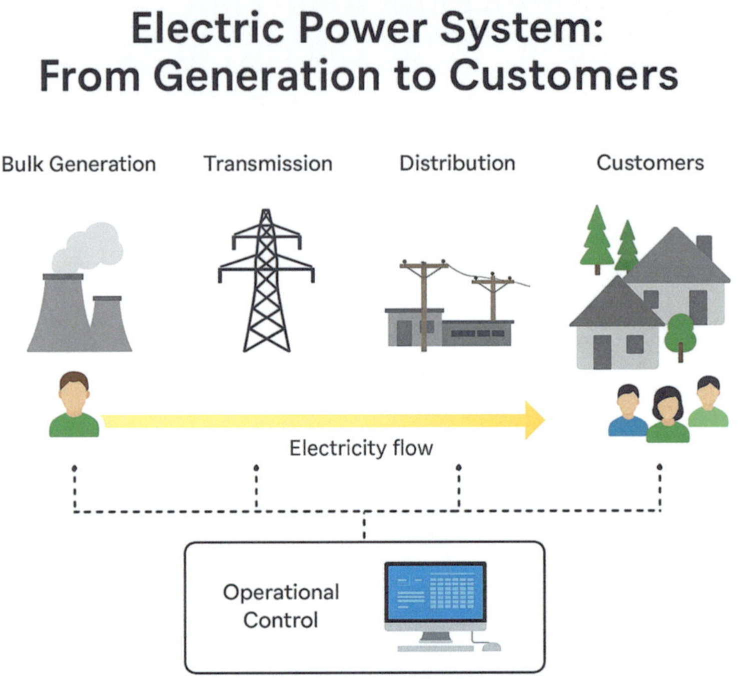

Fig. 1 Traditional energy grid

relying on large power plants burning fossil fuels or harnessing nuclear energy. Figure 1 shows an example of traditional energy grids.

Structure and Operation

- **Generation**: Large power plants generate electricity and then transmit it over long distances.
- **Distribution**: Substations lower the voltage, and then electricity travels to homes and businesses through lower voltage lines.
- **Consumption**: End-users consume electricity, with minimal feedback to the grid.

Traditional grids are characterized by a one-way stream of electricity from the power plants to the consumers. The infrastructure is robust but inflexible, designed to handle predictable, steady demand and supply patterns. The primary challenge with this system is inefficiency; large amounts of energy are lost during transmission, and the grids lack the capability to adapt speedily to changes in energy requests or integrate renewable energy sources effectively.

2.2 Emergence of Smart Grids

The limitations of traditional grids and the increasing need for sustainable energy have led to the creation of smart grids, which use digital communication technology to monitor and adjust electricity distribution in real time, improving efficiency, reliability, and sustainability. Smart grids feature two-way communication between utilities and customers, advanced metering infrastructure (AMI) with detailed usage information, and better integration of renewable energy sources like solar and wind. They enhance reliability by quickly identifying and addressing issues, reducing outages, and supporting energy storage systems and electric vehicles. Overall, smart grids optimize energy usage, reduce costs, and adapt more effectively to diverse energy sources and needs.

2.3 Role of IoT in Energy Systems

The Internet of Things (IoT) helps smart grids work better by linking devices like appliances and cars with sensors and software to gather and share data. In energy systems, IoT allows real-time tracking and control of things like energy use and grid health, predicts when equipment might fail by analyzing data, and adjusts operations based on grid conditions. It also helps add sources of energy like solar panels and home batteries, which improves grid stability.

This integration of IoT into energy systems improves efficiency, control, and interaction between energy producers and consumers, marking a shift from traditional grids, which are inefficient and inflexible, to smart grids that can dynamically respond to changes and incorporate renewable energy sources. As IoT adoption grows, its impact will likely drive further innovations and improvements in energy generation, distribution, and consumption, leading to a more sustainable energy future.

3 Advantages of Edge AI in IOE

3.1 Real-Time Data Processing and Decision-Making

One of the major advantages of Edge AI is its capability to process data in real time. In the context of IoE, this means that energy consumption patterns, production levels, and grid stability metrics can be monitored and analyzed instantaneously. For example, a smart grid equipped with Edge AI can detect fluctuations in energy supply from renewable sources and adjust the distribution of power accordingly, ensuring a balanced and stable grid. This real-time capability is crucial for managing the dynamic and often unpredictable nature of renewable energy.

3.2 Enhanced Data Privacy and Security

Edge AI processes data directly on devices, so there's less need to send sensitive information over networks to central servers. This local processing boosts data privacy and security, which is crucial in the energy sector where data breaches can be very damaging. By cutting down on data transfers, Edge AI also lowers the risk of cyberattacks and makes energy systems more resilient to potential threats.

3.3 Reduced Latency and Bandwidth Usage

Traditional cloud-based AI systems often suffer from latency issues due to the time required to transmit data to and from centralized servers. In contrast, Edge AI eliminates this latency by handling data at the source. This is mainly important for time-sensitive applications within the IoE, such as demand response and predictive maintenance. Additionally, Edge AI reduces bandwidth usage, as less data needs to be communicated over networks. This is advantageous for both cost and efficiency, especially in remote or resource-constrained environments.

3.4 Predictive Maintenance and Fault Detection

Edge AI helps with predictive maintenance by constantly checking the health and performance of energy equipment. It looks at data from sensors to spot patterns or problems that might mean equipment is about to fail. This allows for early fixes, reducing downtime and making equipment last longer. In smart grids, Edge AI can find faults and outages right away, allowing for quick responses and less impact on users

4 Comparative Study of Edge AI for Internet of Energy (IOE)

Edge AI is best for situations where you need quick processing, low delays, and offline access. Cloud AI, on the other hand, works well for tasks that need a lot of computing power and centralized data handling. When it comes to costs, compare the expenses of Edge AI devices with those of Cloud AI services [1]. Figure 2 shows the difference between Cloud AI and Edge AI

Table 1 highlights the key differences between Cloud AI and Edge AI, illustrating how each approach has its own strengths and suitability for various applications.

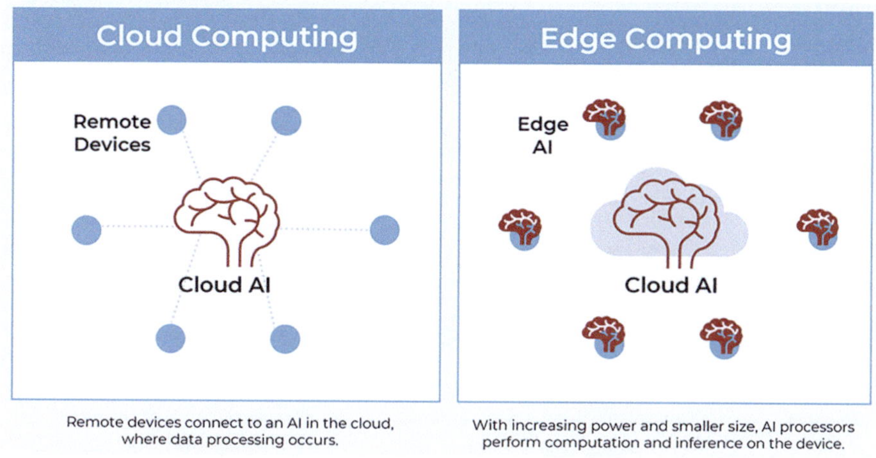

Fig. 2 Edge AI vs. cloud AI

Table 1 Comparing Cloud AI and Edge AI across various aspects

Aspect	Cloud AI	Edge AI
Data processing location	Data is processed in centralized cloud servers	Data is processed locally on edge devices
Response time	Slower response times because of data being sent to and from the cloud	Faster response times because data is processed locally
Bandwidth usage	High bandwidth usage as data is transmitted to the cloud for processing	Low bandwidth usage as data is processed locally, minimizing data transmission
Scalability	Highly scalable with flexible cloud resources; costs can increase with scale	Scalable by adding more edge devices; potential complexity in managing numerous devices
Cost	Ongoing costs for cloud storage, processing, and data transfer; initial low investment	Higher initial investment for edge devices and setup; potential for lower ongoing costs
Computational power	Utilizes powerful, centralized servers; suitable for complex computations	Limited by the computational power of edge devices; optimized for specific tasks
Data security	Data is stored and processed in the cloud, which can be vulnerable to breaches	Data is processed locally, reducing the risk of transmission-based breaches; requires strong device security
Energy consumption	Energy-intensive due to data centers and transmission	Generally, more energy-efficient as data processing happens locally
Connectivity	Requires continuous internet connectivity for data transfer and processing	Can operate independently of continuous internet connectivity, as processing is local

(continued)

Table 1 (continued)

Aspect	Cloud AI	Edge AI
Real-time processing	May not be suitable for real-time applications due to latency in data transfer	Ideal for real-time applications with immediate data processing and response
Maintenance	Managed by cloud providers; regular updates and maintenance handled externally	Requires ongoing maintenance and updates on numerous edge devices
Interoperability	Can integrate with various systems and platforms through APIs and cloud services	Needs standardization and compatibility with other edge devices and systems for interoperability
Data storage	Centralized storage in the cloud; scalable but can be costly	Localized storage on edge devices; limited by device capacity but reduces cloud storage needs
Use cases	Suitable for applications requiring high computational power, large data storage, and complex analytics	Best for applications requiring real-time processing, low latency, and decentralized data handling

5 How Does Edge AI Technology Work?

In Edge AI, a portion of data stored is taken out of the principal data center and moved closer to the data source. Here, machine learning (ML) algorithms are utilized for data processing and analysis within the area in which the data is generated—which could be a smart city, shop floor, or retail store. Internet connectivity is not essential for process completion; real-time decision-making takes place in milliseconds. The process of Edge AI is shown in Fig. 3.

The output of the process (which could be real-time predictions, forecasts, or business insights) that takes place at the edge is sent to the principal data center. In this manner, Edge AI takes data closest to the user's touch points, be it a computer, an Edge server, or an IoT device [2]. For example, when a user communicates something to Alexa or Google, a voice recording is sent to Edge network which is further sent to text through AI, and a response is processed accordingly.

5.1 Life Cycle of an Edge AI Application

To make machines capable of tasks like seeing, detecting objects, driving cars, understanding speech, speaking, or walking, they need to mimic human intelligence. AI uses something called a deep neural network (DNN) to do this. These DNNs are trained by showing them lots of examples of specific questions along with the right answers. The life cycle of the Edge AI application is shown in Fig. 4.

The training process, known as "deep learning," usually happens in data centers or the cloud. This is because it requires a lot of data and the collaboration of data scientists to set up the model correctly. After training, the model becomes an "inference engine," which can then answer real-world questions.

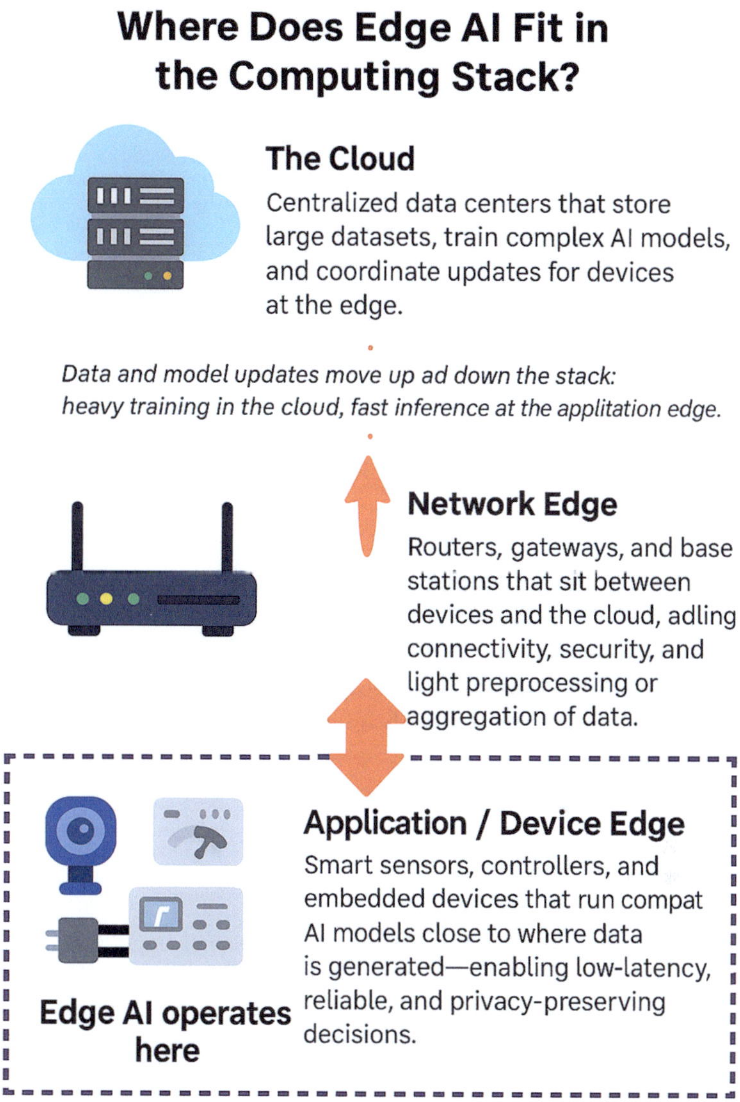

Fig. 3 Working process of Edge AI

In Edge AI, the inference engine runs on devices located in various places like factories, hospitals, cars, satellites, and homes. If the AI encounters a problem it cannot solve, the data from the problem is often sent to the cloud. This data helps retrain the original AI model, which eventually replaces the old inference engine with an updated version. This process of sending data back to improve the model makes the AI smarter over time, as it learns from new challenges and experiences.

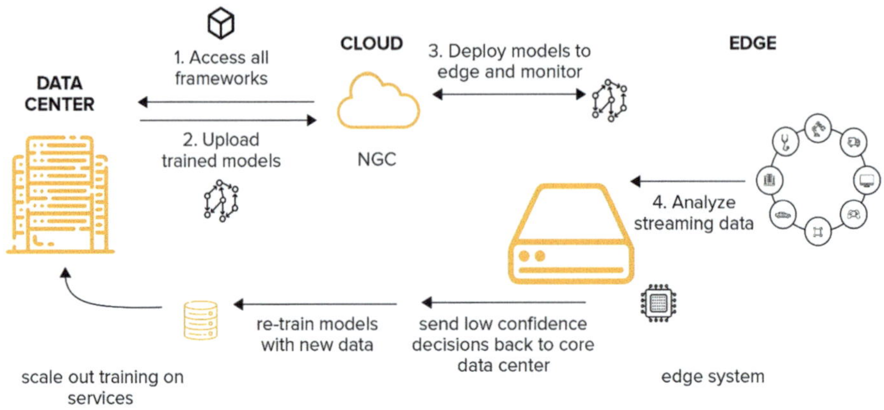

Fig. 4 Life cycle of Edge AI application

6 Role of Edge AI in Renewable Energy Integration

Adding renewable energy sources like solar and wind to the grid is a key part of the Internet of energy (IoE). However, these sources can be unpredictable, which makes managing the grid difficult. Edge AI helps with this by providing real-time data analysis and decision-making.

6.1 Solar Energy Management

Solar energy production is highly dependent on weather conditions, which can change rapidly. Edge AI can analyze data from weather sensors, solar panels, and energy storage systems to predict fluctuations in solar energy production. By integrating this data with information from the grid, Edge AI can optimize the use of solar energy, ensuring that excess power is stored or distributed efficiently.

6.2 Wind Energy Optimization

Similar to solar energy, wind energy is variable and can be difficult to predict. Edge AI can process data from wind turbines and weather stations to forecast wind patterns and adjust the operation of turbines accordingly. This real-time analysis helps in maximizing energy production and minimizing wear and tear on equipment.

7 Edge AI: Real-Time Data Processing

7.1 Importance of Real-Time Data in Energy Systems

In modern energy systems, real-time data processing is crucial because the energy landscape is becoming more complex with sources like solar, wind, and fossil fuels all contributing to the grid. Managing this complexity efficiently requires up-to-date information to ensure stability, optimize performance, and reduce waste. Real-time data helps maintain the balance between electricity supply and demand, preventing blackouts and ensuring a steady supply. It allows for optimizing energy use, reducing waste, and lowering costs. Since renewable energy sources are variable and unpredictable, real-time data helps manage these fluctuations and integrate renewables more effectively. Additionally, providing consumers with real-time data on their energy usage helps them make informed decisions, potentially lowering their energy bills and reducing overall consumption.

7.2 Edge AI for Real-Time Monitoring and Decision-Making

Edge AI means using AI algorithms directly on devices at the edge of the network, like sensors or local servers, instead of relying on central data centers. This enables real-time data handling and decision-making, which is vital for modern energy systems. The reimbursements of Edge AI include reduced latency since processing data locally eliminates delays from sending data to central servers, leading to faster decisions [3]. It enhances security because less data is transmitted over networks, reducing the risk of breaches.

It also offers scalability, with processing spread across many devices rather than one central point, and reliability, as local processing can continue even if the network fails. In energy systems, Edge AI can help manage smart grids by monitoring and managing electricity flows in real time, ensuring efficient distribution and reducing losses. It can also be used for predictive maintenance, analyzing data from apparatus sensors to predict failures and allow timely maintenance. Edge AI supports demand response by enabling devices to adjust their energy use depending on real-time grid circumstances, balancing supply and demand better. Additionally, it helps manage distributed energy resources like solar panels and batteries more efficiently, optimizing their performance and contribution to the grid.

7.3 Case Studies and Applications

Case Study 1: **Smart Grid Management**: In a city with a great perception of renewable energy sources, maintaining grid stability can be challenging. An Edge AI system was deployed to monitor and manage the grid in real time.

Sensors located throughout the grid provided continuous data on electricity flows, while local AI algorithms processed this data to make immediate adjustments. This system improved the efficiency of energy distribution, reduced outages, and facilitated the integration of solar and wind power.

Case Study 2: **Predictive Maintenance in Wind Farms**: Wind turbines are critical for renewable energy, but they are also prone to mechanical failures. A wind farm implemented Edge AI to observe the health of its turbines. Sensors on each turbine collected data on vibrations, temperature, and other indicators of mechanical health. Edge AI processed this data locally, identifying potential issues before they led to failures. This predictive maintenance approach reduced downtime and maintenance costs while increasing the overall energy output of the wind farm.

Case Study 3: **Real-Time Demand Response in Commercial Buildings**: A large commercial building complex aimed to reduce its energy costs and contribute to grid stability. By installing Edge AI-enabled devices, the building's energy management system could respond to real-time data from the grid. During peak demand periods, the system automatically adjusts heating, cooling, and lighting to reduce energy consumption. This real-time demand response not only lowered energy bills but also helped the grid maintain balance during high demand.

Edge AI is revolutionizing real-time data processing in energy systems, by enabling local, immediate decision-making. Real-time data is crucial for balancing supply and demand, integrating renewable energy, and optimizing consumption. Through various applications and case studies, it is evident that Edge AI is a powerful tool in advancing modern energy systems, making them smarter, more responsive, and more sustainable.

8 Enhancing Data Privacy and Security

8.1 Data Privacy Challenges in IoE

The Internet of everything (IoE) links people, processes, data, and things into one big network. While this has many benefits, it also creates major data privacy problems. IoE devices produce a huge amount of data, making it hard to manage and protect. These devices often collect sensitive information like health details, location, and habits. If this data is accessed without permission, it can cause serious privacy issues. Sharing data across different platforms and organizations increases the risk of breaches [4]. IoE devices can also be hacked if they do not have strong security. Plus, meeting data protection laws like GDPR or CCPA is tough because the data is so varied and spread out.

8.2 Security Advantages of Edge AI

Edge AI offers several security benefits that can help address the privacy challenges posed by the Internet of Everything (IoE). By processing data locally on edge devices instead of in centralized data centers, Edge AI enhances data privacy and security in several ways. First, local data processing reduces the need to transmit sensitive information over potentially insecure networks, minimizing the risk of data interception and breaches. Second, by processing data on the edge, only necessary information is shared with central servers, reducing the amount of sensitive data exposed to potential threats. Third, Edge AI can analyze data in real time, enabling faster detection of security threats and anomalies and allowing for immediate responses to potential breaches. Additionally, organizations have more control over their data when it is processed on local devices, allowing for better implementation of security policies and measures. Finally, Edge AI's decentralized nature makes it easier to scale security measures across a wide range of devices and locations without overloading central systems.

8.3 Implementing Secure Edge AI Solutions

To fully benefit from the security advantages of Edge AI, organizations need to implement secure Edge AI solutions using a combination of technological, procedural, and organizational measures. This involves ensuring all edge devices have built-in security features like encryption and secure boot processes, regularly updating and patching devices to protect against vulnerabilities, and using strong authentication methods to protect against unauthorized access.

Data should be encrypted when it is stored and when it is being sent, using strong encryption methods like AES and PKI. Access to data should be tightly controlled, with permissions given based on user roles through role-based access control (RBAC).

AI algorithms should be used to identify anomalies and possible security threats in real time, with machine learning models identifying unusual activities that indicate security breaches. Organizations must also ensure Edge AI solutions comply with relevant data protection regulations, regularly auditing and assessing devices and data processing activities for ongoing compliance. Finally, privacy principles should be incorporated into the plan and development of Edge AI systems from the start, minimizing data collection and processing to only what is necessary for the intended purpose.

8.4 Case Study: Smart Healthcare System

In a smart healthcare system, patient data privacy is paramount. By implementing Edge AI, hospitals can process patient data locally on edge devices within the hospital network. This decreases the need to send sensitive health information over the

Internet, significantly lowering the threat of data breaches. Edge AI can also analyze health data in real time, providing immediate insights to healthcare providers while ensuring that patient data remains secure and private.

Case Study: **Industrial IoT Security**: In an industrial IoT setting, manufacturing plants use numerous connected devices to monitor and control production processes. Implementing Edge AI allows these plants to process data locally, reducing the risk of industrial espionage and cyberattacks. Local processing also enables real-time monitoring of equipment, with AI algorithms detecting and responding to potential security threats instantly, ensuring the continuous and secure operation of the plant.

Enhancing data privacy and security in the era of IoE is critical. Edge AI provides significant advantages by processing data locally, reducing data exposure, and enabling real-time threat detection. Implementing secure Edge AI solutions involves securing devices, encrypting data, controlling access, detecting anomalies, ensuring compliance, and designing for privacy. These measures help defend sensitive information and retain the trust of users, paving the way for a safer and more secure IoE environment.

9 Reducing Latency and Bandwidth Usage

9.1 Latency in Energy Systems

Latency, which is the delay between an action and its response, is very important in energy systems. It affects the efficiency, reliability, and stability of these systems. In traditional energy setups, data from sensors and devices needs to be sent to a central server for processing and analysis, causing delays. High latency can delay the response to changes in energy demand and supply, which can lead to grid instability and even blackouts.

Renewable energy sources like solar and wind are variable and need real-time adjustments, but latency can make it hard to effectively integrate these resources into the grid. Quick adjustments to energy consumption are also necessary to balance the grid, but latency can prevent timely responses, causing inefficiencies. Additionally, delayed detection of equipment failures can lead to longer downtimes and higher maintenance costs.

9.2 How Edge AI Minimizes Latency

Edge AI helps solve latency issues in energy systems by handling data locally on edge devices, close to where the data is generated, instead of sending it to distant central servers. By handling data at the edge, the need to send data to central servers is reduced, which significantly cuts down on transmission time and latency. Edge AI can analyze data and make decisions instantly, allowing energy systems to quickly adapt to changes in demand, supply, and other factors. Additionally, local processing reduces the amount of data transmitted over the network, decreasing congestion and further reducing latency.

In energy systems, Edge AI has several applications. It permits real-time observation and control of smart grids, allowing for instantaneous adjustments to continue stability and efficiency. For renewable energy management, local AI algorithms can rapidly adjust the input from sources like solar and wind based on real-time data, optimizing their integration and usage. Edge AI also helps with predictive maintenance by detecting equipment anomalies and predicting failures in real time, ensuring timely maintenance and reducing downtime.

9.3 Bandwidth Efficiency with Edge AI

In energy systems, efficient bandwidth usage is crucial to ensure reliable communication and avoid network congestion.

Challenges with Bandwidth in Traditional Systems:

- **High Data Volume**: Energy systems produce huge amounts of data that need to be communicated to central servers, consuming significant bandwidth.
- **Network Congestion**: Continuous data transmission can lead to congestion, slowing down the network and affecting system performance.
- **Cost**: High bandwidth usage can be expensive, especially for large-scale energy systems.

9.4 Edge AI Enhances Bandwidth Efficiency

Edge AI reduces the need to transmit large volumes of raw data by processing it locally. This means only essential information and insights are sent to central servers, conserving bandwidth. Edge AI can determine which data is critical and needs to be transmitted and which can be processed and discarded locally, minimizing unnecessary data flow.

Edge AI optimizes performance by processing data locally and transmitting only crucial information, reducing bandwidth usage. Smart meters equipped with Edge AI can analyze energy consumption data onsite and send summarized reports to central servers, minimizing data transmission requirements. For grid monitoring, Edge AI devices can monitor and analyze conditions locally, reporting only significant events or anomalies, thus reducing the continuous flow of data.

Case Study: Smart City Energy Management

In a smart city, many IoT devices are used to monitor and control energy usage in homes, businesses, and public infrastructure [5]. By using Edge AI in these devices, data can be processed locally, allowing for real-time decision-making. For example, smart thermostats can regulate heating and cooling which depends on real-time data, sending only summary reports to the central system. This method not only saves bandwidth but also ensures timely and effective energy management across the city.

Reducing latency and bandwidth usage is crucial for the efficiency and reliability of modern energy systems. Traditional centralized processing models face big challenges due to high latency and high bandwidth consumption. Edge AI provides a powerful solution by processing data locally, which minimizes latency and optimizes bandwidth usage. By allowing real-time decision-making and selective data transmission, Edge AI improves the performance of energy systems, ensuring stability, efficiency, and cost-effectiveness. Various applications and case studies show that Edge AI is essential for the evolution of energy management, leading to smarter and more responsive energy systems.

10 Predictive Maintenance and Fault Detection

10.1 Role of Predictive Maintenance in Energy Systems

Predictive maintenance is a smart way to take care of equipment in energy systems. Instead of waiting for something to break or fixing things on a set schedule, it uses real-time data and advanced tools to predict when equipment might fail. This allows action to be taken before any problems occur.

Benefits of Predictive Maintenance:

- **Reduced Downtime**: By predicting and addressing issues before they lead to equipment failure, predictive maintenance minimizes unplanned downtime.
- **Cost Savings**: Maintenance is performed only when necessary, reducing unnecessary maintenance activities and associated costs.
- **Extended Equipment Life**: Early detection of potential problems allows for timely interventions that can extend the lifespan of equipment.
- **Improved Safety**: Identifying and addressing potential faults before they cause equipment failures can prevent accidents and enhance overall safety.

Applications in Energy Systems:

- **Power Plants**: Predictive maintenance ensures the continuous and efficient operation of turbines, generators, and other critical components.
- **Transmission and Distribution Networks**: Monitoring and maintaining transformers, circuit breakers, and transmission lines help prevent power outages and improve reliability.
- **Renewable Energy Installations**: Wind turbines, solar panels, and other renewable energy equipment benefit from predictive maintenance to optimize performance and minimize downtime.

10.2 Edge AI for Predictive Maintenance

Edge AI enhances predictive maintenance by handling data locally on edge devices, close to the equipment being monitored. This approach enables real-time examination and rapid decision-making, which is crucial for effective predictive maintenance.

Advantages of Edge AI in Predictive Maintenance:

- **Real-Time Monitoring**: Edge AI can continuously monitor the condition of equipment, providing instant insights into its health and performance.
- **Immediate Response**: By analyzing data locally, Edge AI can detect potential issues and trigger maintenance actions immediately, without the delay of transferring data to a central server.
- **Scalability**: Edge AI allows for scalable predictive maintenance solutions, as local processing can be implemented across numerous devices and locations without overloading central systems [6].

Case Study: **Wind Turbine Maintenance**: Wind turbines are critical components of renewable energy systems but are subject to mechanical stress and environmental factors that can lead to failures. By implementing Edge AI, wind turbines can be equipped with sensors to monitor vibrations, temperature, and rotational speed. Local AI algorithms examine this data in real time, detecting anomalies and estimating potential failures. Maintenance teams receive instant alerts, allowing them to address issues before they escalate, thereby reducing downtime and maintenance costs.

11 Fault Detection and Management

Fault detection and management are important for maintaining the consistency and efficiency of energy systems. Faults can arise due to many reasons, including equipment wear and tear, environmental conditions, and operational errors. Effective fault detection and management ensure that faults are identified and addressed promptly, minimizing their impact.

11.1 Edge AI for Fault Detection

Edge AI for fault detection works in several ways. First, it can analyze real-time data to spot unusual patterns, like strange changes in voltage or current, that might indicate a problem. Second, by using past data and machine learning, Edge AI can predict potential issues before they happen, allowing for preventive measures [7]. Lastly, if a fault does occur, Edge AI can automatically isolate the affected part of the system to prevent further problems and keep the overall system stable.

11.2 Fault Management Process

1. **Detection**: Continuous monitoring and real-time analysis to identify faults.
2. **Diagnosis**: Determining the root cause of the fault using AI algorithms.
3. **Alerting**: Notifying maintenance teams or automated systems to take corrective action.

4. **Resolution**: Implementing corrective actions, such as repairs or component replacements.
5. **Post-Mortem Analysis**: Analyzing the fault and response to improve future fault detection and management processes.

Case Study: **Electrical Grid Fault Management**: In an electrical grid, faults such as short circuits, overloads, and equipment failures can disrupt power supply and damage infrastructure. By using Edge AI, sensors placed throughout the grid continuously monitor electrical parameters. When an anomaly is detected, the Edge AI system analyzes the data to determine the nature and location of the fault. Automated systems can then isolate the faulty section, preventing further damage and maintaining grid stability. Maintenance teams receive detailed reports, allowing them to address the issue quickly and efficiently.

12 Integration with Renewable Energy Sources

12.1 Challenges of Renewable Energy Integration

- Adding renewable energy sources like solar and wind to the existing energy grid brings several challenges. These challenges come from the fact that renewable energy is not as predictable as traditional fossil fuel-based energy.
- Key Challenges:
- **Intermittency**: Solar and wind energy are not always available; they depend on the weather and time of day. This makes it hard to keep supply and demand balanced in real time.
- **Grid Stability**: The changing nature of renewables can cause power supply fluctuations, which can destabilize the grid if not managed well.
- **Storage Requirements**: Storing extra energy produced during peak times for use when production is low is essential but difficult to achieve efficiently.
- **Infrastructure Upgrades**: Integrating renewables often requires upgrades to the existing grid infrastructure to handle new energy inputs and maintain reliability.
- **Forecasting**: Accurate prediction of renewable energy production is crucial for planning and balancing the grid but remains a complex task due to the variability of weather conditions.

12.2 Edge AI Solutions for Solar and Wind Energy

Edge AI provides smart solutions for adding solar and wind energy to the grid. By handling data right where the energy is generated, Edge AI can give real-time insights and actions, making renewable energy systems more efficient and reliable.

Advantages of Edge AI in Renewable Energy Integration:

- **Real-Time Monitoring**: Edge AI systems can continuously monitor the performance of solar panels and wind turbines, providing immediate feedback and adjustments to optimize energy production.
- **Predictive Analytics**: Local AI algorithms can analyze weather data and historical performance to forecast energy output, helping grid operatives balance supply and demand more effectively.
- **Energy Storage Management**: Edge AI can manage energy storage systems by predicting when to store excess energy and when to release it, ensuring a steady energy supply even during periods of low production.
- **Dynamic Grid Management**: Edge AI enables dynamic adjustments to the grid, such as switching between different energy sources or adjusting load demands based on real-time data.
- **Fault Detection and Maintenance**: By continuously analyzing data from renewable energy equipment, Edge AI can detect potential faults early and schedule maintenance proactively, reducing downtime and increasing efficiency.

Case Study: **Solar Farm Optimization**: A large solar farm implemented Edge AI to improve its energy production and integration into the grid. Sensors on each solar panel collected data on sunlight exposure, temperature, and energy output. Edge AI algorithms analyzed this data in real time to adjust the angle of the panels for optimal sunlight capture and to detect any performance issues. Additionally, the system predicted daily energy production based on weather forecasts, allowing the grid operator to plan accordingly. This real-time optimization increased the overall energy output and reduced maintenance costs by identifying issues early.

Case Study: **Wind Turbine Efficiency**: A wind farm utilized Edge AI to enhance the performance and reliability of its turbines. Each turbine was equipped with sensors to monitor wind speed, direction, vibration, and energy output. Edge AI processed this data locally to adjust the turbine blades' angles for maximum efficiency based on real-time wind conditions. The system also predicted potential mechanical issues by analyzing vibration patterns, enabling proactive maintenance. As a result, the wind farm experienced fewer breakdowns and improved energy production efficiency.

12.3 Case Studies in Renewable Energy

Case Study: **Smart Microgrid in a Rural Community**: A rural community in a remote area faced challenges with unreliable energy supply. To address this, a smart microgrid driven by solar panels and wind turbines was established, incorporating Edge AI for efficient management. The Edge AI system continuously monitored the energy production from both sources and managed the local energy storage units. During periods of high production, excess energy was stored in batteries, and during low production, the stored energy was utilized.

The system also adjusted local energy consumption based on real-time data, confirming a steady and consistent energy supply for the community. This integration not only provided consistent energy but also reduced the reliance on diesel generators, cutting costs and emissions.

Case Study: **Urban Solar Power Integration**: In a densely populated urban area, integrating rooftop solar panels into the existing grid was challenging due to space constraints and variability in sunlight. Edge AI was implemented to manage the distributed solar energy production from multiple buildings. The system monitored each building's solar panels and predicted their energy output based on weather conditions. It also optimized energy usage within the buildings, adjusting HVAC systems and lighting based on real-time energy availability. Excess energy was fed back into the grid, and Edge AI ensured that this integration did not destabilize the grid. This approach maximized the use of available solar energy, reduced energy costs for building owners, and contributed to a more sustainable urban energy system.

Case Study: **Offshore Wind Farm Management**: An offshore wind farm faced challenges with maintenance and energy integration due to its remote location and harsh environmental conditions. Edge AI was deployed to monitor the wind turbines continuously, analyzing data on wind patterns, turbine performance, and environmental factors. The system predicted maintenance needs and scheduled repairs during calm weather periods to minimize downtime. It also optimized the energy output by adjusting turbine settings based on real-time wind conditions. The Edge AI system communicated with the central grid operator to ensure smooth integration of the variable wind energy, enhancing grid stability and maximizing energy production. This proactive and optimized approach significantly improved the wind farm's efficiency and reliability.

13 Scalability and Interoperability

13.1 Scalability Challenges

Scalability means a system can handle more work as it grows. For the Internet of energy (IoE), scalability is about managing more connected devices, larger amounts of data, and more complex energy systems. This ensures the system can expand and still work well as more devices and data are added.

Key Scalability Challenges:

- **Data Overload**: As more devices are connected to the IoE, the volume of data generated increases exponentially. Managing, storing, and processing this data become a significant challenge.
- **Network Congestion**: High data transmission rates can lead to network congestion, causing delays and reducing the efficiency of real-time applications.

- **Resource Allocation**: Efficiently allocating computational resources to handle varying workloads without over-provisioning or under-provisioning is difficult.
- **System Integration**: Integrating new devices and technologies into existing systems can be complex and costly, especially as the scale increases.
- **Maintenance and Upgrades**: Scaling up systems also means increased maintenance and the need for regular upgrades to ensure reliability and performance.

Example: In a smart city, scaling up the IoE to include more smart meters, sensors, and IoT devices across homes, businesses, and public infrastructure can overwhelm the current infrastructure if not managed properly. The city's energy management system must handle vast amounts of data in real time, ensure reliable network connectivity, and maintain seamless operation as new devices are added.

13.2 Interoperability in IoE

Interoperability is the capability of different systems, devices, and applications to work composed flawlessly, exchanging and using information effectively. In the IoE, interoperability ensures that various energy systems and technologies can communicate and cooperate without compatibility issues.

Importance of Interoperability:

- **Seamless Integration**: Interoperability permits the smooth combination of different energy sources, storage systems, and smart devices into a unified energy management system.
- **Vendor Independence**: It enables the use of devices and systems from different manufacturers, avoiding vendor lock-in and fostering competition and innovation.
- **Data Exchange**: Effective data exchange between systems enhances decision-making, operational efficiency, and overall system performance.
- **Future-Proofing**: Interoperable systems are more adaptable to new technologies and standards, ensuring long-term viability and scalability.

Challenges to Achieving Interoperability:

- **Diverse Standards**: The energy sector has various standards and protocols, making it challenging to achieve universal interoperability.
- **Legacy Systems**: Integrating modern IoT devices with older, legacy systems can be difficult and costly.
- **Security Concerns**: Ensuring secure communication and data exchange between interoperable systems is crucial to prevent cyber threats.

Example: A utility company managing a diverse energy portfolio, including solar, wind, and traditional power plants, needs to integrate these sources into a single management system. Ensuring interoperability among different technologies and standards used by each energy source allows the utility to optimize energy distribution and improve grid stability.

13.3 Edge AI Solutions

Edge AI processes data on devices close to where it is created, offering strong solutions for dealing with scalability and interoperability in the Internet of Energy (IoE) [8].

Edge AI for Scalability:

- **Distributed Processing**: By analyzing data locally on edge devices, Edge AI reduces the load on central servers and networks. This makes it easier to manage large amounts of data efficiently.
- **Real-Time Analysis**: Edge AI can analyze data and make decisions instantly, ensuring quick responses to changes in energy demand and conditions.
- **Resource Optimization**: Edge AI allocates computing power based on current needs, optimizing performance and lowering costs.
- **Modular Expansion**: Edge AI systems can grow by simply adding new edge devices. This allows for easy scaling without needing to redesign the entire system.

Example: In a smart grid, Edge AI can be deployed at various points, such as substations and smart meters, to process data locally. This distributed processing reduces the need for high-bandwidth communication with a central server and allows the grid to scale efficiently as more devices and data sources are added.

13.4 Edge AI for Interoperability

- **Standard Protocols**: Edge AI devices can support multiple communication protocols and standards, facilitating interoperability between different systems and devices.
- **Integration Middleware**: Edge AI can act as middleware, translating data and commands between incompatible systems and ensuring seamless communication and operation.
- **Flexible APIs**: Edge AI systems can provide flexible APIs (application programming interfaces) that allow different devices and applications to interact without compatibility issues.
- **Secure Communication**: Edge AI enhances security by implementing local encryption and authentication protocols, confirming secure data interchange between interoperable systems.

Example: In a renewable energy farm with solar panels and wind turbines from different manufacturers, Edge AI can manage the interoperability of these systems. By supporting various communication protocols and providing a unified interface for data exchange, Edge AI ensures that the different components work together seamlessly to optimize energy production and distribution.

14 Cost and Resource Efficiency

14.1 Cost Considerations of Edge AI

Edge AI processes data on devices close to where it is created, offering several cost benefits compared to traditional centralized methods. However, there are specific costs to consider when setting up Edge AI solutions.

Initial Investment:

- **Hardware Costs**: Deploying Edge AI requires investment in edge devices equipped with sufficient processing power, storage, and connectivity. These devices can range from simple sensors to sophisticated edge servers.
- **Software Development**: Developing and deploying Edge AI applications involve costs related to software development, including the creation of algorithms, data processing pipelines, and user interfaces.
- **Infrastructure Upgrades**: Existing infrastructure may need upgrades to support Edge AI deployment, such as enhanced network connectivity and power supply.

Operational Costs:

- **Maintenance and Updates**: Regular maintenance and software updates are essential to confirm the even operation of Edge AI devices and applications. This includes hardware servicing, firmware updates, and bug fixes.
- **Energy Consumption**: Edge devices consume power, and the cumulative energy cost can be significant, especially in large-scale deployments.
- **Data Management**: While Edge AI reduces data transmission costs, managing and storing locally processed data can still incur expenses.

Cost Savings:

- **Reduced Bandwidth Costs**: Edge AI processes data locally, so less data needs to be sent to central servers, which lowers bandwidth costs.
- **Lower Latency**: Edge AI processes data in real time, which reduces delays and improves efficiency, saving money in industries where quick decisions are important.
- **Decreased Cloud Storage Costs**: With less data being sent to the cloud, cloud storage costs are lower.

Example: A manufacturing plant implementing Edge AI for predictive maintenance of machinery may incur initial costs for purchasing and installing edge devices and developing custom algorithms. However, these costs are offset by savings from reduced machine downtime, lower bandwidth expenses, and decreased reliance on cloud services.

14.2 Resource Efficiency

Resource efficiency is a significant benefit of Edge AI, particularly in how it optimizes the use of computational, network, and energy resources.

Optimized Computational Resources:

- **Distributed Processing**: Edge AI spreads data processing across several devices, preventing the slowdowns that can happen with a central server. This makes better use of computing power.
- **Local Data Processing**: Processing data at the edge lessens the strain on central servers, allowing them to handle other important tasks.

Network Efficiency:

- **Reduced Data Transmission**: By handling data locally, Edge AI cuts down on the amount of data sent over networks. This saves bandwidth and reduces network congestion.
- **Efficient Data Filtering**: Edge devices can sort and process data before sending only the most important information to central servers, making network use more efficient.

Energy Efficiency:

- **Localized Processing**: Edge AI reduces the need to constantly send data to the cloud, which uses a lot of energy. By processing data locally, it saves energy by cutting down on data movement.
- **Power Management**: Edge AI devices can be designed to operate efficiently, with power-saving modes and low-energy components, reducing overall energy consumption.

Example: In a smart agriculture setup, Edge AI can be used to monitor soil moisture, weather conditions, and crop health. By processing this data locally, only essential information is sent to a central server for further analysis, optimizing both network and computational resources. The edge devices can also be powered by solar panels, enhancing energy efficiency.

Cost and resource efficiency are crucial factors in the adoption of Edge AI. While the initial investment in Edge AI can be higher due to hardware and development costs, the long-standing savings in bandwidth, cloud storage, and operational efficiency can be considerable. Edge AI optimizes resource usage by distributing computational loads, reducing network congestion, and conserving energy through localized processing. Compared to Cloud AI, Edge AI offers lower latency, greater reliability, and efficient scalability, making it an attractive solution for real-time, resource-intensive applications. By understanding the cost considerations and resource efficiency benefits, organizations can make knowledgeable decisions about implementing Edge AI to enhance their processes and achieve sustainable growth [9].

15 Future Directions and Challenges

15.1 Emerging Trends in Edge AI

Edge AI, which practices data locally on devices near the source of data generation, is continually evolving. Several emerging trends are shaping its future development and adoption.

Enhanced AI Capabilities:

- **Improved Algorithms**: Advancements in machine learning algorithms are enabling more sophisticated data handling and decision-making at the edge. These improvements allow for more precise predictions, real-time analytics, and adaptive learning.
- **Edge AI Chips**: The development of specialized hardware, such as Edge AI chips, is enhancing the computational power of edge devices while maintaining energy efficiency. These chips are optimized for running AI models and are becoming increasingly common in various applications.

Integration with 5G:

- **Low Latency**: The deployment of 5G networks is significantly reducing latency, enabling faster data transfer between edge devices and central systems.
- **Enhanced Connectivity**: 5G provides higher bandwidth and more reliable connectivity, facilitating the integration of a larger number of edge devices and supporting the scalability of Edge AI systems.

AI at the Edge and Cloud Collaboration:

- **Hybrid Models**: Using both Edge AI and Cloud AI together helps process data more efficiently. Important, time-sensitive data is handled at the edge, while less urgent data goes to the cloud for more detailed analysis and long-term storage.
- **Seamless Integration**: New software frameworks are making it easier to connect Edge AI and Cloud AI, ensuring smooth data flow and consistent system management.

Edge AI in IoT:

- **Smart IoT Devices**: The rise of smart IoT devices is boosting the use of Edge AI. These devices have sensors and AI features that allow them to process data locally, so they do not always need to be connected to the cloud.
- **Interoperability**: Work is being done to create standard protocols and interfaces so different IoT devices can easily work together. This improves how well Edge AI solutions function.

Example: In healthcare, Edge AI is being used to handle data from wearable devices in real time, providing immediate health insights and alerts. This trend is expected to grow with the development of more advanced wearables and improved connectivity through 5G.

15.2 Overcoming Challenges

While Edge AI has many advantages, it also faces some challenges that need to be tackled to fully unlock its potential.

Security and Privacy:

- **Data Protection**: It is crucial to keep data safe and private when processed at the edge. This means protecting it from cyber threats and following data protection laws.

- **Device Security**: Edge devices must be secured against physical tampering and cyberattacks to keep data safe and ensure the system works reliably.

Scalability and Maintenance:
- **Managing Large-Scale Deployments**: As more edge devices are added, managing and maintaining them gets more complicated. We need good solutions for monitoring devices, updating software, and fixing problems.
- **Resource Constraints**: Edge devices often have inadequate computing power and storage, which can be a challenge. Developing efficient algorithms that can operate within these constraints is a significant challenge.

Standardization and Interoperability:
- **Lack of Standards**: The absence of universal standards for Edge AI and IoT devices can lead to compatibility issues. Standardizing protocols and interfaces is necessary for seamless integration and interoperability.
- **Cross-Platform Compatibility**: Ensuring that Edge AI solutions work across different platforms and environments is crucial for widespread adoption.

Energy Efficiency:
- **Power Consumption**: Edge devices need to operate efficiently to conserve energy, especially in remote or resource-constrained environments. Developing low-power AI models and optimizing energy use is essential [10].
- **Sustainable Practices**: Adopting sustainable practices in the plan and deployment of Edge AI systems, such as using renewable energy sources and recyclable materials, is becoming increasingly important.

Example: In smart cities, deploying Edge AI for traffic management involves installing numerous edge devices at various locations. Ensuring these devices are secure, interoperable, and energy-efficient while maintaining them at scale is a significant challenge that needs to be addressed for effective implementation.

15.3 Future Prospects

The future of Edge AI is promising, with numerous opportunities for growth and innovation. As technology improvements and challenges are addressed, Edge AI is likely to play an increasingly important part in various sectors.

Healthcare:
- **Personalized Medicine**: Edge AI will allow real-time health monitoring and personalized treatment plans that depend on data from wearable devices and medical sensors.
- **Remote Diagnostics**: Enhanced connectivity and processing power will support remote diagnostics and telemedicine, improving access to healthcare services.

Industry 4.0:

- **Smart Manufacturing**: Edge AI will advance smart manufacturing by allowing real-time monitoring, predicting when maintenance is needed, and making decisions automatically on the factory floor.
- **Supply Chain Optimization**: Processing data in real time at the edge will improve supply chain management, making it more efficient and lowering costs.

Autonomous Systems:

- **Self-Driving Vehicles**: Edge AI will be crucial for the development of self-driving vehicles, providing real-time processing capabilities for navigation, obstacle detection, and decision-making.
- **Drones and Robotics**: Edge AI will enable more advanced and autonomous operations for drones and robots in various applications, from delivery services to industrial inspections.

Smart Cities:

- **Energy Management**: Edge AI will optimize energy usage in smart cities, balancing stock and request, incorporating renewable energy sources, and improving grid stability.
- **Public Safety**: Real-time data processing at the edge will enhance public safety through advanced surveillance systems, emergency response management, and traffic control.

Retail:

- **Customer Experience**: Edge AI will enhance the retail experience through personalized recommendations, real-time inventory management, and efficient checkout processes.
- **Supply Chain Management**: Real-time data processing will improve supply chain efficiency, reducing waste and optimizing logistics.

The future of Edge AI is optimistic, with emergent trends such as enhanced AI capabilities, integration with 5G, hybrid edge-cloud models, and the proliferation of smart IoT devices driving its evolution. Overcoming challenges related to security, scalability, standardization, and energy efficiency is essential to fully leverage the potential of Edge AI. As these challenges are addressed, Edge AI is expected to revolutionize various industries, from healthcare and manufacturing to autonomous systems and smart cities, offering significant improvements in efficiency and reliability.

16 Conclusion

Edge AI is changing how decisions are made in real time within the Internet of Energy (IoE). By processing data closer to where it is generated, Edge AI decreases the time it takes to create decisions, which is critical for managing energy systems effectively. This means that energy systems can respond faster to changes, such as fluctuations in power demand or unexpected outages, improving their overall performance and reliability.

In addition to reducing delays, Edge AI also enhances privacy and security. Since data is analyzed locally rather than being sent to a central server, there is less risk of sensitive information being intercepted or misused. This local processing also helps protect against cyber threats, as there are fewer points where data can be compromised. Moreover, this approach minimizes the amount of data that needs to be sent, which contributes to more efficient use of network resources.

Looking ahead, Edge AI has the potential to drive significant improvements in the sustainability of energy systems. By enabling more precise and responsive control, it can help optimize energy usage and reduce waste. While there are still challenges to address, such as managing the complexity of distributed systems and ensuring interoperability, the benefits of Edge AI make it a key technology for advancing the future of energy management.

References

1. Yao, Y., Zhuang, X., Zhang, H., & Liu, Y. (2024). Edge AI and cloud AI: A comparative study and future directions. *Journal of Cloud Computing: Advances, Systems and Applications.*
2. Li, J., Liu, S., Xu, X., & Li, X. (2023). A survey on edge AI: From edge computing to edge intelligence. *IEEE Access, 11*, 21034–21051.
3. Chen, Y., & Yang, J. (2024). Cloud computing meets edge computing: The synergy and challenges. *Future Generation Computer Systems, 140*, 477–491.
4. Kumar, A., & Pustokhina, I. (2023). Cloud vs edge AI: A comparative review and performance analysis. *Journal of Computer Networks and Communications.*
5. Zhang, M., Liu, Y., & Yang, F. (2024). Challenges and opportunities in edge AI for smart cities. *IEEE Internet of Things Journal, 11*(2), 1678–1692.
6. Mishra, S., & Ghosh, S. (2024). Edge AI vs. cloud AI: A comparative analysis and future perspectives. *ACM Computing Surveys, 56*(3), 1–28.
7. Kim, H., & Park, S. (2023). Optimizing edge AI for real-time applications: A survey and future directions. *IEEE Transactions on Network and Service Management, 20*(4), 345–362.
8. Gao, X., Zhang, Q., & Liu, J. (2023). Balancing cloud and edge AI: An investigation into hybrid architectures. *Journal of Computer Science and Technology, 38*(1), 55–75.
9. Wang, L., & Xu, S. (2024). Energy-efficient computing in cloud and edge AI systems: A review. *IEEE Transactions on Sustainable Computing, 9*(2), 289–306.
10. Singh, R., & Gupta, M. (2023). Security and privacy in edge AI: A comprehensive review. *Computer Networks, 222*, 108693.

AI-Powered Energy Consumption Prediction and Forecasting

Ghalia Nassreddine, Abir El Abed, Mohamad Nassereddine, and Obada Al-Khatib

1 Introduction

The increase in population has led to a corresponding rise in power consumption. Additionally, advancements in technology are driving an increase in power generation, further exacerbating energy demand. There is a growing global effort to address this rising demand through the utilization of renewable energy sources. This initiative aims to foster the development of smart and sustainable cities that provide advanced technologies, clean energy, and a healthier planet. Such an approach has the potential to replace traditional fossil fuel power generation, significantly reducing carbon dioxide emissions in the atmosphere. Consequently, transitioning to alternative energy sources is imperative. Renewable energy, which harnesses natural resources to provide a consistent and sustainable energy supply while preserving the Earth's resources, is essential for ensuring a sustainable future [1, 2].

Sustainability is commonly defined as the practice of meeting present needs without compromising the ability of future generations to fulfill their own needs. It is widely recognized as crucial for the sustainable progress of societies due to various factors, including economic stability, environmental preservation, and social equity [3]:

G. Nassreddine (✉)
Rafik Hariri University, Mechref, Lebanon

A. El Abed
Lebanese University, Faculty of Engineering, Branch III, Hadath, Lebanon

Lebanese University, Faculty of Public Health, Branch V, Saida, Lebanon

M. Nassereddine · O. Al-Khatib
University of Wollongong in Dubai, Dubai, UAE

P. Raj et al. (eds.), *Artificial Intelligence (AI) for IT Energy Efficiency and Green AI for Environment Sustainability*,
https://doi.org/10.1007/978-3-031-89420-6_27

- Long-Term Resource Availability: Sustainable practices ensure that resources, such as raw materials, energy, and water, are used wisely and preserved for future generations; this approach guarantees long-term availability and prevents resource depletion.
- Environmental Protection: Sustainable development aims to minimize its negative impact on the environment by reducing pollution, greenhouse gas emissions, and habitat destruction. This approach helps safeguard natural resources, biodiversity, and ecosystems, all of which are essential to human well-being.
- Social Equity: Sustainable development fosters justice and equality by ensuring that economic progress benefits every member of society, particularly disadvantaged and vulnerable groups. It emphasizes inclusive progress while considering the needs of future generations.
- Economic Stability: Sustainable practices promote economic stability by encouraging innovation, reducing costs through increased efficiency, and creating new business opportunities in sectors such as green technologies and renewable energy.
- Public Health: Sustainable practices often lead to reduced exposure to harmful chemicals and contaminants, cleaner air and water, and healthier food systems. This not only lowers healthcare costs but also improves overall public health outcomes.
- Global Responsibilities: Sustainable development encourages global cooperation and global responsibilities in our interconnected world, promoting worldwide collaborations to tackle common environmental and social issues, such as poverty and climate change.
- Adaptability to Climate Change: Sustainable development enhances the resilience of communities and economies to extreme weather events and other climate-related challenges, such as rising sea level, by reducing greenhouse gas emissions, and adapting to the impacts of climate change.

Therefore, sustainability ensures that current needs are met without compromising the ability of future generations to meet their own needs, making it essential for the continued progress of society. It integrates social, economic, and environmental aspects to establish a resilient and well-rounded society capable of thriving over time. In contemporary society, there is a significant shift toward the development of sustainable smart cities, where the increasing demand for power is met through clean and sustainable energy sources. This approach not only provides clean energy but also reduces the costs associated with energy generation. Furthermore, it facilitates the integration of energy systems with other sectors, such as waste management and automation [4]. Energy consumption forecasting is crucial for sustainability, as it enhances resource and infrastructure planning, thereby increasing the efficiency of power generation and distribution networks. This enables individuals and enterprises to maximize energy use, successfully integrate renewable energy, and develop strategies for climate mitigation and energy efficiency. Accurate forecasts also improve grid stability, promote sustainable behavioral shifts, and support international efforts to combat climate change by reducing overall energy consumption and endorsing more eco-friendly energy sources [5].

Artificial Intelligence (AI) and its subfields, such as Machine Learning (ML) and deep learning (DL), have gained significant interest due to their excellent performance in various fields, including healthcare, education, and the energy industry [6, 7]. Many researchers have developed energy demand forecasting systems using ML techniques [8]. There are two main classes of forecasting approaches: The first relies on historical data to forecast future values of a variable based on its past values, including autoregressive models and time series analysis [9]. The second class examines the causal relationships between input variables such as social, economic, and climatic factors, and energy demand output. Regression models and artificial neural networks are the two most common causal approaches used to predict energy usage [10]. A revolutionary method of energy demand forecasting using Deep Neural Networks' Long Short-Term Memory (LSTM) capabilities was presented in Marino et al. [11]. Additionally Ghazal [12], proposed an expert energy management system that combines a support vector machine, deep extreme ML techniques, and an Internet of Things-based smart meter.

This chapter explores the impact of artificial intelligence on transforming energy demand prediction and forecasting systems. It discusses various AI techniques, including ML and DL tools, for analyzing historical energy consumption data, recognizing patterns, and generating precise forecasts. It highlights how current AI-driven systems can optimize energy distribution, enhance efficiency, and promote sustainable practices by providing accurate and actionable insights into future energy requirements.

This chapter is organized into six main sections. Section 2 will explore the key factors impacting energy demand. Subsequently, Section 3 will provide a comprehensive review of artificial intelligence models for predicting and forecasting energy demand. Section 3.4 will elaborate on and compare various AI-based methods for forecasting and predicting energy consumption. Section 4 will highlight the advantages of accurate energy demand predictions for sustainability. Finally, the chapter will conclude with a thoughtful summary in Sect. 5.

2 Main Factors Influencing Energy Demand

To effectively forecast and control energy demand and generation, it is crucial to comprehend its dynamics. The amount of energy consumed by users can be influenced by various factors. This section will explore the primary factors impacting energy demand, including population growth, economic development, policy effects, and technological advancements. By examining these key variables, the ability to forecast energy requirements can be improved. Additionally, many strategies can be developed to ensure a stable and sustainable energy future. Figure 1 illustrates the main elements that may influence energy demand [13]:

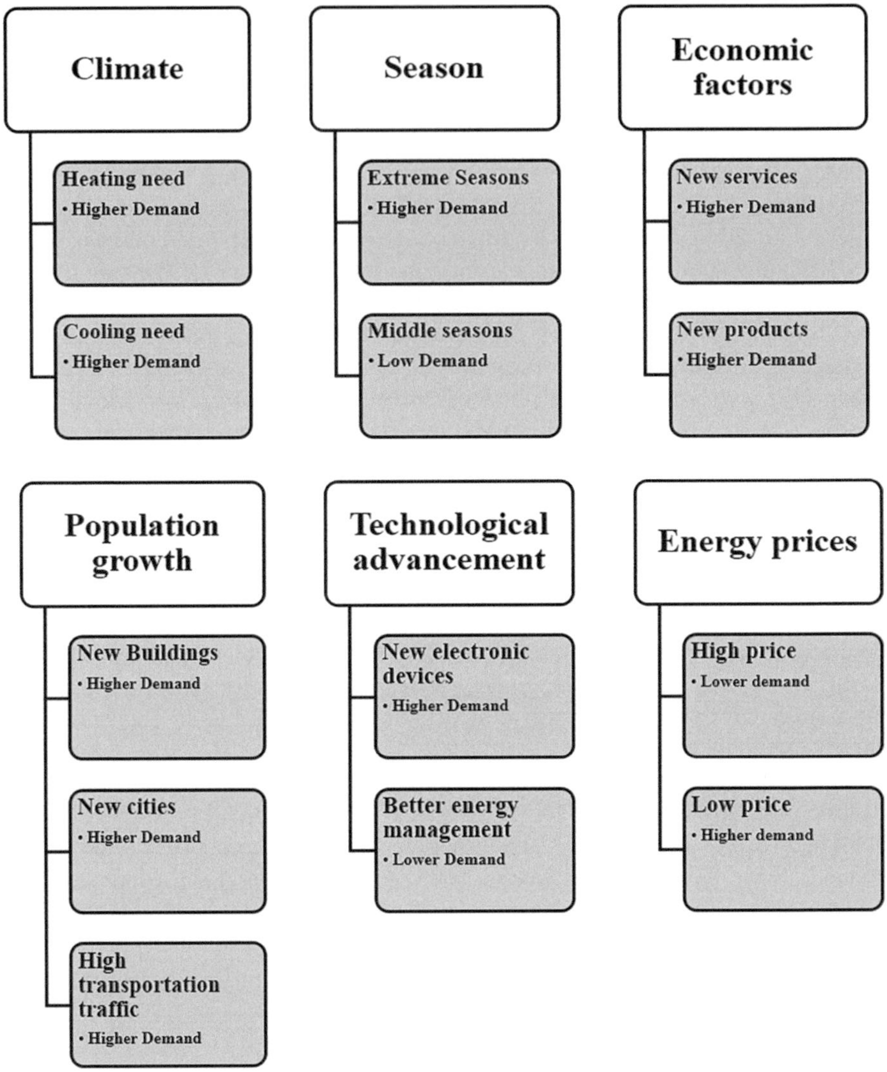

Fig. 1 Main factors influencing energy demand

2.1 *Climate and Weather*

Two types of research study how weather affects energy demand. The first type focuses on forecasting electricity consumption. According to some studies in this area, temperature, humidity, and precipitation are the main weather factors influencing power usage, while wind speed and sunlight are less significant. This makes sense, as extreme temperatures, both hot and cold, can lead to increased energy demand for heating and cooling [14]. The study mentioned in the reference [15] aimed to evaluate the impact of both new and historical data on projecting future energy usage. It focused on educational buildings in Tabriz and Tehran. According to the study, by the year 2080, there will be a 24% reduction in heating energy consumption and a 5% increase in cooling energy consumption in Tehran, compared to 2020. These projections take into account the anticipated changes in seasonal temperatures from 2020 to 2050 to 2080. Additionally, a 19% decrease in heating energy consumption and a 7% increase in cooling energy consumption for Tabriz in 2080 were anticipated.

In the second type of research, when studying energy demand at the regional level, meteorological variables are often considered. For example, a study in Brazil examined the relationship between climate and daily electricity consumption, focusing on heating degree days and cooling degree days, as well as the lag effects between the two. The study found that temperature has a slightly different effect on weekends compared to workdays. Additionally, a model using combined monthly or yearly data may not effectively distinguish between these different scenarios [16].

2.2 *Season*

The impact of seasonal variations on energy demand is significant, mainly due to changes in daylight and temperature. The typical effect of the seasons on energy consumption is as follows [17]:

- In winter, energy consumption increases due to higher heating needs in homes and businesses. Heating systems like furnaces, boilers, and heat pumps use more energy. Additionally, reduced daylight hours lead to increased use of artificial lighting, further raising energy usage.
- In hot weather, energy usage increases due to the operation of air conditioners and fans while longer daylight hours may reduce the need for artificial lighting, they can result in higher usage of the cooling system.
- In spring and fall, there is less need for energy for heating and cooling, but energy use can still be affected by temperature variations and weather patterns. These seasons typically see a balanced use of energy for both cooling and heating, resulting in more moderate overall energy consumption.

2.3 *Economic and Population Growth*

The surge in energy demand fueled by industrialization has led to rapid urbanization and population growth, making energy a vital component of modern life. As the global population swells and consumption patterns evolve, energy consumption continues to soar. Since the 1950s, the world population has experienced a substantial surge, accompanied by a corresponding spike in global energy consumption, particularly in burgeoning economies like China and India [18].

Shaari et al. analyzed the relationship between Malaysia's population, energy usage, and economic growth [19]. The results from the cointegration model showed a long-term relationship between Malaysia's population, energy consumption, and economic growth. The study also revealed how Malaysia's population affects energy usage and how energy consumption drives economic expansion.

Energy and economic growth are closely related. As economies expand, businesses produce more goods and services, leading to a huge increase in energy demand. Furthermore, rising living standards and incomes result in higher consumer spending, which drives up energy demand for heating, cooling, and transportation. Besides, economic growth yields the development and construction of new factories, buildings, and roads, all of which require more energy consumption. Automation in developing economies amplifies energy-intensive activities, while technological advancements and urbanization further contribute to higher energy consumption [20]. Population growth affects energy demand in two main ways [21]:

- Increased Consumption: As the population grows, the overall demand for energy rises due to more people needing it for daily activities such as cooking, heating, cooling, and traveling.
- Greater Residential Energy Use: A larger population living in houses and buildings leads to increased demand for residential energy and a greater need for fuel for heating and electricity.

2.4 Technological Advances

The remarkable progress in technology has revolutionized energy demand, significantly improving efficiency, encouraging the use of renewable energy sources, and facilitating a transition away from fossil fuels to cleaner alternatives. Advanced energy management systems ensure the optimal use and distribution of energy, while automation and demand response technologies provide flexibility in energy consumption. Furthermore, changes in lifestyle and behavior, such as remote work and increased use of digital services, are reshaping energy consumption patterns and contributing to a more balanced and sustainable energy future [22].

Furthermore, these technologies facilitate the integration of critical sectors, such as electricity generation and waste management. This integration has the potential to enhance the sustainability of cities while simultaneously alleviating the waste burden on both national and global scales [1]

2.5 *Energy Prices*

The composition of energy sources and the total energy consumed are directly influenced by energy prices. When the price of a particular energy source rises, its usage typically decreases, while demand for alternative sources increases. However, in the short term, technological limitations restrict how much consumption can be reduced and how feasible it is to switch to different energy sources. Any reduction in energy use or change in energy sources must provide significant financial benefits amid sustained price fluctuations. For instance, when electricity prices surge, people often delay installing heat pumps or replacing inefficient refrigerators. Rising energy costs generally lead to decreased demand and production of goods and services. Industries with lower energy intensity, such as those in the services sectors, tend to experience a relative boost in profitability, while those with higher energy intensity face a decline. Consequently, over time, rising energy prices may contribute to a less energy-intensive industrial structure and an overall reduction in total energy consumption [23].

3 Artificial Intelligence Techniques for Predicting and Forecasting Energy Demand

Predicting and forecasting energy demand is crucial for ensuring sustainability, reducing costs, and optimizing energy systems [24]. Recently, AI techniques have been used to enhance accuracy and efficiency in predicting and forecasting energy demand. In particular, the main subsets of AI, ML, and DL are capable of modeling complex relationships and handling large datasets. This section will first introduce ML and DL, followed by a review of existing prediction and forecasting systems based on these technologies.

3.1 *Machine Learning and Deep Learning*

ML models use existing data to train algorithms, enabling them to make predictions or decisions based on available information without being explicitly programmed. This approach allows systems to improve their performance and efficiency on specific tasks as they are exposed to more data. On the other hand, DL, which is a subset of ML, employs neural networks with multiple layers to automatically learn and extract features from raw data. This method is particularly effective for complex tasks such as image and pattern recognition, due to its ability to model complex patterns and representations [25].

Table 1 illustrates the main advantages and limitations of ML and ML techniques [26]. Figure 2 shows the main categories of ML and DL. According to this figure, ML techniques can be classified into three main categories:

Table 1 Main advantages and disadvantages of Machine learning and Deep Learning techniques

Category	Advantages	Limitations
Supervised Learning	Discerning complex patterns Providing high accuracy in many applications	Need for large amounts of labeled training data Limitations in generalizing to unseen data Susceptibility to overfitting and noisy or biased data
Unsupervised Learning	Require minimal human involvement and can reveal hidden data patterns	Evaluating the performance is challenging due to the lack of labeled data Overfitting in high-dimensional spaces is a concern Careful parameter tuning and feature selection are crucial to avoid unfavorable outcomes
Deep Learning	Automatic feature selections Handle large and complex datasets effectively Uncovering nonlinear relationships in data that are difficult to detect with traditional techniques Effectively manage both structured and unstructured data types Efficiently handle missing data	Training DL models requires high computational costs Overfitting can occur when a model becomes too specialized for the training data, resulting in poor performance on new, unseen data The performance heavily depends on the quality of the training data; noisy, incomplete, or biased data

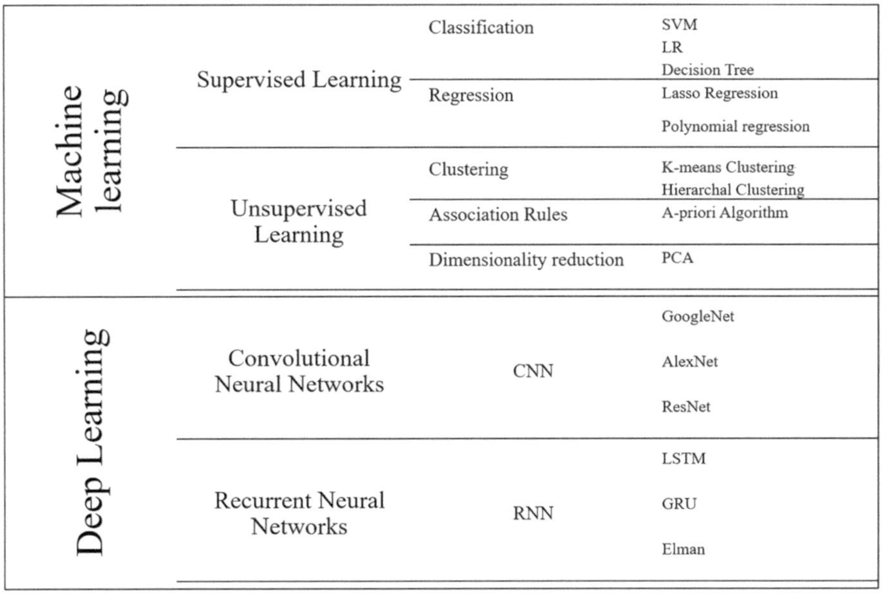

Fig. 2 Main categories and techniques of ML and DL

- **Supervised learning:** This process involves using labeled datasets to train algorithms for making predictions and identifying patterns. It includes uncovering the relationship between the independent (input) variables and the dependent (target) variables. Algorithms in this category can be divided into two main types [27]:
  (a) **Classification:** In this type of algorithm, the target variable is categorical and takes on discrete values. Classification algorithms are widely used in various fields such as finance, healthcare, and fault detection. Examples of algorithms in this category include Support Vector Machine, Logistic Regression, Random Forest, and Gradient Boosting.
  (b) **Regression analysis**: This method is used to predict a continuous output based on one or more input variables. Its goal is to estimate the best-fitting function that enables predictions and helps in understanding trends and patterns within the dataset. The main objective of regression analysis is to model the relationship between dependent and independent variables. Examples of algorithms in this category include Linear Regression, Polynomial Regression, Support Vector Regression, and Lasso Regression.
- **Unsupervised learning:** Methods in this category use ML models to discover relationships within unlabeled data. The main goal is to identify patterns, clusters, or structures in the data. Algorithms in this category can be grouped into three main types [28]:
  (a) **Clustering:** This technique groups unlabeled data based on their similarities or differences, using algorithms to process raw, unclassified data objects into clusters represented by structures or patterns. Examples of algorithms in this category include K-means and Hierarchical clustering. These techniques are widely used in many applications such as Market segmentation, Social network analysis, Medical imaging, and Anomaly detection.
  (b) **Association rules:** These are methods for finding relationships between variables in a dataset. They are frequently applied in market basket analysis to understand product relationships. The A-priori algorithm is an example of a method in this category.
  (c) **Dimensionality reduction:** This technique simplifies high-dimensional datasets by reducing the number of input features while preserving data integrity. It is commonly used in the preprocessing stage. Principle Component Analysis (PCA) is an example of a technique in this category.
- **Deep learning** is an advanced branch of ML that uses multilayered neural networks, known as deep neural networks, to replicate complex decision-making processes similar to those of the human brain. These models typically require less manual intervention because they automatically update their knowledge over time. DL methods can be broadly categorized into two main groups [29]:
  (a) **Convolutional Neural Network (CNN):** CNNs are an extension of artificial neural networks (ANNs) primarily designed to extract features from

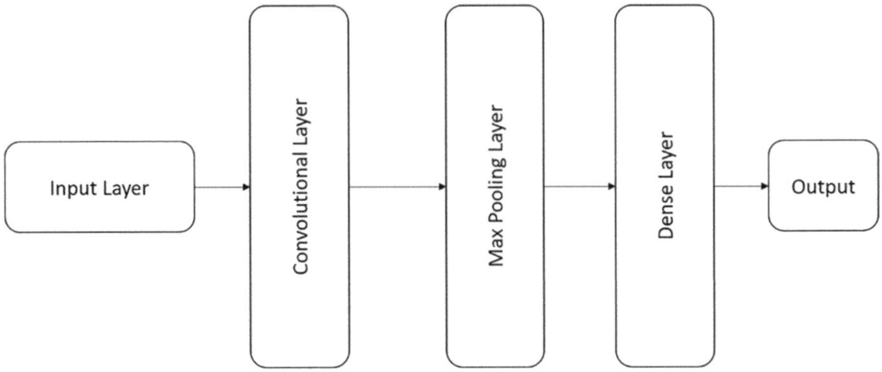

Fig. 3 CNN layers: Convolutional layers use filters to identify features in the input data, creating spatial hierarchies. The max pooling layer reduces the dimensions of the feature maps, aiding in downsampling and enhancing resilience to input variations. In the dense layer, neurons are interconnected, allowing the combination of complex features and producing diverse outputs

grid-like matrix datasets. They are commonly applied to visual datasets such as images or videos, where identifying data patterns is crucial. A CNN consists of multiple layers, including the input layer, convolutional layer, pooling layer, and fully connected layers (see Fig. 3). Convolutional and max pooling layers are used to extract important features from the input variables. Dense layers can project high-dimensional data into a lower-dimensional space while preserving important features. GoogleNet, AlexNet, and ResNet are recent algorithms based on CNNs.

(b) **Recurrent Neural Networks (RNNs):** RNNs are designed to incorporate previous information, making them particularly effective for tasks such as predicting the next word in a sentence. The memory state in RNNs retains crucial sequence information, using consistent parameters for each input, which simplifies the process compared to other network types (see Fig. 4). LSTM is among the most common algorithms built on the RNN methodology. The LSTM architecture includes a memory cell controlled by three gates: the input gate, the forget gate, and the output gate. These gates determine what information is added to, removed from, and output by the memory cell, allowing LSTM networks to selectively retain or discard information and learn long-term dependencies [30].

3.2 Power Demand Prediction and Forecasting System Architecture

Anticipating future electrical power demand is crucial for maintaining a stable energy grid and ensuring efficient energy management. This is where power demand prediction or forecasting systems come into play, offering essential technological

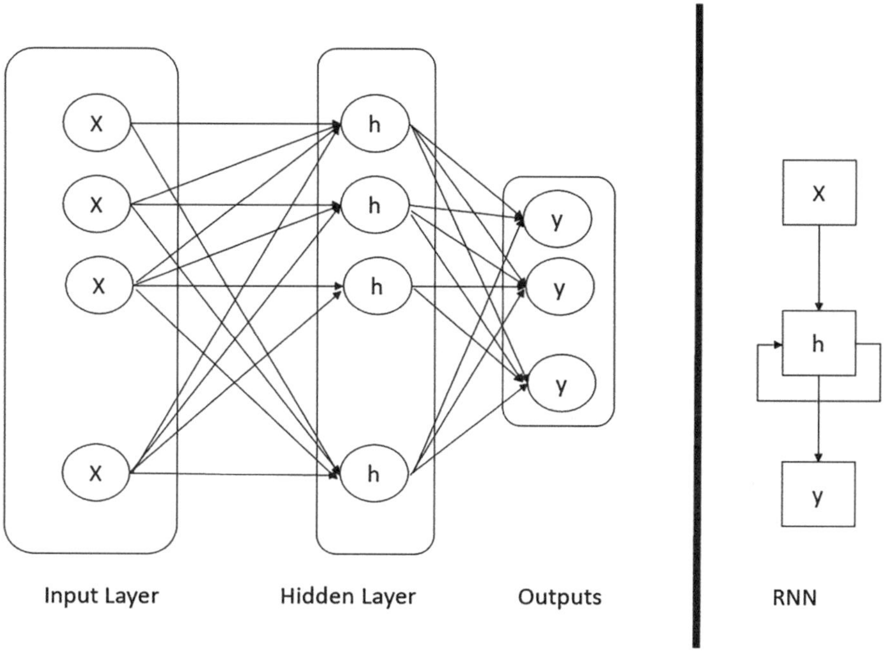

Fig. 4 RNN layers: The input layer processes sequential data one step at a time. The hidden layers capture temporal dependencies and learn patterns by preserving and updating hidden states across sequences. The output layer generates predictions or classifications based on the processed sequence data and the hidden states for each time step or sequence

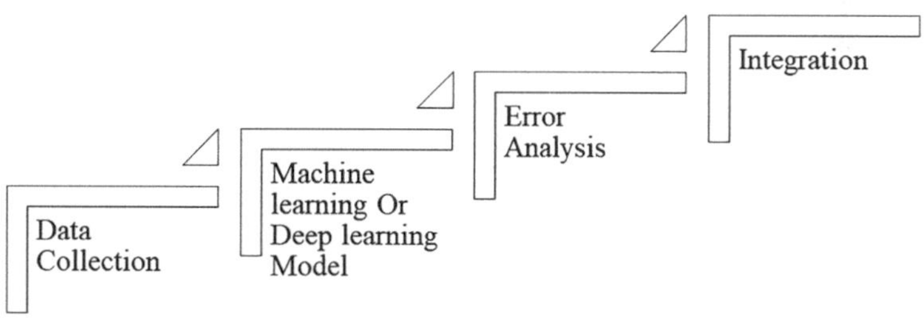

Fig. 5 Main steps of energy demand prediction systems using ML or DL techniques

solutions for meeting energy needs. Figure 5 shows the four main steps of power demand prediction systems.

Data collection: In this phase, two types of data should be gathered:

1. Historical Data: This involves compiling historical information on energy demand, weather trends, economic indicators, and other relevant variables.
2. Real-time Data: This set includes current factors such as temperature, humidity, and calendar events (holidays, workdays).

Let DS be the name of the dataset that covers all collected data. DS can be divided into two subsets:

(a) Features or independent variable, called X: contains historical and real-time data.
(b) Target or dependent variable, called y: represents the current power demand.

ML/DL model: This section employs a variety of ML or DL to analyze the correlations and relationships between different variables, generating accurate estimates for the target variable (y). The system incorporates a wide range of ML and DL approaches, including light gradient boosting, deep neural networks, support vector regressors, and others [31]. Let M represents the ML or DL model. Therefore, the prediction or forecasting phase involves identifying the best model M that produces an accurate estimate of y, referred to $\hat{y}$:

$$\hat{y} = M\left(X\right) \tag{1}$$

Error Analysis: In this phase, the accuracy of the prediction model will be assessed using various metrics [32]:

- ***R*-squared:** This statistic measures the percentage of the dependent variable's variance that is explained by the independent variable(s) in a regression model. Essentially, it indicates how well the data fits the ML or DL model. R^2 is computed using:

$$R^2 = 1 - \frac{\sum_{i=1}^{n}\left(y_i - \hat{y}_i\right)^2}{\sum_{i=1}^{n}\left(y_i - \overline{y}\right)^2} \tag{2}$$

  Where n is the number of collected observations, y_i is target actual value (energy demand) at time step i, $\hat{y}_i$ is target predicted (estimated) value at time step i, and $\overline{y}$ is the mean of the predicted values.
- **MAE (Mean Absolute Error)**: This metric measures the typical magnitude of errors across a set of forecasts, without considering the sign of the errors. It assesses the effectiveness of a regression model and is expressed as the average absolute difference between predicted and actual values. The MAE is given in Eq. (3). It represents the average magnitude of the errors for predictions without considering the direction.

$$\text{MAE} = \frac{1}{n}\sum_{i=1}^{n} y_i - \hat{y}_i \tag{3}$$

- **RMSE (Root Mean Square Error)**: This metric calculates the mean difference between the actual values and the values predicted by a model. It essentially represents the standard deviation of the residuals, which indicates the distance between the data points and the regression line. It is given in Eq. (4).

$$\text{RSME} = \sqrt{\frac{\sum_{i=1}^{n} y_i - \hat{y}_i}{n}} \quad (4)$$

Integration: There are two primary steps in integrating a power demand forecast system into the grid.

- **Energy Management Systems (EMS):** Demand forecasts should be integrated into EMS to improve the coordination of generation, transmission, and distribution. Automated control systems can then use these forecasts to adjust grid operations, such as modifying generation levels or initiating demand response programs.
- **Demand Response Programs:** Demand forecasts should be incorporated into demand response programs to encourage users to adjust their energy usage based on grid conditions. Furthermore, these forecasts can enhance planning and improve customer relations within the programs.

3.3 *Example of ML-Based Electricity Demand Prediction Algorithm*

In this section, a summary of a basic ML-based algorithm for predicting electricity demand will be provided. Algorithm 1 outlines a simplified pseudocode that explains the general process of preparing data, training the model, and making predictions.

Algorithm 1. Simple Algorithm for Electricity Prediction Using ML Technique

- Load Libraries
  - Data Manipulation
  - Machine Learning
  - Data Visualization
- Collect data
  - History of power demand
  - Chronological data
  - Meteorological data
  - Location and building characteristics
- Preprocessing step
  - Handle missing and outliers' values
  - Normalize features
  - Labeling data
  - SPLIT data into features (X) and target (y)
- Feature Selection
  - $X_{best} = \{x_i \in X, \textit{where } x_i \textit{ has a high impact on } y\}$

- Split dataset into training set and test set
  - $X_{best} = \{X_{TRAIN}, X_{test}\}$
  - $y = \{y_{TRAIN}, y_{TEST}\}$
- ML model (M)
  - Initialize
  - Train

    $$M.fit\ (X_{train}, y_{train})$$
- Evaluate Model
  - $\hat{y} = M(X_{TEST})$
  - Calculate performance metrics using Eqs. (2), (3), and (4)
  - Print evaluation metrics
- Prediction
  - Input new ~~data X_{new}~~
  - Compute $y_{new} = M(X_{new})$
  - Print predicted demand

First, libraries such as Pandas, NumPy, Scikit-learn, and Matplotlib in Python are imported to handle data, build models, and visualize data. Next, read the dataset, typically in CSV format. This dataset may contain historical electricity demand data along with important features like location, chronological data, number of buildings, number of rooms per building, and weather conditions. These features should be organized into two main variables:

1. Predictors or independent variables: These include chronological data, weather conditions, and other relevant features.
2. Dependent or target variable: This represents the actual electricity demand.

Given the potentially large number of features, apply a feature selection step using techniques such as correlation, principal component analysis, or regression to improve model performance based on domain knowledge. Next, split the available data into training and test sets for model evaluation. Data preprocessing is necessary to handle missing data, remove duplicates, encode categorical variables, and standardize existing features due to potential data issues.

After that, the data will be ready for regression algorithms such as Linear Regression, Random Forest, or SVR. Train the chosen algorithm using the training data, then use the testing data to make predictions and evaluate the model's performance using metrics like MAE, RMSE, or R^2 scores. If the performance is acceptable, you can use new input to make future predictions of electricity demand. Additionally, you can save the ML model for later use and visualize the results using Matplotlib tools. If the ML model's performance is deemed unacceptable, you can explore an alternative ML model, or perform hyperparameter tuning to improve the current model.

3.4 Existing Prediction and Forecasting Energy Demand Systems

Energy Forecasting (EF) is highly valuable in research on energy efficiency and sustainability. It employs various techniques to predict future patterns in energy distribution, generation, and consumption. Accurate short-term energy demand prediction in buildings and districts is crucial for reducing greenhouse gas emissions. The study in Runge and Saloux [33] explored the use of artificial intelligence techniques to forecast the heating load of a district heating system. It compared prediction and forecasting approaches using various ML and DL algorithms. The study utilized actual weather forecasts from Canadian meteorological sources and data from a Quebec-based district heating system. Hyperparameters for each model were tuned using grid search. The findings indicated that the prediction-based approach was more accurate than the forecasting strategy, with errors not exceeding 16% RMSE and often around 10% RMSE for the best models. All ML models demonstrated good accuracy, with XGBoost and LSTM consistently performing well across different hyperparameters. Additionally, the study found that XGBoost trained significantly faster than LSTM.

As residential and building consumption accounts for half of the world's electrical energy, accurately predicting their electricity usage is essential for the efficient and reliable operation of building microgrids, benefiting both users and utilities. The authors Sekhar and Dahiya [34] proposed a novel optimal hybrid technique that combines bilateral long short-term memory (BiLSTM), gray wolf optimization (GWO), and CNN for building load prediction. GWO was used to determine the optimal parameters for the CNN and BiLSTM algorithms. A one-dimensional CNN was applied to efficiently extract features from time series data. The performance of this technique was tested on four buildings with unique characteristics, using hourly resolution data. The results demonstrated that this method can be effectively applied to other architectures, showing improved accuracy compared to other state-of-the-art technologies for one, two, and three-day forecasts, including ordinary CNN-LSTM, CNN-BiLSTM, optimized BiLSTM, and conventional LSTM and BiLSTM procedures.

The study in Cai et al. [35] proposed an optimal approach for parameter fitting in the SVR model by examining six meta-heuristic optimization algorithms. The investigation of correlation and error parameters revealed that the hybrid model SVR-AEO performs best for simulating the heating and cooling loads of residential buildings. The findings indicated that the R^2 values for the cooling and heating load predictions on the training data were 0.9975 and 0.99955, respectively.

DL algorithms have demonstrated potential in learning load patterns characterized by high volatility and uncertainty. However, many researches have overlooked the empirical assessment of different DL model configurations to improve prediction accuracy. In Morteza et al. [36], the authors explored and modified various deep recurrent neural network (DRNN) architectures to forecast medium- and long-term energy demand. Their goal was to develop a custom DRNN for predicting heating and power usage with a 1-h resolution. They conducted a comprehensive analysis of hyperparameter optimization, which is particularly challenging in DL algorithms due to its empirical nature and reliance on specific applications. Their research involved exploring various configuration variables to identify a set of hyperparameters that enhanced the accuracy of the DRNN model's energy consumption forecast. Their proposed approach was compared with two alternative data-driven models: gradient boosting (GB) regression and support vector regressor (SVR). The suggested model outperformed SVR by 5.4% and GB regression by 7.0% in forecasting energy consumption.

In a study conducted by Shekhar et al. [37], convolutional neural networks were utilized to forecast future power demand. The performance of the models was evaluated using metrics such as mean absolute error, mean square error, root mean square error, and mean constant percentage error. The models were then ranked based on these metrics. The CNN model demonstrated superior accuracy in energy consumption forecasts, with an average absolute percentage deviation of 1.71%, a root mean square error of 0.23%, and a mean absolute error of 0.015%. These results were compared to those of other models. The proposed algorithm was tested using data from the International Renewable Energy Agency on solar power production, wind turbine (WT) power generation, and the "ensue" dataset, which includes hourly power consumption data from the Pennsylvania, New Jersey, and Maryland interconnection.

For many countries, managing energy demand and reducing greenhouse gas emissions are two major challenges. Accurately predicting energy consumption and CO_2 emissions is crucial for addressing these issues. The transportation sector is particularly important due to its substantial contribution to energy use and CO_2 emissions. To improve accuracy in forecasting energy demand and CO_2 emissions in this transportation sector, a hybrid methodology was employed. The approach proposed in Javanmard et al. [38] combined data-driven ML algorithms with a multi-objective mathematical model. Sensitivity analyses were used to evaluate the impact of different energy resource consumption on CO_2 emissions, both individually and collectively. Case studies using national statistics from Canada projected a 34.72% increase in energy demand and a 50.02% increase in CO_2 emissions in the transportation sector between 2019 and 2048. The results also highlighted the varied effects of different energy sources: A 5% increase in energy demand for electricity, gas, oil, and renewable energy would lead to changes in CO_2 emissions of +4.8%, +0.315%, +0.28%, and −0.51%, respectively.

The study in Nazir et al. [39] proposed a Temporal Fusion Transformer model for accurate daily, weekly, and monthly energy consumption forecast. By leveraging primary data sources and advanced batch training techniques, the Temporal Fusion Transformer model outperformed other established models such as LSTM, LSTM interpretable, and Temporal Convolutional Network. With a MSE of 4.09, a RMSE of 2.02, and a MAE of 1.50, the Temporal Fusion Transformer model excels in performance.

Predicting energy and peak power over the long term is crucial for effective power system design. The challenge arises from the fact that utilities often use different forecasting methods to estimate peak power demand and long-term energy consumption. In a study referenced as Zhang et al. [40], a comprehensive long-term forecasting model was proposed, utilizing multiple sequential configurations of the eXtreme Gradient Boosting technique to tackle the issue. The model first forecasted monthly energy consumption one to three years in advance, incorporating external factors such as macroeconomic and meteorological conditions. An input-output sequential technique was utilized, with numerous inputs and outputs that depend on the energy consumption profile. The estimated energy usage was then served as an influencing input in a multivariate model to forecast monthly peak power demand over the same time frame. To address the highly variable nature of peak power consumption, a hybrid direct-recursive sequential arrangement was implemented. By using the predicted energy consumption data to estimate peak power demand, the model achieved improved forecasting accuracy. The model's validity was evaluated using data from the power network of New South Wales (NSW). The study also compared the results with several advanced long-term forecast models to demonstrate the superiority of the proposed long-term forecasting model.

The study conducted by Pierre et al. [41] utilized data from "Compagnie Electrique du Benin" to calculate peak electric power usage during peak hours. The study employed hybrid approaches that combined DL techniques, such as LSTM, gated recurrent unit (GRU), and traditional time series prediction techniques like autoregressive integrated moving average ARIMA to forecast peak electric power consumption. In the methodology, the trend component was modeled using the ARIMA technique, while the fluctuation term was interpreted using DL techniques. The results showed that the hybrid ARIMA-LSTM strategy outperformed the ARIMA-GRU hybrid approach in terms of prediction performance, achieving a RMSE of 7.35 compared to 9.60. It was evident that the hybrid methods outperformed the individual methods, including GRU, LSTM, and ARIMA, with respective RMSE values of 18.11, 18.74, and 49.90.

A new two-stage forecasting approach, consisting of model construction and data preprocessing, was presented in Abbasimehr and Reza Paki [42]. In this approach, input data was preprocessed to extract statistical features, and the significance of each feature was determined using an XGBoost regressor. The forecasting model was then constructed using the original input data along with the extracted features. This approach utilized two cutting-edge DL models, Temporal Convolution Neural Network and Multi-Head Attention, to develop three distinct forecasting models. Two datasets measuring renewable energy consumption were used to evaluate the suggested method. The experiments showed that adding features improved the performance of DL models based on multi-head attention and temporal convolution neural networks.

Accurate forecasting of electricity demand requires reliable, long-term data collection systems supported by smart grid technologies. To effectively anticipate grid demand, a large number of smart meters are necessary to gather sufficient data. However, basic and multi-dimensional power aggregation methods may struggle to capture the relationships between different user categories. As a result, the planning, establishment, and implementation of networks for renewable energy systems, as well as the continuous supply of electricity to users, heavily rely on precise energy demand forecasts. Additionally, the planning of storage devices and their capacity heavily depends on this data. Hour-by-hour forecasting errors can lead to significant financial losses and affect customer satisfaction. By utilizing two forecasting techniques—ARIMA and Artificial Neural Network (ANN)—the study of Tarmanini et al. [43] tried to close the knowledge gap in ML-based techniques for load prediction. The performance of these approaches was compared using Mean Absolute Percentage Error (MAPE). The study was conducted using daily actual load electricity data for 709 randomly selected houses in Ireland over an 18-month period. The outcomes show that for the nonlinear load data, the (ANN) outperforms ARIMA in terms of prediction accuracy. A study by Mounir et al. [44] proposed a forecasting technique using bidirectional LSTM and empirical mode decomposition (EMD) to tackle this issue. EMD proved to be a reliable and powerful tool for time–frequency analysis and signal preprocessing, as it decomposes the time series into components at various resolutions. The proposed model successfully predicted electricity usage for the next 24 h with a 15-min resolution. This was achieved by creating multiple stationary component sequences from the original stochastic electricity usage time series data (IMFs). To predict each Intrinsic Mode Function, a hybrid model BI-LSTM was used. The results of each component's forecast were then combined to provide an overall forecast. Two comparative studies were conducted to justify the choice of the signal processing method and the prediction algorithm. The proposed model demonstrated a minimal MAPE of 0.28% and a value of R_squared closer to 1 compared to other papers.

Energy security can be shaped by examining the factors that influence energy demand and forecasting energy use. The demand for various types of energy utilized in China is estimated in Rao et al. [45]. A two-stage approach called least absolute shrinkage and selection operator-random forest (Lasso-RF) was presented to determine the factors driving energy consumption. Afterward, a support vector regression-compositional data second exponential smoothing (SVR-CDSES) model was created to predict the demand for primary electricity, coal, oil, and natural gas. According to the findings of the empirical investigation, China's energy consumption would be increased at an annualized rate of 2.68% over the next ten years. After an initial rise, the demand for coal is expected to decline annually at a rate of 0.79%. The demand for oil would be expanded at a comparatively steady rate of 3.52% per year. On the other hand, the demand for primary electricity and natural gas would be increased at the fastest annual growth rate of 8.05%, indicating a growing awareness of clean energy.

4 Benefits of Accurately Predicting Energy Demand for Sustainability

Anticipating and forecasting energy consumption is essential for effective energy management. Accurately predicting energy requirements has never been more critical, as demand for sustainable and efficient energy solutions continues to rise.

Table 2 summarizes the main benefits of forecasting energy consumption for sustainability, as detailed in Senyapar et al. [46].

In summary, precise demand forecasting is crucial for optimizing resource utilization and minimizing waste, leading to a more sustainable future. By aligning supply with projected demand, it paves the way for the widespread adoption of renewable energy sources. This strategic approach not only curbs carbon emissions but also reduces our dependence on fossil fuels, significantly impacting on environmental sustainability.

Table 2 Primary advantage of energy consumption forecasting

Benefit	Description
Optimized resource allocation	Precise forecasts of energy demand aid utilities and energy suppliers allocate resources more efficiently, guaranteeing that energy is generated and supplied according to actual needs. This leads to more efficient power generation, positively impacting costs, the environment, and resource utilization
Enhanced integration of renewable energy	Renewable energy sources can be periodic, like solar and wind. Accurate demand forecasts promote efficient planning and the integration of these renewable energy sources into the grid by synchronizing their output with expected consumption trends
Improve grid stability	Load Balancing: Precise forecasts are crucial for ensuring that the grid can manage a balanced load, thus preventing overloading and greatly reducing the likelihood of blackouts or other system malfunctions Integration of Renewables: Despite the unpredictability and variability of renewable energy sources, more accurate demand forecasting is key to successfully integrating them into the system
Decrease environmental impact	Reduced Emissions: By reducing reliance on high-emission power plants, energy generation can be optimized to meet actual demand, which helps to reduce greenhouse gas emissions. Efficient Utilization of Renewables: By minimizing the usage of non-renewable energy sources, accurate demand prediction can help maximize the use of clean energy sources
Improved energy management	Demand Response: Reducing or shifting energy consumption during peak hours is encouraged by demand response programs, which further balance the grid and ease the burden on energy supplies. These programs are made possible by accurate demand projections Demand projections can help utilities and policymakers plan more strategically for investments and infrastructure upgrades, ensuring that resources are allocated where they are most needed
Adaptability and resilience	Adapting to Change: Enhanced forecasting helps in adapting to fluctuations in energy consumption resulting from economic, demographic, or technological advancements Crisis Preparedness: Accurate predictions enable better preparedness for unexpected events or disruptions, such as sudden changes in demand or natural disasters
Customer benefits	Stable Pricing: Enjoy consistent and predictable energy costs with reduced supply volatility Informed Decision-Making: Empower yourself to lower energy expenses by using tools and programs that offer incentives for adjusting energy consumption based on demand forecasts

5 Conclusion

The necessity for a transition from carbon-based energy sources to renewable alternatives has become increasingly more apparent in recent years, driven by the growing demand for electricity resulting from technological advancements and the development of smart cities aimed at preventing environmental degradation. Long-term societal advancement depends on sustainability, which includes resource conservation, environmental protection, social fairness, economic stability, and public health. Forecasting energy consumption is essential to sustainability as it improves grid stability, optimizes resource planning, and efficiently incorporates renewable energy sources. AI, including ML and DL, is transforming energy demand forecasting by providing accurate predictions and improving energy efficiency. This, in turn, is boosting efforts to mitigate climate change and promote sustainable behaviors. This chapter explores the critical factors influencing energy demand. Extreme temperatures drive up the need for heating or cooling, with weather data, including temperature, humidity, and precipitation, playing a pivotal role in energy consumption. Anticipated temperature changes in the future are expected to decrease heating energy usage while increasing cooling energy consumption. Seasonal variations also impact energy needs, with winter and summer driving higher heating and cooling consumption, respectively. Additionally, population growth, economic expansion, industrialization, urbanization, and improved living standards all contribute to increased energy demand. These dynamics are further shaped by energy prices and technological advancements, influencing consumption patterns, energy source preferences, and efficiency. Furthermore, the chapter reviews existing AI-based energy demand forecasting systems and underscores the immense advantages of utilizing such systems for sustainability. Accurate demand forecasting is crucial for maximizing resource utilization, reducing waste, and building a more sustainable future. By balancing supply and demand, it paves the way for the widespread adoption of renewable energy sources. This calculated approach is pivotal for environmental sustainability, leading to reduced carbon emissions and decreased reliance on fossil fuels.

The implementation of AI in energy demand forecasting has the potential to significantly enhance efficiency and precision in the future. Through the integration of diverse data sources and the analysis of intricate data patterns, AI will elevate the accuracy of predictions. Additionally, it will facilitate real-time adjustments through smart grids and Internet of Things devices. Anticipating output and aligning it with demand will optimize the utilization of renewable energy and reduce reliance on fossil fuels. Moreover, by analyzing individual consumption patterns, AI will offer tailored energy solutions and adjust forecasts based on meteorological and climatic data to aid in managing energy demand amid changing environmental conditions.

References

1. Nasserddine, G., Nassereddine, M., & El Arid, A. A. (2023a). Internet of things integration in renewable energy systems. In *Handbook of research on applications of AI, Digital Twin, and Internet of Things for sustainable development* (pp. 159–185). IGI Global.
2. Renna, P., & Materi, S. (2021). A literature review of energy efficiency and sustainability in manufacturing systems. *Applied Sciences, 7366.*
3. Velenturf, A. P. M., & Purnell, P. (2021). Principles for a sustainable circular economy. *Sustainable Production and Consumption*, 1437–1457.
4. Nassereddine, M., Nassreddine, G., & El Arid, A. (2024a). Hybrid photovoltaic and biogas system for stable power system. *Next Energy*, 100172.
5. Khan, I., et al. (2022). Linking energy transitions, energy consumption, and environmental sustainability in OECD countries. *Gondwana Research*, 445–457.
6. Al Kuwaiti, A., et al. (2023). A review of the role of artificial intelligence in healthcare. *Journal of Personalized Medicine*, 951.
7. Nassereddine, M., et al. (2024b). Artificial intelligence applications in renewable power systems. In *Industrial applications of Big Data, AI, and Blockchain* (pp. 26–57). IGI Global.
8. Antonopoulos, I., et al. (2020). Artificial intelligence and machine learning approaches to energy demand-side response: A systematic review. *Renewable and Sustainable Energy Reviews*, 109899.
9. Deb, C., et al. (2017). A review on time series forecasting techniques for building energy consumption. *Renewable and Sustainable Energy Reviews*, 902–924.
10. Román-Portabales, A., López-Nores, M., & Pazos-Arias, J. J. (2021). Systematic review of electricity demand forecast using ANN-based machine learning algorithms. *Sensor*, 4544.
11. Marino, D. L., Amarasinghe, K., & Manic, M. (2016). Building energy load forecasting using deep neural networks. In *IECON 2016-42nd annual conference of the IEEE industrial electronics society*. IEEE.
12. Ghazal, T. M. (2022). Energy demand forecasting using fused machine learning approaches. *Intelligent Automation & Soft Computing*, 539–553.
13. Hasanuzzaman, M., et al. (2020). Energy demand. In *Energy for sustainable development* (pp. 41–87). Academic Press.
14. Szulgowska-Zgrzywa, M., et al. (2020). Impact of users' behavior and real weather conditions on the energy consumption of tenement houses in Wroclaw, Poland: Energy performance gap simulation based on a model calibrated by field measurements. *Energies*, 6707.
15. Reveshti, A. M., Ebrahimpour, A., & Razmara, J. (2023). Investigating the effect of new and old weather data on the energy consumption of buildings affected by global warming in different climates. *International Journal of Thermofluids*, 100377.
16. Trotter, I. M., et al. (2016). Climate change and electricity demand in Brazil: A stochastic approach. *Energy*, 596–604.
17. Cui, Z., et al. (2023). A novel deep learning framework with a COVID-19 adjustment for electricity demand forecasting. *Energy Reports*, 1887–1895.
18. BP Statistical Review of World Energy, 70th ed. (2021). Retrieved 22 August 2024, from https://www.bp.com/en/global/corporate/energy-economics/statistical-review-of-world-energy.html
19. Shaari, M. S., et al. (2023). The linkage between natural gas consumption and industrial output: New evidence based on time series analysis. *Energy*, 129395.
20. Stern, David I., Paul J. Burke, & Stephan B. Bruns (2019). The impact of electricity on economic development: A macroeconomic perspective.
21. Ayad, H., et al. (2023). The impact of economic uncertainty, economic growth and energy consumption on environmental degradation in MENA countries: Fresh insights from multiple thresholds NARDL approach. *Environmental Science and Pollution Research*, 1806–1824.

22. Osabuohien-Irabor, O., & Drapkin, I. M. (2022). The impact of technological innovation on energy consumption in OECD economies: The role of outward foreign direct investment and international trade openness. *International Journal of Energy Economics and Policy*, 317–333.
23. Guan, Y., et al. (2023). Burden of the global energy price crisis on households. *Nature Energy*, 304–316.
24. Rukhiran, M., Boonsong, S., & Netinant, P. (2024). Sustainable optimizing performance and energy efficiency in proof of work Blockchain: A multilinear regression approach. *Sustainability*, 1519.
25. El Samad, M., Nasserddine, G., & Kheir, A. (2023). Introduction to artificial intelligence. In *Artificial intelligence and knowledge processing* (pp. 1–14). CRC Press.
26. Le, T. T., et al. (2024). Harnessing artificial intelligence for data-driven energy predictive analytics: A systematic survey towards enhancing sustainability. *International Journal of Renewable Energy Development*, 270–293.
27. Zoppi, T., et al. (2023). Which algorithm can detect unknown attacks? Comparison of supervised, unsupervised and meta-learning algorithms for intrusion detection. *Computers & Security*, 103107.
28. James, G., et al. (2023). *Unsupervised learning. An introduction to statistical learning: with applications in Python* (pp. 503–556). Springer International Publishing.
29. Taye, M. M. (2023). Understanding of machine learning with deep learning: Architectures, workflow, applications and future directions. *Computers*, 19.
30. Griffis, E. J., et al. (n.d.). Lyapunov-based long short-term memory (Lb-LSTM) neural network-based control. *IEEE Control Systems Letters*.
31. Kapp, S., Choi, J.-K., & Hong, T. (2023). Predicting industrial building energy consumption with statistical and machine-learning models informed by physical system parameters. *Renewable and Sustainable Energy Reviews, 113045*.
32. Nasserddine, G., El Arid, A., & Nasseredine, M. (2023b). Solar PV power prediction system based on machine learning approach. In *IEEE International Conference on Energy Technologies for Future Grids (ETFG)* (pp. 1–7). IEEE.
33. Runge, J., & Saloux, E. (2023). A comparison of prediction and forecasting artificial intelligence models to estimate the future energy demand in a district heating system. *Energy*, 126661.
34. Sekhar, C., & Dahiya, R. (2023). Robust framework based on hybrid deep learning approach for short term load forecasting of building electricity demand. *Energy*, 126660.
35. Cai, W., et al. (2023). Predicting the energy consumption in buildings using the optimized support vector regression model. *Energy*, 127188.
36. Morteza, A., et al. (2023). Deep learning hyperparameter optimization: Application to electricity and heat demand prediction for buildings. *Energy and Buildings*, 113036.
37. Shekhar, H., et al. (2023). Demand side control for energy saving in renewable energy resources using deep learning optimization. *Electric Power Components and Systems*, 2397–2413.
38. Javanmard, M. E., et al. (2023). Forecast energy demand, CO_2 emissions and energy resource impacts for the transportation sector. *Applied Energy*, 120830.
39. Nazir, A., et al. (2023). Forecasting energy consumption demand of customers in smart grid using Temporal Fusion Transformer (TFT). *Results in Engineering*, 100888.
40. Zhang, T., et al. (2023). Long-term energy and peak power demand forecasting based on sequential-XGBoost. *IEEE Transactions on Power Systems*, 3088–3104.
41. Pierre, A. A., et al. (2023). Peak electrical energy consumption prediction by ARIMA, LSTM, GRU, ARIMA-LSTM and ARIMA-GRU approaches. *Energies*, 4739.
42. Abbasimehr, H., & Reza Paki, A. B. (2023). A novel XGBoost-based featurization approach to forecast renewable energy consumption with deep learning models. *Sustainable Computing: Informatics and Systems*, 100863.
43. Tarmanini, C., et al. (2023). Short term load forecasting based on ARIMA and ANN approaches. *Energy Reports*, 550–557.
44. Mounir, N., Ouadi, H., & Jrhilifa, I. (2023). Short-term electric load forecasting using an EMD-BI-LSTM approach for smart grid energy management system. *Energy and Buildings*, 113022.

45. Rao, C., et al. (2023). Energy demand forecasting in China: A support vector regression-compositional data second exponential smoothing model. *Energy*, 125955.
46. Senyapar, D., Nurgul, H., & Aksoz, A. (2024). Empowering sustainability: A consumer-centric analysis based on advanced electricity consumption predictions. *Sustainability*, 2958.

Converging Intelligence and Energy Efficiency in Cloud Environment for Efficient Resource Allocation for Sustainable SME Growth

Deresse Demeke Shallo, Durga Prasad Sharma, and Pethuru Raj Chelliah

1 Introduction

Convergence, the blending of technologies, industries, or markets, is increasingly recognized as a powerful driver of innovation. Convergence creates a fusion framework for decision-making and problem-solving not only for science and technology but also for business and management of technologies. This phenomenon is evident in a wide range of sectors, from information technology. To better understand and harness the potential of convergence, researchers in Technology and Innovation Management have designed its various aspects.

The convergence of Drivers and Patterns area explores the underlying factors that trigger convergence, such as technological advancements, market demands, and policy changes. Researchers investigate how these factors shape the trajectory of convergence and the emergence of new industries and products. The Anticipation of Convergence focuses on the ability of organizations to predict and prepare for future convergence events. By anticipating these shifts, firms can proactively develop strategies to capitalize on emerging opportunities and mitigate potential risks, such as the convergence of AI, ML, and Cloud. Also, the Strategic Reactions to Convergence examines how organizations respond to convergence, whether through internal innovation, external partnerships, or acquisitions. Effective strategic responses can help organizations maintain a competitive edge and adapt to the rapidly changing landscape. In addition, the Converging Products and services such

D. D. Shallo
Arba Minch Institute of Technology, Arba Minch University, Arba Minch, Ethiopia

D. P. Sharma (✉)
Arba Minch University, working under UNDP and MAISM under RTU, Arba Minch, Ethiopia

P. R. Chelliah
Edge AI Division, Reliance Jio Platforms Ltd., Bangalore, India

P. Raj et al. (eds.), *Artificial Intelligence (AI) for IT Energy Efficiency and Green AI for Environment Sustainability*,
https://doi.org/10.1007/978-3-031-89420-6_28

as AI and Cloud investigate the development and commercialization of products that integrate technologies from multiple domains. Researchers explore the challenges and opportunities associated with creating and marketing such innovative products and services [1].

While significant progress has been made in understanding convergence, the field still faces several challenges. One key limitation is the inward-focused nature of much of the research, which often neglects the broader theoretical context. To address this, a more comprehensive framework is needed to guide future research [2, 3].

This article aims to advance the understanding of convergence and its implications for innovation with special reference to AI, ML, and Cloud for Small and Medium Scale Enterprises (SMEs). By exploring emerging areas such as the role of energy efficiency ecosystems, the impact of AI, and ML technologies, and the Cloud computing convergence, both theoretical development and practical applications can be explored for efficient and effective resource allocations in SMEs.

Cloud computing has fundamentally transformed the way businesses operate, providing scalable, on-demand access to a range of computing resources such as storage, servers, and software over the internet. By enabling flexible resource allocation, it has allowed organizations of all sizes particularly small and medium-sized enterprises (SMEs) to enhance their energy efficiency, streamline operations, add on features in a computing environment, and reduce costs. However, as the cloud computing industry grows, concerns over its environmental impact have escalated. Data centers, which are the backbone of cloud services, are substantial energy consumers and contribute significantly to global carbon emissions. This has brought the sustainability of cloud computing into sharp focus, urging businesses to explore green technology solutions within this digital infrastructure [4].

To address these concerns, the convergence of intelligent resource allocation and green technology presents an innovative approach to achieving both operational efficiency and environmental sustainability. By incorporating artificial intelligence (AI) and machine learning (ML) for smart resource management, alongside environmentally conscious practices, cloud providers and users alike can drive significant reductions in energy consumption [5, 6].

Cloud computing has revolutionized the way businesses operate, offering on-demand access to computing resources like storage, servers, and software over the internet. This change in basic assumptions has led to increased efficiency, scalability, and cost savings for organizations of all sizes. However, the rapid growth of cloud computing has also raised concerns about its environmental impact [7]. Data centers, which power cloud services, consume significant amounts of energy and generate substantial carbon emissions. This has prompted a growing focus on green technology in cloud computing, aiming to reduce energy consumption and minimize the industry's carbon footprint.

Several energy efficiency strategies are being employed to make cloud computing more sustainable [8]:

- *Energy-Efficient Data Centers:* Data centers are being designed with energy efficiency in mind, incorporating technologies like advanced cooling systems, high-efficiency servers, and renewable energy sources.
- *Virtualization and Consolidation:* Virtualization allows multiple virtual machines to run on a single physical server, reducing hardware requirements and energy consumption.
- *Renewable Energy Integration:* Data centers are increasingly powered by renewable energy sources like solar and wind power, further reducing their carbon footprint.
- *Intelligent Resource Allocation:* Intervention of AI and machine learning algorithms are used to optimize resource utilization, ensuring that resources are allocated efficiently and minimizing energy waste.

The convergence of cloud computing and green technology offers a promising path toward a more sustainable digital future. By adopting these strategies, businesses can explore the benefits of cloud computing while minimizing their environmental impact. As cloud computing continues to evolve, it is essential to prioritize sustainability and adopt green practices to ensure a positive impact on the environment.

This chapter explores the potential of these technologies to support sustainable growth for SMEs, focusing on strategies for efficient resource allocation and the role of green technology in a cloud computing environment.

2 Brief Overview of Cloud Computing and Its Benefits for SMEs

Cloud computing refers to the delivery of computing services over the Internet. Instead of relying on local servers and data centers, businesses access resources like storage, databases, networking, software, and analytics from remote cloud servers. This shift has transformed how businesses operate, offering numerous benefits, especially for Small and Medium-sized Enterprises (SMEs).

2.1 Key Benefits of Cloud Computing for SMEs

Cloud computing facilitates several key benefits to the users by harnessing the full potential of computing resources. Some of the important benefits are mentioned below:

1. **Cost-Effectiveness:** Cloud computing offers a cost-effective solution by enabling a pay-as-you-go model, where businesses pay only for the resources they use, minimizing upfront investments. It reduces infrastructure and operational costs, as cloud providers handle hardware, maintenance, and updates.

Overall, cloud computing helps businesses reduce costs, improve productivity, and focus on their core operations.

2. **Scalability:** Cloud scalability is the capability of cloud systems to adjust resources (computing power, storage, network capacity) based on demand. It includes two main types:

(**A**) **Vertical Scalability (scaling up)**: Adding power to a single resource, like upgrading a server's CPU or RAM. Ideal for workloads manageable by one powerful machine.

(**B**) **Horizontal Scalability (scaling out):** Adding servers to distribute tasks across multiple machines, suitable for parallel processing. Benefits of cloud scalability include cost-effectiveness, flexibility to adapt to demand changes, resource efficiency, reliability, and a focus on innovation. Cloud scalability enhances agility, efficiency, and cost-effectiveness for businesses.

3. **Improved Productivity:** Cloud computing has transformed business productivity by enabling enhanced collaboration, flexibility and scalability, automation, data security and disaster recovery, and cost savings. These benefits make cloud computing a powerful driver of efficiency, innovation, and productivity for businesses.
4. **Enhanced Security:** Cloud computing offers significant security enhancements compared to traditional on-premises systems. Cloud service providers employ specialized security teams dedicated to threat monitoring and incident management, exploring expertise often inaccessible to smaller businesses. Advanced security infrastructures, including state-of-the-art firewalls and encryption technologies, provide robust defenses against cyber threats. Cloud service providers also adhere to rigorous compliance standards, frequently undergoing audits to meet certifications like ISO 27001, thus ensuring high levels of security and accountability.

Providers maintain updated security patches automatically, securing cloud systems against emerging vulnerabilities. Finally, the shared responsibility model of cloud computing ensures that while providers secure the infrastructure, clients are accountable for safeguarding their data and applications, fostering a comprehensive and resilient approach to cloud security. These features collectively highlight cloud computing as a secure, adaptable, and reliable solution for businesses.

5. **Disaster Recovery and Business Continuity:** Disaster Recovery and Business Continuity (DRBC) are critical elements of a comprehensive IT strategy, and cloud computing provides key advantages in this area.

   (A) **Disaster Recovery:** Cloud solutions facilitate rapid recovery with automated failover mechanisms, significantly reducing downtime and associated costs such as 99.99% uptime for most of the CSPs. The Recovery Time Objective (RTO) is minimized due to redundant cloud infrastructure and geographically distributed data centers, enabling quick restoration of operations post-disaster. Furthermore, frequent backups and data replication in the cloud reduces the Recovery Point Objective (RPO), ensuring minimal data loss.

(B) **Business Continuity**: Cloud-based solutions support uninterrupted business operations by allowing remote access to essential applications and data. Additionally, cloud resources can be scaled according to demand, maintaining continuity during peak periods or following a disruption. Cost-effective pricing models, such as pay-as-you-go, make cloud-based DRBC solutions more affordable than traditional on-premises alternatives.

2.2 Challenges SMEs Facing in Managing Cloud Resources

According to a study [9], small and medium-sized enterprises (SMEs) frequently encounter several challenges in the effective management of cloud resources, which can impact their ability to fully harness cloud computing's potential.

1. **Cost Management:** SMEs often struggle with monitoring and controlling cloud expenses, as fluctuating demand and unexpected usage can lead to cost overruns. Limited budgets make efficient resource allocation crucial, yet complex pricing models and scaling requirements complicate cost prediction and control.
2. **Skill Gaps:** SMEs may lack the specialized skills needed to manage cloud infrastructure, including expertise in cloud architecture, security, and optimization [10]. Recruiting or training personnel to bridge this gap often requires significant investment, which can be challenging for resource-constrained SMEs.
3. **Data Security and Compliance**: While cloud service providers offer robust security measures, SMEs are still responsible for securing their data and ensuring compliance with industry regulations. Navigating the shared responsibility model and addressing compliance requirements such as the General Data Protection Regulation (GDPR) can create significant challenges for SMEs with limited IT staff.
4. **Scalability and Resource Optimization**: Efficiently scaling resources to meet fluctuating demand without over-provisioning remains a challenge. SMEs may lack the monitoring tools and automation capabilities needed to optimize resource allocation, leading to either resource underutilization or insufficient capacity during peak times.
5. **Vendor Lock-In:** Several SMEs experience difficulties in migrating between cloud providers due to compatibility issues and the potentially high cost of data transfer, leading to dependency on a single cloud service provider. This vendor lock-in can limit flexibility and hinder strategic planning.
6. **Performance Monitoring and Management**: Effective performance monitoring of cloud applications is essential but can be complex and resource-intensive. SMEs may struggle to implement real-time monitoring and troubleshooting due to limited access to sophisticated tools or IT expertise. These challenges highlight the need for SMEs to adopt strategies that enhance their cloud management capabilities, ensuring they can efficiently and securely utilize cloud resources to support growth and innovation.

2.3 *Efficient Cloud Resource Allocation*

Intelligent systems, driven by AI and machine learning (ML), offer transformative potential for optimizing resource allocation in cloud environments. These systems can process and analyze enormous amounts of data from cloud infrastructures, identifying patterns and trends that would be challenging, if not impossible, for human administrators to detect. The potential of intelligent systems to optimize resource allocation is a crucial task of efficient and effective resource management for sustainability. Here is how they achieve these benefits:

1. **Optimized Resource Usage:** AI and ML algorithms can monitor cloud resource consumption in real time, predicting usage needs based on historical data and current trends. For example, by analyzing workloads over time, an intelligent system can anticipate peak usage periods and allocate resources accordingly, reducing the risk of under or over-provisioning. This enables cloud systems to deliver necessary resources exactly when they are required [11].
2. **Cost Reduction:** Intelligent systems can suggest or even automate actions that lower costs, such as turning off unused virtual machines, resizing storage or compute instances based on usage, and identifying underutilized resources. By balancing resource allocation dynamically, these systems reduce waste, ensuring that businesses only pay for what they need [12, 13].
3. **Improved Efficiency through Automation:** Intelligent systems use automation to handle routine tasks like scaling, patching, and fault management. Machine learning models can continuously improve from new data, adjusting strategies for greater efficiency without human intervention. Automation also reduces the time spent on manual cloud management, freeing IT personnel to focus on strategic objectives [14].
4. **Enhanced Decision-Making with Big Data Analytics:** By using big data analytics, intelligent systems aggregate and interpret data from diverse sources across the cloud environment. This can provide insights into optimal resource allocation strategies and even predict potential bottlenecks or failures before they occur, allowing for preemptive adjustments.
5. **Adaptive Systems and Continuous Learning:** Intelligent systems evolve by continuously learning from changes in usage patterns, new workloads, and user requirements. This adaptability ensures that resource allocation strategies remain effective, even as conditions and demands shift. Through these capabilities, intelligent systems enable more precise, cost-effective, and efficient management of cloud resources, allowing businesses to maximize the value of their cloud infrastructure while reducing operational overhead [15].

2.4 Intelligent Systems in Cloud Computing Environment

Machine learning (ML), as depicted in Fig. 1, is a field of AI that focuses on developing systems that can automatically learn and improve from experience without being explicitly programmed. It involves using algorithms and statistical models to analyze data, identify patterns, and make decisions or predictions. The key idea is enabling machines to improve their performance as they process more data over time [16].

1. **Self-supervised model:** It is a machine learning method that applies unsupervised learning to problems that typically call for supervised learning. In domains such as computer vision and natural language processing (NLP), where advanced AI models need to be trained on vast quantities of labeled data. The time-consuming annotation of these labeled datasets by human experts can make it prohibitively difficult to collect enough data. However, self-supervised methods can save time and money by eliminating the requirement to manually classify training data in part or in full [17].
2. **A Supervised Learning model:** It is a machine learning task that assigns a function to labeled training data. Supervised learning has two variables: input (I) and output (O). The algorithm's function is to analyze the mapping function from the input variable to the output variable, O = f(I). Supervised learning aims to examine training data and create a full function for mapping new instances. The learning method can accurately generalize class labels based on unobserved cases [18].
3. **Semi-supervised Learning:** The semi-supervised learning deals with performing specific learning tasks utilizing both labeled and unlabeled data [19]. It allows for the utilization of the vast amounts of unlabeled data that are available

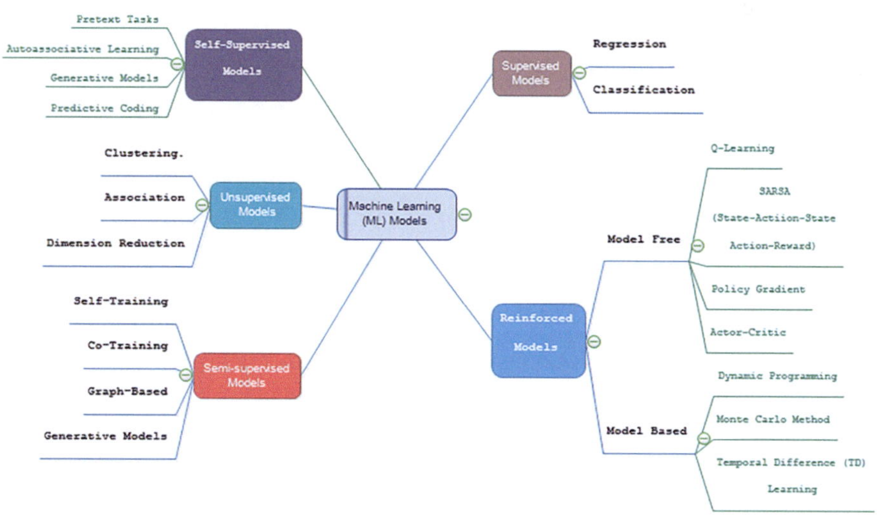

Fig. 1 Taxonomy of machine learning

in many use cases in conjunction with usually smaller collections of labeled data, and it is conceptually located between supervised and unsupervised learning [20].

4. **Unsupervised Machine Learning:** It deals with unlabeled data, meaning the data doesn't have predefined categories or labels [21, 22]. Algorithms in this field discover patterns and structures within the data on their own. Common techniques include clustering (grouping similar data points), dimensionality reduction (simplifying complex data), and anomaly detection (identifying outliers). Unsupervised learning is useful for exploratory data analysis, data compression, and feature engineering.
5. **Reinforced Models:** Reinforcement learning is a machine learning technique that includes training an artificial intelligence agent by repeating behaviors and receiving rewards. As explained by [23], reinforcement learning is the difficulty encountered by an agent that must learn behavior through trial-and-error encounters with a changing environment. There are two main approaches to solving reinforcement-learning problems. The first step is to look through the range of behaviors to identify one that works well in the situation. This approach has been taken by research into genetic algorithms and genetic programming. The second approach is to use statistics and dynamic programming methods to estimate the utility of taking actions in different states of the world. A reinforcement-learning agent experiments in its surroundings, doing actions and receiving rewards when they are correct. Over time, the agent learns which activities will maximize its reward [24].

The integration of Machine Learning (ML) and AI with cloud computing has revolutionized the way businesses operate and innovate [25]. This synergy has led to the development of intelligent cloud solutions that offer enhanced efficiency, scalability, and security.

2.5 Applications of AI/ML in Cloud Computing

As depicted in Fig. 2, AI or ML has many advantages, including predictive analysis, automated response allocation, personalized user experience, optimized data management, and security enhancements. In the cloud environment, over-provisioning is frequently used to prevent quality of service (QoS) degradation and service level agreement (SLA) violations, but it needs intelligent decisions.

1. **Predictive Analytics**: Cloud computing has many advantages, including scalability and high availability. However, achieving these benefits necessitates efficient resource provisioning and scaling to align resources with changing workloads, which can be difficult. Over-provisioning is frequently used to prevent quality of service (QoS) degradation and service level agreement (SLA) violations, but it incurs additional expenses and energy usage.

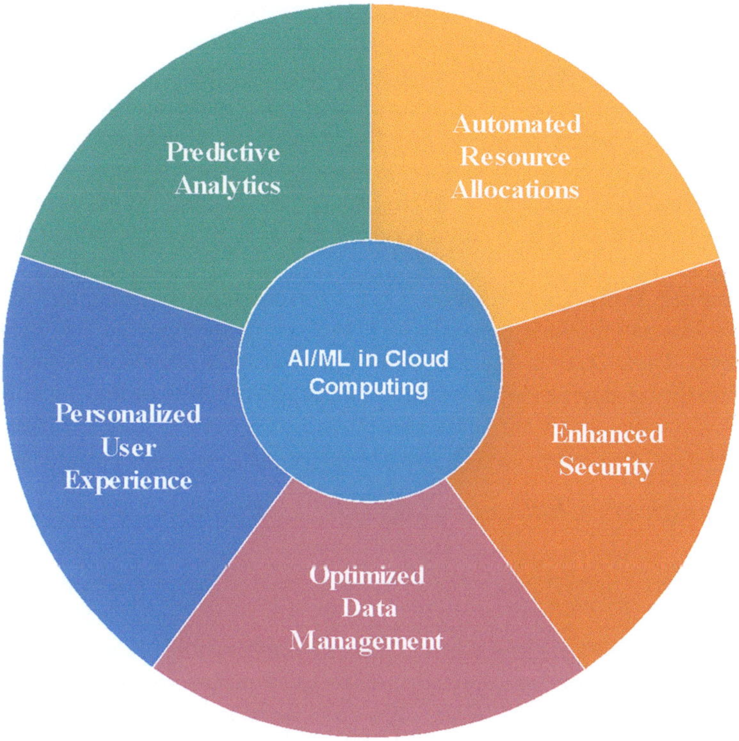

Fig. 2 Applications of AI/ML in Cloud Computing

ML algorithms examine previous data to forecast future resource demands in advance, enabling proactive capacity planning and cost minimization. By finding trends in system logs and performance measurements, ML may forecast possible faults, allowing for timely maintenance and preventing downtime [26]. Workload prediction, made possible by machine learning techniques, presents a feasible option for enhancing efficiency and minimizing operational costs [27].

2. **Automated Resource Allocation:** According to [26], ML algorithms enhance managed cloud systems by enabling dynamic resource allocation, which adjusts computational resources based on real-time demand. These systems detect and resolve issues like server failures and network bottlenecks, reducing disruptions. By predicting workload patterns, ML ensures applications run smoothly, optimizing performance while minimizing costs through efficient resource utilization. This adaptability aligns resources with actual business needs, improving system resilience and responsiveness.
3. **Enhanced Security:** Machine learning in cloud computing security can automate threat detection, reduce manual monitoring, and improve overall security efficacy at a low cost [28]. ML models can identify unusual patterns in network traffic and user behavior, flagging potential security threats. AI-driven systems

can analyze threat intelligence feeds and security logs to proactively protect against cyberattacks.

4. **Personalized User Experiences:** The personalized user experience can be realized using two popularly known techniques, i.e., 1) recommendation systems, and 2) natural language processing (NLP). In a cloud environment, ML algorithms can analyze user behavior and preferences to recommend relevant products, services, or content. Similarly, AI-powered NLP tools enable natural language interactions with cloud services, improving user experience and accessibility [29].
5. **Optimized Data Management:** Data Classification and Tagging: ML algorithms can automatically classify and tag data, facilitating efficient search and retrieval. Data Quality Assurance: AI-powered tools can identify and correct data quality issues, improving data accuracy and reliability [30].

2.6 Significant Challenges and Considerations in AI and ML

AI and ML have several potentials to harness their power in different sectors and enhance the existing system performance and operations. In addition to merits, it has numerous challenges as well. Some of the important challenges are explained below: (1) Data Quality and Quantity: High-quality data is essential for training effective AI/ML models, (2) Model Bias and Fairness: ML models must be carefully designed and trained to avoid biases and discriminatory outcomes, and (3) Ethical Implications: AI and ML raise ethical concerns related to privacy, security, and job displacement. By addressing these challenges and exploring the power of AI and ML, organizations can harness the full potential of cloud computing to drive innovation and achieve sustainable growth.

2.7 Energy-Efficient Technology Convergence with Cloud Computing [31]

Green cloud computing is a sustainable approach to cloud computing that prioritizes energy efficiency and environmental responsibility. By exploring technological advancements and eco-friendly practices, this convergence aims to reduce the carbon footprint associated with digital infrastructure. The use of data centers has been significant in the last two decades, especially cloud data centers. The energy-efficient cloud data centers are designed to minimize energy consumption while maintaining optimal performance.

2.7.1 Key Strategies for Energy-Efficient Data Centers [32]

(A) **Server Virtualization:** Consolidating multiple virtual servers onto fewer physical servers reduces energy consumption and hardware requirements.
(B) **Resource Sharing:** Enabling multiple users to share resources like servers and storage optimizes utilization and reduces energy waste.
(C) **Efficient Cooling Systems**: Using advanced cooling technologies like liquid cooling and air-side economization to reduce energy usage.
(D) **High-Efficiency Servers**: Employing servers with energy-efficient processors and power supplies.
(E) **Power Management**: Implementing power management techniques to reduce energy consumption during idle periods.
(F) **Optimization**: Optimizing data center layout, airflow, and cable management to improve energy efficiency.
(G) **Intelligent Resource Allocation:** Using AI and machine learning to optimize resource allocation and reduce energy waste.
(H) **Renewable Energy Sources:** Utilizing renewable energy sources like solar and wind power to power data centers minimizes reliance on fossil fuels.

2.7.2 Renewable Energy Integration in Cloud Computing

Cloud computing involves utilizing renewable energy sources, such as solar, wind, and hydropower, to power data centers. This helps reduce the carbon footprint of cloud services and promotes sustainable practices. Energy-efficient cloud data centers are designed to minimize energy consumption by renewable energy integration. The following practices are used to explore renewable energy in data centers [33]:

(A) **Direct Power Purchase Agreements (PPAs)**: Cloud providers can directly purchase renewable energy from renewable energy producers.
(B) **Renewable Energy Certificates (RECs)**: By purchasing RECs, cloud providers can support renewable energy projects without directly investing in generation infrastructure.
(C) **On-Site Renewable Energy Generation:** Installing solar panels or wind turbines on data center premises to generate clean energy.
(D) **Energy Storage Systems:** Using batteries or other energy storage technologies to store excess renewable energy for later use, addressing the intermittent nature of renewable sources.
(E) **Demand Response Programs**: Participating in demand response programs to adjust energy consumption based on grid conditions, optimizing energy usage, and supporting renewable energy integration. By adopting these strategies, cloud providers can contribute to a more sustainable future and reduce their reliance on traditional fossil fuel-based energy sources.

2.7.3 Virtualization and Consolidation Techniques in Cloud Computing

Virtualization is a technology that allows multiple operating systems to run concurrently on a single physical machine over the Cloud. This creates virtual machines (VMs), each with its own resources like CPU, memory, and storage. Virtualization enables efficient resource utilization and flexibility in managing server infrastructure.

Consolidation involves combining multiple physical servers into fewer, more powerful servers. This reduces the number of physical servers needed over the cloud, leading to lower energy consumption, reduced cooling costs, and simplified management [34].

2.8 Key Benefits of Virtualization and Consolidation over Cloud Data Centers

(A) *Improved resource utilization: B*y consolidating multiple workloads onto fewer physical servers, resources like CPU, memory, and storage can be utilized more efficiently in the cloud environment.
(B) *Reduced Energy Consumption:* Fewer physical servers mean lower energy consumption for cooling and powering hardware and it is possible over cloud data centers where resources are shared.
(C) *Increased Flexibility:* Virtualization allows for rapid deployment of new servers and applications over the cloud, as well as easy scaling of resources to meet changing demands.
(D) *Enhanced Fault Tolerance*: Virtual machines over the cloud data centers can be easily migrated between physical servers, minimizing downtime in case of hardware failures.
(E) *Simplified Management*: Centralized management of virtualized environments simplifies administration and maintenance tasks in cloud computing data centers without complexity.

By effectively implementing virtualization and consolidation techniques over the cloud, organizations such as SMEs can significantly reduce their operational costs, improve performance, and enhance overall IT efficiency.

2.8.1 Efficient Resource Allocation Strategies over Cloud

(A) *Intelligent Resource Provisioning:* Intelligent Resource Provisioning (IRP) in cloud computing involves the use of AI and machine learning techniques to automate and optimize the allocation of resources like CPU, memory, and storage. This approach aims to improve resource utilization, reduce costs, and enhance the overall performance of cloud services [35].

(B) *Dynamic Resource Allocation:* Dynamic Resource Allocation (DRA) in cloud computing is a technique that allows for the automatic adjustment of resources like CPU, memory, and storage to meet the fluctuating demands of applications and services. This ensures optimal resource utilization and performance while minimizing costs [26, 36].

2.9 Dynamic Resource Allocation (DRA) in Cloud Environment

- *On-demand provisioning:* Resources over the cloud are allocated as needed, eliminating the need for upfront provisioning and reducing idle capacity [37].
- *Auto-scaling:* Resources over the cloud are automatically scaled up or down based on workload requirements, ensuring that applications have the necessary resources to perform optimally [38].
- *Resource optimization:* DRA algorithms analyze workload patterns and resource usage to identify opportunities for optimization, such as consolidating virtual machines or migrating workloads to underloaded servers over the cloud [39].
- *Cost efficiency:* By aligning resource allocation with actual demand, DRA helps reduce cloud computing costs [40].

2.10 Techniques Used in Dynamic Resource Allocation (DRA) in Cloud Environment [41]

- *Predictive analytics*: This can be used by historical data and machine learning to forecast future resource requirements in the cloud environment.
- *Real-time monitoring*: Continuously monitoring resource usage and performance metrics to identify anomalies and trigger adjustments over cloud infrastructure can help in efficient computing.
- *Automation:* Automating the process of allocating, deallocating, and scaling resources over the cloud.
- *Self-healing*: Automatically detecting and resolving resource-related issues over the cloud infrastructure.

  By effectively implementing DRA, cloud providers can offer more flexible, efficient, and cost-effective services to their customers.

2.11 *Predictive Analytics for Resource Forecasting in Cloud Computing*

Predictive analytics, a subfield of data mining, explores historical data to predict future trends and patterns. In the context of cloud computing, it can play a crucial role in resource forecasting. Predictive Analytics can be used in Cloud Computing Resource Forecasting;

1. **Demand Forecasting** [42]: It is very crucial to explore numerous forecasting techniques toward effective resource utilization. Some of the important techniques are explained here.
   (A) *Historical Data Analysis:* By analyzing past usage patterns, such as CPU cycles, memory consumption, and storage usage, predictive models can identify trends and seasonality.
   (B) *External Factors:* Incorporating external factors like economic indicators, industry trends, and social events can enhance forecast accuracy.
   (C) *Machine Learning Techniques:* Advanced machine learning algorithms, such as time series analysis, regression analysis, and neural networks, can be applied to complex data patterns.
2. **Capacity Planning** [27]
   (A) *Optimal Resource Allocation:* Based on predicted demand, the system can automatically allocate resources to different applications or services.
   (B) *Avoiding Over-provisioning*: By accurately forecasting demand, cloud providers can avoid over-provisioning resources, reducing costs.
   (C) *Preventing Outages:* Timely identification of potential resource shortages can prevent service disruptions.
3. **Cost Optimization** [43]
   (A) *Rightsizing Resources:* By predicting future needs, resources can be rightsized to match actual demand, reducing unnecessary costs.
   (B) *Dynamic Pricing:* Predictive analytics can help cloud providers adjust pricing strategies based on anticipated demand fluctuations.

2.12 *Challenges and Considerations in Cloud Resource Allocation Predictions [43]*

- *Data Quality:* Accurate and reliable data is crucial for effective predictive analytics.
- *Model Complexity:* Selecting the appropriate model and tuning its parameters can be complex.

- *External Factors:* Unforeseen events can impact demand and accuracy of forecasts.
- *Continuous Learning:* Models need to be continuously updated and retrained to adapt to changing patterns.

By effectively utilizing predictive analytics, cloud providers can optimize resource allocation, reduce costs, and enhance the overall user experience.

2.13 Self-Healing Systems in Cloud Environment

Self-healing systems are designed to automatically detect, diagnose, and resolve issues without human intervention. In cloud computing, these systems are crucial for maintaining high availability and reliability. The following are the Key Components of Self-Healing Systems [44]:

1. **Continuous Monitoring:**
   - *System Health Monitoring:* Tracking key metrics like CPU usage, memory utilization, disk I/O, and network traffic.
   - *Log Analysis:* Analyzing system logs to identify errors, warnings, and performance bottlenecks.
   - *Threshold-Based Alerts:* Setting predefined thresholds for critical metrics and triggering alerts when these thresholds are exceeded.
2. **Fault Detection and Diagnosis:**
   - *Anomaly Detection:* Using machine learning algorithms to identify abnormal behavior and potential failures.
   - *Root Cause Analysis:* Pinpointing the underlying cause of issues, such as hardware failures, software bugs, or configuration errors.
3. **Automated Remediation:**
   - *Self-Healing Actions:* Automatically taking corrective actions, such as restarting failed services, redeploying applications, or provisioning additional resources.
   - *Rollback Mechanisms:* Implementing rollback strategies to revert to a previous stable state in case of failed remediation attempts.
4. **Learning and Adaptation:**
   - *Machine Learning:* Using machine learning to learn from past incidents and improve future responses.
   - *Adaptive Thresholds:* Adjusting thresholds based on historical data and current system conditions.

2.13.1 Benefits of Self-Healing Systems [44]

- *Increased Uptime*: Reduced downtime and faster recovery from failures.
- *Improved Reliability*: Enhanced system stability and resilience.
- *Reduced Operational Costs*: Automated remediation reduces the need for manual intervention.
- *Enhanced User Experience*: Minimized disruptions and improved service quality.

By implementing self-healing systems, cloud providers can deliver more reliable and efficient services, ultimately improving customer satisfaction.

3 Power-Aware Scheduling in Cloud Computing

Power-aware scheduling is a technique designed to optimize energy consumption in cloud data centers while maintaining performance. It involves intelligently allocating tasks to physical and virtual machines, considering factors like workload intensity, processing power, and energy efficiency.

3.1 Key Techniques Applied for Power-Aware Scheduling

Some key techniques used in power-aware scheduling include:

3.1.1 Dynamic Voltage and Frequency Scaling (DVFS) [45]

This technique adjusts the CPU's voltage and frequency to match the workload and reduces power consumption during idle periods or low-intensity tasks.

- *Task Migration:* This process migrates tasks between physical machines to consolidate workloads and reduce the number of active servers.
- *Server Consolidation:* This technique consolidates multiple virtual machines onto fewer physical servers. In this process, the power-aware task scheduling algorithms are used to schedule tasks based on energy efficiency and performance requirements. This process also prioritizes tasks and allocates resources to minimize energy consumption.
- *Idle Server Power Management: In this technique,* powers down automatically of the idle servers to save energy. It also quickly powers up servers when needed to meet increased demand.

3.2 Benefits of Power-Aware Scheduling

- *Reduced Energy Consumption:* Lower energy consumption leads to significant cost savings and reduced carbon footprint.
- *Improved Performance*: Optimized resource allocation can enhance system performance and responsiveness.
- *Increased Sustainability:* Contributes to a more sustainable IT infrastructure.
- *Enhanced Reliability:* By reducing the number of active servers, it can improve system reliability and fault tolerance.

By implementing power-aware scheduling techniques, cloud providers can significantly reduce their energy consumption and operational costs while maintaining high levels of performance and reliability.

3.3 Load Balancing and Migration Techniques in Cloud Computing

Load balancing is a technique used to distribute incoming traffic across multiple servers to optimize resource utilization, improve performance, and enhance system reliability [45] [46, 47]. This is crucial in cloud environments where workloads can fluctuate significantly.

1. **Common Load-Balancing Techniques** [48]**:** The following load-balancing techniques are used in the cloud computing environment.
    - *Round Robin*: Distributes incoming requests evenly among servers in a circular fashion.
    - *Least Connections:* Directs incoming requests to the server with the fewest active connections.
    - *Least Response Time*: Sends requests to the server with the shortest average response time.
    - *Random:* Distributes requests randomly among servers.
    - Weighted Round Robin: Distributes requests based on server capacity or other weighted factors.
2. **Virtual Machine Migration** [49]**:** which involves moving a running VM from one physical server to another without interrupting its operation. This technique is essential for several reasons:
    - *Load Balancing*: Migrating VMs to less-loaded servers can improve overall system performance.
    - *Fault Tolerance*: Migrating VMs to a different physical server can help prevent downtime in case of hardware failures.
    - *Energy Efficiency:* Consolidating VMs onto fewer physical servers can reduce energy consumption.

3.3.1 Types of VM Migration [49]

- *Live Migration:* Migrates a VM while it is running, minimizing downtime.
- *Pre-Copy Migration*: Migrates a VM by copying its memory image to the target server before shutting it down on the source server.
- *Checkpoint-Restart Migration:* Creates a checkpoint of the VM's state, shuts it down, migrates the checkpoint and VM state, and then restarts the VM on the target server.

3.3.2 Challenges and Considerations

- *Network Bandwidth*: Sufficient network bandwidth is required to transfer large VM images efficiently.
- *Storage Performance*: The performance of the storage system can impact migration speed and downtime.
- *Network Latency:* High network latency can affect the migration process and the performance of migrated VMs.
- *Resource Contention*: Migrating VMs can temporarily increase the load on the target server, potentially impacting other VMs.

By effectively implementing load balancing and VM migration techniques, cloud providers can optimize resource utilization, improve system performance, and enhance fault tolerance.

3.4 Green Data Center and Optimization Techniques

Green data center optimization involves strategies to minimize energy consumption and environmental impact while maximizing operational efficiency. Key strategies for the green data center include:

3.4.1 Energy Efficiency [50] [47]

- *Efficient Cooling Systems*: Implementing advanced cooling technologies like liquid cooling or air-side economizers to reduce energy consumption.
- *High-Efficiency Hardware:* Utilizing energy-efficient servers, storage devices, and networking equipment.
- *Power Usage Effectiveness (PUE):* Monitoring and optimizing PUE, a metric that measures the efficiency of a data center's power usage.
- *Dynamic Power Management:* Adjusting power consumption based on real-time workload demands.

3.4.2 Renewable Energy Integration [51]

- *On-site Generation:* Installing solar panels or wind turbines to generate clean energy.
- *Purchasing Renewable Energy Credits (RECs):* Supporting renewable energy projects and offsetting carbon emissions.
- *Energy Storage Systems:* Using batteries or other storage technologies to store excess renewable energy for later use.

3.4.3 Resource Optimization [52]

- *Virtualization:* Consolidating multiple virtual machines onto fewer physical servers to reduce energy consumption and hardware costs.
- *Workload Consolidation*: Grouping similar workloads onto fewer servers to improve resource utilization.
- *Workload Migration*: Shifting workloads between servers to balance resource utilization and optimize energy consumption.

3.4.4 Sustainable Practices [53]

- *Water Conservation:* Implementing water-saving technologies like air-side economizers and water-efficient cooling systems.
- *Waste Reduction:* Minimizing electronic waste through proper disposal, reuse, and recycling.
- *Sustainable Building Design:* Designing data centers with energy-efficient building materials and practices.

By implementing these strategies, data centers can significantly reduce their environmental impact, lower operational costs, and enhance their sustainability credentials.

3.5 Cloud Cost Management Strategies

Effective cloud cost management is crucial for organizations to maximize the benefits of cloud computing while minimizing expenses. Here are some key strategies that can be explored for the cloud cost optimization [54]:

3.5.1 Rightsizing Resources

- *Identify Over-provisioned Instances*: Use tools to identify instances that are over-sized for their workloads.
- *Rightsized Instances*: Downsize instances to match their actual needs, reducing costs.
- *Utilize Spot Instances*: Leverage spot instances for flexible, cost-effective workloads.

3.5.2 Optimizing Resource Utilization

- *Auto-Scaling*: Automatically scale resources up or down based on demand to avoid over-provisioning.
- *Reservation Purchases:* Utilize Reserved Instances or Savings Plans to lock-in discounted rates for consistent workloads.
- *Shutdown Unused Instances*: Turn off idle instances during off-peak hours or when not in use.

3.5.3 Monitoring and Analysis

- *Cloud Cost Management Tools:* Employ tools to track, analyze, and forecast cloud spending.
- *Identify Cost Anomalies:* Detect unusual spending patterns and investigate their causes.
- *Regular Reviews:* Conduct regular reviews of cloud usage and costs to identify optimization opportunities.

3.6 Best Practices for the Cloud Cost Optimization

By using the following practices, the cost of the cloud resources can be optimized.

- *Tagging Resources*: Properly tag resources to track costs and allocate them to specific projects or departments.
- *Cost Allocation:* Implement a cost allocation model to distribute costs accurately.
- *Regularly Review and Optimize:* Continuously monitor and optimize resource usage to identify and address inefficiencies.
- *Leverage Cloud Provider Tools:* Utilize the cost management tools provided by cloud providers to gain insights and optimize spending.
- *Collaborate with Finance and IT Teams*: Foster collaboration between finance and IT teams to align cloud spending with business objectives.

By implementing these strategies, organizations can significantly reduce their cloud costs while maintaining optimal performance and reliability.

3.7 *Optimal Resource Sizing and Scaling in Cloud Computing*

Optimal resource sizing and scaling are critical for achieving efficient and cost-effective cloud operations. It involves selecting the right instance types and adjusting resource allocations to meet fluctuating demands. Key Strategies for Optimal Resource Sizing and Scaling [55]:

3.7.1 Rightsizing Instances

- *Analyze Workload Patterns*: Understand the workload's CPU, memory, and storage requirements.
- *Choose Appropriate Instance Types:* Select instance types that align with the workload's needs, avoiding over-provisioning.
- *Utilize Instance Families:* Consider instance families with burstable performance for workloads with intermittent spikes.
- *Monitor and Adjust:* Continuously monitor resource utilization and adjust instance sizes as needed.

3.7.2 Auto-Scaling

- *Define Scaling Policies:* Establish rules for automatically scaling resources based on predefined metrics (e.g., CPU utilization, memory usage, network traffic).
- *Implement Scaling Triggers:* Set thresholds to trigger scaling actions, such as adding or removing instances.
- *Cooldown Periods:* Introduce cooldown periods to prevent overreaction to temporary spikes in demand.
- *Monitor and Fine-Tune:* Continuously monitor scaling behavior and adjust policies as needed.

3.7.3 Spot Instances

- *Leverage Spot Instances:* Utilize spot instances for cost-effective workloads that can tolerate interruptions.
- *Set Competitive Bids*: Determine appropriate bid prices to increase the likelihood of instance allocations.
- *Implement Fault Tolerance:* Design applications to handle potential interruptions caused by spot instance terminations.

3.7.4 Scheduling Optimization

- *Task Scheduling:* Use algorithms to optimize task scheduling across resources, considering factors like resource availability, task dependencies, and energy efficiency.
- *Resource Allocation:* Allocate resources efficiently to minimize idle time and maximize utilization.

3.7.5 Monitoring and Analysis

- *Continuous Monitoring:* Monitor resource utilization, performance metrics, and cost metrics.
- *Identify Bottlenecks:* Identify performance bottlenecks and optimize resource allocation accordingly.
- *Analyze Cost Trends*: Track cloud spending and identify opportunities for cost reduction.

3.7.6 Challenges and Considerations

- *Cloud Provider Tools:* Utilize cloud provider tools to automate resource provisioning, scaling, and optimization.
- *Third-Party Tools:* Consider using third-party tools for advanced monitoring, analysis, and optimization.
- *Best Practices*: Follow best practices for cloud resource management, such as tagging resources, using reserved instances, and rightsizing.

By implementing these strategies, organizations can optimize resource utilization, reduce costs, and improve the overall performance of their cloud applications.

3.8 *Case Studies and Real-world Applications:* Cloud Cost Optimization Tools.

Cloud cost optimization tools are essential for managing and reducing cloud expenses. These tools provide insights into spending patterns, identify cost-saving opportunities, and automate cost-saving actions. Here are some popular cloud cost optimization tools:

3.8.1 Cloud Service Providers'-Specific Tools [56]

Following are some of the tools provided by the Cloud Service Providers for cost optimization.

- *AWS Cost Explorer:* This tool provides detailed visibility into AWS usage and costs, helping you identify cost-saving opportunities.
- *Azure Cost Management:* Offers insights into Azure spending, enabling you to optimize resource usage and reduce costs.
- *Google Cloud Billing Report*: Provides detailed reports on Google Cloud Platform usage and costs.

3.8.2 Third-Party Tools for Cloud Cost Optimization [56]

Following are some of the tools provided by the third-party Cloud Service Providers for cost optimization.

- *CloudHealth:* A comprehensive cloud management platform that helps optimize costs, manage governance, and ensure security.
- *CloudChecker:* Provides visibility into cloud usage, identifies cost optimization opportunities, and helps manage cloud security and compliance.
- *CloudZero:* Offers detailed cost allocation and optimization insights, helping you understand the cost drivers of your cloud applications.

3.8.3 Key Features of Cloud Cost Optimization Tools

The following are some of the key features of Cloud Cost Optimization Tools.

- *Cost Allocation:* Accurately allocate costs to specific projects or departments.
- *Anomaly Detection:* Identify unusual spending patterns and potential cost-saving opportunities.
- *Rightsizing Recommendations:* Suggest optimal instance sizes to reduce costs.
- *Reservation Recommendations*: Recommend purchasing Reserved Instances or Savings Plans to lock-in discounts.
- *Spot Instance Optimization*: Help you maximize the use of spot instances to reduce costs.
- *Automated Actions*: Automatically take actions to optimize costs, such as stopping unused instances or rightsizing resources.
- *Custom Reporting:* Create custom reports to analyze spending trends and identify cost-saving opportunities.

By optimizing cloud usage with these tools and best practices, organizations such as SMEs can substantially reduce costs and boost their bottom line.

3.9 Significant Thrust for Greening Cloud Data Centers Through Research and Innovations

The volume of published research on green cloud computing serves as a strong indicator of its importance. As depicted in Fig. 3, the number of papers published annually has steadily increased since 2012, with notable importance after 2018. This trend, observed across renowned publishers such as ACM, IEEE, Elsevier, and Springer, reflects the growing attention and focus on this critical topic. This phenomenon indicates that the Thrust for Greening Cloud Data Centers through Research and Innovations is rapidly evolving.

3.10 Parametric Analysis of the Convergence of AI and Cloud Computing

Table 1 highlights the constructive collaboration (synergy) between AI and Cloud computing, emphasizing their converging potential to drive technological advancements. This technological convergence shows how these technologies complement each other to achieve scalable, intelligent, and energy-efficient computing solutions.

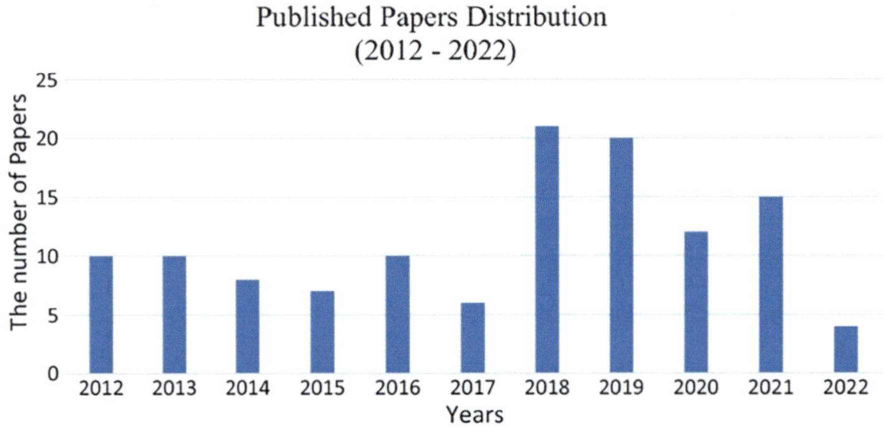

Fig. 3 Significant Research and Innovations for Greening Cloud Data Centers [57]

Table 1 Analysis of the convergence of AI and cloud computing

Aspect	Artificial intelligence (AI)	Cloud computing	Convergence analysis
Core technologies	Machine learning, deep learning, natural language processing, computer vision	Virtualization, containerization, distributed systems, cloud APIs	AI tools and frameworks are increasingly hosted and optimized for cloud environments.
Scalability	Requires significant computational resources for training and inference.	Dynamically scales resources based on demand.	Cloud platforms offer scalability that supports AI workloads like large model training and real-time inference.
Cost efficiency	AI processes can be resource-intensive and costly to implement on-premises.	Offers pay-as-you-go models for cost management.	Cloud computing reduces upfront costs for AI projects by providing resources on demand.
Data storage and management	Relies heavily on large datasets for training and analysis.	Provides vast storage solutions like object storage and data lakes.	Cloud platforms enable enormous storage, processing, and integration of large-scale AI datasets.
Accessibility	AI technologies require specialized expertise to implement and use.	Cloud democratizes access to technologies through easy-to-use interfaces.	Cloud-based AI services (e.g., AWS AI, azure AI, Google AI) make AI accessible to non-experts.
Performance	AI needs high-performance computing (HPC) for tasks like deep learning.	Provides high-performance infrastructure, including GPUs and TPUs.	Cloud environments are optimized for high-performance AI workloads through advanced hardware and services.
Security	AI systems require secure handling of sensitive data.	Implements advanced security protocols and compliance measures.	Cloud ensures secure storage and processing of sensitive AI data with tools like encryption and monitoring.
Interoperability	AI models require compatibility across platforms and tools.	Offers APIs and services for integration with various systems.	Cloud platforms provide seamless integration of AI services with enterprise workflows and applications.
Market trends	Increasing demand for AI solutions in areas like healthcare, finance, and IoT.	Rapid growth in adoption of multi-cloud and hybrid-cloud solutions.	AI and cloud convergence drives innovation in intelligent cloud services and industry-specific solutions.
Future opportunities	Advances in generative AI and autonomous systems.	Growth in edge computing and serverless architectures.	Cloud-AI synergy enables innovations like AI-driven cloud automation, edge AI, and federated learning.

3.11 A Comparison of the Cloud Computing, AI, and Green Computing Convergence

Table 2 describes the comparison of the convergence of Cloud Computing, Artificial Intelligence (AI), and Green Computing.

4 Transforming Small Business Operations Through Smart Resource Allocation over Cloud

For small and medium-sized enterprises (SMEs), success often depends on the ability to optimize limited resources effectively. Intelligent resource allocation is the strategic distribution of resources based on predictive analysis, data-driven insights, and automation has allowed many SMEs to streamline operations, reduce waste, and maximize growth potential.

Case Study-1: Migrating SMEs' application over Cloud
Intelligent Cloud Services: A Game-Changer for SMEs.

Small and medium enterprises (SMEs) can significantly benefit from intelligent cloud services. These services offer a range of advantages, including:

- *Cost Reduction*: By eliminating the need for large upfront investments in on-premises infrastructure, SMEs can reduce operational costs.
- *Enhanced Customer Satisfaction:* Cloud-powered services enable quicker response times, improved service quality, and personalized experiences.
- *Agile and Reliable Services:* Cloud services provide scalability and flexibility, allowing businesses to adapt to changing demands.
- *Robust Security:* Cloud providers invest heavily in security measures to protect sensitive data.
- *Fast Upgrades:* Cloud services are regularly updated with the latest features and security patches.
- *Optimized Expenses:* Cloud services offer pay-as-you-go pricing models, allowing businesses to optimize their spending.

Power of Azure for Data-Driven Insights

By migrating to a cloud platform like Microsoft Azure, SMEs can unlock the power of advanced analytics and machine learning. Azure offers a comprehensive suite of tools, including:

- *Azure Synapse Analytics*: For unified data ingestion, preparation, and analytics.
- *Azure Data Lake:* For scalable data storage and management.
- *Azure Databricks:* For big data processing and machine learning.
- *Azure Machine Learning:* For building, training, and deploying machine learning models.

Table 2 Comparison of the convergence of Cloud Computing, Artificial Intelligence (AI), and Green Computing. This comparison shows how these technologies complement each other to achieve scalable, intelligent, and eco-friendly computing solutions

Parameter	Cloud computing	Artificial intelligence (AI)	Green computing	Convergence analysis
Objective	Provide scalable, cost-effective, and accessible computing.	Automate tasks, enhance efficiency, and enable intelligent decision-making.	Reduce environmental impact and optimize energy usage.	Enables intelligent resource management for sustainability.
Resource efficiency	Dynamically allocates resources based on workload demands.	Optimizes processes through predictive analysis and automation.	Minimizes energy consumption and e-waste.	AI can optimize cloud resource utilization, ensuring greener computing practices.
Energy consumption	Relies on large-scale data centers consuming significant power.	AI workloads like deep learning require high computational power.	Focused on reducing carbon footprints of computing operations.	Cloud providers use AI to optimize energy usage and adopt green technologies like renewable energy.
Scalability	Highly scalable infrastructure for diverse workloads.	Scalable AI models require significant computational resources.	Focus on balancing scalability with energy efficiency.	Scalability is enhanced by AI, while green computing ensures sustainable scaling.
Cost efficiency	Pay-as-you-go models reduce upfront costs.	Reduces operational costs through automation.	Focus on cost reduction by reducing power and hardware usage.	AI-driven cloud systems improve cost efficiency while adhering to green computing principles.
Environmental impact	Large carbon footprint due to energy-intensive data centers.	Indirect impact through energy demands for training and inference.	Direct focus on reducing carbon emissions and e-waste.	AI and cloud together can achieve carbon neutrality through optimization and renewable energy adoption.
Security and compliance	Implements advanced data security protocols.	Handles sensitive data securely for intelligent processing.	Enforces regulations to minimize hazardous waste and energy abuse.	Cloud-AI solutions adhere to green compliance through secure and eco-friendly architectures.

(continued)

Table 2 (continued)

Parameter	Cloud computing	Artificial intelligence (AI)	Green computing	Convergence analysis
Data processing	Centralized and distributed processing of massive datasets.	Processes large datasets for learning and decision-making.	Advocates efficient data management to reduce unnecessary processing.	AI in the cloud enables smarter and more energy-efficient data handling, aligning with green principles.
Infrastructure	Large data centers, virtualized servers, and containerization.	GPUs, TPUs, and other high-performance hardware.	Low-power devices and energy-saving architectures.	AI optimizes infrastructure in cloud systems for greener and more efficient operations.
Future opportunities	Multi-cloud and edge computing for enhanced performance.	Advances in generative AI, autonomous systems, and IoT.	Growth in renewable energy-powered data centers and green tech.	The convergence creates intelligent, sustainable, and efficient solutions for industries worldwide.

Using these tools, SMEs can explore the following benefits:

- Process and analyze large volumes of structured and unstructured data.
- Gain real-time insights into business operations.
- Make data-driven decisions to improve efficiency and profitability.
- Integrate with business intelligence tools like Power BI for comprehensive reporting.
- A cost-effective solution.

Azure offers a flexible pricing model, allowing SMEs to start with minimal upfront costs. In fact, Azure can provide a solution with a monthly cost of $24,051, making it an attractive option for businesses of all sizes. Thus, by embracing cloud technology, SMEs can accelerate their digital transformation, gain a competitive edge, and drive sustainable growth (Fig. 4).

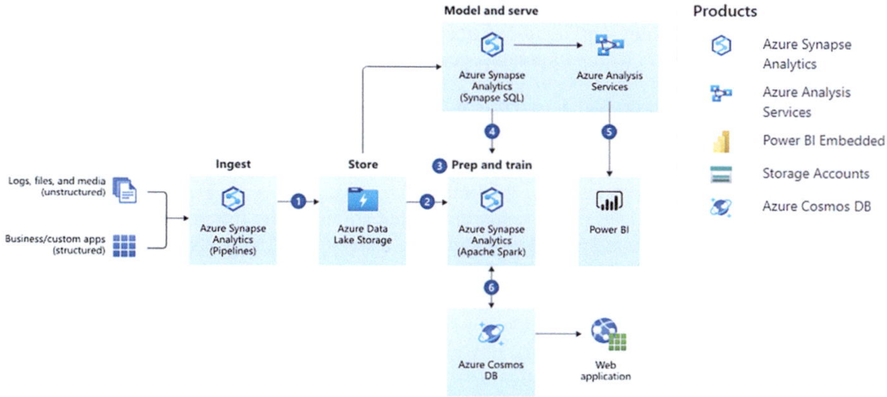

Fig. 4 Microsoft Azure ML integrated web application model [58]

5 Conclusion

In conclusion, the convergence of intelligent resource allocation and green technology offers a powerful solution for SMEs to achieve sustainable growth in the cloud environment. By converging AI, machine learning, Cloud, and environmentally conscious practices, SMEs can optimize resource utilization, reduce energy consumption, and minimize their environmental impact. The case studies and the convergence presented in this chapter article illustrate the practical applications of these technologies. As cloud computing continues to evolve, SMEs must adopt these strategies to remain competitive and contribute to a sustainable future for better Return on Investment (ROI) and remain compatible in the global competitive environment. By embracing intelligent resource allocation and green technology, SMEs can reveal new opportunities, enhance operational efficiency, and minimize their carbon footprint.

References

1. Sick, N., & Bröring, S. (2022). Exploring the research landscape of convergence from a TIM perspective: A review and research agenda. *Technological Forecasting and Social Change, 175*, 121321.
2. Zhao, S., Zeng, D., Li, J., Feng, K., & Wang, Y. (2023). Quantity or quality: The roles of technology and science convergence on firm innovation performance. *Technovation, 126*, 102790.
3. Verma, V. R., Sharma, D. P., & Lamba, C. S. (2018). Stable energy proficient and load balancing based QoS routing in mobile ad-hoc networks: Mobile software based approach. *Malaya Journal of Matematik, S*(1), 79–83.
4. Helali, L., & Omri, M. N. (2021). A survey of data center consolidation in cloud computing systems. *Computer Science Review, 39*, 100366.

5. Mashhadi Moghaddam, S., O'Sullivan, M., Walker, C., Fotuhi Piraghaj, S., & Unsworth, C. P. (2020). Embedding individualized machine learning prediction models for energy efficient VM consolidation within cloud data centers. *Future Generation Computer Systems, 106*, 221–233.
6. Hassannataj Joloudari, J., et al. (2024). *Resource allocation problem and artificial intelligence: the state-of-the-art review (2009–2023) and open research challenges* (Vol. 83., no. 26). Springer US.
7. Al-Sharafi, M. A., Iranmanesh, M., Al-Emran, M., Alzahrani, A. I., Herzallah, F., & Jamil, N. (2023). Determinants of cloud computing integration and its impact on sustainable performance in SMEs: An empirical investigation using the SEM-ANN approach. *Heliyon, 9*(5), e16299.
8. Gill, S. S., & Buyya, R. (2018). A taxonomy and future directions for sustainable cloud computing: 360 degree view. *ACM Computing Surveys, 51*(5), 1.
9. Mohammad, A., & Abbas, Y. (2024). Key challenges of cloud computing resource allocation in small and medium enterprises. *Digital, 4*(2), 372–388.
10. Alemu, M., Adane, A., Singh, B. K., & Sharma, D. P. (2020). Cloud-based outsourcing framework for efficient IT project management practices. *International Journal of Advanced Computer Science and Applications, 11*(9), 153–164.
11. Ashawa, M., Douglas, O., Osamor, J., & Jackie, R. (2022). Improving cloud efficiency through optimized resource allocation technique for load balancing using LSTM machine learning algorithm. *Journal of Cloud Computing, 11*(1).
12. Sobers Smiles David, G., Ramkumar, K., Shanmugavadivu, P., & Eliahim Jeevaraj, P. S. (2021). *Introduction to cloud resource management* (EAI/Springer Innovations in Communication and Computing) (pp. 1–16).
13. Okumura, M., & Fujimura, S. (2019). How do you migrate your system to the cloud environment? An example of conversion from silo type to horizontal type. *Proceedings of the 40th Annual ACM SIGUCCS Conference on User Services*, 51–56.
14. Jia, Z., et al. (2023). The importance of resource awareness in artificial intelligence for healthcare. *Nature Machine Intelligence, 5*(7), 687–698.
15. Aditi Prasad, V. K., Gerogiannis, V. C., Kanavos, A., Dansana, D., & Acharya, B. (2025). Utilizing convolutional neural networks for resource allocation bottleneck analysis in cloud ecosystems. *Cluster Computing, 28*(1).
16. Goyal, P., & Rani, R. (2024). Comparative analysis of machine learning, ensemble learning and deep learning classifiers for Parkinson's disease detection. *SN Computer Science, 5*(1).
17. *What is self-supervised learning?*. [Online]. Available: https://www.ibm.com/topics/self-supervised-learning.
18. N. Thomas Rincy and R. Gupta. (2020). *A survey on machine learning approaches and its techniques*. 2020 IEEE International Students' Conference on Electrical, Electronics and Computer Science (SCEECS).
19. Shan Lee, V. L., Gan, K. H., Tan, T. P., & Abdullah, R. (2019). Semi-supervised learning for sentiment classification using small number of labeled data. *Procedia Computer Science, 161*, 577–584.
20. van Engelen, J. E., & Hoos, H. H. (2020). A survey on semi-supervised learning. *Machine Learning, 109*(2), 373–440.
21. Eckhardt, C. M., et al. (2023). Unsupervised machine learning methods and emerging applications in healthcare. *Knee Surgery, Sports Traumatology, Arthroscopy, 31*(2), 376–381.
22. Lu, H., & Uddin, S. (2024). Unsupervised machine learning for disease prediction: A comparative performance analysis using multiple datasets. *Health and Technology (Berl)., 14*(1), 141–154.
23. Wang, X. S., Wang, R. R., & Cheng, Y. H. (1996). Safe reinforcement learning: A survey. *Journal of Artificial Intelligence Research, 4*(9), 237–285.
24. D. Nelson. (2017). What is reinforcement learning? *Unite.AI*, . [Online]. Retrieved Novomber 18, 2024, from https://www.unite.ai/what-is-reinforcement-learning/.

25. Kanungo, S. (2024). AI-driven resource management strategies for cloud computing systems, services, and applications. *World Journal of Advanced Engineering Technologies and Sciences, 11*(2), 559–566.
26. Inkollu, U. M. R., & Sastry, J. K. R. (2024). AI-driven reinforced optimal cloud resource allocation (ROCRA) for high-speed satellite imagery data processing. *Earth Science Informatics, 17*(2), 1609–1624.
27. Kirchoff, D. F., Meyer, V., Calheiros, R. N., & De Rose, C. A. F. (2024). Evaluating machine learning prediction techniques and their impact on proactive resource provisioning for cloud environments. *The Journal of Supercomputing, 80*(15), 21920–21951.
28. Hasimi, L., Zavantis, D., Shakshuki, E., & Yasar, A. (2024). Cloud computing security and deep learning: An ANN approach. *Procedia Computer Science, 231*(2023), 40–47.
29. M. S. Hasan, F. A. De Oliveira, T. Ledoux, and J. L. Pazat. (2016). *Enabling green energy awareness in interactive cloud application*. In Proceedings of the International Conference on Cloud Computing Technology and Science, CloudCom, pp. 414–422.
30. Ionescu, S. A., & Diaconita, V. (2023). Transforming financial decision-making: The interplay of AI, cloud computing and advanced data management technologies. *International Journal of Computers, Communications & Control, 18*(6), 1–19.
31. W. Kowalski. (2024). *What is green cloud computing and how it benefits organizations*. [Online]. Retrieved Novomber 22, 2024, from https://scand.com/company/blog/what-is-green-cloud-computing-and-how-it-benefits-organizations/.
32. Pandiyan, P., Saravanan, S., Kannadasan, R., Krishnaveni, S., Alsharif, M. H., & Kim, M. K. (2024). A comprehensive review of advancements in green IoT for smart grids: Paving the path to sustainability. *Energy Reports, 11*(April), 5504–5531.
33. Rong, H., Zhang, H., Xiao, S., Li, C., & Hu, C. (2016). Optimizing energy consumption for data centers. *Renewable and Sustainable Energy Reviews, 58*, 674–691.
34. Zolfaghari, R., & Rahmani, A. M. (2020). Virtual machine consolidation in cloud computing systems: Challenges and future trends. *Wireless Personal Communications, 115*(3), 2289–2326.
35. Rawat, P. S., et al. (2025). Efficient virtual machine placement in cloud computing environment using BSO-ANN based hybrid technique. *Alexandria Engineering Journal, 110*, 145–152.
36. V. R. Verma, D. P. Sharma, and C. S. Lamba, *QOS improvement in MANET routing by route optimization through convergence of mobile agent*. In 2018 5th IEEE Uttar Pradesh Section International Conference on Electrical, Electronics and Computer Engineering (UPCON), 2018, pp. 1–6.
37. Okai, S., Uddin, M., Arshad, A., Alsaqour, R., & Shah, A. (2014). Cloud computing adoption model for universities to increase ICT proficiency. *SAGE Open, 4*(3), 1–10.
38. Risco, S., Moltó, G., Naranjo, D. M., & Blanquer, I. (2021). Serverless workflows for containerised applications in the cloud continuum. *Journal of Grid Computing, 19*(3), 30.
39. Bhardwaj, A., & Krishna, C. R. (2021). Virtualization in cloud computing: Moving from hypervisor to containerization—A survey. *Arabian Journal for Science and Engineering, 46*(9), 8585–8601.
40. Sangani, S., Patil, R., & Goudar, R. H. (2024). Efficient algorithm for error optimization and resource prediction to mitigate cost and energy consumption in a cloud environment. *International Journal of Information Technology, 16*(4), 2187–2197.
41. Belgacem, A., Beghdad-Bey, K., Nacer, H., & Bouznad, S. (2020). Efficient dynamic resource allocation method for cloud computing environment. *Cluster Computing, 23*(4), 2871–2889.
42. Ali, T., Khan, H. U., Alarfaj, F. K., & AlReshoodi, M. (2024). Hybrid deep learning and evolutionary algorithms for accurate cloud workload prediction. *Computing, 106*(12), 3905.
43. Seddiki, D., Maldonado Carrascosa, F. J., García Galán, S., Valverde Ibáñez, M., Marciniak, T., & Ruiz Reyes, N. (2024). Enhanced virtual machine migration for energy sustainability optimization in cloud computing through knowledge acquisition. *Computers and Electrical Engineering, 119, no. PA*, 109506.
44. Rouholamini, S. R., Mirabi, M., Farazkish, R., & Sahafi, A. (2024). Proactive self-healing techniques for cloud computing: A systematic review. *Concurrency and Computation: Practice and Experience, 36*.

45. Singh, B. K., Danish, M., & Choudhury, T. (2021). Autonomic resource management in a cloud-based infrastructure environment. *Autonomic Computing in Cloud Resource Management in Industry 4.0.*
46. Mekonnen, D., Megersa, A., Sharma, R. K., & Sharma, D. P. (2022). Designing a component-based throttled load balancing algorithm for cloud data centers. *Mathematical Problems in Engineering, 2022*, 1–12.
47. Getahun, M., Azath, M., Sharma, D. P., Tuni, A., & Adane, A. (2022). Efficient energy utilization algorithm through energy harvesting for heterogeneous clustered wireless sensor network. *Wireless Communications and Mobile Computing, 2022*, 1.
48. Afzal, S., & Kavitha, G. (2019). Load balancing in cloud computing—A hierarchical taxonomical classification. *Journal of Cloud Computing, 8*(1).
49. Smimite, O., & Afdel, K. (2020). Containers placement and migration on cloud system. *International Journal of Computers and Applications, 176*(35), 9–18.
50. Kahouli, B., Hamdi, B., Nafla, A., & Chabaane, N. (2022). Investigating the relationship between ICT, green energy, total factor productivity, and ecological footprint: Empirical evidence from Saudi Arabia. *Energy Strategy Reviews, 42*(February), 100871.
51. J. M. T. I. Jayalath, E. J. A. P. C. Chathumali, K. R. M. Kothalawala, and N. Kuruwitaarachchi. *Green cloud computing: A review on adoption of green-computing attributes and vendor specific implementations.* 2019 International Research Conference on Smart Computing and Systems Engineering (SCSE), pp. 158–164, 2019.
52. N. Ahmad, Q. N. Naveed, and N. Hoda, *Strategy and procedures for migration to the cloud computing.* In 2018 IEEE 5th International Conference on Engineering Technologies and Applied Sciences, ICETAS 2018, 2019, pp. 1–5.
53. Paul, S. G., et al. (2023). A comprehensive review of green computing: Past, present, and future research. *IEEE Access, 11*, 87445–87494.
54. Khan, A. Q., Matskin, M., Prodan, R., Bussler, C., Roman, D., & Soylu, A. (2024). Cost modelling and optimisation for cloud: A graph-based approach. *Journal of Cloud Computing, 13*(1).
55. Mann, Z. Á. (2018). Resource optimization across the cloud stack. *IEEE Transactions on Parallel and Distributed Systems, 29*(1), 169–182.
56. Balobaid, A., & Debnath, D. (2018). *Cloud migration tools: Overview and comparison.* Springer International Publishing.
57. Jahangard, L. R., & Shirmarz, A. (2022). Taxonomy of green cloud computing techniques with environment quality improvement considering: A survey. *International Journal of Energy and Environmental Engineering, 13*(4), 1247–1269.
58. Pricing calculator. [Online]. Retrieved November 24, 2024, from https://azure.microsoft.com/en-us/pricing/calculator/.

Blockchain and AI for Enhanced Threat Detection in Digital Marketing: Maneuvering the Legal Minefield

Madhusudan Narayan, Ashutosh Sharma, and Ashok Kumar Srivastava

1 Introduction

The integration of blockchain and artificial intelligence (AI) into digital marketing represents a significant advancement, enhancing both security and operational efficiency. Blockchain's decentralized ledger provides transparency and immutability, essential for preventing fraud and maintaining data integrity [2]. Concurrently, AI utilizes machine learning and advanced analytics to swiftly identify and address potential threats, optimize targeting strategies, and personalize user experiences [45]. These technologies collectively offer sophisticated solutions to the evolving security challenges in India's digital marketing landscape.

Originally developed for cryptocurrencies, blockchain technology has proven instrumental in various sectors by ensuring data integrity and transparency [23]. In digital marketing, blockchain's immutable ledger plays a critical role in securing digital transactions and maintaining data accuracy [14, 15]. Its ability to provide a transparent record of transactions helps mitigate risks associated with data tampering and fraud. For instance, Zhang and Wang [47] highlight that blockchain's transparency reduces discrepancies and malicious alterations, thereby fostering trust among consumers and marketers. This tamper-proof record supports compliance with stringent regulations like the Digital Personal Data Protection Act (DPDPA) by enabling detailed data tracking and reporting.

AI has revolutionized data analysis and decision-making across industries. In digital marketing, AI technologies, including machine learning and natural language processing (NLP), enable sophisticated analysis of large datasets to identify

M. Narayan (✉) · A. Sharma · A. K. Srivastava
Amity University Jharkhand, Ranchi, India

P. Raj et al. (eds.), *Artificial Intelligence (AI) for IT Energy Efficiency and Green AI for Environment Sustainability*,
https://doi.org/10.1007/978-3-031-89420-6_29

patterns, predict behaviors, and detect anomalies [16, 17]. These capabilities are crucial for enhancing threat detection and management. For example, Narayan [27] demonstrate how machine learning algorithms can effectively flag suspicious activities, reducing the risk of cyberattacks and fraud. This real-time analysis helps maintain consumer trust and ensures marketing operations' security by enabling swift responses to emerging threats [37].

India's regulatory landscape is evolving with the introduction of the Digital Personal Data Protection Act (DPDPA), 2023, which imposes rigorous requirements for data handling, storage, and processing to enhance consumer privacy and ensure ethical data use [30]. This Act establishes the Data Protection Board to oversee compliance and handle grievances, with serve penalties for violations. As of August 2024, the Digital Personal Data Protection Act 2023 (DPDPA) is still under consideration by the Indian Parliament and it has not been enacted into law yet [20].

Blockchain aligns well with the DPDPA's provisions by supporting data security and transparency. Its immutable ledger helps prevent fraudulent activities and unauthorized access, meeting the DPDPA's requirements for detailed audit trails and accountability. Similarly, AI contributes to DPDPA compliance through its advanced analytical capabilities. AI algorithms can detect unusual patterns and threats in real time, supporting proactive data security measures required by the Act [6, 18]. By ensuring AI systems handle data accurately and transparently, organizations can comply with provisions related to data accuracy, access, and correction [30].

Ethical considerations are crucial when deploying blockchain and AI in digital marketing. Blockchain's decentralized nature presents challenges related to data privacy and cross-border data flows, which must be managed to maintain compliance with legal standards [39, 44]. Similarly, AI applications require careful handling to avoid biases and ensure ethical use of consumer data [34]. Transparent AI algorithms are essential for maintaining consumer trust and adhering to ethical guidelines [26, 28, 29].

Case studies from Indian markets illustrate the practical applications of blockchain and AI in digital marketing. Companies like InMobi and Zoho have successfully utilized these technologies to enhance their marketing strategies and address compliance challenges. InMobi's use of AI for refining ad targeting and fraud detection has improved its ability to respond to fraudulent activities in real time [26, 28, 29]. Zoho's implementation of blockchain has increased transparency and reduced fraud in ad transactions [14, 15]. These examples highlight the potential of blockchain and AI to improve threat detection and regulatory compliance while addressing legal and ethical considerations.

Blockchain and AI offer significant opportunities for advancing threat detection and regulatory compliance in digital marketing. In India, these technologies provide innovative solutions to address security challenges and navigate complex regulatory environments. Their effective integration requires a nuanced understanding of legal and ethical issues. By capitalizing blockchain and AI, digital marketers can enhance security measures, improve compliance with regulatory standards, and develop more robust marketing strategies, paving the way for a more secure and compliant digital marketing landscape [12].

This study aims to bridge this gap by examining how blockchain and AI can be effectively integrated to address these key challenges. Specifically, we will investigate their synergistic potential, compliance with regulatory frameworks, ethical considerations, and practical barriers. By providing insights into these areas, this research will highlight the potential impact of integrating these technologies on the Indian digital marketing industry, offering strategies to enhance practices and navigate the evolving legal landscape.

2 Review of Literature

2.1 Specific Features of Blockchain and AI in Digital Marketing

Blockchain technology promotes disintermediation, combats click manipulation, fortifies trust and accountability, guarantees privacy safety, empowers security, and encourages loyalty services that deliver exclusive experiences to their consumers [32, 33]. Blockchain technology is rapidly advancing across various industrial sectors, with both technology giants and startups exploring innovative applications amid a surge in blockchain investments (Fig. 1). Seen as a transformative force, blockchain is set to revolutionize systems in multiple domains. In digital marketing, where big data analytics and artificial intelligence are currently driving change, blockchain is poised to be the next major disruptor (Gupta et al., 2022).

Blockchain technology offers significant benefits in the advertising sector by ensuring transparency in ad transactions [26, 28, 29]. Patel demonstrates that the immutable ledger of blockchain helps combat ad fraud by creating a verifiable and auditable trail of transactions, which aids in verifying ad placements and preventing fraudulent activities. Conversely, AI greatly enhances ad targeting and personalization. Reddy [34] explores how AI utilizes advanced data analytics to analyze user behavior patterns, thereby improving ad delivery and increasing engagement and return on investment (ROI). When addressing customer data management, blockchain plays a pivotal role in securing customer data. By encrypting and decentralizing transactions, blockchain prevents unauthorized access and breaches [34, 35]. Conversely, AI utilizes machine learning algorithms to process large datasets, facilitating personalized marketing strategies. Singh and Kumar [43] demonstrate how AI can predict consumer preferences and trends, which aids in creating targeted campaigns that enhance customer engagement and satisfaction [39, 40]. In supply chain management, blockchain enhances transparency and traceability by maintaining an immutable record of product provenance. Gupta [5, 7, 9] emphasizes how blockchain tracks a product's journey from its origin, ensuring authenticity and reducing counterfeiting. Meanwhile, AI contributes to optimizing logistics and inventory management through predictive analytics. Mehta et al. [21, 22] discuss

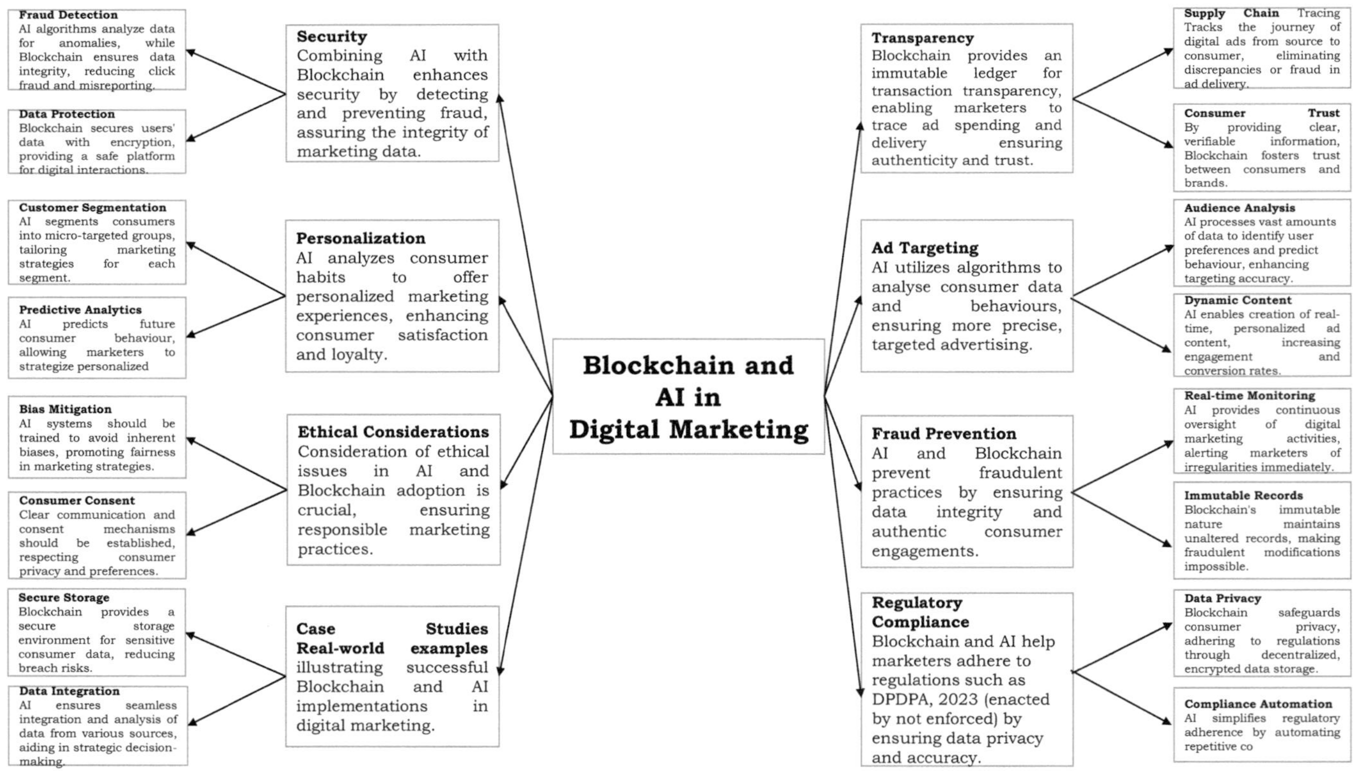

Fig. 1 Blockchain and AI in digital marketing

AI-driven solutions that improve operational efficiency and lower costs by forecasting demand and managing inventory more effectively.

2.2 Comparative Analysis: Benefits and Challenges

The integration of blockchain and AI in digital marketing presents several benefits. Blockchain provides enhanced transparency, security, and trust through its immutable ledger, addressing issues related to fraud and data integrity ([26, 28, 29]; Sharma & Rao, 2024). AI drives efficiency and personalization with its advanced analytics, enabling better targeting and operational optimization [34, 43]. However, the implementation of these technologies comes with challenges. Blockchain faces high implementation costs and technical complexity, which can hinder adoption. Similarly, AI involves complexity and significant costs related to deployment. Mehta et al. [21, 22] highlight the challenges of integrating AI systems, including the need for specialized expertise. Blockchain technology's role in digital marketing and marketing management is set to expand, offering solutions to both current and emerging challenges in Industry 4.0. By eliminating intermediaries and enhancing cybersecurity, blockchain provides a high level of transparency and accountability [31].

2.3 Regulatory Challenges and Compliance

Digital Personal Data Protection Act 2023 (enacted by not enforced yet), introduces strict requirements for data handling that affect digital marketing practices. Blockchain can facilitate compliance by providing transparent records of data transactions, helping meet regulatory demands for data security and accountability (Sharma & Rao, 2024). AI systems, meanwhile, must be designed to adhere to these privacy provisions. Reddy [34] examines how AI can integrate with data protection measures to ensure regulatory compliance. Cross-border data transfers under the DPDPA, 2023, present additional challenges, as the act mandates strict data localization and protection measures. Blockchain's decentralized nature can address these challenges by ensuring secure and compliant data sharing across borders [5, 7, 9]. Ethical considerations also play a crucial role in AI's use, including addressing algorithmic biases and ensuring responsible data handling. Patel [26, 28, 29] emphasizes the need for transparent AI algorithms to prevent misuse and uphold ethical standards.

2.4 Case Studies and Examples

Real-world applications of blockchain and AI showcase their practical benefits. In India, InMobi utilized AI for better ad targeting and fraud detection, leading to a 25% increase in ad engagement and a 20% decrease in fraud [26, 28, 29]. Zoho adopted blockchain to manage ad transactions, enhancing transparency and reducing fraud by 30% [14, 15]. Swiggy used AI to optimize delivery routes and predict customer preferences, improving delivery efficiency by 15% and customer satisfaction by 10% (Reddy & Sharma, 2024). Paytm implemented blockchain for secure financial transactions, reducing fraud by 25%. Tata Consultancy Services (TCS) applied AI analytics to enhance client services, boosting client retention by 20% and project delivery times by 30% [1]. Reliance Industries integrated blockchain in supply chain management, cutting discrepancies by 40%. HDFC Bank adopted AI for credit risk assessment, decreasing loan default rates by 15% [24, 25]. Jio Platforms improved network optimization and customer experience with AI, achieving a 20% increase in network efficiency and a 15% rise in customer satisfaction. Zomato refined its restaurant recommendation engine with AI, resulting in a 30% boost in user engagement and a 25% increase in repeat orders. ICICI Bank used blockchain to streamline trade finance processes, reducing processing time by 35%. HCL Technologies enhanced IT service management with AI, cutting incident resolution time by 25%. Bharat Petroleum tracked its fuel supply chain using blockchain, reducing discrepancies by 30%. Flipkart optimized inventory management with AI, resulting in a 20% reduction in stockouts and a 15% improvement in order fulfillment [14, 15]. Axis Bank used AI for personalized financial advisory services, increasing customer engagement by 25%. Ola improved its ride-matching and dynamic pricing with AI, reducing wait times by 15%. Tech Mahindra applied blockchain to secure IT service contracts, decreasing contract disputes by 25%. Internationally, companies such as IBM and Microsoft have utilized blockchain for supply chain management and AI for customer insights. These examples offer a broader perspective on the global applications and challenges of these technologies [16, 17].

2.5 Emerging Technologies

The impact of emerging technologies like the Internet of Things (IoT) and Augmented Reality (AR) on blockchain and AI integration is significant. IoT can enhance blockchain and AI by providing real-time data for more accurate analytics and decision-making. IoT devices can feed data into blockchain systems for improved traceability and into AI systems for better predictive analytics. AR has the potential to revolutionize digital marketing by offering immersive experiences that can be enhanced by blockchain for secure transactions and AI for personalized interactions. Looking to the future, developments such as decentralized identity management and AI-powered chatbots are expected to further refine digital marketing strategies [47]. As blockchain and AI technologies evolve, their integration into digital marketing will likely become more sophisticated, addressing current limitations and enhancing overall efficiency and compliance [14, 15].

2.6 Research Gap

Despite the growing interest in blockchain and AI for digital marketing, existing research often examines these technologies in isolation. However, there is a significant gap in understanding how to effectively integrate these technologies to address the specific regulatory, legal, and ethical challenges in the Indian context. Detailed studies on their combined application, particularly in relation to India's unique legal requirements, are lacking.

2.7 Statement of Problem

Indian digital marketers are confronting escalating threats, including cyberattacks and data breaches, which necessitate advanced security measures. Blockchain and AI offer promising solutions for enhancing threat detection and ensuring regulatory compliance. However, their integration into digital marketing practices is hindered by India's evolving legal and regulatory environment. This study aims to investigate how blockchain and AI can be effectively combined to bolster security measures and achieve compliance with laws like the DPDPA, 2023, while addressing the unique challenges of the Indian digital marketing sector.

2.8 Research Questions

- How can blockchain and AI be effectively integrated to address key challenges in the Indian digital marketing landscape, such as data privacy, security threats, and consumer trust, while utilizing their combined strengths?
- How can the integration of blockchain and AI address regulatory challenges, including those posed by the DPDPA, 2023 (enacted by not enforced), and other relevant laws in India, to ensure compliance with data privacy and security requirements?
- What are the ethical implications of using AI in digital marketing, particularly concerning bias, transparency, and accountability, and how can blockchain help address these issues?

- What technological limitations, cost constraints, or talent shortages affect the adoption of blockchain and AI in Indian digital marketing, and what strategies can be employed to overcome these barriers?

2.9 Research Methodology

This study employs a qualitative research methodology designed to address the gaps identified in the research gap and the specific challenges outlined in the problem statement. The approach centers on a comprehensive literature review to explore how blockchain and AI technologies can enhance threat detection and regulatory compliance in digital marketing within the Indian context. The research design involves a systematic review of empirical studies, theoretical frameworks, and practical case studies. The aim is to gain a nuanced understanding of how blockchain and AI can be integrated to address security challenges, regulatory compliance issues, and ethical considerations in Indian digital marketing. Data were collected from a range of sources, including scholarly articles, industry reports, regulatory documents, and conference proceedings. Major academic databases such as ScienceDirect, IEEE Xplore, SpringerLink, and Emerald Insight were utilized to ensure a broad and comprehensive search. The literature selected for inclusion focused on relevance to the research questions, publication within the last decade, and peer-reviewed status. Sources lacking empirical evidence or theoretical rigor were excluded to maintain the reliability and depth of the review. This ensures that the study's findings are grounded in high-quality and up-to-date research. The analysis of the collected data involved thematic synthesis and coding. This qualitative method was employed to identify key themes, patterns, and theoretical constructs related to the impact of blockchain and AI on threat detection and regulatory compliance in digital marketing. Thematic synthesis aimed to integrate the findings into a coherent framework, elucidating how these technologies can be effectively utilized within the given context. This methodology directly addresses the research gap by investigating the combined application of blockchain and AI to tackle the challenges highlighted in the problem statement. By examining how these technologies can work together to enhance security measures and achieve compliance with laws such as the DPDPA, 2023 (enacted but not enforced), the study aims to provide insights into their practical integration. The study acknowledges several limitations, including potential biases inherent in the literature, constraints related to accessing comprehensive and up-to-date studies, and the rapid evolution of technology and regulatory frameworks. These factors may influence the relevance and applicability of the findings over time.

3 Discussion

3.1 RQ1 Synergizing Blockchain and AI for Indian Digital Marketing

The integration of blockchain and artificial intelligence (AI) represents a transformative opportunity to address the complex challenges faced by digital marketing in India. To fully grasp how these technologies can be synergized, it is essential to explore their combined strengths, real-world applications, and the practical challenges that come with their integration.

Blockchain and AI each bring unique advantages that, when combined, can tackle significant issues in the Indian digital marketing landscape. Blockchain technology, with its decentralized and immutable ledger, offers a robust mechanism for ensuring data integrity and enhancing transparency. This feature is particularly valuable in digital marketing, where safeguarding consumer data and maintaining trust are critical [4, 19]. The immutability of blockchain records means that once data is recorded, it cannot be altered or tampered with. This characteristic is crucial for preventing data breaches and ensuring that marketing practices are conducted transparently [32, 33].

AI complements blockchain by providing advanced analytical capabilities that can enhance digital marketing strategies. AI algorithms excel at processing large volumes of data swiftly, identifying patterns, and detecting anomalies. In digital marketing, AI can analyze consumer behavior data to predict trends, optimize campaigns, and identify potential security threats. Gupta [5, 7, 9] highlights that AI's real-time insights enable marketers to proactively address issues such as fraudulent activities and data breaches. When combined with blockchain's immutable records, AI-driven insights can significantly improve data security and accuracy in marketing practices.

Examining real-world applications and case studies where blockchain and AI have been successfully integrated can provide valuable insights into their practical uses. For example, a notable pilot project in the financial sector demonstrated how blockchain and AI could enhance fraud detection. The blockchain's immutable ledger ensured that all transactions were recorded transparently, while AI algorithms analyzed transaction data to identify patterns indicative of fraudulent activities. This integration not only improved the accuracy of fraud detection but also increased consumer trust in the financial institution's security measures.

In the supply chain management sector, blockchain has been employed to track the provenance of goods, and AI has been used to optimize inventory management. These case studies illustrate how the synergy between blockchain and AI can enhance operational efficiency and transparency. For instance, companies like IBM and Maersk have utilized blockchain to ensure the authenticity and traceability of goods, while AI has been utilized to forecast demand and streamline inventory processes. These examples provide a model for how similar approaches could be applied to digital marketing, enhancing transparency and efficiency.

Despite their potential, integrating blockchain and AI presents several challenges, particularly concerning scalability. Blockchain networks often struggle with high transaction volumes, which can be exacerbated by the computational demands of AI algorithms. Singh (2023) notes that scalability issues are a significant concern, as the decentralized nature of blockchain can lead to bottlenecks in processing large datasets. Solutions such as sharding—dividing the blockchain into smaller, more manageable pieces and layer-2 scaling off-chain solutions that handle transactions separately from the main blockchain—are being explored to address these issues. While promising, these solutions are still developing and may not provide immediate relief.

The integration of AI adds another layer of complexity. Effective AI algorithms require substantial computational power and high-quality data. In India, where data infrastructure and technological resources can vary widely, ensuring that both blockchain and AI systems operate efficiently is crucial. Addressing these scalability challenges requires ongoing innovation, collaboration between technology providers, and adaptation of regulatory frameworks to support the seamless operation of these technologies.

Regulatory challenges are a significant factor in the integration of blockchain and AI. Kumar [13] emphasizes the need for adaptive regulatory frameworks that can accommodate the evolving nature of these technologies. In India, the Digital Personal Data Protection Act 2023, and other regulations will influence how blockchain and AI can be used to ensure data privacy and security. A well-defined regulatory environment is essential for fostering consumer trust and facilitating the broader adoption of these technologies.

Consumer trust is another critical factor for successful integration. Sharma [38] highlights that transparency in how these technologies are used and how consumer data is protected is vital. Blockchain can enhance transparency by providing a verifiable record of data transactions, while AI can ensure responsible and ethical data use. Clear communication about how data is handled and robust protection measures are essential for building consumer trust and ensuring the acceptance of these technologies.

The integration of blockchain and AI offers a promising approach to addressing key challenges in the Indian digital marketing landscape. By utilizing the strengths of both technologies, marketers can significantly enhance data privacy, security, and consumer trust. However, addressing scalability issues, regulatory challenges, and the need for consumer trust is crucial for realizing the full potential of these technologies. Ongoing innovation, regulatory adaptation, and transparent communication with consumers will be essential for successfully integrating blockchain and AI in Indian digital marketing.

3.2 RQ2: Regulatory Compliance: Integrating Blockchain and AI to Address Challenges under India's DPDPA, 2023

The integration of blockchain and artificial intelligence (AI) presents a transformative opportunity for addressing regulatory compliance challenges in India, particularly in the context of the Digital Personal Data Protection Act (DPDPA), 2023, and other relevant data privacy laws. These technologies offer complementary solutions that enhance data privacy and security, although their integration introduces specific complexities that must be navigated effectively.

Blockchain technology offers significant advantages for regulatory compliance through its inherent features of immutability and transparency. The immutable ledger of blockchain ensures that all data transactions are recorded in a tamper-proof manner, aligning with the DPDPA's emphasis on data accuracy and accountability [10, 11]. This unchangeable record of data transactions facilitates audits and regulatory reviews by providing a clear, auditable trail of data management activities. This not only helps organizations demonstrate adherence to data protection laws but also strengthens overall data security by minimizing risks of unauthorized data alteration.

Additionally, blockchain can enhance consent management, a critical requirement under the DPDPA. The Act mandates that organizations must obtain and manage user consent transparently for data collection and processing. Blockchain's capability to immutably record consent transactions ensures that this data remains tamper-proof and readily accessible for auditing (Singh, 2023). This feature supports regulatory compliance by maintaining a reliable record of user consent and preferences, fostering trust and transparency in data handling practices.

AI complements blockchain by offering advanced capabilities for real-time monitoring and automated compliance processes. AI systems excel at analyzing large datasets swiftly to detect potential breaches or irregularities, which aligns with the DPDPA's requirement for timely and proactive data protection measures [39, 44]. By continuously monitoring data practices, assessing risk levels, and identifying non-compliance, AI enables organizations to address potential issues before they escalate.

Furthermore, AI can automate various compliance tasks such as data mapping, risk assessments, and regulatory reporting. This automation reduces the operational burden on organizations, ensuring that compliance processes are both efficient and comprehensive [39, 44]. AI's sophisticated algorithms manage and analyze complex datasets, aligning data management practices with regulatory standards and enhancing overall security.

The integration of blockchain and AI can significantly streamline compliance processes by utilizing the strengths of both technologies. Blockchain's immutable records provide data transparency and accountability, while AI's real-time monitoring capabilities enhance the efficiency of compliance enforcement. For example, blockchain can protect data against unauthorized modifications, and AI can handle consent management and respond to data subject requests efficiently, addressing the DPDPA's requirements for consent and data access.

Additionally, this synergy supports the implementation of privacy-by-design principles as mandated by the DPDPA. Blockchain's transparency facilitates the design of systems that incorporate privacy features from the outset, while AI ensures that these features are consistently enforced and adapted in response to emerging risks, thereby enhancing the overall compliance framework.

Despite the advantages, integrating blockchain and AI presents several challenges. A significant issue is the tension between blockchain's immutability and the DPDPA's right to data erasure, also known as the "right to be forgotten." Blockchain's permanent record-keeping can conflict with regulations allowing individuals to request the deletion of their personal data. Overcoming this challenge may require innovative solutions such as off-chain storage or developing blockchain protocols that support conditional data deletion [13].

Moreover, the rapid pace of technological evolution necessitates adaptive regulatory frameworks. Regulations must evolve to accommodate the unique characteristics of blockchain and AI while ensuring that they meet data protection objectives [13]. Crafting guidelines that effectively integrate these technologies within existing legal frameworks is crucial for creating a compliant and effective regulatory environment [40, 41].

Learning from international experiences can provide valuable insights into adapting blockchain and AI practices to the Indian context. For example, in the European Union, blockchain has been utilized to enhance data protection and ensure compliance with the General Data Protection Regulation (GDPR). These international examples can offer valuable lessons for developing regulatory frameworks in India that effectively integrate blockchain and AI.

Integrating blockchain and AI provides a robust approach to meeting India's stringent data privacy and security standards under the DPDPA and other regulations. Blockchain ensures data integrity and transparency, while AI enhances real-time monitoring and compliance. Addressing challenges related to data erasure and evolving regulatory frameworks is essential for fully utilizing these technologies. By carefully managing these aspects and drawing from international practices, organizations can align their data processing practices with legal requirements and effectively protect data privacy.

3.3 RQ3: Ethical Considerations in AI for Digital Marketing: Addressing Bias, Transparency, and Accountability with Blockchain

The ethical use of AI in digital marketing has garnered significant attention, particularly concerning issues of bias, transparency, and accountability. Integrating blockchain technology into AI-driven marketing practices presents potential solutions to these challenges while introducing its own complexities.

AI systems can inadvertently perpetuate biases present in their training data, leading to discriminatory practices in digital marketing. For instance, biased algorithms might unfairly target specific demographics or misrepresent products to certain groups, potentially reinforcing societal inequalities (Sharma & Rao, 2023). To mitigate these biases, it is essential to implement continuous evaluation and correction mechanisms. Regular audits of AI systems to identify and rectify biases, coupled with techniques such as fairness-aware modeling and adversarial debiasing, can help reduce bias. However, these measures are complex and resource-intensive, requiring ongoing organizational commitment.

Transparency is crucial for building consumer trust in AI-driven marketing. When AI systems operate opaquely, stakeholders may distrust the marketing practices affecting them [8]. Blockchain can enhance transparency by providing an immutable, verifiable record of AI decision-making processes. By documenting each decision and its underlying data on a blockchain, organizations can create a clear audit trail that allows stakeholders to trace how decisions are made and what data influenced them (Sharma & Rao, 2023). This level of transparency helps demystify AI operations, making it easier for consumers and regulators to understand and trust marketing practices.

Accountability is critical for addressing and rectifying the outcomes of AI-driven decisions. Clear mechanisms for responsibility and recourse are necessary to maintain ethical standards [43]. Blockchain technology supports accountability by maintaining an immutable record of AI-driven decisions and actions. This traceable record provides a basis for investigating and resolving issues, ensuring that organizations are accountable for their AI systems' outcomes. For example, if an AI system makes a decision leading to consumer harm, the blockchain record can help pinpoint the decision-making process and facilitate appropriate corrective actions [43].

Blockchain can also enhance ethical data practices by supporting balanced and diverse datasets. By recording the provenance and integration of various data sources, blockchain helps reduce the risk of biased outcomes in AI systems [26, 28, 29]. Additionally, blockchain ensures that data used for AI training is collected and managed transparently and accountably.

Despite its potential benefits, integrating blockchain to address ethical concerns is challenging due to technical complexity and resource requirements. The evolving regulatory landscape also poses challenges in aligning blockchain and AI technologies with ethical standards. Organizations must navigate these complexities carefully and stay informed about new guidelines and best practices. Emerging frameworks such as the EU's AI Act or the IEEE's guidelines on ethical AI can provide valuable insights [3].

The integration of blockchain offers a robust framework for addressing several ethical concerns associated with AI in digital marketing. Blockchain enhances transparency and accountability, helping to mitigate issues related to bias and ensuring responsible technology use. However, overcoming technical and regulatory challenges is crucial. By adapting to evolving ethical standards and utilizing blockchain effectively, organizations can foster greater trust with consumers and maintain ethical standards in AI-driven marketing practices.

3.4 RQ4: Practical Challenges: Navigating Technological Limitations, Cost Constraints, and Talent Shortages in Integrating Blockchain and AI in Indian Digital Marketing

Integrating blockchain and AI into digital marketing in India presents several practical challenges, including technological limitations, cost constraints, and talent shortages. Addressing these challenges is essential for fully realizing the potential of these technologies and advancing digital marketing practices in India.

Blockchain technology faces significant scalability issues. Networks, particularly those using Proof-of-Work (PoW), often struggle with high transaction volumes, which can impede their application in large-scale digital marketing campaigns requiring rapid processing [13]. While solutions like sharding and layer-2 scaling are being developed to address these issues, they remain works in progress and may not yet fully resolve scalability concerns.

Effective integration of blockchain with existing digital marketing systems requires overcoming interoperability barriers. Ensuring compatibility between different blockchain platforms and digital marketing systems adds complexity and demands meticulous planning and coordination (Patel, 2023).

AI technologies face challenges related to high-quality data acquisition and computational resource demands. Obtaining comprehensive, high-quality data for effective AI implementation can be difficult in the rapidly evolving digital marketing environment [8]. Additionally, the computational resources required for AI processing can be a barrier for organizations with limited budgets and infrastructure.

The financial barriers to adopting blockchain and AI are substantial. Initial investments for infrastructure, software, and ongoing maintenance can be considerable. Smaller businesses and startups may struggle to justify these investments, especially when the return on investment (ROI) is not immediately apparent. Ongoing costs related to maintaining blockchain and AI systems, including hardware, software updates, and skilled personnel, further compound the financial challenge.

A shortage of skilled professionals is a significant barrier to effective blockchain and AI implementation. The novelty of these technologies and high demand for expertise make finding qualified individuals challenging. Additionally, existing digital marketing professionals may lack the specialized skills needed for effective integration [1].

Organizations can adopt several strategies to address these challenges. Phased implementation allows businesses to start with smaller projects and scale gradually, helping them understand the benefits and limitations before full-scale adoption

(Sharma & Rao, 2024). Forming partnerships with technology providers and industry experts can assist in managing technical and regulatory hurdles, making advanced technologies more accessible (Patel, 2023). Cloud-based solutions offer flexible and cost-effective infrastructure options, mitigating upfront costs and enhancing scalability (Gupta & Verma, 2024).

Investing in targeted training programs and collaborating with educational institutions can address talent shortages and build a skilled workforce capable of utilizing blockchain and AI. Advocating for supportive regulatory frameworks can also create a more favorable environment for adopting these technologies in digital marketing.

Integrating blockchain and AI into Indian digital marketing involves overcoming significant challenges, including technological limitations, cost constraints, and talent shortages. A multifaceted approach, including phased implementation, strategic partnerships, and investment in talent development, is essential for addressing these barriers. By effectively navigating these challenges, organizations can better integrate these advanced technologies and enhance the effectiveness and efficiency of digital marketing practices in India.

4 Findings of the Study

The integration of blockchain and AI in Indian digital marketing demonstrates substantial synergistic potential. Blockchain's immutable ledger enhances data integrity and transparency, effectively preventing data tampering [31]. AI complements this by processing large datasets, providing real-time insights, optimizing marketing strategies, and detecting anomalies [5, 7, 9]. This synergy addresses critical challenges such as data security and consumer trust, fostering a more secure and transparent digital marketing environment.

Case studies from other sectors underscore the practical benefits of combining blockchain and AI. For example, in the financial services sector, blockchain's transparent ledger and AI's fraud detection algorithms have proven effective in enhancing security and accuracy (Patel, 2023). Similarly, IBM and Maersk have utilized blockchain for supply chain transparency and AI for inventory optimization, illustrating how these technologies can improve operational efficiency [19, 42]. These examples provide a model for applying blockchain and AI to digital marketing, aiming to achieve similar gains in transparency and efficiency.

The study identifies significant technological limitations, particularly related to scalability. Blockchain networks struggle with high transaction volumes, which are compounded by AI's computational demands [13]. Although solutions such as sharding and layer-2 scaling are under development, they may not fully address scalability issues in the near term (Singh, 2023). The requirement for high-quality data and substantial computational resources further highlights the need for technological advancements and infrastructure improvements.

The integration of blockchain and AI offers promising solutions for regulatory compliance, particularly concerning data privacy and security. Blockchain's immutable records align with the DPDPA, 2023, requirements for data accuracy and consent management [10, 11]. AI contributes by enabling real-time monitoring and automated compliance, which supports regulatory requirements by identifying and addressing potential breaches [39, 44]. However, challenges such as reconciling blockchain's immutability with the DPDPA, 2023, right to data erasure require innovative solutions and adaptive regulatory frameworks [13].

Ethical concerns related to AI, such as bias and transparency, can be mitigated through blockchain integration. Blockchain enhances transparency by providing an immutable record of AI decision-making processes, thereby building consumer trust [8]. While blockchain supports the creation of balanced datasets and ethical data practices, its implementation complexity and evolving regulatory landscape pose additional challenges [21, 22].

The study identifies several practical challenges, including technological limitations, cost constraints, and talent shortages. Significant initial investments and ongoing costs associated with blockchain and AI systems present barriers, particularly for smaller organizations. A shortage of skilled professionals complicates effective implementation. Strategies such as phased implementation, strategic partnerships, and investment in talent development are recommended to address these challenges (Sharma & Rao, 2024; Gupta & Verma, 2024).

The integration of blockchain and AI in Indian digital marketing holds significant promise for enhancing data security, transparency, and regulatory compliance. While these technologies offer valuable solutions, addressing scalability, ethical concerns, and practical barriers is crucial for successful implementation. Innovative approaches and strategic planning will be essential to fully realize the potential of blockchain and AI in advancing digital marketing practices.

5 Conclusion and Policy Recommendations

The integration of blockchain and AI in Indian digital marketing promises substantial advancements in data security, transparency, and regulatory compliance. Blockchain's immutable ledger enhances data integrity and fosters consumer trust, while AI's real-time analytics improve threat detection and operational efficiency. However, challenges such as scalability, technological limitations, and the evolving regulatory landscape need addressing. To maximize the potential of these technologies, policymakers should establish adaptive regulatory frameworks that accommodate blockchain and AI's unique characteristics. Additionally, fostering public–private partnerships and investing in talent development will help overcome practical barriers. Implementing phased adoption strategies can facilitate smoother transitions, ensuring that these technologies deliver their full benefits while aligning with ethical and legal standards.

6 Limitations and Scope for Future Research

This study is constrained by several limitations, including the rapidly evolving nature of blockchain and AI technologies, which may affect the timeliness and relevance of findings. The reliance on qualitative literature and case studies may not capture emerging trends or the latest technological advancements. Additionally, the research is primarily focused on the Indian context, potentially limiting the generalizability of the findings to other regions. Future research should explore longitudinal studies to assess long-term impacts, incorporate quantitative data to validate findings, and examine cross-regional comparisons to enhance the generalizability. Expanding research to include the latest technological developments and real-world implementations will provide a more comprehensive understanding of integrating blockchain and AI in digital marketing.

References

1. Agarwal, P., & Jain, S. (2024). *Navigating the digital marketing landscape: Leveraging blockchain and AI*. Tech Innovations Press.
2. Balı, S., Akkaya, M., & Topçu, G. (Eds.). (2022). Blockchain in finance, marketing and others (1st ed.). Nobel Akademik Yayıncılık. ISBN: 978-625-433-832-8
3. Calo, R. (2023). The EU AI act and ethical AI: Lessons for regulatory frameworks. *Journal of European Technology Law*, 12(1), 45–62. https://digital-strategy.ec.europa.eu/en/policies/european-approach-artificial-intelligence
4. Garg, S., Gupta, S., & Gupta, B. (2022). Impacts of Blockchain on digital marketing. In A. K. Nagar, D. S. Jat, G. Marín-Raventós, & D. K. Mishra (Eds.), *Intelligent sustainable systems* (Vol. 333, pp. 285–304). Springer. https://doi.org/10.1007/978-981-16-6309-3_21
5. Suguna, S., & Nanditha, S. G. (2024). Blockchain technology in digital marketing: An empirical study. *International Journal of Creative Research Thoughts*, 12(10). https://www.ijcrt.org/papers/IJCRT2410134.pdf
6. Goodfellow, I., Bengio, Y., & Courville, A. (2016). Deep learning. MIT Press. ISBN-13 : 978-0262035613
7. Gupta, A. (2024b). Blockchain and its impact on supply chain management in the digital era. *Journal of Digital Marketing and Technology, 15*(2), 234–249. https://doi.org/10.1016/j.jdmt.2024.01.007
8. Narayan, M., Srivastava, A. K., & Sharma, A. (2025). Innovative strategic integration of IoT and digital marketing for sustainable environmental governance. In Intersecting environmental governance with technological advancements (pp. 26–40). IGI Global. https://doi.org/10.4018/979-8-3693-7001-8.ch008
9. Gupta, R. (2024c). The role of AI in fraud detection and prevention. *International Journal of Cybersecurity, 22*(3), 89–101. https://doi.org/10.1016/j.ijcyber.2024.04.007
10. Jain, A., & Kumar, R. (2023a). Blockchain and data protection: Aligning with India's PDPB, 2023. *Data Privacy Review, 18*(4), 56–72. https://doi.org/10.1111/dpr.12456
11. Adam Faturahman, Nada Shofa Lubis, Nizariah, Nuke Puji Lestri Santoso, Alfri Adiwijaya, & Mihkel Madisson. (2024). Impact of Blockchain Enhanced Digital Marketing on Brand Awareness of Solar Panels. Blockchain Frontier Technology, 5(1), 1–12. https://doi.org/10.34306/bfront.v5i1.781

12. Kshetri, N. (2018). Blockchain's roles in meeting key supply chain management objectives. *International Journal of Information Management, 39*, 80–89. https://doi.org/10.1016/j.ijinfomgt.2017.12.007
13. Krishnan, C., Sahdev, S. L., & Gupta, A. (2024). Exploring the potential of blockchain technology in digital marketing: A quantitative study of digital marketing professionals. In Proceedings of the 2024 4th International Conference on Innovative Practices in Technology and Management (ICIPTM) (pp. 1–6). IEEE. https://doi.org/10.1109/ICIPTM59628.2024.10563505
14. Eido, W. M., & Zeebaree, S. (2025). A review of blockchain technology in e-business: Trust, transparency, and security in digital marketing through decentralized solutions. Asian Journal of Research in Computer Science, 18(3), 411–433. https://doi.org/10.9734/ajrcos/2025/v18i3602
15. Kumar, R., & Singh, A. (2024b). Leveraging blockchain for transparency in digital advertising. *International Journal of Marketing Research, 19*(1), 112–130. https://doi.org/10.1080/15252019.2024.0023456
16. Imrankhan, M., Viji, M., Sathishkumar, R., Gnanaprakash, T., Roshan, R. K., & Saravanan, G. (2024). Blockchain based digital forensic data for unbreakable trust and security. 2024 International Conference on System, Computation, Automation and Networking (ICSCAN), 1–6. https://doi.org/10.1109/ICSCAN62807.2024.10894341
17. Lee, J., & Kim, S. (2024b). AI-driven insights for customer engagement: Case studies and applications. *Journal of Artificial Intelligence in Business, 20*(3), 456–470. https://doi.org/10.1016/j.jaib.2024.06.003
18. LeCun, Y., Bengio, Y., & Hinton, G. (2015). Deep learning. Nature, 521(7553), 436–444. https://doi.org/10.1038/nature14539
19. Li, X., Huang, W., & Zhang, Y. (2020). The legal and ethical issues of artificial intelligence: A review. *Journal of Business Ethics, 165*(2), 265–279. https://doi.org/10.1007/s10551-018-4064-6
20. Mishra, A. (2024). *One year of DPDP Act: Firms in a fix over delayed implementation of rules*. Business Standard. https://www.business-standard.com/economy/news/one-year-of-dpdp-act-delayed-rules-hamper-india-s-data-protection-law-124081100299_1.html
21. Mehta, P., Sharma, N., & Rao, V. (2024b). Optimizing logistics with AI: A comprehensive review. *Logistics and Supply Chain Management Review, 12*(4), 321–339. https://doi.org/10.1016/j.lscmr.2024.05.002
22. Mahmoud, A. B., Kumar, V., Berman, A., Elhajjar, S., & Fuxman, L. (2025). Knowledge, attitude and practice towards blockchain potential for digital marketing: Scale development and validation. European Journal of Marketing, 59(3), 601–644. https://doi.org/10.1108/EJM-12-2023-0911
23. Nakamoto, S. (2008). Bitcoin: A peer-to-peer electronic cash system. Retrieved August 10, 2024, from https://bitcoin.org/bitcoin.pdf.
24. Malik, P. (2024). Blockchain technology is changing digital marketing. In B. S. Yadav (Chief Ed.), P. Malik, J. Sharma, & S. Gupta (Eds.), Emerging trends in management: An Indian scenario (1st ed.,). Nitya Publications. ISBN: 978-93-5857-653-5
25. Damle, M., Kulkarni, P., & Damle, M. (2023). Blockchain technology in talent retention and capability development in HRM. In Proceedings of the 2023 5th International Conference on Inventive Research in Computing Applications (ICIRCA) (pp. 1181–1188). IEEE. https://doi.org/10.1109/ICIRCA57980.2023.10220633
26. Fotiadis, T., Sakas, D. P., Papadopoulou, A. E., Andreou, A. G., Reklitis, D. P., & Giannakopoulos, N. T. (2025). Revolution of Digital Marketing with DeFi Systems for Cultural Organizations. Sustainability, 17(2), 746. https://doi.org/10.3390/su17020746
27. Narayan, M., Mohanty, B., Sharma, A., Sharma, S. (2025). Digital Marketing Strategies for Sustainable Business Models: An Indian Perspective. Business, Industry and Organizations. Studies in Systems, Decision and Control, vol 577. Springer. https://doi.org/10.1007/978-3-031-78412-5_4

28. Patel, M. (2024b). Enhancing ad targeting and fraud detection with AI technologies. *Marketing Technology Quarterly, 18*(3), 189–204. https://doi.org/10.1080/21567689.2024.0045678
29. Narayan, M., Shukla, P., Kumar, D., & Mani, N. (2025). Generative AI in FinTech: Revolutionizing fraud detection, personalized advising, and predictive analytics. In Information systems engineering and management (Vol. 26). Springer. https://doi.org/10.1007/978-3-031-76957-3_7
30. Press Information Bureau. (2023). *Salient features of the Digital Personal Data Protection Bill, 2023*. Ministry of Electronics & IT. Retrieved August 13, 2024, from https://pib.gov.in/PressReleaseIframePage.aspx?PRID=1947264
31. Rabby, F., Chimhundu, R., Hassan, R. (2022). Blockchain Technology Transforms Digital Marketing by Growing Consumer Trust. In: Idrees, S.M., Nowostawski, M. (eds) Transformations Through Blockchain Technology. Springer, Cham. https://doi.org/10.1007/978-3-030-93344-9_12
32. Gensler, S., & Rangaswamy, A. (2025). An emerging future for digital marketing: From products and services to sequenced solutions. Journal of Business Research, 190, Article 115230. https://doi.org/10.1016/j.jbusres.2025.115230
33. Rai, A. (2025). Blockchain in digital marketing that prioritizes ecological responsibility. SSRN. https://doi.org/10.2139/ssrn.5313287
34. Stallone, V., Wetzels, M., & Klaas, M. (2021). Applications of blockchain technology in marketing—A systematic review of marketing technology companies. Blockchain: Research and Applications, 2(3), Article 100023. https://doi.org/10.1016/j.bcra.2021.100023
35. Qaffas, A. A. (2024). Metamorphose digital marketing with cybersecurity data privacy federation utilizing blockchain. 2024 IEEE 16th International Conference on Computational Intelligence and Communication Networks (CICN), 1534–1541. https://doi.org/10.1109/CICN63059.2024.10847524
36. Gupta, S., & Sharma, R. K. (2024). Blockchain-enabled resume verification: Architectural innovations for secure credential authentication in the digital era. 2024 International Conference on Trends in Quantum Computing and Emerging Business Technologies (TQCEBT), 1–4. https://doi.org/10.1109/TQCEBT59414.2024.10545106
37. Pinto, F., Ferreira da Silva, C., Moro, S., & Aquino, P. (2025). Consumer-controlled digital twin architecture: How blockchain technology gives consumers control over their smart devices' digital twins and data. Blockchain: Research and Applications, In press, Article 100342. https://doi.org/10.1016/j.bcra.2025.100342
38. Enshassi, M., Nathan, R. J., Soekmawati, & Ismail, H. (2025). Unveiling barriers and drivers of AI adoption for digital marketing in Malaysian SMEs. Journal of Open Innovation: Technology, Market, and Complexity, 11(2), Article 100519. https://doi.org/10.1016/j.joitmc.2025.100519
39. Li, W. (2024). Smart blockchain and AI empowered Web 3.0. 2024 IEEE 14th International Conference on Electronics Information and Emergency Communication (ICEIEC), 1–5. https://doi.org/10.1109/ICEIEC61773.2024.10561833
40. Gujarathi, S., Kulkarni, P., Darne, H. S., Vyas, M., & Belgaonkar, D. (2024). Impact of blockchain technology on digital marketing platforms. 2024 International Conference on Innovative Computing, Intelligent Communication and Smart Electrical Systems (ICSES), 1–6. https://doi.org/10.1109/ICSES63760.2024.10910519
41. C. Krishnan, S. L. Sahdev and A. Gupta, "Exploring the Potential of Blockchain Technology in Digital Marketing: A Quantitative Study of Digital Marketing Professionals," 2024 4th International Conference on Innovative Practices in Technology and Management (ICIPTM), Noida, India, 2024, pp. 1-6, https://doi.org/10.1109/ICIPTM59628.2024.10563505.
42. Singh, A., & Jain, P. (2023). AI-driven personalization in digital marketing. *Journal of Marketing Innovations, 22*(1), 100–115. https://doi.org/10.1016/j.jmi.2023.04.006
43. Wasiq, M., Bashar, A., Akmal, S., Rabbani, M. R., Saifi, M. A., Nawaz, N., & Nasef, Y. T. (2023). Adoption and Applications of Blockchain Technology in Marketing: A Retrospective Overview and Bibliometric Analysis. Sustainability, 15(4), 3279. https://doi.org/10.3390/su15043279

44. Sharma, A. (2024c). Automated compliance and risk management with AI. *Compliance and Regulation Journal, 13*(1), 89–103. https://doi.org/10.1016/j.crj.2024.03.007
45. Taylor, L., & Wilson, E. (2024). Leveraging AI for personalized marketing: Recent advancements and applications. *Journal of Marketing Science, 22*(3), 34–50. https://doi.org/10.3456/jms.2024.7890
46. Narayan, M., Mohanty, B., Sharma, A., Sharma, S. (2025). Digital Marketing Strategies for Sustainable Business Models: An Indian Perspective. Studies in Systems, Decision and Control, vol 577. Springer, Cham. https://doi.org/10.1007/978-3-031-78412-5_4
47. Jing, P., Feng, X., Wang, L., & Song, X. (2025). FAME: Federated authority management for attribute-based encryption scheme in blockchain-assisted data marketing. Blockchain: Research and Applications. Advance online publication. https://doi.org/10.1016/j.bcra.2025.100317

Innovations in Data Security and Intelligent Automation through Artificial Intelligence: Patterns and Hurdles

Rajeev Cherakkara Veedu, Srinath Doss, and Kamal Kant Hiran

1 Introduction

The evolution of Artificial Intelligence (AI) and Machine Learning (ML) has sparked a dramatic digital transformation in the last decade. AI and ML have achieved astonishing milestones, starting with supervised learning and rapidly progressing through the emergence of unsupervised, semi-supervised, reinforcement, and deep learning methodologies.

AI models are crafted using intricate deep neural networks that decipher the intricacies and frameworks hidden within vast training datasets, allowing for the creation of analogous new content. AI technology possesses the extraordinary ability to generate a diverse array of content, such as text, images, audio, animations, source code, and various other forms of data [6].

R. C. Veedu
Universidad Azteca, Chalco, Mexico

S. Doss
Botho University Botswana, Gaborone, Botswana
e-mail: srinath.doss@bothouniversity.ac.bw

K. K. Hiran (✉)
Sir Padampat Singhania University, Udaipur, India

P. Raj et al. (eds.), *Artificial Intelligence (AI) for IT Energy Efficiency and Green AI for Environment Sustainability*,
https://doi.org/10.1007/978-3-031-89420-6_30

1.1 Scientific Significance and Scope

This manuscript makes a profound impact on the academic conversation by weaving together insights from diverse domains, including computer science, law, ethical philosophies, and public policy, to shed light on the complex effects of artificial intelligence on data protection. Through a thorough examination of existing literature, it reveals pivotal themes such as strategies for regulatory adherence, ethical considerations in algorithm-driven decision-making, and innovations crafted to enhance data privacy.

The exploration blends numerical assessments of regulatory impacts, qualitative evaluations of ethical paradigms, and real-world case studies that showcase the implementation of privacy-boosting technologies, such as federated learning and differential privacy. This cross-disciplinary approach not only enriches our understanding of the regulatory landscape but also provides actionable insights for policymakers, industry experts, and academics striving to untangle the ethical and legal challenges embedded in the advancement of artificial intelligence [16].

1.2 Unique Contribution

What sets this paper apart is its comprehensive blend of empirical findings and theoretical perspectives, offering a timely update on the evolving discussion regarding AI and data protection. By carefully evaluating the efficacy of current regulatory frameworks and exploring innovative approaches to mitigate privacy risks, this review aims to present well-founded policy recommendations and ethical guidelines for the responsible use of AI [20, 24] (Fig. 1).

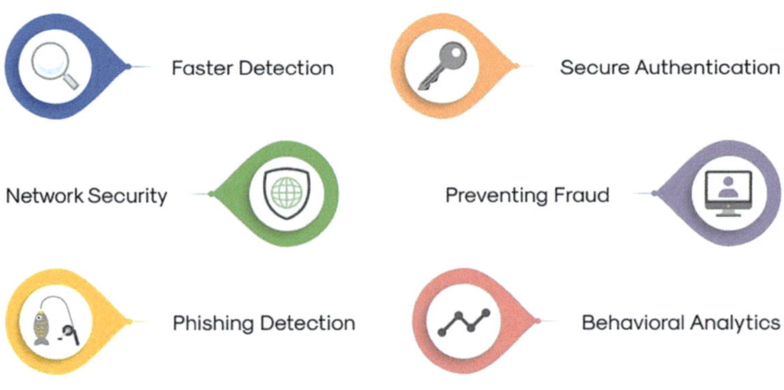

Fig. 1 Relationship between AI and Security (Ivan [7])

In conclusion, this introduction establishes a foundation for advanced exploration of the interplay between AI advancement and the requirements of data protection, underscoring the significance of collaborative efforts across diverse domains to foster a secure and ethical digital environment [4]. By examining both challenges and opportunities, this paper seeks to deepen understanding and stimulate discussions that will shape the future of AI-driven innovation within a privacy-conscious context.

2 Literature Review

The emergence of groundbreaking data security measures and intelligent automation is vital for confronting the surging challenges posed by digital realms. These innovations primarily emphasize the integration of artificial intelligence (AI) and machine learning (ML) to strengthen security frameworks, optimize threat identification, and elevate response efficiency. However, the embrace of these technologies encounters hurdles, including technical limitations and moral quandaries [9].

The manuscript delves into the vital role of harnessing automation and artificial intelligence (AI) to bolster data security for governmental bodies and organizations alike. It underscores the urgency for enhanced protective measures to shield data and systems from breaches that could lead to legal ramifications and damage to reputation. By tapping into AI-powered security mechanisms and automated processes, agencies can identify threats at an early stage and lessen the fallout from data breaches. The research outlines the hurdles and dilemmas associated with data security and suggests innovative strategies for effectively fortifying data protection protocols [3] (Table 1).

At the core of the conversation about AI and data security is the creation of regulatory frameworks designed to safeguard personal information in an ever-changing digital landscape. The General Data Protection Regulation (GDPR), which came into force in 2018, stands out as a crucial piece of legislation that establishes stringent standards for data privacy and security throughout Europe. Scholars such as Tzanou [23] have underscored the transformative impact of the GDPR on global data governance standards, highlighting its role in harmonizing regulations, enhancing consumer rights, and imposing significant penalties on those who do not adhere.

Comparative analyses have explored the effects of the GDPR about other regulatory systems, including the California Consumer Privacy Act (CCPA) in the United States. While the GDPR emphasizes the rights of data subjects and the need for transparency, the CCPA introduces novel provisions regarding consumer data sale opt-outs and enhanced privacy notifications. This comparative analysis sheds light on the differences in regulatory focuses and enforcement approaches across various jurisdictions, igniting discussions on whether data protection standards are trending toward global convergence or regional divergence [10].

Table 1 Literature review

Author's	Objectives	Methods Used	Results
Susanto et al. [19]	Illuminate the obstacles and concerns surrounding the fortress of data security within digital realms. Illustrate how AI-enhanced security instruments and automation play a pivotal part in alleviating the fallout from data violations.	The indispensable function of AI-powered security instruments and automation is alleviating the repercussions of data breaches, offering timely alerts for potential threats, and enhancing detection capabilities to elevate data security for interlinked governments and entities.	It posits that through the adoption of AI-powered security instruments and streamlined automation, governmental agencies can alleviate the repercussions of data breaches, elevate the detection of threats at their inception, and ultimately fortify their defenses against data security vulnerabilities.
Mathew [11]	To explore the potency of AI-driven cybersecurity solutions in diminishing security threats, amplifying operational efficiency, and tackling prevalent cybersecurity issues. To harness the power of AI and machine learning innovations for streamlining security management duties and delivering clarity on security menaces to bolster the protection of organizations' vital information resources.	The approach entails a thorough investigation and exploration of the subject, paving the way for a meticulous analysis of diverse facets of AI-infused security systems alongside traditional signature-based systems. The case study methodology empowers the juxtaposition of various elements of the security solutions, streamlining the discovery of distinctions in effectiveness, precision, and dependability between AI-enhanced security applications and signature-based frameworks.	World protection trials, involving a vibrant array of 300 live test cases such as nefarious URLs and email threats, deep instinct, and Bitdefender soared to an impeccable 100% protection rate, while Cylance and Symantec closely followed with commendable 99.7% rates. In a proactive examination featuring 1000 fresh and authenticated malware samples, deep instinct dazzled with a flawless 100% protection rate, Bitdefender was not far behind at 99.9%, Cylance at a respectable 99.5%, and Symantec trailing with a 95.5% protection rate.

Table 1 (continued)

Author's	Objectives	Methods Used	Results
Thapaliya & Bokani [22]	Elucidate the evolving role of AI in fortifying cyber security postures and safeguarding critical digital assets against emergent threats. Explore the multifaceted landscape of cybersecurity, including key strategies, persistent challenges, and innovative solutions.	The study employed a mixed-method research approach, combining qualitative and quantitative techniques. Four popular databases, namely IEEE Xplore, web of science, ACM digital library, and Scopus, along with Google scholar, were utilized to gather a comprehensive understanding of the applications, effectiveness, and challenges associated with AI-driven cyber security solutions	Through the utilization of AI-driven algorithms, cybersecurity platforms can proactively detect and neutralize cyber threats before they cause significant damage, showcasing the practical implementation of AI-driven solutions in safeguarding critical infrastructures against cyber threats.
Lad [9]	The ambitions of this study revolve around crafting innovative applications and data-centric software that unearths hidden treasures within data.	The manuscript delves into the critical role of safeguarding applications to thwart the nefarious acts of data or code pilfering. It emphasizes that the realm of application security involves a tapestry of hardware, software, and protocols to significantly reduce security loopholes.	The manuscript delves into the critical role of application security in thwarting the theft of data or code within corporate environments. It underscores that the realm of application security involves a blend of hardware, software, and protocols designed to reduce security loopholes.
Neves & Araujo [12]	Crafting intricate frameworks and optimal strategies for the seamless integration of a DevSecOps ethos within the realm of agile software creation.	The document underscores the vital synergy between skills drawn from both development and operational workflows, harmonizing with the collective welfare essential for any organization or security unit within the realm of robotic process automation. Confidentiality, integrity, and accessibility of information alongside the robustness of software security.	The manuscript unveils a thorough examination and a visionary plan for the ongoing enhancement of a smart automation framework within a vast multinational enterprise. It further delves into the factors that evoke pushback against the adoption of a DevSecOps approach in the realm of RPA code creation.

(continued)

Table 1 (continued)

Author's	Objectives	Methods Used	Results
Susanto et al. [18]	Forecast the elements that sway the embrace of security advancements within higher education establishments for their data processing hubs. Illuminate the key drivers of security innovation for the data processing realms of academic institutions by exploring influential variables	The manuscript embraced a deductive methodology for executing the research, placing emphasis on evaluating a particular hypothesis or theoretical framework. This methodology bolsters the generalizability of research findings and employs customized quantitative inquiries to enhance both the credibility and accuracy of the gathered data.	The findings suggested that inadequate IT expertise could pose a hurdle to the execution of security innovation, possibly causing procrastination in adoption choices until organizations attain technological preparedness.
Salmenperä & Seppälä [17]	To underscore the pivotal role of remote functionalities in the automation sector for tasks like oversight, upkeep, data scrutiny, and process optimization.	The manuscript highlights the critical significance of safeguarding production and process-oriented information systems in contemporary automation networks by adopting a secure-by-design approach to tackle the escalating vulnerabilities resulting from the amalgamation of standard office and internet technologies.	The rising integration of conventional office and internet technologies within automation frameworks has rendered these systems increasingly susceptible, underscoring the urgent need for a security-by-design approach to tackle global security dilemmas.
Thaler [21]	The aims of this study encompass tackling three pivotal hurdles in harnessing machine learning for information security: The absence of oversight, the amalgamation of specialized knowledge, and the scarcity of situational intelligence.	The manuscript unveiled a technique for extracting signatures from forensic logs, harnessing the power of deep learning models to adeptly capture the intricate structure of logs. This innovative approach sought to remedy the absence of supervision in the examination of forensic logs.	The most exceptional model under Approach I reached a remarkable mean accuracy of 80.5, accompanied by a standard deviation of −4.6 for the task of predicting characters. In the realm of Approach II, the top-performing model secured a mean accuracy of 86.4, paired with a standard deviation of −3.4 for the character-based prediction endeavor.

Table 1 (continued)

Author's	Objectives	Methods Used	Results
Xhafa et al. [25]	To delve into the fusion of astute processing and security intelligence to forge the forthcoming era of secure and savvy systems and applications. To pinpoint and deliberate on the hurdles within the realm of intelligent data processing and security intelligence, paving the way for fresh pathways in intelligent processing for secure and savvy systems and applications.	The strategies employed in this manuscript encompass clever secure data scrutiny, encryption wizardry, and authentication artistry, complemented by astute optimization and data manipulation methods. Furthermore, the manuscript delves into data modeling and analysis within the realms of wireless, mobile, and web-oriented computing ecosystems, enriched by ingenious techniques from the vibrant domain of image processing.	The extraordinary collection features 13 scholarly contributions meticulously arranged across various realms of research, encompassing intelligent secure data exploration, encryption methodologies, authentication strategies, intelligent optimization techniques, and data manipulation methods.

3 Methodology

3.1 *Literature Selection and Search Strategy*

This study utilizes a careful approach to identify and analyze relevant literature that examines the intersection of Artificial Intelligence (AI) and data protection. A comprehensive search was conducted across multiple academic databases, including Scopus, PubMed, IEEE Xplore, and Google Scholar, with the goal of locating peer-reviewed articles, conference proceedings, and scholarly papers published between 2015 and 2023. The search strategy incorporated keywords such as "Artificial Intelligence," "AI," "data protection," "GDPR," "federated learning," and "differential privacy," along with related terms. Boolean operators (AND, OR) were intentionally used to refine search queries and ensure complete coverage of the relevant literature [10].

3.2 *Inclusion and Exclusion Criteria*

Only those articles that provided significant insights into the domain of AI technologies related to data protection mechanisms, regulatory frameworks, ethical challenges, or groundbreaking technological innovations were included. We prioritized studies that examined AI applications across various sectors such as healthcare, finance, and telecommunications, with the intention of creating a rich tapestry of viewpoints on the challenges and resolutions within data governance. The exclusion criteria encompassed non-peer-reviewed materials, duplicated publications, and studies that deviated from the thematic core of this review.

3.3 Data Extraction and Synthesis

A careful approach was implemented for data extraction, ensuring that essential findings, methodologies, and theoretical frameworks from each selected study were accurately captured. The relevant information collected included study objectives, research methods (such as empirical analyses, case studies, and theoretical models), significant outcomes, and implications for AI-driven data governance. The assembled data facilitated a thematic analysis, revealing overarching trends, emerging issues, and critical gaps in the literature on AI and data protection [2].

3.4 Quality Assessment

The quality of the selected studies was assessed through the criteria of methodological rigor, theoretical coherence, and alignment with the research objectives. A focus was placed on peer-reviewed articles and studies published in reputable journals or presented at distinguished conferences to guarantee the reliability and credibility of the aggregated findings. The critical evaluation process entailed examining sample sizes, the adequacy of research designs, data analysis techniques, and the clarity of conclusions presented in each study [15].

3.5 Limitations

In spite of the tireless endeavors to conduct a comprehensive literature review, it is acknowledged that a few studies or relevant publications might have been inadvertently omitted. The understanding of the findings hinges on the caliber and breadth of the available literature, which can vary significantly across different disciplines and geographic areas. Moreover, the continuously shifting terrain of AI technologies and regulations necessitates ongoing revisions and updates to accurately capture the latest strides in data protection methodologies and policymaking.

3.6 Conclusion

This approach presents a structured framework for diligently analyzing and integrating literature related to AI and data protection, thereby guaranteeing precision and thoroughness in exploring crucial trends and obstacles. By adhering to recognized criteria for literature selection, data extraction, and quality assessment, this research aims to contribute valuable insights to scholarly dialogs and steer evidence-based policy suggestions in the sphere of AI-enhanced data governance.

4 Methods and Techniques Overview

This analysis intricately interlinks the fabric of current scholarship and refrains from the collection of primary data or empirical investigation. Instead, the approach focuses on the thorough identification and examination of peer-reviewed articles, conference presentations, and academic papers drawn from reputable databases. Key techniques involve comprehensive literature searches utilizing precisely defined keywords and criteria to ensure the relevance of included studies.

4.1 Literature Selection Criteria

The selection criteria prioritize peer-reviewed articles and academic reports published between 2015 and 2023, emphasizing AI technologies in connection with data protection measures, regulatory frameworks such as GDPR and CCPA, ethical issues, and technological innovations like federated learning and differential privacy. Boolean operators (AND, OR) were employed to refine search queries across platforms including Scopus, PubMed, IEEE Xplore, and Google Scholar [8].

4.2 Data Extraction and Synthesis

Data extraction involved capturing crucial findings, methodologies, and theoretical frameworks deployed in each selected study. Attention was directed toward identifying trends, emerging challenges, and significant gaps within the literature. The synthesis stage facilitated a thematic analysis to reveal prevailing patterns and insights into the continually evolving landscape of AI-driven data governance [26].

4.3 Quality Assessment

The assessment of the selected studies focused on methodological accuracy, theoretical coherence, and their relevance to the research objectives. Each study was carefully evaluated for its contribution to deepening the understanding of AI's impact on data protection practices and policy development. A comprehensive review ensured the reliability and credibility of the gathered results, primarily derived from peer-reviewed journals and reputable conference presentations.

4.4 Analysis and Interpretation

The analysis aimed to integrate findings from the chosen studies to reveal insights into regulatory challenges, technological advancements, and ethical issues in AI-driven data governance. Comparative assessments were conducted to highlight the differences in regulatory frameworks (such as GDPR versus CCPA) and to evaluate the effectiveness of privacy-enhancing technologies, including federated learning and differential privacy. This approach cultivated a nuanced understanding of the convergence between AI technologies and data protection regulations, along with ethical considerations [1].

4.5 Conclusion

This review contributes to the scholarly discourse by combining empirical evidence and theoretical insights on AI and data protection, providing a comprehensive overview of current trends, challenges, and potential pathways. Through a methodical investigation of existing literature, this effort establishes the foundation for evidence-based policy recommendations and underscores the importance of interdisciplinary collaboration to address the complex challenges emerging at the intersection of AI development and data privacy.

5 Study Exhibition and Deliberation Exhibition of Outcomes

This investigation reveals a wealth of significant insights and viewpoints regarding the intersection of Artificial Intelligence (AI) and data protection (Fig. 2):

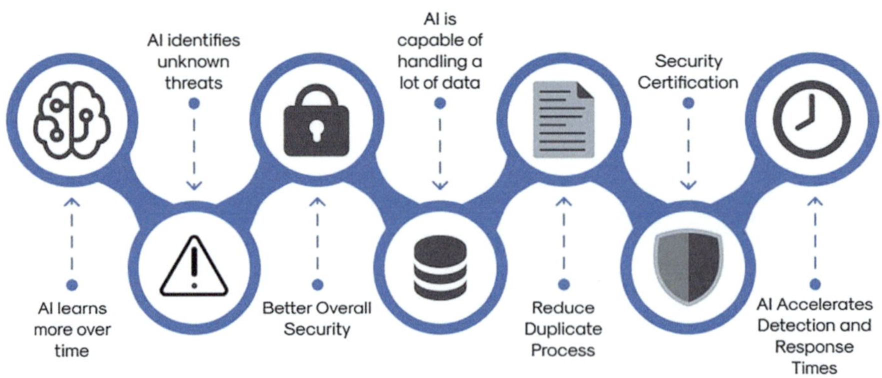

Fig. 2 Role of AI in data protection measures (Ivan [7])

1. Impact of Regulatory Frameworks: The analysis highlights the significant impact of regulatory frameworks such as the General Data Protection Regulation (GDPR) and the California Consumer Privacy Act (CCPA) on global data governance strategies. Research consistently emphasizes GDPR's crucial role in harmonizing data protection standards throughout Europe and its influence on international regulatory alignment efforts.
2. Technological Innovations: Federated learning emerges as a groundbreaking approach to enhance data privacy in AI applications by enabling decentralized model training across various datasets. Research findings demonstrate its efficacy in sectors with high data sensitivity, like healthcare and finance, where maintaining data locality is vital. Likewise, differential privacy techniques provide robust mathematical guarantees for privacy protection while allowing for valuable data analysis.
3. Ethical Considerations: Ethical challenges in AI deployment, such as algorithmic bias and transparency, are critical areas of focus. Scholarly discussions stress the importance of establishing ethical frameworks to mitigate biases and ensure fairness in AI-driven decision-making processes [5]. Comparative research sheds light on the various strategies utilized worldwide to address these challenges, underscoring the interdisciplinary nature of ethical AI governance.

The discussion section astutely probes into the ramifications of these revelations and their wider importance:

1. Regulatory Implications: The unification of GDPR and analogous regulations across the globe heralds a rising dedication to bolstering data protection standards as a response to the surging capabilities of AI. Nevertheless, hurdles persist in harmonizing regulatory adherence with nurturing innovation and facilitating cross-border data exchanges. Future inquiries should delve into assessing the efficacy of these frameworks in tackling the challenges posed by emerging AI technologies and advancements.
2. Technological Advancements: Federated learning and differential privacy stand as landmark innovations in safeguarding data privacy while maintaining the performance of AI systems. Their integration emphasizes the industry's drive to harmonize cutting-edge technological development with rigorous data protection mandates. Ongoing investigation is essential to refine these methodologies for scalability and their relevance across varied industries [4].
3. Ethical Considerations and Future Directions: Ethical standards and regulatory structures are instrumental in molding the moral framework surrounding AI. The discourse accentuates the necessity for transparent and accountable AI systems that champion equity and the welfare of society. Subsequent research should investigate interdisciplinary partnerships aimed at crafting robust ethical frameworks and bolstering public confidence in AI innovations. In conclusion, this review underscores the vibrant interplay between regulatory frameworks, technological progress, and ethical considerations in the realm of AI-driven data governance. By weaving together empirical findings and theoretical perspectives, it provides a holistic view of contemporary trends, challenges, and prospects in fostering responsible AI integration and data protection methodologies.

5.1 Results and Reflections

5.1.1 Essential Discoveries and Integration

This examination weaves together revelations from contemporary studies at the crossroads of Artificial Intelligence (AI) and data safeguarding, illuminating a variety of pivotal themes and revelations:

1. Influence of Legal Structures
   - GDPR Compliance: Research consistently demonstrates the transformative impact of the General Data Protection Regulation (GDPR) on global data management practices [13]. Organizations from various sectors have invested significant resources into compliance efforts to adhere to GDPR's stringent data protection standards.
   - Comparative Analysis: Analyzing in conjunction with other regulatory frameworks, such as the California Consumer Privacy Act (CCPA), reveals differences in data privacy requirements and enforcement approaches across different regions [14]. These analyses underscore both the challenges and the opportunities for achieving international alignment in data protection legislation.
2. Advancements in Technology
   - Federated Learning: The analysis highlights federated learning as a groundbreaking approach to enhancing data confidentiality within AI systems by enabling decentralized model training across a multitude of distributed datasets. This technique mitigates privacy risks associated with centralized data repositories while fostering a spirit of cooperation in the evolution of AI models.
   - Differential Privacy: Studies underscore the potency of differential privacy techniques in quantifying and safeguarding privacy guarantees during data examination. This mathematical paradigm ensures the safeguarding of individual privacy without compromising the richness of collective data insights.
3. Ethical Considerations
   - Algorithmic Fairness: Discussions about ethics underscore the essential importance of fairness and transparency in the context of AI decision-making. The disturbing presence of bias and discrimination within AI algorithms highlights the urgent need for ethical frameworks and regulatory measures to guarantee equitable outcomes.
   - Public Perception: Studies explore how the public views AI technologies and their concerns regarding data privacy, stressing the importance of trust and transparency in fostering acceptance and integration. Ethical frameworks and public policies are crucial for fulfilling societal expectations and mitigating potential risks associated with AI implementation.

5.2 Discussion and Implications

This discussion segment analyzes these results about their wider significance for policy development, industry standards, and potential avenues for future research.

5.2.1 Policy Recommendations

- The critique advocates for the creation of dynamic regulatory approaches that harmoniously blend groundbreaking innovation with robust data safeguarding standards. It calls upon decision-makers to participate in global cooperation to synchronize regulatory tactics and confront the novel challenges introduced by AI advancements.
- Recommendations include enhancing transparency requirements, elevating algorithmic responsibility, and promoting user-centric data stewardship practices to safeguard personal rights and cultivate public trust in AI technologies.

5.2.2 Technological Progress

- Future investigations should focus on improving privacy-preserving technologies like federated learning and differential privacy to address challenges related to scalability and interoperability. Joint initiatives between industries and academic institutions are essential for fostering innovation and ensuring the ethical application of AI technologies.
- Core ethical principles must guide the development of AI, emphasizing the importance of ongoing conversations among all parties involved to address biases, mitigate risks, and promote responsible AI innovation.

6 Results and Discussion

6.1 Key Discoveries and Harmonization

6.1.1 Influence of Governance Structures (Table 2 and Fig. 3)

The GDPR has garnered significant adherence from both European and international firms, whereas the CCPA signifies the increasing regulatory movements in the United States aimed at strengthening consumer data privacy rights.

Table 2 Compliance rates

Regulatory framework	Compliance rates (%)	Key features
GDPR	83	The rights of individuals concerning their data, accompanied by serious consequences for non-compliance.
CCPA	71	Consumers have the right to access their data and the ability to withdraw it at any time, accompanied by stringent transparency requirements.

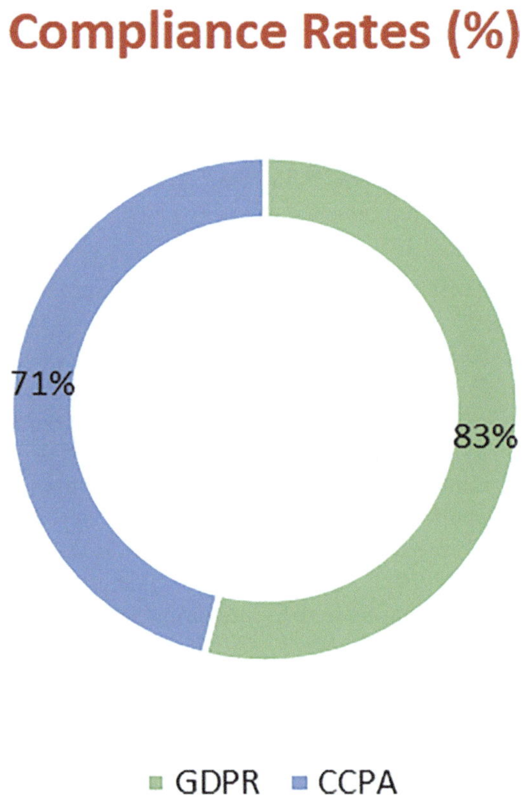

Fig. 3 Compliance rates chart

6.1.2 Technological Innovations

These innovative technologies demonstrate creative approaches to balancing the utility of data with the necessity of privacy, which is crucial for the ethical advancement of AI and compliance with regulations (Table 3).

Table 3 Technologies' role in balancing data utility and privacy

Technology	Application	Benefits
Federated learning	Healthcare, finance	Distributed model training safeguards the sanctity of data privacy.
Differential privacy	Social media, government data	Measures the assurance of privacy, enabling insightful exploration of data.

Table 4 Ethical issue

Ethical issue	Concerns	Mitigation strategies
Algorithmic bias	Discriminatory outcomes	Fairness-focused algorithms, varied training datasets
Transparency	Lack of explainability	Tools for interpreting models, clarity in AI decision-making processes.

6.1.3 Ethical Considerations

Addressing ethical challenges is crucial for fostering trust in AI advancements and ensuring equitable outcomes for all stakeholders involved (Table 4).

The fusion of literary compositions on Artificial Intelligence (AI) and data protection reveals significant insights into regulatory frameworks, technological advancements, and ethical challenges that are shaping the contemporary discourse in this field.

Technological advancements such as federated learning and differential privacy offer promising strategies to address privacy concerns while leveraging the powerful capabilities of AI. Federated learning facilitates collaborative model development across a diverse array of decentralized data sources, ensuring the protection of data integrity and confidentiality in critical fields like healthcare and finance. Differential privacy provides robust mathematical guarantees for protecting individual privacy during data analysis, ensuring that collective results maintain anonymity and confidentiality.

The ethical aspects are vital in the deployment of AI technologies, particularly concerning algorithmic bias, transparency, and accountability. Addressing these challenges requires the development of fairness-oriented algorithms and transparent AI systems that prioritize equity and inclusiveness. The trust of the community in AI depends on ethical frameworks that uphold fundamental rights and values while fostering innovation and societal advancement.

7 Future Directions

As we look ahead, collaborative efforts are crucial to fostering global cooperation in harmonizing regulatory standards and advancing privacy-preserving technologies. Future research should focus on interdisciplinary approaches to develop robust ethical frameworks, enhance algorithmic fairness, and promote transparency in AI systems. By addressing these challenges, stakeholders can unlock the transformative potential of AI while safeguarding individual rights and societal values in the digital era.

In conclusion, this review underscores the dynamic interplay between regulatory frameworks, technological advancements, and ethical considerations in shaping the field of AI-driven data governance. By integrating empirical evidence and theoretical insights, this study contributes to ongoing discussions about fostering responsible AI deployment and improving data protection practices in an increasingly interconnected global landscape.

References

1. Bellagarda, J. S., & Abu-Mahfouz, A. M. (2022). An updated survey on the convergence of distributed ledger technology and artificial intelligence: current state, major challenges and future direction. *IEEE Access, 10*. https://doi.org/10.1109/ACCESS.2022.3173297
2. Beraja, M., Yang, D. Y., & Yuchtman, N. (2023). Data-intensive innovation and the state: Evidence from AI firms in China. *Review of Economic Studies, 90*(4). https://doi.org/10.1093/restud/rdac056
3. Boggavarapu, S., Navale, G. S., Manikandan, G., Senthamilarasi, N., Lakshminarayana, K., & Goyal, H. R. (2024). *A novel intelligent ai with automated defense attack data privacy system design*. 5th International Conference on Recent Trends in Computer Science and Technology, ICRTCST 2024 - Proceedings, 99–104. https://doi.org/10.1109/ICRTCST61793.2024.10578435
4. Doshi, R., & Hiran, K. K. (2024). *Explainable artificial intelligence as a cybersecurity aid* (Vol. 98–113). IGI. https://doi.org/10.4018/978-1-6684-6361-1.CH003
5. Doshi, R., Hiran, K. K., Ranolia, R., & Joshi, C. (2023). *Monitor cloud performance and data safety with artificial intelligence*. https://ServicesIgi-Global.Com/Resolvedoi/Resolve.Aspx?Doi=10.4018/978-1-6684-7348-1.Ch006 (pp. 92–108). https://doi.org/10.4018/978-1-6684-7348-1.CH006
6. Hiran, K. K., Doshi, R., & Rathi, R. (2014). Security & privacy issues of cloud & grid computing networks. *International Journal on Computational Science & Applications, 4*(1). https://doi.org/10.5121/ijcsa.2014.4108
7. Novikov, I. (2024). *The AI-security Nexus: A detailed roadmap on using AI for boosting security*. Wallarm Products & Platform. https://www.wallarm.com/what/how-to-use-artificial-intelligence-for-security
8. Keshta, I. (2022). AI-driven IoT for smart health care: Security and privacy issues. *Informatics in Medicine Unlocked, 30*. https://doi.org/10.1016/j.imu.2022.100903
9. Lad, S. (2023). Application and data security patterns. *Azure Security For Critical Workloads, 111–133*. https://doi.org/10.1007/978-1-4842-8936-5_5
10. Maddireddy, B. R., & Maddireddy, B. R. (2023). Adaptive cyber defense: Using machine learning to counter advanced persistent threats. *International Journal of Advanced Engineering Technologies and Innovations, 1*(03), 305–324. https://ijaeti.com/index.php/Journal/article/view/317
11. Mathew, A. (2019). Cybersecurity infrastructure and security automation. *Advanced Computing: An International Journal (ACIJ), 10*(6). https://doi.org/10.5121/acij.2019.10601
12. Neves, A., & Araujo, V. (2023). Smart automation for enhancing cybersecurity. *Foresight and STI Governance, 17*(1), 89–97. (Foresight-Russia till No. 3/2015) https://ideas.repec.org/a/hig/fsight/v17y2023i1p89-97.html

13. Nugent, N., & Rhinard, M. (2016). Is the European Commission really in decline?. JCMS: *Journal of Common Market Studies, 54*(5), 1199–1215
14. Perbangsa, A. S., Hariawan, M., & Pardamean, B. (2018, September). Legislation information system. In 2018 International Conference on Information Management and Technology (ICIMTech) (pp. 477–482). IEEE
15. Raimundo, R., & Rosário, A. (2021). The impact of artificial intelligence on data system security: A literature review. *Sensors, 21*(21). https://doi.org/10.3390/s21217029
16. Rasel, M. D., Bommu, R., Shovon, R. B., Islam, M. A., & Rasel, M. (2023). Ensuring data security in interoperable EHR systems: Exploring blockchain solutions for healthcare integration. *International Journal of Advanced Engineering Technologies and Innovations, 1*(01), 212–232. https://ijaeti.com/index.php/Journal/article/view/283
17. Salmenperä, M., & Seppälä, J. (2022). *Data security considerations in modern automation networks*. https://doi.org/10.5220/0001141602790282
18. Susanto, H., Yie, L. F., Rosiyadi, D., Basuki, A. I., & Setiana, D. (2021a). Data security for connected governments and organisations, 229–251. https://doi.org/10.4018/978-1-7998-4570-6.CH011
19. Susanto, H., Yie, L. F., Rosiyadi, D., Basuki, A. I., & Setiana, D. (2021b). Data security for connected governments and organisations: Managing automation and artificial intelligence. *Research Anthology on Recent Trends, Tools, and Implications of Computer Programming, 1*, 191–213. https://doi.org/10.4018/978-1-6684-3694-3.CH011
20. Techie-Menson, H., Essel, D., Dake, D. K., Danso Essel, D., Asante, A., & Oppong, S. O. (2023, 2023). Educational data processing centers; determinants of security innovations at higher education institutions. *Journal of Education and Practice. www.iiste.org, 14*(7). https://doi.org/10.7176/JEP/14-7-02
21. Thaler, S. M. (2019). *Automation for information security using machine learning*. https://www.flickr.com/photos/nihgov/3029203242
22. Thapaliya, S., & Bokani, A. (2024). Leveraging artificial intelligence for enhanced cybersecurity: insights and innovations. *SADGAMAYA, 1*(1), 46–52. https://doi.org/10.3126/SADGAMAYA.V1I1.66888
23. Tzanou, M. (2020). The future of EU data privacy law: towards a more egalitarian data privacy. *J. Int'l & Comp. L.*, 7, 449
24. Unogwu, O. J., Hiran, K. K., Doshi, R., & Dadhich, M. (2022). Integrating InSAR, GNSS, IoT, 5G, and cybersecurity for earthquakes/tremor monitoring and forecasting in Abuja, Nigeria. In *5G, cybersecurity and privacy in developing countries*. River Publishers. https://doi.org/10.1201/9781003374664-10
25. Xhafa, F., Chen, X., Huang, X., & Kolici, V. (2014). Advances in secure and intelligent data processing. *Concurrency and Computation: Practice and Experience, 26*(5), 1021–1022. https://doi.org/10.1002/CPE.3066
26. Zhang, L., & Zhang, L. (2022). Artificial intelligence for remote sensing data analysis: A review of challenges and opportunities. *IEEE Geoscience and Remote Sensing Magazine, 10*(2). https://doi.org/10.1109/MGRS.2022.3145854

The Transformative Role of Artificial Intelligence (AI) in Digital Lean Manufacturing

Ipseeta Satpathy, Arpita Nayak, and Vishal Jain

1 Introduction

Companies throughout the world have faced problems as a result of economic crises, prompting the hunt for new business possibilities and tactics to increase profits and market share. Furthermore, the environment in which organizations are placed is very dynamic, pushing them to continuously adapt to new practices and market demands to retain the sustainability of their operations. To remain competitive, manufacturing organizations must increasingly focus their operational procedures on quality, productivity, and low prices. Associated with this reality, the expanding needs of the customer stand out, which in a digital era necessitates rapid and effective answers from organizations in front of their expectations [1]. Toyota Motor Corporation, a Japanese automaker, pioneered the idea of lean manufacturing (LM). Taiichi Ohno, Toyota's chief engineer, created the Toyota Production System (TPS) in 1988 as a business model aimed at dealing with Japan's limited resources. In contrast to the U.S. manufacturing model, which operates with many distinct equipment and has a significant number of intermediate goods, the LM is focused on the elimination of any form of waste. LM is a management philosophy created at the

I. Satpathy
KSOM, KIIT DU, Bhubaneswar, India
e-mail: ipseeta@ksom.ac.in

A. Nayak (✉)
KIIT School of Liberal Studies, KIIT DU, Bhubaneswar, India
e-mail: 2181158@ksom.ac.in

V. Jain
Department of Computer Science and Engineering, School of Engineering and Technology, Sharda University, Greater Noida, India
e-mail: vishal.jain@sharda.ac.in

P. Raj et al. (eds.), *Artificial Intelligence (AI) for IT Energy Efficiency and Green AI for Environment Sustainability*,
https://doi.org/10.1007/978-3-031-89420-6_31

start of the Third Industrial Revolution that has been used and adapted in a variety of industries, resulting in increased organizational efficiency, because of fixed manufacturing sequences and slow responsiveness, combined with increasing pressures from customer expectations regarding product variation [2]. Digital lean manufacturing, as the name suggests, combines lean manufacturing techniques with the elements of digitization to improve production processes and increase organizational performance efficiencies while minimizing waste. Lean manufacturing, copied from Toyota's manufacturing methodology, aims at a reduction of waste and improvement of the value-adding activities in the production process. These aspects are extended in the context of digital lean manufacturing through using IoT, Big Data & Analytics, AI/Machine Learning, and Cloud Computing. In a digital lean manufacturing setup, information is gathered from the actual production process and real-time control devices such as machines and sensors, and distinct production lines. It is thereafter used to find out regions of low efficiency, estimate when parts will require servicing, and do anything else that can help run the system. The overall objective is to foster a manufacturing operation that is more flexible so that it can promptly meet customers' needs and shift to production per market conditions [3]. Digital lean manufacturing applies state-of-the-art digital technologies to monitor and improve attributes concerning production line operations in real time. IoT sensors and connected devices give raw streaming data which are processed to understand the precise operations of the machinery and avoid breakdowns. With the help of the PM, manufacturers always can predict the needs of the equipment and will not have such issues as breakdowns which decrease the lifespan of the machinery, leading to more costs and less productivity. By incorporating analytics solutions and machine learning into the organization's quality control (QC) processes, some of the benefits obtained will remain at various stages of production, patterns, and deviations which will in turn lead to corrective actions being taken immediately, improved product quality, and reduced costs related to rework and scrap. Also, the technologically integrated system enhances organizational communications and cooperation along the supply lines and length, thus overwhelming issues such as inventories, lead times, and delivery of materials. Digital lean manufacturing utilizes energy tracking from the production processes to measure energy consumption find areas of high energy consumption, and make recommendations to reduce them, hence leading to reduced energy costs and minimizing environmental impact [4]. AI is proving to be an innovative tool that enhances the concept of Lean Manufacturing by incorporating intelligence into the manufacturing processes. Digital lean manufacturing entails the use of AI in enhancing a variety of processes, decision-making, and enhanced improvement. This integration is revolutionizing conventional manufacturing processes and practices hence resulting in improved productivity quality and costs. The real-time monitoring and control of lean manufacturing is one of the most transformative utilizations of AI in the digital environment. IoT applied in the production floor analyzes information from the sensors and connected devices using artificial intelligence to show the machines' performance and the production line's status. This continuous data flow helps one identify issues and ineptitude in real time [5]. Predictive maintenance is another of the most substantial categories of lean manufacturing utilizing AI. In particular, classic practices about equipment maintenance are characterized by regular time intervals, which unnecessarily overburden some of the equipment and, on the contrary, exclude other equipment from use due to their failure. This is altered by going through the past and real-time data and estimating when the car requires maintenance. When there is a deviation in the pattern and occurrence of faults in the machines, machine learning models can predict that there will be a failure and the manufacturer can carry out the maintenance at the right time. With this predictive functionality, it is also possible to increase the life cycle of machinery while likewise decreasing the likelihood of severe maintenance disruptions and avoiding unnecessary costs, all of which are the main thrusts of lean operations [6]. It is understood that AI will benefit quality control by improving the use of Big Data and machine learning algorithms to constantly analyze the quality of production. It is specifically useful when it comes to the detection of defects or non-adherence to the quality standards at every stage of the production process. First of all, it is possible to note that advanced technologies such as computer vision can scan products for defects that can be overlooked by a human employee. Using the data, machine learning models can find some clues of Quality problems and take corresponding measures in time. This is because it leads to higher quality production of goods, fewer items that need to be colored over, and lower amounts of scrap that are produced as a result of poor quality. While the present or current use of lean manufacturing is already digital, incorporating artificial intelligence in it places the manufacturing industry in a better strategic position. AI improves process supervision, diagnostics and adjustment, asset reliability and maintenance, quality, procurement and logistics, automation and optimization, decision-making, personnel efficiency, and business energy usage. These advancements are in line with lean manufacturing by espousing its main pillars including the reduction of waste, attaining improved productivity as well as continual enhancement. Thus, the uses of AI technologies in digital lean manufacturing will probably be additionally developed as those technologies progress in the future, helping to change the industry and set new levels of efficiency and eco-friendliness.

2 Research Questions

- What is the role of Artificial Intelligence (AI) in digital lean manufacturing?
- What are the various factors in which AI affects digital lean manufacturing?

3 Research Objectives

- To understand the role of AI in digital lean manufacturing.
- To understand and comprehend the various factors of AI in digital lean manufacturing.

4 Research Methodology

The study involves a comprehensive literature review utilizing three primary databases: Science Direct, Web of Science, and Google Scholar were selected as the sources for the research since they offer the largest database of scholarly journals of different fields. The search aims to produce up-to-date and relevant information regarding the role of AI in digital lean manufacturing. The first level of conducting the research includes identifying the accurate keywords and phrases to search the articles in these databases. Keywords include Digital Lean Manufacturing, AI, predictive maintenance, Supply chain optimization, and quality control and inspection. These include consideration according to the type of study, the relevance of the articles, years of publication, and theoretically determined inclusion criteria. The articles to be reviewed should be relevant to the mentioned study area, published in the last 5 years. Newspaper articles and other articles that are irrelevant to the study's focus are eliminated, such as research papers in other fields. The objective is to find the most relevant papers that devote their content to the systematic evaluation of the effect of AI on Digital lean manufacturing. Out of these papers, the analysis of 39 papers has been considered after applying the inclusion and exclusion criteria. These selected publications are then carefully reviewed to identify the main findings, methods, and theories related to green manufacturing and its influence on the three study areas of interest. The findings derived from these publications are then summarized to identify shared themes, tenets, and views on the effect of AI on digital lean manufacturing.

5 Artificial Intelligence (AI)-Powered Predictive Maintenance for Digital Lean Manufacturing

Predictive maintenance is one of the revolutionary domains that incorporates artificial intelligence to define the field of digital lean manufacturing. By using AI, companies that operate in the manufacturing industry can ensure a higher level of equipment reliability and, therefore, increase production rates. Such a preventive measure follows the lean manufacturing concept since it eliminates the prolonged time that some equipment may spend immobile so that it can be serviced, means less spending on repairs, and increases the longevity cycle of the equipment. Concerning predictive maintenance, AI works through the incorporation of sensors and other IoT gadgets. They are fit on numerous machinery and equipment in the production line and gather large amounts of data frequently on several parameters like temperature, vibration, pressure as well as speed. This data is then fed to centralized systems where AI algorithms are applied in real time. The problem that may occur is the sheer number of sensor data that needs to be processed and dealt with, but as scholars showed AI is very capable of handling big datasets and finding correlations that might go unnoticed by operators [7]. Big data analytics is among the cornerstones of AI for predictive maintenance. Another use is that of an analytical model for comparing current and historical data, which if different may suggest pending problems or general wear and tear on machinery in a plant. For example, a small change in vibration or rising temperature may not be very significant on its own but may indicate the onset of some issue when compared with other readings. AI facilitates a manufacturer to arrange maintenance activities as soon as such small changes are prompt, thereby preventing a minor problem from worsening and leading to a failure. Predictive maintenance can be considered a machine learning application in turn, as this approach helps the system enhance its predictive capacity on the job. This field involves using the predictive streak usually associated with machine learning algorithms where the systems extract information on historical cases and study what factors have led to equipment malfunction. These algorithms are then able to use this knowledge to predict future failures with greater and greater accuracy. More data is collected and analyzed over and above the period, and the machine learning models get more refined making the predictions more accurate [8]. Indeed, the application of AI in making predictions for lean manufacturing has the following benefits: By identifying the probable equipment problems early enough, the manufacturers are in a better position to avoid the occurrence of time-consuming breakdowns. This is especially true in lean manufacturing where just-in-time production is prompted by the smooth running of the manufacturing assets. Such a situation greatly affects the extensive production process since it slows down and increases the costs, which in turn will lower the satisfaction levels of customers. Besides, the application of AI in maintaining the systems lowers the costs of maintenance while increasing the machine's uptime. The conventional practice of maintenance entails the fixating of certain periods at which the equipment is to undergo servicing even if they are not worn out. This can result in regular maintenance interventions and swapping of various parts that are still viable in utterance to their probable life lovemaking. While preventive maintenance schedules all the equipment in a given system regardless of how often it is required, predictive maintenance focuses only on the equipment that requires attention, hence efficiency and effectiveness in the use of resources and minimization of avoidable expenses. This not only reduces on costs of maintenance but also ensures that machinery is only worked on when there is a need to be worked on [9]. AI utilized in predictive maintenance increases the efficiency and sustainability of manufacturing processes. Minimally, manufacturers can work on making their machinery run with improved efficiency and with minimal wastage. The applicant believes that technically superior equipment should consume less energy and produce a few defects, and this view is consistent with lean manufacturing. The schedules for the future failures of the equipment are also predicted, which help to avoid emergency conditions and the corresponding environmental consequences of the express delivery of replacement components. It also improves the decision of maintenance teams through the incorporation of AI in the predictive maintenance function. Maintenance personnel can reach efficient decisions since AI systems are capable of analyzing data and presenting AID insights and recommendations. This eliminates guesswork and makes the strategies to be implemented in carrying out of maintenance to be more efficient. Furthermore, through the implementation of AI, many human operators' time is saved from busy work such as monitoring and analyzing data [10]. As AI technology expands, the latter will have great potential in the sphere of predictive maintenance. For instance, various strategies like deep learning and neural network integration can improve the quality and effectiveness of the predictive models. These technologies help to clarify the relationships between variables in the data and can uncover nonlinear phenomena and tendencies connected to the equipment behavior and possible failure modes. Also, the incorporation of AI with augmented reality (AR) and virtual reality (VR) must help the maintenance teams to have efficient visual instructions and directions while executing the maintenance [11]. Edge computing is one of the latest trends that are used in conjunction with AI and predictive maintenance. Edge computing enables data to be processed closer to where it originated, hence the machinery and the equipment in the manufacturing floor. This saves time and analyzes the data to be real time, thus in the instances where there is a problem, we can be in a position to act right away. Therefore, by analyzing data locally, edge computing makes predictive maintenance systems more prompt hence any issue detected is worked on before it becomes a much bigger problem to handle [12]. AI-supported predictive maintenance expands that skill and uses complex prediction algorithms and digital replicas of machinery. A digital twin is an informational copy of a physical object or a parameter of the production process, which is complemented by the contemporary information of its real counterpart. It can be seen that by creating the digital twins of manufacturing equipment, AI can simulate situations and even predict potential failures with remarkable accuracy. Maintenance simulations offer vital clues to machinery conditions, hence helping the maintenance teams in the formulation of perfect maintenance strategies for their machinery. Another advantage is the fact that the implementation of digital twins also makes it easier to do a foresight analysis or a simulation of certain contingencies that may occur on the shop floor to enable the management to determine the best way to schedule the maintenance and resource distribution plans. AI-based predictive maintenance is also being approved by the addition of augmented reality (AR) and virtual reality (VR) technologies. AR and VR training assists the maintenance personnel to see a live model of the machinery and also get immediate training during the performance of his/her task. For instance, AR can place information on real-life objects—items that require attention and notice areas that require maintenance, plus a sequence on how to fix the particular object. VR can be applied in training maintenance staff since it creates a platform for maintenance activities to be conducted virtually before physically executing them on the tools. This integration enhances the element of quality of maintenance plans besides increasing the efficiency in organization productivity [13].With the help of modern technologies, AI initiates the creation of Self-Maintenance Systems when machines are capable of diagnosing their problems and sometimes can fix them independently. Autonomous maintenance utilizes robots and drones that have sensors and diagnostic tools that are special for the completion of the job. Such systems can watch mechanical equipment, and even diagnose problems or decide if a problem needs repair or adjustment [14]. The use of AI-based systems to support predictive maintenance is considered one of the aspects that help achieve sustainability and minimize environmental impacts in lean manufacturing. In achieving this, maintenance schedules are improved and made optimal while monitoring that the machinery is well functioning, cuts out wastage, and saves energy. In addition, predictive maintenance also helps in the elongation of the useful life of the machinery to avoid frequent replacement and therefore lowering the environmental implications of manufacturing or procuring new equipment, and/or disposal of the old ones. Also, AI can specify potential areas where energy conservation and process improvement are possible, which will advance the idea of making manufacturing more sustainable [4]. Digital lean manufacturing is extremely beneficial in the use of AI in proactive maintenance and in using data to provide advanced estimations for maintenance requirements. With the help of sensors that interconnect IoT devices, big data processing, and machine learning, AI can forecast equipment failures and suggest proper maintenance activities. This in turn helps in avoiding regular or rather frequent breakdowns of the machines and also cuts down on expenses incurred in the process of carrying out maintenance, thus boosting the universal efficiency and durability of the systems. However, with advancements in the use of AI technology, the application of the technique in predictive maintenance will continue to be valuable in the quest of attaining the goals of lean manufacturing, aiming at more improvement throughout the manufacturing industry.

6 Enhancing Quality Control and Inspection in Digital Lean Manufacturing with Artificial Intelligence (AI)

Quality control is a procedure that ensures that goods are free of faults by using best practices and adhering strictly to standards and specifications. Companies that use strict quality control (QC) techniques may better manage costs, eliminate waste, and protect themselves from brand harm and liability, which are common consequences of faulty products. The key to maintaining quality control programs is to strictly adhere to standardized methods. Any deviation from the procedure is likely to result in defective goods as well as other types of waste. Collecting data on part quality may be useful for monitoring production processes and ensuring that standardized work and work instructions are consistent with producing quality components while minimizing waste [15]. QC initiatives have a substantial influence on cost and on-time delivery. Without effective quality control, firms will create more scrap and incur greater labor expenses due to additional handling and rework. This error also disrupts the supply chain's performance and accuracy, as planners must buy extra material to account for scrap loss. QC also serves as the final line of defense for client satisfaction. Customer returns are reduced when quality control is correct, and brand impression improves. This helps to ensure a high percentage of recurring consumers. Robust QC evaluates a company's procedures and guarantees that production systems function properly. Defect detection may reveal a malfunctioning process in the manufacturing chain. Changing the process using formal process improvement approaches such as Lean will help enhance performance and reduce [16]. AI learns from numerous data files obtained from sensors and production equipment to detect patterns and possible correlations which may imply the existence of quality problems. Based on identified patterns, AI can anticipate that certain parts are more prone to have defects and make manufacturers eliminate possible causes. Machine learning algorithms can be trained on the baseline values of operational parameters of the machinery and the processes. These concepts can be marked as abnormalities if variations from the norms are found in such cases. This early detection is crucial in avoiding having a quality problem act as a cause of widespread problems, thus improving the homogeneity of products [17]. The use of AI algorithms can also be very effective in identifying the types of defects that are disastrous to the business and their frequency to aid the quality control teams. Thus, it is valuable for carrying out relevant corrective actions and constant improvement activities. There are Machine learning techniques that are available with AI systems to study the patterns of the reported defects and correlate them with the causes like equipment failures, process variations, and material variations. In this way, by identifying the actual sources of failure, the manufacturers can take some steps to eradicate the causes affecting the quality of products [18]. Introducing AI can result in constant supervision of the production processes, which provides suggestions for the operators. It provides a real-time result which means that if the standard is not met, changes can be made immediately to ensure that substandard goods are not produced. Inspection criteria of AI systems can be adjusted when new information arrives and the systems can learn from it. This makes the quality control process change with time and conditions under which production is carried out to meet the set quality standards regardless of the processes or products that are involved [19]. The architectural plans of specifications of AI systems have integrated documentation for stored results of the inspection and related images and data to every inspected item. Such documentation is important for documentation and records, and it facilitates traceability, compliance as well as quality audits, which help prove that a certain product is of quality. The large amounts of data gathered and processed by the AI systems contain important information about individuals to be used in decision-making. Quality managers can use this data in ways that seek to understand the patterns, analyze the results of the implemented quality control measure, and where necessary, make the right decisions that would enhance the identified processes [20]. Quality assurance executed by artificial intelligence reduces operational loss due to the increased ratio of detecting defective products that later have to be either fixed or disposed of. This tallies with the lean manufacturing model's objective of reducing or eliminating any wasteful activities in the use of resources and helps in bringing about sustained enhancement as constant information on the processes and quality problems can be made available within the organization. These may help the manufacturers to apply incremental adjustments iteratively, so that, they improve both the quality of their products and the production processes [21]. AI-based technologies help the workers because these tools give them feedback and support at the time of carrying out inspections. Examples of applications are augmented reality (AR) systems, which can place digital information on the physical products and mark the areas to be attended, as well as enhance the accuracy of the inspection. AI systems have the additional advantage of being able to perform dangerous inspection activities that may prove fatal to the human inspector's life. In other words, by assuming some monotonous or risky operations, AI makes workers concentrate on more creative and worthwhile tasks, thus improving the safety level in the work environment [22]. In the area of digital lean manufacturing, the basics of quality control and inspection improve significantly due to the involvement of AI technology which provides better tools for accuracy, efficiency, and quality. Therefore, utilizing AI for inspection results in additional benefits to manufacturers through predicting the development of quality issues, making classifications on defects, real-time performance, and direction toward continuous improvement under lean production. The IT application of quality control not only minimizes waste and costs but also utilizes employees and increases workplace safety to make a better manufacturing community [23]. The application of Artificial Intelligence to computer vision systems is already making a forceful entry into the aspect of quality inspection. This employs the latest image processing techniques where the images or the video feed of a product are analyzed in real time by these systems. Compared to other approaches, AI systems are capable of detecting surface cracks, discoloration, or differences in the dimensions of an item that human inspectors might not discern. This high level of precision makes it possible to provide the customer with only superior quality products and this in turn minimizes the incidence of returns and ultimately improves customer satisfaction [24]. AI speeds up the process of inspection and excludes human errors at the same time. The methods involving the use of workers to inspect the tiles are slow and involve errors that are inherent in manual processes. AI systems on the other hand are capable of processing thousands of images per minute and can give quick feedback and save much time for quality control. Having a rapid inspection capability benefits the organization because it is possible to not only identify the defect quickly but also eradicate it to ensure that it does not progress further and get transferred to more complex and irreversible production stages with substandard products. Big data and AI algorithms are very helpful when it comes to the classification of the type of defect and simplify the process of analysis of the root causes. Comprehending the causes and frequencies of the specific types of defects allows for the detection of problems in the manufacturing process. Targeted in this case means that precise corrective actions are taken, which ensures the improvement of the product quality [25]. AI systems enable the enhancement of the existing workflows by offering feedback loops. Checks and audits generate information that evaluates trends and patterns for manufacturers to make the right decisions. For instance, if there are anomalies that are recurrent and that are commonly observed in items produced from a certain facilitating line, then AI plays a crucial role in highlighting such a challenge or issue that might be with the machinery, process, or materials. This ongoing analysis helps to apply one of the lean principles, namely, continuous improvement, which implies that production processes should change and progress over time [26]. Applying AI in conjunction with AR and VR boosts the support of the personnel at the time of inspections. It can also add virtual information on top of the tangible products, orient the inspector throughout the inspection processes, and point out certain problems. VR can be applied in training, enabling the workers to train in how to perform the inspection procedures in the Olympics before performing them on the products. This not only increases the efficiency and effectiveness of the inspection but also cuts training time therefore costs [27]. Automated self-inspection robots are constantly being used more frequently in manufacturing areas. Some of these robots are endoscope models with sensors and computer vision that can move through production floors and inspect on their own without human guidance. They can reach the areas that are difficult to access and inspect, repeat the same procedure with equal efficiency, and work in regions that require special safety measures, thus increasing the efficacy and safety of inspection activities. In the big manufacturing plants, flying drones which are also AI-enabled are being used for aerial inspection. These drones are also useful for scanning large areas for defects or for inspecting equipment or products from directions that are difficult for an inspector to access. This capability is highly effective in long fields of structure, like wind turbine blades or storage tanks that require some periods and could be risky to inspect physically [28].

7 Transforming Supply Chain Optimization with Artificial Intelligence in Digital Lean Manufacturing

Supply chain management (SCM) involves the management of committing a firm's resources for the procurement, creation, and delivery of goods and services to the customer in the most effective manner. It entails the proper coordination of the manufacture and distribution of products, as well as the supply chain of the information and finances from the raw material suppliers up to the final consumer. Some of the objectives include the maintenance of low overheads, achieving high levels of customer satisfaction, and being able to act promptly to changes in the market environment. The goals of today's supply chain management, commonly crossing national borders and involving a host of players, can be attained only with sophisticated methods and technology [29]. AI becomes a central element in SCM because it helps enhance forecasting, decision-making, and reduces the number of routine tasks in the process. In satisfying digital lean manufacturing objectives, AI's impact is especially noteworthy when considered in light of AI solutions the most important of which adhere to the principles of lean manufacturing of eliminating waste and enhancing processes. One of the most important parts of supply chain optimization is effectively forecasting demand. Traditional forecasting approaches frequently fail owing to their reliance on past data and incapacity to respond to unexpected market developments. AI, on the other hand, uses machine learning algorithms to analyze massive volumes of data from multiple sources, such as market trends, social media, and economic indicators, to provide more accurate demand projections. This enhanced forecasting allows businesses to maintain ideal inventory levels, lowering the risk of overstocking and stockouts. Companies that ensure inventory is aligned with actual demand can reduce holding costs and improve customer satisfaction through timely delivery [30]. Visibility is improved in the supply chain because real-time tracking of goods within the supply chain is supported by AI. Using sensors within IoT and AI, the real-time location of the shipment can be determined along with the environmental conditions and probable issues that may occur in the supply chain. Such visibility makes it possible for organizations to nip some risks in the bud, hence improving the efficiency and reliability of supply chain processes. The application of AI can include data analysis from various phases of the supply chain to locate ineffective areas or problems. The information given by AI then assists businesses in making smarter decisions and cutting out unnecessary time-wasting [31]. It is critical in digital lean manufacturing to ensure equipment uptime. Unexpected equipment breakdowns can cause major production delays and higher expenses. AI-driven predictive maintenance predicts when equipment may break based on sensor data and prior maintenance records. By spotting trends and abnormalities, AI can inform maintenance personnel to do preventative maintenance, decreasing downtime and increasing machinery lifespan. Product quality assurance is another key part of supply chain optimization. AI-powered quality control systems utilize computer vision and machine learning algorithms to evaluate items for flaws and deviations from quality specifications. These devices may function faster and more accurately than human inspectors, assuring constant product quality and eliminating waste from faulty items [32]. A resilient supply chain requires effective supplier relationship management. AI may use data from supplier performance, market trends, and risk considerations to assess and pick the best suppliers. AI-powered platforms may improve communication and cooperation with suppliers, ensuring that all parties are aligned and working toward the same goals. Logistics and transportation can also be made efficient by using AI in such areas as the best path to take, the flow of traffic, and fuel consumption. Some of the things that complex models can identify include the best route of transport for consignments to bring down delivery cycles and transportation prices. Additionally, the concept of dynamic navigation can be applied to route planning, meaning that the found routes may be constantly updated depending on such conditions as traffic jams or any other occurrences [33]. AI uses predictive ML techniques with complex sales data from various sources, which include sales history, market factors, customers' behavior, and other regulating factors like economic status and weather conditions. The older models of demand forecasting bear several inconveniences and one of these is that they use little data set and linear models, hence they cannot capture the finest interactions and patterns. On the other hand, AI systems can analyze and understand large chunks of data and identify intricate patterns and connections that human beings or less sophisticated algorithms cannot notice [34]. Flexibility, adaptability, and precise demand forecasting are AI's great benefits that can change supply chain management. Machine learning models gradually learn from the new data inputs by making better predictive models over time. This adaptive capability is top-notch, especially in the contemporary marketplace where changes in consumer trends and market conditions can be frequent. For instance, a retail company that employs AI in demand forecasting will observe a spike in the trends in the number of searches and posts on social media about a particular gadget. The AI system can quickly process this information and calculate a tendency that corresponds to a rise in the demand for this gadget; therefore, the company can minimize the cases of missed sales and precarious overstocking [34]. AI in demand forecasting assists manufacturers and retailers in minimizing their stock-holding costs. Forecasts help a business to ensure it has the right stock in the right quantities guessing the market demands so that it does not end up with goods it cannot sell or rather it does not end up without stock of goods it can sell. Overstocking means investing capital in slow-moving or non-moving stocks; it also means having to provide for warehousing costs and having to take the risk if the stocks become obsolete. On the other hand, stockouts cost time and result in reduced sales, angry customers, and damage to the brand image AI can level the playing field thus allowing companies to achieve the desired middle ground. For instance, AI applied in a fashion retailer company can be in a position to forecast which types of clothes will be fashionable the next season and hence order appropriate quantities to meet this demand without incurring unnecessary stock during the next season's end [35]. Hyperautomation applies the use of AI in automating specific and sometimes complicated tasks in handling orders, management of inventories as well as planning of deliveries. Thus, this well-coordinated deployment of one or more technologies yields a coherent and effective supply chain system. AI always improves these operations increase productivity and help minimize errors that may have been made. In lean manufacturing, work with AI to increase the efficiency and safety of human and robotic employees. When applied along with other lean values like Jidoka, which means automation that incorporates the human aspect, AI holds the key to enhancing the quality as well as productivity in manufacturing. AI solutions for supervising and identifying the potentially problematic situations which, in turn, are present in case human operators can dedicate their attention to higher-level decisions. The interaction of skillful AI and human skills results in improved decision-making processes and innovations [29]. Supply chain and the application of lean manufacturing will depend much on the future growth of artificial intelligence and digitalization. Previous trends have been diversified with features such as AI use for carbon neutrality, fortifying smart factories, designing the next generation of organizational learning. That is why businesses that take AI early will have competitive advantages that will help them become more efficient, sustainable, and innovative [36]. Supply chain optimization therefore plays a vital role in digital lean manufacturing, which is made possible through AI, in demand forecasting, visibility, maintenance, quality assurance, suppliers, and logistics. AI when applied in businesses enables organizations to operate more optimally, cuts down on expenses, and offers high-end services to its clients. With time going forward, there are enhanced uses and developments of AI in SCM, where advanced optimization and creativity will be enhanced.

8 Conclusion

The importance of AI in lean manufacturing in the digital world is evident due to its contribution to improving the predictive maintenance, processes, quality control, and supply chain. Given this, it can parse big amounts of data and offer decision-making information that stands for reducing waste, improving the business, and attaining optimal efficiency in line with lean principles. Therefore, with the advances in AI, the flood of benefits on lean manufacturing is likely to increase resulting in further advancements in the manufacturing sector. Manufacturers have to adapt to the new conditions of the rapidly growing market, and applying AI allows for increasing productivity, product quality, and sustainability.

References

1. Valamede, L. S., & Akkari, A. C. S. (2020). Lean 4.0: A new holistic approach for the integration of lean manufacturing tools and digital technologies. *International Journal of Mathematical, Engineering and Management Sciences, 5*(5), 851.
2. Powell, D., et al. (2024). A digital lean world: From digital lean manufacturing to lean 4.0. In *A research agenda for lean management* (pp. 215–226). Edward Elgar Publishing.
3. Romero, David, et al. (2018). *Digital lean cyber-physical production systems: the emergence of digital lean manufacturing and the significance of digital waste*. Advances in Production Management Systems. Production Management for Data-Driven, Intelligent, Collaborative, and Sustainable Manufacturing: IFIP WG 5.7 International Conference, APMS 2018, Seoul, Korea, august 26–30, 2018, Proceedings, Part I. Springer International Publishing.
4. Medyński, D., et al. (2023). Digital standardization of lean manufacturing tools according to industry 4.0 concept. *Applied Sciences, 13*(10), 6259.
5. Sharma, A. K., & Pinca-Bretotean, C. (2023). Artificial intelligence in lean manufacturing paradigm: A review. In *E3S web of conferences* (Vol. 391). EDP Sciences.
6. Shahin, M., et al. (2024). Improving operations through a lean AI paradigm: A view to an AI-aided lean manufacturing via versatile convolutional neural network. *The International Journal of Advanced Manufacturing Technology*, 1–77.
7. Chomklin, A., Jaiyen, S., & Wattanakitrungroj, N. (2023). A survey of AI techniques based on predictive maintenance in lean manufacturing. *Science, Technology, and Social Sciences Procedia, 4*(2023), CiM03.
8. Perico, Paolo, and Juliette Mattioli. (2020). *Empowering process and control in lean 4.0 with artificial intelligence*. 2020 third international conference on artificial intelligence for industries (AI4I). IEEE.
9. Boute, R. N., Gijsbrechts, J., & Van Mieghem, J. A. (2022). Digital lean operations: Smart automation and artificial intelligence in financial services. In *Innovative technology at the interface of finance and operations* (Vol. I, pp. 175–188).
10. Kliestik, T., et al. (2023). Artificial intelligence-based predictive maintenance, time-sensitive networking, and big data-driven algorithmic decision-making in the economics of industrial internet of things. *Oeconomia Copernicana, 14*(4), 1097–1138.
11. Fordal, J. M., et al. (2023). Application of sensor data based predictive maintenance and artificial neural networks to enable industry 4.0. *Advances in Manufacturing, 11*(2), 248–263.

12. Shahin, M., et al. (2023). Using machine learning and deep learning algorithms for downtime minimization in manufacturing systems: An early failure detection diagnostic service. *The International Journal of Advanced Manufacturing Technology, 128*(9–10), 3857–3883.
13. Berg, Philip, and Axel Dannevig. (2023). *How artificial intelligence can be used to improve lean manufacturing and production processes. A case study of Hennig Olsen*. MS thesis. University of Agder.
14. Mehta, Arvinder Kour, et al. (2023). *AI & Lean Management Principles Based Pharmaceutical Manufacturing Processes*. 2023 10th IEEE Uttar Pradesh Section International Conference on Electrical, Electronics and Computer Engineering (UPCON). Vol. 10. IEEE .
15. Powell, Daryl, et al. 2018. *Towards digital lean cyber-physical production systems: Industry 4.0 technologies as enablers of leaner production*. Advances in Production Management Systems. Smart Manufacturing for Industry 4.0: IFIP WG 5.7 International Conference, APMS 2018, Seoul, Korea, August 26-30, 2018, Proceedings, Part II. Springer International Publishing.
16. Arey, D., Le, C. H., & Gao, J. (2021). Lean industry 4.0: A digital value stream approach to process improvement. *Procedia Manufacturing, 54*, 19–24.
17. Broday, E. E. (2022). The evolution of quality: From inspection to quality 4.0. *International Journal of Quality and Service Sciences, 14*(3), 368–382.
18. Sini, Francesca. 2020. Lean management in additive manufacturing: A methodological proposal for quality control. Diss. Politecnico di Torino
19. Dutta, G., et al. (2021). Digitalization priorities of quality control processes for SMEs: A conceptual study in perspective of industry 4.0 adoption. *Journal of Intelligent Manufacturing, 32*(6), 1679–1698.
20. Hussamadin, R., Jansson, G., & Mukkavaara, J. (2023). Digital quality control system—A tool for reliable on-site inspection and documentation. *Buildings, 13*(2), 358.
21. Ghani, Mohammad Kashif. "Digital tools of lean manufacturing." (2018).
22. Sundaram, S., & Zeid, A. (2023). Artificial intelligence-based smart quality inspection for manufacturing. *Micromachines, 14*(3), 570.
23. Eleftheriadis, Ragnhild J., and Odd Myklebust. 2016. *A quality pathway to digitalization in manufacturing thru zero defect manufacturing practices*. 6th International Workshop of Advanced Manufacturing and Automation. Atlantis Press.
24. Marsden, P. (2019). *Digital quality management in construction*. Routledge.
25. Babic, M., Farahani, M. A., & Wuest, T. (2021). Image based quality inspection in smart manufacturing systems: A literature review. *Procedia CIRP, 103*, 262–267.
26. Endrigo Sordan, J., et al. (2024). How industry 4.0, artificial intelligence and augmented reality can boost digital lean six sigma. *Total Quality Management & Business Excellence, 35*, 1–25.
27. Pourhejazy, P., & Kwon, O. K. (2016). The new generation of operations research methods in supply chain optimization: A review. *Sustainability, 8*(10), 1033.
28. Scholz, J., et al. (2018). Digital technologies for forest supply chain optimization: Existing solutions and future trends. *Environmental Management, 62*, 1108–1133.
29. Tay, H. L., & Loh, H. S. (2022). Digital transformations and supply chain management: A lean six sigma perspective. *Journal of Asia Business Studies, 16*(2), 340–353.
30. Wang, X., et al. (2022). Impact of digital technology on supply chain efficiency in manufacturing industry. In *Digital transformation in industry: Digital twins and new business models* (pp. 347–371). Springer International Publishing.
31. Rahamneh, A., et al. (2023). The effect of digital supply chain on lean manufacturing: A structural equation modelling approach. *Uncertain Supply Chain Management, 11*(1), 391–402.
32. Buer, Sven-Vegard. Investigating the relationship between lean manufacturing and industry 4.0. (2020).
33. Sah, B. P., et al. (2024). The integration of industry 4.0 and lean technologies in manufacturing industries: A systematic literature review. *International Journal of Management Information Systems and Data Science, 1*(3), 14–25.

34. Turconi, G., et al. (2022). A literature review on lean manufacturing in the industry 4.0: From integrated systems to IoT and smart factories. In *IoT and cloud computing for societal good* (pp. 181–194).
35. Devereaux, D. (2019). *Smaller manufacturers get lean with artificial intelligence*. National Institute of Standards and Technology (NIST). [Online]. Available: https://www.nist.gov/blogs/manufacturing-innovation-blog/smaller-manufacturers-get-lean-artificial-intelligence
36. Haddud, A., & Khare, A. (2020). Digitalizing supply chains potential benefits and impact on lean operations. *International Journal of Lean Six Sigma, 11*(4), 731–765.

Green Digital Finance and its Impact on Sustainable Development: Empirical Evidence from India

M. Shuaib Ahmed, M. Mohammed Mustafa, and Shaply Abdul Kareem

1 Introduction

Over the past few decades, India's economy has experienced significant growth, propelled by the implementation of new government economic policies aimed at ensuring proper utilization of environmental resources. Consequently, there is a growing recognition of the critical need to conserve ecosystems and mitigate various adverse impacts on human health stemming from the overexploitation of forest resources, leading to pollution and environmental degradation [1]. Given that the rural population in India heavily relies on local environmental resources for their livelihoods, current efforts to promote a green ecosystem are increasingly tailored to address their specific needs. Notably, fossil fuels have garnered increased attention recently, particularly in meeting approximately 60 different energy demands in India [2].

Consequently, the widespread use of fossil fuels has adversely impacted the ecology, releasing substantial pollutants into the atmosphere. Many scientists are actively researching strategies to mitigate the pace of environmental degradation [3]. Furthermore, the Indian government has been enacting policies to promote green practices and foster robust economic growth, aiming to set sustainable objectives for the nation. The sustainable growth of the global economy significantly

M. Shuaib Ahmed (✉)
School of Management, C. Abdul Hakeem College of Engineering and Technology, Melvisharam, India

M. Mohammed Mustafa
Department of AI&DS, Sri Eshwar College of Engineering, Coimbatore, India

S. A. Kareem
School of Commerce, Presidency University, Bangalore, India

P. Raj et al. (eds.), *Artificial Intelligence (AI) for IT Energy Efficiency and Green AI for Environment Sustainability*,
https://doi.org/10.1007/978-3-031-89420-6_32

relies on employing green financing mechanisms to fund diverse projects while ensuring alignment with these established criteria [4].

The concept of "green finance"primarily encompasses two key areas: finance and the environment. The company's product has gained widespread recognition and has significantly contributed to remarkable advancements across various sectors worldwide [5]. To facilitate the transition towards a low-carbon economy, India issued $6.11 billion in green bonds in 2021 and secured an additional $3.3 billion in 2022. Green finance offers numerous advantages, as it aims to furnish the financial sector with operational, investment, fund financing, and ancillary services, while also offering perspectives for projects. However, in India, carbon finance is emerging belatedly yet rapidly, a pivotal development for establishing a robust carbon trading market [5, 6]. While green finance has gained traction in emerging economies, its precise contribution to high-quality economic development remains uncertain. Additionally, Fintech in India has demonstrated exceptional performance in recent years since its inception [7]. The adoption of financial technology among both consumers and businesses has surged. The evolution of financial technology is expected to play a pivotal role in the country's transition towards a sustainable financial system and achieving high-quality economic growth [8]. It offers avenues for retail investors to explore and invest in products aligned with environmental sustainability goals and high-quality economic development. Nonetheless, a gap persists between the supply and demand of funds in renewable energy, necessitating urgent efforts to overcome this challenge [9].

It fosters the advancement of green finance by mitigating information disparities among investors and stakeholders, enhancing efficiency, appraising natural assets, and advocating for sustainable lifestyles to bolster high-quality economic growth [9, 10]. Green finance is intrinsically linked to both high-quality and sustainable development, serving as critical catalysts for economic progress. Moreover, it embodies the characteristic of sustainable expansion, providing a framework for closely aligned planning with high-quality sustainable development [11, 12]. Additionally, it introduces a pivotal element of economic growth, laying the groundwork for orchestrated prosperity. Green finance involves the engagement of both public and private financial frameworks in ecological business models aimed at initiatives such as clean energy, enhancing water and air quality, and reducing disposable plastics [12, 13]. Furthermore, green finance serves as a conduit, bridging the gaps between investors, lenders, consumers, business owners, and all stakeholders.

This study makes several contributions to the literature on green finance, enriching understanding across various dimensions. First, it provides clarity on the subject of green finance, building upon previous scholarly inquiries to explore the relationship between green finance and enhanced economic growth [14]. While prior research primarily concentrated on individual environmental aspects of economic development, this study delves into how green finance influences comprehensive and superior economic progress. Second, it addresses a recent call for information concerning the role of Fintech in advancing green financing. It was noted that there is insufficient research on how Fintech affects the relationship between green financing and economic development [15, 16]. Thus, Fintech was regarded as a moderating variable, prompting an investigation into its subtle influence on the link between green finance and high-quality economic growth. Additionally, the introduction of this study offers a genuine understanding of the developmental impact of green finance [16]. The potential findings regarding Fintech mentioned above could offer significant insights for stakeholders, enabling them to reshape economic development by advancing in green finance and financial technology [13, 16].

2 Literature Review

In line with green finance principles, financial resources are directed towards environmental initiatives and policies that foster the advancement of a more sustainable economy and projects for sustainable progress. Particularly, financial institutions prioritize environmental considerations in their funding and investment deliberations, channeling capital towards green enterprises [18, 19]. This approach acknowledges the adverse environmental impacts and the necessary measures to achieve a balance between natural resource exploitation and economic growth. Unlike past financial objectives solely focused on maximizing profits, this strategy considers the environmental implications, representing a shift towards more sustainable practices [13, 17]. Green finance plays a pivotal role in bridging the gap necessary for ensuring sustainable economic development [18]. By facilitating more efficient resource allocation and environmental protection, it helps steer resources towards environmentally friendly endeavors [19, 20]. Research indicates that green finance contributes to high-quality economic development by positively impacting ecological conditions, economic efficiency, and structural frameworks [11]. Fintech, an amalgamation of financial services delivered through technology, offers ample opportunities for organizations in the financial sector to streamline operations and devise more effective solutions [21]. However, the extent of its impact on establishing a high-quality and sustainable economic framework remains a subject of ongoing research, given its nascent stage and evolving implications. Achieving high-quality economic development involves the synergy of numerous factors [20, 22].

Kishor et al. conducted a study examining the utilization of digital finance, particularly Fintech, for fostering financial inclusion and sustainable development [6, 10, 18, 23–34]. The authors commence by emphasizing the Sustainable Development Goals (SDGs) outlined by the United Nations in 2015, with a particular focus on the 3Ps (People, Planet, and Profit) aimed at achieving sustainable development [24]. They posit that while banking has traditionally been perceived as an environmentally benign industry, the current magnitude of banking activities has led to increased carbon emissions attributable to energy consumption, paper waste, and the absence of green buildings. Consequently, the authors advocate for the indispensability of digital finance in advancing sustainable development [23].

Yao and Yang explore the progression and hurdles of digital finance within China. The authors offer an overview of its historical trajectory, accentuating pivotal milestones such as the emergence of Alipay and WeChat Pay [23]. Additionally, they delve into the regulatory framework governing digital finance in China, outlining the roles of entities like the People's Bank of China (PBOC) and other governmental bodies [25]. Furthermore, the study delineates various challenges and risks inherent in the expansion of digital finance in China, encompassing concerns regarding data privacy and security, systemic risks within the financial sector, and potential disruptions to traditional financial institutions. The authors underscore the imperative for on-going regulatory vigilance and policy direction to foster the sustainable and accountable growth of digital finance in China [23, 25].

Studies reveal that countries have extensively exploited natural resources to sustain economic growth, particularly due to the absence of technological innovation and energy efficiency in traditional industries, resulting in severe pollution. However, transitioning to economic growth driven by renewable energy production could yield enduring benefits for both the environment and the economy. The growth of high-quality economic development is recognized as multifaceted, influenced by factors including technological innovation, organizational dynamics, financial considerations, and political contexts [25].

The study focuses on three key objectives. First, it aims to comprehend the concept of digital finance, exploring its definition, key components, and underlying technologies. Second, it seeks to elucidate the diverse roles of digital finance in promoting environmental sustainability, such as facilitating green investments, supporting carbon tracking, and enabling sustainable financial practices. Finally, the study examines the effects of digital finance on environmental sustainability, analyzing both its positive contributions, like advancing eco-friendly initiatives and potential negative impacts. By addressing these objectives, the study aims to provide a comprehensive understanding of how digital finance intersects with and influences sustainability efforts.

3 Role of Digital Finance in Environmental Sustainability

India has actively promoted inclusive finance and embraced digital technologies to enhance traditional financial systems. This initiative has led to the broadening of inclusive financial services, benefiting a wider spectrum of the economy.

Digital inclusive finance holds the potential to mitigate some of the constraints associated with traditional finance, thereby catalyzing research and development in green technologies [18]. However, it is imperative to explore whether the expansion of digital finance influences green innovation. Only through addressing such inquiries can governments effectively foster green technologies via the advancement of digital finance. Given the rapid advancement of digital finance, it is reasonable to anticipate its pivotal role in the future financial landscape. In India's pursuit of sustainable development and improved environmental quality, prioritizing innovation and ecological civilization is paramount [26].

Green innovation is pivotal for realizing an innovation-driven development strategy, aligning with the new development concept and fostering harmonious coexistence between humans and the natural environment. Consequently, it is crucial to investigate the influence of digital finance on green innovation and identify the factors that shape this influence [27] (Fig. 1).

In our perspective, digital finance fosters green innovation through several avenues.

Firstly, according to Rao et al., Yao and Yang, it plays a vital role in alleviating information asymmetry between financial institutions and firms engaged in green technology projects, thereby enhancing financial support for innovative green enterprises and easing their financing constraints [36, 37].

Secondly, according to Tang et al. and Kishor et al., there exists a positive correlation between digital finance and heightened green innovation, as digital finance can reduce costs associated with financial transactions and operations [34, 35].

Thirdly, it can be deduced that digital finance positively impacts green innovation by accelerating the green transformation of traditional industries. Therefore, digital finance contributes significantly to the advancement of green innovation.

Fourthly, following the confirmation of digital finance's impact on green innovation, attention should be directed towards specific aspects of digital inclusive finance, such as its coverage, depth, and utilization, to scrutinize their effects on green innovation.

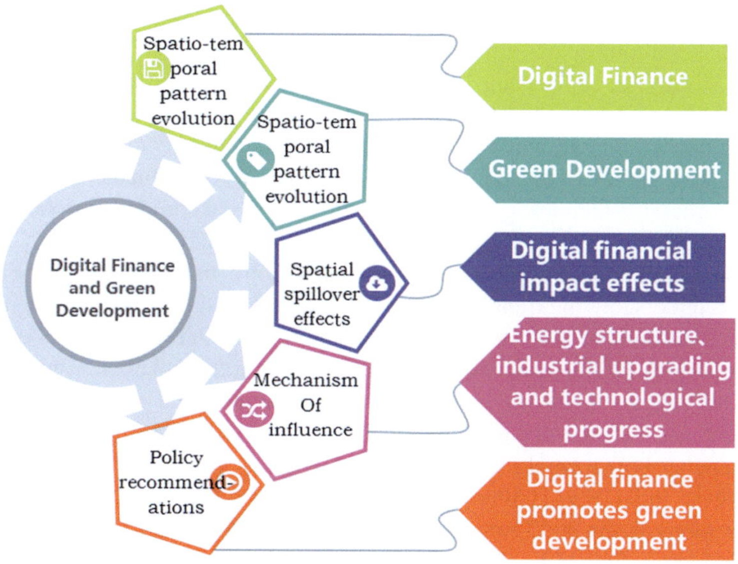

Fig. 1 Digital finance and green development

4 Significance of Green Finance

Green finance holds significance for three primary reasons, some of the key practical areas discussed below:

Promotes sustainable development: One of the key rationales behind the importance of green finance is its role in fostering sustainable development, by facilitating the flow of financial instruments and related services, it supports the establishment and implementation of sustainable business models, investments, projects, and procedures across economic, environmental, and social domains [28].

Stimulates innovation in banks: Another pivotal aspect underscoring the importance of green finance is its ability to drive innovation within banking institutions. By mitigating risks associated with environmental policy violations, it helps banks avoid potential penalties that could diminish profits and hinder repayment capabilities [6]. This, in turn, safeguards banks and lenders from heightened risks by Alkuwaiti and Nobanee.

Emphasis on holistic organizational value creation: Following the 2008 financial crisis, there has been a notable shift in organizational focus towards not only economic objectives but also environmental and social considerations, alongside creating value for investors stated by Cen, Tao, and He, Renke.

5 Results and Discussion

The global community has embraced the United Nations Sustainable Development Goals (UN SDGs), comprising 17 goals and 169 targets aimed at attaining environmental sustainability by 2030. A prominent area of emphasis within these goals is financial inclusion facilitated by digital finance, which has become a focal point for policymakers globally [29]. The objective is to achieve sustainable financial inclusion by harnessing technology and embracing a multi-stakeholder approach (Fig. 2).

FinTech companies leverage their technological capabilities to drive financial inclusion effectively. Initiatives like FinTech4FI by the Alliance for Financial Inclusion underscore the transformative potential of digital financial strategies [29]. Presently, sustainable finance, financial inclusion, and FinTech rank among the top policy priorities for numerous national governments and regulators. For instance, in India, FinTech has facilitated credit access for 80% of adults and enabled 350 million individuals to acquire their first-ever bank accounts. Moreover, digital footprints generate both formal and informal data, shaping various aspects of society [30].

However, India faces challenges in environmental sustainability, as evidenced by its last ranking in the 2022 Environmental Performance Index (EPI) with a score of 18.9. To address this, India must embrace financial technologies that bolster digital finance, aligning with the demands of environmental sustainability [31]. As India

Fig. 2 United Nations Sustainable Development Goals (UN SDGs)

pursues high-quality development, green development emerges as a shared objective, with innovation serving as the driving force for sustainable progress [32]. Green innovation, characterized by technology advancement and environmental preservation, is pivotal for achieving environmental sustainability [33]. This entails leveraging technology-driven innovation in the financial sector to realize energy savings, emissions reduction, and effective environmental management. By prioritizing green innovation in the financial domain, India can make significant strides towards achieving its environmental sustainability goals [10, 33].

6 Limitations of the Study

This research paper acknowledges certain limitations, primarily stemming from inadequate data accessibility, which complicates the analysis of heterogeneous issues that ultimately impact the research model. Green finance encompasses various themes such as green bonds, green insurance, and green credit, yet the lack of comprehensive information sharing poses challenges in obtaining relevant statistics. Consequently, our examination cannot fully ascertain the significance of numerous green financing strategies on ecological sustainability.

Academic research is an ongoing process, and efforts to gather pertinent data and conduct future studies on the impacts of different green financing methods on a sustainable economy remain imperative. While our focus lies on environmental sustainability, future investigations may explore how green financing influences corporate environmental behavior in other spheres to enrich the findings. It is important to note that CO2 emissions represent only a fraction of the complex environmental contamination issues faced by developing nations.

7 Conclusion

The current research paper utilized panel data from Indian states spanning from 2013 to 2023. It formulated a conceptual framework to analyze the role of financial technology and green finance in achieving the objectives of sustainable economic growth. Five parameters were selected to assess the magnitude of green finance, which were then combined using a General Principal Component Analysis Model (GPCAM). Concurrently, the researcher established parameters for sustainable economic growth based on three dimensions: environmental quality protection, economic efficiency, and economic system.

The study's original findings revealed that green finance significantly contributes to sustainable economic growth, exerting a considerable impact on all three dimensions: environmental quality protection, economic efficiency, and the economic system. These results analytically validate the beneficial effects of green finance. Furthermore, financial technology facilitates a positive and substantial influence of green finance on environmental quality protection and economic system characteristics. However, Fintech does not regulate the relationship between green finance and its economic impact. This underscores the areas where green finance, complemented by Fintech, has potential for growth and development in India.

Digital finance, an emerging financial system spurred by the rapid advancement of digital information technology and financial development, holds significant potential in enhancing the accessibility of financial resources for green enterprises. By enabling financial institutions to accurately assess borrower creditworthiness and monitor the flow and utilization of financial resources, digital finance facilitates the provision of financial assistance to green enterprises typically excluded from traditional financial services. Consequently, digital finance affords green enterprises the opportunity to allocate greater financial capital towards research and development projects cantered on green technologies, thereby elevating the level of green innovation.

Furthermore, digital finance has the capacity to reduce the costs associated with financial operations and transactions, a factor often correlated with increased levels of green innovation. Serving as a fusion of digital information technology and financial advancement, digital finance extends support to enterprises involved in green innovation by facilitating access to financial services. It empowers green enterprises to efficiently and inexpensively connect with suitable lenders, thereby diminishing the financing costs of innovative green ventures and alleviating pressure on their green technological innovation endeavors. Ultimately, the support provided by digital finance is poised to elevate the level of green innovation. Lastly, digital finance has the potential to instigate the transformation of green development and industrial upgrading, thereby fostering green innovation. This transformation paves the way for the emergence of novel green technologies and practices, ultimately contributing to sustainable development.

References

1. Akomea-Frimpong, I., Adeabah, D., Ofosu, D., & Tenakwah, E. J. (2022). A review of studies on green finance of banks, research gaps and future directions. *Journal of Sustainable Finance & Investment, 12*(4), 1241–1264.
2. Haddad, C., & Hornuf, L. (2019). The emergence of the global fintech market: Economic and technological determinants. *Small Business Economics, 53*(1), 81–105.
3. Varghese, G. (2017). Inflationary effects of oil price shocks in Indian economy. *Journal of Public Affairs, 17*(3), e1614.
4. Zhou, G., Zhu, J., & Luo, S. (2022). The impact of fintech innovation on green growth in China: Mediating effect of green finance. *Ecological Economics, 193*, 107308.
5. Geddes, A., Schmidt, T. S., & Steffen, B. (2018). The multiple roles of state investment banks in low-carbon energy finance: An analysis of Australia, the UK and Germany. *Energy Policy, 115*, 158–170.
6. Gu, W., Wang, J., Hua, X., & Liu, Z. (2021). Entrepreneurship and high-quality economic development: Based on the triple bottom line of sustainable development. *International Entrepreneurship and Management Journal, 17*, 1–27.
7. Herring, H., & Roy, R. (2007). Technological innovation, energy efficient design and the rebound effect. *Technovation, 27*(4), 194–203.
8. Jin, J., & Han, L. (2018). Assessment of Chinese green funds: Performance and industry allocation. *Journal of Cleaner Production, 171*, 1084–1093.
9. Park, J., Konana, P., Gu, B., Kumar, A., & Raghunathan, R. (2013). Information valuation and confirmation bias in virtual communities: Evidence from stock message boards. *Information Systems Research, 24*(4), 1050–1067.
10. Zhao, J., Zhao, Z., & Zhang, H. (2021). The impact of growth, energy and financial development on environmental pollution in China: New evidence from a spatial econometric analysis. *Energy Economics, 93*, 104506.
11. Riti, J. S., Song, D., Shu, Y., & Kamah, M. (2017). Decoupling CO2 emission and economic growth in China: Is there consistency in estimation results in analyzing environmental Kuznets curve? *Journal of Cleaner Production, 166*, 1448–1461.
12. Aldieri, L., Brahmi, M., Bruno, B., & Vinci, C. P. (2021). Circular economy business models: The complementarities with sharing economy and eco-innovations investments. *Sustainability, 13*(22), 12438.
13. Ma, L., Long, H., Chen, K., Tu, S., Zhang, Y., & Liao, L. (2019). Green growth efficiency of Chinese cities and its spatio-temporal pattern. *Resources, Conservation and Recycling, 146*, 441–451.
14. Lee, I., & Shin, Y. J. (2018). Fintech: Ecosystem, business models, investment decisions, and challenges. *Business Horizons, 61*(1), 35–46.
15. Anshari, M., Almunawar, M. N., Masri, M., & Hamdan, M. (2019). Digital marketplace and FinTech to support agriculture sustainability. *Energy Procedia, 156*, 234–238.
16. Missaoui, I., Brahmi, M., & BenRajeb, J. (2018). Quantitative relationship between corruption and development of the Tunisian stock market. *Public and Municipal Finance, 7*(2), 39–47.
17. Sreenu, N. (2022). Impact of FDI, crude oil price and economic growth on CO2 emission in India:-symmetric and asymmetric analysis through ARDL and non-linear ARDL approach. *Environmental Science and Pollution Research, 29*(28), 42452–42465.
18. Nenavath, S. (2022). Impact of fintech and green finance on environmental quality protection in India: By applying the semi-parametric difference-in-differences (SDID). *Renewable Energy, 193*, 913–919.
19. Sreenu, N. (2018). The effects of oil price shock on the Indian economy- A study. *The Indian Economic Journal, 66*(1–2), 190–202.

20. Qiu, S., Wang, Z., & Geng, S. (2021). How do environmental regulation and foreign investment behavior affect green productivity growth in the industrial sector? An empirical test based on Chinese provincial panel data. *Journal of Environmental Management, 287*, 112282.
21. Shaikh, E., Tunio, M. N., Khoso, W. M., Brahmi, M., & Rasool, S. (2022). The COVID-19 pandemic overlaps entrepreneurial activities and triggered new challenges: A review study. In *Managing Human Resources in SMEs and Start-ups: International Challenges and Solutions* (pp. 155–182).
22. Hafner, S., Jones, A., Anger-Kraavi, A., & Pohl, J. (2020). Closing the green finance gap–a systems perspective. *Environmental Innovation and Societal Transitions, 34*, 26–60.
23. Solarin, S. A., Al-Mulali, U., & Ozturk, I. (2017). Validating the environmental Kuznets curve hypothesis in India and China: The role of hydroelectricity consumption. *Renewable and Sustainable Energy Reviews, 80*, 1578–1587.
24. Salari, M., Javid, R. J., & Noghanibehambari, H. (2021). The nexus between CO2 emissions, energy consumption, and economic growth in the US. *Economic Analysis and Policy, 69*, 182–194.
25. Song, Z., Storesletten, K., & Zilibotti, F. (2011). Growing like china. *American Economic Review, 101*(1), 196–233.
26. Muganyi, T., Yan, L., & Sun, H. P. (2021). Green finance, fintech and environmental protection: Evidence from China. *Environmental Science and Ecotechnology, 7*, 100107.
27. Xu, T. (2018). Investigating environmental Kuznets curve in China–aggregation bias and policy implications. *Energy Policy, 114*, 315–322.
28. Tawiah, V., Zakari, A., & Adedoyin, F. F. (2021). Determinants of green growth in developed and developing countries. *Environmental Science and Pollution Research, 28*(29), 39227–39242.
29. Yang, X., Zhang, J., Ren, S., & Ran, Q. (2021). Can the new energy demonstration city policy reduce environmental pollution? Evidence from a quasi-natural experiment in China. *Journal of Cleaner Production, 287*, 125015.
30. Zhou, X., Tang, X., & Zhang, R. (2020). Impact of green finance on economic development and environmental quality: A study based on provincial panel data from China. *Environmental Science and Pollution Research, 27*, 19915–19932.
31. Wang, Y., & Zhi, Q. (2016). The role of green finance in environmental protection: Two aspects of market mechanism and policies. *Energy Procedia, 104*, 311–316.
32. Yang, Y., Su, X., & Yao, S. (2021). Nexus between green finance, fintech, and high-quality economic development: Empirical evidence from China. *Resources Policy, 74*, 102445.
33. Arabeche, Z., Soudani, A., Brahmi, M., Aldieri, L., Vinci, C. P., & Abdelli, M. E. A. (2022). Entrepreneurial orientation, organizational culture and business performance in SMEs: Evidence from emerging economy. *Sustainability, 14*(9), 5160.
34. Kishor, K., Bansal, S. K., & Kumar, R. (2024). The role of Fintech in promoting financial inclusion to achieve sustainable development: An integrated bibliometric analysis and systematic literature review. *Journal of the Knowledge Economy, 16*, 1–29.
35. Tang, H., Tong, M., & Chen, Y. (2024). Green investor behavior and corporate green innovation: Evidence from Chinese listed companies. *Journal of Environmental Management, 366*, 121691.
36. Rao, H., Chen, D., Shen, F., & Shen, Y. (2022). Can green bonds stimulate green innovation in enterprises? Evidence from China. *Sustainability, 14*(23), 15631.
37. Yang, L., & Yuan, H. (2024). Supply chain financing and Total factor productivity: Evidence from listed manufacturing companies in China. *Journal of International Development.*

Correction to: Artificial Intelligence (AI) for IT Energy Efficiency and Green AI for Environment Sustainability

Pethuru Raj, Durga Prasad Sharma, Pushan Kumar Dutta, B. V. V. Siva Prasad, and P. Beaulah Soundarabai

Correction to:
P. Raj et al. (eds.), *Artificial Intelligence (AI) for IT Energy Efficiency and Green AI for Environment Sustainability*, https://doi.org/10.1007/978-3-031-89420-6

In the original version of the book, the following post-publication corrections have been incorporated:

The author name "Jitendra Chandnani" has been changed to "Chirag Jitendra Chandnani" in the front matter and in the chapter "Artificial Intelligence for Energy-Efficient and Sustainable Smart Grids".

In the chapter "Empowering Sustainability: Assessing the Impact and Viability of Renewable Energy Loans in Accelerating Green Transitions", the affiliation "Lakshmi Narayan College of Technology" of author "Vinita Saxena" has been changed to "Lakshmi Narain College of Technology".

The author name "Priyanka" has been changed to "Priyanka Deswal" in the front matter and in the chapter "Exploring the Influence of Workplace Spirituality on Job Performance: A Study in Haryana, India".

In the chapter "Blockchain and AI for Enhanced Threat Detection in Digital Marketing: Maneuvering the Legal Minefield", the affiliation "Hensard University, Toru-Orua, Bayelsa, Nigeria" of author "Ashok Kumar Srivastava" has been changed to "Amity University Jharkhand, Ranchi, India".

The updated version of this book can be found at:
https://doi.org/10.1007/978-3-031-89420-6

P. Raj et al. (eds.), *Artificial Intelligence (AI) for IT Energy Efficiency and Green AI for Environment Sustainability*,
https://doi.org/10.1007/978-3-031-89420-6_33

Index

P. Raj et al. (eds.), *Artificial Intelligence (AI) for IT Energy Efficiency and Green AI for Environment Sustainability*,
https://doi.org/10.1007/978-3-031-89420-6

F

G

MIX
Papier aus verantwortungsvollen Quellen
Paper from responsible sources
FSC® C105338

If you have any concerns about our products,
you can contact us on
ProductSafety@springernature.com

In case Publisher is established outside the EU,
the EU authorized representative is:
Springer Nature Customer Service Center GmbH
Europaplatz 3, 69115 Heidelberg, Germany

Printed by Libri Plureos GmbH
in Hamburg, Germany